Alexander Apelblat
Bessel and Related Functions

Also of interest

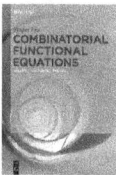

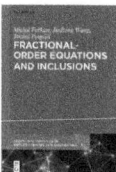

Alexander Apelblat

Bessel and Related Functions

Mathematical Operations with Respect to the Order
Volume 2: Numerical Results

DE GRUYTER

Author
Prof. Dr. Alexander Apelblat
Ben-Gurion University of the Negev
Department of Chemical Engineering
Beer Sheva, Israel
apelblat@bgu.ac.il

ISBN 978-3-11-068163-5
e-ISBN (PDF) 978-3-11-068247-2
e-ISBN (EPUB) 978-3-11-068347-9

Library of Congress Control Number: 2020933852

Bibliographic information published by the Deutsche Nationalbibliothek
The Deutsche Nationalbibliothek lists this publication in the Deutsche Nationalbibliografie;
detailed bibliographic data are available on the Internet at http://dnb.dnb.de.

www.degruyter.com

To Ira and Yoram

Contents

1 Numerical Results – Tabulation of First, Second and Third Derivatives with Respect to the Order of the Bessel and Modified Bessel Functions

The earliest short table of the first and second derivatives with respect to the order of the Bessel function $J_v(t)$ was published in 1935 by Airey [42, 43, for list of References see Part 1]. He calculated $\partial J_v(t)/\partial v$ and $\partial^2 J_v(t)/\partial v^2$ with the argument of the Bessel function at steps $t = 1, 2, 3, \ldots, 20$, but with the order v in short ranges, $t - 1 \leq v \leq t + 1$. The table is constructed with 0.1 intervals and four places of decimals. Airey also mentioned, in references, the existence of his tabulation from 1928, where values of the first derivatives are reported with respect to the order for v equals $\pm 1/2$ and $3/2$. Lee and Radosevich [55] in 1960 calculated to four places the first derivatives with respect to the order for v equals $k + 1/4$, $k + 1/3$, $k + 2/3$ and $k + 3/4$, $k = 0, 1, 2, 3, 4$ and $1 \leq t \leq 15$ in 0.5 intervals. With the same precision, Erber and Gordon [49] in 1963 reported the first derivatives of the modified Bessel function of the first kind $\partial I_v(t)/\partial v$ for $\pm 1/3$ with $0 \leq t \leq 1$ in 0.01 intervals, and with $1 \leq t \leq 5$ in 0.05 intervals. From the literature search, it is evident that only a small number of tabulations of derivatives with respect to the order of the Bessel functions exist and these tables have rather accidental, sporadic character.

In the case of the integral Bessel functions, even functions themselves are tabulated in a very limited form. In the Abramowitz–Stegun handbook [9], only the Bessel functions with zero order are tabulated. Their integrands are $(1 - J_0(t))/t$, $(1 - I_0(t))/t$, $Y_0(t)/t$ and $K_0(t)/t$, and the integration limits vary in $0 \leq t \leq 5$ region in 0.1 intervals.

The Kelvin functions $\mathrm{ber}_v(t)$, $\mathrm{bei}_v(t)$, $\mathrm{ker}_v(t)$ and $\mathrm{kei}_v(t)$ and their derivatives with respect to the argument t were tabulated by McLachlan [22] for integer orders $n = 1$–5 and arguments 0–10, mainly for the zero order.

Considering this unfortunate situation in the literature, I decided to prepare in systematic way a set of tables, where the first, the second and the third derivatives are given with respect to the order of the Bessel functions, the modified Bessel functions, the Struve functions, the modified Struve functions and the Anger and Weber functions (Chapter 2). In the case of the integral Bessel functions (Chapter 4) and the Kelvin functions (Chapter 3), these functions, their first and second derivatives with respect to the order are tabulated. The orders cover $0 \leq v \leq 5$ range, with 0.05 intervals. The arguments of functions lie in $0.05 \leq t \leq 5$ range with 0.05 intervals for $t \leq 1.0$, and with 1.0 intervals for $t > 1.0$. These ranges of v and t give the most characteristic patterns of computed derivatives and functions as can be observed from graphical representations presented in this book. Values of derivatives in tables are presented in the form to preserve six significant figures coming from computations. Numerical calculations were performed by using MATHEMATICA program.

https://doi.org/10.1515/9783110682472-001

Derivatives with respect to the order were mainly calculated by determination of integrals in the integral representations of Bessel functions. In other cases, calculations were performed directly, by applying the central-difference formulas of order $O(h^4)$ with $h = 0.001$, with the Bessel and related functions taken from MATHEMATICA. The consistency of both methods of calculations was checked and was found to be very satisfactory. The reliability of calculations was also confirmed by comparison of results presented in Tables 1.1–1.6, reported by Airey [42, 43], Lee and Radosevich [55] and Erber and Gordon [49].

1.1 First, Second and Third Derivatives with Respect to the Order of the Bessel Function of the First Kind

Changes in values of the first, second and third derivatives with respect to the order of the Bessel function of the first kind are limited, and occur only when v and t are small. As can be observed in the following figures, the functional form of these derivatives is rather simple.

Using MATHEMATICA program, the derivatives with respect to the order of the Bessel function of the first kind were determined by applying the following integral representations:

$$\frac{\partial J_v(z)}{\partial v} = \frac{1}{\pi} \int_0^\pi t \sin(z \sin t - v t) \, dt$$

$$+ \frac{1}{\pi} \int_0^\infty e^{-z \sinh t - vt} [t \sin(\pi v) - \pi \cos(\pi v)] \, dt \tag{1.1.1}$$

$$\frac{\partial^2 J_v(z)}{\partial v^2} = -\frac{1}{\pi} \int_0^\pi t^2 \cos(z \sin t - v t) \, dt$$

$$+ \frac{1}{\pi} \int_0^\infty e^{-z \sinh t - vt} [(\pi^2 - t^2) \sin(\pi v) + 2\pi t \cos(\pi v)] \, dt \tag{1.1.2}$$

$$\frac{\partial^3 J_v(z)}{\partial v^3} = -\frac{1}{\pi} \int_0^\pi t^3 \sin(z \sin t - v t) \, dt$$

$$+ \frac{1}{\pi} \int_0^\infty e^{-z \sinh t - vt} [(t^3 - 3\pi^2 t) \sin(\pi v) + (\pi^3 - 3\pi t^2) \cos(\pi v)] \, dt \tag{1.1.3}$$

Similar to the first derivatives, the second derivatives with respect to the order of the Bessel functions of the first kind $\partial^2 J_v(t)/\partial v^2$ vary for small values of the order and argument, and they tend to be zero for larger values of v and t (Figures 1.1 and 1.2).

$$\frac{\partial J_\nu(t)}{\partial \nu}$$

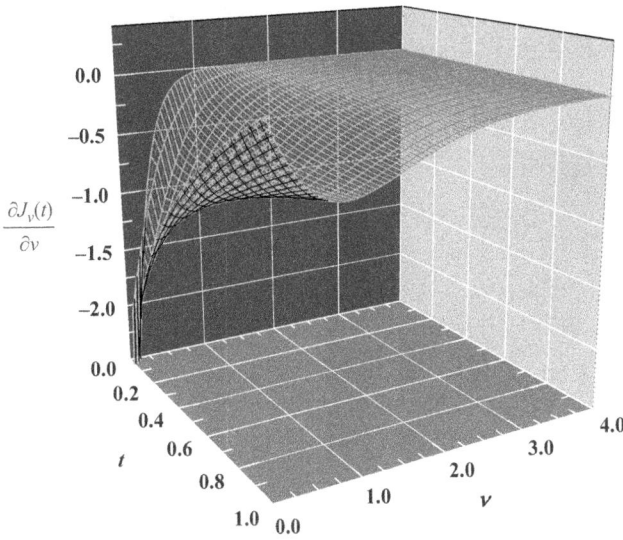

Figure 1.1: First derivatives with respect to the order of the Bessel functions of the first kind $\partial J_\nu(t)/\partial \nu$ as a function of ν and t.

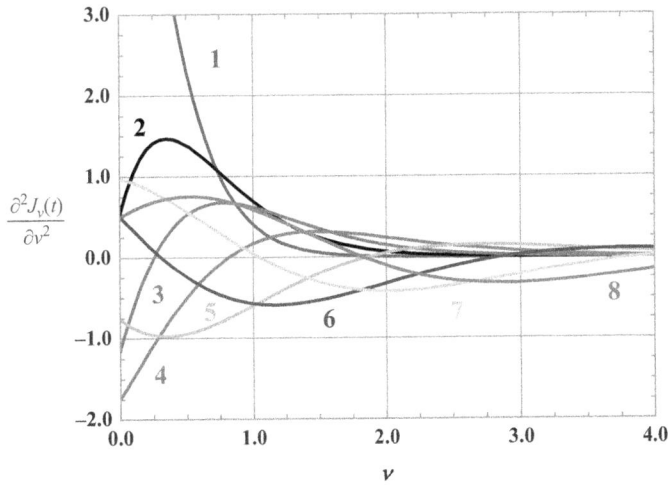

$$\frac{\partial^2 J_\nu(t)}{\partial \nu^2}$$

Figure 1.2: Second derivatives of the Bessel function of the first kind with respect to the order ν as a function of ν, at constant values of argument t.

$1 - t = 0.05$; $2 - t = 0.25$; $3 - t = 0.50$; $4 - t = 1.0$; $5 - t = 2.0$; $6 - t = 3.0$; $7 - t = 4.0$; $8 - t = 5.0$.

Third derivatives with respect to the order of the Bessel functions of the first kind $\partial^3 J_\nu(t)/\partial \nu^3$, presented in Figures 1.3 and 1.4, show a more oscillatory character than lower derivatives, but essentially their behaviour is similar. It is worthwhile to note that with increasing value of the order v, for small arguments t (Figure 1.4) they have nearly zero value for $v > 1$ (Figure 1.5).

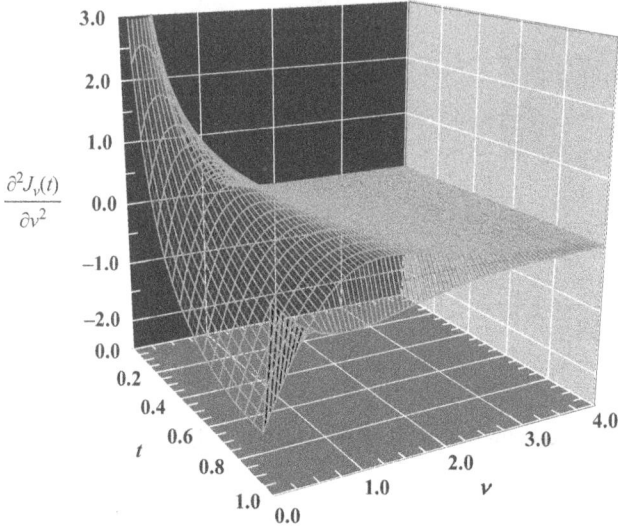

Figure 1.3: Second derivatives with respect to the order of the Bessel functions of the first kind $\partial^2 J_v(t)/\partial v^2$ as a function of v and t.

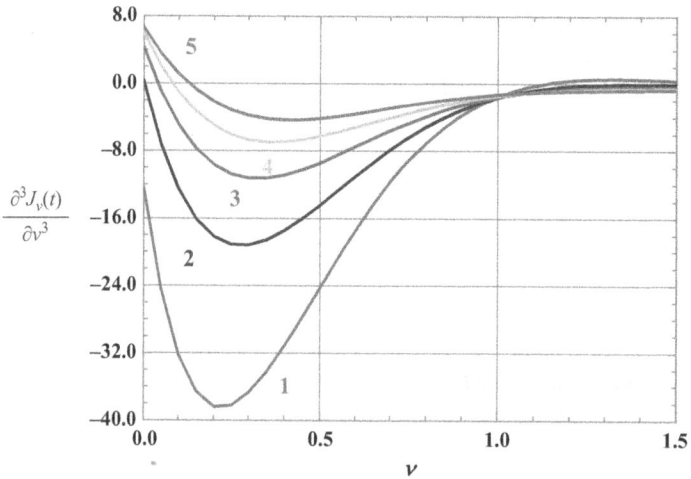

Figure 1.4: Third derivatives of the Bessel function of the first kind with respect to the order v as a function of v, at constant values of argument t.

$1 - t = 0.05$; $2 - t = 0.10$; $3 - t = 0.15$; $4 - t = 0.20$; $5 - t = 0.25$.

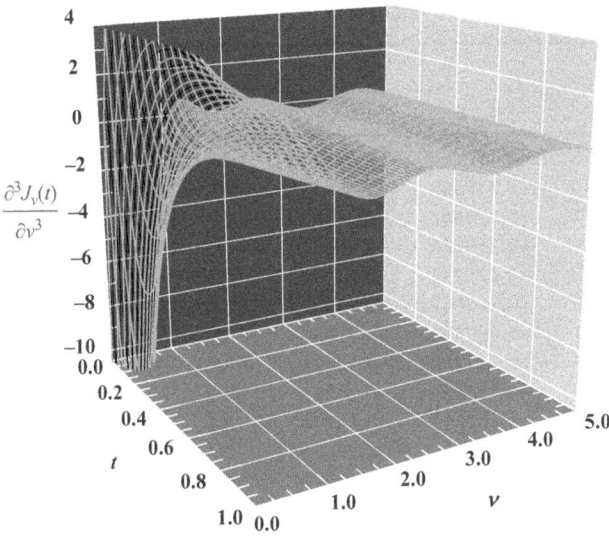

Figure 1.5: Third derivatives with respect to the order of the Bessel functions of the first kind $\partial^3 J_\nu(t)/\partial\nu^3$ as a function of v and t.

Table 1.1: First derivatives with respect to the order of the Bessel functions of the first kind $\partial J_\nu(t)/\partial\nu$.

v	t					
	0.05	0.10	0.15	0.20	0.25	0.30
0.00	-4.72096	-1.99750	-1.996130	-1.69820	-1.46331	-1.26806
0.05	-3.78340	-1.80104	-1.873580	-1.62828	-1.42946	-1.26051
0.10	-3.01634	-1.61483	-1.744890	-1.54762	-1.38264	-1.23899
0.15	-2.39338	-1.44049	-1.613850	-1.45970	-1.32595	-1.20618
0.20	-1.89076	-1.27898	-1.483450	-1.36743	-1.26205	-1.16450
0.25	-1.48762	-1.13068	-1.355980	-1.27320	-1.19325	-1.11605
0.30	-1.16602	-0.995593	-1.233180	-1.17894	-1.12147	-1.06266
0.35	-0.910702	-0.873385	-1.116290	-1.08620	-1.04832	-1.00593
0.40	-0.708930	-0.763515	-1.006140	-0.996157	-0.975117	-0.947201
0.45	-0.550136	-0.665289	-0.903265	-0.909710	-0.902934	-0.887600
0.50	-0.425648	-0.577920	-0.807907	-0.827506	-0.832612	-0.828068
0.55	-0.328408	-0.500567	-0.720118	-0.749979	-0.764802	-0.769372
0.60	-0.252709	-0.432376	-0.639786	-0.677393	-0.699987	-0.712125
0.65	-0.193967	-0.372500	-0.566679	-0.609870	-0.638509	-0.656808
0.70	-0.148520	-0.320119	-0.500478	-0.547420	-0.580590	-0.603783
0.75	-0.113459	-0.274452	-0.440804	-0.489962	-0.526352	-0.553314

Table 1.1 (continued)

			t			
v	0.05	0.10	0.15	0.20	0.25	0.30
0.80	−0.086483	−0.234766	−0.387237	−0.437351	−0.475839	−0.505578
0.85	−0.065782	−0.200384	−0.339340	−0.389386	−0.429027	−0.460680
0.90	−0.049934	−0.170681	−0.296667	−0.345836	−0.385841	−0.418668
0.95	−0.037829	−0.145091	−0.258776	−0.306440	−0.346165	−0.379538
1.00	−0.028605	−0.123100	−0.225239	−0.270927	−0.309855	−0.343247
1.05	−0.021591	−0.104249	−0.195642	−0.239019	−0.276743	−0.309722
1.10	−0.016268	−0.088127	−0.169598	−0.210438	−0.246649	−0.278866
1.15	−0.012236	−0.074370	−0.146740	−0.184910	−0.219384	−0.250562
1.20	−0.009189	−0.062656	−0.126729	−0.162171	−0.194753	−0.224681
1.25	−0.006889	−0.052701	−0.109252	−0.141970	−0.172565	−0.201088
1.30	−0.005157	−0.044259	−0.094025	−0.124066	−0.152629	−0.179640
1.35	−0.003855	−0.037113	−0.080785	−0.108237	−0.134762	−0.160194
1.40	−0.002877	−0.031075	−0.069299	−0.094272	−0.118786	−0.142608
1.45	−0.002144	−0.025982	−0.059353	−0.081978	−0.104535	−0.126743
1.50	−0.001596	−0.021693	−0.050758	−0.071178	−0.091848	−0.112461
1.55	−0.001186	−0.018088	−0.043345	−0.061708	−0.080579	−0.099634
1.60	−0.000881	−0.015062	−0.036961	−0.053421	−0.070588	−0.088137
1.65	−0.000653	−0.012527	−0.031474	−0.046182	−0.061746	−0.077853
1.70	−0.000483	−0.010405	−0.026766	−0.039869	−0.053937	−0.068671
1.75	−0.000358	−0.008632	−0.022732	−0.034373	−0.047052	−0.060489
1.80	−0.000264	−0.007152	−0.019282	−0.029596	−0.040992	−0.053210
1.85	−0.000195	−0.005920	−0.016335	−0.025451	−0.035666	−0.046746
1.90	−0.000144	−0.004894	−0.013821	−0.021859	−0.030993	−0.041015
1.95	−0.000106	−0.004041	−0.011681	−0.018752	−0.026900	−0.035942
2.00	−0.000078	−0.003334	−0.009860	−0.016068	−0.023320	−0.031459
2.05	−0.000057	−0.002747	−0.008314	−0.013752	−0.020192	−0.027502
2.10	−0.000042	−0.002261	−0.007003	−0.011756	−0.017464	−0.024015
2.15	−0.000031	−0.001859	−0.005892	−0.010040	−0.015088	−0.020947
2.20	−0.000023	−0.001527	−0.004952	−0.008564	−0.013021	−0.018251
2.25	−0.000016	−0.001254	−0.004158	−0.007298	−0.011225	−0.015884
2.30	−0.000012	−0.001028	−0.003487	−0.006213	−0.009667	−0.013810
2.35	−0.000009	−0.000842	−0.002922	−0.005283	−0.008317	−0.011995
2.40	−0.000006	−0.000689	−0.002446	−0.004489	−0.007148	−0.010407
2.45	−0.000005	−0.000563	−0.002046	−0.003810	−0.006137	−0.009021
2.50	−0.000003	−0.000460	−0.001709	−0.003231	−0.005265	−0.007812
2.55	−0.000002	−0.000376	−0.001427	−0.002737	−0.004512	−0.006758
2.60	−0.000002	−0.000306	−0.001190	−0.002317	−0.003863	−0.005841
2.65	−0.000001	−0.000250	−0.000991	−0.001959	−0.003304	−0.005044
2.70	−0.000001	−0.000203	−0.000825	−0.001655	−0.002824	−0.004352
2.75	−0.000001	−0.000165	−0.000687	−0.001397	−0.002412	−0.003751
2.80	−0.000001	−0.000134	−0.000571	−0.001178	−0.002058	−0.003231

Table 1.1 (continued)

v	0.05	0.10	0.15	0.20	0.25	0.30
2.85		−0.000109	−0.000474	−0.000993	−0.001754	−0.002780
2.90		−0.000089	−0.000393	−0.000836	−0.001494	−0.002391
2.95		−0.000072	−0.000326	−0.000704	−0.001272	−0.002054
3.00		−0.000058	−0.000270	−0.000592	−0.001081	−0.001763
3.05		−0.000047	−0.000224	−0.000497	−0.000919	−0.001512
3.10		−0.000038	−0.000185	−0.000417	−0.000780	−0.001296
3.15		−0.000031	−0.000153	−0.000350	−0.000662	−0.001110
3.20		−0.000025	−0.000126	−0.000293	−0.000561	−0.000950
3.25		−0.000020	−0.000104	−0.000246	−0.000475	−0.000812
3.30		−0.000016	−0.000086	−0.000206	−0.000402	−0.000694
3.35		−0.000013	−0.000071	−0.000172	−0.000340	−0.000593
3.40		−0.000011	−0.000058	−0.000144	−0.000288	−0.000506
3.45		−0.000008	−0.000048	−0.000120	−0.000243	−0.000431
3.50		−0.000007	−0.000039	−0.000100	−0.000205	−0.000367
3.55		−0.000005	−0.000032	−0.000084	−0.000173	−0.000313
3.60		−0.000004	−0.000027	−0.000070	−0.000146	−0.000266
3.65		−0.000004	−0.000022	−0.000058	−0.000123	−0.000226
3.70		−0.000003	−0.000018	−0.000048	−0.000104	−0.000192
3.75		−0.000002	−0.000015	−0.000040	−0.000087	−0.000163
3.80		−0.000002	−0.000012	−0.000033	−0.000073	−0.000139
3.85		−0.000001	−0.000010	−0.000028	−0.000062	−0.000118
3.90		−0.000001	−0.000008	−0.000023	−0.000052	−0.000100
3.95		−0.000001	−0.000007	−0.000019	−0.000043	−0.000084
4.00		−0.000001	−0.000005	−0.000016	−0.000036	−0.000071
4.05		−0.000001	−0.000004	−0.000013	−0.000031	−0.000060
4.10			−0.000004	−0.000011	−0.000026	−0.000051
4.15			−0.000003	−0.000009	−0.000021	−0.000043
4.20			−0.000002	−0.000007	−0.000018	−0.000036
4.25			−0.000002	−0.000006	−0.000015	−0.000031
4.30			−0.000002	−0.000005	−0.000013	−0.000026
4.35			−0.000001	−0.000004	−0.000010	−0.000022
4.40			−0.000001	−0.000003	−0.000009	−0.000018
4.45			−0.000001	−0.000003	−0.000007	−0.000016
4.50			−0.000001	−0.000002	−0.000006	−0.000013
4.55			−0.000001	−0.000002	−0.000005	−0.000011
4.60				−0.000002	−0.000004	−0.000009
4.65				−0.000001	−0.000004	−0.000008
4.70				−0.000001	−0.000003	−0.000007
4.75				−0.000001	−0.000002	−0.000005
4.80				−0.000001	−0.000002	−0.000005
4.85				−0.000001	−0.000002	−0.000004

Table 1.1 (continued)

				t		
v	0.05	0.10	0.15	0.20	0.25	0.30
4.90				−0.000001	−0.000001	−0.000003
4.95					−0.000001	−0.000003
5.00					−0.000001	−0.000002

				t		
v	0.35	0.40	0.45	0.50	0.55	0.60
0.00	−1.10005	−0.951941	−0.819058	−0.698248	−0.587300	−0.484606
0.05	−1.11247	−0.979932	−0.859402	−0.748505	−0.645560	−0.549336
0.10	−1.11058	−0.993665	−0.885798	−0.785281	−0.690905	−0.601783
0.15	−1.09670	−0.995147	−0.899956	−0.810022	−0.724550	−0.642951
0.20	−1.07296	−0.986240	−0.903497	−0.824126	−0.747687	−0.673849
0.25	−1.04128	−0.968656	−0.897936	−0.828926	−0.761477	−0.695479
0.30	−1.00335	−0.943950	−0.884679	−0.825676	−0.767033	−0.708818
0.35	−0.960691	−0.913525	−0.865015	−0.815552	−0.765413	−0.714809
0.40	−0.914607	−0.878634	−0.840118	−0.799637	−0.757609	−0.714355
0.45	−0.866239	−0.840385	−0.811046	−0.778928	−0.744548	−0.708305
0.50	−0.816560	−0.799750	−0.778744	−0.754328	−0.727083	−0.697458
0.55	−0.766395	−0.757572	−0.744052	−0.726655	−0.705997	−0.682556
0.60	−0.716431	−0.714577	−0.707703	−0.696637	−0.682000	−0.664280
0.65	−0.667234	−0.671378	−0.670340	−0.664921	−0.655733	−0.643255
0.70	−0.619259	−0.628491	−0.632510	−0.632076	−0.627768	−0.620047
0.75	−0.572865	−0.586341	−0.594684	−0.598594	−0.598611	−0.595165
0.80	−0.528329	−0.545274	−0.557254	−0.564902	−0.568707	−0.569065
0.85	−0.485851	−0.505561	−0.520546	−0.531362	−0.538445	−0.542148
0.90	−0.445572	−0.467414	−0.484824	−0.498279	−0.508158	−0.514769
0.95	−0.407576	−0.430989	−0.450298	−0.465905	−0.478129	−0.487235
1.00	−0.371904	−0.396392	−0.417129	−0.434444	−0.448598	−0.459810
1.05	−0.338561	−0.363691	−0.385438	−0.404059	−0.419762	−0.432719
1.10	−0.307518	−0.332919	−0.355308	−0.374877	−0.391780	−0.406150
1.15	−0.278724	−0.304080	−0.326791	−0.346988	−0.364780	−0.380260
1.20	−0.252109	−0.277155	−0.299912	−0.320458	−0.338859	−0.355174
1.25	−0.227587	−0.252103	−0.274672	−0.295324	−0.314089	−0.330993
1.30	−0.205062	−0.228872	−0.251056	−0.271606	−0.290518	−0.307790
1.35	−0.184430	−0.207394	−0.229032	−0.249304	−0.268176	−0.285623
1.40	−0.165582	−0.187594	−0.208557	−0.228402	−0.247073	−0.264526
1.45	−0.148409	−0.169390	−0.189575	−0.208873	−0.227209	−0.244522
1.50	−0.132799	−0.152698	−0.172027	−0.190680	−0.208569	−0.225617
1.55	−0.118643	−0.137428	−0.155845	−0.173779	−0.191128	−0.207807
1.60	−0.105833	−0.123491	−0.140961	−0.158118	−0.174854	−0.191078
1.65	−0.094266	−0.110800	−0.127301	−0.143642	−0.159710	−0.175407
1.70	−0.083842	−0.099266	−0.114793	−0.130292	−0.145650	−0.160767
1.75	−0.074466	−0.088807	−0.103364	−0.118009	−0.132628	−0.147122

Table 1.1 (continued)

v	0.35	0.40	0.45	0.50	0.55	0.60
1.80	−0.066048	−0.079340	−0.092942	−0.106730	−0.120595	−0.134436
1.85	−0.058505	−0.070786	−0.083456	−0.096395	−0.109498	−0.122667
1.90	−0.051756	−0.063072	−0.074839	−0.086943	−0.099285	−0.111771
1.95	−0.045728	−0.056127	−0.067024	−0.078315	−0.089905	−0.101705
2.00	−0.040353	−0.049884	−0.059950	−0.070453	−0.081305	−0.092424
2.05	−0.035567	−0.044283	−0.053557	−0.063301	−0.073435	−0.083882
2.10	−0.031312	−0.039263	−0.047788	−0.056806	−0.066245	−0.076034
2.15	−0.027534	−0.034773	−0.042590	−0.050917	−0.059687	−0.068836
2.20	−0.024186	−0.030761	−0.037915	−0.045586	−0.053716	−0.062246
2.25	−0.021221	−0.027182	−0.033715	−0.040767	−0.048286	−0.056221
2.30	−0.018600	−0.023994	−0.029948	−0.036417	−0.043357	−0.050722
2.35	−0.016286	−0.021157	−0.026573	−0.032496	−0.038889	−0.045711
2.40	−0.014245	−0.018637	−0.023554	−0.028967	−0.034844	−0.041150
2.45	−0.012448	−0.016400	−0.020857	−0.025795	−0.031187	−0.037005
2.50	−0.010866	−0.014418	−0.018450	−0.022946	−0.027885	−0.033243
2.55	−0.009477	−0.012662	−0.016305	−0.020392	−0.024908	−0.029834
2.60	−0.008257	−0.011110	−0.014396	−0.018105	−0.022227	−0.026747
2.65	−0.007188	−0.009739	−0.012698	−0.016059	−0.019816	−0.023957
2.70	−0.006251	−0.008530	−0.011190	−0.014231	−0.017649	−0.021437
2.75	−0.005432	−0.007464	−0.009852	−0.012599	−0.015705	−0.019165
2.80	−0.004716	−0.006525	−0.008666	−0.011145	−0.013962	−0.017118
2.85	−0.004091	−0.005700	−0.007617	−0.009850	−0.012402	−0.015275
2.90	−0.003546	−0.004974	−0.006689	−0.008697	−0.011007	−0.013619
2.95	−0.003070	−0.004338	−0.005869	−0.007673	−0.009760	−0.012132
3.00	−0.002657	−0.003779	−0.005145	−0.006764	−0.008647	−0.010799
3.05	−0.002297	−0.003290	−0.004507	−0.005958	−0.007654	−0.009603
3.10	−0.001984	−0.002862	−0.003945	−0.005244	−0.006771	−0.008534
3.15	−0.001713	−0.002488	−0.003450	−0.004611	−0.005984	−0.007577
3.20	−0.001478	−0.002161	−0.003015	−0.004052	−0.005284	−0.006722
3.25	−0.001274	−0.001875	−0.002633	−0.003558	−0.004663	−0.005959
3.30	−0.001097	−0.001627	−0.002297	−0.003122	−0.004112	−0.005278
3.35	−0.000944	−0.001410	−0.002003	−0.002737	−0.003623	−0.004672
3.40	−0.000812	−0.001221	−0.001745	−0.002398	−0.003190	−0.004132
3.45	−0.000698	−0.001056	−0.001519	−0.002099	−0.002806	−0.003652
3.50	−0.000599	−0.000913	−0.001322	−0.001836	−0.002467	−0.003226
3.55	−0.000514	−0.000789	−0.001149	−0.001605	−0.002168	−0.002847
3.60	−0.000441	−0.000682	−0.000999	−0.001402	−0.001903	−0.002511
3.65	−0.000378	−0.000588	−0.000867	−0.001224	−0.001670	−0.002213
3.70	−0.000324	−0.000507	−0.000752	−0.001068	−0.001464	−0.001949
3.75	−0.000277	−0.000437	−0.000652	−0.000931	−0.001283	−0.001715
3.80	−0.000237	−0.000377	−0.000565	−0.000811	−0.001123	−0.001509
3.85	−0.000203	−0.000324	−0.000489	−0.000706	−0.000983	−0.001326
3.90	−0.000173	−0.000279	−0.000424	−0.000615	−0.000859	−0.001165

Table 1.1 (continued)

				t		
v	0.35	0.40	0.45	0.50	0.55	0.60
3.95	−0.000148	−0.000240	−0.000366	−0.000535	−0.000751	−0.001023
4.00	−0.000126	−0.000206	−0.000317	−0.000465	−0.000656	−0.000897
4.05	−0.000108	−0.000177	−0.000274	−0.000403	−0.000572	−0.000787
4.10	−0.000092	−0.000152	−0.000236	−0.000350	−0.000499	−0.000689
4.15	−0.000078	−0.000130	−0.000204	−0.000304	−0.000435	−0.000604
4.20	−0.000066	−0.000112	−0.000176	−0.000263	−0.000379	−0.000528
4.25	−0.000057	−0.000096	−0.000151	−0.000228	−0.000330	−0.000462
4.30	−0.000048	−0.000082	−0.000130	−0.000198	−0.000287	−0.000404
4.35	−0.000041	−0.000070	−0.000112	−0.000171	−0.000250	−0.000353
4.40	−0.000035	−0.000060	−0.000097	−0.000148	−0.000217	−0.000308
4.45	−0.000029	−0.000051	−0.000083	−0.000128	−0.000189	−0.000269
4.50	−0.000025	−0.000044	−0.000071	−0.000111	−0.000164	−0.000234
4.55	−0.000021	−0.000037	−0.000061	−0.000095	−0.000142	−0.000204
4.60	−0.000018	−0.000032	−0.000053	−0.000082	−0.000123	−0.000178
4.65	−0.000015	−0.000027	−0.000045	−0.000071	−0.000107	−0.000155
4.70	−0.000013	−0.000023	−0.000039	−0.000061	−0.000093	−0.000135
4.75	−0.000011	−0.000020	−0.000033	−0.000053	−0.000080	−0.000117
4.80	−0.000009	−0.000017	−0.000028	−0.000046	−0.000069	−0.000102
4.85	−0.000008	−0.000014	−0.000024	−0.000039	−0.000060	−0.000089
4.90	−0.000007	−0.000012	−0.000021	−0.000034	−0.000052	−0.000077
4.95	−0.000006	−0.000010	−0.000018	−0.000029	−0.000045	−0.000067
5.00	−0.000005	−0.000009	−0.000015	−0.000025	−0.000039	−0.000058

				t		
v	0.65	0.70	0.75	0.80	0.85	0.90
0.00	−0.388974	−0.299496	−0.215470	−0.136349	−0.061695	0.008841
0.05	−0.458917	−0.373603	−0.292854	−0.216246	−0.143444	−0.074180
0.10	−0.517249	−0.436793	−0.360025	−0.286638	−0.216392	−0.149096
0.15	−0.564785	−0.489714	−0.417483	−0.347891	−0.280785	−0.216047
0.20	−0.602368	−0.533059	−0.465782	−0.400435	−0.336942	−0.275250
0.25	−0.630853	−0.567544	−0.505516	−0.444749	−0.385237	−0.326982
0.30	−0.651089	−0.593901	−0.537304	−0.481349	−0.426088	−0.371574
0.35	−0.663911	−0.612860	−0.561777	−0.510772	−0.459947	−0.409393
0.40	−0.670128	−0.625140	−0.579569	−0.533571	−0.487285	−0.440838
0.45	−0.670517	−0.631444	−0.591308	−0.550299	−0.508588	−0.466327
0.50	−0.665814	−0.632447	−0.597606	−0.561510	−0.524347	−0.486290
0.55	−0.656716	−0.628793	−0.599058	−0.567742	−0.535049	−0.501164
0.60	−0.643869	−0.621094	−0.596232	−0.569522	−0.541176	−0.511384
0.65	−0.627876	−0.609921	−0.589667	−0.567354	−0.543194	−0.517378
0.70	−0.609289	−0.595808	−0.579873	−0.561720	−0.541556	−0.519568
0.75	−0.588611	−0.579246	−0.567326	−0.553075	−0.536692	−0.518358

Table 1.1 (continued)

				t		
ν	0.65	0.70	0.75	0.80	0.85	0.90
0.80	−0.566299	−0.560687	−0.552466	−0.541846	−0.529013	−0.514138
0.85	−0.542764	−0.540543	−0.535701	−0.528431	−0.518904	−0.507278
0.90	−0.518371	−0.519185	−0.517403	−0.513198	−0.506726	−0.498129
0.95	−0.493443	−0.496945	−0.497910	−0.496487	−0.492815	−0.487020
1.00	−0.468264	−0.474122	−0.477527	−0.478606	−0.477479	−0.474255
1.05	−0.443081	−0.450977	−0.456527	−0.459835	−0.461002	−0.460120
1.10	−0.418103	−0.427740	−0.435153	−0.440427	−0.443642	−0.444875
1.15	−0.393512	−0.404608	−0.413618	−0.420606	−0.425633	−0.428758
1.20	−0.369456	−0.381753	−0.392110	−0.400572	−0.407183	−0.411986
1.25	−0.346060	−0.359318	−0.370789	−0.380501	−0.388481	−0.394755
1.30	−0.323424	−0.337423	−0.349794	−0.360546	−0.369689	−0.377239
1.35	−0.301625	−0.316168	−0.329241	−0.340837	−0.350955	−0.359593
1.40	−0.280722	−0.295631	−0.309226	−0.321488	−0.332402	−0.341956
1.45	−0.260758	−0.275873	−0.289829	−0.302593	−0.314139	−0.324446
1.50	−0.241759	−0.256940	−0.271111	−0.284229	−0.296258	−0.307167
1.55	−0.223741	−0.238865	−0.253121	−0.266459	−0.278835	−0.290208
1.60	−0.206706	−0.221666	−0.235894	−0.249333	−0.261932	−0.273644
1.65	−0.190647	−0.205353	−0.219454	−0.232889	−0.245600	−0.257537
1.70	−0.175552	−0.189926	−0.203815	−0.217153	−0.229879	−0.241937
1.75	−0.161400	−0.175378	−0.188982	−0.202142	−0.214796	−0.226885
1.80	−0.148163	−0.161694	−0.174952	−0.187867	−0.200373	−0.212410
1.85	−0.135813	−0.148855	−0.161717	−0.174328	−0.186621	−0.198536
1.90	−0.124315	−0.136837	−0.149262	−0.161521	−0.173546	−0.185277
1.95	−0.113634	−0.125613	−0.137570	−0.149437	−0.161148	−0.172642
2.00	−0.103731	−0.115152	−0.126618	−0.138061	−0.149420	−0.160632
2.05	−0.094567	−0.105422	−0.116380	−0.127377	−0.138351	−0.149244
2.10	−0.086104	−0.096390	−0.106830	−0.117363	−0.127930	−0.138473
2.15	−0.078301	−0.088022	−0.097939	−0.107996	−0.118137	−0.128307
2.20	−0.071120	−0.080282	−0.089677	−0.099252	−0.108955	−0.118733
2.25	−0.064521	−0.073135	−0.082012	−0.091104	−0.100361	−0.109734
2.30	−0.058467	−0.066547	−0.074915	−0.083525	−0.092333	−0.101293
2.35	−0.052923	−0.060484	−0.068352	−0.076487	−0.084847	−0.093388
2.40	−0.047851	−0.054912	−0.062295	−0.069963	−0.077878	−0.086000
2.45	−0.043220	−0.049800	−0.056713	−0.063924	−0.071401	−0.079106
2.50	−0.038996	−0.045116	−0.051575	−0.058344	−0.065390	−0.072684
2.55	−0.035149	−0.040830	−0.046854	−0.053194	−0.059822	−0.066710
2.60	−0.031649	−0.036914	−0.042521	−0.048448	−0.054670	−0.061163
2.65	−0.028471	−0.033341	−0.038551	−0.044081	−0.049911	−0.056019
2.70	−0.025586	−0.030084	−0.034917	−0.040068	−0.045521	−0.051255
2.75	−0.022973	−0.027120	−0.031595	−0.036385	−0.041475	−0.046850
2.80	−0.020607	−0.024425	−0.028562	−0.033009	−0.037753	−0.042782

Table 1.1 (continued)

				t		
v	0.65	0.70	0.75	0.80	0.85	0.90
2.85	−0.018468	−0.021977	−0.025797	−0.029918	−0.034333	−0.039029
2.90	−0.016537	−0.019757	−0.023278	−0.027092	−0.031194	−0.035573
2.95	−0.014794	−0.017746	−0.020986	−0.024511	−0.028316	−0.032393
3.00	−0.013224	−0.015926	−0.018904	−0.022157	−0.025681	−0.029471
3.05	−0.011811	−0.014280	−0.017013	−0.020011	−0.023270	−0.026789
3.10	−0.010540	−0.012794	−0.015299	−0.018057	−0.021068	−0.024330
3.15	−0.009398	−0.011453	−0.013746	−0.016281	−0.019059	−0.022078
3.20	−0.008373	−0.010244	−0.012341	−0.014668	−0.017226	−0.020018
3.25	−0.007454	−0.009156	−0.011071	−0.013204	−0.015558	−0.018135
3.30	−0.006631	−0.008177	−0.009923	−0.011876	−0.014039	−0.016416
3.35	−0.005894	−0.007297	−0.008888	−0.010674	−0.012659	−0.014848
3.40	−0.005235	−0.006506	−0.007954	−0.009586	−0.011406	−0.013420
3.45	−0.004646	−0.005797	−0.007114	−0.008602	−0.010269	−0.012119
3.50	−0.004121	−0.005162	−0.006357	−0.007714	−0.009238	−0.010937
3.55	−0.003652	−0.004593	−0.005677	−0.006912	−0.008305	−0.009862
3.60	−0.003235	−0.004084	−0.005066	−0.006189	−0.007461	−0.008887
3.65	−0.002863	−0.003628	−0.004518	−0.005538	−0.006698	−0.008002
3.70	−0.002532	−0.003221	−0.004026	−0.004952	−0.006008	−0.007200
3.75	−0.002238	−0.002858	−0.003585	−0.004425	−0.005386	−0.006474
3.80	−0.001976	−0.002534	−0.003190	−0.003951	−0.004825	−0.005818
3.85	−0.001745	−0.002246	−0.002837	−0.003526	−0.004319	−0.005224
3.90	−0.001539	−0.001989	−0.002521	−0.003144	−0.003864	−0.004688
3.95	−0.001357	−0.001760	−0.002239	−0.002802	−0.003455	−0.004204
4.00	−0.001195	−0.001556	−0.001988	−0.002496	−0.003087	−0.003767
4.05	−0.001052	−0.001376	−0.001763	−0.002221	−0.002756	−0.003374
4.10	−0.000926	−0.001215	−0.001563	−0.001976	−0.002459	−0.003020
4.15	−0.000814	−0.001073	−0.001385	−0.001756	−0.002193	−0.002701
4.20	−0.000715	−0.000946	−0.001226	−0.001560	−0.001955	−0.002415
4.25	−0.000628	−0.000834	−0.001085	−0.001385	−0.001741	−0.002157
4.30	−0.000552	−0.000735	−0.000959	−0.001229	−0.001550	−0.001926
4.35	−0.000484	−0.000647	−0.000848	−0.001090	−0.001379	−0.001719
4.40	−0.000424	−0.000570	−0.000749	−0.000966	−0.001226	−0.001533
4.45	−0.000372	−0.000501	−0.000661	−0.000856	−0.001089	−0.001366
4.50	−0.000326	−0.000441	−0.000583	−0.000758	−0.000968	−0.001217
4.55	−0.000285	−0.000387	−0.000515	−0.000670	−0.000859	−0.001083
4.60	−0.000249	−0.000340	−0.000453	−0.000593	−0.000762	−0.000964
4.65	−0.000218	−0.000298	−0.000399	−0.000524	−0.000675	−0.000857
4.70	−0.000191	−0.000262	−0.000352	−0.000463	−0.000599	−0.000762
4.75	−0.000166	−0.000230	−0.000309	−0.000409	−0.000530	−0.000677
4.80	−0.000145	−0.000201	−0.000272	−0.000361	−0.000469	−0.000601
4.85	−0.000127	−0.000176	−0.000239	−0.000318	−0.000415	−0.000533
4.90	−0.000111	−0.000154	−0.000210	−0.000280	−0.000367	−0.000473

Table 1.1 (continued)

			t			
v	0.65	0.70	0.75	0.80	0.85	0.90
4.95	-0.000096	-0.000135	-0.000185	-0.000247	-0.000324	-0.000419
5.00	-0.000084	-0.000118	-0.000162	-0.000218	-0.000287	-0.000371

			t			
v	0.95	1.0	2.0	3.0	4.0	5.0
0.00	0.075540	0.138634	0.801696	0.591955	-0.026611	-0.484618
0.05	-0.008240	0.054550	0.762004	0.615226	0.021489	-0.458634
0.10	-0.084601	-0.022786	0.719669	0.634155	0.068344	-0.430381
0.15	-0.153586	-0.093333	0.675192	0.648836	0.113706	-0.400104
0.20	-0.215321	-0.157133	0.629047	0.659388	0.157351	-0.368051
0.25	-0.269996	-0.214297	0.581687	0.665957	0.199078	-0.334472
0.30	-0.317858	-0.264997	0.533535	0.668704	0.238710	-0.299617
0.35	-0.359200	-0.309451	0.484987	0.667811	0.276091	-0.263733
0.40	-0.394347	-0.347921	0.436409	0.663473	0.311091	-0.227061
0.45	-0.423655	-0.380700	0.388137	0.655898	0.343599	-0.189840
0.50	-0.447495	-0.408104	0.340475	0.645303	0.373529	-0.152300
0.55	-0.466252	-0.430467	0.293700	0.631911	0.400815	-0.114665
0.60	-0.480317	-0.448135	0.248055	0.615950	0.425410	-0.077146
0.65	-0.490080	-0.461459	0.203757	0.597654	0.447289	-0.039949
0.70	-0.495928	-0.470792	0.160993	0.577252	0.466444	-0.003267
0.75	-0.498238	-0.476484	0.119923	0.554977	0.482884	0.032721
0.80	-0.497377	-0.478876	0.080681	0.531057	0.496636	0.067843
0.85	-0.493699	-0.478302	0.043375	0.505715	0.507740	0.101944
0.90	-0.487540	-0.475084	0.008090	0.479171	0.516250	0.134879
0.95	-0.479220	-0.469528	-0.025109	0.451636	0.522235	0.166518
1.00	-0.469039	-0.461929	-0.056181	0.423313	0.525773	0.196744
1.05	-0.457278	-0.452560	-0.085100	0.394399	0.526952	0.225451
1.10	-0.444197	-0.441681	-0.111860	0.365080	0.525872	0.252551
1.15	-0.430039	-0.429534	-0.136469	0.335531	0.522636	0.277963
1.20	-0.415024	-0.416340	-0.158951	0.305918	0.517358	0.301625
1.25	-0.399353	-0.402306	-0.179340	0.276395	0.510155	0.323482
1.30	-0.383209	-0.387620	-0.197683	0.247107	0.501149	0.343495
1.35	-0.366757	-0.372453	-0.214036	0.218187	0.490465	0.361636
1.40	-0.350142	-0.356958	-0.228464	0.189755	0.478231	0.377888
1.45	-0.333495	-0.341275	-0.241037	0.161921	0.464575	0.392243
1.50	-0.316930	-0.325526	-0.251833	0.134786	0.449628	0.404707
1.55	-0.300547	-0.309821	-0.260934	0.108437	0.433518	0.415293
1.60	-0.284430	-0.294254	-0.268426	0.082951	0.416373	0.424023
1.65	-0.268654	-0.278909	-0.274396	0.058398	0.398320	0.430929
1.70	-0.253278	-0.263857	-0.278934	0.034833	0.379482	0.436050
1.75	-0.238354	-0.249156	-0.282131	0.012305	0.359980	0.439432
1.80	-0.223923	-0.234859	-0.284078	-0.009147	0.339932	0.441127
1.85	-0.210015	-0.221004	-0.284865	-0.029493	0.319451	0.441194

Table 1.1 (continued)

				t		
v	0.95	1.0	2.0	3.0	4.0	5.0
1.90	−0.196655	−0.207626	−0.284581	−0.048710	0.298646	0.439696
1.95	−0.183860	−0.194748	−0.283315	−0.066784	0.277621	0.436702
2.00	−0.171640	−0.182390	−0.281152	−0.083706	0.256475	0.432284
2.05	−0.160000	−0.170563	−0.278175	−0.099477	0.235303	0.426517
2.10	−0.148939	−0.159273	−0.274465	−0.114102	0.214193	0.419479
2.15	−0.138454	−0.148526	−0.270098	−0.127590	0.193228	0.411250
2.20	−0.128537	−0.138316	−0.265149	−0.139958	0.172486	0.401913
2.25	−0.119176	−0.128639	−0.259689	−0.151226	0.152039	0.391550
2.30	−0.110359	−0.119488	−0.253784	−0.161419	0.131952	0.380244
2.35	−0.102070	−0.110850	−0.247498	−0.170564	0.112286	0.368080
2.40	−0.094291	−0.102712	−0.240891	−0.178693	0.093096	0.355141
2.45	−0.087005	−0.095061	−0.234018	−0.185839	0.074431	0.341508
2.50	−0.080191	−0.087878	−0.226931	−0.192038	0.056335	0.327264
2.55	−0.073829	−0.081148	−0.219680	−0.197329	0.038846	0.312489
2.60	−0.067899	−0.074851	−0.212309	−0.201751	0.021999	0.297260
2.65	−0.062380	−0.068970	−0.204860	−0.205346	0.005820	0.281654
2.70	−0.057250	−0.063484	−0.197369	−0.208154	−0.009665	0.265745
2.75	−0.052490	−0.058376	−0.189873	−0.210217	−0.024437	0.249604
2.80	−0.048078	−0.053625	−0.182401	−0.211580	−0.038483	0.233299
2.85	−0.043994	−0.049213	−0.174983	−0.212283	−0.051790	0.216896
2.90	−0.040219	−0.045120	−0.167644	−0.212370	−0.064353	0.200458
2.95	−0.036735	−0.041330	−0.160406	−0.211882	−0.076167	0.184044
3.00	−0.033521	−0.037823	−0.153290	−0.210861	−0.087234	0.167710
3.05	−0.030562	−0.034583	−0.146311	−0.209347	−0.097556	0.151509
3.10	−0.027840	−0.031592	−0.139487	−0.207381	−0.107139	0.135490
3.15	−0.025338	−0.028835	−0.132829	−0.204999	−0.115994	0.119699
3.20	−0.023042	−0.026297	−0.126348	−0.202241	−0.124130	0.104180
3.25	−0.020936	−0.023961	−0.120054	−0.199143	−0.131563	0.088970
3.30	−0.019008	−0.021816	−0.113952	−0.195738	−0.138307	0.074106
3.35	−0.017243	−0.019846	−0.108050	−0.192061	−0.144379	0.059620
3.40	−0.015630	−0.018040	−0.102351	−0.188144	−0.149800	0.045542
3.45	−0.014157	−0.016386	−0.096857	−0.184017	−0.154588	0.031897
3.50	−0.012813	−0.014872	−0.091571	−0.179709	−0.158767	0.018708
3.55	−0.011588	−0.013488	−0.086492	−0.175248	−0.162358	0.005995
3.60	−0.010473	−0.012223	−0.081620	−0.170659	−0.165385	−0.006226
3.65	−0.009457	−0.011069	−0.076953	−0.165968	−0.167871	−0.017939
3.70	−0.008535	−0.010017	−0.072490	−0.161196	−0.169842	−0.029134
3.75	−0.007697	−0.009058	−0.068227	−0.156365	−0.171322	−0.039803
3.80	−0.006936	−0.008186	−0.064160	−0.151495	−0.172337	−0.049938
3.85	−0.006246	−0.007392	−0.060287	−0.146604	−0.172911	−0.059536
3.90	−0.005621	−0.006671	−0.056601	−0.141711	−0.173069	−0.068595
3.95	−0.005056	−0.006016	−0.053098	−0.136829	−0.172837	−0.077116
4.00	−0.004544	−0.005422	−0.049773	−0.131974	−0.172240	−0.085100

Table 1.1 (continued)

			t			
v	0.95	1.0	2.0	3.0	4.0	5.0
4.05	−0.004081	−0.004883	−0.046621	−0.127158	−0.171301	−0.092552
4.10	−0.003663	−0.004395	−0.043636	−0.122394	−0.170045	−0.099477
4.15	−0.003286	−0.003954	−0.040811	−0.117693	−0.168495	−0.105883
4.20	−0.002946	−0.003554	−0.038142	−0.113062	−0.166674	−0.111779
4.25	−0.002640	−0.003193	−0.035621	−0.108512	−0.164605	−0.117174
4.30	−0.002364	−0.002867	−0.033244	−0.104050	−0.162310	−0.122079
4.35	−0.002115	−0.002572	−0.031004	−0.099682	−0.159808	−0.126507
4.40	−0.001892	−0.002307	−0.028895	−0.095414	−0.157121	−0.130471
4.45	−0.001691	−0.002068	−0.026912	−0.091251	−0.154268	−0.133984
4.50	−0.001510	−0.001852	−0.025048	−0.087196	−0.151267	−0.137062
4.55	−0.001349	−0.001658	−0.023298	−0.083253	−0.148137	−0.139718
4.60	−0.001203	−0.001484	−0.021657	−0.079425	−0.144894	−0.141970
4.65	−0.001073	−0.001327	−0.020118	−0.075713	−0.141556	−0.143833
4.70	−0.000956	−0.001186	−0.018677	−0.072120	−0.138136	−0.145323
4.75	−0.000852	−0.001059	−0.017329	−0.068645	−0.134650	−0.146457
4.80	−0.000759	−0.000946	−0.016068	−0.065289	−0.131112	−0.147252
4.85	−0.000675	−0.000844	−0.014890	−0.062051	−0.127534	−0.147725
4.90	−0.000600	−0.000752	−0.013790	−0.058933	−0.123929	−0.147892
4.95	−0.000534	−0.000671	−0.012764	−0.055932	−0.120308	−0.147770
5.00	−0.000474	−0.000598	−0.011808	−0.053047	−0.116681	−0.147377

Table 1.2: Second derivatives with respect to the order of the Bessel functions of the first kind $\partial^2 J_\nu(t)/\partial \nu^2$.

			t			
v	0.05	0.10	0.15	0.20	0.25	0.30
0.00	8.02736	4.17671	2.36010	1.26439	0.524520	−0.008416
0.05	7.38071	4.13885	2.52650	1.51839	0.817778	0.300277
0.10	6.70037	4.01641	2.60847	1.69643	1.04474	0.551842
0.15	6.01777	3.83350	2.62316	1.81072	1.21409	0.752412
0.20	5.35506	3.60969	2.58533	1.87220	1.33393	0.907921
0.25	4.72710	3.36064	2.50753	1.89061	1.41171	1.02401
0.30	4.14320	3.09865	2.40036	1.87451	1.45420	1.10598
0.35	3.60842	2.83321	2.27259	1.83132	1.46750	1.15874
0.40	3.12472	2.57148	2.13147	1.76743	1.45701	1.18678
0.45	2.69183	2.31868	1.98283	1.68828	1.42752	1.19417
0.50	2.30790	2.07846	1.83133	1.59845	1.38319	1.18457
0.55	1.97008	1.85322	1.68061	1.50175	1.32761	1.16123
0.60	1.67492	1.64438	1.53345	1.40129	1.26386	1.12703

Table 1.2 (continued)

			t			
v	0.05	0.10	0.15	0.20	0.25	0.30
0.65	1.41862	1.45258	1.39190	1.29960	1.19454	1.08445
0.70	1.19732	1.27788	1.25742	1.19869	1.12183	1.03567
0.75	1.00722	1.11990	1.13098	1.10010	1.04752	0.982521
0.80	0.844683	0.977963	1.01316	1.00503	0.973078	0.926583
0.85	0.706302	0.851174	0.904218	0.914322	0.899658	0.869163
0.90	0.588960	0.738508	0.804179	0.828575	0.828180	0.811345
0.95	0.489824	0.638867	0.712875	0.748161	0.759336	0.754010
1.00	0.406358	0.551132	0.629999	0.673277	0.693636	0.697862
1.05	0.336313	0.474190	0.555148	0.603975	0.631435	0.643452
1.10	0.277710	0.406967	0.487854	0.540198	0.572956	0.591196
1.15	0.228820	0.348440	0.427606	0.481801	0.518317	0.541400
1.20	0.188144	0.297650	0.373876	0.428578	0.467550	0.494271
1.25	0.154389	0.253709	0.326132	0.380275	0.420618	0.449936
1.30	0.126447	0.215804	0.283845	0.336608	0.377428	0.408459
1.35	0.103369	0.183194	0.246513	0.297276	0.337851	0.369844
1.40	0.084353	0.155212	0.213651	0.261967	0.301724	0.334058
1.45	0.068716	0.131261	0.184806	0.230371	0.268866	0.301029
1.50	0.055885	0.110807	0.159552	0.202180	0.239080	0.270661
1.55	0.045376	0.093380	0.137498	0.177098	0.212164	0.242839
1.60	0.036786	0.078562	0.118285	0.154841	0.187913	0.217432
1.65	0.029777	0.065989	0.101583	0.135141	0.166124	0.194303
1.70	0.024068	0.055342	0.087097	0.117744	0.146598	0.173309
1.75	0.019425	0.046343	0.074559	0.102417	0.129143	0.154304
1.80	0.015657	0.038750	0.063727	0.088943	0.113576	0.137144
1.85	0.012602	0.032355	0.054388	0.077121	0.099725	0.121688
1.90	0.010130	0.026978	0.046351	0.066771	0.087426	0.107799
1.95	0.008133	0.022464	0.039446	0.057725	0.076529	0.095346
2.00	0.006521	0.018681	0.033523	0.049834	0.066891	0.084203
2.05	0.005222	0.015516	0.028453	0.042963	0.058385	0.074254
2.10	0.004177	0.012870	0.024118	0.036990	0.050889	0.065386
2.15	0.003338	0.010663	0.020418	0.031806	0.044297	0.057498
2.20	0.002664	0.008824	0.017264	0.027314	0.038508	0.050494
2.25	0.002123	0.007294	0.014580	0.023427	0.033434	0.044285
2.30	0.001691	0.006022	0.012299	0.020069	0.028992	0.038790
2.35	0.001345	0.004966	0.010363	0.017173	0.025110	0.033935
2.40	0.001069	0.004091	0.008722	0.014678	0.021722	0.029652
2.45	0.000848	0.003367	0.007333	0.012531	0.018770	0.025879
2.50	0.000673	0.002768	0.006158	0.010686	0.016201	0.022560
2.55	0.000533	0.002273	0.005167	0.009104	0.013968	0.019645
2.60	0.000422	0.001865	0.004330	0.007747	0.012030	0.017087
2.65	0.000334	0.001529	0.003626	0.006586	0.010350	0.014847
2.70	0.000264	0.001252	0.003033	0.005593	0.008895	0.012887
2.75	0.000208	0.001024	0.002534	0.004746	0.007638	0.011174

Table 1.2 (continued)

			t			
v	0.05	0.10	0.15	0.20	0.25	0.30
2.80	0.000164	0.000837	0.002116	0.004023	0.006551	0.009679
2.85	0.000129	0.000684	0.001765	0.003407	0.005614	0.008376
2.90	0.000102	0.000558	0.001471	0.002882	0.004807	0.007242
2.95	0.000080	0.000455	0.001225	0.002436	0.004111	0.006255
3.00	0.000063	0.000370	0.001019	0.002057	0.003513	0.005397
3.05	0.000049	0.000301	0.000847	0.001736	0.003000	0.004653
3.10	0.000039	0.000245	0.000703	0.001464	0.002559	0.004008
3.15	0.000030	0.000199	0.000583	0.001233	0.002181	0.003450
3.20	0.000024	0.000162	0.000484	0.001038	0.001857	0.002966
3.25	0.000019	0.000131	0.000401	0.000873	0.001581	0.002549
3.30	0.000015	0.000106	0.000332	0.000733	0.001344	0.002188
3.35	0.000011	0.000086	0.000274	0.000616	0.001142	0.001877
3.40	0.000009	0.000070	0.000227	0.000517	0.000969	0.001608
3.45	0.000007	0.000056	0.000187	0.000433	0.000822	0.001377
3.50	0.000005	0.000046	0.000155	0.000363	0.000697	0.001179
3.55	0.000004	0.000037	0.000127	0.000304	0.000590	0.001008
3.60	0.000003	0.000030	0.000105	0.000254	0.000499	0.000861
3.65	0.000003	0.000024	0.000087	0.000212	0.000422	0.000735
3.70	0.000002	0.000019	0.000071	0.000177	0.000357	0.000627
3.75	0.000002	0.000016	0.000059	0.000148	0.000301	0.000535
3.80	0.000001	0.000013	0.000048	0.000124	0.000254	0.000456
3.85	0.000001	0.000010	0.000040	0.000103	0.000214	0.000388
3.90	0.000001	0.000008	0.000032	0.000086	0.000181	0.000330
3.95	0.000001	0.000006	0.000027	0.000071	0.000152	0.000281
4.00		0.000005	0.000022	0.000059	0.000128	0.000238
4.05		0.000004	0.000018	0.000049	0.000108	0.000202
4.10		0.000003	0.000015	0.000041	0.000091	0.000172
4.15		0.000003	0.000012	0.000034	0.000076	0.000146
4.20		0.000002	0.000010	0.000028	0.000064	0.000123
4.25		0.000002	0.000008	0.000023	0.000054	0.000105
4.30		0.000001	0.000007	0.000019	0.000045	0.000089
4.35		0.000001	0.000005	0.000016	0.000038	0.000075
4.40		0.000001	0.000004	0.000013	0.000032	0.000063
4.45		0.000001	0.000004	0.000011	0.000026	0.000053
4.50		0.000001	0.000003	0.000009	0.000022	0.000045
4.55			0.000002	0.000008	0.000018	0.000038
4.60			0.000002	0.000006	0.000015	0.000032
4.65			0.000002	0.000005	0.000013	0.000027
4.70			0.000001	0.000004	0.000011	0.000023
4.75			0.000001	0.000003	0.000009	0.000019
4.80			0.000001	0.000003	0.000007	0.000016
4.85			0.000001	0.000002	0.000006	0.000014
4.90			0.000001	0.000002	0.000005	0.000011

Table 1.2 (continued)

			t			
v	0.05	0.10	0.15	0.20	0.25	0.30
4.95				0.000002	0.000004	0.000010
5.00				0.000001	0.000004	0.000008

			t			
v	0.35	0.40	0.45	0.50	0.55	0.60
0.00	−0.408281	−0.716458	−0.958174	−1.14979	−1.30244	−1.42399
0.05	−0.097115	−0.410317	−0.661555	−0.865433	−1.03201	−1.16846
0.10	0.165197	−0.145693	−0.399980	−0.610415	−0.785878	−0.932782
0.15	0.382884	0.080204	−0.171761	−0.383898	−0.563878	−0.717297
0.20	0.560218	0.270355	0.025069	−0.184716	−0.365472	−0.521973
0.25	0.701413	0.427835	0.192670	−0.011447	−0.189840	−0.346477
0.30	0.810555	0.555733	0.333315	0.137506	−0.035939	−0.190227
0.35	0.891539	0.657094	0.449328	0.263863	0.097438	−0.052445
0.40	0.948035	0.734864	0.543032	0.369409	0.211615	0.067942
0.45	0.983456	0.791858	0.616708	0.455955	0.307990	0.171529
0.50	1.00094	0.830731	0.672562	0.525299	0.387997	0.259867
0.55	1.00337	0.853957	0.712698	0.579201	0.453077	0.333965
0.60	0.993305	0.863823	0.739101	0.619356	0.504657	0.394994
0.65	0.973078	0.862420	0.753624	0.647382	0.544123	0.444123
0.70	0.944737	0.851643	0.757984	0.664804	0.572809	0.482503
0.75	0.910087	0.833196	0.753752	0.673045	0.591987	0.511248
0.80	0.870702	0.808600	0.742358	0.673425	0.602852	0.531426
0.85	0.827940	0.779195	0.725090	0.667154	0.606520	0.544052
0.90	0.782959	0.746160	0.703101	0.655335	0.604027	0.550082
0.95	0.736742	0.710515	0.677413	0.638962	0.596323	0.550406
1.00	0.690107	0.673140	0.648928	0.618930	0.584272	0.545849
1.05	0.643729	0.634784	0.618430	0.596032	0.568659	0.537168
1.10	0.598156	0.596077	0.586597	0.570971	0.550188	0.525054
1.15	0.553823	0.557542	0.554012	0.544359	0.529485	0.510133
1.20	0.511067	0.519610	0.521166	0.516729	0.507106	0.492967
1.25	0.470139	0.482625	0.488472	0.488540	0.483537	0.474058
1.30	0.431221	0.446858	0.456268	0.460180	0.459201	0.453851
1.35	0.394431	0.412517	0.424832	0.431977	0.434464	0.432736
1.40	0.359837	0.379753	0.394380	0.404202	0.409636	0.411053
1.45	0.327463	0.348671	0.365084	0.377076	0.384981	0.389097
1.50	0.297299	0.319334	0.337068	0.350778	0.360717	0.367118
1.55	0.269309	0.291771	0.310421	0.325445	0.337023	0.345331
1.60	0.243431	0.265987	0.285199	0.301180	0.314044	0.323911
1.65	0.219588	0.241958	0.261434	0.278058	0.291891	0.303004
1.70	0.197692	0.219648	0.239130	0.256128	0.270651	0.282728
1.75	0.177644	0.199001	0.218278	0.235416	0.250385	0.263175
1.80	0.159339	0.179955	0.198851	0.215932	0.231133	0.244416
1.85	0.142671	0.162435	0.180809	0.197668	0.212920	0.226500

Table 1.2 (continued)

			t			
v	0.35	0.40	0.45	0.50	0.55	0.60
1.90	0.127531	0.146364	0.164105	0.180605	0.195753	0.209463
1.95	0.113813	0.131662	0.148683	0.164714	0.179628	0.193323
2.00	0.101411	0.118243	0.134483	0.149959	0.164532	0.178088
2.05	0.090224	0.106025	0.121441	0.136294	0.150439	0.163754
2.10	0.080152	0.094925	0.109491	0.123673	0.137322	0.150310
2.15	0.071103	0.084862	0.098567	0.112045	0.125145	0.137738
2.20	0.062988	0.075757	0.088603	0.101355	0.113868	0.126012
2.25	0.055725	0.067536	0.079532	0.091551	0.103450	0.115103
2.30	0.049234	0.060126	0.071291	0.082577	0.093847	0.104979
2.35	0.043445	0.053459	0.063818	0.074380	0.085015	0.095606
2.40	0.038288	0.047470	0.057054	0.066906	0.076907	0.086945
2.45	0.033703	0.042101	0.050942	0.060105	0.069480	0.078961
2.50	0.029632	0.037294	0.045428	0.053927	0.062688	0.071613
2.55	0.026023	0.032996	0.040463	0.048325	0.056488	0.064865
2.60	0.022828	0.029161	0.035998	0.043252	0.050839	0.058679
2.65	0.020003	0.025743	0.031989	0.038666	0.045700	0.053018
2.70	0.017509	0.022700	0.028395	0.034528	0.041033	0.047845
2.75	0.015310	0.019996	0.025177	0.030797	0.036800	0.043127
2.80	0.013374	0.017595	0.022300	0.027440	0.032968	0.038830
2.85	0.011670	0.015467	0.019731	0.024423	0.029502	0.034923
2.90	0.010174	0.013583	0.017440	0.021715	0.026373	0.031375
2.95	0.008861	0.011916	0.015400	0.019288	0.023551	0.028157
3.00	0.007710	0.010444	0.013586	0.017115	0.021010	0.025243
3.05	0.006703	0.009145	0.011973	0.015172	0.018724	0.022608
3.10	0.005822	0.008001	0.010542	0.013437	0.016671	0.020228
3.15	0.005052	0.006993	0.009274	0.011889	0.014829	0.018082
3.20	0.004380	0.006107	0.008151	0.010510	0.013178	0.016147
3.25	0.003794	0.005328	0.007157	0.009282	0.011701	0.014406
3.30	0.003284	0.004645	0.006279	0.008191	0.010379	0.012842
3.35	0.002840	0.004046	0.005505	0.007222	0.009199	0.011437
3.40	0.002454	0.003521	0.004821	0.006362	0.008146	0.010176
3.45	0.002119	0.003062	0.004219	0.005599	0.007208	0.009047
3.50	0.001828	0.002661	0.003690	0.004924	0.006372	0.008037
3.55	0.001576	0.002310	0.003224	0.004327	0.005628	0.007133
3.60	0.001358	0.002004	0.002815	0.003800	0.004968	0.006326
3.65	0.001169	0.001738	0.002456	0.003334	0.004381	0.005605
3.70	0.001005	0.001505	0.002141	0.002923	0.003861	0.004963
3.75	0.000864	0.001303	0.001865	0.002560	0.003400	0.004391
3.80	0.000742	0.001127	0.001623	0.002241	0.002991	0.003882
3.85	0.000637	0.000974	0.001412	0.001961	0.002630	0.003430
3.90	0.000546	0.000842	0.001227	0.001714	0.002311	0.003028
3.95	0.000468	0.000726	0.001066	0.001497	0.002029	0.002671
4.00	0.000401	0.000627	0.000925	0.001307	0.001780	0.002354

Table 1.2 (continued)

			t			
v	0.35	0.40	0.45	0.50	0.55	0.60
4.05	0.000343	0.000540	0.000803	0.001140	0.001561	0.002074
4.10	0.000294	0.000465	0.000696	0.000994	0.001367	0.001825
4.15	0.000251	0.000401	0.000603	0.000866	0.001197	0.001606
4.20	0.000214	0.000345	0.000522	0.000753	0.001048	0.001412
4.25	0.000183	0.000296	0.000451	0.000656	0.000916	0.001240
4.30	0.000156	0.000255	0.000390	0.000570	0.000800	0.001089
4.35	0.000133	0.000219	0.000337	0.000495	0.000699	0.000955
4.40	0.000114	0.000188	0.000291	0.000430	0.000610	0.000838
4.45	0.000097	0.000161	0.000251	0.000373	0.000532	0.000734
4.50	0.000082	0.000138	0.000217	0.000324	0.000464	0.000643
4.55	0.000070	0.000118	0.000187	0.000280	0.000404	0.000562
4.60	0.000060	0.000101	0.000161	0.000243	0.000352	0.000492
4.65	0.000051	0.000087	0.000139	0.000210	0.000306	0.000430
4.70	0.000043	0.000074	0.000119	0.000182	0.000266	0.000376
4.75	0.000036	0.000063	0.000103	0.000157	0.000231	0.000328
4.80	0.000031	0.000054	0.000088	0.000136	0.000201	0.000286
4.85	0.000026	0.000046	0.000076	0.000117	0.000174	0.000250
4.90	0.000022	0.000039	0.000065	0.000101	0.000151	0.000218
4.95	0.000019	0.000034	0.000056	0.000088	0.000131	0.000190
5.00	0.000016	0.000029	0.000048	0.000075	0.000114	0.000165

			t			
v	0.65	0.70	0.75	0.80	0.85	0.90
0.00	−1.52020	−1.59536	−1.65278	−1.69505	−1.72423	−1.74197
0.05	−1.28009	−1.37090	−1.44399	−1.50180	−1.54628	−1.57904
0.10	−1.05590	−1.15887	−1.24452	−1.31511	−1.37246	−1.41807
0.15	−0.848357	−0.960294	−1.05564	−1.13644	−1.20434	−1.26070
0.20	−0.657837	−0.775873	−0.878303	−0.966918	−1.04319	−1.10833
0.25	−0.484397	−0.605981	−0.713128	−0.807379	−0.890005	−0.962069
0.30	−0.327838	−0.450734	−0.560493	−0.658408	−0.745552	−0.822825
0.35	−0.187743	−0.310020	−0.420545	−0.520368	−0.610364	−0.691278
0.40	−0.063528	−0.183545	−0.293237	−0.393421	−0.484785	−0.567920
0.45	0.045530	−0.070861	−0.178358	−0.277562	−0.368987	−0.453074
0.50	0.140253	0.028600	−0.075565	−0.172646	−0.262994	−0.346915
0.55	0.221536	0.115499	0.015595	−0.078405	−0.166705	−0.249488
0.60	0.290318	0.190565	0.095659	0.005524	−0.079913	−0.160725
0.65	0.347566	0.254573	0.165226	0.079581	−0.002324	−0.080466
0.70	0.394251	0.308327	0.224943	0.144265	0.066425	−0.008469
0.75	0.431335	0.352643	0.275486	0.200120	0.126755	0.055567
0.80	0.459758	0.388334	0.317546	0.247719	0.179126	0.111998
0.85	0.480430	0.416201	0.351816	0.287651	0.224026	0.161217
0.90	0.494219	0.437024	0.378985	0.320514	0.261966	0.203645
0.95	0.501947	0.451551	0.399725	0.346904	0.293460	0.239724

Table 1.2 (continued)

v	0.65	0.70	0.75	0.80	0.85	0.90
1.00	0.504388	0.460497	0.414688	0.367404	0.319029	0.269903
1.05	0.502263	0.464538	0.424499	0.382586	0.339185	0.294638
1.10	0.496238	0.464307	0.429752	0.393000	0.354429	0.314379
1.15	0.486925	0.460395	0.431006	0.399169	0.365248	0.329572
1.20	0.474883	0.453345	0.428787	0.401593	0.372109	0.340649
1.25	0.460617	0.443658	0.423581	0.400739	0.375455	0.348026
1.30	0.444582	0.431791	0.415838	0.397045	0.375708	0.352102
1.35	0.427184	0.418157	0.405972	0.390918	0.373263	0.353256
1.40	0.408783	0.403126	0.394357	0.382731	0.368486	0.351846
1.45	0.389696	0.387031	0.381336	0.372829	0.361720	0.348205
1.50	0.370200	0.370167	0.367213	0.361524	0.353278	0.342646
1.55	0.350533	0.352792	0.352263	0.349099	0.343448	0.335456
1.60	0.330900	0.335132	0.336728	0.335808	0.332492	0.326901
1.65	0.311474	0.317383	0.320821	0.321877	0.320646	0.317223
1.70	0.292399	0.299714	0.304729	0.307509	0.308122	0.306641
1.75	0.273795	0.282265	0.288613	0.292880	0.295110	0.295356
1.80	0.255759	0.265155	0.272611	0.278144	0.281777	0.283544
1.85	0.238364	0.248483	0.256841	0.263435	0.268272	0.271367
1.90	0.221670	0.232326	0.241400	0.248869	0.254723	0.258963
1.95	0.205717	0.216748	0.226368	0.234541	0.241242	0.246457
2.00	0.190534	0.201795	0.211810	0.220532	0.227923	0.233956
2.05	0.176136	0.187501	0.197778	0.206909	0.214847	0.221554
2.10	0.162530	0.173890	0.184310	0.193726	0.202081	0.209330
2.15	0.149713	0.160973	0.171434	0.181024	0.189681	0.197351
2.20	0.137674	0.148755	0.159167	0.168834	0.177689	0.185673
2.25	0.126397	0.137233	0.147520	0.157180	0.166141	0.174342
2.30	0.115863	0.126398	0.136496	0.146075	0.155062	0.163393
2.35	0.106046	0.116237	0.126092	0.135527	0.144471	0.152854
2.40	0.096918	0.106732	0.116298	0.125538	0.134377	0.142748
2.45	0.088451	0.097860	0.107104	0.116105	0.124788	0.133087
2.50	0.080612	0.089599	0.098493	0.107218	0.115703	0.123880
2.55	0.073371	0.081923	0.090447	0.098869	0.107119	0.115132
2.60	0.066693	0.074806	0.082945	0.091041	0.099028	0.106842
2.65	0.060547	0.068218	0.075964	0.083718	0.091419	0.099006
2.70	0.054900	0.062132	0.069481	0.076883	0.084281	0.091616
2.75	0.049720	0.056521	0.063471	0.070516	0.077597	0.084662
2.80	0.044977	0.051355	0.057911	0.064595	0.071353	0.078134
2.85	0.040640	0.046607	0.052776	0.059099	0.065529	0.072018
2.90	0.036681	0.042250	0.048040	0.054008	0.060108	0.066298
2.95	0.033072	0.038259	0.043681	0.049298	0.055071	0.060958
3.00	0.029786	0.034607	0.039673	0.044948	0.050398	0.055983
3.05	0.026800	0.031271	0.035994	0.040938	0.046070	0.051356
3.10	0.024088	0.028228	0.032623	0.037246	0.042067	0.047058

Table 1.2 (continued)

			t			
v	0.65	0.70	0.75	0.80	0.85	0.90
3.15	0.021630	0.025456	0.029537	0.033851	0.038372	0.043074
3.20	0.019404	0.022933	0.026716	0.030734	0.034965	0.039385
3.25	0.017390	0.020640	0.024141	0.027877	0.031828	0.035975
3.30	0.015571	0.018559	0.021794	0.025261	0.028944	0.032827
3.35	0.013930	0.016673	0.019656	0.022868	0.026296	0.029925
3.40	0.012451	0.014964	0.017712	0.020683	0.023868	0.027253
3.45	0.011119	0.013419	0.015945	0.018689	0.021643	0.024796
3.50	0.009921	0.012023	0.014342	0.016873	0.019608	0.022540
3.55	0.008844	0.010763	0.012890	0.015220	0.017749	0.020470
3.60	0.007878	0.009627	0.011574	0.013717	0.016052	0.018575
3.65	0.007012	0.008604	0.010384	0.012351	0.014504	0.016840
3.70	0.006236	0.007684	0.009309	0.011113	0.013095	0.015254
3.75	0.005542	0.006856	0.008338	0.009990	0.011813	0.013805
3.80	0.004921	0.006113	0.007463	0.008974	0.010648	0.012484
3.85	0.004366	0.005446	0.006675	0.008055	0.009590	0.011281
3.90	0.003871	0.004848	0.005965	0.007224	0.008631	0.010185
3.95	0.003430	0.004313	0.005326	0.006474	0.007761	0.009189
4.00	0.003037	0.003834	0.004753	0.005798	0.006974	0.008283
4.05	0.002686	0.003406	0.004238	0.005189	0.006262	0.007461
4.10	0.002375	0.003023	0.003776	0.004640	0.005618	0.006716
4.15	0.002098	0.002682	0.003362	0.004146	0.005037	0.006041
4.20	0.001853	0.002377	0.002992	0.003702	0.004513	0.005430
4.25	0.001635	0.002106	0.002660	0.003304	0.004041	0.004877
4.30	0.001441	0.001864	0.002364	0.002946	0.003616	0.004378
4.35	0.001270	0.001649	0.002099	0.002625	0.003233	0.003926
4.40	0.001118	0.001458	0.001863	0.002338	0.002889	0.003519
4.45	0.000984	0.001288	0.001652	0.002081	0.002579	0.003153
4.50	0.000865	0.001138	0.001464	0.001851	0.002302	0.002822
4.55	0.000761	0.001004	0.001297	0.001645	0.002053	0.002525
4.60	0.000668	0.000885	0.001148	0.001461	0.001829	0.002257
4.65	0.000587	0.000780	0.001016	0.001297	0.001629	0.002017
4.70	0.000515	0.000687	0.000898	0.001151	0.001450	0.001801
4.75	0.000451	0.000605	0.000793	0.001020	0.001290	0.001607
4.80	0.000395	0.000532	0.000701	0.000904	0.001147	0.001433
4.85	0.000346	0.000468	0.000618	0.000801	0.001019	0.001277
4.90	0.000303	0.000411	0.000545	0.000709	0.000905	0.001138
4.95	0.000265	0.000361	0.000481	0.000627	0.000803	0.001013
5.00	0.000232	0.000317	0.000424	0.000554	0.000712	0.000901

			t			
v	0.95	1.00	2.00	3.00	4.00	5.00
0.00	−1.74967	−1.74845	−0.763993	0.509350	0.972687	0.495369
0.05	−1.60140	−1.61447	−0.821966	0.421736	0.950402	0.543185

Table 1.2 (continued)

v	0.95	1.00	2.00	3.00	4.00	5.00
				t		
0.10	−1.45317	−1.47881	−0.869750	0.335734	0.922949	0.586134
0.15	−1.30667	−1.34321	−0.907771	0.251892	0.890780	0.624135
0.20	−1.16336	−1.20917	−0.936509	0.170697	0.854357	0.657145
0.25	−1.02447	−1.07798	−0.956488	0.092580	0.814145	0.685161
0.30	−0.890998	−0.950733	−0.968268	0.017915	0.770609	0.708213
0.35	−0.763746	−0.828322	−0.972430	−0.052979	0.724211	0.726363
0.40	−0.643333	−0.711469	−0.969573	−0.119835	0.675407	0.739706
0.45	−0.530210	−0.600734	−0.960306	−0.182442	0.624639	0.748364
0.50	−0.424681	−0.496532	−0.945236	−0.240634	0.572341	0.752485
0.55	−0.326919	−0.399147	−0.924968	−0.294293	0.518928	0.752237
0.60	−0.236981	−0.308749	−0.900094	−0.343344	0.464800	0.747813
0.65	−0.154827	−0.225403	−0.871194	−0.387756	0.410335	0.739421
0.70	−0.080331	−0.149087	−0.838826	−0.427530	0.355892	0.727284
0.75	−0.013295	−0.079703	−0.803528	−0.462707	0.301807	0.711638
0.80	0.046535	−0.017089	−0.765809	−0.493356	0.248393	0.692731
0.85	0.099462	0.038971	−0.726154	−0.519576	0.195939	0.670818
0.90	0.145823	0.088735	−0.685017	−0.541490	0.144710	0.646158
0.95	0.185983	0.132497	−0.642821	−0.559242	0.094946	0.619017
1.00	0.220328	0.170575	−0.599958	−0.572999	0.046862	0.589659
1.05	0.249252	0.203305	−0.556788	−0.582939	0.000650	0.558352
1.10	0.273156	0.231038	−0.513641	−0.589257	−0.043525	0.525357
1.15	0.292441	0.254127	−0.470813	−0.592157	−0.085519	0.490935
1.20	0.307501	0.272929	−0.428571	−0.591853	−0.125211	0.455341
1.25	0.318722	0.287797	−0.387151	−0.588562	−0.162505	0.418822
1.30	0.326479	0.299078	−0.346760	−0.582507	−0.197323	0.381618
1.35	0.331132	0.307109	−0.307579	−0.573913	−0.229610	0.343962
1.40	0.333022	0.312215	−0.269759	−0.563003	−0.259329	0.306074
1.45	0.332474	0.314707	−0.233429	−0.550001	−0.286460	0.268165
1.50	0.329793	0.314880	−0.198693	−0.535124	−0.311004	0.230435
1.55	0.325265	0.313014	−0.165635	−0.518587	−0.332973	0.193070
1.60	0.319154	0.309370	−0.134315	−0.500600	−0.352397	0.156247
1.65	0.311705	0.304194	−0.104777	−0.481364	−0.369318	0.120127
1.70	0.303144	0.297710	−0.077047	−0.461074	−0.383791	0.084860
1.75	0.293675	0.290130	−0.051136	−0.439914	−0.395881	0.050582
1.80	0.283485	0.281643	−0.027040	−0.418063	−0.405663	0.017417
1.85	0.272741	0.272425	−0.004743	−0.395686	−0.413220	−0.014526
1.90	0.261594	0.262633	0.015782	−0.372942	−0.418643	−0.045150
1.95	0.250179	0.252412	0.034573	−0.349978	−0.422030	−0.074368
2.00	0.238614	0.241887	0.051674	−0.326931	−0.423483	−0.102111
2.05	0.227003	0.231173	0.067139	−0.303927	−0.423108	−0.128316
2.10	0.215436	0.220370	0.081025	−0.281083	−0.421016	−0.152936
2.15	0.203992	0.209566	0.093397	−0.258505	−0.417317	−0.175934
2.20	0.192737	0.198837	0.104321	−0.236289	−0.412125	−0.197282

Table 1.2 (continued)

			t			
v	0.95	1.00	2.00	3.00	4.00	5.00
2.25	0.181727	0.188248	0.113869	−0.214521	−0.405555	−0.216965
2.30	0.171008	0.177856	0.122112	−0.193278	−0.397720	−0.234975
2.35	0.160618	0.167707	0.129126	−0.172627	−0.388733	−0.251315
2.40	0.150587	0.157840	0.134983	−0.152628	−0.378706	−0.265995
2.45	0.140938	0.148285	0.139760	−0.133330	−0.367748	−0.279034
2.50	0.131688	0.139068	0.143529	−0.114775	−0.355966	−0.290458
2.55	0.122848	0.130208	0.146365	−0.096999	−0.343465	−0.300299
2.60	0.114424	0.121717	0.148339	−0.080028	−0.330345	−0.308597
2.65	0.106419	0.113605	0.149521	−0.063882	−0.316703	−0.315393
2.70	0.098832	0.105876	0.149979	−0.048577	−0.302632	−0.320738
2.75	0.091658	0.098531	0.149779	−0.034120	−0.288222	−0.324685
2.80	0.084889	0.091568	0.148982	−0.020516	−0.273557	−0.327289
2.85	0.078518	0.084983	0.147651	−0.007761	−0.258716	−0.328612
2.90	0.072532	0.078768	0.145841	0.004149	−0.243775	−0.328715
2.95	0.066920	0.072916	0.143607	0.015226	−0.228805	−0.327662
3.00	0.061669	0.067415	0.141002	0.025482	−0.213870	−0.325521
3.05	0.056763	0.062255	0.138072	0.034936	−0.199033	−0.322357
3.10	0.052188	0.057423	0.134865	0.043605	−0.184348	−0.318240
3.15	0.047928	0.052906	0.131421	0.051514	−0.169869	−0.313238
3.20	0.043969	0.048690	0.127780	0.058685	−0.155641	−0.307419
3.25	0.040294	0.044762	0.123979	0.065146	−0.141706	−0.300850
3.30	0.036888	0.041108	0.120051	0.070923	−0.128103	−0.293600
3.35	0.033737	0.037714	0.116028	0.076045	−0.114866	−0.285733
3.40	0.030824	0.034565	0.111937	0.080542	−0.102023	−0.277316
3.45	0.028136	0.031648	0.107805	0.084443	−0.089600	−0.268412
3.50	0.025658	0.028949	0.103654	0.087781	−0.077619	−0.259082
3.55	0.023376	0.026456	0.099506	0.090584	−0.066098	−0.249385
3.60	0.021278	0.024155	0.095378	0.092884	−0.055052	−0.239380
3.65	0.019352	0.022035	0.091289	0.094712	−0.044492	−0.229121
3.70	0.017584	0.020083	0.087253	0.096098	−0.034427	−0.218662
3.75	0.015965	0.018288	0.083282	0.097071	−0.024862	−0.208053
3.80	0.014482	0.016640	0.079388	0.097661	−0.015799	−0.197342
3.85	0.013127	0.015127	0.075581	0.097896	−0.007240	−0.186574
3.90	0.011889	0.013741	0.071868	0.097805	0.000817	−0.175793
3.95	0.010759	0.012472	0.068256	0.097413	0.008377	−0.165038
4.00	0.009728	0.011311	0.064752	0.096748	0.015444	−0.154346
4.05	0.008790	0.010250	0.061359	0.095833	0.022028	−0.143753
4.10	0.007937	0.009282	0.058081	0.094693	0.028136	−0.133290
4.15	0.007161	0.008399	0.054920	0.093351	0.033780	−0.122986
4.20	0.006456	0.007594	0.051878	0.091828	0.038970	−0.112870
4.25	0.005816	0.006861	0.048955	0.090146	0.043720	−0.102964
4.30	0.005236	0.006195	0.046152	0.088323	0.048043	−0.093292
4.35	0.004711	0.005590	0.043468	0.086379	0.051954	−0.083872

Table 1.2 (continued)

			t			
v	0.95	1.00	2.00	3.00	4.00	5.00
4.40	0.004235	0.005040	0.040902	0.084330	0.055468	−0.074722
4.45	0.003805	0.004541	0.038453	0.082194	0.058600	−0.065857
4.50	0.003416	0.004089	0.036118	0.079984	0.061366	−0.057290
4.55	0.003065	0.003679	0.033896	0.077717	0.063782	−0.049031
4.60	0.002749	0.003309	0.031784	0.075404	0.065866	−0.041090
4.65	0.002463	0.002973	0.029778	0.073058	0.067633	−0.033474
4.70	0.002206	0.002670	0.027876	0.070690	0.069100	−0.026188
4.75	0.001974	0.002397	0.026075	0.068311	0.070284	−0.019236
4.80	0.001766	0.002150	0.024372	0.065930	0.071201	−0.012621
4.85	0.001579	0.001927	0.022762	0.063555	0.071868	−0.006341
4.90	0.001410	0.001727	0.021242	0.061195	0.072299	−0.000399
4.95	0.001259	0.001546	0.019809	0.058856	0.072511	0.005210
5.00	0.001124	0.001383	0.018460	0.056546	0.072519	0.010487

Table 1.3: Third derivatives with respect to the order of the Bessel functions of the first kind $\partial^3 J_v(t)/\partial v^3$.

			t			
v	0.05	0.10	0.15	0.20	0.25	0.30
0.00	−12.3308	0.274815	4.298970	5.92714	6.58867	6.78657
0.05	−24.3781	−7.181620	−0.944008	2.01838	3.57730	4.42020
0.10	−32.1583	−12.5099	−4.89302	−1.03274	1.16154	2.47870
0.15	−36.5692	−16.0638	−7.72808	−3.32808	−0.71934	0.92503
0.20	−38.3831	−18.1687	−9.62305	−4.97049	−2.12952	−0.28199
0.25	−38.2525	−19.1176	−10.7414	−6.06029	−3.13381	−1.18550
0.30	−36.7202	−19.1692	−11.2332	−6.69264	−3.79553	−1.82880
0.35	−34.2295	−18.5476	−11.2336	−6.95570	−4.17477	−2.25397
0.40	−31.1363	−17.4442	−10.8620	−6.92939	−4.32725	−2.50076
0.45	−27.7204	−16.0186	−10.2217	−6.68472	−4.30345	−2.60584
0.50	−24.1966	−14.4019	−9.40056	−6.28354	−4.14821	−2.60225
0.55	−20.7250	−12.6992	−8.47159	−5.77858	−3.90046	−2.51911
0.60	−17.4206	−10.9925	−7.49418	−5.21384	−3.59328	−2.38150
0.65	−14.3612	−9.34358	−6.51529	−4.62507	−3.25407	−2.21050
0.70	−11.5952	−7.79726	−5.57078	−4.04047	−2.90491	−2.02330
0.75	−9.147520	−6.38369	−4.68679	−3.48143	−2.56294	−1.83348
0.80	−7.025020	−5.12104	−3.88110	−2.96331	−2.24086	−1.65128
0.85	−5.221180	−4.01774	−3.16446	−2.49628	−1.94747	−1.48395
0.90	−3.719730	−3.07451	−2.54180	−2.08611	−1.68819	−1.33618
0.95	−2.497740	−2.28615	−2.01337	−1.73489	−1.46560	−1.21038

Table 1.3 (continued)

				t		
v	0.05	0.10	0.15	0.20	0.25	0.30
1.00	−1.528070	−1.64311	−1.57579	−1.44184	−1.27995	−1.10718
1.05	−0.781361	−1.13278	−1.22296	−1.20387	−1.12968	−1.02571
1.10	−0.227582	−0.740638	−0.946877	−1.01626	−1.01183	−0.963994
1.15	0.162838	−0.451171	−0.738317	−0.873155	−0.922466	−0.919255
1.20	0.418121	−0.248615	−0.587436	−0.768019	−0.857065	−0.888214
1.25	0.564534	−0.117552	−0.484252	−0.694043	−0.810800	−0.867337
1.30	0.625976	−0.043363	−0.419042	−0.644462	−0.778819	−0.853052
1.35	0.623743	−0.012568	−0.382646	−0.612821	−0.756463	−0.841938
1.40	0.576412	−0.013049	−0.366697	−0.593175	−0.739438	−0.830859
1.45	0.499837	−0.034197	−0.363780	−0.580236	−0.723943	−0.817083
1.50	0.407227	−0.066976	−0.367532	−0.569478	−0.706765	−0.798357
1.55	0.309273	−0.103932	−0.372686	−0.557189	−0.685329	−0.772950
1.60	0.214332	−0.139150	−0.375078	−0.540491	−0.657719	−0.739682
1.65	0.128635	−0.168168	−0.371611	−0.517327	−0.622679	−0.697913
1.70	0.056509	−0.187869	−0.360191	−0.486421	−0.579580	−0.647525
1.75	0.000620	−0.196344	−0.339644	−0.447217	−0.528373	−0.588878
1.80	−0.037793	−0.192748	−0.309611	−0.399804	−0.469528	−0.522762
1.85	−0.058680	−0.177142	−0.270437	−0.344824	−0.403960	−0.450327
1.90	−0.062957	−0.150332	−0.223048	−0.283379	−0.332949	−0.373014
1.95	−0.052287	−0.113717	−0.168832	−0.216928	−0.258054	−0.292484
2.00	−0.028886	−0.069138	−0.109517	−0.147192	−0.181026	−0.210541
2.05	0.004662	−0.018730	−0.047057	−0.076050	−0.103729	−0.129053
2.10	0.045566	0.035203	0.016481	−0.005453	−0.028055	−0.049890
2.15	0.090975	0.090310	0.079010	0.062664	0.044147	0.025151
2.20	0.138094	0.144292	0.138514	0.126455	0.111138	0.094397
2.25	0.184297	0.194998	0.193125	0.184232	0.171349	0.156357
2.30	0.227204	0.240491	0.241183	0.234516	0.223422	0.209757
2.35	0.264746	0.279110	0.281286	0.276083	0.266254	0.253580
2.40	0.295216	0.309509	0.312330	0.307993	0.299023	0.287083
2.45	0.317294	0.330685	0.333530	0.329612	0.321200	0.309813
2.50	0.330062	0.341995	0.344429	0.340619	0.332560	0.321614
2.55	0.333005	0.343151	0.344904	0.341006	0.333178	0.322618
2.60	0.325999	0.334216	0.335150	0.331064	0.323415	0.313234
2.65	0.309285	0.315576	0.315663	0.311369	0.303900	0.294131
2.70	0.283438	0.287919	0.287213	0.282750	0.275506	0.266205
2.75	0.249325	0.252193	0.250807	0.246257	0.239310	0.230553
2.80	0.208064	0.209563	0.207654	0.203127	0.196567	0.188436
2.85	0.160967	0.161374	0.159116	0.154737	0.148661	0.141240
2.90	0.109497	0.109094	0.106671	0.102565	0.097067	0.090433
2.95	0.055208	0.054269	0.051859	0.048147	0.043312	0.037527
3.00	−0.000304	−0.001523	−0.003759	−0.006971	−0.011075	−0.015963
3.05	−0.055447	−0.056723	−0.058645	−0.061273	−0.064595	−0.068561
3.10	−0.108687	−0.109831	−0.111328	−0.113310	−0.115823	−0.118861

Table 1.3 (continued)

			t			
v	0.05	0.10	0.15	0.20	0.25	0.30
3.15	−0.158585	−0.159449	−0.160439	−0.161741	−0.163441	−0.165568
3.20	−0.203836	−0.204310	−0.204745	−0.205359	−0.206267	−0.207522
3.25	−0.243300	−0.243318	−0.243181	−0.243124	−0.243281	−0.243725
3.30	−0.276029	−0.275561	−0.274863	−0.274178	−0.273648	−0.273359
3.35	−0.301285	−0.300335	−0.299117	−0.297867	−0.296732	−0.295804
3.40	−0.318553	−0.317158	−0.315480	−0.313747	−0.312104	−0.310642
3.45	−0.327547	−0.325766	−0.323711	−0.321591	−0.319546	−0.317667
3.50	−0.328208	−0.326123	−0.323784	−0.321384	−0.319053	−0.316878
3.55	−0.320703	−0.318408	−0.315890	−0.313322	−0.310822	−0.308474
3.60	−0.305404	−0.303003	−0.300415	−0.297794	−0.295244	−0.292842
3.65	−0.282881	−0.280480	−0.277932	−0.275369	−0.272883	−0.270543
3.70	−0.253873	−0.251578	−0.249176	−0.246776	−0.244461	−0.242290
3.75	−0.219267	−0.217179	−0.215020	−0.212881	−0.210832	−0.208926
3.80	−0.180071	−0.178280	−0.176450	−0.174657	−0.172958	−0.171399
3.85	−0.137380	−0.135964	−0.134538	−0.133161	−0.131881	−0.130736
3.90	−0.092351	−0.091373	−0.090408	−0.089502	−0.088693	−0.088014
3.95	−0.046165	−0.045672	−0.045209	−0.044812	−0.044511	−0.044331
4.00	−0.000002	−0.000023	−0.000087	−0.000219	−0.000443	−0.000779
4.05	0.044990	0.044445	0.043849	0.043184	0.042435	0.041583
4.10	0.087727	0.086664	0.085546	0.084364	0.083103	0.081752
4.15	0.127209	0.125652	0.124040	0.122368	0.120628	0.118809
4.20	0.162545	0.160532	0.158468	0.156351	0.154176	0.151934
4.25	0.192972	0.190557	0.188096	0.185589	0.183035	0.180427
4.30	0.217870	0.215116	0.212323	0.209495	0.206628	0.203720
4.35	0.236771	0.233753	0.230703	0.227626	0.224522	0.221387
4.40	0.249372	0.246169	0.242942	0.239697	0.236435	0.233152
4.45	0.255530	0.252226	0.248906	0.245577	0.242238	0.238888
4.50	0.255266	0.251946	0.248619	0.245289	0.241957	0.238621
4.55	0.248758	0.245507	0.242255	0.239006	0.235762	0.232519
4.60	0.236332	0.233231	0.230134	0.227046	0.223966	0.220892
4.65	0.218452	0.215576	0.212708	0.209852	0.207007	0.204171
4.70	0.195700	0.193117	0.190544	0.187985	0.185438	0.182902
4.75	0.168764	0.166532	0.164311	0.162104	0.159909	0.157725
4.80	0.138414	0.136580	0.134758	0.132948	0.131149	0.129359
4.85	0.105483	0.104084	0.102695	0.101316	0.099945	0.098580
4.90	0.070843	0.069903	0.068970	0.068044	0.067122	0.066203
4.95	0.035387	0.034917	0.034450	0.033986	0.033523	0.033057
5.00			−0.000002	−0.000005	−0.000013	−0.000028

			t			
v	0.35	0.40	0.45	0.50	0.55	0.60
0.00	6.73939	6.55519	6.29225	5.98406	5.65074	5.30483
0.05	4.85781	5.04992	5.08626	5.02058	4.88672	4.70683

Table 1.3 (continued)

v	0.35	0.40	0.45	0.50	0.55	0.60
0.10	3.28372	3.76834	4.04246	4.17325	4.20384	4.16304
0.15	1.99451	2.69695	3.15322	3.43825	3.60075	3.67371
0.20	0.96344	1.81842	2.40752	2.80880	3.07358	3.23691
0.25	0.16137	1.11298	1.79199	2.27603	2.61660	2.84917
0.30	−0.44181	0.559541	1.29190	1.82973	2.22287	2.50591
0.35	−0.87600	0.136719	0.891924	1.45900	1.88475	2.20190
0.40	−1.16996	−0.176428	0.576832	1.15283	1.59438	1.93166
0.45	−1.35060	−0.399764	0.331980	0.900494	1.34404	1.68976
0.50	−1.44253	−0.551632	0.143723	0.691939	1.12642	1.47105
0.55	−1.46765	−0.648532	−0.000295	0.518016	0.934895	1.27086
0.60	−1.44503	−0.704928	−0.110959	0.370659	0.763627	1.08515
0.65	−1.39085	−0.733162	−0.197574	0.242987	0.607683	0.910544
0.70	−1.31843	−0.743448	−0.267839	0.129342	0.463073	0.744426
0.75	−1.23838	−0.743940	−0.327882	0.025277	0.326744	0.584898
0.80	−1.15878	−0.740858	−0.382338	−0.072499	0.196543	0.430769
0.85	−1.08543	−0.738657	−0.434473	−0.166211	0.071145	0.281488
0.90	−1.02208	−0.740218	−0.486329	−0.257138	−0.050042	0.137074
0.95	−0.970794	−0.747068	−0.538894	−0.345736	−0.166976	−0.001978
1.00	−0.932150	−0.759601	−0.592272	−0.431787	−0.279109	−0.134796
1.05	−0.905582	−0.777303	−0.645872	−0.514544	−0.385501	−0.260235
1.10	−0.889635	−0.798970	−0.698576	−0.592876	−0.484945	−0.376976
1.15	−0.882221	−0.822914	−0.748909	−0.665404	−0.576082	−0.483628
1.20	−0.880854	−0.847149	−0.795198	−0.730633	−0.657509	−0.578819
1.25	−0.882854	−0.869566	−0.835704	−0.787068	−0.727874	−0.661281
1.30	−0.885527	−0.888072	−0.868753	−0.833310	−0.785967	−0.729915
1.35	−0.886315	−0.900721	−0.892828	−0.868149	−0.830784	−0.783855
1.40	−0.882914	−0.905810	−0.906659	−0.890625	−0.861582	−0.822511
1.45	−0.873370	−0.901954	−0.909281	−0.900078	−0.877925	−0.845600
1.50	−0.856139	−0.888141	−0.900077	−0.896185	−0.879702	−0.853167
1.55	−0.830131	−0.863760	−0.878802	−0.878972	−0.867143	−0.845585
1.60	−0.794721	−0.828617	−0.845586	−0.848819	−0.840810	−0.823553
1.65	−0.749752	−0.782926	−0.800930	−0.806444	−0.801587	−0.788074
1.70	−0.695508	−0.727286	−0.745679	−0.752882	−0.750651	−0.740430
1.75	−0.632678	−0.662648	−0.680990	−0.689447	−0.689438	−0.682145
1.80	−0.562309	−0.590269	−0.608287	−0.617695	−0.619599	−0.614941
1.85	−0.485751	−0.511660	−0.529210	−0.539365	−0.542953	−0.540693
1.90	−0.404588	−0.428525	−0.445559	−0.456338	−0.461437	−0.461379
1.95	−0.320573	−0.342699	−0.359241	−0.370571	−0.377050	−0.379027
2.00	−0.235564	−0.256088	−0.272202	−0.284047	−0.291800	−0.295662
2.05	−0.151449	−0.170606	−0.186378	−0.198719	−0.207656	−0.213258
2.10	−0.070089	−0.088114	−0.103635	−0.116461	−0.126490	−0.133693
2.15	0.006746	−0.010367	−0.025722	−0.039013	−0.050043	−0.058698
2.20	0.077426	0.061033	0.045779	0.032053	0.020123	0.010172

Table 1.3 (continued)

			t			
ν	0.35	0.40	0.45	0.50	0.55	0.60
2.25	0.140511	0.124680	0.109485	0.095370	0.082657	0.071574
2.30	0.194786	0.179400	0.164249	0.149812	0.136444	0.124406
2.35	0.239291	0.224272	0.209180	0.194506	0.180624	0.167817
2.40	0.273341	0.258652	0.243660	0.228853	0.214611	0.201227
2.45	0.296537	0.282178	0.267349	0.252529	0.238090	0.224327
2.50	0.308769	0.294769	0.280191	0.265488	0.251021	0.237077
2.55	0.310209	0.296626	0.282399	0.267952	0.253626	0.239697
2.60	0.301298	0.288208	0.274444	0.260394	0.246376	0.232652
2.65	0.282728	0.270220	0.257034	0.243521	0.229971	0.216630
2.70	0.255411	0.243581	0.231088	0.218244	0.205312	0.192518
2.75	0.220453	0.209396	0.197704	0.185651	0.173472	0.161371
2.80	0.179114	0.168920	0.158127	0.146972	0.135663	0.124380
2.85	0.132774	0.123521	0.113710	0.103543	0.093199	0.082839
2.90	0.082889	0.074643	0.065879	0.056768	0.047462	0.038104
2.95	0.030958	0.023764	0.016092	0.008083	-0.000135	-0.008437
3.00	-0.021524	-0.027641	-0.034198	-0.041082	-0.048186	-0.055405
3.05	-0.073101	-0.078136	-0.083578	-0.089338	-0.095328	-0.101459
3.10	-0.122391	-0.126361	-0.130711	-0.135371	-0.140270	-0.145334
3.15	-0.168118	-0.171064	-0.174366	-0.177971	-0.181824	-0.185863
3.20	-0.209145	-0.211126	-0.213444	-0.216060	-0.218932	-0.222010
3.25	-0.244491	-0.245587	-0.247002	-0.248713	-0.250687	-0.252883
3.30	-0.273357	-0.273661	-0.274272	-0.275176	-0.276349	-0.277759
3.35	-0.295136	-0.294756	-0.294673	-0.294880	-0.295362	-0.296092
3.40	-0.309422	-0.308475	-0.307817	-0.307447	-0.307355	-0.307521
3.45	-0.316015	-0.314626	-0.313517	-0.312694	-0.312151	-0.311873
3.50	-0.314919	-0.313213	-0.311781	-0.310632	-0.309762	-0.309162
3.55	-0.306333	-0.304437	-0.302808	-0.301457	-0.300384	-0.299581
3.60	-0.290641	-0.288678	-0.286974	-0.285541	-0.284383	-0.283493
3.65	-0.268398	-0.266483	-0.264819	-0.263418	-0.262285	-0.261416
3.70	-0.240308	-0.238547	-0.237027	-0.235761	-0.234754	-0.234005
3.75	-0.207202	-0.205689	-0.204406	-0.203366	-0.202575	-0.202032
3.80	-0.170014	-0.168829	-0.167863	-0.167126	-0.166626	-0.166363
3.85	-0.129757	-0.128965	-0.128378	-0.128007	-0.127858	-0.127932
3.90	-0.087490	-0.087140	-0.086979	-0.087020	-0.087266	-0.087722
3.95	-0.044294	-0.044417	-0.044715	-0.045196	-0.045867	-0.046732
4.00	-0.001245	-0.001857	-0.002626	-0.003562	-0.004670	-0.005955
4.05	0.040614	0.039516	0.038277	0.036889	0.035346	0.033644
4.10	0.080298	0.078731	0.077040	0.075217	0.073257	0.071154
4.15	0.116902	0.114896	0.112784	0.110558	0.108211	0.105740
4.20	0.149618	0.147220	0.144730	0.142144	0.139454	0.136655
4.25	0.177759	0.175022	0.172211	0.169318	0.166337	0.163263
4.30	0.200765	0.197755	0.194685	0.191547	0.188337	0.185048
4.35	0.218217	0.215006	0.211746	0.208433	0.205060	0.201622

Table 1.3 (continued)

			t			
v	0.35	0.40	0.45	0.50	0.55	0.60
4.40	0.229844	0.226507	0.223133	0.219717	0.216252	0.212734
4.45	0.235523	0.232137	0.228725	0.225280	0.221798	0.218272
4.50	0.235277	0.231921	0.228547	0.225149	0.221722	0.218259
4.55	0.229276	0.226026	0.222764	0.219485	0.216183	0.212852
4.60	0.217821	0.214749	0.211670	0.208578	0.205468	0.202334
4.65	0.201341	0.198512	0.195680	0.192838	0.189981	0.187104
4.70	0.180372	0.177846	0.175317	0.172780	0.170230	0.167661
4.75	0.155547	0.153372	0.151195	0.149010	0.146812	0.144596
4.80	0.127574	0.125790	0.124003	0.122207	0.120397	0.118569
4.85	0.097218	0.095855	0.094486	0.093106	0.091710	0.090294
4.90	0.065283	0.064358	0.063424	0.062476	0.061510	0.060521
4.95	0.032587	0.032107	0.031615	0.031105	0.030573	0.030016
5.00	−0.000053	−0.000092	−0.000147	−0.000224	−0.000326	−0.000458

			t			
v	0.65	0.70	0.75	0.80	0.85	0.90
0.00	4.95434	4.60451	4.25877	3.91944	3.58804	3.26561
0.05	4.49585	4.26414	4.01904	3.76584	3.50844	3.24971
0.10	4.07071	3.94090	3.78378	3.60689	3.41586	3.21499
0.15	3.68013	3.63653	3.55502	3.44463	3.31227	3.16324
0.20	3.32353	3.35133	3.33354	3.28016	3.19887	3.09566
0.25	2.99898	3.08444	3.11925	3.11387	3.07637	3.01308
0.30	2.70363	2.83425	2.91144	2.94566	2.94507	2.91610
0.35	2.43405	2.59869	2.70900	2.77515	2.80510	2.80520
0.40	2.18657	2.37543	2.51065	2.60182	2.65654	2.68084
0.45	1.95751	2.16217	2.31511	2.42520	2.49948	2.54361
0.50	1.74340	1.95672	2.12123	2.24493	2.33421	2.39421
0.55	1.54113	1.75723	1.92813	2.06089	2.16117	2.23358
0.60	1.34806	1.56220	1.73522	1.87321	1.98110	2.06290
0.65	1.16213	1.37057	1.54228	1.68235	1.79497	1.88357
0.70	0.981808	1.18176	1.34945	1.48907	1.60406	1.69725
0.75	0.806171	0.995614	1.15727	1.29444	1.40986	1.50580
0.80	0.634834	0.812444	0.966606	1.09980	1.21411	1.31130
0.85	0.467914	0.632932	0.778621	0.906724	1.01873	1.11593
0.90	0.305960	0.458091	0.594728	0.716966	0.825765	0.921979
0.95	0.149880	0.289192	0.416522	0.532400	0.637328	0.731778
1.00	0.000848	0.127690	0.245709	0.354959	0.455549	0.547624
1.05	−0.139781	−0.024859	0.084036	0.186565	0.282510	0.371739
1.10	−0.270560	−0.166870	−0.066778	0.029067	0.120181	0.206208
1.15	−0.390035	−0.296811	−0.205108	−0.115817	−0.029626	0.052927
1.20	−0.496826	−0.413265	−0.329487	−0.246556	−0.165315	−0.086437
1.25	−0.589697	−0.514997	−0.438657	−0.361857	−0.285545	−0.210493
1.30	−0.667614	−0.601000	−0.531612	−0.460696	−0.389265	−0.318151

Table 1.3 (continued)

v	0.65	0.70	0.75	0.80	0.85	0.90
			t			
1.35	-0.729797	-0.670541	-0.607639	-0.542357	-0.475738	-0.408647
1.40	-0.775754	-0.723188	-0.666335	-0.606446	-0.544563	-0.481557
1.45	-0.805309	-0.758832	-0.707627	-0.652903	-0.595675	-0.536802
1.50	-0.818611	-0.777690	-0.731772	-0.682002	-0.629349	-0.574640
1.55	-0.816136	-0.780306	-0.739353	-0.694343	-0.646184	-0.595657
1.60	-0.798677	-0.767533	-0.731262	-0.690833	-0.647086	-0.600747
1.65	-0.767321	-0.740517	-0.708672	-0.672660	-0.633240	-0.591078
1.70	-0.723425	-0.700658	-0.673014	-0.641262	-0.606078	-0.568066
1.75	-0.668571	-0.649581	-0.625930	-0.598285	-0.567240	-0.533330
1.80	-0.604532	-0.589086	-0.569235	-0.545545	-0.518529	-0.488654
1.85	-0.533221	-0.521108	-0.504870	-0.484978	-0.461867	-0.435937
1.90	-0.456641	-0.447663	-0.434854	-0.418597	-0.399249	-0.377148
1.95	-0.376837	-0.370803	-0.361236	-0.348436	-0.332692	-0.314283
2.00	-0.295842	-0.292561	-0.286043	-0.276512	-0.264193	-0.249311
2.05	-0.215634	-0.214912	-0.211238	-0.204772	-0.195681	-0.184139
2.10	-0.138085	-0.139719	-0.138676	-0.135056	-0.128975	-0.120562
2.15	-0.064921	-0.068701	-0.070062	-0.069054	-0.065750	-0.060237
2.20	0.002311	-0.003394	-0.006922	-0.008279	-0.007501	-0.004639
2.25	0.062280	0.054879	0.049433	0.045967	0.044482	0.044955
2.30	0.113889	0.105030	0.097920	0.092614	0.089138	0.087493
2.35	0.156297	0.146224	0.137710	0.130835	0.125645	0.122164
2.40	0.188928	0.177886	0.168231	0.160060	0.153436	0.148400
2.45	0.211472	0.199707	0.189173	0.179976	0.172196	0.165885
2.50	0.223889	0.211642	0.200483	0.190526	0.181860	0.174549
2.55	0.226393	0.213897	0.202359	0.191897	0.182606	0.174558
2.60	0.219439	0.206915	0.195230	0.184504	0.174835	0.166300
2.65	0.203701	0.191356	0.179741	0.168974	0.159155	0.150366
2.70	0.180050	0.168072	0.156722	0.146118	0.136359	0.127527
2.75	0.149521	0.138076	0.127166	0.116906	0.107392	0.098707
2.80	0.113284	0.102514	0.092195	0.082433	0.073325	0.064950
2.85	0.072606	0.062630	0.053024	0.043891	0.035322	0.027397
2.90	0.028820	0.019728	0.010933	0.002532	-0.005391	-0.012756
2.95	-0.016713	-0.024856	-0.032770	-0.040366	-0.047562	-0.054284
3.00	-0.062639	-0.069796	-0.076786	-0.083528	-0.089946	-0.095967
3.05	-0.107645	-0.113803	-0.119853	-0.125719	-0.131331	-0.136620
3.10	-0.150488	-0.155659	-0.160776	-0.165770	-0.170573	-0.175123
3.15	-0.190025	-0.194245	-0.198459	-0.202605	-0.206621	-0.210447
3.20	-0.225238	-0.228562	-0.231925	-0.235269	-0.238537	-0.241674
3.25	-0.255256	-0.257757	-0.260336	-0.262941	-0.265519	-0.268020
3.30	-0.279368	-0.281134	-0.283012	-0.284956	-0.286918	-0.288849
3.35	-0.297041	-0.298171	-0.299442	-0.300814	-0.302243	-0.303683
3.40	-0.307921	-0.308522	-0.309290	-0.310188	-0.311176	-0.312212
3.45	-0.311840	-0.312025	-0.312397	-0.312924	-0.313568	-0.314293

Table 1.3 (continued)

			t			
v	0.65	0.70	0.75	0.80	0.85	0.90
3.50	−0.308815	−0.308697	−0.308783	−0.309043	−0.309443	−0.309950
3.55	−0.299036	−0.298729	−0.298637	−0.298733	−0.298988	−0.299371
3.60	−0.282862	−0.282474	−0.282308	−0.282341	−0.282546	−0.282895
3.65	−0.260805	−0.260436	−0.260294	−0.260356	−0.260601	−0.261001
3.70	−0.233508	−0.233252	−0.233222	−0.233399	−0.233763	−0.234290
3.75	−0.201734	−0.201670	−0.201829	−0.202195	−0.202750	−0.203471
3.80	−0.166333	−0.166531	−0.166945	−0.167562	−0.168366	−0.169337
3.85	−0.128229	−0.128742	−0.129463	−0.130380	−0.131479	−0.132745
3.90	−0.088386	−0.089255	−0.090320	−0.091572	−0.093001	−0.094590
3.95	−0.047790	−0.049038	−0.050470	−0.052080	−0.053856	−0.055787
4.00	−0.007418	−0.009055	−0.010864	−0.012838	−0.014969	−0.017246
4.05	0.031781	0.029758	0.027578	0.025245	0.022767	0.020152
4.10	0.068908	0.066518	0.063985	0.061313	0.058507	0.055574
4.15	0.103141	0.100413	0.097558	0.094576	0.091473	0.088252
4.20	0.133745	0.130721	0.127584	0.124333	0.120972	0.117506
4.25	0.160093	0.156824	0.153454	0.149984	0.146414	0.142749
4.30	0.181676	0.178219	0.174674	0.171040	0.167318	0.163509
4.35	0.198113	0.194532	0.190873	0.187137	0.183323	0.179430
4.40	0.209157	0.205518	0.201812	0.198038	0.194194	0.190281
4.45	0.214696	0.211068	0.207382	0.203636	0.199827	0.195956
4.50	0.214754	0.211205	0.207605	0.203952	0.200243	0.196477
4.55	0.209486	0.206081	0.202632	0.199135	0.195588	0.191987
4.60	0.199170	0.195972	0.192734	0.189452	0.186124	0.182745
4.65	0.184200	0.181264	0.178292	0.175279	0.172222	0.169117
4.70	0.165068	0.162444	0.159787	0.157090	0.154349	0.151563
4.75	0.142356	0.140086	0.137783	0.135441	0.133056	0.130625
4.80	0.116715	0.114832	0.112915	0.110959	0.108959	0.106913
4.85	0.088851	0.087378	0.085869	0.084320	0.082726	0.081085
4.90	0.059503	0.058453	0.057365	0.056235	0.055059	0.053833
4.95	0.029427	0.028803	0.028139	0.027430	0.026674	0.025866
5.00	−0.000624	−0.000828	−0.001075	−0.001369	−0.001713	−0.002111

			t			
v	0.95	1.00	2.00	3.00	4.00	5.00
0.00	2.95280	2.65006	−1.263870	−1.764470	−0.391066	1.004240
0.05	2.99182	2.73644	−0.973283	−1.675720	−0.453478	0.941866
0.10	3.00762	2.79634	−0.694715	−1.578940	−0.509397	0.875735
0.15	3.00172	2.83100	−0.431260	−1.476620	−0.559579	0.805913
0.20	2.97524	2.84139	−0.185579	−1.371090	−0.604759	0.732510
0.25	2.92906	2.82836	0.040116	−1.264460	−0.645626	0.655689
0.30	2.86392	2.79270	0.244086	−1.158660	−0.682805	0.575672
0.35	2.78053	2.73526	0.425062	−1.055330	−0.716840	0.492748
0.40	2.67965	2.65698	0.582250	−0.955885	−0.748182	0.407272

Table 1.3 (continued)

			t			
v	0.95	1.00	2.00	3.00	4.00	5.00
0.45	2.56218	2.55899	0.715314	−0.861467	−0.777182	0.319669
0.50	2.42917	2.44261	0.824358	−0.772945	−0.804082	0.230435
0.55	2.28193	2.30938	0.909903	−0.690926	−0.829021	0.140128
0.60	2.12196	2.16108	0.972847	−0.615766	−0.852031	0.049366
0.65	1.95101	1.99971	1.014430	−0.547583	−0.873048	−0.041185
0.70	1.77104	1.82750	1.036190	−0.486276	−0.891916	−0.130814
0.75	1.58423	1.64681	1.039920	−0.431552	−0.908406	−0.218784
0.80	1.39289	1.46019	1.027590	−0.382949	−0.922222	−0.304341
0.85	1.19944	1.27025	1.001350	−0.339869	−0.933024	−0.386726
0.90	1.00637	1.07965	0.963402	−0.301603	−0.940438	−0.465194
0.95	0.816199	0.891017	0.916024	−0.267370	−0.944077	−0.539024
1.00	0.631356	0.706935	0.861473	−0.236340	−0.943559	−0.607533
1.05	0.454190	0.529850	0.801950	−0.207669	−0.938521	−0.670094
1.10	0.286887	0.362037	0.739559	−0.180527	−0.928640	−0.726145
1.15	0.131426	0.205551	0.676267	−0.154125	−0.913643	−0.775199
1.20	−0.010460	0.062185	0.613867	−0.127735	−0.893322	−0.816859
1.25	−0.137327	−0.066562	0.553953	−0.100717	−0.867549	−0.850822
1.30	−0.248043	−0.179511	0.497897	−0.072532	−0.836279	−0.876888
1.35	−0.341807	−0.275826	0.446831	−0.042754	−0.799562	−0.894962
1.40	−0.418166	−0.355016	0.401638	−0.011084	−0.757542	−0.905058
1.45	−0.477015	−0.416941	0.362947	0.022647	−0.710464	−0.907297
1.50	−0.518586	−0.461804	0.331139	0.058478	−0.658666	−0.901908
1.55	−0.543444	−0.490136	0.306351	0.096316	−0.602581	−0.889218
1.60	−0.552455	−0.502773	0.288490	0.135948	−0.542729	−0.869652
1.65	−0.546765	−0.500827	0.277252	0.177045	−0.479704	−0.843719
1.70	−0.527762	−0.485652	0.272142	0.219177	−0.414171	−0.812006
1.75	−0.497039	−0.458805	0.272502	0.261827	−0.346845	−0.775162
1.80	−0.456348	−0.422005	0.277537	0.304409	−0.278485	−0.733892
1.85	−0.407560	−0.377086	0.286347	0.346283	−0.209874	−0.688937
1.90	−0.352616	−0.325953	0.297955	0.386781	−0.141805	−0.641062
1.95	−0.293477	−0.270535	0.311341	0.425217	−0.075065	−0.591045
2.00	−0.232087	−0.212739	0.325472	0.460917	−0.010420	−0.539658
2.05	−0.170323	−0.154413	0.339332	0.493233	0.051400	−0.487653
2.10	−0.109956	−0.097299	0.351952	0.521560	0.109717	−0.435755
2.15	−0.052617	−0.043003	0.362430	0.545357	0.163916	−0.384642
2.20	0.000235	0.007039	0.369960	0.564158	0.213458	−0.334938
2.25	0.047342	0.051586	0.373848	0.577588	0.257891	−0.287204
2.30	0.087663	0.089612	0.373529	0.585371	0.296853	−0.241926
2.35	0.120394	0.120320	0.368578	0.587337	0.330086	−0.199515
2.40	0.144972	0.143152	0.358717	0.583429	0.357429	−0.160295
2.45	0.161078	0.157790	0.343821	0.573704	0.378831	−0.124506
2.50	0.168637	0.164152	0.323915	0.558329	0.394341	−0.092302

Table 1.3 (continued)

			t			
v	0.95	1.00	2.00	3.00	4.00	5.00
2.55	0.167806	0.162388	0.299171	0.537584	0.404107	−0.063750
2.60	0.158959	0.152857	0.269900	0.511848	0.408375	−0.038837
2.65	0.142670	0.136119	0.236538	0.481595	0.407474	−0.017471
2.70	0.119690	0.112903	0.199639	0.447382	0.401817	0.000512
2.75	0.090920	0.084089	0.159849	0.409835	0.391879	0.015336
2.80	0.057380	0.050674	0.117893	0.369637	0.378196	0.027283
2.85	0.020184	0.013745	0.074555	0.327509	0.361343	0.036679
2.90	−0.019497	−0.025552	0.030654	0.284199	0.341930	0.043884
2.95	−0.060466	−0.066046	−0.012976	0.240459	0.320578	0.049284
3.00	−0.101528	−0.106570	−0.055507	0.197034	0.297914	0.053274
3.05	−0.141525	−0.145988	−0.096137	0.154645	0.274552	0.056256
3.10	−0.179361	−0.183229	−0.134112	0.113972	0.251085	0.058619
3.15	−0.214027	−0.217306	−0.168743	0.075642	0.228067	0.060736
3.20	−0.244626	−0.247342	−0.199419	0.040212	0.206006	0.062952
3.25	−0.270392	−0.272586	−0.225625	0.008168	0.185357	0.065577
3.30	−0.290703	−0.292431	−0.246949	−0.020094	0.166506	0.068877
3.35	−0.305091	−0.306422	−0.263094	−0.044270	0.149773	0.073073
3.40	−0.313256	−0.314264	−0.273879	−0.064151	0.135400	0.078329
3.45	−0.315059	−0.315827	−0.279245	−0.079629	0.123555	0.084760
3.50	−0.310527	−0.311138	−0.279251	−0.090692	0.114325	0.092418
3.55	−0.299849	−0.300386	−0.274075	−0.097426	0.107721	0.101305
3.60	−0.283358	−0.283902	−0.264003	−0.100009	0.103680	0.111365
3.65	−0.261529	−0.262155	−0.249422	−0.098701	0.102071	0.122492
3.70	−0.234956	−0.235733	−0.230808	−0.093841	0.102697	0.134532
3.75	−0.204338	−0.205324	−0.208717	−0.085833	0.105306	0.147290
3.80	−0.170457	−0.171702	−0.183765	−0.075137	0.109598	0.160535
3.85	−0.134157	−0.135698	−0.156616	−0.062255	0.115236	0.174010
3.90	−0.096324	−0.098186	−0.127965	−0.047719	0.121853	0.187434
3.95	−0.057859	−0.060055	−0.098520	−0.032078	0.129066	0.200517
4.00	−0.019658	−0.022189	−0.068986	−0.015885	0.136485	0.212962
4.05	0.017410	0.014554	−0.040049	0.000320	0.143723	0.224479
4.10	0.052522	0.049363	−0.012360	0.016017	0.150405	0.234789
4.15	0.084923	0.081492	0.013478	0.030719	0.156181	0.243632
4.20	0.113938	0.110277	0.036925	0.043988	0.160732	0.250774
4.25	0.138992	0.135149	0.057513	0.055435	0.163779	0.256017
4.30	0.159617	0.155646	0.074857	0.064738	0.165089	0.259199
4.35	0.175462	0.171421	0.088660	0.071638	0.164479	0.260200
4.40	0.186298	0.182248	0.098717	0.075951	0.161823	0.258946
4.45	0.192021	0.188025	0.104923	0.077568	0.157051	0.255410
4.50	0.192652	0.188769	0.107264	0.076453	0.150150	0.249614
4.55	0.188332	0.184621	0.105824	0.072646	0.141169	0.241627
4.60	0.179315	0.175832	0.100772	0.066258	0.130208	0.231563

Table 1.3 (continued)

			t			
v	0.95	1.00	2.00	3.00	4.00	5.00
4.65	0.165962	0.162755	0.092365	0.057466	0.117423	0.219582
4.70	0.148727	0.145840	0.080929	0.046507	0.103016	0.205881
4.75	0.128145	0.125614	0.066858	0.033672	0.087230	0.190692
4.80	0.104817	0.102668	0.050600	0.019296	0.070345	0.174277
4.85	0.079392	0.077645	0.032643	0.003749	0.052670	0.156920
4.90	0.052554	0.051219	0.013505	-0.012573	0.034531	0.138922
4.95	0.025002	0.024081	-0.006281	-0.029260	0.016268	0.120595
5.00	-0.002566	-0.003082	-0.026181	-0.045898	-0.001776	0.102250

1.2 First, Second and Third Derivatives with Respect to the Order of the Bessel Function of the Second Kind

First three derivatives with respect to the order of the Bessel function of the second kind were determined directly using the central-difference formulas of order $O(h^4)$ with $h = 10^{-3}$. The Bessel functions of the second kind $Y_v(t)$ were taken from MATHEMATICA program:

$$\frac{\partial Y_v(t)}{\partial v} = \frac{- Y_{v+2h}(t) + 8 Y_{v+h}(t) - 8 Y_{v-h}(t) + Y_{v-2h}(t)}{12h} \qquad (1.2.1)$$

$$\frac{\partial^2 Y_v(t)}{\partial v^2} = \frac{- Y_{v+2h}(t) + 16 Y_{v+h}(t) - 30 Y_v(t) + 16 Y_{v-h}(t) - Y_{v-2h}(t)}{12h^2} \qquad (1.2.2)$$

$$\frac{\partial^3 Y_v(t)}{\partial v^3} = \frac{- Y_{v+3h}(t) + 8 Y_{v+2h}(t) - 13 Y_{v+h}(t)}{8h^3} + \frac{13 Y_{v-h}(t) - 8 Y_{v-2h}(t) + Y_{v-3h}(t)}{8h^3} \qquad (1.2.3)$$

As shown in Figure 1.6, derivatives of the Bessel function of the second kind with respect to the order have a very large negative values for large v and t.

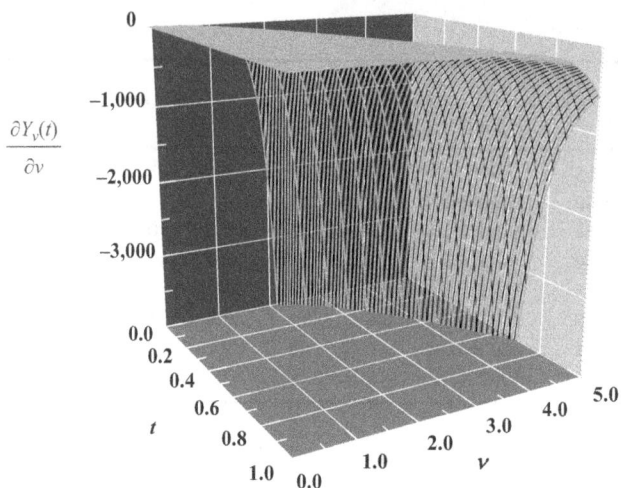

Figure 1.6: First derivatives with respect to the order of the Bessel functions of the second kind $\partial Y_v(t)/\partial v$ as a function of v and t.

Behaviour of second and third derivatives with respect to the order of the Bessel functions of the first kind is very similar to that of the first derivatives (Figures 1.7 and 1.8).

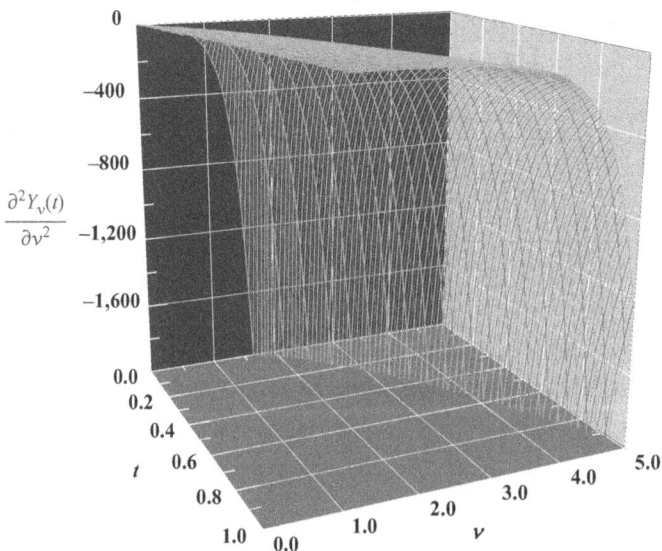

Figure 1.7: Second derivatives with respect to the order of the Bessel functions of the second kind $\partial^2 Y_v(t)/\partial v^2$ as a function of v and t.

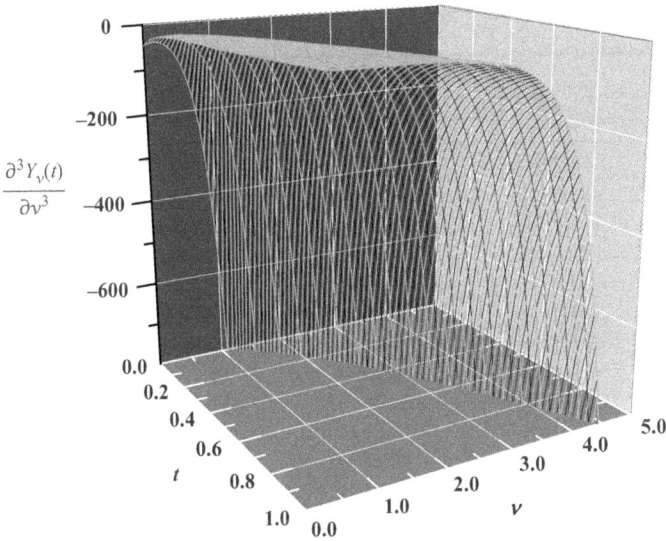

Figure 1.8: Third derivatives with respect to the order of the Bessel functions of the second kind $\partial^3 Y_\nu(t)/\partial \nu^3$ as a function of v and t.

Table 1.4: First derivatives with respect to the order of the Bessel functions of the second kind $\partial Y_\nu(t)/\partial \nu$.

			t		
v	0.05	0.10	0.15	0.20	0.25
0.00	−1.56981	−1.56687	−1.56197	−1.55513	−1.54635
0.05	−1.59324	−1.47130	−1.43504	−1.42024	−1.41246
0.10	−1.72222	−1.44010	−1.35319	−1.31869	−1.30384
0.15	−1.95152	−1.46997	−1.31462	−1.24965	−1.22033
0.20	−2.27936	−1.55864	−1.31787	−1.21234	−1.16165
0.25	−2.70757	−1.70498	−1.36188	−1.20610	−1.12752
0.30	−3.24175	−1.90913	−1.44612	−1.23047	−1.11768
0.35	−3.89151	−2.17261	−1.57067	−1.28524	−1.13196
0.40	−4.67089	−2.49844	−1.73628	−1.37056	−1.17036
0.45	−5.59890	−2.89127	−1.94448	−1.48695	−1.23307
0.50	−6.70020	−3.35756	−2.19764	−1.63538	−1.32050
0.55	−8.00604	−3.90584	−2.49908	−1.81736	−1.43340
0.60	−9.55549	−4.54691	−2.85319	−2.03493	−1.57281
0.65	−11.3969	−5.29422	−3.26553	−2.29081	−1.74017
0.70	−13.5898	−6.16425	−3.74303	−2.58845	−1.93735
0.75	−16.2075	−7.17705	−4.29415	−2.93209	−2.16670
0.80	−19.3397	−8.35680	−4.92909	−3.32691	−2.43114
0.85	−23.0965	−9.73258	−5.66009	−3.77916	−2.73417
0.90	−27.6129	−11.3393	−6.50172	−4.29626	−3.08002

Table 1.4 (continued)

			t		
v	0.05	0.10	0.15	0.20	0.25
0.95	−33.0547	−13.2187	−7.47131	−4.88702	−3.47370
1.00	−39.6255	−15.4208	−8.58932	−5.56182	−3.92111
1.05	−47.5756	−18.0056	−9.87999	−6.33287	−4.42920
1.10	−57.2132	−21.0447	−11.3719	−7.21447	−5.00606
1.15	−68.9181	−24.6240	−13.0989	−8.22339	−5.66116
1.20	−83.1591	−28.8465	−15.1008	−9.37924	−6.40549
1.25	−100.515	−33.8357	−17.4247	−10.7050	−7.25187
1.30	−121.703	−39.7399	−20.1263	−12.2274	−8.21513
1.35	−147.609	−46.7375	−23.2713	−13.9778	−9.31254
1.40	−179.332	−55.0428	−26.9376	−15.9930	−10.5641
1.45	−218.239	−64.9141	−31.2172	−18.3159	−11.9931
1.50	−266.024	−76.6626	−36.2194	−20.9968	−13.6265
1.55	−324.797	−90.6641	−42.0734	−24.0946	−15.4957
1.60	−397.185	−107.372	−48.9330	−27.6784	−17.6371
1.65	−486.463	−127.334	−56.9805	−31.8293	−20.0931
1.70	−596.716	−151.215	−66.4328	−36.6426	−22.9131
1.75	−733.050	−179.817	−77.5480	−42.2302	−26.1544
1.80	−901.847	−214.114	−90.6332	−48.7237	−29.8841
1.85	−1,111.10	−255.287	−106.055	−56.2781	−34.1800
1.90	−1,370.81	−304.772	−124.249	−65.0760	−39.1334
1.95	−1,693.54	−364.312	−145.738	−75.3324	−44.8504
2.00	−2,095.04	−436.029	−171.145	−87.3013	−51.4554
2.05	−2,595.12	−522.504	−201.213	−101.282	−59.0935
2.10	−3,218.68	−626.885	−236.836	−117.630	−67.9348
2.15	−3,997.05	−753.009	−279.079	−136.763	−78.1783
2.20	−4,969.74	−905.561	−329.222	−159.176	−90.0573
2.25	−6,186.54	−1,090.26	−388.800	−185.457	−103.845
2.30	−7,710.29	−1,314.11	−459.654	−216.300	−119.863
2.35	−9,620.41	−1,585.66	−543.996	−252.531	−138.487
2.40	−12,017.2	−1,915.39	−644.487	−295.129	−160.161
2.45	−15,027.8	−2,316.14	−764.325	−345.254	−185.406
2.50	−18,812.9	−2,803.67	−907.361	−404.289	−214.833
2.55	−23,576.4	−3,397.28	−1,078.24	−473.875	−249.165
2.60	−29,576.7	−4,120.73	−1,282.54	−555.967	−289.250
2.65	−37,142.1	−5,003.17	−1,527.03	−652.892	−336.092
2.70	−46,689.3	−6,080.49	−1,819.85	−767.423	−390.872
2.75	−58,748.5	−7,396.84	−2,170.83	−902.867	−454.985
2.80	−73,993.7	−9,006.61	−2,591.88	−1,063.17	−530.079
2.85	−93,283.6	−10,976.8	−3,097.39	−1,253.04	−618.103
2.90	−117,712	−13,390.2	−3,704.78	−1,478.10	−721.359
2.95	−148,674	−16,348.8	−4,435.16	−1,745.09	−842.574
3.00	−187,948	−19,978.6	−5,314.09	−2,062.04	−984.975
3.05	−237,809	−24,435.5	−6,372.61	−2,438.58	−1,152.39
3.10	−301,159	−29,912.0	−7,648.36	−2,886.26	−1,349.34

Table 1.4 (continued)

			t		
v	0.05	0.10	0.15	0.20	0.25
3.15	−381,712	−36,646.6	−9,187.07	−3,418.89	−1,581.22
3.20	−484,219	−44,934.5	−11,044.3	−4,053.05	−1,854.41
3.25	−614,763	−55,141.6	−13,287.5	−4,808.61	−2,176.48
3.30	−781,141	−67,721.5	−15,999.0	−5,709.45	−2,556.44
3.35	−993,344	−83,236.9	−19,278.8	−6,784.25	−3,005.01
3.40	−1.2642E+6	−102,387	−23,248.7	−8,067.46	−3,534.91
3.45	−1.6102	−126,038	−28,057.2	−9,600.53	−4,161.32
3.50	−2.0524	−155,271	−33,885.4	−11,433.3	−4,902.29
3.55	−2.6180	−191,426	−40,954.4	−13,625.9	−5,779.34
3.60	−3.3421	−236,172	−49,533.9	−16,250.5	−6,818.12
3.65	−4.2695	−291,590	−59,953.5	−19,394.4	−8,049.24
3.70	−5.4582	−360,268	−72,616.1	−23,162.7	−9,509.20
3.75	−6.9829	−445,436	−88,014.4	−27,682.1	−11,241.6
3.80	−8.9399	−551,121	−106,751	−33,105.9	−13,298.6
3.85	−1.1453E+7	−682,348	−129,564	−39,618.9	−15,742.5
3.90	−1.4683	−845,394	−157,358	−47,444.6	−18,647.6
3.95	−1.8837	−1.0481E+6	−191,240	−56,853.4	−22,103.3
4.00	−2.4182	−1.3003	−232,570	−68,172.0	−26,216.1
4.05	−3.1064	−1.6142	−283,013	−81,796.2	−31,113.9
4.10	−3.9931	−2.0051	−344,618	−98,205.1	−36,949.8
4.15	−5.1362	−2.4923	−419,896	−117,979	−43,907.5
4.20	−6.6107	−3.0999	−511,935	−141,823	−52,207.3
4.25	−8.5139	−3.8580	−624,534	−170,589	−62,113.5
4.30	−1.0972E+8	−4.8044	−762,360	−205,313	−73,943.5
4.35	−1.4148	−5.9866	−931,162	−247,253	−88,078.8
4.40	−1.8255	−7.4641	−1.13802E+6	−297,935	−104,978
4.45	−2.3567	−9.3119	−1.39164	−359,215	−125,191
4.50	−3.0445	−1.1624E+7	−1.70277	−433,349	−149,382
4.55	−3.9351	−1.4518	−2.08466	−523,080	−178,348
4.60	−5.0894	−1.8144	−2.55364	−631,747	−213,050
4.65	−6.5859	−2.2687	−3.12989	−763,415	−254,646
4.70	−8.5273	−2.8385	−3.83831	−923,034	−304,529
4.75	−1.1047E+9	−3.5533	−4.70966	−1.11664E+6	−364,382
4.80	−1.4320	−4.4505	−5.78195	−1.35158	−436,232
4.85	−1.8572	−5.5773	−7.10221	−1.63682	−522,528
4.90	−2.4099	−6.9931	−8.72859	−1.98332	−626,224
4.95	−3.1289	−8.7731	−1.07331E+7	−2.40442	−750,890
5.00	−4.0645	−1.1012E+8	−1.32047	−2.91644	−900,839

			t		
v	0.30	0.35	0.40	0.45	0.50
0.00	−1.53565	−1.52306	−1.50859	−1.49228	−1.47414
0.05	−1.40677	−1.40109	−1.39446	−1.38639	−1.37663

Table 1.4 (continued)

			t		
v	0.30	0.35	0.40	0.45	0.50
0.10	−1.29723	−1.29392	−1.29153	−1.28878	−1.28498
0.15	−1.20731	−1.20212	−1.20057	−1.20038	−1.20024
0.20	−1.13712	−1.12610	−1.12220	−1.12193	−1.12324
0.25	−1.08669	−1.06609	−1.05683	−1.05399	−1.05465
0.30	−1.05596	−1.02225	−1.00479	−0.997023	−0.995037
0.35	−1.04490	−0.994671	−0.966289	−0.951348	−0.944812
0.40	−1.05351	−0.983441	−0.941505	−0.917229	−0.904322
0.45	−1.08188	−0.988657	−0.930586	−0.894878	−0.873842
0.50	−1.13024	−1.01047	−0.933684	−0.884479	−0.853600
0.55	−1.19897	−1.04912	−0.950985	−0.886216	−0.843802
0.60	−1.28865	−1.10495	−0.982732	−0.900291	−0.844646
0.65	−1.40008	−1.17846	−1.02925	−0.926946	−0.856341
0.70	−1.53435	−1.27029	−1.09096	−0.966482	−0.879124
0.75	−1.69284	−1.38131	−1.16841	−1.01928	−0.913278
0.80	−1.87728	−1.51258	−1.26232	−1.08581	−0.959145
0.85	−2.08979	−1.66546	−1.37354	−1.16667	−1.017140
0.90	−2.33293	−1.84156	−1.50316	−1.26259	−1.087790
0.95	−2.60976	−2.04283	−1.65245	−1.37444	−1.171690
1.00	−2.92389	−2.2716	−1.82298	−1.50329	−1.269590
1.05	−3.27957	−2.5306	−2.01656	−1.65038	−1.382380
1.10	−3.68176	−2.8230	−2.23534	−1.81721	−1.511110
1.15	−4.13624	−3.1527	−2.48184	−2.00551	−1.657000
1.20	−4.64972	−3.5238	−2.75895	−2.21728	−1.821480
1.25	−5.22999	−3.9414	−3.07006	−2.45488	−2.006210
1.30	−5.88603	−4.4113	−3.41903	−2.72098	−2.213110
1.35	−6.62823	−4.9402	−3.81034	−3.01870	−2.444380
1.40	−7.46859	−5.5357	−4.24911	−3.35157	−2.702550
1.45	−8.42095	−6.2066	−4.74122	−3.72366	−2.990510
1.50	−9.50128	−6.9630	−5.29339	−4.13962	−3.311560
1.55	−10.7280	−7.8166	−5.91332	−4.60474	−3.669460
1.60	−12.1225	−8.7806	−6.60982	−5.12508	−4.068490
1.65	−13.7092	−9.8705	−7.39296	−5.70752	−4.513510
1.70	−15.5166	−11.104	−8.27426	−6.35992	−5.010070
1.75	−17.5776	−12.501	−9.26688	−7.09123	−5.564430
1.80	−19.9300	−14.085	−10.3859	−7.91164	−6.183750
1.85	−22.6178	−15.882	−11.6486	−8.83277	−6.876120
1.90	−25.6919	−17.924	−13.0746	−9.86787	−7.650760
1.95	−29.2113	−20.246	−14.6867	−11.0321	−8.518130
2.00	−33.2443	−22.888	−16.5107	−12.3425	−9.490110
2.05	−37.8703	−25.899	−18.5763	−13.8190	−10.5802
2.10	−43.1812	−29.331	−20.9176	−15.4839	−11.8038
2.15	−49.2842	−33.248	−23.5738	−17.3628	−13.1784
2.20	−56.3036	−37.722	−26.5897	−19.4852	−14.7238

Table 1.4 (continued)

			t		
v	0.30	0.35	0.40	0.45	0.50
2.25	−64.3842	−42.838	−30.0171	−21.8844	−16.4628
2.30	−73.6943	−48.690	−33.9153	−24.5990	−18.4210
2.35	−84.4304	−55.392	−38.3527	−27.6727	−20.6281
2.40	−96.8211	−63.074	−43.4078	−31.1559	−23.1174
2.45	−111.133	−71.885	−49.1714	−35.1064	−25.9274
2.50	−127.678	−81.999	−55.7479	−39.5902	−29.1016
2.55	−146.820	−93.619	−63.2577	−44.6832	−32.6902
2.60	−168.984	−106.98	−71.8400	−50.4725	−36.7501
2.65	−194.667	−122.35	−81.6553	−57.0584	−41.3466
2.70	−224.451	−140.05	−92.8890	−64.5559	−46.5546
2.75	−259.017	−160.45	−105.756	−73.0974	−52.4594
2.80	−299.164	−183.98	−120.504	−82.8352	−59.1592
2.85	−345.826	−211.13	−137.420	−93.9448	−66.7662
2.90	−400.101	−242.48	−156.837	−106.628	−75.4093
2.95	−463.277	−278.72	−179.139	−121.118	−85.2363
3.00	−536.868	−320.63	−204.775	−137.684	−96.4168
3.05	−622.649	−369.13	−234.262	−156.636	−109.146
3.10	−722.712	−425.31	−268.202	−178.331	−123.647
3.15	−839.515	−490.41	−307.293	−203.183	−140.179
3.20	−975.952	−565.91	−352.348	−231.671	−159.037
3.25	−1,135.43	−653.53	−404.309	−264.347	−180.563
3.30	−1,321.97	−755.27	−464.274	−301.851	−205.149
3.35	−1,540.30	−873.51	−533.522	−344.924	−233.249
3.40	−1,796.01	−1,011.0	−613.540	−394.424	−265.384
3.45	−2,095.70	−1,170.9	−706.060	−451.345	−302.155
3.50	−2,447.15	−1,357.2	−813.104	−516.839	−344.259
3.55	−2,859.57	−1,574.1	−937.026	−592.244	−392.496
3.60	−3,343.84	−1,827.0	−1,080.58	−679.112	−447.794
3.65	−3,912.82	−2,122.0	−1,246.97	−779.245	−511.223
3.70	−4,581.74	−2,466.2	−1,439.94	−894.737	−584.020
3.75	−5,368.64	−2,868.3	−1,663.88	−1,028.02	−667.619
3.80	−6,294.89	−3,338.0	−1,923.91	−1,181.93	−763.677
3.85	−7,385.80	−3,887.3	−2,226.02	−1,359.75	−874.113
3.90	−8,671.42	−4,529.9	−2,577.22	−1,565.32	−1,001.15
3.95	−10,187.4	−5,282.0	−2,985.72	−1,803.11	−1,147.37
4.00	−11,975.9	−6,162.9	−3,461.13	−2,078.30	−1,315.75
4.05	−14,087.4	−7,195.2	−4,014.73	−2,396.97	−1,509.77
4.10	−16,581.4	−8,405.7	−4,659.73	−2,766.19	−1,733.44
4.15	−19,528.9	−9,825.7	−5,411.63	−3,194.20	−1,991.44
4.20	−23,014.4	−11,492.5	−6,288.63	−3,690.64	−2,289.19
4.25	−27,138.2	−13,450.2	−7,312.09	−4,266.74	−2,633.01
4.30	−32,019.9	−15,750.6	−8,507.09	−4,935.65	−3,030.22
4.35	−37,802.0	−18,455.3	−9,903.14	−5,712.71	−3,489.36

Table 1.4 (continued)

			t		
v	0.30	0.35	0.40	0.45	0.50
4.40	−44,654.2	−21,636.8	−11,534.9	−6,615.90	−4,020.34
4.45	−52,778.7	−25,381.3	−13,443.2	−7,666.21	−4,634.73
4.50	−62,416.8	−29,790.7	−15,675.9	−8,888.23	−5,345.98
4.55	−73,856.6	−34,985.7	−18,289.7	−10,310.7	−6,169.78
4.60	−87,441.7	−41,109.3	−21,351.1	−11,967.5	−7,124.41
4.65	−103,583	−48,331.2	−24,938.4	−13,898.0	−8,231.18
4.70	−122,770	−56,852.7	−29,144.2	−16,148.5	−9,514.98
4.75	−145,590	−66,912.5	−34,077.5	−18,773.5	−11,004.8
4.80	−172,744	−78,794.2	−39,867.0	−21,836.7	−12,734.6
4.85	−205,071	−92,834.6	−46,664.4	−25,412.9	−14,744.0
4.90	−243,575	−109,434	−54,649.2	−29,590.1	−17,079.2
4.95	−289,458	−129,068	−64,033.1	−34,471.5	−19,794.3
5.00	−344,162	−152,302	−75,066.6	−40,178.5	−22,952.7

			t		
v	0.55	0.60	0.65	0.70	0.75
0.00	−1.45423	−1.43257	−1.40921	−1.38419	−1.35755
0.05	−1.36505	−1.35159	−1.33624	−1.31901	−1.29993
0.10	−1.27973	−1.27278	−1.26402	−1.25337	−1.24081
0.15	−1.19936	−1.19728	−1.19368	−1.18839	−1.18130
0.20	−1.12487	−1.12603	−1.12621	−1.12507	−1.12239
0.25	−1.05700	−1.05986	−1.06246	−1.06428	−1.06496
0.30	−0.996376	−0.999439	−1.003150	−1.006770	−1.00979
0.35	−0.943506	−0.945340	−0.948900	−0.953202	−0.957549
0.40	−0.898800	−0.898034	−0.900229	−0.904130	−0.908838
0.45	−0.862596	−0.857910	−0.857574	−0.860031	−0.864160
0.50	−0.835170	−0.825294	−0.821303	−0.821310	−0.823954
0.55	−0.816759	−0.800459	−0.791725	−0.788313	−0.788596
0.60	−0.807577	−0.783642	−0.769108	−0.761336	−0.758411
0.65	−0.807827	−0.775058	−0.753684	−0.740634	−0.733677
0.70	−0.817719	−0.774910	−0.745665	−0.726433	−0.714640
0.75	−0.837485	−0.783408	−0.745253	−0.718940	−0.701518
0.80	−0.867390	−0.800774	−0.752651	−0.718354	−0.694512
0.85	−0.907745	−0.827255	−0.768072	−0.724869	−0.693813
0.90	−0.958924	−0.863141	−0.791750	−0.738690	−0.699610
0.95	−1.02137	−0.908767	−0.823953	−0.760040	−0.712098
1.00	−1.09563	−0.964532	−0.864985	−0.789164	−0.731488
1.05	−1.18234	−1.03091	−0.915205	−0.826344	−0.758008
1.10	−1.28224	−1.10844	−0.975031	−0.871904	−0.791919
1.15	−1.39623	−1.19780	−1.04495	−0.926219	−0.833516
1.20	−1.52533	−1.29973	−1.12554	−0.989725	−0.883140
1.25	−1.67076	−1.41513	−1.21746	−1.06293	−0.941180
1.30	−1.83391	−1.54502	−1.32148	−1.14641	−1.00809

Table 1.4 (continued)

			t		
ν	0.55	0.60	0.65	0.70	0.75
1.35	−2.01637	−1.69059	−1.43848	−1.24085	−1.08439
1.40	−2.21998	−1.85319	−1.56950	−1.34703	−1.17068
1.45	−2.44685	−2.03439	−1.71570	−1.46583	−1.26764
1.50	−2.69938	−2.23596	−1.87841	−1.59827	−1.37606
1.55	−2.98029	−2.45991	−2.05915	−1.74551	−1.49682
1.60	−3.29269	−2.70856	−2.25966	−1.90887	−1.63096
1.65	−3.64010	−2.98449	−2.48189	−2.08983	−1.77961
1.70	−4.02653	−3.29067	−2.72805	−2.29010	−1.94408
1.75	−4.45651	−3.63044	−3.00066	−2.51157	−2.12584
1.80	−4.93516	−4.00757	−3.30254	−2.75640	−2.32655
1.85	−5.46831	−4.42633	−3.63689	−3.02702	−2.54809
1.90	−6.06255	−4.89154	−4.00731	−3.32617	−2.79255
1.95	−6.72533	−5.40866	−4.41786	−3.65691	−3.06231
2.00	−7.46510	−5.98384	−4.87311	−4.02271	−3.36000
2.05	−8.29143	−6.62402	−5.37823	−4.42746	−3.68863
2.10	−9.21518	−7.33707	−5.93902	−4.87554	−4.05154
2.15	−10.2486	−8.13186	−6.56202	−5.37187	−4.45248
2.20	−11.4057	−9.01842	−7.25463	−5.92198	−4.89567
2.25	−12.7023	−10.0081	−8.02515	−6.53210	−5.38585
2.30	−14.1562	−11.1137	−8.88298	−7.20923	−5.92831
2.35	−15.7880	−12.3498	−9.83869	−7.96123	−6.52903
2.40	−17.6206	−13.7327	−10.9042	−8.79697	−7.19469
2.45	−19.6806	−15.2812	−12.0931	−9.72644	−7.93281
2.50	−21.9977	−17.0163	−13.4206	−10.7609	−8.75181
2.55	−24.6061	−18.9619	−14.9038	−11.9129	−9.66120
2.60	−27.5446	−21.1451	−16.5623	−13.1968	−10.6716
2.65	−30.8572	−23.5968	−18.4180	−14.6288	−11.7951
2.70	−34.5943	−26.3519	−20.4959	−16.2268	−13.0451
2.75	−38.8133	−29.4500	−22.8242	−18.0116	−14.4368
2.80	−43.5797	−32.9365	−25.4349	−20.0061	−15.9872
2.85	−48.9682	−36.8625	−28.3642	−22.2365	−17.7158
2.90	−55.0642	−41.2865	−31.6531	−24.7325	−19.6441
2.95	−61.9651	−46.2750	−35.3482	−27.5274	−21.7967
3.00	−69.7826	−51.9037	−39.5025	−30.6591	−24.2012
3.05	−78.6442	−58.2589	−44.1760	−34.1704	−26.8887
3.10	−88.6958	−65.4391	−49.4369	−38.1098	−29.8945
3.15	−100.105	−73.5565	−55.3629	−42.5322	−33.2583
3.20	−113.062	−82.7391	−62.0421	−47.5000	−37.0252
3.25	−127.788	−93.1333	−69.5750	−53.0838	−41.2458
3.30	−144.534	−104.906	−78.0759	−59.3638	−45.9778
3.35	−163.589	−118.249	−87.6748	−66.4311	−51.2862
3.40	−185.284	−133.380	−98.5203	−74.3890	−57.2449
3.45	−210.000	−150.548	−110.78	−83.3551	−63.9372

Table 1.4 (continued)

			t		
v	0.55	0.60	0.65	0.70	0.75
3.50	−238.176	−170.041	−124.65	−93.4629	−71.4579
3.55	−270.314	−192.186	−140.35	−104.864	−79.9144
3.60	−306.992	−217.358	−158.13	−117.733	−89.4284
3.65	−348.878	−245.987	−178.27	−132.264	−100.138
3.70	−396.736	−278.567	−201.11	−148.684	−112.201
3.75	−451.452	−315.665	−227.02	−167.247	−125.794
3.80	−514.042	−357.929	−256.43	−188.244	−141.122
3.85	−585.681	−406.107	−289.83	−212.008	−158.414
3.90	−667.721	−461.056	−327.79	−238.918	−177.932
3.95	−761.724	−523.762	−370.93	−269.407	−199.976
4.00	−869.495	−595.358	−420.02	−303.968	−224.883
4.05	−993.115	−677.150	−475.88	−343.167	−253.043
4.10	−1,134.99	−770.638	−539.49	−387.650	−284.895
4.15	−1,297.91	−877.551	−611.96	−438.154	−320.942
4.20	−1,485.09	−999.881	−694.57	−495.523	−361.757
4.25	−1,700.25	−1,139.92	−788.78	−560.725	−407.996
4.30	−1,947.70	−1,300.32	−896.29	−634.866	−460.403
4.35	−2,232.44	−1,484.14	−1,019.03	−719.214	−519.832
4.40	−2,560.26	−1,694.89	−1,159.21	−815.221	−587.257
4.45	−2,937.85	−1,936.64	−1,319.42	−924.554	−663.792
4.50	−3,373.00	−2,214.10	−1,502.59	−1,049.12	−750.707
4.55	−3,874.73	−2,532.68	−1,712.11	−1,191.12	−849.461
4.60	−4,453.50	−2,898.68	−1,951.91	−1,353.05	−961.717
4.65	−5,121.48	−3,319.34	−2,226.47	−1,537.82	−1,089.38
4.70	−5,892.78	−3,803.05	−2,541.00	−1,748.74	−1,234.64
4.75	−6,783.82	−4,359.55	−2,901.47	−1,989.62	−1,400.00
4.80	−7,813.66	−5,000.08	−3,314.80	−2,264.85	−1,588.31
4.85	−9,004.50	−5,737.68	−3,788.95	−2,579.47	−1,802.87
4.90	−10,382.1	−6,587.44	−4,333.12	−2,939.28	−2,047.46
4.95	−11,976.6	−7,566.87	−4,957.95	−3,350.97	−2,326.38
5.00	−13,823.0	−8,696.28	−5,675.70	−3,822.22	−2,644.61

			t		
v	0.80	0.85	0.90	0.95	1.0
0.00	−1.32935	−1.29963	−1.26846	−1.23588	−1.20197
0.05	−1.27904	−1.25638	−1.23202	−1.20600	−1.17839
0.10	−1.22634	−1.20999	−1.19179	−1.17178	−1.15001
0.15	−1.17234	−1.16150	−1.14878	−1.13418	−1.11775
0.20	−1.11802	−1.11188	−1.10392	−1.09412	−1.08248
0.25	−1.06425	−1.06200	−1.05808	−1.05243	−1.04501
0.30	−1.01183	−1.01264	−1.01203	−1.00988	−1.00610
0.35	−0.961439	−0.964506	−0.966483	−0.967175	−0.966442
0.40	−0.913698	−0.918228	−0.922068	−0.924948	−0.926667

Table 1.4 (continued)

v	0.80	0.85	0.90	0.95	1.0
			t		
0.45	-0.869140	-0.874359	-0.879355	-0.883777	-0.887358
0.50	-0.828230	-0.833385	-0.838848	-0.844179	-0.849039
0.55	-0.791372	-0.795735	-0.800994	-0.806616	-0.812183
0.60	-0.758915	-0.761779	-0.766185	-0.771495	-0.777212
0.65	-0.731163	-0.731846	-0.734767	-0.739182	-0.744505
0.70	-0.708381	-0.706217	-0.707043	-0.709996	-0.714397
0.75	-0.690802	-0.685145	-0.683281	-0.684221	-0.687188
0.80	-0.678637	-0.668851	-0.663716	-0.662109	-0.663142
0.85	-0.672078	-0.657537	-0.648561	-0.643881	-0.642496
0.90	-0.671309	-0.651388	-0.638007	-0.629740	-0.625461
0.95	-0.676514	-0.650581	-0.632234	-0.619870	-0.612228
1.00	-0.687879	-0.655292	-0.631410	-0.614441	-0.602973
1.05	-0.705600	-0.665698	-0.635705	-0.613616	-0.597859
1.10	-0.729895	-0.681987	-0.645286	-0.617555	-0.597043
1.15	-0.761003	-0.704361	-0.660334	-0.626422	-0.600678
1.20	-0.799197	-0.733045	-0.681037	-0.640384	-0.608920
1.25	-0.844785	-0.768288	-0.707605	-0.659622	-0.621927
1.30	-0.898122	-0.810375	-0.740270	-0.684331	-0.639872
1.35	-0.959616	-0.859629	-0.779295	-0.714729	-0.662938
1.40	-1.02973	-0.916418	-0.824976	-0.751057	-0.691327
1.45	-1.10900	-0.981165	-0.877649	-0.793589	-0.725265
1.50	-1.19804	-1.05435	-0.937696	-0.842633	-0.765004
1.55	-1.29755	-1.13652	-1.00555	-0.898538	-0.810829
1.60	-1.40830	-1.22829	-1.08171	-0.961700	-0.863060
1.65	-1.53121	-1.33038	-1.16673	-1.03257	-0.922059
1.70	-1.66728	-1.44358	-1.26126	-1.11165	-0.988236
1.75	-1.81767	-1.56879	-1.36599	-1.19952	-1.06205
1.80	-1.98367	-1.70704	-1.48175	-1.29682	-1.14403
1.85	-2.16673	-1.85947	-1.60944	-1.40427	-1.23475
1.90	-2.36849	-2.02736	-1.75008	-1.52271	-1.33488
1.95	-2.59080	-2.21217	-1.90483	-1.65304	-1.44516
2.00	-2.83571	-2.41551	-2.07496	-1.79630	-1.56640
2.05	-3.10555	-2.63921	-2.26192	-1.95363	-1.69955
2.10	-3.40290	-2.88530	-2.46732	-2.12632	-1.84563
2.15	-3.73069	-3.15605	-2.69297	-2.31582	-2.00580
2.20	-4.09215	-3.45401	-2.94087	-2.52372	-2.18136
2.25	-4.49095	-3.78204	-3.21327	-2.75182	-2.37374
2.30	-4.93117	-4.14332	-3.51269	-3.00211	-2.58454
2.35	-5.41739	-4.54141	-3.84193	-3.27683	-2.81557
2.40	-5.95473	-4.98029	-4.20411	-3.57846	-3.06879
2.45	-6.54894	-5.46441	-4.60274	-3.90977	-3.34644
2.50	-7.20645	-5.99876	-5.04170	-4.27383	-3.65096
2.55	-7.93450	-6.58891	-5.52535	-4.67410	-3.98512

Table 1.4 (continued)

			t		
v	0.80	0.85	0.90	0.95	1.0
2.60	−8.74117	−7.24109	−6.05856	−5.11440	−4.35195
2.65	−9.63553	−7.96227	−6.64675	−5.59901	−4.75486
2.70	−10.6278	−8.76027	−7.29599	−6.13269	−5.19763
2.75	−11.7294	−9.64384	−8.01305	−6.72076	−5.68447
2.80	−12.9532	−10.6228	−8.80552	−7.36916	−6.22007
2.85	−14.3135	−11.7081	−9.68188	−8.08449	−6.80966
2.90	−15.8268	−12.9120	−10.6516	−8.87415	−7.45906
2.95	−17.5111	−14.2485	−11.7253	−9.74639	−8.17476
3.00	−19.3871	−15.7330	−12.9149	−10.7104	−8.96400
3.05	−21.4778	−17.3829	−14.2336	−11.7766	−9.83484
3.10	−23.8093	−19.2178	−15.6965	−12.9564	−10.7963
3.15	−26.4109	−21.2597	−17.3201	−14.2627	−11.8584
3.20	−29.3156	−23.5333	−19.1234	−15.7099	−13.0324
3.25	−32.5609	−26.0663	−21.1273	−17.3143	−14.3309
3.30	−36.1886	−28.8902	−23.3554	−19.0938	−15.7678
3.35	−40.2463	−32.0401	−25.8343	−21.0688	−17.3588
3.40	−44.7876	−35.5556	−28.5938	−23.2620	−19.1215
3.45	−49.8730	−39.4816	−31.6675	−25.6988	−21.0754
3.50	−55.5711	−43.8683	−35.0929	−28.4079	−23.2426
3.55	−61.9592	−48.7726	−38.9126	−31.4213	−25.6476
3.60	−69.1250	−54.2587	−43.1742	−34.7750	−28.3180
3.65	−77.1675	−60.3989	−47.9314	−38.5096	−31.2846
3.70	−86.1990	−67.2750	−53.2449	−42.6704	−34.5821
3.75	−96.3466	−74.9793	−59.1827	−47.3087	−38.2494
3.80	−107.755	−83.6162	−65.8219	−52.4820	−42.3301
3.85	−120.586	−93.3039	−73.2491	−58.2550	−46.8730
3.90	−135.027	−104.176	−81.5622	−64.7005	−51.9333
3.95	−151.286	−116.383	−90.8719	−71.9006	−57.5727
4.00	−169.604	−130.097	−101.303	−79.9478	−63.8606
4.05	−190.251	−145.511	−112.996	−88.9463	−70.8752
4.10	−213.535	−162.845	−126.111	−99.0136	−78.7044
4.15	−239.807	−182.348	−140.828	−110.282	−87.4471
4.20	−269.464	−204.302	−157.350	−122.902	−97.2146
4.25	−302.961	−229.028	−175.909	−137.041	−108.132
4.30	−340.812	−256.890	−196.766	−152.891	−120.342
4.35	−383.605	−288.300	−220.216	−170.667	−134.003
4.40	−432.009	−323.727	−246.595	−190.613	−149.294
4.45	−486.785	−363.705	−276.283	−213.004	−166.419
4.50	−548.804	−408.840	−309.711	−238.152	−185.606
4.55	−619.055	−459.820	−347.367	−266.409	−207.114
4.60	−698.669	−517.431	−389.806	−298.175	−231.235
4.65	−788.937	−582.564	−437.658	−333.902	−258.298
4.70	−891.331	−656.237	−491.638	−374.101	−288.675
4.75	−1,007.54	−739.607	−552.559	−419.354	−322.790

Table 1.4 (continued)

			t		
v	0.80	0.85	0.90	0.95	1.0
4.80	−1,139.47	−833.993	−621.344	−470.319	−361.117
4.85	−1,289.34	−940.900	−699.044	−527.742	−404.198
4.90	−1,459.65	−1,062.04	−786.852	−592.471	−452.642
4.95	−1,653.29	−1,199.38	−886.129	−665.466	−507.140
5.00	−1,873.53	−1,355.14	−998.421	−747.821	−568.478

			t		
v	1.5	2.0	3.0	4.0	5.0
0.00	−0.803977	−0.351687	0.408489	0.623841	0.278968
0.05	−0.830077	−0.404731	0.358356	0.619232	0.313776
0.10	−0.849903	−0.453060	0.307641	0.611034	0.346047
0.15	−0.863866	−0.496663	0.256703	0.599438	0.375659
0.20	−0.872398	−0.535571	0.205882	0.584645	0.402515
0.25	−0.875944	−0.569857	0.155493	0.566871	0.426536
0.30	−0.874956	−0.599624	0.105830	0.546338	0.447669
0.35	−0.869888	−0.625010	0.057160	0.523279	0.465878
0.40	−0.861193	−0.646175	0.009727	0.497928	0.481150
0.45	−0.849315	−0.663304	−0.036249	0.470524	0.493489
0.50	−0.834689	−0.676601	−0.080576	0.441305	0.502918
0.55	−0.817737	−0.686284	−0.123082	0.410512	0.509478
0.60	−0.798866	−0.692585	−0.163622	0.378378	0.513224
0.65	−0.778465	−0.695743	−0.202075	0.345137	0.514226
0.70	−0.756906	−0.696004	−0.238339	0.311015	0.512569
0.75	−0.734538	−0.693619	−0.272337	0.276232	0.508349
0.80	−0.711694	−0.688839	−0.304011	0.240999	0.501673
0.85	−0.688683	−0.681914	−0.333323	0.205522	0.492657
0.90	−0.665794	−0.673093	−0.360253	0.169993	0.481426
0.95	−0.643296	−0.662618	−0.384800	0.134597	0.468113
1.00	−0.621439	−0.650729	−0.406976	0.099505	0.452857
1.05	−0.600452	−0.637657	−0.426810	0.064881	0.435799
1.10	−0.580546	−0.623625	−0.444343	0.030873	0.417086
1.15	−0.561918	−0.608849	−0.459630	−0.002381	0.396868
1.20	−0.544744	−0.593535	−0.472735	−0.034754	0.375295
1.25	−0.529189	−0.577878	−0.483733	−0.066134	0.352518
1.30	−0.515403	−0.562066	−0.492706	−0.096418	0.328688
1.35	−0.503525	−0.546275	−0.499744	−0.125517	0.303953
1.40	−0.493683	−0.530671	−0.504945	−0.153353	0.278462
1.45	−0.485997	−0.515411	−0.508409	−0.179860	0.252358
1.50	−0.480581	−0.500641	−0.510243	−0.204982	0.225781
1.55	−0.477540	−0.486498	−0.510554	−0.228676	0.198868
1.60	−0.476980	−0.473111	−0.509455	−0.250908	0.171751
1.65	−0.479001	−0.460600	−0.507057	−0.271654	0.144556
1.70	−0.483704	−0.449077	−0.503474	−0.290900	0.117402

Table 1.4 (continued)

			t		
v	1.5	2.0	3.0	4.0	5.0
1.75	−0.491194	−0.438644	−0.498820	−0.308639	0.090406
1.80	−0.501575	−0.429401	−0.493206	−0.324877	0.063674
1.85	−0.514959	−0.421439	−0.486744	−0.339622	0.037309
1.90	−0.531464	−0.414843	−0.479544	−0.352894	0.011404
1.95	−0.551216	−0.409696	−0.471713	−0.364719	−0.013953
2.00	−0.574353	−0.406075	−0.463356	−0.375126	−0.038681
2.05	−0.601024	−0.404056	−0.454576	−0.384153	−0.062706
2.10	−0.631394	−0.403710	−0.445471	−0.391842	−0.085963
2.15	−0.665645	−0.405111	−0.436137	−0.398240	−0.108392
2.20	−0.703976	−0.408329	−0.426666	−0.403397	−0.129941
2.25	−0.746610	−0.413436	−0.417147	−0.407367	−0.150565
2.30	−0.793790	−0.420505	−0.407664	−0.410207	−0.170226
2.35	−0.845791	−0.429613	−0.398299	−0.411977	−0.188890
2.40	−0.902912	−0.440839	−0.389128	−0.412737	−0.206532
2.45	−0.965487	−0.454266	−0.380226	−0.412552	−0.223133
2.50	−1.03389	−0.469984	−0.371662	−0.411486	−0.238680
2.55	−1.10852	−0.488087	−0.363503	−0.409603	−0.253163
2.60	−1.18983	−0.508679	−0.355812	−0.406970	−0.266580
2.65	−1.27833	−0.531872	−0.348649	−0.403652	−0.278933
2.70	−1.37456	−0.557788	−0.342072	−0.399715	−0.290230
2.75	−1.47913	−0.586558	−0.336134	−0.395224	−0.300482
2.80	−1.59271	−0.618329	−0.330887	−0.390244	−0.309704
2.85	−1.71604	−0.653259	−0.326380	−0.384837	−0.317916
2.90	−1.84994	−0.691522	−0.322661	−0.379067	−0.325140
2.95	−1.99529	−0.733309	−0.319775	−0.372995	−0.331404
3.00	−2.15309	−0.778829	−0.317766	−0.366680	−0.336736
3.05	−2.32443	−0.828311	−0.316678	−0.360180	−0.341167
3.10	−2.51050	−0.882008	−0.316551	−0.353552	−0.344733
3.15	−2.71262	−0.940194	−0.317429	−0.346851	−0.347468
3.20	−2.93224	−1.00317	−0.319352	−0.340129	−0.349411
3.25	−3.17096	−1.07127	−0.322361	−0.333438	−0.350601
3.30	−3.43052	−1.14486	−0.326499	−0.326826	−0.351079
3.35	−3.71288	−1.22432	−0.331809	−0.320341	−0.350886
3.40	−4.02016	−1.31010	−0.338335	−0.314028	−0.350066
3.45	−4.35470	−1.40267	−0.346123	−0.307931	−0.348660
3.50	−4.71909	−1.50254	−0.355221	−0.302091	−0.346712
3.55	−5.11618	−1.61029	−0.365679	−0.296549	−0.344265
3.60	−5.54910	−1.72653	−0.377552	−0.291343	−0.341365
3.65	−6.02132	−1.85193	−0.390895	−0.286509	−0.338052
3.70	−6.53665	−1.98725	−0.405771	−0.282083	−0.334372
3.75	−7.09930	−2.13329	−0.422243	−0.278099	−0.330365
3.80	−7.71391	−2.29093	−0.440381	−0.274589	−0.326076
3.85	−8.38560	−2.46114	−0.460262	−0.271585	−0.321543

Table 1.4 (continued)

			t		
v	1.5	2.0	3.0	4.0	5.0
3.90	−9.12003	−2.64496	−0.481967	−0.269116	−0.316810
3.95	−9.92344	−2.84356	−0.505582	−0.267213	−0.311914
4.00	−10.8027	−3.05819	−0.531204	−0.265904	−0.306896
4.05	−11.7656	−3.29024	−0.558934	−0.265218	−0.301792
4.10	−12.8203	−3.54120	−0.588885	−0.265182	−0.296640
4.15	−13.9763	−3.81272	−0.621177	−0.265824	−0.291475
4.20	−15.2440	−4.10660	−0.655941	−0.267170	−0.286332
4.25	−16.6346	−4.42482	−0.693318	−0.269250	−0.281245
4.30	−18.1608	−4.76952	−0.733461	−0.272089	−0.276246
4.35	−19.8367	−5.14306	−0.776537	−0.275717	−0.271366
4.40	−21.6777	−5.54804	−0.822726	−0.280162	−0.266635
4.45	−23.7010	−5.98726	−0.872222	−0.285455	−0.262083
4.50	−25.9257	−6.46383	−0.925238	−0.291625	−0.257737
4.55	−28.3729	−6.98114	−0.982000	−0.298705	−0.253626
4.60	−31.0660	−7.54291	−1.04276	−0.306729	−0.249773
4.65	−34.0310	−8.15321	−1.10778	−0.315731	−0.246206
4.70	−37.2969	−8.81651	−1.17735	−0.325749	−0.242948
4.75	−40.8956	−9.53771	−1.25179	−0.336823	−0.240022
4.80	−44.8628	−10.3222	−1.33144	−0.348995	−0.237452
4.85	−49.2380	−11.1758	−1.41666	−0.362308	−0.235259
4.90	−54.0654	−12.1051	−1.50785	−0.376812	−0.233464
4.95	−59.3937	−13.1172	−1.60544	−0.392557	−0.232089
5.00	−65.2775	−14.2199	−1.70990	−0.409598	−0.231154

Table 1.5: Second derivatives with respect to the order of the Bessel functions of the second kind $\partial^2 Y_v(t)/\partial v^2$.

			t		
v	0.05	0.10	0.15	0.20	0.25
0.00	0.639167	2.58204	3.00261	3.03613	2.93053
0.05	−1.54729	1.25504	2.08143	2.36175	2.42478
0.10	−3.59451	0.003844	1.19863	1.70325	1.92061
0.15	−5.57135	−1.191080	0.348929	1.06102	1.42095
0.20	−7.54834	−2.35171	−0.475212	0.433527	0.927114
0.25	−9.59875	−3.50221	−1.28330	−0.182439	0.439003
0.30	−11.8002	−4.66891	−2.08657	−0.791568	−0.044739
0.35	−14.2368	−5.88045	−2.89789	−1.39987	−0.526587
0.40	−17.0017	−7.16813	−3.73171	−2.01460	−1.01004
0.45	−20.2006	−8.56643	−4.60418	−2.64417	−1.49953
0.50	−23.9551	−10.1137	−5.53326	−3.29819	−2.00041

Table 1.5 (continued)

			t		
v	0.05	0.10	0.15	0.20	0.25
0.55	−28.4083	−11.8533	−6.53897	−3.98753	−2.51892
0.60	−33.7300	−13.8342	−7.64376	−4.72437	−3.06221
0.65	−40.1240	−16.1129	−8.87294	−5.52242	−3.63839
0.70	−47.8367	−18.7551	−10.2552	−6.39708	−4.25663
0.75	−57.1681	−21.8372	−11.8233	−7.36574	−4.92724
0.80	−68.4845	−25.4493	−13.6149	−8.44810	−5.66183
0.85	−82.2351	−29.6977	−15.6735	−9.66657	−6.47352
0.90	−98.9721	−34.7090	−18.0495	−11.0468	−7.37716
0.95	−119.376	−40.6335	−20.8019	−12.6181	−8.38959
1.00	−144.286	−47.6513	−23.9993	−14.4144	−9.53001
1.05	−174.740	−55.9778	−27.7227	−16.4749	−10.8203
1.10	−212.024	−65.8715	−32.0671	−18.8449	−12.2857
1.15	−257.729	−77.6432	−37.1447	−21.5772	−13.9549
1.20	−313.833	−91.6670	−43.0880	−24.7332	−15.8612
1.25	−382.790	−108.394	−50.0539	−28.3847	−18.0431
1.30	−467.654	−128.368	−58.2281	−32.6157	−20.5448
1.35	−572.228	−152.246	−67.8313	−37.5244	−23.4180
1.40	−701.250	−180.823	−79.1255	−43.2262	−26.7223
1.45	−860.634	−215.062	−92.4225	−49.8564	−30.5271
1.50	−1,057.76	−256.128	−108.093	−57.5742	−34.9133
1.55	−1,301.87	−305.435	−126.580	−66.5669	−39.9750
1.60	−1,604.51	−364.699	−148.409	−77.0551	−45.8221
1.65	−1,980.15	−436.004	−174.212	−89.2989	−52.5827
1.70	−2,446.92	−521.884	−204.738	−103.605	−60.4068
1.75	−3,027.60	−625.424	−240.889	−120.336	−69.4694
1.80	−3,750.79	−750.380	−283.738	−139.920	−79.9757
1.85	−4,652.42	−901.330	−334.574	−162.863	−92.1659
1.90	−5,777.73	−1,083.86	−394.941	−189.765	−106.322
1.95	−7,183.69	−1,304.79	−466.690	−221.336	−122.773
2.00	−8,942.11	−1,572.44	−552.042	−258.417	−141.908
2.05	−11,143.6	−1,897.01	−653.665	−302.006	−164.183
2.10	−13,902.5	−2,290.96	−774.768	−353.287	−190.131
2.15	−17,363.4	−2,769.56	−919.206	−413.666	−220.384
2.20	−21,709.1	−3,351.52	−1,091.62	−484.817	−255.683
2.25	−27,171.1	−4,059.77	−1,297.61	−568.726	−296.900
2.30	−34,042.4	−4,922.49	−1,543.90	−667.760	−345.064
2.35	−42,694.7	−5,974.26	−1,838.63	−784.737	−401.388
2.40	−53,599.6	−7,257.60	−2,191.61	−923.013	−467.305
2.45	−67,355.7	−8,824.81	−2,614.69	−1,086.59	−544.505
2.50	−84,723.8	−10,740.2	−3,122.18	−1,280.25	−634.984
2.55	−106,671	−13,083.2	−3,731.40	−1,509.69	−741.103
2.60	−134,430	−15,951.4	−4,463.30	−1,781.73	−865.654
2.65	−169,567	−19,465.4	−5,343.27	−2,104.50	−1,011.94

Table 1.5 (continued)

			t		
v	0.05	0.10	0.15	0.20	0.25
2.70	−214,080	−23,773.9	−6,402.02	−2,487.74	−1,183.88
2.75	−270,519	−29,060.7	−7,676.86	−2,943.12	−1,386.11
2.80	−342,136	−35,552.8	−9,212.98	−3,484.59	−1,624.13
2.85	−433,084	−43,531.0	−11,065.3	−4,128.87	−1,904.46
2.90	−548,673	−53,342.8	−13,300.4	−4,896.02	−2,234.85
2.95	−695,692	−65,418.5	−15,999.5	−5,810.09	−2,624.48
3.00	−882,828	−80,291.2	−19,261.0	−6,899.95	−3,084.29
3.05	−1.1212E+6	−98,621.7	−23,204.8	−8,200.29	−3,627.27
3.10	−1.4251	−121,230	−27,977.0	−9,752.77	−4,268.88
3.15	−1.8128	−149,134	−33,755.5	−11,607.5	−5,027.51
3.20	−2.3077	−183,597	−40,757.1	−13,824.8	−5,925.07
3.25	−2.9399	−226,192	−49,246.2	−16,477.1	−6,987.67
3.30	−3.7483	−278,870	−59,545.5	−19,651.9	−8,246.43
3.35	−4.7824	−344,064	−72,049.3	−23,454.4	−9,738.48
3.40	−6.1062	−424,799	−87,238.9	−28,011.7	−11,508.1
3.45	−7.8022	−524,846	−105,703	−33,476.7	−13,608.3
3.50	−9.9764	−648,903	−128,161	−40,034.5	−16,102.1
3.55	−1.2765E+7	−802,829	−155,494	−47,908.1	−19,065.2
3.60	−1.6345	−993,938	−188,781	−57,367.1	−22,587.8
3.65	−2.0943	−1.2314E+6	−229,342	−68,737.7	−26,778.1
3.70	−2.6853	−1.5265	−278,797	−82,413.9	−31,765.4
3.75	−3.4454	−1.8936	−339,133	−98,872.8	−37,704.8
3.80	−4.4235	−2.3505	−412,784	−118,692	−44,781.9
3.85	−5.6830	−2.9194	−502,743	−142,571	−53,219.2
3.90	−7.3057	−3.6284	−612,683	−171,357	−63,283.8
3.95	−9.3978	−4.5123	−747,117	−206,078	−75,296.1
4.00	−1.2097E+8	−5.6150	−911,596	−247,981	−89,640.7
4.05	−1.5580	−6.9914	−1.1130E+6	−298,579	−10,6779
4.10	−2.0079	−8.7105	−1.3596	−359,708	−127,268
4.15	−2.5893	−1.0859E+7	−1.6618	−433,600	−151,772
4.20	−3.3410	−1.3544	−2.0324	−522,968	−181,097
4.25	−4.3135	−1.6904	−2.4870	−631,109	−216,206
4.30	−5.5723	−2.1109	−3.0451	−762,033	−258,264
4.35	−7.2026	−2.6376	−3.7305	−920,623	−308,670
4.40	−9.3153	−3.2975	−4.5726	−1.11282E+6	−369,112
4.45	−1.2055E+9	−4.1247	−5.6079	−1.34587	−441,625
4.50	−1.5608	−5.1623	−6.8813	−1.62859	−528,660
4.55	−2.0220	−6.4645	−8.4484	−1.97174	−633,179
4.60	−2.6209	−8.0994	−1.0378E+7	−2.38845	−758,753
4.65	−3.3991	−1.0153	−1.2755	−2.89471	−909,696
4.70	−4.4107	−1.2735E+8	−1.5684	−3.51008	−1.09122
4.75	−5.7265	−1.5981	−1.9296	−4.25843	−1.30962E+6
4.80	−7.4386	−2.0065	−2.3752	−5.16892	−1.57252

Table 1.5 (continued)

			t		
v	0.05	0.10	0.15	0.20	0.25
4.85	−9.6676	−2.5205	−2.9252	−6.27722	−1.88913
4.90	−1.2571E+10	−3.1678	−3.6042	−7.62691	−2.27058
4.95	−1.6355	−3.9833	−4.4432	−9.27134	−2.73039
5.00	−2.1288	−5.0113	−5.4801	−1.12758E+7	−3.28487

			t		
v	0.30	0.35	0.40	0.45	0.50
0.00	2.76807	2.58211	2.38793	2.19297	2.00106
0.05	2.38555	2.29373	2.17367	2.03842	1.89539
0.10	1.99525	1.99130	1.94119	1.86290	1.76702
0.15	1.60119	1.67926	1.69497	1.67072	1.62000
0.20	1.206090	1.36099	1.43862	1.46553	1.45786
0.25	0.811476	1.03888	1.17496	1.25031	1.28359
0.30	0.417830	0.714428	0.906073	1.02744	1.09969
0.35	0.024669	0.388334	0.633341	0.798701	0.908156
0.40	−0.369356	0.060545	0.357515	0.565339	0.710534
0.45	−0.766387	−0.269671	0.078763	0.328097	0.507933
0.50	−1.16931	−0.603679	−0.203298	0.087246	0.301054
0.55	−1.58173	−0.943438	−0.489559	−0.157391	0.090209
0.60	−2.00793	−1.29148	−0.781394	−0.406406	−0.124651
0.65	−2.45292	−1.65088	−1.08064	−0.660788	−0.343925
0.70	−2.92242	−2.02530	−1.38960	−0.921912	−0.568342
0.75	−3.42292	−2.41893	−1.71100	−1.191530	−0.798953
0.80	−3.96177	−2.83661	−2.04807	−1.471760	−1.03712
0.85	−4.54724	−3.28378	−2.40447	−1.76511	−1.28453
0.90	−5.18866	−3.76660	−2.78440	−2.07447	−1.54317
0.95	−5.89653	−4.29202	−3.19261	−2.40314	−1.81537
1.00	−6.68276	−4.86785	−3.63445	−2.75489	−2.10378
1.05	−7.56081	−5.50289	−4.11591	−3.13392	−2.41143
1.10	−8.54597	−6.20710	−4.64378	−3.54500	−2.74174
1.15	−9.65564	−6.99170	−5.22567	−3.99348	−3.09855
1.20	−10.9097	−7.86943	−5.87016	−4.48535	−3.48618
1.25	−12.3309	−8.85474	−6.58692	−5.02737	−3.90947
1.30	−13.9452	−9.96406	−7.38692	−5.62713	−4.37385
1.35	−15.7827	−11.2161	−8.28254	−6.29318	−4.88542
1.40	−17.8776	−12.6324	−9.28785	−7.03516	−5.45102
1.45	−20.2698	−14.2372	−10.4188	−7.86398	−6.07837
1.50	−23.0049	−16.0587	−11.6936	−8.79198	−6.77613
1.55	−26.1359	−18.1290	−13.1330	−9.83313	−7.55407
1.60	−29.7240	−20.4850	−14.7605	−11.0033	−8.42325
1.65	−33.8399	−23.1691	−16.6033	−12.3205	−9.39615
1.70	−38.5660	−26.2304	−18.6923	−13.8053	−10.4869
1.75	−43.9975	−29.7250	−21.0628	−15.4809	−11.7115

Table 1.5 (continued)

			t		
v	0.30	0.35	0.40	0.45	0.50
1.80	−50.2451	−33.7179	−23.7555	−17.3741	−13.0881
1.85	−57.4373	−38.2843	−26.8169	−19.5153	−14.6375
1.90	−65.7239	−43.5106	−30.3006	−21.9391	−16.3829
1.95	−75.2788	−49.4970	−34.2680	−24.6854	−18.3513
2.00	−86.3050	−56.3595	−38.7900	−27.7995	−20.5729
2.05	−99.0388	−64.2322	−43.9478	−31.3335	−23.0825
2.10	−113.756	−73.2706	−49.8354	−35.3470	−25.9196
2.15	−130.778	−83.6550	−56.5610	−39.9086	−29.1293
2.20	−150.480	−95.5943	−64.2491	−45.0966	−32.7633
2.25	−173.302	−109.331	−73.0438	−51.0013	−36.8806
2.30	−199.757	−125.148	−83.1114	−57.7263	−41.5486
2.35	−230.445	−143.372	−94.6439	−65.3906	−46.8445
2.40	−266.070	−164.384	−107.864	−74.1314	−52.8567
2.45	−307.454	−188.628	−123.027	−84.1063	−59.6866
2.50	−355.563	−216.619	−140.433	−95.4971	−67.4503
2.55	−411.530	−248.960	−160.424	−108.513	−76.2812
2.60	−476.681	−286.350	−183.401	−123.396	−86.3321
2.65	−552.576	−329.607	−209.827	−140.423	−97.7788
2.70	−641.048	−379.685	−240.238	−159.917	−110.823
2.75	−744.249	−437.697	−275.257	−182.249	−125.697
2.80	−864.712	−504.944	−315.610	−207.847	−142.668
2.85	−1,005.42	−582.945	−362.138	−237.207	−162.042
2.90	−1,169.87	−673.480	−415.817	−270.904	−184.173
2.95	−1,362.21	−778.627	−477.787	−309.599	−209.470
3.00	−1,587.31	−900.822	−549.372	−354.063	−238.400
3.05	−1,850.91	−1,042.92	−632.112	−405.184	−271.507
3.10	−2,159.78	−1,208.25	−727.804	−463.995	−309.413
3.15	−2,521.95	−1,400.75	−838.542	−531.691	−352.841
3.20	−2,946.85	−1,625.00	−966.767	−609.661	−402.622
3.25	−3,445.66	−1,886.41	−1,115.33	−699.514	−459.718
3.30	−4,031.59	−2,191.30	−1,287.55	−803.121	−525.242
3.35	−4,720.26	−2,547.12	−1,487.31	−922.655	−600.477
3.40	−5,530.19	−2,962.62	−1,719.15	−1,060.64	−686.913
3.45	−6,483.26	−3,448.09	−1,988.38	−1,220.02	−786.269
3.50	−7,605.43	−4,015.63	−2,301.19	−1,404.20	−900.540
3.55	−8,927.47	−4,679.50	−2,664.85	−1,617.16	−1,032.04
3.60	−10,485.9	−5,456.47	−3,087.85	−1,863.55	−1,183.43
3.65	−12,323.9	−6,366.33	−3,580.15	−2,148.74	−1,357.83
3.70	−14,493.0	−7,432.37	−4,153.40	−2,479.03	−1,558.84
3.75	−17,054.1	−8,682.11	−4,821.28	−2,861.76	−1,790.63
3.80	−20,079.9	−10,148.0	−5,599.83	−3,305.48	−2,058.06
3.85	−23,656.5	−11,868.3	−6,507.84	−3,820.18	−2,366.77
3.90	−27,886.5	−13,888.2	−7,567.42	−4,417.51	−2,723.30
3.95	−32,891.9	−16,261.3	−8,804.49	−5,111.12	−3,135.28

Table 1.5 (continued)

			t		
v	0.30	0.35	0.40	0.45	0.50
4.00	−38,818.1	−19,050.6	−10,249.5	−5,916.90	−3,611.57
4.05	−45,838.0	−22,330.9	−11,938.4	−6,853.48	−4,162.46
4.10	−54,157.9	−26,190.7	−13,913.2	−7,942.64	−4,799.98
4.15	−64,023.6	−30,734.5	−16,223.4	−9,209.84	−5,538.10
4.20	−75,728.3	−36,086.2	−18,927.5	−10,684.9	−6,393.09
4.25	−89,621.7	−42,392.8	−22,094.1	−12,402.8	−7,383.96
4.30	−10,6121	−49,828.2	−25,804.2	−14,404.5	−8,532.84
4.35	−125,726	−58,598.7	−30,153.0	−16,737.9	−9,865.54
4.40	−149,032	−68,949.2	−35,253.1	−19,459.3	−11,412.2
4.45	−176,750	−81,170.1	−41,237.1	−22,634.8	−13,208.1
4.50	−209,733	−95,606.4	−48,261.5	−26,341.7	−15,294.2
4.55	−249,000	−112,668	−56,511.0	−30,671.1	−17,718.6
4.60	−295,769	−132,841	−66,203.8	−35,729.9	−20,537.4
4.65	−351,500	−156,705	−77,597.6	−41,643.5	−23,816.3
4.70	−417,942	−184,947	−90,997.2	−48,559.6	−27,632.2
4.75	−497,191	−218,388	−106,763	−56,651.7	−32,074.7
4.80	−591,758	−258,000	−125,320	−66,124.0	−37,249.3
4.85	−704,655	−304,945	−147,174	−77,216.9	−43,279.1
4.90	−839,498	−360,605	−172,921	−90,213.3	−50,308.5
4.95	−1.00062E+6	−426,627	−203,268	−105,446	−58,506.9
5.00	−1.19324	−504,975	−239,053	−123,309	−68,072.6

			t		
v	0.55	0.60	0.65	0.70	0.75
0.00	1.81414	1.63320	1.45868	1.29074	1.12938
0.05	1.74900	1.60195	1.45598	1.31219	1.17131
0.10	1.66015	1.54656	1.42910	1.30973	1.18983
0.15	1.55135	1.47043	1.38110	1.28608	1.18734
0.20	1.42593	1.37666	1.31481	1.24381	1.16614
0.25	1.28680	1.26801	1.23283	1.18531	1.12841
0.30	1.13645	1.14691	1.13748	1.11276	1.07618
0.35	0.976986	1.01547	1.03080	1.02813	1.01130
0.40	0.810108	0.875476	0.914594	0.933182	0.935468
0.45	0.637154	0.728383	0.790384	0.829441	0.850189
0.50	0.459108	0.575356	0.659436	0.718231	0.756795
0.55	0.276616	0.417266	0.522766	0.600652	0.656431
0.60	0.090005	0.254704	0.381146	0.477596	0.550063
0.65	−0.100701	0.087993	0.235112	0.349750	0.438477
0.70	−0.295770	−0.082800	0.084976	0.217600	0.322285
0.75	−0.495750	−0.257853	−0.069170	0.081440	0.201928
0.80	−0.701459	−0.437583	−0.227443	−0.058623	0.077681
0.85	−0.913983	−0.622640	−0.390166	−0.202663	−0.050342
0.90	−1.13467	−0.813901	−0.557860	−0.350934	−0.182187

Table 1.5 (continued)

			t		
v	0.55	0.60	0.65	0.70	0.75
0.95	−1.36515	−1.012470	−0.731246	−0.503860	−0.318052
1.00	−1.60731	−1.219690	−0.911239	−0.662039	−0.458287
1.05	−1.86333	−1.437120	−1.09896	−0.826239	−0.603396
1.10	−2.13570	−1.666580	−1.29571	−0.997403	−0.754029
1.15	−2.42722	−1.910130	−1.50304	−1.17665	−0.910990
1.20	−2.74107	−2.170140	−1.72268	−1.36528	−1.07524
1.25	−3.08078	−2.449230	−1.95662	−1.56478	−1.24789
1.30	−3.45034	−2.750370	−2.20709	−1.77685	−1.43024
1.35	−3.85421	−3.076880	−2.47660	−2.00341	−1.62374
1.40	−4.29737	−3.432480	−2.76797	−2.24659	−1.83004
1.45	−4.78543	−3.821310	−3.08431	−2.50879	−2.05100
1.50	−5.32465	−4.248040	−3.42915	−2.79270	−2.28868
1.55	−5.92208	−4.717840	−3.80639	−3.10130	−2.54541
1.60	−6.58564	−5.236560	−4.22040	−3.43794	−2.82376
1.65	−7.32425	−5.810710	−4.67607	−3.80633	−3.12660
1.70	−8.14796	−6.447640	−5.17886	−4.21061	−3.45712
1.75	−9.06809	−7.155570	−5.73489	−4.65543	−3.81890
1.80	−10.0975	−7.943770	−6.35103	−5.14596	−4.21590
1.85	−11.2505	−8.822680	−7.03497	−5.68798	−4.65255
1.90	−12.5437	−9.804060	−7.79534	−6.28797	−5.13379
1.95	−13.9956	−10.9012	−8.64186	−6.95319	−5.66513
2.00	−15.6271	−12.1290	−9.58547	−7.69178	−6.25275
2.05	−17.4623	−13.5046	−10.6385	−8.51288	−6.90353
2.10	−19.5283	−15.0470	−11.8148	−9.42673	−7.62522
2.15	−21.8559	−16.7779	−13.1300	−10.4449	−8.42646
2.20	−24.4803	−18.7221	−14.6019	−11.5803	−9.31698
2.25	−27.4414	−20.9073	−16.2503	−12.8477	−10.3077
2.30	−30.7847	−23.3653	−18.0980	−14.2634	−11.4108
2.35	−34.5622	−26.1320	−20.1704	−15.8462	−12.6402
2.40	−38.8330	−29.2483	−22.4966	−17.6168	−14.0112
2.45	−43.6647	−32.7607	−25.1093	−19.5992	−15.5414
2.50	−49.1345	−36.7219	−28.0457	−21.8199	−17.2505
2.55	−55.3304	−41.1922	−31.3480	−24.3094	−19.1605
2.60	−62.3533	−46.2400	−35.0641	−27.1017	−21.2965
2.65	−70.3183	−51.9435	−39.2483	−30.2358	−23.6867
2.70	−79.3573	−58.3916	−43.9624	−33.7554	−26.3630
2.75	−89.6213	−65.6859	−49.2766	−37.7104	−29.3613
2.80	−101.283	−73.9422	−55.2708	−42.1572	−32.7223
2.85	−114.541	−83.2927	−62.0359	−47.1597	−36.4921
2.90	−129.622	−93.8888	−69.6753	−52.7904	−40.7226
2.95	−146.788	−105.903	−78.3068	−59.1320	−45.4729
3.00	−166.336	−119.533	−88.0647	−66.2779	−50.8097
3.05	−188.612	−135.004	−99.1021	−74.3347	−56.8086
3.10	−214.009	−152.576	−111.594	−83.4232	−63.5554

Table 1.5 (continued)

			t		
v	0.55	0.60	0.65	0.70	0.75
3.15	−242.982	−172.544	−125.739	−93.6812	−71.1473
3.20	−276.052	−195.248	−141.765	−105.265	−79.6946
3.25	−313.819	−221.076	−159.931	−118.354	−89.3227
3.30	−356.976	−250.475	−180.536	−133.150	−100.174
3.35	−406.317	−283.956	−203.917	−149.885	−112.409
3.40	−462.759	−322.106	−230.464	−168.823	−126.213
3.45	−527.359	−365.599	−260.620	−190.265	−141.795
3.50	−601.336	−415.210	−294.895	−214.554	−159.391
3.55	−686.096	−471.828	−333.871	−242.083	−179.273
3.60	−783.260	−536.478	−378.214	−273.300	−201.748
3.65	−894.704	−610.336	−428.690	−308.715	−227.167
3.70	−1,022.59	−694.756	−486.176	−348.913	−255.930
3.75	−1,169.42	−791.299	−551.678	−394.564	−288.493
3.80	−1,338.09	−901.759	−626.351	−446.432	−325.376
3.85	−1,531.94	−1,028.21	−711.521	−505.392	−367.171
3.90	−1,754.85	−1,173.03	−808.710	−572.447	−414.557
3.95	−2,011.29	−1,338.98	−919.670	−648.745	−468.304
4.00	−2,306.46	−1,529.22	−1,046.41	−735.600	−529.298
4.05	−2,646.37	−1,747.43	−1,191.25	−834.520	−598.546
4.10	−3,038.00	−1,997.83	−1,356.84	−947.235	−677.203
4.15	−3,489.42	−2,285.31	−1,546.26	−1,075.73	−766.589
4.20	−4,010.03	−2,615.52	−1,763.03	−1,222.27	−868.211
4.25	−4,610.70	−2,994.97	−2,011.22	−1,389.49	−983.800
4.30	−5,304.08	−3,431.23	−2,295.50	−1,580.37	−1,115.33
4.35	−6,104.85	−3,933.03	−2,621.28	−1,798.37	−1,265.08
4.40	−7,030.07	−4,510.48	−2,994.79	−2,047.46	−1,435.63
4.45	−8,099.57	−5,175.28	−3,423.21	−2,332.19	−1,629.97
4.50	−9,336.42	−5,941.00	−3,914.83	−2,657.80	−1,851.51
4.55	−10,767.5	−6,823.35	−4,479.23	−3,030.33	−2,104.16
4.60	−12,423.9	−7,840.55	−5,127.47	−3,456.74	−2,392.43
4.65	−14,342.1	−9,013.72	−5,872.33	−3,945.01	−2,721.47
4.70	−16,564.5	−10,367.4	−6,728.58	−4,504.38	−3,097.22
4.75	−19,140.3	−11,930.0	−7,713.31	−5,145.46	−3,526.48
4.80	−22,127.2	−13,734.5	−8,846.27	−5,880.50	−4,017.07
4.85	−25,592.2	−15,819.4	−10,150.3	−6,723.63	−4,578.01
4.90	−29,613.5	−18,229.2	−11,652.0	−7,691.15	−5,219.64
4.95	−34,282.5	−21,015.7	−13,381.9	−8,801.88	−5,953.87
5.00	−39,705.8	−24,239.2	−15,375.6	−10,077.5	−6,794.41

			t		
v	0.80	0.85	0.90	0.95	1.0
0.00	0.974515	0.826013	0.683725	0.547499	0.417183
0.05	1.03382	0.900043	0.770187	0.644398	0.522771

Table 1.5 (continued)

			t		
v	0.80	0.85	0.90	0.95	1.0
0.10	1.07040	0.952152	0.835643	0.721283	0.609391
0.15	1.08633	0.984145	0.881627	0.779433	0.678085
0.20	1.08367	0.997816	0.909692	0.820174	0.729966
0.25	1.06438	0.994916	0.921384	0.844853	0.766192
0.30	1.03031	0.977129	0.918214	0.854812	0.787938
0.35	0.983196	0.946057	0.901644	0.851369	0.796380
0.40	0.924650	0.903199	0.873062	0.835797	0.792673
0.45	0.856125	0.849947	0.833778	0.809315	0.777937
0.50	0.778933	0.787572	0.785006	0.773068	0.753243
0.55	0.694232	0.717218	0.727860	0.728126	0.719605
0.60	0.603025	0.639901	0.663348	0.675471	0.677969
0.65	0.506164	0.556507	0.592365	0.615995	0.629205
0.70	0.404344	0.467787	0.515697	0.550494	0.574108
0.75	0.298114	0.374363	0.434012	0.479665	0.513386
0.80	0.187874	0.276726	0.347866	0.404106	0.447664
0.85	0.073878	0.175239	0.257702	0.324318	0.377480
0.90	-0.043757	0.070139	0.163849	0.240701	0.303285
0.95	-0.165056	-0.038459	0.066528	0.153553	0.225439
1.00	-0.290176	-0.150565	-0.034151	0.063080	0.144219
1.05	-0.419410	-0.266305	-0.138185	-0.030613	0.059812
1.10	-0.553183	-0.385925	-0.245678	-0.127515	-0.027681
1.15	-0.692053	-0.509789	-0.356837	-0.227710	-0.118243
1.20	-0.836713	-0.638378	-0.471978	-0.331377	-0.211943
1.25	-0.987991	-0.772295	-0.591517	-0.438787	-0.308936
1.30	-1.14686	-0.912261	-0.715983	-0.550309	-0.409459
1.35	-1.31443	-1.05912	-0.846007	-0.666405	-0.513836
1.40	-1.49198	-1.21386	-0.982338	-0.787636	-0.622478
1.45	-1.68095	-1.37759	-1.12584	-0.914661	-0.735881
1.50	-1.88294	-1.55157	-1.27749	-1.04825	-0.854635
1.55	-2.09977	-1.73722	-1.43841	-1.18927	-0.979422
1.60	-2.33344	-1.93612	-1.60984	-1.33871	-1.11102
1.65	-2.58621	-2.15004	-1.79320	-1.49769	-1.25032
1.70	-2.86056	-2.38094	-1.99003	-1.66745	-1.39831
1.75	-3.15927	-2.63102	-2.20207	-1.84938	-1.55610
1.80	-3.48543	-2.90269	-2.43125	-2.04502	-1.72494
1.85	-3.84247	-3.19865	-2.67972	-2.25608	-1.90620
1.90	-4.23422	-3.52192	-2.94984	-2.48446	-2.10141
1.95	-4.66493	-3.87580	-3.24426	-2.73226	-2.31226
2.00	-5.13935	-4.26401	-3.56588	-3.00182	-2.54064
2.05	-5.66278	-4.69067	-3.91797	-3.29571	-2.78860
2.10	-6.24112	-5.16037	-4.30413	-3.61681	-3.05846
2.15	-6.88100	-5.67821	-4.72836	-3.96830	-3.35276

Table 1.5 (continued)

			t		
v	0.80	0.85	0.90	0.95	1.0
2.20	−7.58979	−6.24992	−5.19513	−4.35369	−3.67431
2.25	−8.37577	−6.88184	−5.70940	−4.77691	−4.02624
2.30	−9.24821	−7.58110	−6.27670	−5.24232	−4.41202
2.35	−10.2175	−8.35564	−6.90320	−5.75475	−4.83549
2.40	−11.2952	−9.21436	−7.59579	−6.31960	−5.30092
2.45	−12.4946	−10.1672	−8.36214	−6.94287	−5.81305
2.50	−13.8302	−11.2254	−9.21084	−7.63125	−6.37716
2.55	−15.3186	−12.4013	−10.1515	−8.39221	−6.99913
2.60	−16.9783	−13.7090	−11.1948	−9.23408	−7.68548
2.65	−18.8303	−15.1643	−12.3529	−10.1662	−8.44350
2.70	−20.8981	−16.7849	−13.6392	−11.1988	−9.28131
2.75	−23.2083	−18.5905	−15.0687	−12.3438	−10.2080
2.80	−25.7906	−20.6036	−16.6584	−13.6139	−11.2336
2.85	−28.6789	−22.8492	−18.4274	−15.0239	−12.3694
2.90	−31.9112	−25.3557	−20.3969	−16.5899	−13.6282
2.95	−35.5304	−28.1548	−22.5909	−18.3304	−15.0239
3.00	−39.5851	−31.2825	−25.0363	−20.2657	−16.5724
3.05	−44.1301	−34.7791	−27.7633	−22.4188	−18.2913
3.10	−49.2272	−38.6902	−30.8061	−24.8155	−20.2004
3.15	−54.9468	−43.0673	−34.2030	−27.4849	−22.3218
3.20	−61.3679	−47.9683	−37.9970	−30.4593	−24.6804
3.25	−68.5805	−53.4588	−42.2367	−33.7753	−27.3041
3.30	−76.6862	−59.6129	−46.9769	−37.4740	−30.2240
3.35	−85.8002	−66.5140	−52.2793	−41.6016	−33.4752
3.40	−96.0531	−74.2568	−58.2135	−46.2102	−37.0971
3.45	−107.593	−82.9483	−64.8580	−51.3581	−41.1338
3.50	−120.588	−92.7094	−72.3014	−57.1113	−45.6350
3.55	−135.229	−103.677	−80.6438	−63.5441	−50.6566
3.60	−151.731	−116.007	−89.9984	−70.7400	−56.2613
3.65	−170.342	−129.874	−100.493	−78.7935	−62.5196
3.70	−191.341	−145.477	−112.271	−87.8110	−69.5113
3.75	−215.044	−163.043	−125.498	−97.9125	−77.3256
3.80	−241.815	−182.828	−140.357	−109.234	−86.0634
3.85	−272.063	−205.122	−157.058	−121.927	−95.8383
3.90	−306.256	−230.256	−175.837	−136.166	−106.778
3.95	−344.928	−258.604	−196.964	−152.146	−119.028
4.00	−388.684	−290.591	−220.742	−170.088	−132.750
4.05	−438.218	−326.702	−247.516	−190.242	−148.127
4.10	−494.316	−367.487	−277.678	−212.890	−165.369
4.15	−557.879	−413.572	−311.671	−238.353	−184.708
4.20	−629.932	−465.668	−349.998	−266.993	−206.410
4.25	−711.645	−524.586	−393.232	−299.219	−230.774

Table 1.5 (continued)

			t		
v	0.80	0.85	0.90	0.95	1.0
4.30	−804.356	−591.249	−442.022	−335.499	−258.138
4.35	−909.590	−666.708	−497.106	−376.357	−288.884
4.40	−1,029.09	−752.161	−559.323	−422.393	−323.445
4.45	−1,164.86	−848.973	−629.627	−474.285	−362.312
4.50	−1,319.17	−958.703	−709.102	−532.801	−406.038
4.55	−1,494.63	−1,083.13	−798.984	−598.816	−455.253
4.60	−1,694.23	−1,224.27	−900.678	−673.322	−510.667
4.65	−1,921.38	−1,384.45	−1,015.78	−757.446	−573.088
4.70	−2,180.00	−1,566.31	−1,146.13	−852.468	−643.431
4.75	−2,474.57	−1,772.87	−1,293.78	−959.844	−722.732
4.80	−2,810.22	−2,007.58	−1,461.12	−1,081.23	−812.169
4.85	−3,192.85	−2,274.40	−1,650.84	−1,218.51	−913.078
4.90	−3,629.22	−2,577.82	−1,866.02	−1,373.83	−1,026.98
4.95	−4,127.06	−2,923.02	−2,110.19	−1,549.63	−1,155.59
5.00	−4,695.28	−3,315.91	−2,387.35	−1,748.68	−1,300.87

			t		
v	1.5	2.0	3.0	4.0	5.0
0.00	−0.587371	−1.10774	−1.10774	−0.055120	0.720597
0.05	−0.457925	−1.01384	−1.01384	−0.128675	0.671228
0.10	−0.336475	−0.919277	−0.919277	−0.198600	0.619201
0.15	−0.223484	−0.824940	−0.824940	−0.264594	0.564971
0.20	−0.119281	−0.731645	−0.731645	−0.326404	0.508988
0.25	−0.024067	−0.640126	−0.640126	−0.383824	0.451688
0.30	0.062066	−0.551029	−0.551029	−0.436694	0.393496
0.35	0.139125	−0.464922	−0.464922	−0.484895	0.334823
0.40	0.207205	−0.382293	−0.382293	−0.528350	0.276060
0.45	0.266473	−0.303556	−0.303556	−0.567023	0.217579
0.50	0.317162	−0.229051	−0.229051	−0.600913	0.159731
0.55	0.359557	−0.159054	−0.159054	−0.630054	0.102845
0.60	0.393984	−0.093776	−0.093776	−0.654511	0.047227
0.65	0.420805	−0.033368	−0.033368	−0.674378	−0.006843
0.70	0.440404	0.022071	0.022071	−0.689775	−0.059109
0.75	0.453184	0.072490	0.072490	−0.700847	−0.109339
0.80	0.459556	0.117883	0.117883	−0.707758	−0.157328
0.85	0.459934	0.158285	0.158285	−0.710692	−0.202895
0.90	0.454731	0.193764	0.193764	−0.709847	−0.245885
0.95	0.444349	0.224418	0.224418	−0.705435	−0.286166
1.00	0.429180	0.250371	0.250371	−0.697678	−0.323633
1.05	0.409598	0.271771	0.271771	−0.686807	−0.358200
1.10	0.385959	0.288780	0.288780	−0.673057	−0.389807

Table 1.5 (continued)

			t		
v	1.5	2.0	3.0	4.0	5.0
1.15	0.358595	0.301576	0.301576	−0.656670	−0.418415
1.20	0.327813	0.310348	0.310348	−0.637888	−0.444003
1.25	0.293893	0.315291	0.315291	−0.616954	−0.466572
1.30	0.257085	0.316605	0.316605	−0.594108	−0.486140
1.35	0.217611	0.314491	0.314491	−0.569589	−0.502743
1.40	0.175659	0.309150	0.309150	−0.543630	−0.516432
1.45	0.131388	0.300778	0.300778	−0.516459	−0.527271
1.50	0.084922	0.289568	0.289568	−0.488297	−0.535340
1.55	0.036352	0.275704	0.275704	−0.459356	−0.540729
1.60	−0.014266	0.259363	0.259363	−0.429841	−0.543538
1.65	−0.066908	0.240709	0.240709	−0.399946	−0.543878
1.70	−0.121585	0.219897	0.219897	−0.369856	−0.541868
1.75	−0.178346	0.197068	0.197068	−0.339746	−0.537632
1.80	−0.237272	0.172352	0.172352	−0.309779	−0.531300
1.85	−0.298482	0.145861	0.145861	−0.280107	−0.523008
1.90	−0.362131	0.117694	0.117694	−0.250871	−0.512894
1.95	−0.428413	0.087934	0.087934	−0.222200	−0.501099
2.00	−0.497559	0.056649	0.056649	−0.194213	−0.487764
2.05	−0.569842	0.023888	0.023888	−0.167018	−0.473033
2.10	−0.645576	−0.010316	−0.010316	−0.140709	−0.457048
2.15	−0.725117	−0.045947	−0.045947	−0.115373	−0.439948
2.20	−0.808868	−0.083007	−0.083007	−0.091083	−0.421873
2.25	−0.897282	−0.121518	−0.121518	−0.067906	−0.402959
2.30	−0.990859	−0.161518	−0.161518	−0.045894	−0.383339
2.35	−1.09016	−0.203066	−0.203066	−0.025095	−0.363142
2.40	−1.19580	−0.246240	−0.246240	−0.005544	−0.342492
2.45	−1.30845	−0.291137	−0.291137	0.012731	−0.321510
2.50	−1.42887	−0.337877	−0.337877	0.029709	−0.300310
2.55	−1.55787	−0.386600	−0.386600	0.045377	−0.279002
2.60	−1.69636	−0.437467	−0.437467	0.059730	−0.257690
2.65	−1.84534	−0.490663	−0.490663	0.072768	−0.236472
2.70	−2.00588	−0.546399	−0.546399	0.084498	−0.215439
2.75	−2.17917	−0.604908	−0.604908	0.094930	−0.194678
2.80	−2.36654	−0.666452	−0.666452	0.104081	−0.174268
2.85	−2.56941	−0.731320	−0.731320	0.111971	−0.154284
2.90	−2.78937	−0.799832	−0.799832	0.118626	−0.134791
2.95	−3.02813	−0.872338	−0.872338	0.124072	−0.115853
3.00	−3.28761	−0.949224	−0.949224	0.128339	−0.097523
3.05	−3.56990	−1.03091	−1.03091	0.131460	−0.079852
3.10	−3.87729	−1.11786	−1.11786	0.133470	−0.062883
3.15	−4.21232	−1.21058	−1.21058	0.134404	−0.046654
3.20	−4.57776	−1.30962	−1.30962	0.134299	−0.031199

Table 1.5 (continued)

			t		
v	1.5	2.0	3.0	4.0	5.0
3.25	−4.97670	−1.41558	−1.41558	0.133193	−0.016544
3.30	−5.41249	−1.52911	−1.52911	0.131122	−0.002712
3.35	−5.88888	−1.65092	−1.65092	0.128126	0.010278
3.40	−6.40997	−1.78180	−1.78180	0.124240	0.022413
3.45	−6.98029	−1.92260	−1.92260	0.119502	0.033684
3.50	−7.60484	−2.07423	−2.07423	0.113947	0.044086
3.55	−8.28916	−2.23771	−2.23771	0.107608	0.053616
3.60	−9.03935	−2.41413	−2.41413	0.100520	0.062276
3.65	−9.86216	−2.60471	−2.60471	0.092711	0.070072
3.70	−10.7651	−2.81075	−2.81075	0.084213	0.077010
3.75	−11.7563	−3.03368	−3.03368	0.075050	0.083100
3.80	−12.8450	−3.27508	−3.27508	0.065248	0.088356
3.85	−14.0414	−3.53664	−3.53664	0.054829	0.092792
3.90	−15.3566	−3.82025	−3.82025	0.043811	0.096424
3.95	−16.8030	−4.12795	−4.12795	0.032211	0.099269
4.00	−18.3945	−4.46198	−4.46198	0.020042	0.101348
4.05	−20.1462	−4.82479	−4.82479	0.007315	0.102680
4.10	−22.0752	−5.21908	−5.21908	−0.005964	0.103287
4.15	−24.2003	−5.64778	−5.64778	−0.019792	0.103189
4.20	−26.5422	−6.11412	−6.11412	−0.034167	0.102410
4.25	−29.1243	−6.62165	−6.62165	−0.049094	0.100972
4.30	−31.9722	−7.17425	−7.17425	−0.064579	0.098896
4.35	−35.1145	−7.77617	−7.77617	−0.080635	0.096205
4.40	−38.5831	−8.43209	−8.43209	−0.097276	0.092922
4.45	−42.4134	−9.14715	−9.14715	−0.114521	0.089067
4.50	−46.6446	−9.92700	−9.92700	−0.132393	0.084661
4.55	−51.3206	−10.7778	−10.7778	−0.150922	0.079725
4.60	−56.4901	−11.7065	−11.7065	−0.170138	0.074277
4.65	−62.2073	−12.7204	−12.7204	−0.190079	0.068335
4.70	−68.5328	−13.8279	−13.8279	−0.210787	0.061916
4.75	−75.5338	−15.0380	−15.0380	−0.232309	0.055036
4.80	−83.2854	−16.3607	−16.3607	−0.254698	0.047709
4.85	−91.8713	−17.8070	−17.8070	−0.278010	0.039948
4.90	−101.385	−19.3891	−19.3891	−0.302310	0.031764
4.95	−111.930	−21.1202	−21.1202	−0.327668	0.023167
5.00	−123.624	−23.0151	−23.0151	−0.354159	0.014163

Table 1.6: Third derivatives with respect to the order of the Bessel functions of the second kind $\partial^3 Y_\nu(t)/\partial \nu^3$.

			t		
ν	0.05	0.10	0.15	0.20	0.25
0.00	−45.5748	−27.4145	−18.830	−13.6325	−10.1027
0.05	−42.1093	−25.7213	−18.027	−13.3336	−10.1119
0.10	−40.0089	−24.3917	−17.303	−13.0053	−10.0452
0.15	−39.2995	−23.4785	−16.710	−12.6891	−9.93684
0.20	−40.0237	−23.0272	−16.288	−12.4213	−9.81694
0.25	−42.2507	−23.0806	−16.073	−12.2330	−9.71207
0.30	−46.0870	−23.6828	−16.100	−12.1522	−9.64589
0.35	−51.6864	−24.8829	−16.401	−12.2042	−9.63970
0.40	−59.2614	−26.7388	−17.006	−12.4129	−9.71306
0.45	−69.0959	−29.3208	−17.952	−12.8019	−9.88439
0.50	−81.5606	−32.7153	−19.277	−13.3953	−10.1716
0.55	−97.1314	−37.0293	−21.026	−14.2187	−10.5926
0.60	−116.413	−42.3947	−23.250	−15.3005	−11.1661
0.65	−140.165	−48.9741	−26.013	−16.6726	−11.9119
0.70	−169.341	−56.9668	−29.387	−18.3717	−12.8520
0.75	−205.131	−66.6164	−33.463	−20.4403	−14.0109
0.80	−249.016	−78.2199	−38.346	−22.9287	−15.4164
0.85	−302.840	−92.1385	−44.165	−25.8955	−17.1007
0.90	−368.897	−108.811	−51.072	−29.4105	−19.1009
0.95	−450.037	−128.769	−59.250	−33.5555	−21.4602
1.00	−549.809	−152.658	−68.917	−38.4278	−24.2292
1.05	−672.629	−181.259	−80.335	−44.1418	−27.4669
1.10	−823.998	−215.521	−93.815	−50.8332	−31.2424
1.15	−1,010.77	−256.590	−109.73	−58.6620	−35.6367
1.20	−1,241.51	−305.859	−128.524	−67.8175	−40.7446
1.25	−1,526.90	−365.019	−150.727	−78.5231	−46.6773
1.30	−1,880.29	−436.119	−176.975	−91.0431	−53.5649
1.35	−2,318.39	−521.652	−208.024	−105.690	−61.5602
1.40	−2,862.13	−624.645	−244.781	−122.832	−70.8420
1.45	−3,537.73	−748.784	−288.330	−142.906	−81.6202
1.50	−4,378.09	−898.554	−339.967	−166.429	−94.1410
1.55	−5,424.53	−1,079.42	−401.247	−194.012	−108.693
1.60	−6,728.95	−1,298.04	−474.032	−226.380	−125.616
1.65	−8,356.67	−1,562.56	−560.556	−264.391	−145.307
1.70	−10,389.9	−1,882.89	−663.499	−309.062	−168.236
1.75	−12,932.2	−2,271.18	−786.080	−361.600	−194.951
1.80	−16,114.2	−2,742.26	−932.167	−423.441	−226.100
1.85	−20,100.9	−3,314.29	−1,106.41	−496.285	−262.445
1.90	−25,100.4	−4,009.52	−1,314.41	−582.159	−304.883
1.95	−31,376.1	−4,855.22	−1,562.91	−683.468	−354.469
2.00	−39,261.1	−5,884.82	−1,860.03	−803.078	−412.451
2.05	−49,177.1	−7,139.39	−2,215.57	−944.403	−480.298

Table 1.6 (continued)

v	0.05	0.10	0.15	0.20	0.25
			t		
2.10	−61,658.7	−8,669.35	−2,641.33	−1,111.51	−559.744
2.15	−77,383.8	−10,536.7	−3,151.60	−1,309.24	−652.838
2.20	−97,212.4	−12,817.6	−3,763.61	−1,543.38	−762.001
2.25	−122,237	−15,606.0	−4,498.20	−1,820.85	−890.095
2.30	−153,847	−19,017.5	−5,380.59	−2,149.88	−1,040.51
2.35	−193,809	−23,194.5	−6,441.29	−2,540.34	−1,217.24
2.40	−244,370	−28,312.7	−7,717.26	−3,004.02	−1,425.05
2.45	−308,395	−34,589.2	−9,253.32	−3,555.03	−1,669.56
2.50	−389,536	−42,291.6	−11,103.8	−4,210.27	−1,957.44
2.55	−492,448	−51,751.1	−13,334.6	−4,989.98	−2,296.60
2.60	−623,077	−63,376.9	−16,025.8	−5,918.44	−2,696.45
2.65	−789,017	−77,675.6	−19,274.7	−7,024.74	−3,168.13
2.70	−999,974	−95,274.4	−23,199.5	−8,343.83	−3,724.92
2.75	−1.26836E+6	−116,950	−27,943.9	−9,917.65	−4,382.57
2.80	−1.61007	−143,666	−33,683.2	−11,796.6	−5,159.84
2.85	−2.04545	−176,616	−40,630.3	−14,041.3	−6,079.07
2.90	−2.60059	−217,284	−49,045.1	−16,724.6	−7,166.83
2.95	−3.30893	−267,512	−59,244.1	−19,934.2	−8,454.82
3.00	−4.21339	−329,587	−71,613.6	−23,775.7	−9,980.79
3.05	−5.36908	−406,355	−86,625.1	−28,376.3	−11,789.8
3.10	−6.84681	−501,355	−104,854	−33,889.4	−13,935.6
3.15	−8.73760	−618,994	−127,004	−40,500.0	−16,482.4
3.20	−1.11585E+7	−764,760	−153,936	−48,431.1	−19,506.9
3.25	−1.42604	−945,492	−186,700	−57,952.3	−23,100.6
3.30	−1.82372	−1.169720E+6	−226,584	−69,389.0	−27,373.3
3.35	−2.33391	−1.448070	−275,164	−83,134.3	−32,455.9
3.40	−2.98886	−1.793840	−334,372	−99,664.0	−38,505.5
3.45	−3.83019	−2.223610	−406,574	−119,553	−45,709.9
3.50	−4.91160	−2.758100	−494,674	−143,498	−54,294.3
3.55	−6.30249	−3.423220	−602,231	−172,341	−64,528.5
3.60	−8.09249	−4.251390	−733,620	−207,105	−76,736.2
3.65	−1.03975	−5.283170	−894,209	−249,026	−91,305.8
3.70	−1.33676	−6.569370	−1.09060E+6	−299,607	−108,703
3.75	−1.71969	−8.173620	−1.33090	−360,667	−129,488
3.80	−2.21368	−1.017570E+7	−1.62509	−434,419	−154,334
3.85	−2.85132	−1.267570	−1.98546	−523,546	−184,047
3.90	−3.67484	−1.579910	−2.42711	−631,311	−219,602
3.95	−4.73905	−1.970350	−2.96868	−761,677	−262,167
4.00	−6.11507	−2.458680	−3.63312	−919,465	−313,149
4.05	−7.89524	−3.069780	−4.44873	−1.11054E+6	−374,245
4.10	−1.01995E+8	−3.834910	−5.45042	−1.34204	−447,496
4.15	−1.31839	−4.793410	−6.68126	−1.62266	−535,364
4.20	−1.70512	−5.994770	−8.19444	−1.96300	−640,815

Table 1.6 (continued)

			t		
v	0.05	0.10	0.15	0.20	0.25
4.25	−2.20652	−7.501290	−1.00557E+7	−2.37594	−767,430
4.30	−2.85694	−9.391450	−1.23461	−2.87724	−919,528
4.35	−3.70112	−1.176420E+8	−1.51662	−3.48609	−1.10232E+6
4.40	−4.79735	−1.474410	−1.86400	−4.22592	−1.32212
4.45	−6.22162	−1.848850	−2.29213	−5.12533	−1.58652
4.50	−8.07302	−2.319590	−2.82001	−6.21925	−1.90473
4.55	−1.04809E+9	−2.911660	−3.47120	−7.55036	−2.28788
4.60	−1.36140	−3.656730	−4.27489	−9.17086	−2.74941
4.65	−1.76928	−4.594750	−5.26726	−1.11446E+7	−3.30563
4.70	−2.30055	−5.776270	−6.49315	−1.35495	−3.97626
4.75	−2.99286	−7.265190	−8.00823	−1.64813	−4.78518
4.80	−3.08955	−9.142380	−9.88157	−2.00569	−5.76134
4.85	−5.07278	−1.151020E+9	−1.21989E+8	−2.44196	−6.93984
4.90	−6.60918	−1.449810	−1.50668	−2.97451	−8.36323
4.95	−8.61511	−1.827050	−1.86176	−3.62486	−1.00831E+7
5.00	−1.12353E+10	−2.303520	−2.30159	−4.41942	−1.21622

			t		
v	0.30	0.35	0.40	0.45	0.50
0.00	−7.53848	−5.59201	−4.06836	−2.84880	−1.85636
0.05	−7.74396	−5.92485	−4.48373	−3.31634	−2.35489
0.10	−7.85475	−6.15762	−4.80050	−3.69030	−2.76661
0.15	−7.89859	−6.31262	−5.03645	−3.98470	−3.10247
0.20	−7.90039	−6.41027	−5.20819	−4.21295	−3.37320
0.25	−7.88249	−6.46921	−5.33119	−4.38779	−3.58914
0.30	−7.86498	−6.50645	−5.41974	−4.52118	−3.76025
0.35	−7.86601	−6.53749	−5.48708	−4.62432	−3.89596
0.40	−7.90215	−6.57653	−5.54539	−4.70770	−4.00523
0.45	−7.98873	−6.63668	−5.60600	−4.78107	−4.09648
0.50	−8.14026	−6.73017	−5.67943	−4.85359	−4.17766
0.55	−8.37073	−6.86861	−5.77560	−4.93386	−4.25629
0.60	−8.69408	−7.06323	−5.90395	−5.03002	−4.33945
0.65	−9.12452	−7.32511	−6.07364	−5.14986	−4.43394
0.70	−9.67699	−7.66550	−6.29370	−5.30096	−4.54625
0.75	−10.3676	−8.09604	−6.57320	−5.49078	−4.68274
0.80	−11.2140	−8.62913	−6.92153	−5.72682	−4.84966
0.85	−12.2359	−9.27816	−7.34852	−6.01676	−5.05329
0.90	−13.4557	−10.0579	−7.86470	−6.36863	−5.30006
0.95	−14.8990	−10.9848	−8.48155	−6.79091	−5.59659
1.00	−16.5950	−12.0776	−9.21167	−7.29278	−5.94982
1.05	−18.5777	−13.3573	−10.0695	−7.88438	−6.36753
1.10	−20.8864	−14.8482	−11.0707	−8.57673	−6.85764
1.15	−23.5667	−16.5782	−12.2334	−9.38231	−7.42918

Table 1.6 (continued)

v	0.30	0.35	0.40	0.45	0.50
1.20	−26.6715	−18.5791	−13.5783	−10.3151	−8.09211
1.25	−30.2624	−20.8880	−15.1289	−11.3909	−8.85755
1.30	−34.4111	−23.5475	−16.9122	−12.6278	−9.73800
1.35	−39.2010	−26.6070	−18.9594	−14.0462	−10.7476
1.40	−44.7293	−30.1236	−21.3060	−15.6695	−11.9023
1.45	−51.1089	−34.1634	−23.9931	−17.5245	−13.2204
1.50	−58.4718	−38.8029	−27.0680	−19.6419	−14.7224
1.55	−66.9714	−44.1309	−30.5852	−22.0569	−16.4322
1.60	−76.7866	−50.2500	−34.6075	−24.8097	−18.3765
1.65	−88.1263	−57.2793	−39.2074	−27.9467	−20.5863
1.70	−101.234	−65.3571	−44.4685	−31.5212	−23.0967
1.75	−116.393	−74.6436	−50.4872	−35.5940	−25.9481
1.80	−133.937	−85.3246	−57.3748	−40.2355	−29.1865
1.85	−154.251	−97.6162	−65.2601	−45.5264	−32.8647
1.90	−177.790	−111.769	−74.2916	−51.5595	−37.0432
1.95	−205.083	−128.075	−84.6410	−58.4415	−41.7914
2.00	−236.750	−146.872	−96.5071	−66.2955	−47.1886
2.05	−273.516	−168.554	−110.120	−75.2630	−53.3259
2.10	−316.232	−193.582	−125.745	−85.5071	−60.3078
2.15	−365.892	−222.487	−143.691	−97.2157	−68.2543
2.20	−423.666	−255.894	−164.314	−110.606	−77.3030
2.25	−490.922	−294.526	−188.028	−125.927	−87.6122
2.30	−569.270	−339.230	−215.314	−143.468	−99.3636
2.35	−660.597	−390.993	−246.727	−163.562	−112.766
2.40	−767.124	−450.967	−282.914	−186.593	−128.060
2.45	−891.460	−520.498	−324.626	−213.008	−145.522
2.50	−1,036.68	−601.158	−372.735	−243.319	−165.470
2.55	−1,206.38	−694.787	−428.254	−278.122	−188.272
2.60	−1,404.84	−803.536	−492.364	−318.105	−214.349
2.65	−1,637.06	−929.922	−566.438	−364.066	−244.189
2.70	−1,908.96	−1,076.90	−652.073	−416.929	−278.354
2.75	−2,227.50	−1,247.91	−751.132	−477.764	−317.492
2.80	−2,600.93	−1,447.02	−865.784	−547.812	−362.351
2.85	−3,038.94	−1,678.97	−998.561	−628.515	−413.795
2.90	−3,553.03	−1,949.33	−1,152.41	−721.544	−472.824
2.95	−4,156.76	−2,264.65	−1,330.79	−828.841	−540.592
3.00	−4,866.17	−2,632.62	−1,537.71	−952.660	−618.433
3.05	−5,700.25	−3,062.26	−1,777.88	−1,095.63	−707.894
3.10	−6,681.47	−3,564.20	−2,056.78	−1,260.78	−810.763
3.15	−7,836.44	−4,150.92	−2,380.85	−1,451.68	−929.110
3.20	−9,196.70	−4,837.13	−2,757.61	−1,672.45	−1,065.33
3.25	−10,799.6	−5,640.13	−3,195.85	−1,927.90	−1,222.22
3.30	−12,689.5	−6,580.30	−3,705.87	−2,223.62	−1,402.99

Table 1.6 (continued)

			t		
v	0.30	0.35	0.40	0.45	0.50
3.35	−14,919.1	−7,681.68	−4,299.74	−2,566.14	−1,611.39
3.40	−17,550.6	−8,972.58	−4,991.62	−2,963.08	−1,851.75
3.45	−20,658.5	−10,486.4	−5,798.08	−3,423.31	−2,129.13
3.50	−24,330.7	−12,262.6	−6,738.60	−3,957.20	−2,449.39
3.55	−28,672.2	−14,347.8	−7,836.02	−4,576.83	−2,819.32
3.60	−33,807.4	−16,796.8	−9,117.15	−5,296.35	−3,246.84
3.65	−39,884.8	−19,674.8	−10,613.5	−6,132.25	−3,741.17
3.70	−47,081.0	−23,058.5	−12,362.1	−7,103.85	−4,313.00
3.75	−55,606.2	−27,038.8	−14,406.4	−8,233.72	−4,974.82
3.80	−65,710.9	−31,723.1	−16,797.8	−9,548.28	−5,741.15
3.85	−77,694.0	−37,238.9	−19,596.3	−11,078.4	−6,628.90
3.90	−91,911.6	−43,736.8	−22,873.0	−12,860.4	−7,657.79
3.95	−108,789	−51,395.4	−26,711.3	−14,936.6	−8,850.83
4.00	−128,833	−60,426.4	−31,209.7	−17,356.7	−10,234.8
4.05	−152,650	−71,080.8	−36,484.2	−20,179.0	−11,841.1
4.10	−180,963	−83,656.5	−42,671.4	−23,471.9	−13,706.1
4.15	−214,638	−98,506.9	−49,932.9	−27,315.6	−15,872.6
4.20	−254,709	−116,052	−58,458.9	−31,804.2	−18,390.4
4.25	−302,414	−136,789	−68,474.4	−37,048.3	−21,317.9
4.30	−359,232	−161,312	−80,244.7	−43,177.8	−24,723.0
4.35	−426,937	−190,323	−94,083.5	−50,345.5	−28,685.5
4.40	−507,651	−224,661	−110,362	−58,730.7	−33,298.7
4.45	−603,919	−265,322	−129,518	−68,544.6	−38,671.8
4.50	−718,789	−313,491	−152,071	−80,035.7	−44,932.4
4.55	−855,917	−370,580	−178,634	−93,496.3	−52,230.4
4.60	−1.01969E+6	−438,271	−209,934	−10,9271	−60,741.3
4.65	−1.21537	−518,567	−246,832	−127,765	−70,670.8
4.70	−1.44928	−613,857	−290,347	−149,456	−82,260.2
4.75	−1.72901	−726,990	−341,689	−174,908	−95,792.4
4.80	−2.06367	−861,363	−402,289	−204,785	−111,600
4.85	−2.46424	−1.02103E+6	−473,848	−239,871	−130,072
4.90	−2.94388	−1.21084	−558,382	−281,092	−151,668
4.95	−3.51845	−1.43657	−658,285	−329,539	−176,926
5.00	−4.20704	−1.70512	−776,399	−386,502	−206,477

			t		
v	0.55	0.60	0.65	0.70	0.75
0.00	−1.03875	−0.359062	0.209586	0.687271	1.08933
0.05	−1.55312	−0.878283	−0.306380	0.180555	0.596335
0.10	−1.98828	−1.32611	−0.758704	−0.269998	0.152379
0.15	−2.35260	−1.70881	−1.15189	−0.667439	−0.244359
0.20	−2.65453	−2.03290	−1.49084	−1.01529	−0.596253

Table 1.6 (continued)

			t		
ν	0.55	0.60	0.65	0.70	0.75
0.25	−2.90243	−2.30502	−1.78069	−1.31743	−0.906109
0.30	−3.10445	−2.53178	−2.02674	−1.57800	−1.17706
0.35	−3.26850	−2.71974	−2.23434	−1.80129	−1.41246
0.40	−3.40217	−2.87528	−2.40881	−1.99168	−1.61586
0.45	−3.51270	−3.00458	−2.55538	−2.15354	−1.79085
0.50	−3.60700	−3.11360	−2.67915	−2.29122	−1.94110
0.55	−3.69160	−3.20804	−2.78508	−2.40897	−2.07024
0.60	−3.77275	−3.29338	−2.87795	−2.51097	−2.18187
0.65	−3.85639	−3.37484	−2.96237	−2.60126	−2.27953
0.70	−3.94821	−3.45745	−3.04278	−2.68376	−2.36668
0.75	−4.05371	−3.54605	−3.12345	−2.76228	−2.44668
0.80	−4.17827	−3.64532	−3.20855	−2.84049	−2.52281
0.85	−4.32720	−3.75988	−3.30210	−2.92198	−2.59827
0.90	−4.50583	−3.89426	−3.40808	−3.01026	−2.67618
0.95	−4.71957	−4.05303	−3.53041	−3.10877	−2.75959
1.00	−4.97402	−4.24082	−3.67305	−3.22094	−2.85152
1.05	−5.27508	−4.46242	−3.83998	−3.35021	−2.95499
1.10	−5.62899	−4.72282	−4.03535	−3.50007	−3.07301
1.15	−6.04253	−5.02732	−4.26344	−3.67410	−3.20864
1.20	−6.52307	−5.38161	−4.52879	−3.87602	−3.36502
1.25	−7.07875	−5.79186	−4.83624	−4.10977	−3.54542
1.30	−7.71860	−6.26484	−5.19103	−4.37952	−3.75327
1.35	−8.45271	−6.80801	−5.59887	−4.68976	−3.99219
1.40	−9.29245	−7.42970	−6.06605	−5.04537	−4.26607
1.45	−10.2506	−8.13920	−6.59950	−5.45170	−4.57915
1.50	−11.3416	−8.94694	−7.20697	−5.91463	−4.93600
1.55	−12.5820	−9.86467	−7.89710	−6.44069	−5.34168
1.60	−13.9902	−10.9057	−8.67959	−7.03715	−5.80176
1.65	−15.5876	−12.0849	−9.56536	−7.71213	−6.32242
1.70	−17.3983	−13.4195	−10.5667	−8.47475	−6.91056
1.75	−19.4497	−14.9286	−11.6975	−9.33522	−7.57385
1.80	−21.7732	−16.6343	−12.9736	−10.3051	−8.32094
1.85	−24.4045	−18.5613	−14.4126	−11.3973	−9.16148
1.90	−27.3845	−20.7380	−16.0347	−12.6266	−10.1064
1.95	−30.7595	−23.1967	−17.8628	−14.0094	−11.1678
2.00	−34.5829	−25.9739	−19.9227	−15.5646	−12.3597
2.05	−38.9153	−29.1114	−22.2439	−17.3133	−13.6975
2.10	−43.8262	−32.6567	−24.8598	−19.2794	−15.1988
2.15	−49.3949	−36.6639	−27.8083	−21.4901	−16.8834
2.20	−55.7123	−41.1949	−31.1325	−23.9763	−18.7737
2.25	−62.8822	−46.3200	−34.8813	−26.7727	−20.8951
2.30	−71.0236	−52.1195	−39.1105	−29.9190	−23.2762
2.35	−80.2726	−58.6851	−43.8835	−33.4598	−25.9493

Table 1.6 (continued)

			t		
v	0.55	0.60	0.65	0.70	0.75
2.40	−90.7853	−66.1214	−49.2723	−37.4463	−28.9511
2.45	−102.741	−74.5480	−55.3591	−41.9360	−32.3232
2.50	−116.343	−84.1015	−62.2374	−46.9946	−36.1124
2.55	−131.829	−94.9378	−70.0139	−52.6967	−40.3720
2.60	−149.468	−107.236	−78.8098	−59.1269	−45.1624
2.65	−169.570	−121.199	−88.7639	−66.3816	−50.5519
2.70	−192.492	−137.063	−100.034	−74.5703	−56.6182
2.75	−218.642	−155.093	−112.801	−83.8175	−63.4492
2.80	−248.492	−175.598	−127.270	−94.2653	−71.1449
2.85	−282.582	−198.928	−143.677	−106.075	−79.8188
2.90	−321.537	−225.487	−162.290	−119.431	−89.5997
2.95	−366.072	−255.737	−183.417	−134.542	−100.634
3.00	−417.016	−290.209	−207.409	−151.649	−113.089
3.05	−475.319	−329.511	−234.669	−171.023	−127.153
3.10	−542.080	−374.344	−265.657	−192.977	−143.042
3.15	−618.564	−425.510	−300.899	−217.864	−161.001
3.20	−706.233	−483.935	−341.001	−246.092	−181.310
3.25	−806.772	−550.682	−386.653	−278.124	−204.287
3.30	−922.130	−626.971	−438.650	−314.489	−230.294
3.35	−1,054.55	−714.213	−497.901	−355.793	−259.745
3.40	−1,206.65	−814.025	−565.452	−402.730	−293.111
3.45	−1,381.42	−928.276	−642.501	−456.093	−330.930
3.50	−1,582.34	−1,059.12	−730.426	−516.789	−373.817
3.55	−1,813.44	−1,209.03	−830.809	−585.858	−422.472
3.60	−2,079.38	−1,380.87	−945.468	−664.492	−477.696
3.65	−2,385.56	−1,577.94	−1,076.49	−754.055	−540.405
3.70	−2,738.24	−1,804.06	−1,226.29	−856.114	−611.644
3.75	−3,144.66	−2,063.62	−1,397.63	−972.463	−692.611
3.80	−3,613.23	−2,361.70	−1,593.70	−1,105.16	−784.673
3.85	−4,153.71	−2,704.18	−1,818.16	−1,256.58	−889.398
3.90	−4,777.43	−3,097.86	−2,075.25	−1,429.43	−1,008.58
3.95	−5,497.52	−3,550.58	−2,369.83	−1,626.84	−1,144.27
4.00	−6,329.26	−4,071.44	−2,707.53	−1,852.38	−1,298.83
4.05	−7,290.39	−4,670.96	−3,094.83	−2,110.19	−1,474.95
4.10	−8,401.54	−5,361.32	−3,539.19	−2,405.00	−1,675.72
4.15	−9,686.69	−6,156.63	−4,049.26	−2,742.27	−1,904.70
4.20	−11,173.7	−7,073.26	−4,635.01	−3,128.28	−2,165.96
4.25	−12,895.2	−8,130.16	−5,307.93	−3,570.27	−2,464.17
4.30	−14,888.8	−9,349.32	−6,081.35	−4,076.55	−2,804.70
4.35	−17,198.7	−10,756.2	−6,970.64	−4,656.72	−3,193.71
4.40	−19,876.0	−12,380.6	−7,993.58	−5,321.85	−3,638.29
4.45	−22,980.8	−14,256.6	−9,170.76	−6,084.68	−4,146.59
4.50	−26,582.5	−16,424.4	−10,526.0	−6,959.93	−4,727.97

Table 1.6 (continued)

			t		
v	0.55	0.60	0.65	0.70	0.75
4.55	−30,762.7	−18,930.2	−12,086.9	−7,964.58	−5,393.21
4.60	−35,616.2	−21,828.0	−13,885.3	−9,118.21	−6,154.70
4.65	−41,253.8	−25,180.5	−15,958.2	−10,443.5	−7,026.72
4.70	−47,804.8	−29,060.6	−18,348.6	−11,966.5	−8,025.71
4.75	−55,420.4	−33,553.1	−21,106.1	−13,717.4	−9,170.60
4.80	−64,277.1	−38,756.9	−24,288.4	−15,731.2	−1,0483.2
4.85	−74,581.5	−44,786.8	−27,962.4	−18,048.3	−11,988.7
4.90	−86,574.9	−51,777.0	−32,205.7	−20,715.2	−13,716.1
4.95	−100,540	−59,883.4	−37,108.4	−23,786.1	−15,698.8
5.00	−116,806	−69,287.9	−42,775.2	−27,323.6	−17,975.5

			t		
v	0.80	0.85	0.90	0.95	1.0
0.00	1.42769	1.71181	1.94928	2.14628	2.30790
0.05	0.951741	1.25536	1.51413	1.73369	1.91875
0.10	0.518135	0.834998	1.10923	1.34599	1.54953
0.15	0.126047	0.450727	0.735312	0.984436	1.20195
0.20	−0.225914	0.101972	0.392460	0.649705	0.877167
0.25	−0.539629	−0.212356	0.080257	0.341939	0.575788
0.30	−0.817363	−0.493774	−0.202164	0.060831	0.297991
0.35	−1.06167	−0.744128	−0.456028	−0.194306	0.043551
0.40	−1.27534	−0.965534	−0.682869	−0.424492	−0.188094
0.45	−1.46128	−1.16030	−0.884449	−0.630996	−0.397780
0.50	−1.62253	−1.33090	−1.06272	−0.815311	−0.586587
0.55	−1.76216	−1.47987	−1.21976	−0.979090	−0.755779
0.60	−1.88326	−1.60984	−1.35775	−1.12412	−0.906778
0.65	−1.98888	−1.72343	−1.47893	−1.25227	−1.04112
0.70	−2.08207	−1.82329	−1.58558	−1.36547	−1.16042
0.75	−2.16580	−1.91203	−1.67997	−1.46570	−1.26637
0.80	−2.24298	−1.99224	−1.76439	−1.55496	−1.36070
0.85	−2.31646	−2.06645	−1.84108	−1.63523	−1.44515
0.90	−2.38903	−2.13716	−1.91229	−1.70849	−1.52147
0.95	−2.46343	−2.20682	−1.98019	−1.77672	−1.59143
1.00	−2.54235	−2.27784	−2.04697	−1.84186	−1.65676
1.05	−2.62843	−2.35258	−2.11475	−1.90582	−1.71919
1.10	−2.72435	−2.43343	−2.18565	−1.97053	−1.78046
1.15	−2.83277	−2.52273	−2.26178	−2.03787	−1.84226
1.20	−2.95638	−2.62286	−2.34523	−2.10971	−1.90628
1.25	−3.09798	−2.73621	−2.43812	−2.18794	−1.97422
1.30	−3.26042	−2.86525	−2.54260	−2.27448	−2.04778
1.35	−3.44673	−3.01253	−2.66088	−2.37123	−2.12867
1.40	−3.66010	−3.18073	−2.79523	−2.48019	−2.21864
1.45	−3.90392	−3.37265	−2.94802	−2.60339	−2.31946

Table 1.6 (continued)

			t		
v	0.80	0.85	0.90	0.95	1.0
1.50	−4.18187	−3.59131	−3.12177	−2.74296	−2.43297
1.55	−4.49795	−3.83992	−3.31913	−2.90115	−2.56110
1.60	−4.85651	−4.12198	−3.54296	−3.08033	−2.70586
1.65	−5.26238	−4.44130	−3.79635	−3.28304	−2.86938
1.70	−5.72085	−4.80208	−4.08264	−3.51202	−3.05393
1.75	−6.23784	−5.20892	−4.40552	−3.77025	−3.26197
1.80	−6.81993	−5.66692	−4.76901	−4.06097	−3.49615
1.85	−7.47443	−6.18176	−5.17756	−4.38772	−3.75933
1.90	−8.20960	−6.75975	−5.63612	−4.75443	−4.05469
1.95	−9.03463	−7.40794	−6.15015	−5.16541	−4.38567
2.00	−9.95987	−8.13423	−6.72575	−5.62546	−4.75611
2.05	−10.9970	−8.94743	−7.36978	−6.13992	−5.17019
2.10	−12.1590	−9.85746	−8.08979	−6.71468	−5.63262
2.15	−13.4606	−10.8755	−8.89432	−7.35638	−6.14860
2.20	−14.9186	−12.0139	−9.79291	−8.07240	−6.72389
2.25	−16.5515	−13.2868	−10.7963	−8.87098	−7.36496
2.30	−18.3804	−14.7100	−11.9163	−9.76135	−8.07898
2.35	−20.4293	−16.3013	−13.1667	−10.7539	−8.87398
2.40	−22.7249	−18.0806	−14.5623	−11.8601	−9.75896
2.45	−25.2975	−20.0706	−16.1203	−13.0930	−10.7439
2.50	−28.1815	−22.2966	−17.8598	−14.4672	−11.8401
2.55	−31.4157	−24.7873	−19.8022	−15.9990	−13.0602
2.60	−35.0436	−27.5749	−21.9717	−17.7067	−14.4181
2.65	−39.1150	−30.6960	−24.3956	−19.6111	−15.9297
2.70	−43.6857	−34.1917	−27.1046	−21.7352	−17.6128
2.75	−48.8193	−38.1083	−30.1331	−24.1050	−19.4872
2.80	−54.5874	−42.4984	−33.5200	−26.7499	−21.5751
2.85	−61.0713	−47.4211	−37.3093	−29.7027	−23.9016
2.90	−68.3632	−52.9434	−41.5503	−33.0005	−26.4948
2.95	−76.5673	−59.1411	−46.2988	−36.6850	−29.3863
3.00	−85.8021	−66.0998	−51.6179	−40.8032	−32.6115
3.05	−96.2018	−73.9162	−57.5785	−45.4078	−36.2102
3.10	−107.919	−82.7002	−64.2609	−50.5585	−40.2273
3.15	−121.126	−92.5758	−71.7557	−56.3224	−44.7133
3.20	−136.019	−103.684	−80.1654	−62.7752	−49.7247
3.25	−152.822	−116.183	−89.6058	−70.0025	−55.3256
3.30	−171.789	−130.255	−100.208	−78.1004	−61.5876
3.35	−193.206	−146.105	−112.120	−87.1778	−68.5920
3.40	−217.403	−163.964	−125.509	−97.3577	−76.4299
3.45	−244.753	−184.097	−140.566	−108.779	−85.2041
3.50	−275.679	−206.803	−157.505	−121.597	−95.0307
3.55	−310.666	−232.422	−176.570	−135.992	−106.040
3.60	−350.264	−261.341	−198.037	−152.161	−118.381

Table 1.6 (continued)

			t		
v	0.80	0.85	0.90	0.95	1.0
3.65	−395.102	−294.000	−222.220	−170.333	−132.219
3.70	−445.894	−330.897	−249.473	−190.765	−147.743
3.75	−503.457	−372.602	−280.199	−213.746	−165.165
3.80	−568.722	−419.760	−314.856	−239.605	−184.726
3.85	−642.751	−473.107	−353.964	−268.716	−206.696
3.90	−726.758	−533.483	−398.112	−301.501	−231.384
3.95	−822.128	−601.841	−447.970	−338.439	−259.137
4.00	−930.444	−679.270	−504.302	−380.072	−290.347
4.05	−1,053.52	−767.009	−567.973	−427.018	−325.459
4.10	−1,193.41	−866.474	−639.971	−479.975	−364.979
4.15	−1,352.50	−979.279	−721.415	−539.738	−409.476
4.20	−1,533.49	−1,107.26	−813.585	−607.209	−459.597
4.25	−1,739.48	−12,52.53	−917.934	−683.410	−516.075
4.30	−1,974.02	−1,417.48	−1,036.12	−769.509	−579.743
4.35	−2,241.18	−1,604.86	−1,170.03	−866.826	−651.544
4.40	−2,545.60	−1,817.80	−1,321.82	−976.869	−732.546
4.45	−2,892.64	−2,059.88	−1,493.94	−1,101.35	−823.967
4.50	−3,288.42	−2,335.20	−1,689.19	−1,242.22	−927.185
4.55	−3,739.96	−2,648.46	−1,910.76	−1,401.69	−1,043.77
4.60	−4,255.32	−3,005.01	−2,162.31	−1,582.29	−1,175.49
4.65	−4,843.76	−3,411.00	−2,448.00	−1,786.91	−1,324.39
4.70	−5,515.89	−3,873.46	−2,772.59	−2,018.82	−1,492.75
4.75	−6,283.92	−4,400.44	−3,141.51	−2,281.75	−1,683.20
4.80	−7,161.88	−5,001.20	−3,560.98	−2,579.99	−1,898.72
4.85	−8,165.87	−5,686.29	−4,038.10	−2,918.38	−2,142.69
4.90	−9,314.44	−6,467.88	−4,581.00	−3,302.48	−2,418.97
4.95	−10,628.9	−7,359.87	−5,198.98	−3,738.62	−2,731.96
5.00	−12,133.8	−8,378.26	−5,902.68	−4,234.05	−3,086.66

			t		
v	1.5	2.0	3.0	4.0	5.0
0.00	2.66507	1.86473	−0.384443	−1.50515	−0.957719
0.05	2.51075	1.88773	−0.232363	−1.43588	−1.01549
0.10	2.34571	1.89189	−0.088227	−1.36011	−1.06409
0.15	2.17281	1.87890	0.047339	−1.27881	−1.10363
0.20	1.99465	1.85044	0.173839	−1.19292	−1.13428
0.25	1.81360	1.80819	0.290891	−1.10336	−1.15630
0.30	1.63175	1.75380	0.398233	−1.01102	−1.16998
0.35	1.45094	1.68886	0.495707	−0.916752	−1.17564
0.40	1.27281	1.61492	0.583257	−0.821349	−1.17365
0.45	1.09871	1.53344	0.660913	−0.725575	−1.16443
0.50	0.929811	1.44582	0.728792	−0.630143	−1.14840
0.55	0.767068	1.35336	0.787081	−0.535705	−1.12602

Table 1.6 (continued)

			t		
v	1.5	2.0	3.0	4.0	5.0
0.60	0.611231	1.25728	0.836036	−0.442874	−1.09778
0.65	0.462883	1.15871	0.875964	−0.352198	−1.06417
0.70	0.322431	1.05866	0.907227	−0.264176	−1.02569
0.75	0.190134	0.958074	0.930226	−0.179255	−0.982833
0.80	0.066121	0.857784	0.945397	−0.097816	−0.936116
0.85	−0.049619	0.758526	0.953198	−0.020204	−0.886039
0.90	−0.157185	0.660955	0.954123	0.053297	−0.833099
0.95	−0.256791	0.565620	0.948655	0.122442	−0.777811
1.00	−0.348723	0.473011	0.937308	0.186912	−0.720578
1.05	−0.433349	0.383519	0.920592	0.247008	−0.661904
1.10	−0.511095	0.297440	0.899009	0.302171	−0.602234
1.15	−0.582442	0.215039	0.873066	0.352499	−0.541989
1.20	−0.647911	0.136487	0.843257	0.397968	−0.481552
1.25	−0.708069	0.061898	0.810062	0.438599	−0.421288
1.30	−0.763505	−0.008671	0.773945	0.474434	−0.361566
1.35	−0.814835	−0.075214	0.735351	0.505548	−0.302718
1.40	−0.862693	−0.137776	0.694707	0.532048	−0.245038
1.45	−0.907732	−0.196442	0.652418	0.554058	−0.188798
1.50	−0.950614	−0.251338	0.608864	0.571731	−0.134256
1.55	−0.992014	−0.302625	0.564400	0.585227	−0.081627
1.60	−1.03261	−0.350491	0.519357	0.594732	−0.031116
1.65	−1.07310	−0.395154	0.474041	0.600441	0.017108
1.70	−1.11417	−0.436854	0.428731	0.602559	0.062894
1.75	−1.15653	−0.475854	0.383676	0.601298	0.106124
1.80	−1.20090	−0.512433	0.339106	0.596883	0.146695
1.85	−1.24801	−0.546884	0.295224	0.589535	0.184523
1.90	−1.29860	−0.579515	0.252204	0.579483	0.219561
1.95	−1.35344	−0.610646	0.210202	0.566945	0.251753
2.00	−1.41332	−0.640604	0.169350	0.552162	0.280820
2.05	−1.47905	−0.669735	0.129752	0.535346	0.307644
2.10	−1.55148	−0.698368	0.091491	0.516715	0.331318
2.15	−1.63150	−0.726875	0.054640	0.496490	0.352199
2.20	−1.72006	−0.755609	0.019243	0.474870	0.370335
2.25	−1.81815	−0.784945	−0.014671	0.452058	0.385774
2.30	−1.92681	−0.815253	−0.047091	0.428243	0.398596
2.35	−2.04720	−0.846925	−0.078018	0.403613	0.408877
2.40	−2.18053	−0.880353	−0.107471	0.378337	0.416711
2.45	−2.32810	−0.915944	−0.135481	0.352580	0.422195
2.50	−2.49135	−0.954121	−0.162094	0.326494	0.425434
2.55	−2.67182	−0.995315	−0.187364	0.300223	0.426537
2.60	−2.87120	−1.039980	−0.211358	0.273899	0.425625
2.65	−3.09132	−1.088580	−0.234153	0.247643	0.422811
2.70	−3.33422	−1.141620	−0.255830	0.221568	0.418212

Table 1.6 (continued)

			t		
ν	1.5	2.0	3.0	4.0	5.0
2.75	−3.60209	−1.199610	−0.276484	0.195770	0.411965
2.80	−3.89737	−1.263110	−0.296215	0.170343	0.404173
2.85	−4.22273	−1.332680	−0.315127	0.145367	0.394974
2.90	−4.58113	−1.408950	−0.333332	0.120908	0.384506
2.95	−4.97582	−1.492580	−0.350949	0.097024	0.372840
3.00	−5.41037	−1.584320	−0.368101	0.073775	0.360163
3.05	−5.88876	−1.684790	−0.384914	0.051190	0.346532
3.10	−6.41535	−1.794960	−0.401518	0.029316	0.332088
3.15	−6.99501	−1.915660	−0.418054	0.008164	0.316939
3.20	−7.63307	−2.047850	−0.434656	−0.012237	0.301198
3.25	−8.33549	−2.192580	−0.451477	−0.031894	0.284931
3.30	−9.10883	−2.350970	−0.468655	−0.050794	0.268270
3.35	−9.96037	−2.524260	−0.486351	−0.068945	0.251292
3.40	−10.8982	−2.713790	−0.504720	−0.086360	0.234089
3.45	−11.9312	−2.921030	−0.523923	−0.103052	0.216736
3.50	−13.0694	−3.147590	−0.544131	−0.119051	0.199315
3.55	−14.3237	−3.395230	−0.565513	−0.134377	0.181893
3.60	−15.7063	−3.665860	−0.588253	−0.149068	0.164536
3.65	−17.2309	−3.961580	−0.612537	−0.163162	0.147303
3.70	−18.9125	−4.284730	−0.638557	−0.176694	0.130248
3.75	−20.7678	−4.637810	−0.666521	−0.189721	0.113420
3.80	−22.8154	−5.023620	−0.696641	−0.202282	0.096856
3.85	−25.0760	−5.445210	−0.729136	−0.214433	0.080607
3.90	−27.5726	−5.905930	−0.764246	−0.226226	0.064720
3.95	−30.3306	−6.409470	−0.802225	−0.237736	0.049162
4.00	−33.3786	−6.959860	−0.843261	−0.248995	0.034040
4.05	−36.7481	−7.561630	−0.887827	−0.260085	0.019301
4.10	−40.4744	−8.219600	−0.936048	−0.271069	0.005012
4.15	−44.5968	−8.939200	−0.988281	−0.282020	−0.008838
4.20	−49.1588	−9.726380	−1.04487	−0.293004	−0.022252
4.25	−54.2091	−10.5877	−1.10619	−0.304094	−0.035216
4.30	−59.8020	−11.5303	−1.17262	−0.315370	−0.047737
4.35	−65.9980	−12.5621	−1.24458	−0.326908	−0.059811
4.40	−72.8644	−13.6919	−1.32252	−0.338786	−0.071451
4.45	−80.4765	−14.9294	−1.40693	−0.351092	−0.082675
4.50	−88.9183	−16.2851	−1.49832	−0.363911	−0.093487
4.55	−98.2834	−17.7709	−1.59727	−0.377330	−0.103909
4.60	−108.677	−19.3995	−1.70438	−0.391443	−0.113959
4.65	−120.214	−21.1853	−1.82030	−0.406343	−0.123659
4.70	−133.028	−23.1441	−1.94577	−0.422138	−0.133034
4.75	−147.262	−25.2932	−2.08155	−0.438921	−0.142111
4.80	−163.081	−27.6519	−2.22847	−0.456805	−0.150918
4.85	−180.666	−30.2413	−2.38746	−0.475904	−0.159484

Table 1.6 (continued)

v	1.5	2.0	3.0	4.0	5.0
			t		
4.90	−200.222	−33.0849	−2.55949	−0.496334	−0.167844
4.95	−221.977	−36.2087	−2.74566	−0.518220	−0.176007
5.00	−246.187	−39.6412	−2.94710	−0.541710	−0.184104

1.3 First, Second and Third Derivatives with Respect to the Order of the Modified Bessel Function of the First Kind

Derivatives with respect to the order of the modified Bessel function of the first kind were determined using MATHEMATICA program by applying the following integral representations:

$$\frac{\partial I_v(z)}{\partial v} = \frac{1}{\pi}\int_0^\pi t\,e^{z\,\cos t}\,\sin(v\,t)\,dt + \frac{1}{\pi}\int_0^\infty e^{-z\,\cosh t\,-\,vt}[t\,\sin(\pi v) - \pi\,\cos(\pi v)]\,dt$$

(1.3.1)

$$\frac{\partial^2 I_v(z)}{\partial v^2} = -\frac{1}{\pi}\int_0^\pi t^2 e^{z\,\cos t}\,\cos(v\,t)\,dt$$
$$+ \frac{1}{\pi}\int_0^\infty e^{-z\,\cosh t\,-\,vt}[(\pi^2 - t^2)\sin(\pi v) + 2\pi t\,\cos(\pi v)]\,dt$$

(1.3.2)

$$\frac{\partial^3 I_v(z)}{\partial v^3} = \frac{1}{\pi}\int_0^\pi t^3 e^{z\,\cos t}\,\sin(v\,t)\,dt$$
$$- \frac{1}{\pi}\int_0^\infty e^{-z\,\cosh t\,-\,vt}[(t^3 - 3\pi^2 t)\sin(v\,t) + (\pi^3 - 3\pi t^2)\,\cos(v\,t)]\,dt$$

(1.3.3)

The form of derivatives with respect to the order of the modified Bessel function of the first kind differ from those of the first kind $\partial J_v(t)/\partial v$ and higher derivatives (Tables 1.7–1.12). Figures 1.9–1.17 show that these derivatives tend quickly to zero for small values of the order v and argument t.

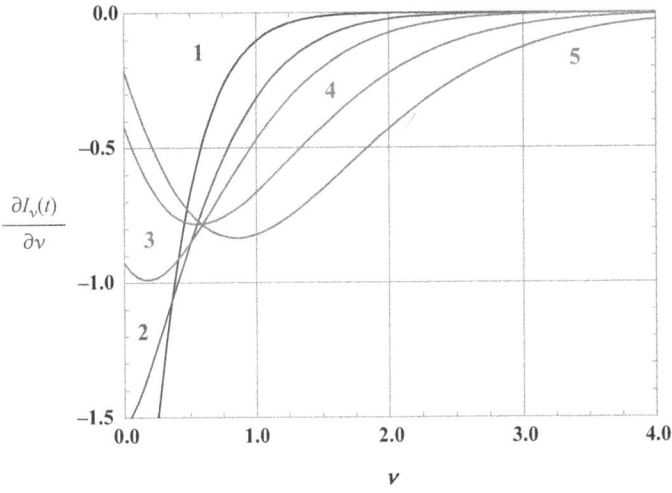

Figure 1.9: First derivatives of the modified Bessel function of the first kind with respect to the order v as a function of v, at constant values of argument t.

$1 - t = 0.05$; $2 - t = 0.25$; $3 - t = 0.50$; $4 - t = 1.0$; $5 - t = 1.50$.

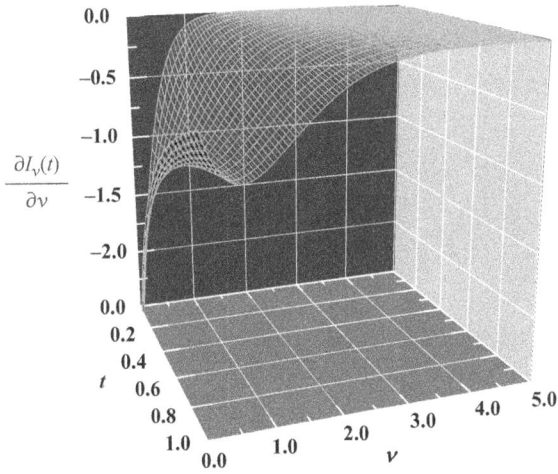

Figure 1.10: First derivatives with respect to the order of the modified Bessel functions of the first kind $\partial I_v(t)/\partial v$ as a function of v and t.

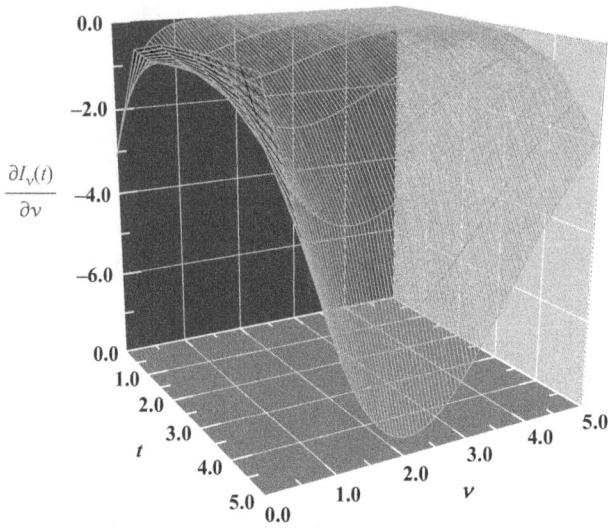

Figure 1.11: First derivatives with respect to the order of the modified Bessel functions of the first kind $\partial I_\nu(t)/\partial \nu$ as a function of ν and t.

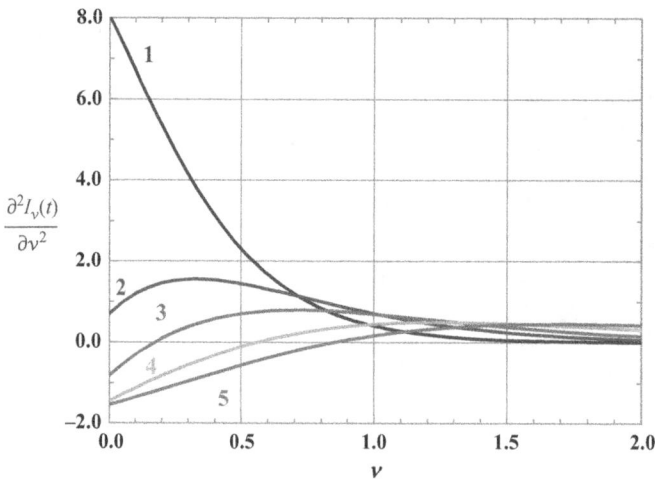

Figure 1.12: Second derivatives of the modified Bessel function of the first kind with respect to the order ν as a function of ν, at constant values of argument t.

$1 - t = 0.05$; $2 - t = 0.25$; $3 - t = 0.50$; $4 - t = 1.0$; $5 - t = 1.50$.

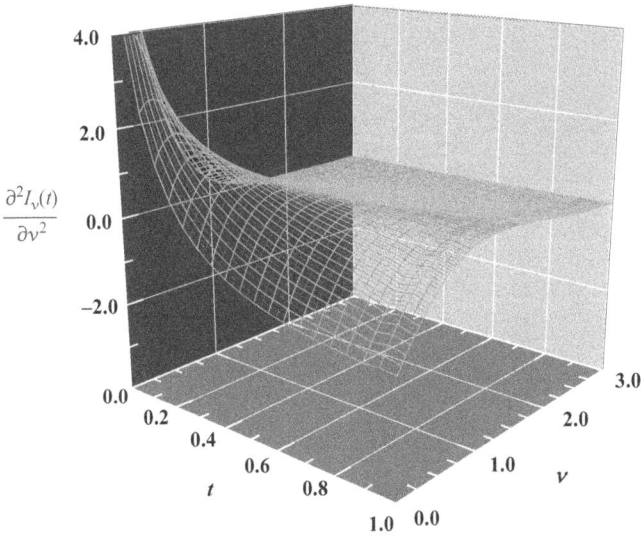

Figure 1.13: Second derivatives with respect to the order of the modified Bessel functions of the first kind $\partial^2 I_\nu(t)/\partial \nu^2$ as a function of ν and t.

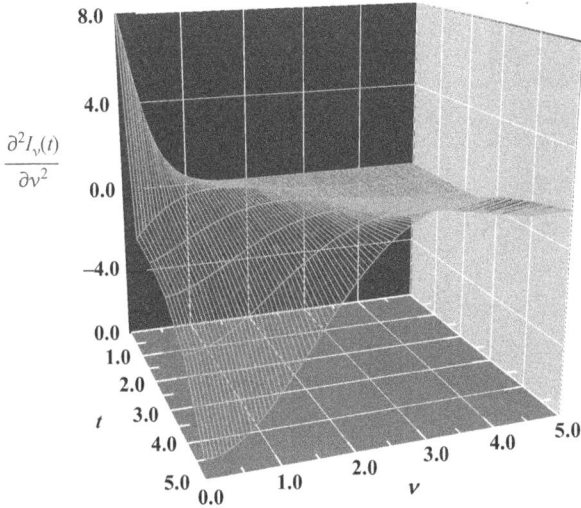

Figure 1.14: Second derivatives with respect to the order of the modified Bessel functions of the first kind $\partial^2 I_\nu(t)/\partial \nu^2$ as a function of ν and t.

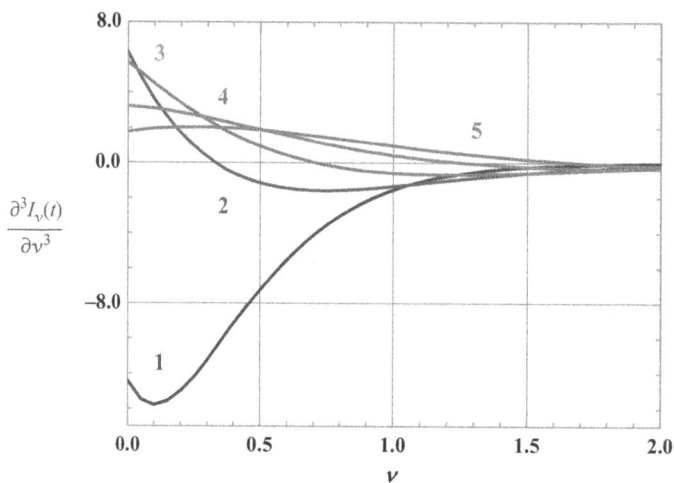

Figure 1.15: Third derivatives of the modified Bessel function of the first kind with respect to the order v as a function of v, at constant values of argument t.
$1 - t = 0.05;\ 2 - t = 0.25;\ 3 - t = 0.50;\ 4 - t = 1.0;\ 5 - t = 1.50.$

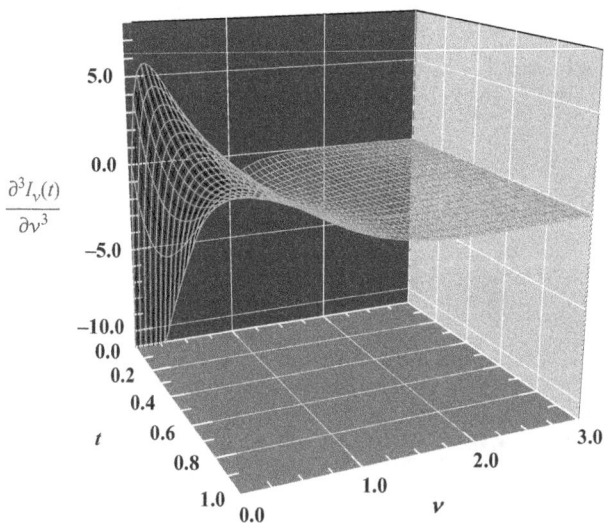

Figure 1.16: Third derivatives with respect to the order of the modified Bessel functions of the first kind $\partial^3 I_v(t)/\partial v^3$ as a function of v and t.

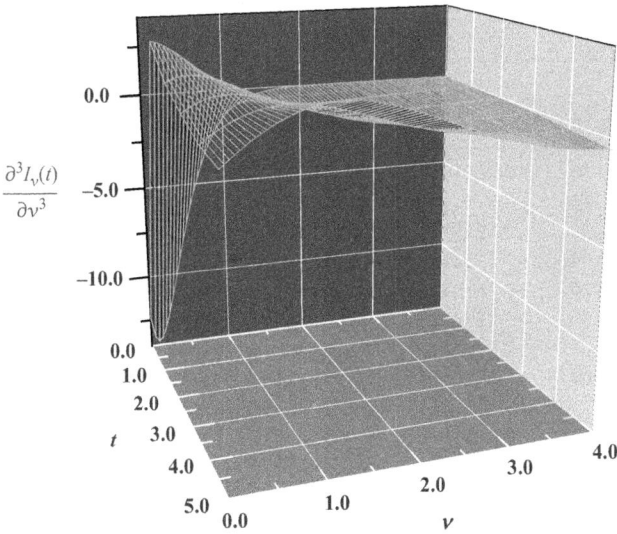

Figure 1.17: Third derivatives with respect to the order of the modified Bessel functions of the first kind $\partial^3 I_\nu(t)/\partial \nu^3$ as a function of ν and t.

Table 1.7: First derivatives with respect to the order of the modified Bessel functions of the first kind $\partial I_\nu(t)/\partial \nu$.

ν	t					
	0.05	0.10	0.15	0.20	0.25	0.30
0.00	−3.11423	−2.42707	−2.03003	−1.75270	−1.54151	−1.37246
0.05	−2.72789	−2.21621	−1.90302	−1.67636	−1.49928	−1.35468
0.10	−2.37503	−2.00982	−1.77041	−1.58995	−1.44485	−1.32373
0.15	−2.05648	−1.81148	−1.63593	−1.49689	−1.38126	−1.28230
0.20	−1.77176	−1.62365	−1.50251	−1.40004	−1.31115	−1.23273
0.25	−1.51945	−1.44794	−1.37242	−1.30174	−1.23675	−1.17709
0.30	−1.29754	−1.28525	−1.24732	−1.20388	−1.15993	−1.11718
0.35	−1.10367	−1.13596	−1.12844	−1.10795	−1.08227	−1.05453
0.40	−0.935324	−1.00002	−1.01657	−1.01510	−1.00504	−0.990455
0.45	−0.789928	−0.877101	−0.912191	−0.926182	−0.929260	−0.926030
0.50	−0.664979	−0.766626	−0.815540	−0.841806	−0.855739	−0.862159
0.55	−0.558088	−0.667890	−0.726636	−0.762376	−0.785089	−0.799567
0.60	−0.467031	−0.580091	−0.645344	−0.688124	−0.717755	−0.738830
0.65	−0.389763	−0.502378	−0.571412	−0.619146	−0.654049	−0.680392
0.70	−0.324436	−0.433884	−0.504503	−0.555427	−0.594162	−0.624581
0.75	−0.269390	−0.373754	−0.444221	−0.496865	−0.538191	−0.571630
0.80	−0.223156	−0.321160	−0.390135	−0.443293	−0.486152	−0.521687
0.85	−0.184440	−0.275316	−0.341795	−0.394496	−0.437999	−0.474830
0.90	−0.152112	−0.235482	−0.298744	−0.350224	−0.393636	−0.431080

Table 1.7 (continued)

v			t			
	0.05	0.10	0.15	0.20	0.25	0.30
0.95	−0.125190	−0.200976	−0.260532	−0.310204	−0.352930	−0.390413
1.00	−0.102828	−0.171171	−0.226721	−0.274153	−0.315719	−0.352764
1.05	−0.084298	−0.145495	−0.196892	−0.241779	−0.281820	−0.318041
1.10	−0.068979	−0.123434	−0.170650	−0.212797	−0.251040	−0.286129
1.15	−0.056344	−0.104524	−0.147625	−0.186925	−0.223176	−0.256896
1.20	−0.045944	−0.088353	−0.127473	−0.163890	−0.198026	−0.230199
1.25	−0.037401	−0.074556	−0.109877	−0.143434	−0.175386	−0.205890
1.30	−0.030397	−0.062809	−0.094549	−0.125313	−0.155058	−0.183814
1.35	−0.024667	−0.052827	−0.081224	−0.109297	−0.136851	−0.163820
1.40	−0.019986	−0.044362	−0.069666	−0.095172	−0.120582	−0.145753
1.45	−0.016170	−0.037197	−0.059661	−0.082743	−0.106077	−0.129468
1.50	−0.013064	−0.031144	−0.051015	−0.071826	−0.093171	−0.114821
1.55	−0.010539	−0.026038	−0.043559	−0.062257	−0.081712	−0.101676
1.60	−0.008491	−0.021740	−0.037140	−0.053886	−0.071558	−0.089902
1.65	−0.006832	−0.018126	−0.031623	−0.046575	−0.062576	−0.079376
1.70	−0.005490	−0.015093	−0.026890	−0.040201	−0.054647	−0.069986
1.75	−0.004406	−0.012552	−0.022835	−0.034653	−0.047658	−0.061622
1.80	−0.003532	−0.010425	−0.019367	−0.029832	−0.041508	−0.054186
1.85	−0.002828	−0.008648	−0.016406	−0.025650	−0.036106	−0.047585
1.90	−0.002262	−0.007166	−0.013880	−0.022027	−0.031368	−0.041737
1.95	−0.001807	−0.005930	−0.011730	−0.018893	−0.027219	−0.036562
2.00	−0.001441	−0.004903	−0.009901	−0.016186	−0.023591	−0.031991
2.05	−0.001149	−0.004048	−0.008348	−0.013851	−0.020423	−0.027958
2.10	−0.000915	−0.003339	−0.007030	−0.011840	−0.017660	−0.024406
2.15	−0.000728	−0.002752	−0.005915	−0.010110	−0.015254	−0.021282
2.20	−0.000578	−0.002265	−0.004971	−0.008623	−0.013162	−0.018537
2.25	−0.000459	−0.001862	−0.004173	−0.007347	−0.011345	−0.016129
2.30	−0.000364	−0.001530	−0.003500	−0.006254	−0.009768	−0.014020
2.35	−0.000288	−0.001256	−0.002933	−0.005318	−0.008402	−0.012173
2.40	−0.000228	−0.001029	−0.002455	−0.004518	−0.007220	−0.010560
2.45	−0.000180	−0.000843	−0.002053	−0.003834	−0.006198	−0.009151
2.50	−0.000143	−0.000690	−0.001715	−0.003251	−0.005316	−0.007923
2.55	−0.000112	−0.000564	−0.001431	−0.002754	−0.004555	−0.006853
2.60	−0.000089	−0.000461	−0.001194	−0.002330	−0.003900	−0.005922
2.65	−0.000070	−0.000376	−0.000995	−0.001971	−0.003335	−0.005113
2.70	−0.000055	−0.000307	−0.000828	−0.001665	−0.002850	−0.004410
2.75	−0.000043	−0.000250	−0.000689	−0.001405	−0.002434	−0.003801
2.80	−0.000034	−0.000203	−0.000572	−0.001185	−0.002076	−0.003273
2.85	−0.000027	−0.000165	−0.000475	−0.000999	−0.001769	−0.002816
2.90	−0.000021	−0.000135	−0.000394	−0.000841	−0.001507	−0.002421
2.95	−0.000016	−0.000109	−0.000327	−0.000707	−0.001282	−0.002079
3.00	−0.000013	−0.000089	−0.000271	−0.000595	−0.001090	−0.001784
3.05	−0.000010	−0.000072	−0.000224	−0.000500	−0.000926	−0.001530
3.10	−0.000008	−0.000058	−0.000185	−0.000419	−0.000787	−0.001311

Table 1.7 (continued)

			t			
v	0.05	0.10	0.15	0.20	0.25	0.30
3.15	−0.000006	−0.000047	−0.000153	−0.000352	−0.000667	−0.001123
3.20	−0.000005	−0.000038	−0.000127	−0.000295	−0.000566	−0.000961
3.25	−0.000004	−0.000031	−0.000104	−0.000247	−0.000479	−0.000821
3.30	−0.000003	−0.000025	−0.000086	−0.000207	−0.000406	−0.000702
3.35	−0.000002	−0.000020	−0.000071	−0.000173	−0.000343	−0.000599
3.40	−0.000002	−0.000016	−0.000058	−0.000144	−0.000290	−0.000511
3.45	−0.000001	−0.000013	−0.000048	−0.000121	−0.000245	−0.000436
3.50	−0.000001	−0.000011	−0.000040	−0.000101	−0.000207	−0.000371
3.55	−0.000001	−0.000008	−0.000033	−0.000084	−0.000174	−0.000316
3.60	−0.000001	−0.000007	−0.000027	−0.000070	−0.000147	−0.000269
3.65	−0.000001	−0.000005	−0.000022	−0.000058	−0.000124	−0.000229
3.70		−0.000004	−0.000018	−0.000048	−0.000104	−0.000194
3.75		−0.000004	−0.000015	−0.000040	−0.000088	−0.000165
3.80		−0.000003	−0.000012	−0.000034	−0.000074	−0.000140
3.85		−0.000002	−0.000010	−0.000028	−0.000062	−0.000119
3.90		−0.000002	−0.000008	−0.000023	−0.000052	−0.000101
3.95		−0.000001	−0.000007	−0.000019	−0.000044	−0.000085
4.00		−0.000001	−0.000005	−0.000016	−0.000037	−0.000072
4.05		−0.000001	−0.000004	−0.000013	−0.000031	−0.000061
4.10		−0.000001	−0.000004	−0.000011	−0.000026	−0.000052
4.15		−0.000001	−0.000003	−0.000009	−0.000022	−0.000044
4.20			−0.000002	−0.000007	−0.000018	−0.000037
4.25			−0.000002	−0.000006	−0.000015	−0.000031
4.30			−0.000002	−0.000005	−0.000013	−0.000026
4.35			−0.000001	−0.000004	−0.000011	−0.000022
4.40			−0.000001	−0.000003	−0.000009	−0.000019
4.45			−0.000001	−0.000003	−0.000007	−0.000016
4.50			−0.000001	−0.000002	−0.000006	−0.000013
4.55			−0.000001	−0.000002	−0.000005	−0.000011
4.60				−0.000002	−0.000004	−0.000009
4.65				−0.000001	−0.000004	−0.000008
4.70				−0.000001	−0.000003	−0.000007
4.75				−0.000001	−0.000002	−0.000006
4.80				−0.000001	−0.000002	−0.000005
4.85				−0.000001	−0.000002	−0.000004
4.90				−0.000001	−0.000001	−0.000003
4.95					−0.000001	−0.000003
5.00					−0.000001	−0.000002

			t			
v	0.35	0.40	0.45	0.50	0.55	0.60
0.00	−1.23271	−1.11453	−1.01291	−0.924419	−0.846568	−0.777522
0.05	−1.23316	−1.12899	−1.03834	−0.958573	−0.887750	−0.824416

Table 1.7 (continued)

				t		
v	0.35	0.40	0.45	0.50	0.55	0.60
0.10	−1.22014	−1.13000	−1.05059	−0.979930	−0.916594	−0.859471
0.15	−1.19595	−1.11959	−1.05138	−0.989978	−0.934374	−0.883774
0.20	−1.16269	−1.09959	−1.04234	−0.990139	−0.942332	−0.898403
0.25	−1.12224	−1.07169	−1.02499	−0.981759	−0.941659	−0.904411
0.30	−1.07627	−1.03744	−1.00072	−0.966097	−0.933491	−0.902814
0.35	−1.02625	−0.998190	−0.970801	−0.944323	−0.918893	−0.894581
0.40	−0.973439	−0.955175	−0.936380	−0.917509	−0.898859	−0.880628
0.45	−0.918949	−0.909464	−0.898490	−0.886632	−0.874308	−0.861812
0.50	−0.863710	−0.861992	−0.858045	−0.852575	−0.846082	−0.838930
0.55	−0.808504	−0.813565	−0.815851	−0.816127	−0.814945	−0.812713
0.60	−0.753982	−0.764869	−0.772609	−0.777989	−0.781585	−0.783832
0.65	−0.700670	−0.716483	−0.728924	−0.738775	−0.746617	−0.752892
0.70	−0.648989	−0.668885	−0.685312	−0.699023	−0.710586	−0.720436
0.75	−0.599263	−0.622466	−0.642206	−0.659193	−0.673967	−0.686950
0.80	−0.551736	−0.577535	−0.599964	−0.619677	−0.637178	−0.652862
0.85	−0.506580	−0.534335	−0.558880	−0.580806	−0.600572	−0.618546
0.90	−0.463905	−0.493044	−0.519185	−0.542851	−0.564454	−0.584326
0.95	−0.423770	−0.453789	−0.481058	−0.506033	−0.529074	−0.550479
1.00	−0.386191	−0.416650	−0.444632	−0.470526	−0.494642	−0.517241
1.05	−0.351151	−0.381669	−0.409999	−0.436463	−0.461325	−0.484805
1.10	−0.318599	−0.348855	−0.377216	−0.403943	−0.429253	−0.453332
1.15	−0.288467	−0.318190	−0.346309	−0.373029	−0.398525	−0.422948
1.20	−0.260665	−0.289633	−0.317281	−0.343763	−0.369213	−0.393751
1.25	−0.235093	−0.263127	−0.290113	−0.316157	−0.341361	−0.365815
1.30	−0.211640	−0.238601	−0.264767	−0.290209	−0.314995	−0.339188
1.35	−0.190188	−0.215970	−0.241195	−0.265898	−0.290119	−0.313903
1.40	−0.170618	−0.195147	−0.219334	−0.243188	−0.266726	−0.289971
1.45	−0.152809	−0.176036	−0.199116	−0.222035	−0.244791	−0.267392
1.50	−0.136639	−0.158538	−0.180464	−0.202385	−0.224283	−0.246152
1.55	−0.121991	−0.142556	−0.163300	−0.184178	−0.205159	−0.226227
1.60	−0.108750	−0.127990	−0.147541	−0.167347	−0.187371	−0.207584
1.65	−0.096805	−0.114743	−0.133104	−0.151826	−0.170864	−0.190185
1.70	−0.086050	−0.102719	−0.119905	−0.137542	−0.155581	−0.173985
1.75	−0.076384	−0.091828	−0.107864	−0.124426	−0.141462	−0.158934
1.80	−0.067714	−0.081980	−0.096899	−0.112405	−0.128446	−0.144982
1.85	−0.059949	−0.073092	−0.086933	−0.101409	−0.116469	−0.132074
1.90	−0.053007	−0.065084	−0.077892	−0.091370	−0.105470	−0.120155
1.95	−0.046812	−0.057881	−0.069702	−0.082219	−0.095388	−0.109171
2.00	−0.041290	−0.051413	−0.062297	−0.073894	−0.086161	−0.099067
2.05	−0.036377	−0.045613	−0.055612	−0.066331	−0.077733	−0.089787
2.10	−0.032012	−0.040420	−0.049587	−0.059472	−0.070045	−0.081279
2.15	−0.028139	−0.035779	−0.044163	−0.053261	−0.063045	−0.073492
2.20	−0.024707	−0.031635	−0.039289	−0.047645	−0.056680	−0.066375

Table 1.7 (continued)

			t			
v	0.35	0.40	0.45	0.50	0.55	0.60
2.25	−0.021670	−0.027940	−0.034915	−0.042575	−0.050901	−0.059881
2.30	−0.018987	−0.024651	−0.030995	−0.038003	−0.045662	−0.053963
2.35	−0.016619	−0.021727	−0.027486	−0.033886	−0.040919	−0.048578
2.40	−0.014532	−0.019130	−0.024349	−0.030185	−0.036631	−0.043685
2.45	−0.012694	·−0.016827	−0.021549	−0.026860	−0.032759	−0.039245
2.50	−0.011078	−0.014786	−0.019052	−0.023878	−0.029267	−0.035221
2.55	−0.009658	−0.012981	−0.016828	−0.021207	−0.026122	−0.031579
2.60	−0.008413	−0.011385	−0.014850	−0.018816	−0.023293	−0.028286
2.65	−0.007321	−0.009977	−0.013092	−0.016680	−0.020750	−0.025313
2.70	−0.006366	−0.008734	−0.011532	−0.014773	−0.018468	−0.022631
2.75	−0.005530	−0.007640	−0.010149	−0.013072	−0.016423	−0.020215
2.80	−0.004800	−0.006677	−0.008924	−0.011556	−0.014591	−0.018041
2.85	−0.004162	−0.005830	−0.007840	−0.010208	−0.012952	−0.016087
2.90	−0.003607	−0.005087	−0.006881	−0.009009	−0.011487	−0.014332
2.95	−0.003122	−0.004434	−0.006035	−0.007944	−0.010180	−0.012758
3.00	−0.002701	−0.003862	−0.005289	−0.007000	−0.009013	−0.011348
3.05	−0.002335	−0.003362	−0.004631	−0.006163	−0.007974	−0.010085
3.10	−0.002017	−0.002923	−0.004052	−0.005421	−0.007050	−0.008955
3.15	−0.001740	−0.002540	−0.003542	−0.004765	−0.006227	−0.007946
3.20	−0.001501	−0.002206	−0.003095	−0.004185	−0.005496	−0.007045
3.25	−0.001293	−0.001914	−0.002701	−0.003673	−0.004848	−0.006241
3.30	−0.001114	−0.001659	−0.002356	−0.003222	−0.004272	−0.005525
3.35	−0.000958	−0.001438	−0.002054	−0.002823	−0.003762	−0.004888
3.40	−0.000824	−0.001245	−0.001789	−0.002472	−0.003311	−0.004320
3.45	−0.000708	−0.001077	−0.001557	−0.002164	−0.002912	−0.003816
3.50	−0.000608	−0.000931	−0.001354	−0.001892	−0.002559	−0.003369
3.55	−0.000522	−0.000804	−0.001177	−0.001654	−0.002247	−0.002972
3.60	−0.000447	−0.000695	−0.001020	−0.001444	−0.001972	−0.002619
3.65	−0.000383	−0.000599	−0.000888	−0.001260	−0.001729	−0.002307
3.70	−0.000328	−0.000517	−0.000770	−0.001099	−0.001516	−0.002031
3.75	−0.000281	−0.000445	−0.000667	−0.000958	−0.001327	−0.001787
3.80	−0.000240	−0.000383	−0.000578	−0.000834	−0.001162	−0.001571
3.85	−0.000205	−0.000330	−0.000501	−0.000726	−0.001016	−0.001380
3.90	−0.000175	−0.000284	−0.000433	−0.000632	−0.000888	−0.001212
3.95	−0.000150	−0.000244	−0.000374	−0.000549	−0.000776	−0.001063
4.00	−0.000128	−0.000209	−0.000324	−0.000477	−0.000677	−0.000932
4.05	−0.000109	−0.000180	−0.000279	−0.000414	−0.000591	−0.000817
4.10	−0.000093	−0.000154	−0.000241	−0.000360	−0.000515	−0.000716
4.15	−0.000079	−0.000132	−0.000208	−0.000312	−0.000449	−0.000627
4.20	−0.000067	−0.000113	−0.000179	−0.000270	−0.000391	−0.000548
4.25	−0.000057	−0.000097	−0.000155	−0.000234	−0.000341	−0.000479
4.30	−0.000049	−0.000083	−0.000133	−0.000203	−0.000296	−0.000419
4.35	−0.000041	−0.000071	−0.000115	−0.000175	−0.000258	−0.000366

Table 1.7 (continued)

			t			
v	0.35	0.40	0.45	0.50	0.55	0.60
4.40	−0.000035	−0.000061	−0.000099	−0.000152	−0.000224	−0.000319
4.45	−0.000030	−0.000052	−0.000085	−0.000131	−0.000194	−0.000278
4.50	−0.000025	−0.000044	−0.000073	−0.000113	−0.000169	−0.000243
4.55	−0.000021	−0.000038	−0.000062	−0.000098	−0.000146	−0.000212
4.60	−0.000018	−0.000032	−0.000054	−0.000084	−0.000127	−0.000184
4.65	−0.000015	−0.000028	−0.000046	−0.000073	−0.000110	−0.000160
4.70	−0.000013	−0.000023	−0.000039	−0.000063	−0.000095	−0.000140
4.75	−0.000011	−0.000020	−0.000034	−0.000054	−0.000082	−0.000121
4.80	−0.000009	−0.000017	−0.000029	−0.000047	−0.000071	−0.000105
4.85	−0.000008	−0.000014	−0.000025	−0.000040	−0.000062	−0.000092
4.90	−0.000007	−0.000012	−0.000021	−0.000034	−0.000053	−0.000080
4.95	−0.000006	−0.000010	−0.000018	−0.000030	−0.000046	−0.000069
5.00	−0.000005	−0.000009	−0.000016	−0.000025	−0.000040	−0.000060

			t			
v	0.65	0.70	0.75	0.80	0.85	0.90
0.00	−0.715873	−0.660520	−0.610582	−0.565347	−0.524226	−0.486730
0.05	−0.767445	−0.715946	−0.669201	−0.626621	−0.587718	−0.552081
0.10	−0.807694	−0.760570	−0.717532	−0.678111	−0.641913	−0.608605
0.15	−0.837544	−0.795171	−0.756227	−0.720354	−0.687250	−0.656652
0.20	−0.857927	−0.820550	−0.785970	−0.753932	−0.724214	−0.696624
0.25	−0.869768	−0.837516	−0.807467	−0.779454	−0.753332	−0.728967
0.30	−0.873974	−0.846876	−0.821430	−0.797548	−0.775151	−0.754160
0.35	−0.871421	−0.849419	−0.828568	−0.808848	−0.790235	−0.772701
0.40	−0.862949	−0.845913	−0.829578	−0.813984	−0.799154	−0.785101
0.45	−0.849355	−0.837093	−0.825139	−0.813578	−0.802475	−0.791877
0.50	−0.831387	−0.823658	−0.815900	−0.808234	−0.800756	−0.793540
0.55	−0.809742	−0.806269	−0.802483	−0.798533	−0.794539	−0.790598
0.60	−0.785064	−0.785541	−0.785474	−0.785030	−0.784348	−0.783541
0.65	−0.757944	−0.762047	−0.765423	−0.768253	−0.770685	−0.772845
0.70	−0.728920	−0.736314	−0.742843	−0.748695	−0.754025	−0.758965
0.75	−0.698477	−0.708821	−0.718205	−0.726817	−0.734815	−0.742334
0.80	−0.667050	−0.680004	−0.691941	−0.703045	−0.713472	−0.723357
0.85	−0.635025	−0.650256	−0.664447	−0.677773	−0.690386	−0.702417
0.90	−0.602741	−0.619926	−0.636077	−0.651358	−0.665913	−0.679869
0.95	−0.570493	−0.589324	−0.607150	−0.624123	−0.640380	−0.656039
1.00	−0.538538	−0.558721	−0.577949	−0.596362	−0.614085	−0.631228
1.05	−0.507093	−0.528351	−0.548723	−0.568333	−0.587295	−0.605709
1.10	−0.476340	−0.498418	−0.519690	−0.540267	−0.560250	−0.579729
1.15	−0.446430	−0.469091	−0.491038	−0.512366	−0.533163	−0.553512
1.20	−0.417486	−0.440514	−0.462926	−0.484803	−0.506222	−0.527253
1.25	−0.389603	−0.412803	−0.435490	−0.457730	−0.479589	−0.501128
1.30	−0.362854	−0.386052	−0.408839	−0.431273	−0.453406	−0.475289

Table 1.7 (continued)

			t			
v	**0.65**	**0.70**	**0.75**	**0.80**	**0.85**	**0.90**
1.35	-0.337292	-0.360331	-0.383065	-0.405537	-0.427791	-0.449869
1.40	-0.312951	-0.335696	-0.358238	-0.380610	-0.402846	-0.424981
1.45	-0.289850	-0.312182	-0.334411	-0.356559	-0.378653	-0.400719
1.50	-0.267992	-0.289812	-0.311622	-0.333438	-0.355278	-0.377163
1.55	-0.247372	-0.268593	-0.289895	-0.311284	-0.332773	-0.354376
1.60	-0.227972	-0.248526	-0.269242	-0.290124	-0.311176	-0.332409
1.65	-0.209768	-0.229598	-0.249666	-0.269971	-0.290514	-0.311301
1.70	-0.192727	-0.211789	-0.231160	-0.250832	-0.270804	-0.291077
1.75	-0.176812	-0.195075	-0.213708	-0.232702	-0.252051	-0.271755
1.80	-0.161982	-0.179424	-0.197290	-0.215570	-0.234255	-0.253344
1.85	-0.148192	-0.164800	-0.181879	-0.199418	-0.217408	-0.235844
1.90	-0.135394	-0.151163	-0.167445	-0.184225	-0.201495	-0.219250
1.95	-0.123541	-0.138474	-0.153953	-0.169964	-0.186498	-0.203550
2.00	-0.112583	-0.126688	-0.141366	-0.156604	-0.172393	-0.188727
2.05	-0.102470	-0.115761	-0.129646	-0.144113	-0.159153	-0.174762
2.10	-0.093153	-0.105649	-0.118753	-0.132456	-0.146749	-0.161629
2.15	-0.084584	-0.096306	-0.108646	-0.121596	-0.135150	-0.149304
2.20	-0.076715	-0.087688	-0.099284	-0.111497	-0.124321	-0.137755
2.25	-0.069501	-0.079752	-0.090626	-0.102120	-0.114230	-0.126954
2.30	-0.062896	-0.072454	-0.082632	-0.093428	-0.104840	-0.116869
2.35	-0.056857	-0.065753	-0.075261	-0.085383	-0.096117	-0.107467
2.40	-0.051345	-0.059609	-0.068476	-0.077948	-0.088026	-0.098716
2.45	-0.046320	-0.053983	-0.062237	-0.071086	-0.080532	-0.090582
2.50	-0.041744	-0.048839	-0.056510	-0.064762	-0.073601	-0.083033
2.55	-0.037584	-0.044142	-0.051259	-0.058942	-0.067199	-0.076037
2.60	-0.033805	-0.039857	-0.046451	-0.053593	-0.061294	-0.069562
2.65	-0.030378	-0.035955	-0.042053	-0.048683	-0.055854	-0.063577
2.70	-0.027273	-0.032404	-0.038037	-0.044181	-0.050849	-0.058052
2.75	-0.024463	-0.029178	-0.034372	-0.040058	-0.046249	-0.052958
2.80	-0.021923	-0.026249	-0.031033	-0.036288	-0.042028	-0.048267
2.85	-0.019629	-0.023593	-0.027993	-0.032843	-0.038158	-0.043952
2.90	-0.017561	-0.021188	-0.025229	-0.029699	-0.034614	-0.039989
2.95	-0.015697	-0.019012	-0.022719	-0.026834	-0.031372	-0.036351
3.00	-0.014019	-0.017045	-0.020441	-0.024224	-0.028410	-0.033016
3.05	-0.012510	-0.015269	-0.018376	-0.021850	-0.025707	-0.029962
3.10	-0.011155	-0.013667	-0.016507	-0.019693	-0.023241	-0.027169
3.15	-0.009939	-0.012223	-0.014816	-0.017735	-0.020996	-0.024616
3.20	-0.008848	-0.010924	-0.013288	-0.015958	-0.018952	-0.022286
3.25	-0.007871	-0.009755	-0.011908	-0.014349	-0.017094	-0.020160
3.30	-0.006997	-0.008704	-0.010664	-0.012892	-0.015406	-0.018223
3.35	-0.006215	-0.007761	-0.009542	-0.011575	-0.013875	-0.016460
3.40	-0.005517	-0.006915	-0.008532	-0.010384	-0.012487	-0.014857
3.45	-0.004893	-0.006157	-0.007624	-0.009309	-0.011229	-0.013400

Table 1.7 (continued)

			t			
v	0.65	0.70	0.75	0.80	0.85	0.90
3.50	−0.004337	−0.005478	−0.006807	−0.008339	−0.010091	−0.012077
3.55	−0.003841	−0.004870	−0.006074	−0.007465	−0.009061	−0.010877
3.60	−0.003400	−0.004327	−0.005415	−0.006678	−0.008131	−0.009789
3.65	−0.003007	−0.003842	−0.004825	−0.005970	−0.007292	−0.008804
3.70	−0.002658	−0.003409	−0.004296	−0.005334	−0.006535	−0.007913
3.75	−0.002348	−0.003023	−0.003823	−0.004762	−0.005852	−0.007107
3.80	−0.002073	−0.002678	−0.003400	−0.004248	−0.005237	−0.006379
3.85	−0.001829	−0.002372	−0.003021	−0.003788	−0.004684	−0.005722
3.90	−0.001612	−0.002099	−0.002683	−0.003375	−0.004187	−0.005130
3.95	−0.001420	−0.001856	−0.002381	−0.003006	−0.003740	−0.004595
4.00	−0.001251	−0.001641	−0.002112	−0.002675	−0.003338	−0.004114
4.05	−0.001101	−0.001449	−0.001872	−0.002379	−0.002978	−0.003681
4.10	−0.000968	−0.001279	−0.001659	−0.002114	−0.002655	−0.003292
4.15	−0.000851	−0.001129	−0.001469	−0.001878	−0.002366	−0.002942
4.20	−0.000747	−0.000995	−0.001300	−0.001667	−0.002107	−0.002627
4.25	−0.000656	−0.000877	−0.001149	−0.001479	−0.001875	−0.002345
4.30	−0.000576	−0.000773	−0.001016	−0.001312	−0.001668	−0.002092
4.35	−0.000505	−0.000680	−0.000897	−0.001163	−0.001483	−0.001865
4.40	−0.000442	−0.000598	−0.000792	−0.001030	−0.001317	−0.001662
4.45	−0.000387	−0.000526	−0.000699	−0.000912	−0.001170	−0.001480
4.50	−0.000339	−0.000462	−0.000616	−0.000806	−0.001038	−0.001317
4.55	−0.000297	−0.000406	−0.000543	−0.000713	−0.000921	−0.001172
4.60	−0.000260	−0.000356	−0.000478	−0.000630	−0.000816	−0.001042
4.65	−0.000227	−0.000313	−0.000421	−0.000557	−0.000723	−0.000926
4.70	−0.000198	−0.000274	−0.000371	−0.000491	−0.000641	−0.000822
4.75	−0.000173	−0.000240	−0.000326	−0.000434	−0.000567	−0.000730
4.80	−0.000151	−0.000210	−0.000287	−0.000382	−0.000502	−0.000648
4.85	−0.000132	−0.000184	−0.000252	−0.000337	−0.000443	−0.000574
4.90	−0.000115	−0.000161	−0.000221	−0.000297	−0.000392	−0.000509
4.95	−0.000100	−0.000141	−0.000194	−0.000262	−0.000346	−0.000451
5.00	−0.000087	−0.000123	−0.000170	−0.000230	−0.000306	−0.000399

			t			
v	0.95	1.0	2.0	3.0	4.0	5.0
0.00	−0.452447	−0.421024	−0.113894	−0.034740	−0.011160	−0.003691
0.05	−0.519362	−0.489263	−0.193810	−0.139336	−0.179060	−0.312908
0.10	−0.577902	−0.549556	−0.270916	−0.242763	−0.346264	−0.621444
0.15	−0.628335	−0.602103	−0.344898	−0.344649	−0.512338	−0.928708
0.20	−0.670990	−0.647165	−0.415483	−0.444640	−0.676861	−1.23411
0.25	−0.706243	−0.685053	−0.482438	−0.542405	−0.839418	−1.53709
0.30	−0.734505	−0.716118	−0.545570	−0.637631	−0.999612	−1.83706
0.35	−0.756215	−0.740746	−0.604725	−0.730031	−1.15706	−2.13348
0.40	−0.771830	−0.759340	−0.659787	−0.819340	−1.31139	−2.42581

Table 1.7 (continued)

v	0.95	1.0	2.0	3.0	4.0	5.0
				t		
0.45	-0.781817	-0.772324	-0.710673	-0.905317	-1.46225	-2.71352
0.50	-0.786648	-0.780126	-0.757333	-0.987747	-1.60931	-2.99610
0.55	-0.786790	-0.783180	-0.799750	-1.06644	-1.75226	-3.27307
0.60	-0.782703	-0.781915	-0.837934	-1.14122	-1.89081	-3.54397
0.65	-0.774838	-0.776755	-0.871922	-1.21196	-2.02469	-3.80834
0.70	-0.763629	-0.768111	-0.901774	-1.27853	-2.15364	-4.06576
0.75	-0.749489	-0.756383	-0.927575	-1.34084	-2.27745	-4.31584
0.80	-0.732816	-0.741951	-0.949425	-1.39882	-2.39591	-4.55820
0.85	-0.713982	-0.725180	-0.967445	-1.45242	-2.50884	-4.79250
0.90	-0.693336	-0.706413	-0.981769	-1.50162	-2.61608	-5.01841
0.95	-0.671205	-0.685972	-0.992543	-1.54640	-2.71750	-5.23564
1.00	-0.647889	-0.664159	-0.999927	-1.58677	-2.81300	-5.44393
1.05	-0.623666	-0.641250	-1.00409	-1.62278	-2.90247	-5.64303
1.10	-0.598788	-0.617504	-1.00519	-1.65447	-2.98586	-5.83273
1.15	-0.573485	-0.593153	-1.00343	-1.68189	-3.06311	-6.01286
1.20	-0.547962	-0.568412	-0.998971	-1.70514	-3.13422	-6.18325
1.25	-0.522404	-0.543473	-0.992008	-1.72430	-3.19916	-6.34378
1.30	-0.496973	-0.518506	-0.982722	-1.73949	-3.25797	-6.49434
1.35	-0.471814	-0.493665	-0.971298	-1.75080	-3.31066	-6.63488
1.40	-0.447049	-0.469085	-0.957916	-1.75839	-3.35731	-6.76532
1.45	-0.422785	-0.444882	-0.942756	-1.76236	-3.39797	-6.88567
1.50	-0.399114	-0.421157	-0.925993	-1.76288	-3.43273	-6.99592
1.55	-0.376111	-0.397995	-0.907796	-1.76009	-3.46170	-7.09610
1.60	-0.353836	-0.375470	-0.888332	-1.75415	-3.48498	-7.18626
1.65	-0.332339	-0.353640	-0.867758	-1.74520	-3.50272	-7.26648
1.70	-0.311657	-0.332552	-0.846228	-1.73343	-3.51504	-7.33685
1.75	-0.291817	-0.312242	-0.823887	-1.71898	-3.52210	-7.39751
1.80	-0.272837	-0.292738	-0.800873	-1.70202	-3.52406	-7.44857
1.85	-0.254727	-0.274057	-0.777319	-1.68273	-3.52108	-7.49021
1.90	-0.237488	-0.256210	-0.753346	-1.66127	-3.51336	-7.52259
1.95	-0.221117	-0.239200	-0.729072	-1.63779	-3.50106	-7.54591
2.00	-0.205605	-0.223025	-0.704604	-1.61247	-3.48439	-7.56038
2.05	-0.190937	-0.207677	-0.680043	-1.58546	-3.46354	-7.56621
2.10	-0.177094	-0.193144	-0.655483	-1.55693	-3.43870	-7.56364
2.15	-0.164056	-0.179409	-0.631009	-1.52702	-3.41008	-7.55292
2.20	-0.151799	-0.166453	-0.606699	-1.49589	-3.37789	-7.53430
2.25	-0.140295	-0.154254	-0.582625	-1.46368	-3.34232	-7.50806
2.30	-0.129517	-0.142787	-0.558850	-1.43053	-3.30358	-7.47447
2.35	-0.119435	-0.132027	-0.535434	-1.39658	-3.26189	-7.43380
2.40	-0.110020	-0.121946	-0.512427	-1.36195	-3.21744	-7.38636
2.45	-0.101240	-0.112515	-0.489875	-1.32678	-3.17044	-7.33244
2.50	-0.093066	-0.103707	-0.467817	-1.29119	-3.12108	-7.27233
2.55	-0.085465	-0.095492	-0.446288	-1.25527	-3.06957	-7.20634

Table 1.7 (continued)

			t			
v	0.95	1.0	2.0	3.0	4.0	5.0
2.60	−0.078407	−0.087841	−0.425317	−1.21915	−3.01610	−7.13476
2.65	−0.071863	−0.080724	−0.404928	−1.18292	−2.96087	−7.05792
2.70	−0.065802	−0.074113	−0.385141	−1.14668	−2.90405	−6.97611
2.75	−0.060197	−0.067980	−0.365971	−1.11052	−2.84583	−6.88963
2.80	−0.055019	−0.062298	−0.347430	−1.07453	−2.78639	−6.79880
2.85	−0.050241	−0.057040	−0.329526	−1.03877	−2.72589	−6.70391
2.90	−0.045838	−0.052179	−0.312264	−1.00332	−2.66450	−6.60527
2.95	−0.041785	−0.047691	−0.295645	−0.968254	−2.60239	−6.50316
3.00	−0.038058	−0.043553	−0.279667	−0.933623	−2.53970	−6.39787
3.05	−0.034634	−0.039740	−0.264328	−0.899485	−2.47658	−6.28971
3.10	−0.031493	−0.036231	−0.249622	−0.865891	−2.41317	−6.17893
3.15	−0.028614	−0.033006	−0.235541	−0.832886	−2.34961	−6.06582
3.20	−0.025977	−0.030043	−0.222074	−0.800509	−2.28602	−5.95065
3.25	−0.023565	−0.027326	−0.209211	−0.768796	−2.22252	−5.83368
3.30	−0.021360	−0.024835	−0.196940	−0.737779	−2.15923	−5.71515
3.35	−0.019347	−0.022554	−0.185246	−0.707483	−2.09626	−5.59532
3.40	−0.017511	−0.020467	−0.174114	−0.677932	−2.03370	−5.47442
3.45	−0.015838	−0.018560	−0.163530	−0.649144	−1.97165	−5.35269
3.50	−0.014314	−0.016818	−0.153477	−0.621135	−1.91019	−5.23034
3.55	−0.012927	−0.015229	−0.143938	−0.593916	−1.84942	−5.10759
3.60	−0.011667	−0.013780	−0.134896	−0.567497	−1.78940	−4.98464
3.65	−0.010522	−0.012461	−0.126333	−0.541882	−1.73020	−4.86168
3.70	−0.009483	−0.011260	−0.118233	−0.517076	−1.67189	−4.73891
3.75	−0.008541	−0.010168	−0.110576	−0.493077	−1.61452	−4.61650
3.80	−0.007688	−0.009176	−0.103346	−0.469885	−1.55815	−4.49461
3.85	−0.006915	−0.008276	−0.096525	−0.447494	−1.50283	−4.37341
3.90	−0.006216	−0.007458	−0.090096	−0.425899	−1.44858	−4.25305
3.95	−0.005584	−0.006718	−0.084040	−0.405091	−1.39546	−4.13366
4.00	−0.005013	−0.006047	−0.078342	−0.385060	−1.34348	−4.01539
4.05	−0.004498	−0.005439	−0.072984	−0.365796	−1.29269	−3.89835
4.10	−0.004033	−0.004890	−0.067951	−0.347286	−1.24310	−3.78265
4.15	−0.003614	−0.004394	−0.063226	−0.329515	−1.19473	−3.66842
4.20	−0.003237	−0.003945	−0.058795	−0.312469	−1.14759	−3.55574
4.25	−0.002897	−0.003540	−0.054641	−0.296132	−1.10170	−3.44471
4.30	−0.002591	−0.003175	−0.050751	−0.280487	−1.05705	−3.33540
4.35	−0.002317	−0.002846	−0.047110	−0.265518	−1.01367	−3.22790
4.40	−0.002070	−0.002550	−0.043706	−0.251207	−0.971541	−3.12227
4.45	−0.001849	−0.002283	−0.040524	−0.237535	−0.930667	−3.01858
4.50	−0.001650	−0.002043	−0.037553	−0.224483	−0.891042	−2.91687
4.55	−0.001472	−0.001827	−0.034781	−0.212033	−0.852660	−2.81720
4.60	−0.001312	−0.001633	−0.032195	−0.200167	−0.815510	−2.71961
4.65	−0.001169	−0.001459	−0.029786	−0.188864	−0.779580	−2.62412
4.70	−0.001041	−0.001303	−0.027542	−0.178105	−0.744857	−2.53077

Table 1.7 (continued)

			t			
v	0.95	1.0	2.0	3.0	4.0	5.0
4.75	−0.000927	−0.001163	−0.025454	−0.167872	−0.711325	−2.43959
4.80	−0.000825	−0.001037	−0.023511	−0.158146	−0.678965	−2.35058
4.85	−0.000733	−0.000925	−0.021706	−0.148907	−0.647761	−2.26377
4.90	−0.000652	−0.000824	−0.020029	−0.140137	−0.617690	−2.17916
4.95	−0.000579	−0.000734	−0.018472	−0.131818	−0.588733	−2.09675
5.00	−0.000514	−0.000653	−0.017028	−0.123932	−0.560866	−2.01655

Table 1.8: Second derivatives with respect to the order of the modified Bessel functions of the first kind $\partial^2 I_v(t)/\partial v^2$.

			t			
v	0.05	0.10	0.15	0.20	0.25	0.30
0.00	8.04768	4.23192	2.45498	1.40004	0.700029	0.204756
0.05	7.39753	4.18635	2.61009	1.64005	0.977469	0.496675
0.10	6.71426	4.05717	2.68192	1.80521	1.18957	0.732161
0.15	6.02922	3.86840	2.68753	1.90771	1.34506	0.917435
0.20	5.36447	3.63950	2.64160	1.95846	1.45202	1.05849
0.25	4.73482	3.38605	2.55662	1.96714	1.51791	1.16100
0.30	4.14952	3.12026	2.44307	1.94223	1.54947	1.23028
0.35	3.61359	2.85156	2.30969	1.89111	1.55275	1.27123
0.40	3.12894	2.58702	2.16362	1.82011	1.53313	1.28833
0.45	2.69526	2.33182	2.01064	1.73461	1.49533	1.28565
0.50	2.31069	2.08955	1.85533	1.63910	1.44347	1.26679
0.55	1.97235	1.86257	1.70130	1.53735	1.38109	1.23497
0.60	1.67676	1.65225	1.55124	1.43242	1.31121	1.19303
0.65	1.42011	1.45919	1.40718	1.32676	1.23640	1.14341
0.70	1.19853	1.28342	1.27052	1.22235	1.15876	1.08824
0.75	1.00820	1.12454	1.14219	1.12068	1.08005	1.02932
0.80	0.845466	0.981842	1.02274	1.02290	1.00168	0.96816
0.85	0.706933	0.854413	0.912396	0.929815	0.924766	0.906048
0.90	0.589466	0.741208	0.811148	0.841987	0.850187	0.844012
0.95	0.490230	0.641116	0.718805	0.759755	0.778597	0.782896
1.00	0.406684	0.553002	0.635039	0.683285	0.710469	0.723367
1.05	0.336575	0.475743	0.559426	0.612602	0.646125	0.665938
1.10	0.277919	0.408256	0.491480	0.547625	0.585759	0.610993
1.15	0.228987	0.349508	0.430676	0.488187	0.529460	0.558805
1.20	0.188277	0.298533	0.376471	0.434062	0.477236	0.509553
1.25	0.154495	0.254439	0.328323	0.384978	0.429026	0.463337
1.30	0.126531	0.216407	0.285693	0.340637	0.384718	0.420194

Table 1.8 (continued)

			t			
v	0.05	0.10	0.15	0.20	0.25	0.30
1.35	0.103437	0.183691	0.248070	0.300723	0.344164	0.380110
1.40	0.084406	0.155622	0.214962	0.264913	0.307185	0.343026
1.45	0.068759	0.131598	0.185908	0.232886	0.273584	0.308855
1.50	0.055918	0.111084	0.160478	0.204324	0.243152	0.277482
1.55	0.045403	0.093607	0.138275	0.178925	0.215675	0.248778
1.60	0.036807	0.078749	0.118936	0.156396	0.190936	0.222597
1.65	0.029793	0.066142	0.102128	0.136462	0.168725	0.198791
1.70	0.024081	0.055467	0.087553	0.118867	0.148833	0.177203
1.75	0.019436	0.046445	0.074939	0.103370	0.131061	0.157680
1.80	0.015665	0.038834	0.064045	0.089750	0.115222	0.140068
1.85	0.012609	0.032423	0.054653	0.077805	0.101135	0.124218
1.90	0.010135	0.027034	0.046572	0.067349	0.088634	0.109986
1.95	0.008137	0.022510	0.039629	0.058214	0.077561	0.097235
2.00	0.006524	0.018718	0.033676	0.050247	0.067774	0.085833
2.05	0.005225	0.015546	0.028580	0.043311	0.059138	0.075659
2.10	0.004179	0.012895	0.024223	0.037284	0.051532	0.066597
2.15	0.003339	0.010683	0.020505	0.032053	0.044845	0.058540
2.20	0.002665	0.008840	0.017337	0.027522	0.038975	0.051390
2.25	0.002124	0.007307	0.014640	0.023602	0.033830	0.045054
2.30	0.001692	0.006033	0.012349	0.020216	0.029329	0.039451
2.35	0.001346	0.004975	0.010404	0.017296	0.025396	0.034502
2.40	0.001069	0.004098	0.008756	0.014781	0.021965	0.030137
2.45	0.000849	0.003373	0.007361	0.012618	0.018976	0.026294
2.50	0.000673	0.002773	0.006182	0.010759	0.016375	0.022916
2.55	0.000533	0.002277	0.005186	0.009164	0.014116	0.019949
2.60	0.000422	0.001868	0.004346	0.007798	0.012155	0.017347
2.65	0.000334	0.001531	0.003639	0.006629	0.010456	0.015069
2.70	0.000264	0.001254	0.003043	0.005629	0.008985	0.013076
2.75	0.000208	0.001026	0.002543	0.004775	0.007713	0.011335
2.80	0.000164	0.000839	0.002123	0.004047	0.006615	0.009817
2.85	0.000129	0.000685	0.001771	0.003427	0.005668	0.008493
2.90	0.000102	0.000559	0.001476	0.002899	0.004852	0.007341
2.95	0.000080	0.000455	0.001229	0.002451	0.004149	0.006339
3.00	0.000063	0.000371	0.001022	0.002069	0.003545	0.005469
3.05	0.000049	0.000302	0.000849	0.001746	0.003027	0.004714
3.10	0.000039	0.000246	0.000705	0.001472	0.002582	0.004060
3.15	0.000030	0.000199	0.000585	0.001240	0.002200	0.003494
3.20	0.000024	0.000162	0.000485	0.001043	0.001873	0.003004
3.25	0.000019	0.000131	0.000402	0.000877	0.001594	0.002580
3.30	0.000015	0.000106	0.000333	0.000737	0.001355	0.002214
3.35	0.000011	0.000086	0.000275	0.000619	0.001151	0.001899
3.40	0.000009	0.000070	0.000227	0.000519	0.000977	0.001627
3.45	0.000007	0.000056	0.000188	0.000435	0.000829	0.001394

Table 1.8 (continued)

			t			
v	0.05	0.10	0.15	0.20	0.25	0.30
3.50	0.000005	0.000046	0.000155	0.000365	0.000702	0.001192
3.55	0.000004	0.000037	0.000128	0.000305	0.000595	0.001019
3.60	0.000003	0.000030	0.000105	0.000255	0.000503	0.000871
3.65	0.000003	0.000024	0.000087	0.000213	0.000426	0.000743
3.70	0.000002	0.000019	0.000071	0.000178	0.000360	0.000634
3.75	0.000002	0.000016	0.000059	0.000149	0.000304	0.000541
3.80	0.000001	0.000013	0.000048	0.000124	0.000256	0.000461
3.85	0.000001	0.000010	0.000040	0.000103	0.000216	0.000392
3.90	0.000001	0.000008	0.000033	0.000086	0.000182	0.000334
3.95	0.000001	0.000007	0.000027	0.000072	0.000153	0.000284
4.00		0.000006	0.000022	0.000060	0.000129	0.000241
4.05		0.000004	0.000018	0.000050	0.000109	0.000204
4.10		0.000003	0.000015	0.000041	0.000091	0.000174
4.15		0.000003	0.000012	0.000034	0.000077	0.000147
4.20		0.000002	0.000010	0.000028	0.000064	0.000125
4.25		0.000002	0.000008	0.000024	0.000054	0.000106
4.30		0.000001	0.000006	0.000020	0.000045	0.000089
4.35		0.000001	0.000005	0.000016	0.000038	0.000076
4.40		0.000001	0.000004	0.000013	0.000032	0.000064
4.45		0.000001	0.000004	0.000011	0.000027	0.000054
4.50		0.000001	0.000003	0.000009	0.000022	0.000046
4.55			0.000002	0.000008	0.000019	0.000038
4.60			0.000002	0.000006	0.000016	0.000032
4.65			0.000002	0.000005	0.000013	0.000027
4.70			0.000001	0.000004	0.000011	0.000023
4.75			0.000001	0.000004	0.000009	0.000019
4.80			0.000001	0.000003	0.000008	0.000016
4.85			0.000001	0.000002	0.000005	0.000014
4.90			0.000001	0.000002	0.000005	0.000012
4.95				0.000002	0.000004	0.000010
5.00				0.000001	0.000004	0.000008

			t			
v	0.35	0.40	0.45	0.50	0.55	0.60
0.00	−0.160477	−0.437631	−0.652332	−0.821219	−0.955614	−1.06351
0.05	0.133765	−0.147824	−0.370783	−0.550061	−0.695964	−0.815859
0.10	0.379503	0.100417	−0.124764	−0.309188	−0.462037	−0.589956
0.15	0.581113	0.310085	0.087666	−0.097463	−0.253317	−0.385764
0.20	0.742980	0.484326	0.268690	0.086539	−0.068972	−0.202922
0.25	0.869402	0.626345	0.420645	0.244461	0.092080	−0.040798
0.30	0.964521	0.739337	0.545950	0.378090	0.231107	0.101448
0.35	1.03227	0.826421	0.647048	0.489300	0.349509	0.224822
0.40	1.07634	0.890601	0.726353	0.580007	0.448776	0.330447

Table 1.8 (continued)

v	t					
	0.35	0.40	0.45	0.50	0.55	0.60
0.45	1.10015	0.934727	0.786218	0.652124	0.530440	0.419529
0.50	1.10683	0.961476	0.828899	0.707533	0.596051	0.493322
0.55	1.09923	0.973330	0.856533	0.748054	0.647142	0.553099
0.60	1.07991	0.972572	0.871128	0.775432	0.685209	0.600132
0.65	1.05117	0.961281	0.874546	0.791314	0.711698	0.635673
0.70	1.01501	0.941334	0.868501	0.797245	0.727982	0.660933
0.75	0.973206	0.914410	0.854555	0.794657	0.735360	0.677076
0.80	0.927295	0.882001	0.834123	0.784868	0.735045	0.685206
0.85	0.878594	0.845417	0.808473	0.769082	0.728162	0.686363
0.90	0.828222	0.805801	0.778732	0.748388	0.715745	0.681517
0.95	0.777122	0.764141	0.745896	0.723762	0.698737	0.671565
1.00	0.726076	0.721280	0.710835	0.696076	0.677992	0.657331
1.05	0.675720	0.677931	0.674303	0.666100	0.654277	0.639565
1.10	0.626568	0.634692	0.636946	0.634508	0.628275	0.618948
1.15	0.579020	0.592051	0.599315	0.601886	0.600591	0.596089
1.20	0.533382	0.550405	0.561871	0.568737	0.571756	0.571532
1.25	0.489876	0.510069	0.524993	0.535489	0.542230	0.545758
1.30	0.448655	0.471283	0.489002	0.502504	0.512409	0.519190
1.35	0.409811	0.434227	0.454112	0.470079	0.482633	0.492192
1.40	0.373388	0.399025	0.420547	0.438459	0.453184	0.465082
1.45	0.339389	0.365758	0.388439	0.407837	0.424300	0.438128
1.50	0.307783	0.334465	0.357887	0.378364	0.396172	0.411556
1.55	0.278514	0.305156	0.328958	0.350154	0.368955	0.385554
1.60	0.251504	0.277812	0.301686	0.323286	0.342767	0.360275
1.65	0.226661	0.252395	0.276079	0.297812	0.317698	0.335839
1.70	0.203882	0.228849	0.252127	0.273761	0.293811	0.312342
1.75	0.183055	0.207105	0.229798	0.251138	0.271145	0.289854
1.80	0.164065	0.187083	0.209051	0.229935	0.249723	0.268423
1.85	0.146795	0.168700	0.189832	0.210127	0.229549	0.248080
1.90	0.131126	0.151865	0.172078	0.191679	0.210612	0.228840
1.95	0.116944	0.136486	0.155721	0.174547	0.192892	0.210705
2.00	0.104135	0.122471	0.140691	0.158681	0.176359	0.193663
2.05	0.092591	0.109727	0.126911	0.144024	0.160977	0.177698
2.10	0.082208	0.098163	0.114306	0.130517	0.146701	0.162781
2.15	0.072887	0.087691	0.102802	0.118099	0.133484	0.148880
2.20	0.064535	0.078228	0.092324	0.106705	0.121276	0.135957
2.25	0.057064	0.069691	0.082799	0.096275	0.110025	0.123973
2.30	0.050393	0.062004	0.074156	0.086744	0.099678	0.112883
2.35	0.044447	0.055095	0.066329	0.078053	0.090181	0.102642
2.40	0.039154	0.048894	0.059253	0.070141	0.081481	0.093204
2.45	0.034451	0.043339	0.052866	0.062952	0.073525	0.084523
2.50	0.030277	0.038369	0.047111	0.056430	0.066263	0.076552
2.55	0.026579	0.033930	0.041933	0.050523	0.059646	0.069248
2.60	0.023307	0.029971	0.037281	0.045182	0.053625	0.062564

Table 1.8 (continued)

			t			
v	0.35	0.40	0.45	0.50	0.55	0.60
2.65	0.020415	0.026445	0.033108	0.040360	0.048157	0.056459
2.70	0.017864	0.023308	0.029370	0.036012	0.043197	0.050892
2.75	0.015615	0.020522	0.026026	0.032097	0.038705	0.045822
2.80	0.013635	0.018050	0.023039	0.028578	0.034643	0.041212
2.85	0.011895	0.015860	0.020374	0.025418	0.030975	0.037026
2.90	0.010367	0.013922	0.017999	0.022585	0.027667	0.033231
2.95	0.009026	0.012210	0.015886	0.020048	0.024687	0.029794
3.00	0.007852	0.010697	0.014007	0.017778	0.022006	0.026686
3.05	0.006824	0.009363	0.012339	0.015750	0.019598	0.023879
3.10	0.005925	0.008189	0.010859	0.013941	0.017436	0.021346
3.15	0.005140	0.007155	0.009548	0.012328	0.015499	0.019065
3.20	0.004455	0.006246	0.008388	0.010892	0.013764	0.017011
3.25	0.003859	0.005448	0.007363	0.009614	0.012213	0.015165
3.30	0.003339	0.004748	0.006457	0.008480	0.010827	0.013508
3.35	0.002887	0.004134	0.005658	0.007472	0.009590	0.012021
3.40	0.002494	0.003597	0.004954	0.006579	0.008487	0.010689
3.45	0.002153	0.003127	0.004334	0.005788	0.007505	0.009496
3.50	0.001857	0.002717	0.003788	0.005088	0.006631	0.008429
3.55	0.001601	0.002358	0.003309	0.004469	0.005854	0.007477
3.60	0.001379	0.002045	0.002888	0.003923	0.005164	0.006626
3.65	0.001186	0.001773	0.002519	0.003440	0.004552	0.005868
3.70	0.001020	0.001535	0.002195	0.003015	0.004009	0.005193
3.75	0.000877	0.001329	0.001911	0.002640	0.003529	0.004592
3.80	0.000753	0.001149	0.001663	0.002310	0.003104	0.004057
3.85	0.000646	0.000993	0.001446	0.002020	0.002728	0.003582
3.90	0.000554	0.000857	0.001257	0.001765	0.002395	0.003160
3.95	0.000475	0.000740	0.001091	0.001541	0.002102	0.002786
4.00	0.000407	0.000638	0.000947	0.001345	0.001844	0.002455
4.05	0.000348	0.000550	0.000821	0.001173	0.001616	0.002161
4.10	0.000298	0.000474	0.000712	0.001022	0.001415	0.001902
4.15	0.000254	0.000408	0.000616	0.000890	0.001238	0.001672
4.20	0.000217	0.000351	0.000533	0.000774	0.001083	0.001469
4.25	0.000186	0.000301	0.000461	0.000674	0.000947	0.001290
4.30	0.000158	0.000259	0.000399	0.000585	0.000827	0.001132
4.35	0.000135	0.000222	0.000344	0.000509	0.000722	0.000993
4.40	0.000115	0.000191	0.000297	0.000441	0.000630	0.000870
4.45	0.000098	0.000164	0.000257	0.000383	0.000549	0.000762
4.50	0.000083	0.000140	0.000221	0.000332	0.000479	0.000667
4.55	0.000071	0.000120	0.000191	0.000288	0.000417	0.000584
4.60	0.000060	0.000103	0.000164	0.000249	0.000363	0.000510
4.65	0.000051	0.000088	0.000141	0.000216	0.000315	0.000446
4.70	0.000043	0.000075	0.000122	0.000187	0.000274	0.000389

Table 1.8 (continued)

			t			
v	0.35	0.40	0.45	0.50	0.55	0.60
4.75	0.000037	0.000064	0.000105	0.000161	0.000238	0.000340
4.80	0.000031	0.000055	0.000090	0.000139	0.000207	0.000296
4.85	0.000027	0.000047	0.000077	0.000120	0.000180	0.000258
4.90	0.000022	0.000040	0.000066	0.000104	0.000156	0.000225
4.95	0.000019	0.000034	0.000057	0.000090	0.000135	0.000196
5.00	0.000016	0.000029	0.000049	0.000077	0.000117	0.000171

			t			
v	0.65	0.70	0.75	0.80	0.85	0.90
0.00	−1.15071	−1.22156	−1.27934	−1.32662	−1.36540	−1.42353
0.05	−0.915164	−0.997965	−1.06740	−1.12593	−1.17551	−1.25376
0.10	−0.697889	−0.789603	−0.868024	−0.935461	−0.993767	−1.088720
0.15	−0.499224	−0.597097	−0.682053	−0.756220	−0.821319	−0.929702
0.20	−0.319178	−0.420751	−0.510028	−0.588931	−0.659026	−0.777757
0.25	−0.157487	−0.260597	−0.352219	−0.434054	−0.507502	−0.633721
0.30	−0.013660	−0.116432	−0.208661	−0.291822	−0.367141	−0.498223
0.35	0.112975	0.012138	−0.079193	−0.162268	−0.238138	−0.371708
0.40	0.223233	0.125663	0.036510	−0.045258	−0.120521	−0.254459
0.45	0.318035	0.224817	0.138910	0.059486	−0.014173	−0.146609
0.50	0.398373	0.310368	0.228575	0.152356	0.081148	−0.048169
0.55	0.465292	0.383150	0.306154	0.233840	0.165785	0.040964
0.60	0.519858	0.444044	0.372362	0.304499	0.240162	0.120984
0.65	0.563140	0.493954	0.427950	0.364949	0.304770	0.192165
0.70	0.596197	0.533795	0.473698	0.415843	0.360148	0.254843
0.75	0.620062	0.564474	0.510394	0.457859	0.406868	0.309406
0.80	0.635731	0.586879	0.538826	0.491686	0.445528	0.356282
0.85	0.644154	0.601874	0.559768	0.518012	0.476734	0.395927
0.90	0.646232	0.610285	0.573973	0.537520	0.501098	0.428815
0.95	0.642808	0.612898	0.582167	0.550875	0.519222	0.455432
1.00	0.634668	0.610455	0.585044	0.558717	0.531699	0.476268
1.05	0.622537	0.603648	0.583259	0.561665	0.539102	0.491810
1.10	0.607081	0.593121	0.577430	0.560302	0.541982	0.502535
1.15	0.588905	0.579467	0.568129	0.555181	0.540865	0.508911
1.20	0.568555	0.563229	0.555890	0.546817	0.536247	0.511388
1.25	0.546524	0.544900	0.541200	0.535690	0.528595	0.510398
1.30	0.523249	0.524925	0.524508	0.522244	0.518345	0.506350
1.35	0.499115	0.503703	0.506218	0.506885	0.505900	0.499633
1.40	0.474462	0.481589	0.486696	0.489983	0.491631	0.490609
1.45	0.449584	0.458896	0.466267	0.471874	0.475877	0.479618
1.50	0.424734	0.435898	0.445222	0.452859	0.458949	0.466975
1.55	0.400127	0.412833	0.423816	0.433207	0.441125	0.452967
1.60	0.375945	0.389906	0.402273	0.413156	0.422655	0.437861
1.65	0.352337	0.367289	0.380787	0.392917	0.403762	0.421897

Table 1.8 (continued)

			t			
v	0.65	0.70	0.75	0.80	0.85	0.90
1.70	0.329425	0.345129	0.359522	0.372671	0.384642	0.405292
1.75	0.307305	0.323544	0.338618	0.352577	0.365469	0.388242
1.80	0.286051	0.302632	0.318194	0.332770	0.346393	0.370923
1.85	0.265719	0.282469	0.298344	0.313362	0.327542	0.353487
1.90	0.246344	0.263113	0.279146	0.294448	0.309027	0.336071
1.95	0.227951	0.244607	0.260661	0.276105	0.290939	0.318792
2.00	0.210547	0.226978	0.242932	0.258394	0.273354	0.301753
2.05	0.194133	0.210242	0.225993	0.241362	0.256334	0.285040
2.10	0.178698	0.194404	0.209863	0.225044	0.239926	0.268726
2.15	0.164223	0.179461	0.194553	0.209464	0.224168	0.252871
2.20	0.150684	0.165401	0.180064	0.194636	0.209086	0.237524
2.25	0.138052	0.152206	0.166390	0.180564	0.194695	0.222722
2.30	0.126293	0.139855	0.153520	0.167249	0.181007	0.208496
2.35	0.115372	0.128319	0.141436	0.154682	0.168022	0.194866
2.40	0.105251	0.117570	0.130117	0.142850	0.155736	0.181846
2.45	0.095889	0.107575	0.119537	0.131737	0.144141	0.169441
2.50	0.087246	0.098299	0.109670	0.121322	0.133223	0.157655
2.55	0.079282	0.089708	0.100486	0.111582	0.122966	0.146484
2.60	0.071957	0.081765	0.091954	0.102492	0.113349	0.135919
2.65	0.065230	0.074435	0.084043	0.094024	0.104351	0.125951
2.70	0.059063	0.067682	0.076720	0.086150	0.095949	0.116564
2.75	0.053419	0.061471	0.069953	0.078842	0.088116	0.107743
2.80	0.048260	0.055766	0.063709	0.072070	0.080828	0.099469
2.85	0.043552	0.050536	0.057959	0.065805	0.074058	0.091722
2.90	0.039262	0.045747	0.052670	0.060019	0.067779	0.084482
2.95	0.035358	0.041368	0.047813	0.054682	0.061964	0.077726
3.00	0.031810	0.037370	0.043359	0.049768	0.056589	0.071433
3.05	0.028589	0.033725	0.039280	0.045249	0.051626	0.065579
3.10	0.025670	0.030405	0.035549	0.041099	0.047050	0.060142
3.15	0.023027	0.027386	0.032141	0.037293	0.042837	0.055100
3.20	0.020637	0.024643	0.029033	0.033806	0.038964	0.050431
3.25	0.018478	0.022155	0.026200	0.030617	0.035407	0.046112
3.30	0.016530	0.019900	0.023623	0.027703	0.032144	0.042123
3.35	0.014774	0.017858	0.021279	0.025044	0.029156	0.038443
3.40	0.013194	0.016012	0.019152	0.022619	0.026422	0.035053
3.45	0.011773	0.014345	0.017222	0.020412	0.023923	0.031933
3.50	0.010496	0.012840	0.015474	0.018405	0.021642	0.029065
3.55	0.009350	0.011484	0.013891	0.016580	0.019561	0.026432
3.60	0.008322	0.010263	0.012460	0.014925	0.017666	0.024017
3.65	0.007401	0.009164	0.011168	0.013424	0.015942	0.021805
3.70	0.006578	0.008177	0.010002	0.012064	0.014374	0.019780
3.75	0.005841	0.007290	0.008950	0.010833	0.012950	0.017929
3.80	0.005183	0.006495	0.008003	0.009721	0.011659	0.016238

Table 1.8 (continued)

			t			
v	0.65	0.70	0.75	0.80	0.85	0.90
3.85	0.004596	0.005782	0.007151	0.008716	0.010488	0.014696
3.90	0.004072	0.005143	0.006385	0.007809	0.009427	0.013289
3.95	0.003606	0.004572	0.005697	0.006992	0.008468	0.012009
4.00	0.003190	0.004061	0.005079	0.006255	0.007601	0.010844
4.05	0.002821	0.003605	0.004525	0.005592	0.006817	0.009784
4.10	0.002492	0.003198	0.004029	0.004996	0.006110	0.008822
4.15	0.002201	0.002835	0.003585	0.004461	0.005473	0.007949
4.20	0.001942	0.002511	0.003187	0.003980	0.004899	0.007158
4.25	0.001712	0.002223	0.002832	0.003548	0.004382	0.006440
4.30	0.001509	0.001967	0.002515	0.003161	0.003917	0.005791
4.35	0.001329	0.001739	0.002231	0.002815	0.003499	0.005204
4.40	0.001170	0.001537	0.001979	0.002505	0.003123	0.004673
4.45	0.001029	0.001357	0.001754	0.002228	0.002787	0.004194
4.50	0.000904	0.001197	0.001553	0.001980	0.002484	0.003761
4.55	0.000794	0.001056	0.001375	0.001758	0.002214	0.003371
4.60	0.000698	0.000931	0.001216	0.001561	0.001971	0.003020
4.65	0.000612	0.000820	0.001075	0.001385	0.001754	0.002703
4.70	0.000537	0.000722	0.000950	0.001227	0.001560	0.002419
4.75	0.000470	0.000635	0.000839	0.001088	0.001387	0.002163
4.80	0.000412	0.000558	0.000740	0.000963	0.001232	0.001933
4.85	0.000361	0.000491	0.000653	0.000852	0.001094	0.001726
4.90	0.000316	0.000431	0.000576	0.000754	0.000971	0.001540
4.95	0.000276	0.000378	0.000507	0.000667	0.000861	0.001374
5.00	0.000241	0.000332	0.000447	0.000589	0.000763	0.001225

			t			
v	0.95	1.0	2.0	3.0	4.0	5.0
0.00	−1.39727	−1.44521	−1.62410	−2.10105	−3.36204	−6.18718
0.05	−1.21769	−1.28477	−1.57134	−2.08150	−3.35250	−6.17950
0.10	−1.04444	−1.12762	−1.51187	−2.05434	−3.33421	−6.15996
0.15	−0.878758	−0.975123	−1.44651	−2.01992	−3.30736	−6.12865
0.20	−0.721608	−0.828382	−1.37610	−1.97863	−3.27215	−6.08571
0.25	−0.573731	−0.688306	−1.30145	−1.93091	−3.22882	−6.03134
0.30	−0.435649	−0.555611	−1.22333	−1.87719	−3.17765	−5.96576
0.35	−0.307696	−0.430836	−1.14250	−1.81793	−3.11895	−5.88924
0.40	−0.190037	−0.314358	−1.059690	−1.75363	−3.05305	−5.80207
0.45	−0.082689	−0.206416	−0.975570	−1.68475	−2.98033	−5.70459
0.50	0.014455	−0.107118	−0.890783	−1.61180	−2.90116	−5.59716
0.55	0.101608	−0.016463	−0.805926	−1.53527	−2.81594	−5.48019
0.60	0.179076	0.065647	−0.721551	−1.45566	−2.72511	−5.35409
0.65	0.247234	0.139392	−0.638164	−1.37344	−2.62910	−5.21931
0.70	0.306515	0.205022	−0.556222	−1.28911	−2.52834	−5.07632
0.75	0.357397	0.262842	−0.476139	−1.20312	−2.42330	−4.92562

Table 1.8 (continued)

v			t			
	0.95	1.0	2.0	3.0	4.0	5.0
0.80	0.400389	0.313202	−0.398280	−1.11594	−2.31443	−4.76771
0.85	0.436020	0.356490	−0.322965	−1.02800	−2.20220	−4.60311
0.90	0.464831	0.393116	−0.250473	−0.939714	−2.08707	−4.43237
0.95	0.487367	0.423511	−0.181037	−0.851493	−1.96948	−4.25601
1.00	0.504168	0.448113	−0.114852	−0.763712	−1.84991	−4.07461
1.05	0.515764	0.467367	−0.052072	−0.676729	−1.72878	−3.88870
1.10	0.522671	0.481713	0.007183	−0.590876	−1.60655	−3.69885
1.15	0.525385	0.491587	0.062828	−0.506464	−1.48364	−3.50562
1.20	0.524381	0.497414	0.114809	−0.423778	−1.36046	−3.30957
1.25	0.520109	0.499604	0.163103	−0.343078	−1.23740	−3.11124
1.30	0.512994	0.498551	0.207712	−0.264601	−1.11487	−2.91117
1.35	0.503432	0.494631	0.248664	−0.188558	−0.993211	−2.70991
1.40	0.491793	0.488201	0.286006	−0.115135	−0.872788	−2.50797
1.45	0.478417	0.479594	0.319808	−0.044496	−0.753928	−2.30588
1.50	0.463617	0.469123	0.350155	0.023219	−0.636943	−2.10412
1.55	0.447679	0.457080	0.377148	0.087893	−0.522125	−1.90318
1.60	0.430862	0.443732	0.400900	0.149430	−0.409744	−1.70353
1.65	0.413398	0.429325	0.421536	0.207756	−0.300050	−1.50562
1.70	0.395496	0.414085	0.439188	0.262817	−0.193273	−1.30986
1.75	0.377342	0.398214	0.453997	0.314576	−0.089623	−1.11669
1.80	0.359099	0.381898	0.466109	0.363017	0.010716	−0.926469
1.85	0.340909	0.365300	0.475673	0.408137	0.107574	−0.739578
1.90	0.322896	0.348567	0.482842	0.449951	0.200806	−0.556358
1.95	0.305166	0.331827	0.487768	0.488489	0.290287	−0.377131
2.00	0.287807	0.315194	0.490606	0.523793	0.375909	−0.202195
2.05	0.270896	0.298764	0.491506	0.555916	0.457585	−0.031826
2.10	0.254491	0.282623	0.490619	0.584925	0.535247	0.133725
2.15	0.238643	0.266840	0.488092	0.610895	0.608842	0.294229
2.20	0.223389	0.251474	0.484070	0.633910	0.678339	0.449481
2.25	0.208756	0.236576	0.478690	0.654062	0.743718	0.599298
2.30	0.194764	0.222181	0.472089	0.671449	0.804977	0.743520
2.35	0.181425	0.208322	0.464395	0.686175	0.862131	0.882010
2.40	0.168743	0.195019	0.455733	0.698348	0.915204	1.01465
2.45	0.156718	0.182288	0.446220	0.708083	0.964237	1.14135
2.50	0.145343	0.170137	0.435970	0.715493	1.00928	1.26204
2.55	0.134608	0.158571	0.425087	0.720698	1.05040	1.37666
2.60	0.124500	0.147587	0.413672	0.723815	1.08767	1.48518
2.65	0.115001	0.137180	0.401819	0.724965	1.12116	1.58758
2.70	0.106094	0.127341	0.389615	0.724268	1.15098	1.68387
2.75	0.097756	0.118059	0.377140	0.721842	1.17722	1.77407
2.80	0.089967	0.109320	0.364471	0.717806	1.19998	1.85821
2.85	0.082702	0.101106	0.351676	0.712277	1.21937	1.93634

Table 1.8 (continued)

			t			
v	0.95	1.0	2.0	3.0	4.0	5.0
2.90	0.075937	0.093401	0.338819	0.705368	1.23553	2.00854
2.95	0.069649	0.086186	0.325958	0.697192	1.24855	2.07488
3.00	0.063813	0.079440	0.313145	0.687858	1.25858	2.13545
3.05	0.058404	0.073144	0.300428	0.677471	1.26573	2.19036
3.10	0.053399	0.067275	0.287850	0.666135	1.27014	2.23973
3.15	0.048774	0.061814	0.275447	0.653949	1.27194	2.28367
3.20	0.044505	0.056739	0.263254	0.641007	1.27126	2.32233
3.25	0.040571	0.052030	0.251299	0.627401	1.26823	2.35585
3.30	0.036950	0.047665	0.239608	0.613217	1.26299	2.38436
3.35	0.033621	0.043626	0.228203	0.598540	1.25567	2.40805
3.40	0.030564	0.039892	0.217101	0.583448	1.24641	2.42705
3.45	0.027761	0.036444	0.206318	0.568014	1.23532	2.44155
3.50	0.025192	0.033266	0.195865	0.552310	1.22254	2.45170
3.55	0.022842	0.030338	0.185753	0.536402	1.20819	2.45769
3.60	0.020694	0.027644	0.175987	0.520351	1.19240	2.45969
3.65	0.018732	0.025168	0.166572	0.504215	1.17528	2.45788
3.70	0.016943	0.022895	0.157511	0.488048	1.15696	2.45244
3.75	0.015312	0.020811	0.148805	0.471899	1.13754	2.44355
3.80	0.013828	0.018901	0.140452	0.455814	1.11714	2.43138
3.85	0.012477	0.017153	0.132450	0.439836	1.09585	2.41612
3.90	0.011251	0.015555	0.124794	0.424003	1.07380	2.39795
3.95	0.010137	0.014095	0.117481	0.408349	1.05106	2.37704
4.00	0.009126	0.012763	0.110504	0.392907	1.02774	2.35356
4.05	0.008211	0.011548	0.103856	0.377705	1.00392	2.32770
4.10	0.007382	0.010442	0.097529	0.362767	0.979700	2.29961
4.15	0.006632	0.009434	0.091515	0.348117	0.955147	2.26946
4.20	0.005955	0.008518	0.085805	0.333774	0.930343	2.23742
4.25	0.005343	0.007686	0.080390	0.319756	0.905361	2.20364
4.30	0.004790	0.006930	0.075261	0.306075	0.880270	2.16828
4.35	0.004292	0.006244	0.070406	0.292745	0.855133	2.13148
4.40	0.003843	0.005623	0.065817	0.279776	0.830012	2.09340
4.45	0.003439	0.005060	0.061484	0.267175	0.804963	2.05417
4.50	0.003076	0.004551	0.057395	0.254949	0.780039	2.01393
4.55	0.002749	0.004090	0.053541	0.243102	0.755287	1.97281
4.60	0.002455	0.003673	0.049912	0.231636	0.730754	1.93093
4.65	0.002192	0.003297	0.046498	0.220553	0.706480	1.88842
4.70	0.001955	0.002958	0.043289	0.209852	0.682502	1.84539
4.75	0.001743	0.002652	0.040275	0.199532	0.658857	1.80195
4.80	0.001553	0.002376	0.037447	0.189590	0.635573	1.75820
4.85	0.001383	0.002128	0.034796	0.180023	0.612681	1.71424
4.90	0.001231	0.001905	0.032312	0.170826	0.590204	1.67018
4.95	0.001095	0.001704	0.029987	0.161993	0.568165	1.62609
5.00	0.000974	0.001523	0.027813	0.153519	0.546583	1.58206

Table 1.9: Third derivatives with respect to the order of the modified Bessel functions of the first kind $\partial^3 I_\nu(t)/\partial\nu^3$.

			t			
ν	0.05	0.10	0.15	0.20	0.25	0.30
0.00	−12.4072	0.110158	4.06137	5.63582	6.26299	6.44487
0.05	−13.4564	−1.835720	2.20843	4.00906	4.86544	5.25305
0.10	−13.7706	−3.25156	0.721652	2.63818	3.64780	4.18719
0.15	−13.5582	−4.23465	−0.448325	1.49817	2.59859	3.24377
0.20	−12.9830	−4.86941	−1.34733	0.563988	1.70492	2.41728
0.25	−12.1724	−5.22815	−2.01719	−0.188678	0.953132	1.70084
0.30	−11.2233	−5.37211	−2.49558	−0.782993	0.329327	1.08666
0.35	−10.2080	−5.35253	−2.81595	−1.24062	−0.180259	0.566367
0.40	−9.17922	−5.21178	−3.00770	−1.58151	−0.588934	0.131386
0.45	−8.17428	−4.98448	−3.09634	−1.82381	−0.909342	−0.226865
0.50	−7.21827	−4.69856	−3.10381	−1.98383	−1.15331	−0.516762
0.55	−6.32706	−4.37620	−3.04874	−2.07604	−1.33176	−0.746344
0.60	−5.50951	−4.03476	−2.94682	−2.11320	−1.45464	−0.923200
0.65	−4.76936	−3.68755	−2.81108	−2.10639	−1.53096	−1.05441
0.70	−4.10665	−3.34455	−2.65224	−2.06518	−1.56872	−1.14650
0.75	−3.51883	−3.01300	−2.47899	−1.99773	−1.57504	−1.20541
0.80	−3.00171	−2.69796	−2.29827	−1.91094	−1.55614	−1.23652
0.85	−2.55004	−2.40272	−2.11554	−1.81058	−1.51743	−1.24464
0.90	−2.15807	−2.12920	−1.93501	−1.70141	−1.46355	−1.23402
0.95	−1.81984	−1.87829	−1.75982	−1.58732	−1.39848	−1.20840
1.00	−1.52951	−1.65003	−1.59226	−1.47141	−1.32556	−1.17107
1.05	−1.28148	−1.44392	−1.43389	−1.35617	−1.24756	−1.12482
1.10	−1.07050	−1.25902	−1.28571	−1.24348	−1.16680	−1.07207
1.15	−0.891773	−1.09412	−1.14827	−1.13479	−1.08513	−1.01486
1.20	−0.740922	−0.947826	−1.02173	−1.03114	−1.00405	−0.954903
1.25	−0.614044	−0.818664	−0.906023	−0.933216	−0.924729	−0.893613
1.30	−0.507679	−0.705128	−0.800834	−0.841481	−0.848059	−0.832148
1.35	−0.418781	−0.605732	−0.705722	−0.756158	−0.774701	−0.771435
1.40	−0.344697	−0.519041	−0.620137	−0.677306	−0.705122	−0.712207
1.45	−0.283127	−0.443696	−0.543467	−0.604848	−0.639623	−0.655027
1.50	−0.232089	−0.378424	−0.475063	−0.538611	−0.578376	−0.600312
1.55	−0.189885	−0.322053	−0.414264	−0.478341	−0.521442	−0.548357
1.60	−0.155068	−0.273508	−0.360415	−0.423738	−0.468799	−0.499358
1.65	−0.126409	−0.231817	−0.312875	−0.374464	−0.420360	−0.453421
1.70	−0.102869	−0.196105	−0.271035	−0.330160	−0.375983	−0.410589
1.75	−0.083573	−0.165588	−0.234318	−0.290461	−0.335493	−0.370845
1.80	−0.067788	−0.139572	−0.202183	−0.255001	−0.298686	−0.334129
1.85	−0.054898	−0.117441	−0.174131	−0.223421	−0.265344	−0.300351
1.90	−0.044392	−0.098656	−0.149703	−0.195377	−0.235239	−0.269390
1.95	−0.035845	−0.082744	−0.128480	−0.170537	−0.208137	−0.241111
2.00	−0.028901	−0.069290	−0.110083	−0.148590	−0.183809	−0.215365

Table 1.9 (continued)

			t			
v	0.05	0.10	0.15	0.20	0.25	0.30
2.05	−0.023271	−0.057937	−0.094168	−0.129247	−0.162029	−0.191996
2.10	−0.018712	−0.048374	−0.080429	−0.112236	−0.142579	−0.170845
2.15	−0.015027	−0.040332	−0.068591	−0.097308	−0.125253	−0.151753
2.20	−0.012052	−0.033581	−0.058411	−0.084236	−0.109852	−0.134563
2.25	−0.009654	−0.027923	−0.049670	−0.072810	−0.096194	−0.119123
2.30	−0.007724	−0.023188	−0.042180	−0.062844	−0.084106	−0.105285
2.35	−0.006172	−0.019231	−0.035772	−0.054165	−0.073428	−0.092912
2.40	−0.004927	−0.015930	−0.030297	−0.046621	−0.064014	−0.081870
2.45	−0.003928	−0.013180	−0.025628	−0.040075	−0.055730	−0.072035
2.50	−0.003128	−0.010892	−0.021652	−0.034403	−0.048453	−0.063294
2.55	−0.002489	−0.008991	−0.018271	−0.029498	−0.042071	−0.055537
2.60	−0.001978	−0.007413	−0.015400	−0.025260	−0.036483	−0.048666
2.65	−0.001570	−0.006106	−0.012965	−0.021606	−0.031598	−0.042591
2.70	−0.001245	−0.005023	−0.010903	−0.018459	−0.027334	−0.037227
2.75	−0.000987	−0.004129	−0.009159	−0.015752	−0.023617	−0.032499
2.80	−0.000781	−0.003390	−0.007685	−0.013428	−0.020383	−0.028338
2.85	−0.000618	−0.002780	−0.006442	−0.011434	−0.017572	−0.024682
2.90	−0.000488	−0.002278	−0.005395	−0.009726	−0.015132	−0.021472
2.95	−0.000385	−0.001865	−0.004513	−0.008264	−0.013017	−0.018660
3.00	−0.000304	−0.001525	−0.003772	−0.007015	−0.011186	−0.016198
3.05	−0.000239	−0.001246	−0.003149	−0.005949	−0.009602	−0.014047
3.10	−0.000188	−0.001018	−0.002627	−0.005040	−0.008235	−0.012168
3.15	−0.000148	−0.000830	−0.002189	−0.004266	−0.007055	−0.010530
3.20	−0.000116	−0.000676	−0.001823	−0.003607	−0.006039	−0.009104
3.25	−0.000091	−0.000551	−0.001517	−0.003047	−0.005164	−0.007863
3.30	−0.000072	−0.000448	−0.001261	−0.002572	−0.004412	−0.006785
3.35	−0.000056	−0.000364	−0.001047	−0.002169	−0.003766	−0.005850
3.40	−0.000044	−0.000296	−0.000869	−0.001828	−0.003211	−0.005038
3.45	−0.000035	−0.000240	−0.000720	−0.001539	−0.002736	−0.004336
3.50	−0.000027	−0.000195	−0.000597	−0.001294	−0.002329	−0.003728
3.55	−0.000021	−0.000158	−0.000494	−0.001088	−0.001981	−0.003203
3.60	−0.000016	−0.000128	−0.000408	−0.000914	−0.001684	−0.002749
3.65	−0.000013	−0.000103	−0.000338	−0.000767	−0.001430	−0.002357
3.70	−0.000010	−0.000084	−0.000279	−0.000643	−0.001214	−0.002020
3.75	−0.000008	−0.000067	−0.000230	−0.000539	−0.001029	−0.001730
3.80	−0.000006	−0.000054	−0.000190	−0.000451	−0.000872	−0.001480
3.85	−0.000005	−0.000044	−0.000156	−0.000377	−0.000738	−0.001265
3.90	−0.000004	−0.000035	−0.000129	−0.000315	−0.000624	−0.001081
3.95	−0.000003	−0.000029	−0.000106	−0.000264	−0.000528	−0.000923
4.00	−0.000002	−0.000023	−0.000087	−0.000220	−0.000446	−0.000787
4.05	−0.000002	−0.000018	−0.000072	−0.000184	−0.000376	−0.000671
4.10	−0.000001	−0.000015	−0.000059	−0.000153	−0.000317	−0.000571
4.15	−0.000001	−0.000012	−0.000048	−0.000127	−0.000268	−0.000486

Table 1.9 (continued)

			t			
v	0.05	0.10	0.15	0.20	0.25	0.30
4.20	−0.000001	−0.000010	−0.000040	−0.000106	−0.000225	−0.000414
4.25	−0.000001	−0.000008	−0.000032	−0.000088	−0.000190	−0.000352
4.30		−0.000006	−0.000027	−0.000073	−0.000160	−0.000299
4.35		−0.000005	−0.000022	−0.000061	−0.000134	−0.000254
4.40		−0.000004	−0.000018	−0.000051	−0.000113	−0.000215
4.45		−0.000003	−0.000015	−0.000042	−0.000095	−0.000182
4.50		−0.000003	−0.000012	−0.000035	−0.000079	−0.000154
4.55		−0.000002	−0.000010	−0.000029	−0.000067	−0.000131
4.60		−0.000002	−0.000008	−0.000024	−0.000056	−0.000111
4.65		−0.000001	−0.000006	−0.000020	−0.000047	−0.000094
4.70		−0.000001	−0.000005	−0.000016	−0.000039	−0.000079
4.75		−0.000001	−0.000004	−0.000014	−0.000033	−0.000067
4.80		−0.000001	−0.000003	−0.000011	−0.000027	−0.000056
4.85		−0.000001	−0.000003	−0.000009	−0.000023	−0.000048
4.90			−0.000002	−0.000008	−0.000019	−0.000040
4.95			−0.000002	−0.000006	−0.000016	−0.000034
5.00			−0.000002	−0.000005	−0.000013	−0.000028

			t			
v	0.35	0.40	0.45	0.50	0.55	0.60
0.00	6.39856	6.23057	5.99765	5.73179	5.45174	5.16870
0.05	5.38533	5.37086	5.26970	5.11710	4.93466	4.73600
0.10	4.45882	4.56879	4.57755	4.52164	4.42418	4.30024
0.15	3.62025	3.82844	3.92695	3.95210	3.92742	3.86865
0.20	2.86878	3.15189	3.32182	3.41349	3.45001	3.44718
0.25	2.20192	2.53955	2.76446	2.90937	2.99630	3.04059
0.30	1.61597	1.99052	2.25591	2.44204	2.56945	2.65263
0.35	1.10634	1.50287	1.79607	2.01276	2.17163	2.28608
0.40	0.667834	1.07384	1.38401	1.62189	1.80415	1.94295
0.45	0.294900	0.700149	1.01811	1.26907	1.46760	1.62451
0.50	−0.018199	0.378078	0.696231	0.953350	1.16195	1.33144
0.55	−0.277212	0.103688	0.415841	0.673334	0.886673	1.06389
0.60	−0.487783	−0.127082	0.174164	0.427271	0.640855	0.821629
0.65	−0.655354	−0.318305	−0.031735	0.213162	0.423263	0.604037
0.70	−0.785099	−0.473983	−0.204866	0.028838	0.232432	0.410242
0.75	−0.881873	−0.597990	−0.348238	−0.127967	0.066733	0.239154
0.80	−0.950181	−0.694016	−0.464810	−0.259562	−0.075576	0.089526
0.85	−0.994159	−0.765540	−0.557444	−0.368248	−0.196295	−0.040001
0.90	−1.01757	−0.815805	−0.628873	−0.456282	−0.297244	−0.150853
0.95	−1.02382	−0.847802	−0.681676	−0.525845	−0.380229	−0.244491
1.00	−1.01592	−0.864267	−0.718266	−0.579020	−0.447018	−0.322383
1.05	−0.996587	−0.867682	−0.740874	−0.617777	−0.499313	−0.385977

Table 1.9 (continued)

				t		
v	0.35	0.40	0.45	0.50	0.55	0.60
1.10	−0.968169	−0.860277	−0.751548	−0.643955	−0.538742	−0.436684
1.15	−0.932732	−0.844043	−0.752154	−0.659264	−0.566839	−0.475861
1.20	−0.892058	−0.820741	−0.744374	−0.665274	−0.585043	−0.504805
1.25	−0.847674	−0.791919	−0.729719	−0.663419	−0.594689	−0.524735
1.30	−0.800872	−0.758921	−0.709529	−0.654997	−0.597005	−0.536794
1.35	−0.752737	−0.722911	−0.684989	−0.641177	−0.593115	−0.542045
1.40	−0.704169	−0.684883	−0.657136	−0.622999	−0.584038	−0.541464
1.45	−0.655902	−0.645677	−0.626870	−0.601386	−0.570694	−0.535947
1.50	−0.608526	−0.605999	−0.594966	−0.577150	−0.553904	−0.526305
1.55	−0.562505	−0.566430	−0.562084	−0.550998	−0.534398	−0.513273
1.60	−0.518196	−0.527446	−0.528781	−0.523542	−0.512819	−0.497508
1.65	−0.475860	−0.489424	−0.495522	−0.495307	−0.489731	−0.479594
1.70	−0.435682	−0.452661	−0.462688	−0.466735	−0.465622	−0.460047
1.75	−0.397779	−0.417382	−0.430589	−0.438202	−0.440912	−0.439322
1.80	−0.362214	−0.383751	−0.399469	−0.410014	−0.415960	−0.417811
1.85	−0.329004	−0.351877	−0.369515	−0.382425	−0.391065	−0.395855
1.90	−0.298130	−0.321827	−0.340869	−0.355634	−0.366480	−0.373743
1.95	−0.269543	−0.293633	−0.313630	−0.329799	−0.342409	−0.351721
2.00	−0.243174	−0.267294	−0.287859	−0.305038	−0.319018	−0.329993
2.05	−0.218933	−0.242786	−0.263591	−0.281435	−0.296437	−0.308728
2.10	−0.196721	−0.220065	−0.240835	−0.259049	−0.274764	−0.288061
2.15	−0.176430	−0.199073	−0.219579	−0.237909	−0.254071	−0.268099
2.20	−0.157944	−0.179739	−0.199795	−0.218030	−0.234407	−0.248925
2.25	−0.141150	−0.161986	−0.181444	−0.199405	−0.215802	−0.230599
2.30	−0.125929	−0.145729	−0.164473	−0.182017	−0.198266	−0.213161
2.35	−0.112168	−0.130881	−0.148824	−0.165835	−0.181799	−0.196637
2.40	−0.099755	−0.117354	−0.134434	−0.150822	−0.166388	−0.181037
2.45	−0.088581	−0.105059	−0.121236	−0.136932	−0.152009	−0.166362
2.50	−0.078544	−0.093909	−0.109160	−0.124115	−0.138633	−0.152601
2.55	−0.069545	−0.083819	−0.098136	−0.112319	−0.126224	−0.139736
2.60	−0.061493	−0.074706	−0.088094	−0.101487	−0.114742	−0.127742
2.65	−0.054301	−0.066491	−0.078966	−0.091563	−0.104143	−0.116590
2.70	−0.047887	−0.059100	−0.070686	−0.082490	−0.094381	−0.106247
2.75	−0.042178	−0.052463	−0.063189	−0.074213	−0.085412	−0.096677
2.80	−0.037104	−0.046511	−0.056412	−0.066676	−0.077186	−0.087842
2.85	−0.032601	−0.041184	−0.050299	−0.059825	−0.069658	−0.079702
2.90	−0.028612	−0.036424	−0.044792	−0.053610	−0.062782	−0.072218
2.95	−0.025082	−0.032176	−0.039840	−0.047981	−0.056512	−0.065351
3.00	−0.021963	−0.028391	−0.035394	−0.042891	−0.050805	−0.059062
3.05	−0.019211	−0.025023	−0.031408	−0.038297	−0.045619	−0.053311
3.10	−0.016787	−0.022031	−0.027840	−0.034155	−0.040914	−0.048062
3.15	−0.014653	−0.019376	−0.024651	−0.030427	−0.036653	−0.043279
3.20	−0.012777	−0.017023	−0.021803	−0.027076	−0.032798	−0.038928

Table 1.9 (continued)

			t			
v	0.35	0.40	0.45	0.50	0.55	0.60
3.25	-0.011131	-0.014941	-0.019265	-0.024069	-0.029317	-0.034974
3.30	-0.009687	-0.013101	-0.017005	-0.021373	-0.026178	-0.031388
3.35	-0.008422	-0.011476	-0.014995	-0.018960	-0.023350	-0.028140
3.40	-0.007316	-0.010043	-0.013209	-0.016803	-0.020807	-0.025202
3.45	-0.006349	-0.008780	-0.011626	-0.014877	-0.018522	-0.022548
3.50	-0.005506	-0.007670	-0.010222	-0.013159	-0.016473	-0.020153
3.55	-0.004770	-0.006694	-0.008980	-0.011628	-0.014636	-0.017995
3.60	-0.004129	-0.005837	-0.007882	-0.010267	-0.012992	-0.016053
3.65	-0.003571	-0.005085	-0.006912	-0.009056	-0.011521	-0.014307
3.70	-0.003085	-0.004427	-0.006056	-0.007981	-0.010209	-0.012739
3.75	-0.002664	-0.003850	-0.005301	-0.007028	-0.009037	-0.011333
3.80	-0.002298	-0.003346	-0.004637	-0.006183	-0.007994	-0.010074
3.85	-0.001981	-0.002905	-0.004053	-0.005436	-0.007065	-0.008946
3.90	-0.001707	-0.002521	-0.003539	-0.004775	-0.006238	-0.007938
3.95	-0.001469	-0.002186	-0.003088	-0.004191	-0.005504	-0.007038
4.00	-0.001264	-0.001893	-0.002693	-0.003675	-0.004852	-0.006235
4.05	-0.001086	-0.001639	-0.002346	-0.003220	-0.004275	-0.005519
4.10	-0.000933	-0.001418	-0.002042	-0.002820	-0.003763	-0.004882
4.15	-0.000800	-0.001226	-0.001777	-0.002467	-0.003309	-0.004314
4.20	-0.000686	-0.001059	-0.001545	-0.002157	-0.002909	-0.003810
4.25	-0.000588	-0.000914	-0.001342	-0.001885	-0.002555	-0.003362
4.30	-0.000504	-0.000788	-0.001165	-0.001646	-0.002242	-0.002965
4.35	-0.000431	-0.000680	-0.001010	-0.001436	-0.001966	-0.002613
4.40	-0.000369	-0.000585	-0.000876	-0.001252	-0.001723	-0.002301
4.45	-0.000315	-0.000504	-0.000759	-0.001091	-0.001509	-0.002025
4.50	-0.000269	-0.000434	-0.000657	-0.000950	-0.001321	-0.001780
4.55	-0.000230	-0.000373	-0.000568	-0.000826	-0.001155	-0.001564
4.60	-0.000196	-0.000320	-0.000491	-0.000718	-0.001010	-0.001374
4.65	-0.000167	-0.000275	-0.000425	-0.000624	-0.000882	-0.001206
4.70	-0.000142	-0.000236	-0.000367	-0.000542	-0.000770	-0.001057
4.75	-0.000121	-0.000202	-0.000316	-0.000470	-0.000671	-0.000927
4.80	-0.000103	-0.000173	-0.000273	-0.000408	-0.000585	-0.000812
4.85	-0.000088	-0.000149	-0.000235	-0.000354	-0.000510	-0.000711
4.90	-0.000075	-0.000127	-0.000203	-0.000306	-0.000444	-0.000622
4.95	-0.000063	-0.000109	-0.000174	-0.000265	-0.000386	-0.000543
5.00	-0.000054	-0.000093	-0.000150	-0.000230	-0.000336	-0.000475

			t			
v	0.65	0.70	0.75	0.80	0.85	0.90
0.00	4.88947	4.61813	4.35706	4.10757	3.87028	3.64536
0.05	4.52993	4.32220	4.11657	3.91552	3.72063	3.53287
0.10	4.15995	4.01020	3.85578	3.70001	3.54521	3.39301

Table 1.9 (continued)

			t			
v	0.65	0.70	0.75	0.80	0.85	0.90
0.15	3.78666	3.68905	3.58127	3.46725	3.34986	3.23119
0.20	3.41608	3.36468	3.29882	3.22279	3.13982	3.05235
0.25	3.05321	3.04217	3.01344	2.97151	2.91981	2.86100
0.30	2.70212	2.72575	2.72943	2.71767	2.69400	2.66116
0.35	2.36602	2.41889	2.45038	2.46494	2.46603	2.45642
0.40	2.04738	2.12437	2.17930	2.21640	2.23904	2.24991
0.45	1.74799	1.84436	1.91860	1.97465	2.01571	2.04436
0.50	1.46905	1.58046	1.67016	1.74178	1.79829	1.84210
0.55	1.21127	1.33377	1.43542	1.51946	1.58861	1.64507
0.60	0.974916	1.10500	1.21539	1.30897	1.38813	1.45490
0.65	0.759902	0.894473	1.01074	1.11121	1.19799	1.27286
0.70	0.565837	0.702192	0.821810	0.926814	1.01902	1.09998
0.75	0.392089	0.527915	0.648673	0.756122	0.851787	0.936998
0.80	0.237834	0.371182	0.491180	0.599245	0.696628	0.784436
0.85	0.102104	0.231363	0.348995	0.456101	0.553674	0.642611
0.90	−0.016178	0.107692	0.221626	0.326437	0.422883	0.511664
0.95	−0.118158	−0.000693	0.108463	0.209869	0.304069	0.391582
1.00	−0.205026	−0.094729	0.008802	0.105901	0.196922	0.282224
1.05	−0.277989	−0.175396	−0.078129	0.013952	0.101036	0.183338
1.10	−0.338249	−0.243697	−0.153149	−0.066624	0.015924	0.094586
1.15	−0.386987	−0.300644	−0.217102	−0.136516	−0.058958	0.015557
1.20	−0.425350	−0.347235	−0.270845	−0.196444	−0.124200	−0.054215
1.25	−0.454439	−0.384447	−0.315233	−0.247141	−0.180418	−0.115238
1.30	−0.475300	−0.413225	−0.351105	−0.289344	−0.228249	−0.168046
1.35	−0.488919	−0.434474	−0.379279	−0.323783	−0.268331	−0.213195
1.40	−0.496220	−0.449050	−0.400543	−0.351169	−0.301303	−0.251244
1.45	−0.498058	−0.457761	−0.415646	−0.372193	−0.327791	−0.282757
1.50	−0.495222	−0.461361	−0.425301	−0.387518	−0.348404	−0.308286
1.55	−0.488434	−0.460550	−0.430175	−0.397773	−0.363731	−0.328375
1.60	−0.478351	−0.455971	−0.430891	−0.403553	−0.374332	−0.343548
1.65	−0.465564	−0.448213	−0.428026	−0.405416	−0.380741	−0.354308
1.70	−0.450607	−0.437813	−0.422111	−0.403882	−0.383459	−0.361134
1.75	−0.433951	−0.425255	−0.413631	−0.399429	−0.382956	−0.364480
1.80	−0.416016	−0.410971	−0.403029	−0.392502	−0.379667	−0.364773
1.85	−0.397170	−0.395350	−0.390703	−0.383502	−0.373997	−0.362409
1.90	−0.377733	−0.378735	−0.377010	−0.372797	−0.366314	−0.357758
1.95	−0.357982	−0.361425	−0.362270	−0.360718	−0.356957	−0.351161
2.00	−0.338154	−0.343685	−0.346765	−0.347561	−0.346233	−0.342928
2.05	−0.318448	−0.325740	−0.330742	−0.333592	−0.334418	−0.333345
2.10	−0.299034	−0.307785	−0.314420	−0.319043	−0.321759	−0.322670
2.15	−0.280049	−0.289985	−0.297983	−0.304121	−0.308478	−0.311135
2.20	−0.261604	−0.272478	−0.281594	−0.289005	−0.294770	−0.298947
2.25	−0.243787	−0.255376	−0.265387	−0.273851	−0.280806	−0.286293

Table 1.9 (continued)

			t			
v	0.65	0.70	0.75	0.80	0.85	0.90
2.30	-0.226668	-0.238773	-0.249476	-0.258790	-0.266735	-0.273334
2.35	-0.210295	-0.222740	-0.233955	-0.243936	-0.252686	-0.260215
2.40	-0.194702	-0.207334	-0.218900	-0.229382	-0.238769	-0.247061
2.45	-0.179911	-0.192595	-0.204370	-0.215205	-0.225079	-0.233979
2.50	-0.165930	-0.178551	-0.190411	-0.201467	-0.211691	-0.221060
2.55	-0.152759	-0.165220	-0.177056	-0.188218	-0.198670	-0.208381
2.60	-0.140390	-0.152608	-0.164327	-0.175495	-0.186068	-0.196009
2.65	-0.128808	-0.140714	-0.152239	-0.163325	-0.173923	-0.183993
2.70	-0.117991	-0.129531	-0.140795	-0.151725	-0.162267	-0.172378
2.75	-0.107915	-0.119045	-0.129996	-0.140705	-0.151120	-0.161194
2.80	-0.098552	-0.109239	-0.119832	-0.130269	-0.140497	-0.150467
2.85	-0.089871	-0.100091	-0.110292	-0.120415	-0.130404	-0.140213
2.90	-0.081841	-0.091576	-0.101360	-0.111134	-0.120844	-0.130443
2.95	-0.074426	-0.083669	-0.093018	-0.102416	-0.111813	-0.121161
3.00	-0.067595	-0.076341	-0.085243	-0.094247	-0.103304	-0.112368
3.05	-0.061312	-0.069563	-0.078012	-0.086608	-0.095305	-0.104059
3.10	-0.055544	-0.063306	-0.071301	-0.079482	-0.087804	-0.096228
3.15	-0.050258	-0.057541	-0.065085	-0.072846	-0.080785	-0.088863
3.20	-0.045421	-0.052238	-0.059337	-0.066680	-0.074231	-0.081952
3.25	-0.041003	-0.047368	-0.054032	-0.060961	-0.068122	-0.075481
3.30	-0.036974	-0.042903	-0.049144	-0.055666	-0.062439	-0.069434
3.35	-0.033304	-0.038815	-0.044647	-0.050771	-0.057162	-0.063793
3.40	-0.029967	-0.035079	-0.040516	-0.046255	-0.052271	-0.058541
3.45	-0.026936	-0.031669	-0.036728	-0.042093	-0.047744	-0.053659
3.50	-0.024187	-0.028561	-0.033259	-0.038265	-0.043561	-0.049129
3.55	-0.021697	-0.025732	-0.030087	-0.034748	-0.039702	-0.044933
3.60	-0.019445	-0.023160	-0.027190	-0.031522	-0.036147	-0.041051
3.65	-0.017410	-0.020826	-0.024547	-0.028567	-0.032877	-0.037465
3.70	-0.015573	-0.018708	-0.022140	-0.025864	-0.029872	-0.034158
3.75	-0.013918	-0.016791	-0.019951	-0.023394	-0.027116	-0.031112
3.80	-0.012427	-0.015056	-0.017961	-0.021140	-0.024591	-0.028310
3.85	-0.011087	-0.013489	-0.016155	-0.019086	-0.022280	-0.025735
3.90	-0.009882	-0.012074	-0.014518	-0.017215	-0.020167	-0.023374
3.95	-0.008801	-0.010799	-0.013035	-0.015515	-0.018239	-0.021209
4.00	-0.007832	-0.009650	-0.011694	-0.013970	-0.016481	-0.019228
4.05	-0.006964	-0.008616	-0.010482	-0.012568	-0.014879	-0.017417
4.10	-0.006187	-0.007687	-0.009388	-0.011298	-0.013421	-0.015763
4.15	-0.005492	-0.006852	-0.008402	-0.010148	-0.012097	-0.014254
4.20	-0.004872	-0.006103	-0.007513	-0.009107	-0.010894	-0.012878
4.25	-0.004319	-0.005432	-0.006713	-0.008167	-0.009803	-0.011626
4.30	-0.003825	-0.004832	-0.005993	-0.007318	-0.008814	-0.010488
4.35	-0.003386	-0.004294	-0.005347	-0.006553	-0.007919	-0.009453

Table 1.9 (continued)

			t			
v	0.65	0.70	0.75	0.80	0.85	0.90
4.40	−0.002994	−0.003813	−0.004767	−0.005863	−0.007110	−0.008514
4.45	−0.002647	−0.003384	−0.004246	−0.005242	−0.006378	−0.007663
4.50	−0.002337	−0.003001	−0.003780	−0.004683	−0.005718	−0.006892
4.55	−0.002063	−0.002660	−0.003363	−0.004181	−0.005122	−0.006193
4.60	−0.001820	−0.002355	−0.002989	−0.003730	−0.004585	−0.005562
4.65	−0.001604	−0.002084	−0.002656	−0.003325	−0.004101	−0.004991
4.70	−0.001413	−0.001844	−0.002358	−0.002963	−0.003666	−0.004476
4.75	−0.001244	−0.001629	−0.002092	−0.002638	−0.003275	−0.004011
4.80	−0.001094	−0.001439	−0.001854	−0.002347	−0.002924	−0.003592
4.85	−0.000962	−0.001270	−0.001643	−0.002087	−0.002608	−0.003215
4.90	−0.000845	−0.001121	−0.001455	−0.001854	−0.002326	−0.002875
4.95	−0.000742	−0.000988	−0.001287	−0.001647	−0.002072	−0.002570
5.00	−0.000651	−0.000870	−0.001139	−0.001461	−0.001845	−0.002296

			t			
v	0.95	1.0	2.0	3.0	4.0	5.0
0.00	3.43267	3.23191	0.982423	0.312509	0.102611	0.034400
0.05	3.35282	3.18076	1.12514	0.468239	0.278765	0.272507
0.10	3.24453	3.10051	1.25108	0.617108	0.451958	0.508999
0.15	3.11278	2.99577	1.36037	0.758411	0.621388	0.743015
0.20	2.96223	2.87086	1.45330	0.891540	0.786298	0.973716
0.25	2.79712	2.72979	1.53029	1.01598	0.945984	1.20029
0.30	2.62133	2.57623	1.59188	1.13132	1.09979	1.42197
0.35	2.43833	2.41355	1.63873	1.23723	1.24713	1.63800
0.40	2.25122	2.24476	1.67155	1.33348	1.38746	1.84768
0.45	2.06273	2.07257	1.69114	1.41990	1.52029	2.05035
0.50	1.87523	1.89935	1.69835	1.49643	1.64522	2.24539
0.55	1.69074	1.72718	1.69408	1.56308	1.76187	2.43223
0.60	1.51098	1.55783	1.67923	1.61991	1.86995	2.61034
0.65	1.33737	1.39282	1.65474	1.66706	1.96921	2.77926
0.70	1.17105	1.23339	1.62155	1.70474	2.05947	2.93855
0.75	1.01292	1.08057	1.58057	1.73318	2.14060	3.08784
0.80	0.863650	0.935145	1.53273	1.75269	2.21253	3.22682
0.85	0.723720	0.797732	1.47892	1.76362	2.27522	3.35522
0.90	0.593424	0.668757	1.42000	1.76633	2.32872	3.47282
0.95	0.472902	0.548494	1.35680	1.76124	2.37309	3.57944
1.00	0.362161	0.437081	1.29011	1.74879	2.40846	3.67499
1.05	0.261091	0.334535	1.22068	1.72943	2.43498	3.75939
1.10	0.169486	0.240770	1.14922	1.70364	2.45286	3.83261
1.15	0.087060	0.155612	1.07639	1.67189	2.46233	3.89470
1.20	0.013461	0.078818	1.00279	1.63468	2.46366	3.94572
1.25	−0.051713	0.010084	0.928981	1.59251	2.45715	3.98578

Table 1.9 (continued)

v	0.95	1.0	2.0	3.0	4.0	5.0
			t			
1.30	−0.108905	−0.050944	0.855478	1.54587	2.44312	4.01505
1.35	−0.158582	−0.104651	0.782742	1.49524	2.42191	4.03371
1.40	−0.201231	−0.151452	0.711186	1.44112	2.39389	4.04200
1.45	−0.237349	−0.191778	0.641179	1.38396	2.35945	4.04017
1.50	−0.267434	−0.226071	0.573041	1.32424	2.31897	4.02853
1.55	−0.291980	−0.254775	0.507051	1.26240	2.27287	4.00740
1.60	−0.311471	−0.278335	0.443446	1.19885	2.22155	3.97712
1.65	−0.326380	−0.297185	0.382421	1.13402	2.16543	3.93806
1.70	−0.337158	−0.311751	0.324137	1.06829	2.10493	3.89063
1.75	−0.344240	−0.322444	0.268718	1.00203	2.04048	3.83522
1.80	−0.348038	−0.329659	0.216259	0.935581	1.97249	3.77227
1.85	−0.348940	−0.333770	0.166820	0.869273	1.90136	3.70221
1.90	−0.347310	−0.335133	0.120439	0.803404	1.82752	3.62550
1.95	−0.343487	−0.334083	0.077126	0.738252	1.75135	3.54258
2.00	−0.337785	−0.330931	0.036869	0.674071	1.67324	3.45393
2.05	−0.330492	−0.325968	−0.000363	0.611093	1.59358	3.36001
2.10	−0.321873	−0.319462	−0.034619	0.549526	1.51271	3.26127
2.15	−0.312169	−0.311659	−0.065965	0.489556	1.43100	3.15820
2.20	−0.301599	−0.302785	−0.094478	0.431347	1.34878	3.05125
2.25	−0.290357	−0.293044	−0.120248	0.375045	1.26637	2.94088
2.30	−0.278619	−0.282622	−0.143374	0.320770	1.18407	2.82755
2.35	−0.266541	−0.271684	−0.163965	0.268627	1.10216	2.71168
2.40	−0.254261	−0.260378	−0.182132	0.218702	1.02091	2.59373
2.45	−0.241898	−0.248837	−0.197995	0.171061	0.940585	2.47410
2.50	−0.229558	−0.237177	−0.211675	0.125755	0.861404	2.35321
2.55	−0.217329	−0.225497	−0.223296	0.082820	0.783585	2.23146
2.60	−0.205289	−0.213886	−0.232982	0.042275	0.707326	2.10922
2.65	−0.193502	−0.202420	−0.240859	0.004128	0.632806	1.98687
2.70	−0.182020	−0.191161	−0.247050	−0.031626	0.560188	1.86476
2.75	−0.170887	−0.180164	−0.251678	−0.065006	0.489615	1.74321
2.80	−0.160138	−0.169472	−0.254862	−0.096039	0.421215	1.62255
2.85	−0.149798	−0.159121	−0.256719	−0.124760	0.355101	1.50308
2.90	−0.139887	−0.149138	−0.257364	−0.151217	0.291368	1.38507
2.95	−0.130418	−0.139544	−0.256904	−0.175462	0.230096	1.26879
3.00	−0.121398	−0.130356	−0.255446	−0.197555	0.171352	1.15448
3.05	−0.112830	−0.121581	−0.253090	−0.217560	0.115188	1.04237
3.10	−0.104713	−0.113226	−0.249932	−0.235549	0.061641	0.932671
3.15	−0.097043	−0.105291	−0.246063	−0.251596	0.010738	0.825569
3.20	−0.089811	−0.097775	−0.241570	−0.265779	−0.037507	0.721237
3.25	−0.083008	−0.090671	−0.236533	−0.278178	−0.083092	0.619828
3.30	−0.076621	−0.083973	−0.231028	−0.288875	−0.126024	0.521482
3.35	−0.070637	−0.077670	−0.225127	−0.297955	−0.166321	0.426317

Table 1.9 (continued)

v	t					
	0.95	1.0	2.0	3.0	4.0	5.0
3.40	−0.065042	−0.071751	−0.218895	−0.305502	−0.204008	0.334437
3.45	−0.059819	−0.066203	−0.212394	−0.311601	−0.239122	0.245931
3.50	−0.054952	−0.061012	−0.205680	−0.316337	−0.271704	0.160872
3.55	−0.050425	−0.056164	−0.198806	−0.319794	−0.301804	0.079316
3.60	−0.046221	−0.051644	−0.191818	−0.322056	−0.329476	0.001308
3.65	−0.042322	−0.047436	−0.184760	−0.323204	−0.354782	−0.073122
3.70	−0.038712	−0.043526	−0.177672	−0.323319	−0.377787	−0.143958
3.75	−0.035374	−0.039896	−0.170588	−0.322479	−0.398562	−0.211194
3.80	−0.032292	−0.036533	−0.163540	−0.320760	−0.417180	−0.274837
3.85	−0.029450	−0.033420	−0.156557	−0.318237	−0.433718	−0.334906
3.90	−0.026833	−0.030543	−0.149664	−0.314979	−0.448255	−0.391427
3.95	−0.024425	−0.027887	−0.142882	−0.311057	−0.460873	−0.444438
4.00	−0.022214	−0.025439	−0.136230	−0.306535	−0.471653	−0.493984
4.05	−0.020185	−0.023185	−0.129724	−0.301477	−0.480681	−0.540122
4.10	−0.018325	−0.021112	−0.123380	−0.295943	−0.488041	−0.582911
4.15	−0.016623	−0.019208	−0.117207	−0.289990	−0.493816	−0.622421
4.20	−0.015066	−0.017460	−0.111216	−0.283671	−0.498093	−0.658726
4.25	−0.013643	−0.015859	−0.105414	−0.277039	−0.500955	−0.691906
4.30	−0.012345	−0.014393	−0.099807	−0.270140	−0.502485	−0.722047
4.35	−0.011162	−0.013051	−0.094399	−0.263021	−0.502766	−0.749238
4.40	−0.010084	−0.011826	−0.089193	−0.255723	−0.501879	−0.773571
4.45	−0.009104	−0.010707	−0.084190	−0.248286	−0.499904	−0.795143
4.50	−0.008212	−0.009687	−0.079390	−0.240746	−0.496917	−0.814053
4.55	−0.007403	−0.008757	−0.074793	−0.233138	−0.492994	−0.830402
4.60	−0.006668	−0.007911	−0.070397	−0.225492	−0.488210	−0.844292
4.65	−0.006002	−0.007142	−0.066200	−0.217838	−0.482635	−0.855827
4.70	−0.005399	−0.006442	−0.062198	−0.210201	−0.476339	−0.865112
4.75	−0.004853	−0.005807	−0.058387	−0.202606	−0.469387	−0.872252
4.80	−0.004359	−0.005231	−0.054763	−0.195074	−0.461845	−0.877352
4.85	−0.003913	−0.004709	−0.051322	−0.187625	−0.453772	−0.880515
4.90	−0.003510	−0.004237	−0.048058	−0.180277	−0.445229	−0.881846
4.95	−0.003147	−0.003809	−0.044966	−0.173044	−0.436271	−0.881448
5.00	−0.002819	−0.003422	−0.042040	−0.165942	−0.426950	−0.879422

1.4 First, Second and Third Derivatives with Respect to the Order of the Modified Bessel Function of the Second Kind

Using MATHEMATICA program, derivatives with respect to the order of the modified Bessel function of the second kind were determined by applying the following integral representations:

$$\frac{\partial K_v(z)}{\partial v} = \int_0^\infty t e^{-z \cosh t} \sinh(vt)\, dt \tag{1.4.1}$$

$$\frac{\partial^2 K_v(z)}{\partial v^2} = \int_0^\infty t^2 e^{-z \cosh t} \cosh(vt)\, dt \tag{1.4.2}$$

$$\frac{\partial^3 K_v(z)}{\partial v^3} = \int_0^\infty t^3 e^{-z \cosh t} \sinh(vt)\, dt \tag{1.4.3}$$

In calculated range of the order v and the argument t, contrary to other derivatives with respect to the order of the Bessel functions, the derivatives of the modified Bessel functions of the second kind have complex form with distinct extremal values (Figures 1.18–1.20).

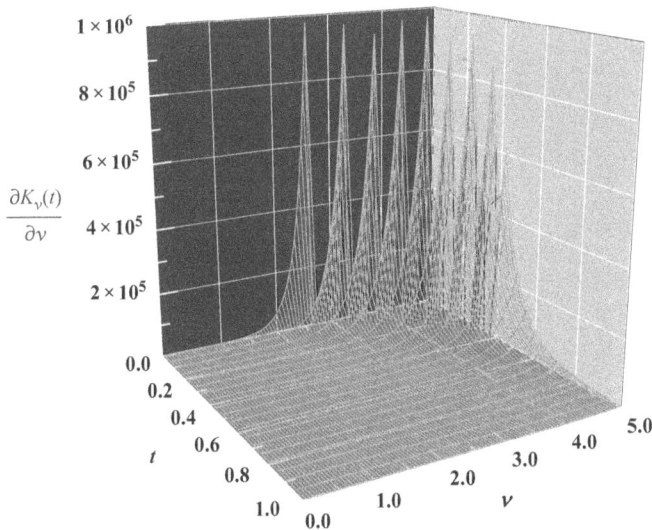

Figure 1.18: First derivatives with respect to the order of the modified Bessel functions of the second kind $\partial K_v(t)/\partial v$ as a function of v and t.

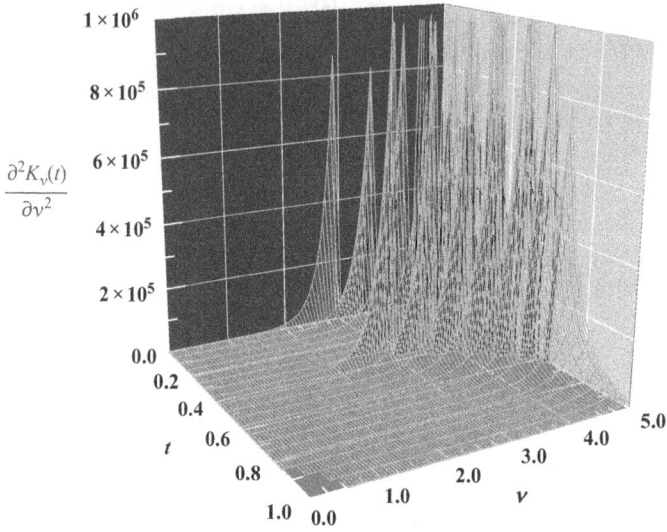

Figure 1.19: Second derivatives with respect to the order of the modified Bessel functions of the second kind $\partial^2 K_\nu(t)/\partial \nu^2$ as a function of ν and t.

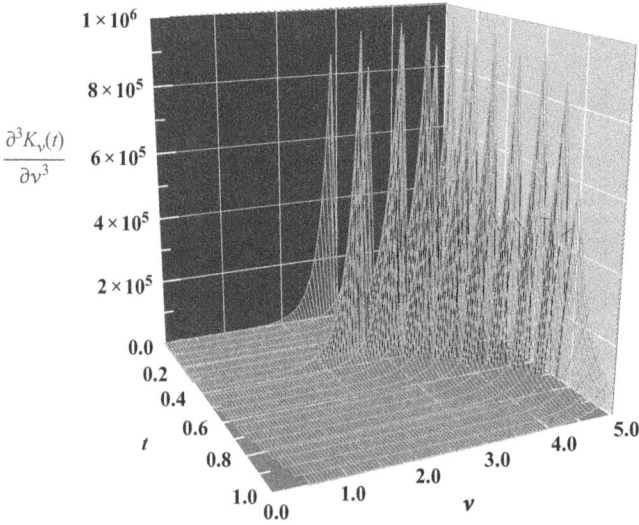

Figure 1.20: Third derivatives with respect to the order of the modified Bessel functions of the second kind $\partial^3 K_\nu(t)/\partial \nu^3$ as a function of ν and t.

Table 1.10: First derivatives with respect to the order of the modified Bessel functions of the second kind $\partial K_\nu(t)/\partial \nu$.

ν				t		
	0.05	0.10	0.15	0.20	0.25	0.30
0.00	0.00000	0.00000	0.00000	0.00000	0.00000	0.00000
0.05	7.23717	2.93400	1.61066	1.01184	0.68734	0.491650
0.10	14.7175	5.94176	3.25494	2.04205	1.38582	0.990532
0.15	22.6939	9.09928	4.96733	3.10940	2.10678	1.504010
0.20	31.4390	12.4871	6.78402	4.23347	2.86204	2.039720
0.25	41.2563	16.1927	8.74387	5.43512	3.66408	2.605680
0.30	52.4925	20.3132	10.8894	6.73694	4.52632	3.210480
0.35	65.5517	24.9587	13.2678	8.16375	5.46339	3.863430
0.40	80.9128	30.2551	15.9324	9.74321	6.49143	4.574700
0.45	99.1494	36.3489	18.9438	11.5064	7.62845	5.355550
0.50	120.955	43.4113	22.3719	13.4886	8.89464	6.218550
0.55	147.172	51.6443	26.2974	15.7302	10.3129	7.177790
0.60	178.831	61.2870	30.8141	18.2776	11.9093	8.249200
0.65	217.196	72.6243	36.0321	21.1841	13.7135	9.450810
0.70	263.821	85.9957	42.0799	24.5118	15.7597	10.8032
0.75	320.623	101.808	49.1091	28.3326	18.0871	12.3298
0.80	389.966	120.548	57.2982	32.7304	20.7412	14.0575
0.85	474.779	142.802	66.8581	37.8030	23.7743	16.0170
0.90	578.687	169.274	78.0381	43.6647	27.2470	18.2440
0.95	706.188	200.812	91.1335	50.4492	31.2298	20.7790
1.00	862.869	238.441	106.495	58.3134	35.8043	23.6692
1.05	1,055.68	283.395	124.537	67.4412	41.0656	26.9688
1.10	1,293.25	337.169	145.756	78.0489	47.1243	30.7406
1.15	1,586.38	401.569	170.739	90.3903	54.1092	35.0569
1.20	1,948.48	478.785	200.186	104.764	62.1707	40.0018
1.25	2,396.34	571.469	234.931	121.524	71.4840	45.6725
1.30	2,950.93	682.839	275.970	141.082	82.2540	52.1817
1.35	3,638.48	816.803	324.491	163.931	94.7201	59.6603
1.40	4,491.83	978.111	381.913	190.646	109.162	68.2602
1.45	5,552.16	1,172.54	449.933	221.912	125.909	78.1579
1.50	6,871.12	1,407.12	530.581	258.534	145.343	89.5586
1.55	8,513.54	1,690.42	626.289	301.468	167.915	102.701
1.60	10,560.90	2,032.88	739.971	351.845	194.154	117.864
1.65	13,115.8	2,447.25	875.120	411.003	224.679	135.370
1.70	16,307.3	2,949.10	1,035.93	480.531	260.218	155.598
1.75	20,297.9	3,557.43	1,227.43	562.312	301.627	178.988
1.80	25,292.8	4,295.52	1,455.68	658.582	349.911	206.053
1.85	31,550.7	5,191.83	1,727.95	771.997	406.256	237.396
1.90	39,398.8	6,281.25	2,052.99	905.715	472.056	273.716
1.95	49,250.2	7,606.50	2,441.36	1,063.49	548.954	315.836
2.00	61,628.1	9,220.03	2,905.76	1,249.80	638.886	364.714
2.05	77,194.5	11,186.2	3,461.52	1,469.95	744.138	421.476
2.10	96,788.7	13,584.1	4,127.11	1,730.30	867.407	487.436

Table 1.10 (continued)

			t			
v	0.05	0.10	0.15	0.20	0.25	0.30
2.15	121,475	16,510.8	4,924.88	2,038.42	1,011.88	564.138
2.20	152,603	20,085.9	5,881.78	2,403.32	1,181.32	653.392
2.25	191,890	24,456.7	7,030.44	2,835.78	1,380.18	757.319
2.30	241,515	29,804.2	8,410.31	3,348.69	1,613.73	878.414
2.35	304,252	36,352.0	10,069.1	3,957.42	1,888.21	1,019.60
2.40	383,631	44,375.6	12,064.8	4,680.39	2,210.99	1,184.33
2.45	484,148	54,215.1	14,467.4	5,539.63	2,590.84	1,376.63
2.50	611,537	66,290.6	17,362.0	6,561.52	3,038.15	1,601.28
2.55	773,110	81,121.1	20,851.7	7,777.65	3,565.22	1,863.87
2.60	978,200	99,348.7	25,062.0	9,225.94	4,186.68	2,171.01
2.65	1.238E+6	121,768	30,145.0	10,951.8	4,919.91	2,530.48
2.70	1.569	149,361	36,285.8	13,009.9	5,785.55	2,951.44
2.75	1.991	183,348	43,709.6	15,465.6	6,808.16	3,444.72
2.80	2.528	225,239	52,690.4	18,397.7	8,016.95	4,023.09
2.85	3.211	276,908	63,562.0	21,900.9	9,446.68	4,701.63
2.90	4.083	340,681	76,731.0	26,088.9	11,138.8	5,498.16
2.95	5.195	419,445	92,693.2	31,098.8	13,142.6	6,433.75
3.00	6.615	516,791	112,053	37,095.6	15,517.0	7,533.29
3.05	8.423	637,180	135,550	44,278.0	18,332.1	8,826.29
3.10	1.075E+7	786,167	164,084	52,885.7	21,671.7	10,347.6
3.15	1.372	970,661	198,757	63,207.7	25,635.8	12,138.6
3.20	1.752	1.200E+6	240,917	75,592.7	30,344.0	14,248.3
3.25	2.239	1.483	292,210	90,461.7	35,939.1	16,734.7
3.30	2.864	1.834	354,652	108,323	42,591.8	19,666.8
3.35	3.665	2.271	430,713	129,792	50,506.6	23,126.2
3.40	4.693	2.813	523,417	155,612	59,928.1	27,210.1
3.45	6.014	3.487	636,470	186,681	71,149.2	32,033.8
3.50	7.712	4.326	774,421	224,088	84,521.1	37,734.2
3.55	9.896	5.369	942,848	269,151	100,465	44,474.3
3.60	1.271E+8	6.668	1.149 E+6	323,467	119,484	52,447.9
3.65	1.633	8.287	1.400	388,971	142,186	61,885.7
3.70	2.099	1.030E+7	1.708	468,009	169,296	73,062.2
3.75	2.700	1.282	2.084	563,431	201,688	86,304.4
3.80	3.476	1.596	2.545	678,691	240,411	102,002
3.85	4.477	1.988	3.109	817,989	286,727	120,620
3.90	5.770	2.478	3.801	986,425	342,150	142,713
3.95	7.441	3.091	4.649	1.190E+6	408,508	168,941
4.00	9.602	3.857	5.690	1.437	487,995	200,094
4.05	1.240E+9	4.816	6.968	1.736	583,257	237,115
4.10	1.602	6.016	8.534	2.097	697,480	281,130
4.15	2.070	7.520	1.047E+7	2.536	834,507	333,486
4.20	2.678	9.405	1.284	3.068	998,967	395,793
4.25	3.465	1.177E+8	1.575	3.714	1.196E+6	469,976

Table 1.10 (continued)

			t			
v	0.05	0.10	0.15	0.20	0.25	0.30
4.30	4.486	1.474	1.934	4.498	1.434	558,341
4.35	5.812	1.846	2.376	5.450	1.719	663,647
4.40	7.533	2.313	2.920	6.607	2.062	789,200
4.45	9.770	2.901	3.591	8.013	2.474	938,961
4.50	1.268E+10	3.639	4.418	9.724	2.971	1.118E+6
4.55	1.646	4.568	5.438	1.1806E+7	3.569	1.331
4.60	2.138	5.737	6.698	1.434	4.289	1.586
4.65	2.778	7.209	8.253	1.743	5.157	1.890
4.70	3.613	9.063	1.017E+8	2.119	6.203	2.254
4.75	4.700	1.140E+9	1.255	2.578	7.466	2.690
4.80	6.117	1.434	1.548	3.137	8.990	3.211
4.85	7.966	1.806	1.912	3.819	1.083E+7	3.834
4.90	1.038E+11	2.275	2.361	4.653	1.305	4.581
4.95	1.353	2.867	2.917	5.670	1.574	5.476
5.00	1.764	3.615	3.607	6.913	1.898	6.548

			t			
v	0.35	0.40	0.45	0.50	0.55	0.60
0.00	0.00000	0.00000	0.00000	0.00000	0.00000	0.00000
0.05	0.364897	0.278468	0.217199	0.172421	0.138880	0.113240
0.10	0.734729	0.560431	0.436947	0.346744	0.279207	0.227600
0.15	1.11452	0.849440	0.661829	0.524899	0.422449	0.344214
0.20	1.50944	1.14915	0.894507	0.708865	0.570110	0.464243
0.25	1.92494	1.46340	1.13776	0.900706	0.723753	0.588893
0.30	2.36681	1.79624	1.39451	1.10259	0.885020	0.719424
0.35	2.84127	2.15201	1.66789	1.31684	1.05565	0.857174
0.40	3.35511	2.53542	1.96128	1.54592	1.23751	1.00357
0.45	3.91579	2.95163	2.27835	1.79253	1.43262	1.16014
0.50	4.53157	3.40632	2.62313	2.05961	1.64317	1.32856
0.55	5.21166	3.90576	3.00008	2.35040	1.87155	1.51065
0.60	5.96640	4.45698	3.41411	2.66845	2.12041	1.70838
0.65	6.80743	5.06784	3.87074	3.01773	2.39266	1.92397
0.70	7.74791	5.74717	4.37611	3.40265	2.69155	2.15983
0.75	8.80278	6.50496	4.93714	3.82814	3.02068	2.41865
0.80	9.98904	7.35249	5.56160	4.29973	3.38408	2.70344
0.85	11.3261	8.30254	6.25828	4.82363	3.78626	3.01753
0.90	12.8360	9.36965	7.03710	5.40684	4.23227	3.36465
0.95	14.5441	10.5703	7.90929	6.05723	4.72778	3.74899
1.00	16.4794	11.9235	8.88760	6.78371	5.27919	4.17522
1.05	18.6751	13.4506	9.98651	7.59636	5.89370	4.64862
1.10	21.1696	15.1761	11.2225	8.50657	6.57942	5.17509
1.15	24.0066	17.1282	12.6142	9.52727	7.34551	5.76129
1.20	27.2366	19.3391	14.1830	10.6731	8.20235	6.41471

Table 1.10 (continued)

			t			
v	0.35	0.40	0.45	0.50	0.55	0.60
1.25	30.9179	21.8454	15.9533	11.9607	9.16165	7.14381
1.30	35.1175	24.6893	17.9528	13.4091	10.2367	7.95812
1.35	39.9126	27.9193	20.2131	15.0396	11.4424	8.86842
1.40	45.3924	31.5909	22.7704	16.8768	12.7960	9.88685
1.45	51.6600	35.7679	25.6662	18.9485	14.3167	11.0272
1.50	58.8343	40.5236	28.9478	21.2865	16.0265	12.3050
1.55	67.0530	45.9423	32.6694	23.9270	17.9502	13.7378
1.60	76.4753	52.1212	36.8931	26.9112	20.1163	15.3457
1.65	87.2855	59.1720	41.6901	30.2863	22.5569	17.1512
1.70	99.6972	67.2234	47.1420	34.1061	25.3087	19.1800
1.75	113.958	76.4241	53.3424	38.4320	28.4134	21.4612
1.80	130.355	86.9452	60.3990	43.3345	31.9185	24.0278
1.85	149.222	98.9846	68.4351	48.8939	35.8782	26.9172
1.90	170.945	112.770	77.5929	55.2023	40.3543	30.1722
1.95	195.974	128.567	88.0357	62.3651	45.4172	33.8412
2.00	224.831	146.679	99.9512	70.5032	51.1472	37.9793
2.05	258.125	167.459	113.556	79.7549	57.6364	42.6491
2.10	296.564	191.316	129.099	90.2791	64.9895	47.9222
2.15	340.970	218.722	146.867	102.258	73.3265	53.8799
2.20	392.304	250.227	167.192	115.901	82.7846	60.6149
2.25	451.686	286.465	190.456	131.448	93.5207	68.2328
2.30	520.419	328.173	217.098	149.176	105.714	76.8544
2.35	600.028	376.206	247.630	169.402	119.572	86.6172
2.40	692.291	431.557	282.638	192.491	135.328	97.6784
2.45	799.286	495.380	322.802	218.864	153.255	110.218
2.50	923.441	569.015	368.910	249.005	173.660	124.440
2.55	1,067.60	654.021	421.870	283.472	196.902	140.580
2.60	1,235.08	752.212	482.737	322.908	223.387	158.905
2.65	1,429.77	865.698	552.730	368.053	253.585	179.725
2.70	1,656.23	996.939	633.264	419.764	288.036	203.388
2.75	1,919.81	1,148.80	725.977	479.027	327.358	230.300
2.80	2,226.75	1,324.62	832.771	546.982	372.265	260.921
2.85	2,584.41	1,528.29	955.851	624.946	423.576	295.780
2.90	3,001.39	1,764.36	1,097.78	714.441	482.237	335.484
2.95	3,487.83	2,038.14	1,261.53	817.226	549.334	380.730
3.00	4,055.61	2,355.81	1,450.56	935.338	626.120	432.318
3.05	4,718.71	2,724.63	1,668.90	1,071.13	714.039	491.166
3.10	5,493.54	3,153.05	1,921.20	1,227.34	814.756	558.328
3.15	6,399.44	3,650.98	2,212.93	1,407.12	930.194	635.018
3.20	7,459.17	4,230.00	2,550.40	1,614.13	1,062.57	722.631
3.25	8,699.50	4,903.68	2,941.00	1,852.63	1,214.44	822.770
3.30	10,152.0	5,687.90	3,393.32	2,127.53	1,388.77	937.282
3.35	11,853.9	6,601.28	3,917.39	2,444.56	1,588.98	1,068.29
3.40	13,849.0	7,665.64	4,524.88	2,810.35	1,819.01	1,218.25

Table 1.10 (continued)

			t			
v	0.35	0.40	0.45	0.50	0.55	0.60
3.45	16,189.1	8,906.57	5,229.44	3,232.61	2,083.44	1,389.99
3.50	18,935.3	10,354.1	6,046.98	3,720.29	2,387.56	1,586.74
3.55	22,159.7	12,043.5	6,996.08	4,283.80	2,737.49	1,812.27
3.60	25,947.4	14,016.1	8,098.47	4,935.26	3,140.33	2,070.91
3.65	30,399.3	16,320.6	9,379.52	5,688.74	3,604.29	2,367.65
3.70	35,634.3	19,014.1	10,868.9	6,560.64	4,138.90	2,708.26
3.75	41,793.4	22,163.9	12,601.3	7,570.05	4,755.19	3,099.40
3.80	49,043.2	25,849.0	14,617.5	8,739.19	5,465.99	3,548.78
3.85	57,580.9	30,162.5	16,964.8	10,094.0	6,286.14	4,065.31
3.90	67,640.5	35,213.9	19,699.1	11,664.6	7,232.92	4,659.27
3.95	79,498.8	41,132.4	22,885.6	13,486.3	8,326.35	5,342.58
4.00	93,484.2	48,069.9	26,600.8	15,600.2	9,589.73	6,129.03
4.05	109,986	56,205.8	30,934.4	18,054.2	11,050.1	7,034.59
4.10	129,466	65,751.5	35,991.8	20,904.4	12,739.0	8,077.74
4.15	152,473	76,956.3	41,896.3	24,216.1	14,692.9	9,279.91
4.20	179,659	90,114.9	48,793.0	28,065.9	16,954.5	10,665.9
4.25	211,795	105,575	56,852.4	32,543.0	19,573.4	12,264.6
4.30	249,802	123,746	66,274.3	37,752.1	22,607.2	14,109.4
4.35	294,773	145,115	77,294.3	43,815.4	26,123.3	16,239.0
4.40	348,007	170,255	90,188.8	50,876.0	30,199.9	18,698.4
4.45	411,051	199,844	105,283	59,101.4	34,928.6	21,540.0
4.50	485,746	234,685	122,961	68,688.0	40,415.9	24,824.4
4.55	574,284	275,727	143,672	79,865.6	46,786.1	28,622.3
4.60	679,277	324,097	167,948	92,903.7	54,184.3	33,015.6
4.65	803,836	381,125	196,415	108,118	62,779.9	38,099.8
4.70	951,671	448,391	229,809	125,880	72,770.9	43,986.0
4.75	1.127E+6	527,765	269,000	146,625	84,388.5	50,803.2
4.80	1.336	621,468	315,012	170,862	97,902.8	58,702.1
4.85	1.584	732,130	369,057	199,191	113,630	67,857.6
4.90	1.878	862,876	432,562	232,317	131,939	78,474.1
4.95	2.229	1.017E+6	507,211	271,068	153,263	90,789.3
5.00	2.645	1.200	594,998	316,415	178,108	105,081

			t			
v	0.65	0.70	0.75	0.80	0.85	0.90
0.00	0.00000	0.00000	0.00000	0.00000	0.00000	0.00000
0.05	0.093304	0.077575	0.065010	0.054860	0.046584	0.039776
0.10	0.187486	0.155848	0.130578	0.110174	0.093537	0.079856
0.15	0.283436	0.235523	0.197273	0.166398	0.141234	0.120548
0.20	0.382062	0.317322	0.265669	0.224000	0.190056	0.162164
0.25	0.484308	0.401990	0.336366	0.283463	0.240395	0.205027
0.30	0.591157	0.490306	0.409983	0.345286	0.292659	0.249472
0.35	0.703649	0.583089	0.487175	0.409998	0.347275	0.295847

Table 1.10 (continued)

			t			
v	0.65	0.70	0.75	0.80	0.85	0.90
0.40	0.822893	0.681210	0.568634	0.478154	0.404696	0.344523
0.45	0.950076	0.785603	0.655101	0.550347	0.465400	0.395891
0.50	1.08649	0.897272	0.747372	0.627216	0.529903	0.450368
0.55	1.23352	1.01731	0.846307	0.709445	0.598757	0.508405
0.60	1.39270	1.14690	0.952841	0.797781	0.672560	0.570486
0.65	1.56571	1.28734	1.06800	0.893032	0.751963	0.637136
0.70	1.75439	1.44007	1.19289	0.996086	0.837673	0.708929
0.75	1.96079	1.60666	1.32876	1.10792	0.930467	0.786487
0.80	2.18718	1.78885	1.47695	1.22959	1.03120	0.870495
0.85	2.43608	1.98857	1.63897	1.36228	1.14079	0.961702
0.90	2.71029	2.20797	1.81647	1.50730	1.26029	1.06093
0.95	3.01295	2.44944	2.01131	1.66608	1.39083	1.16910
1.00	3.34756	2.71563	2.22553	1.84022	1.53367	1.28720
1.05	3.71804	3.00951	2.46141	2.03151	1.69021	1.41634
1.10	4.12878	3.33441	2.72149	2.24190	1.86199	1.55776
1.15	4.58471	3.69402	3.00861	2.47359	2.05073	1.71281
1.20	5.09136	4.09251	3.32593	2.72903	2.25834	1.88299
1.25	5.65495	4.53453	3.67699	3.01095	2.48695	2.06999
1.30	6.28248	5.02531	4.06576	3.32239	2.73893	2.27566
1.35	6.98182	5.57070	4.49667	3.66675	3.01690	2.50207
1.40	7.76184	6.17730	4.97468	4.04781	3.32381	2.75152
1.45	8.63254	6.85250	5.50536	4.46984	3.66294	3.02657
1.50	9.60519	7.60463	6.09495	4.93759	4.03796	3.33008
1.55	10.6925	8.44305	6.75048	5.45637	4.45295	3.66522
1.60	11.9089	9.37832	7.47980	6.03215	4.91248	4.03554
1.65	13.2706	10.4223	8.29177	6.67161	5.42168	4.44502
1.70	14.7959	11.5884	9.19633	7.38225	5.98626	4.89806
1.75	16.5056	12.8918	10.2047	8.17247	6.61263	5.39960
1.80	18.4231	14.3495	11.3294	9.05172	7.30797	5.95517
1.85	20.5752	15.9806	12.5847	10.031	8.08032	6.57095
1.90	22.9919	17.8071	13.9864	11.121	8.93870	7.25383
1.95	25.7073	19.8534	15.5527	12.336	9.89323	8.01155
2.00	28.7602	22.1473	17.3038	13.692	10.955	8.85276
2.05	32.1945	24.7202	19.2627	15.204	12.138	9.78716
2.10	36.0601	27.6077	21.4551	16.893	13.454	10.826
2.15	40.4135	30.8501	23.9102	18.779	14.922	11.980
2.20	45.3190	34.4930	26.6610	20.887	16.558	13.265
2.25	50.8498	38.5879	29.7448	23.244	18.383	14.695
2.30	57.0890	43.1936	33.2036	25.880	20.420	16.287
2.35	64.1310	48.3765	37.0852	28.832	22.695	18.061
2.40	72.0836	54.2119	41.4434	32.138	25.236	20.039
2.45	81.0692	60.7855	46.3392	35.841	28.077	22.245
2.50	91.2274	68.1946	51.8419	39.993	31.254	24.706
2.55	102.717	76.5496	58.0298	44.650	34.808	27.453

Table 1.10 (continued)

			t			
ν	0.65	0.70	0.75	0.80	0.85	0.90
2.60	115.720	85.9760	64.9916	49.876	38.787	30.522
2.65	130.443	96.6168	72.8281	55.743	43.244	33.950
2.70	147.122	108.634	81.6536	62.333	48.237	37.783
2.75	166.027	122.214	91.5978	69.739	53.835	42.070
2.80	187.465	137.566	102.808	78.065	60.113	46.868
2.85	211.788	154.930	115.451	87.431	67.157	52.238
2.90	239.399	174.580	129.718	97.972	75.066	58.253
2.95	270.758	196.828	145.825	109.840	83.948	64.992
3.00	306.390	222.029	164.017	123.209	93.928	72.548
3.05	346.899	250.590	184.574	138.275	105.147	81.021
3.10	392.974	282.972	207.815	155.263	117.766	90.528
3.15	445.404	319.707	234.103	174.425	131.963	101.201
3.20	505.097	361.397	263.850	196.051	147.946	113.186
3.25	573.090	408.735	297.528	220.468	165.945	126.652
3.30	650.575	462.509	335.673	248.049	186.225	141.788
3.35	738.917	523.624	378.899	279.217	209.085	158.808
3.40	839.687	593.115	427.903	314.455	234.863	177.956
3.45	954.686	672.164	483.483	354.312	263.946	199.507
3.50	1,085.98	762.129	546.550	399.415	296.771	223.772
3.55	1,235.96	864.562	618.146	450.474	333.837	251.105
3.60	1,407.35	981.244	699.458	508.303	375.708	281.908
3.65	1,603.30	1,114.22	791.847	573.827	423.029	316.635
3.70	1,827.43	1,265.82	896.866	648.103	476.530	355.802
3.75	2,083.91	1,438.75	1016.30	732.336	537.045	399.996
3.80	2,377.54	1,636.08	1152.17	827.900	605.523	449.883
3.85	2,713.85	1,861.36	1,306.83	936.367	683.043	506.219
3.90	3,099.21	2,118.66	1,482.93	1,059.53	770.835	569.864
3.95	3,540.97	2,412.65	1,683.54	1,199.44	870.304	641.796
4.00	4,047.60	2,748.72	1,912.16	1,358.44	983.046	723.126
4.05	4,628.89	3,133.04	2,172.83	1,539.21	1,110.89	815.120
4.10	5,296.12	3,572.73	2,470.14	1,744.81	1,255.91	919.218
4.15	6,062.31	4,075.99	2,809.39	1,978.75	1,420.48	1,037.06
4.20	6,942.53	4,652.24	3,196.67	2,245.05	1,607.32	1,170.51
4.25	7,954.18	5,312.33	3,638.94	2,548.31	1,819.52	1,321.70
4.30	9,117.34	6,068.80	4,144.23	2,893.78	2,060.63	1,493.05
4.35	10,455.3	6,936.05	4,721.75	3,287.52	2,334.68	1,687.33
4.40	11,994.9	7,930.72	5,382.09	3,736.43	2,646.30	1,907.69
4.45	13,767.3	9,071.99	6,137.42	4,248.45	3,000.77	2,157.73
4.50	15,808.6	10,382.0	7,001.76	4,832.68	3,404.16	2,441.55
4.55	18,160.4	11,886.3	7,991.22	5,499.57	3,863.39	2,763.85
4.60	20,871.1	13,614.4	9,124.38	6,261.12	4,386.38	3,129.97
4.65	23,996.8	15,600.3	10,422.6	7,131.07	4,982.22	3,546.04
4.70	27,602.3	17,883.5	11,910.5	8,125.26	5,661.31	4,019.04
4.75	31,763.1	20,509.4	13,616.4	9,261.86	6,435.57	4,556.96

Table 1.10 (continued)

				t		
v	0.65	0.70	0.75	0.80	0.85	0.90
4.80	36,566.3	23,530.7	15,573.1	10,561.8	7,318.68	5,168.95
4.85	42,113.6	27,008.3	17,818.3	12,049.0	8,326.32	5,865.45
4.90	48,522.5	31,012.5	20,395.5	13,751.1	9,476.46	6,658.44
4.95	55,929.7	35,625.0	23,354.8	15,700.1	10,789.7	7,561.59
5.00	64,494.2	40,940.0	26,754.3	17,932.3	12,289.9	8,590.60

				t		
v	0.95	1.0	2.0	3.0	4.0	5.0
0.00	0.00000	0.00000	0.00000	0.00000	0.00000	0.00000
0.05	0.034134	0.029425	0.002362	0.000506	0.000126	0.000034
0.10	0.068519	0.059060	0.004731	0.001013	0.000251	0.000068
0.15	0.103411	0.089116	0.007113	0.001522	0.000377	0.000102
0.20	0.139068	0.119810	0.009516	0.002035	0.000504	0.000136
0.25	0.175757	0.151363	0.011945	0.002551	0.000632	0.000170
0.30	0.213753	0.184004	0.014409	0.003073	0.000760	0.000205
0.35	0.253346	0.217972	0.016914	0.003601	0.000890	0.000239
0.40	0.294839	0.253519	0.019467	0.004136	0.001021	0.000275
0.45	0.338552	0.290911	0.022076	0.004680	0.001154	0.000310
0.50	0.384831	0.330432	0.024749	0.005233	0.001289	0.000346
0.55	0.434042	0.372384	0.027493	0.005798	0.001426	0.000382
0.60	0.486581	0.417092	0.030318	0.006374	0.001565	0.000419
0.65	0.542877	0.464910	0.033232	0.006964	0.001706	0.000456
0.70	0.603394	0.516217	0.036244	0.007568	0.001851	0.000494
0.75	0.668641	0.571428	0.039363	0.008188	0.001998	0.000533
0.80	0.739169	0.630995	0.042600	0.008826	0.002149	0.000572
0.85	0.815586	0.695411	0.045966	0.009482	0.002303	0.000613
0.90	0.898556	0.765218	0.049471	0.010159	0.002461	0.000654
0.95	0.988811	0.841008	0.053127	0.010858	0.002623	0.000695
1.00	1.08716	0.923433	0.056947	0.011580	0.002790	0.000738
1.05	1.19448	1.01321	0.060944	0.012327	0.002961	0.000782
1.10	1.31177	1.11114	0.065133	0.013102	0.003138	0.000827
1.15	1.44010	1.21808	0.069527	0.013906	0.003319	0.000873
1.20	1.58068	1.33500	0.074145	0.014740	0.003507	0.000920
1.25	1.73483	1.46298	0.079001	0.015608	0.003700	0.000969
1.30	1.90404	1.60319	0.084116	0.016512	0.003900	0.001019
1.35	2.08994	1.75693	0.089508	0.017453	0.004107	0.001070
1.40	2.29435	1.92567	0.095198	0.018434	0.004321	0.001123
1.45	2.51928	2.11100	0.101208	0.019458	0.004543	0.001178
1.50	2.76699	2.31470	0.107563	0.020528	0.004772	0.001234
1.55	3.03997	2.53877	0.114289	0.021647	0.005011	0.001292
1.60	3.34101	2.78540	0.121413	0.022818	0.005258	0.001352
1.65	3.67319	3.05704	0.128964	0.024043	0.005516	0.001414
1.70	4.03999	3.35640	0.136975	0.025327	0.005783	0.001478

Table 1.10 (continued)

v	t 0.95	1.0	2.0	3.0	4.0	5.0
1.75	4.44524	3.68653	0.145480	0.026674	0.006061	0.001544
1.80	4.89323	4.05078	0.154516	0.028087	0.006351	0.001613
1.85	5.38877	4.45291	0.164123	0.029570	0.006652	0.001684
1.90	5.93719	4.89711	0.174343	0.031129	0.006967	0.001758
1.95	6.54447	5.38805	0.185223	0.032768	0.007295	0.001834
2.00	7.21730	5.93092	0.196813	0.034491	0.007637	0.001913
2.05	7.96312	6.53152	0.209166	0.036305	0.007994	0.001995
2.10	8.79030	7.19635	0.222340	0.038214	0.008367	0.002081
2.15	9.70817	7.93263	0.236399	0.040227	0.008757	0.002169
2.20	10.7272	8.74846	0.251409	0.042348	0.009164	0.002262
2.25	11.8591	9.65286	0.267445	0.044584	0.009590	0.002358
2.30	13.1169	10.6559	0.284584	0.046944	0.010036	0.002457
2.35	14.5154	11.7690	0.302913	0.049436	0.010503	0.002561
2.40	16.0711	13.0047	0.322523	0.052067	0.010992	0.002669
2.45	17.8025	14.3772	0.343517	0.054847	0.011504	0.002782
2.50	19.7302	15.9023	0.366000	0.057785	0.012041	0.002899
2.55	21.8777	17.5978	0.390093	0.060892	0.012605	0.003021
2.60	24.2710	19.4836	0.415921	0.064180	0.013195	0.003149
2.65	26.9395	21.5821	0.443622	0.067659	0.013815	0.003282
2.70	29.9163	23.9182	0.473348	0.071343	0.014466	0.003420
2.75	33.2386	26.5202	0.505259	0.075245	0.015150	0.003565
2.80	36.9480	29.4194	0.539532	0.079380	0.015868	0.003717
2.85	41.0916	32.6513	0.576359	0.083763	0.016622	0.003874
2.90	45.7223	36.2558	0.615948	0.088411	0.017415	0.004040
2.95	50.8998	40.2775	0.658524	0.093342	0.018250	0.004212
3.00	56.6910	44.7666	0.704332	0.098575	0.019127	0.004392
3.05	63.1718	49.7799	0.753640	0.104130	0.020051	0.004581
3.10	70.4273	55.3807	0.806738	0.110030	0.021023	0.004778
3.15	78.5538	61.6408	0.863940	0.116299	0.022047	0.004985
3.20	87.6598	68.6406	0.925591	0.122961	0.023126	0.005201
3.25	97.8677	76.4711	0.992064	0.130044	0.024262	0.005427
3.30	109.316	85.2345	1.06377	0.137578	0.025460	0.005664
3.35	122.161	95.0461	1.14114	0.145593	0.026723	0.005912
3.40	136.578	106.036	1.22467	0.154123	0.028055	0.006173
3.45	152.769	118.350	1.31489	0.163205	0.029460	0.006445
3.50	170.957	132.155	1.41236	0.172877	0.030943	0.006731
3.55	191.399	147.637	1.51772	0.183182	0.032509	0.007031
3.60	214.383	165.007	1.63164	0.194164	0.034162	0.007346
3.65	240.236	184.503	1.75488	0.205872	0.035908	0.007676
3.70	269.328	206.395	1.88825	0.218359	0.037753	0.008023
3.75	302.078	230.986	2.03263	0.231680	0.039702	0.008386
3.80	338.963	258.621	2.18900	0.245896	0.041763	0.008769
3.85	380.519	289.689	2.35843	0.261071	0.043943	0.009170
3.90	427.359	324.630	2.54207	0.277277	0.046249	0.009592

Table 1.10 (continued)

			t			
v	0.95	1.0	2.0	3.0	4.0	5.0
3.95	480.174	363.944	2.74120	0.294588	0.048688	0.010035
4.00	539.752	408.194	2.95720	0.313087	0.051270	0.010501
4.05	606.985	458.020	3.19159	0.332860	0.054004	0.010991
4.10	682.887	514.146	3.44604	0.354003	0.056898	0.011506
4.15	768.609	577.395	3.72237	0.376618	0.059964	0.012048
4.20	865.459	648.697	4.02258	0.400815	0.063213	0.012618
4.25	974.925	729.108	4.34885	0.426713	0.066656	0.013219
4.30	1,098.70	819.828	4.70359	0.454440	0.070307	0.013851
4.35	1,238.70	922.216	5.08941	0.484136	0.074178	0.014516
4.40	1,397.13	1,037.82	5.50921	0.515949	0.078284	0.015217
4.45	1,576.47	1,168.39	5.96614	0.550041	0.082640	0.015956
4.50	1,779.55	1,315.92	6.46367	0.586588	0.087264	0.016734
4.55	2,009.62	1,482.68	7.00561	0.625779	0.092172	0.017554
4.60	2,270.36	1,671.24	7.59615	0.667817	0.097385	0.018419
4.65	2,565.95	1,884.53	8.23987	0.712924	0.102921	0.019331
4.70	2,901.20	2,125.88	8.94183	0.761339	0.108803	0.020293
4.75	3,281.55	2,399.09	9.70757	0.813320	0.115054	0.021309
4.80	3,713.23	2,708.48	10.5432	0.869147	0.121699	0.022380
4.85	4,203.35	3,058.95	11.4554	0.929125	0.128764	0.023511
4.90	4,760.03	3,456.11	12.4515	0.993579	0.136279	0.024704
4.95	5,392.52	3,906.34	13.5398	1.06287	0.144274	0.025965
5.00	6,111.42	4,416.91	14.7290	1.13737	0.152782	0.027297

Table 1.11: Second derivatives with respect to the order of the modified Bessel functions of the second kind $\partial^2 K_v(t)/\partial v^2$.

			t			
v	0.05	0.10	0.15	0.20	0.25	0.30
0.00	14.3812	7.94802	5.32473	3.88756	2.98369	2.36692
0.05	14.5616	8.02121	5.36493	3.91282	3.00085	2.37920
0.10	15.1089	8.24265	5.48636	3.98905	3.05261	2.41621
0.15	16.0416	8.61788	5.69156	4.11764	3.13981	2.47850
0.20	17.3911	9.15641	5.98483	4.30093	3.26386	2.56698
0.25	19.2033	9.87188	6.37234	4.54228	3.42679	2.68297
0.30	21.5403	10.7826	6.86231	4.84611	3.63127	2.82820
0.35	24.4829	11.9119	7.46516	5.21805	3.88067	3.00482
0.40	28.1338	13.2893	8.19385	5.66503	4.17912	3.21551
0.45	32.6221	14.9507	9.06417	6.19544	4.53163	3.46345

Table 1.11 (continued)

			t			
v	0.05	0.10	0.15	0.20	0.25	0.30
0.50	38.1082	16.9403	10.0952	6.81932	4.94412	3.75243
0.55	44.7910	19.3114	11.3096	7.54862	5.42362	4.08691
0.60	52.9159	22.1282	12.7348	8.39743	5.97838	4.47208
0.65	62.7857	25.4683	14.4027	9.38236	6.61801	4.91400
0.70	74.7730	29.4245	16.3518	10.5229	7.35375	5.41967
0.75	89.3372	34.1085	18.6271	11.8417	8.19866	5.99722
0.80	107.044	39.6540	21.2820	13.3657	9.16791	6.65600
0.85	128.591	46.2217	24.3797	15.1260	10.2791	7.40683
0.90	154.840	54.0044	27.9948	17.1592	11.5527	8.26216
0.95	186.852	63.2334	32.2154	19.5079	13.0123	9.23635
1.00	225.944	74.1870	37.1459	22.2221	14.6855	10.3460
1.05	273.741	87.1995	42.9095	25.3602	16.6042	11.6101
1.10	332.257	102.673	49.6525	28.9908	18.8053	13.0507
1.15	403.990	121.093	57.5478	33.1940	21.3320	14.6932
1.20	492.042	143.044	66.8007	38.0637	24.2342	16.5669
1.25	600.265	169.229	77.6546	43.7102	27.5699	18.7055
1.30	733.452	200.501	90.3986	50.2627	31.4068	21.1481
1.35	897.570	237.889	105.376	57.8732	35.8236	23.9398
1.40	1,100.06	282.637	122.996	66.7202	40.9118	27.1328
1.45	1,350.20	336.252	143.744	77.0136	46.7782	30.7874
1.50	1,659.58	400.559	168.199	89.0005	53.5474	34.9736
1.55	2,042.69	477.775	197.053	102.972	61.3647	39.7723
1.60	2,517.67	570.586	231.128	119.271	70.4000	45.2774
1.65	3,107.23	682.260	271.407	138.303	80.8515	51.5977
1.70	3,839.84	816.767	319.067	160.544	92.9513	58.8597
1.75	4,751.25	978.942	375.511	186.560	106.971	67.2102
1.80	5,886.34	1,174.67	442.421	217.017	123.228	76.8197
1.85	7,301.54	1,411.13	521.812	252.704	142.096	87.8867
1.90	9,067.87	1,697.07	616.095	294.554	164.012	100.642
1.95	11,274.7	2,043.18	728.165	343.676	189.487	115.355
2.00	14,034.9	2,462.52	861.497	401.379	219.124	132.338
2.05	17,490.6	2,971.05	1,020.26	469.219	253.630	151.958
2.10	21,821.4	3,588.32	1,209.48	549.043	293.838	174.639
2.15	27,254.2	4,338.26	1,435.17	643.045	340.724	200.880
2.20	34,076.0	5,250.19	1,704.61	753.830	395.442	231.262
2.25	42,650.2	6,360.09	2,026.54	884.501	459.346	266.464
2.30	53,437.0	7,712.11	2,411.51	1,038.75	534.037	307.280
2.35	67,019.8	9,360.49	2,872.22	1,220.97	621.401	354.640
2.40	84,139.0	11,371.9	3,424.03	1,436.39	723.662	409.632
2.45	105,735	13,828.3	4,085.47	1,691.27	843.450	473.531
2.50	133,001	16,830.6	4,878.96	1,993.06	983.870	547.833
2.55	167,457	20,503.2	5,831.58	2,350.66	1,148.59	634.291

Table 1.11 (continued)

| v | \multicolumn{6}{c}{t} |
	0.05	0.10	0.15	0.20	0.25	0.30
2.60	211,036	24,999.2	6,976.12	2,774.69	1,341.96	734.962
2.65	266,199	30,507.8	8,352.30	3,277.87	1,569.12	852.263
2.70	336,084	37,262.1	10,008.2	3,875.41	1,836.16	989.034
2.75	424,693	45,550.3	12,002.2	4,585.48	2,150.29	1,148.61
2.80	537,131	55,728.5	14,405.1	5,429.90	2,520.08	1,334.92
2.85	679,922	68,237.0	17,302.7	6,434.76	2,955.66	1,552.59
2.90	861,400	83,620.7	20,799.5	7,631.38	3,469.11	1,807.05
2.95	1.092E+6	102,555	25,022.3	9,057.33	4,074.73	2,104.72
3.00	1.386	125,874	30,125.4	10,757.7	4,789.54	2,453.16
3.05	1.760	154,617	36,296.5	12,786.6	5,633.77	2,861.27
3.10	2.237	190,068	43,764.3	15,209.2	6,631.48	3,339.59
3.15	2.846	233,825	52,807.1	18,103.7	7,811.33	3,900.52
3.20	3.623	287,869	63,764.6	21,564.3	9,207.45	4,558.76
3.25	4.616	354,664	77,050.8	25,704.4	10,860.5	5,331.63
3.30	5.885	437,276	93,171.2	30,660.3	12,819.0	6,239.65
3.35	7.509	539,518	112,743	36,596.7	15,140.7	7,307.09
3.40	9.587	666,135	136,519	43,712.0	17,894.7	8,562.67
3.45	1.225E+7	823,042	165,423	52,245.3	21,163.5	10,040.4
3.50	1.566	1.017E+6	200,581	62,485.6	25,045.4	11,780.6
3.55	2.004	1.259	243,372	74,781.6	29,658.4	13,831.1
3.60	2.566	1.559	295,487	89,554.8	35,143.1	16,248.6
3.65	3.288	1.931	358,994	107,315	41,668.1	19,100.3
3.70	4.216	2.394	436,429	128,677	49,435.1	22,466.1
3.75	5.410	2.970	530,904	154,388	58,685.6	26,440.9
3.80	6.946	3.687	646,234	185,350	69,709.3	31,137.5
3.85	8.923	4.579	787,106	222,657	82,853.2	36,689.8
3.90	1.147E+8	5.691	959,273	267,634	98,533.6	43,257.3
3.95	1.476	7.078	1.170E+6	321,888	117,250	51,029.9
4.00	1.899	8.807	1.427	387,368	139,603	60,233.4
4.05	2.446	1.097E+7	1.743	466,439	166,312	71,136.9
4.10	3.153	1.366	2.129	561,975	198,244	84,061.3
4.15	4.066	1.703	2.602	677,464	236,440	99,388.8
4.20	5.246	2.125	3.183	817,149	282,152	11,7576
4.25	6.773	2.652	3.895	986,185	336,886	139,166
4.30	8.750	3.311	4.769	1.191E+6	402,458	164,809
4.35	1.131E+9	4.138	5.843	1.439	481,053	195,282
4.40	1.463	5.173	7.162	1.739	575,304	231,511
4.45	1.893	6.471	8.784	2.104	688,385	274,605
4.50	2.451	8.099	1.078E+7	2.546	824,125	325,889
4.55	3.175	1.014E+8	1.323	3.082	987,144	386,950
4.60	4.116	1.271	1.626	3.733	1.183E+6	459,684
4.65	5.337	1.593	1.998	4.525	1.418	546,366

Table 1.11 (continued)

| | | | | t | | | |
v	0.05	0.10	0.15	0.20	0.25	0.30
4.70	6.926	1.998	2.457	5.488	1.702	649,718
4.75	8.992	2.507	3.023	6.658	2.042	773,001
4.80	1.168E+10	3.148	3.721	8.082	2.453	920,129
4.85	1.518	3.955	4.583	9.816	2.947	1.096E+6
4.90	1.974	4.970	5.647	1.193E+7	3.542	1.306
4.95	2.568	6.250	6.962	1.450	4.259	1.556
5.00	3.343	7.863	8.587	1.763	5.125	1.856

| | | | | t | | | |
v	0.35	0.40	0.45	0.50	0.55	0.60
0.00	1.92259	1.58980	1.33314	1.13062	0.967852	0.835045
0.05	1.93170	1.59675	1.33856	1.13493	0.971321	0.837873
0.10	1.95916	1.61770	1.35490	1.14789	0.981764	0.846387
0.15	2.00534	1.65291	1.38234	1.16966	0.999290	0.860671
0.20	2.07087	1.70283	1.42122	1.20048	1.02408	0.880866
0.25	2.15663	1.76807	1.47197	1.24068	1.05640	0.907172
0.30	2.26380	1.84948	1.53522	1.29072	1.09658	0.939853
0.35	2.39385	1.94808	1.61170	1.35115	1.14506	0.979235
0.40	2.54858	2.06514	1.70234	1.42265	1.20234	1.02571
0.45	2.73015	2.20217	1.80822	1.50603	1.26903	1.07976
0.50	2.94108	2.36094	1.93063	1.60224	1.34586	1.14193
0.55	3.18437	2.54354	2.07107	1.71239	1.43364	1.21285
0.60	3.46349	2.75238	2.23126	1.83774	1.53335	1.29325
0.65	3.78245	2.99023	2.41319	1.97975	1.64608	1.38398
0.70	4.14589	3.26030	2.61915	2.14010	1.77306	1.48599
0.75	4.55915	3.56626	2.85173	2.32069	1.91573	1.60035
0.80	5.02836	3.91229	3.11392	2.52369	2.07570	1.72828
0.85	5.56056	4.30321	3.40909	2.75154	2.25479	1.87118
0.90	6.16384	4.74450	3.74111	3.00704	2.45506	2.03059
0.95	6.84746	5.24240	4.11436	3.29334	2.67884	2.20826
1.00	7.62203	5.80407	4.53381	3.61403	2.92877	2.40618
1.05	8.49973	6.43765	5.00513	3.97315	3.20781	2.62657
1.10	9.49452	7.15243	5.53474	4.37529	3.51932	2.87193
1.15	10.6224	7.95904	6.12997	4.82565	3.86709	3.14508
1.20	11.9017	8.86957	6.79911	5.33010	4.25539	3.44918
1.25	13.3536	9.89786	7.55163	5.89532	4.68903	3.78781
1.30	15.0021	11.0597	8.39826	6.52885	5.17348	4.16498
1.35	16.8752	12.3732	9.35125	7.23926	5.71488	4.58522
1.40	19.0048	13.8590	10.4245	8.03625	6.32018	5.05362
1.45	21.4276	15.5407	11.6339	8.93085	6.99726	5.57593
1.50	24.1859	17.4454	12.9976	9.93554	7.75501	6.15863

Table 1.11 (continued)

			t			
v	0.35	0.40	0.45	0.50	0.55	0.60
1.55	27.3284	19.6041	14.5360	11.0645	8.60348	6.80901
1.60	30.9113	22.0523	16.2729	12.3340	9.55407	7.53533
1.65	34.9991	24.8308	18.2349	13.7622	10.6197	8.34688
1.70	39.6665	27.9862	20.4528	15.3700	11.8149	9.25417
1.75	44.9996	31.5723	22.9616	17.1812	13.1564	10.2691
1.80	51.0979	35.6506	25.8013	19.2228	14.6630	11.4050
1.85	58.0762	40.2921	29.0178	21.5256	16.3559	12.6772
1.90	66.0676	45.5781	32.6635	24.1247	18.2594	14.1029
1.95	75.2258	51.6026	36.7985	27.0601	20.4011	15.7014
2.00	85.7287	58.4734	41.4916	30.3774	22.8122	17.4948
2.05	97.7828	66.3149	46.8217	34.1289	25.5284	19.5081
2.10	111.627	75.2705	52.8795	38.3741	28.5902	21.7698
2.15	127.539	85.5056	59.7687	43.1811	32.0438	24.3118
2.20	145.839	97.2110	67.6088	48.6276	35.9416	27.1707
2.25	166.903	110.607	76.5369	54.8029	40.3435	30.3880
2.30	191.164	125.949	86.7106	61.8088	45.3179	34.0107
2.35	219.126	143.530	98.3115	69.7621	50.9427	38.0924
2.40	251.377	163.691	111.548	78.7967	57.3067	42.6940
2.45	288.601	186.827	126.661	89.0659	64.5116	47.8849
2.50	331.592	213.392	143.927	100.746	72.6735	53.7438
2.55	381.278	243.917	163.666	114.038	81.9249	60.3608
2.60	438.740	279.014	186.246	129.176	92.4177	67.8382
2.65	505.239	319.393	212.092	146.424	104.325	76.2929
2.70	582.246	365.880	241.695	166.090	117.847	85.8580
2.75	671.480	419.431	275.621	188.527	133.210	96.6858
2.80	774.948	481.159	314.527	214.138	150.676	108.950
2.85	894.999	552.356	359.169	243.391	170.543	122.848
2.90	1,034.38	634.527	410.426	276.825	193.156	138.608
2.95	1,196.30	729.420	469.311	315.057	218.908	156.489
3.00	1,384.51	839.072	537.000	358.802	248.251	176.787
3.05	1,603.44	965.853	614.855	408.885	281.705	199.842
3.10	1,858.25	1,112.53	704.456	466.256	319.868	226.042
3.15	2,154.98	1,282.32	807.632	532.012	363.426	255.833
3.20	2,500.75	1,478.98	926.510	607.421	413.169	289.726
3.25	2,903.90	1,706.89	1,063.56	693.950	470.007	328.305
3.30	3,374.23	1,971.18	1,221.64	793.291	534.987	372.242
3.35	3,923.26	2,277.83	1,404.08	907.404	609.315	422.308
3.40	4,564.51	2,633.83	1,614.77	1,038.56	694.382	479.387
3.45	5,313.90	3,047.34	1,858.19	1,189.38	791.789	544.496
3.50	6,190.18	3,527.92	2,139.59	1,362.91	903.387	618.803
3.55	7,215.39	4,086.77	2,465.08	1,562.67	1,031.31	703.650
3.60	8,415.53	4,736.99	2,841.75	1,792.75	1,178.02	800.581

Table 1.11 (continued)

			t			
v	0.35	0.40	0.45	0.50	0.55	0.60
3.65	9,821.19	5,493.91	3,277.90	2,057.90	1,346.36	911.375
3.70	11,468.5	6,375.52	3,783.17	2,363.60	1,539.62	1,038.08
3.75	13,400.0	7,402.92	4,368.84	2,716.25	1,761.61	1,183.04
3.80	15,666.0	8,600.83	5,048.03	3,123.27	2,016.72	1,348.99
3.85	18,325.8	9,998.28	5,836.10	3,593.26	2,310.03	1,539.04
3.90	21,449.4	11,629.4	6,750.96	4,136.26	2,647.44	1,756.81
3.95	25,119.8	13,534.1	7,813.54	4,763.90	3,035.76	2,006.45
4.00	29,434.7	15,759.6	9,048.32	5,489.75	3,482.90	2,292.78
4.05	34,510.1	18,361.1	10,483.9	6,329.59	3,998.02	2,621.34
4.10	40,482.9	21,403.6	12,153.8	7,301.77	4,591.72	2,998.53
4.15	47,515.5	24,963.8	14,097.2	8,427.71	5,276.33	3,431.76
4.20	55,799.8	29,131.8	16,360.0	9,732.34	6,066.14	3,929.59
4.25	65,563.7	34,013.6	18,995.9	11,244.7	6,977.73	4,501.90
4.30	77,077.1	39,734.4	22,067.9	12,998.8	8,030.38	5,160.15
4.35	90,659.9	46,441.5	25,649.9	15,034.2	9,246.48	5,917.58
4.40	106,692	54,308.7	29,828.5	17,397.0	10,652.0	6,789.54
4.45	125,624	63,541.1	34,705.4	20,141.2	12,277.3	7,793.79
4.50	147,991	74,380.5	40,399.9	23,329.8	14,157.5	8,950.92
4.55	174,429	87,112.8	47,052.0	27,036.5	16,333.6	10,284.8
4.60	205,694	102,075	54,826.5	31,347.3	18,853.3	11,823.1
4.65	242,683	119,667	63,916.6	36,363.1	21,772.0	13,597.9
4.70	286,465	140,358	74,550.0	42,201.6	25,154.5	15,646.4
4.75	338,311	164,707	86,994.0	49,000.8	29,076.2	18,012.0
4.80	399,734	193,373	101,564	56,922.2	33,624.9	20,744.7
4.85	472,537	227,136	118,629	66,155.1	38,903.3	23,903.1
4.90	558,866	266,921	138,627	76,921.2	45,030.9	27,554.8
4.95	661,278	313,820	162,071	89,480.5	52,147.4	31,778.8
5.00	782,825	369,132	189,567	10,4138	60,415.9	36,666.7

			t			
v	0.65	0.70	0.75	0.80	0.85	0.90
0.00	0.725604	0.633647	0.556382	0.490726	0.434541	0.386159
0.05	0.727635	0.635585	0.558006	0.492097	0.435705	0.387153
0.10	0.734649	0.641416	0.562893	0.496220	0.439205	0.390142
0.15	0.746413	0.651193	0.571083	0.503130	0.445071	0.395149
0.20	0.763038	0.665004	0.582648	0.512883	0.453347	0.402212
0.25	0.784680	0.682973	0.597688	0.525561	0.464102	0.411386
0.30	0.811545	0.705263	0.616333	0.541268	0.477419	0.422741
0.35	0.843889	0.732078	0.638746	0.560137	0.493407	0.436365
0.40	0.882023	0.763661	0.665122	0.582326	0.512193	0.452364
0.45	0.926311	0.800303	0.695693	0.608020	0.533931	0.470863
0.50	0.977184	0.842343	0.730729	0.637438	0.558797	0.492005

Table 1.11 (continued)

			t			
v	0.65	0.70	0.75	0.80	0.85	0.90
0.55	1.03514	0.890170	0.770541	0.670831	0.586994	0.515959
0.60	1.10074	0.944233	0.815486	0.708484	0.618755	0.542913
0.65	1.17464	1.00504	0.865969	0.750724	0.654343	0.573083
0.70	1.25757	1.07317	0.922448	0.797918	0.694056	0.606712
0.75	1.35037	1.14928	0.985441	0.850479	0.738228	0.644072
0.80	1.45398	1.23410	1.05553	0.908873	0.787235	0.685469
0.85	1.56946	1.32846	1.13337	0.973621	0.841496	0.731242
0.90	1.69801	1.43328	1.21968	1.04531	0.901479	0.781772
0.95	1.84097	1.54962	1.31530	1.12458	0.967709	0.837484
1.00	1.99984	1.67864	1.42114	1.21217	1.04077	0.898848
1.05	2.17632	1.82164	1.53822	1.30889	1.12130	0.966388
1.10	2.37231	1.98011	1.66768	1.41564	1.21004	1.040690
1.15	2.58995	2.15566	1.81082	1.53343	1.30779	1.12239
1.20	2.83163	2.35016	1.96905	1.66339	1.41543	1.21222
1.25	3.10004	2.56564	2.14397	1.80678	1.53397	1.31097
1.30	3.39819	2.80442	2.33737	1.96498	1.66451	1.41953
1.35	3.72948	3.06908	2.55125	2.13956	1.80830	1.53888
1.40	4.09771	3.36251	2.78782	2.33227	1.96669	1.67012
1.45	4.50717	3.68795	3.04959	2.54503	2.14121	1.81446
1.50	4.96267	4.04904	3.33934	2.78001	2.33358	1.97325
1.55	5.46960	4.44985	3.66019	3.03964	2.54568	2.14799
1.60	6.03407	4.89496	4.01562	3.32660	2.77962	2.34036
1.65	6.66291	5.38950	4.40954	3.64391	3.03775	2.55220
1.70	7.36385	5.93923	4.84634	3.99495	3.32271	2.78559
1.75	8.14557	6.55063	5.33091	4.38346	3.63741	3.04282
1.80	9.01787	7.23098	5.86875	4.81368	3.98512	3.32645
1.85	9.99179	7.98846	6.46602	5.29030	4.36949	3.63934
1.90	11.0798	8.83229	7.12965	5.81860	4.79458	3.98466
1.95	12.2960	9.77283	7.86740	6.40448	5.26496	4.36596
2.00	13.6562	10.8218	8.68800	7.05457	5.78570	4.78721
2.05	15.1783	11.9922	9.60125	7.77628	6.36249	5.25279
2.10	16.8828	13.2990	10.6182	8.57793	7.00169	5.76766
2.15	18.7925	14.7589	11.7511	9.46884	7.71042	6.33729
2.20	20.9333	16.3907	13.0141	10.4595	8.49667	6.96785
2.25	23.3348	18.2157	14.4228	11.5616	9.36935	7.66621
2.30	26.0301	20.2580	15.9948	12.7885	10.3385	8.44004
2.35	29.0571	22.5447	17.7502	14.1549	11.4153	9.29795
2.40	32.4584	25.1065	19.7113	15.6776	12.6124	10.2496
2.45	36.2826	27.9782	21.9035	17.3753	13.9439	11.3056
2.50	40.5848	31.1990	24.3553	19.2692	15.4257	12.4783
2.55	45.4275	34.8134	27.0991	21.3831	17.0756	13.7810
2.60	50.8818	38.8718	30.1712	23.7437	18.9136	15.2289
2.65	57.0282	43.4312	33.6128	26.3813	20.9623	16.8391

Table 1.11 (continued)

v				t		
	0.65	**0.70**	**0.75**	**0.80**	**0.85**	**0.90**
2.70	63.9587	48.5564	37.4705	29.3300	23.2469	18.6307
2.75	71.7775	54.3207	41.7968	32.6282	25.7961	20.6250
2.80	80.6035	60.8073	46.6514	36.3192	28.6418	22.8462
2.85	90.5721	68.1108	52.1015	40.4520	31.8201	25.3213
2.90	101.837	76.3384	58.2235	45.0819	35.3719	28.0807
2.95	114.574	85.6121	65.1040	50.2713	39.3429	31.1586
3.00	128.984	96.0704	72.8408	56.0908	43.7850	34.5935
3.05	145.295	107.871	81.5450	62.6204	48.7564	38.4287
3.10	163.767	121.193	91.3429	69.9504	54.3230	42.7129
3.15	184.698	136.241	102.377	78.1830	60.5592	47.5009
3.20	208.428	153.246	114.811	87.4340	67.5491	52.8548
3.25	235.346	172.474	128.828	97.8346	75.3874	58.8442
3.30	265.895	194.227	144.638	109.533	84.1816	65.5478
3.35	300.584	218.848	162.480	122.699	94.0529	73.0544
3.40	339.994	246.729	182.624	137.523	105.14	81.4641
3.45	384.790	278.318	205.380	154.222	117.59	90.8900
3.50	435.734	314.127	231.097	173.042	131.60	101.460
3.55	493.699	354.739	260.177	194.262	147.34	113.318
3.60	559.687	400.821	293.073	218.202	165.06	126.628
3.65	634.843	453.135	330.307	245.221	185.00	141.575
3.70	720.486	512.553	372.469	275.730	207.47	158.366
3.75	818.125	580.072	420.234	310.197	232.77	177.238
3.80	929.495	656.834	474.372	349.153	261.30	198.460
3.85	1,056.59	744.146	535.764	393.202	293.48	222.334
3.90	1,201.70	843.504	605.412	443.034	329.78	249.204
3.95	1,367.45	956.623	684.465	499.434	370.75	279.459
4.00	1,556.87	1,085.47	774.233	563.296	417.03	313.540
4.05	1,773.45	1,232.30	876.215	635.640	469.31	351.949
4.10	2,021.19	1,399.70	992.126	717.630	528.40	395.254
4.15	2,304.71	1,590.63	1,123.93	810.593	595.23	444.100
4.20	2,629.33	1,808.51	1,273.86	916.044	670.82	499.220
4.25	3,001.16	2,057.25	1,444.51	1,035.71	756.38	561.446
4.30	3,427.27	2,341.35	1,638.81	1,171.58	853.26	631.725
4.35	3,915.80	2,665.98	1,860.14	1,325.90	962.99	711.133
4.40	4,476.15	3,037.08	2,112.36	1,501.25	1,087.35	800.893
4.45	5,119.17	3,461.49	2,399.93	1,700.59	1,228.34	902.397
4.50	5,857.36	3,947.09	2,727.92	1,927.30	1,388.24	1,017.23
4.55	6,705.20	4,502.93	3,102.19	2,185.23	1,569.68	1,147.19
4.60	7,679.38	5,139.44	3,529.44	2,478.83	1,775.64	1,294.34
4.65	8,799.23	5,868.66	4,017.37	2,813.15	2,009.53	1,461.02
4.70	10,087.1	6,704.43	4,574.85	3,194.01	2,275.25	1,649.90
4.75	11,568.7	7,662.72	5,212.05	3,628.05	2,577.25	1,864.01

Table 1.11 (continued)

			t			
v	0.65	0.70	0.75	0.80	0.85	0.90
4.80	13,274.1	8,761.95	5,940.67	4,122.91	2,920.62	2,106.83
4.85	15,237.7	10,023.4	6,774.17	4,687.35	3,311.19	2,382.31
4.90	17,499.8	11,471.6	7,728.03	5,331.39	3,755.63	2,694.98
4.95	20,106.6	13,134.8	8,820.09	6,066.57	4,261.56	3,049.99
5.00	23,112.0	15,045.8	10,070.8	6,906.12	4,837.71	3,453.23

			t			
v	0.95	1.0	2.0	3.0	4.0	5.0
0.00	0.344266	0.307811	0.047222	0.010119	0.002510	0.000677
0.05	0.345119	0.308546	0.047288	0.010129	0.002512	0.000677
0.10	0.347684	0.310757	0.047489	0.010160	0.002518	0.000678
0.15	0.351980	0.314459	0.047825	0.010211	0.002528	0.000681
0.20	0.358038	0.319679	0.048297	0.010284	0.002542	0.000684
0.25	0.365904	0.326455	0.048907	0.010377	0.002560	0.000688
0.30	0.375635	0.334834	0.049658	0.010492	0.002582	0.000693
0.35	0.387306	0.344877	0.050553	0.010629	0.002609	0.000699
0.40	0.401002	0.356657	0.051597	0.010788	0.002639	0.000705
0.45	0.416826	0.370260	0.052794	0.010969	0.002674	0.000713
0.50	0.434900	0.385784	0.054148	0.011174	0.002714	0.000722
0.55	0.455358	0.403343	0.055667	0.011403	0.002758	0.000731
0.60	0.478359	0.423068	0.057357	0.011656	0.002806	0.000742
0.65	0.504079	0.445104	0.059225	0.011935	0.002859	0.000754
0.70	0.532717	0.469617	0.061280	0.012241	0.002918	0.000766
0.75	0.564497	0.496791	0.063531	0.012574	0.002981	0.000780
0.80	0.599669	0.526833	0.065989	0.012935	0.003049	0.000795
0.85	0.638512	0.559971	0.068665	0.013326	0.003123	0.000811
0.90	0.681337	0.596464	0.071570	0.013748	0.003202	0.000828
0.95	0.728489	0.636593	0.074719	0.014202	0.003287	0.000847
1.00	0.780353	0.680675	0.078127	0.014691	0.003378	0.000866
1.05	0.837355	0.729059	0.081808	0.015215	0.003475	0.000887
1.10	0.899967	0.782132	0.085781	0.015776	0.003579	0.000910
1.15	0.968715	0.840323	0.090065	0.016377	0.003690	0.000933
1.20	1.04418	0.904106	0.094680	0.017019	0.003807	0.000958
1.25	1.12701	0.974007	0.099649	0.017705	0.003932	0.000985
1.30	1.21792	1.05061	0.104995	0.018436	0.004065	0.001013
1.35	1.31769	1.13455	0.110746	0.019216	0.004206	0.001043
1.40	1.42722	1.22655	0.116929	0.020048	0.004355	0.001074
1.45	1.54747	1.32739	0.123576	0.020934	0.004513	0.001108
1.50	1.67952	1.43796	0.130720	0.021877	0.004681	0.001143
1.55	1.82459	1.55920	0.138397	0.022881	0.004858	0.001180
1.60	1.98399	1.69221	0.146647	0.023949	0.005046	0.001219
1.65	2.15921	1.83816	0.155514	0.025086	0.005244	0.001260

Table 1.11 (continued)

v	0.95	1.0	t 2.0	3.0	4.0	5.0
1.70	2.35189	1.99838	0.165042	0.026295	0.005454	0.001303
1.75	2.56385	2.17431	0.175282	0.027582	0.005676	0.001348
1.80	2.79712	2.36760	0.186289	0.028950	0.005910	0.001396
1.85	3.05396	2.58002	0.198121	0.030406	0.006158	0.001447
1.90	3.33688	2.81359	0.210844	0.031954	0.006420	0.001500
1.95	3.64866	3.07051	0.224526	0.033601	0.006697	0.001555
2.00	3.99242	3.35326	0.239242	0.035352	0.006989	0.001614
2.05	4.37161	3.66457	0.255075	0.037216	0.007298	0.001676
2.10	4.79009	4.00748	0.272114	0.039198	0.007624	0.001740
2.15	5.25215	4.38539	0.290455	0.041308	0.007969	0.001808
2.20	5.76259	4.80207	0.310202	0.043553	0.008333	0.001880
2.25	6.32675	5.26171	0.331470	0.045942	0.008718	0.001955
2.30	6.95059	5.76900	0.354383	0.048486	0.009125	0.002035
2.35	7.64078	6.32914	0.379075	0.051194	0.009555	0.002118
2.40	8.40475	6.94794	0.405693	0.054078	0.010009	0.002206
2.45	9.25082	7.63189	0.434395	0.057150	0.010489	0.002298
2.50	10.1883	8.38820	0.465355	0.060422	0.010997	0.002395
2.55	11.2275	9.22495	0.498761	0.063909	0.011534	0.002496
2.60	12.3801	10.1512	0.534818	0.067626	0.012101	0.002604
2.65	13.6592	11.1769	0.573750	0.071587	0.012702	0.002716
2.70	15.0792	12.3133	0.615801	0.075812	0.013337	0.002835
2.75	16.6566	13.5731	0.661235	0.080318	0.014008	0.002959
2.80	18.4095	14.9703	0.710341	0.085125	0.014719	0.003091
2.85	20.3586	16.5206	0.763435	0.090254	0.015472	0.003229
2.90	22.5268	18.2416	0.820860	0.095728	0.016268	0.003374
2.95	24.9399	20.1531	0.882991	0.101573	0.017111	0.003527
3.00	27.6270	22.2772	0.950237	0.107815	0.018004	0.003688
3.05	30.6205	24.6385	1.02304	0.114483	0.018950	0.003858
3.10	33.9571	27.2649	1.10190	0.121607	0.019951	0.004036
3.15	37.6778	30.1876	1.18734	0.129220	0.021013	0.004224
3.20	41.8288	33.4413	1.27994	0.137360	0.022138	0.004422
3.25	46.4621	37.0655	1.38035	0.146064	0.023331	0.004631
3.30	51.6362	41.1040	1.48924	0.155374	0.024595	0.004851
3.35	57.4170	45.6064	1.60740	0.165334	0.025936	0.005083
3.40	63.8785	50.6282	1.73564	0.175995	0.027358	0.005327
3.45	71.1043	56.2320	1.87488	0.187407	0.028867	0.005585
3.50	79.1887	62.4881	2.02611	0.199627	0.030469	0.005856
3.55	88.2376	69.4755	2.19042	0.212716	0.032168	0.006143
3.60	98.3709	77.2833	2.36902	0.226739	0.033973	0.006445
3.65	109.724	86.0116	2.56320	0.241769	0.035889	0.006764
3.70	122.448	95.7735	2.77440	0.257881	0.037925	0.007101
3.75	136.717	106.696	3.00420	0.275157	0.040087	0.007456
3.80	152.725	118.923	3.25431	0.293689	0.042385	0.007831

Table 1.11 (continued)

v	0.95	1.0	2.0	3.0	4.0	5.0
3.85	170.691	132.615	3.52663	0.313571	0.044827	0.008227
3.90	190.865	147.956	3.82323	0.334910	0.047424	0.008645
3.95	213.527	165.151	4.14640	0.357817	0.050185	0.009087
4.00	238.995	184.433	4.49862	0.382416	0.053122	0.009553
4.05	267.630	206.064	4.88264	0.408838	0.056247	0.010047
4.10	299.838	230.340	5.30148	0.437226	0.059572	0.010568
4.15	336.082	257.597	5.75846	0.467736	0.063112	0.011119
4.20	376.884	288.213	6.25720	0.500536	0.066880	0.011701
4.25	422.838	322.618	6.80172	0.535806	0.070892	0.012318
4.30	474.615	361.296	7.39642	0.573745	0.075166	0.012969
4.35	532.978	404.795	8.04614	0.614564	0.079720	0.013659
4.40	598.792	453.738	8.75621	0.658496	0.084573	0.014389
4.45	673.039	508.826	9.53250	0.705791	0.089745	0.015161
4.50	756.835	570.859	10.3815	0.756720	0.095260	0.015979
4.55	851.444	640.739	11.3102	0.811579	0.101141	0.016844
4.60	958.308	719.490	12.3266	0.870686	0.107414	0.017761
4.65	1,079.06	808.274	13.4392	0.934388	0.114107	0.018733
4.70	1,215.56	908.410	14.6576	1.00306	0.121249	0.019762
4.75	1,369.93	1,021.39	15.9922	1.07712	0.128873	0.020852
4.80	1,544.57	1,148.92	17.4547	1.15700	0.137014	0.022008
4.85	1,742.23	1,292.92	19.0578	1.24318	0.145707	0.023234
4.90	1,966.02	1,455.59	20.8156	1.33620	0.154993	0.024533
4.95	2,219.49	1,639.42	22.7436	1.43661	0.164915	0.025912
5.00	2,506.71	1,847.23	24.8591	1.54504	0.175519	0.027374

Table 1.12: Third derivatives with respect to the order of the modified Bessel functions of the second kind $\partial^3 K_v(t)/\partial v^3$.

v	0.05	0.10	0.15	0.20	0.25	0.30
0.00	0.000000	0.000000	0.000000	0.000000	0.000000	0.000000
0.05	7.23717	2.93400	1.61066	1.01184	0.68734	0.49165
0.10	14.7175	5.94176	3.25494	2.04205	1.38582	0.99053
0.15	22.6939	9.09928	4.96733	3.10940	2.10678	1.50401
0.20	31.4390	12.4871	6.78402	4.23347	2.86204	2.03972
0.25	41.2563	16.1927	8.74387	5.43512	3.66408	2.60568
0.30	52.4925	20.3132	10.8894	6.73694	4.52632	3.21048
0.35	65.5517	24.9587	13.2678	8.16375	5.46339	3.86343

Table 1.12 (continued)

				t		
v	0.05	0.10	0.15	0.20	0.25	0.30
0.40	80.9128	30.2551	15.9324	9.74321	6.49143	4.57470
0.45	99.1494	36.3489	18.9438	11.5064	7.62845	5.35555
0.50	120.955	43.4113	22.3719	13.4886	8.89464	6.21855
0.55	147.172	51.6443	26.2974	15.7302	10.3129	7.17779
0.60	178.831	61.2870	30.8141	18.2776	11.9093	8.24920
0.65	217.196	72.6243	36.0321	21.1841	13.7135	9.45081
0.70	263.821	85.9957	42.0799	24.5118	15.7597	10.8032
0.75	320.623	101.808	49.1091	28.3326	18.0871	12.3298
0.80	389.966	120.548	57.2982	32.7304	20.7412	14.0575
0.85	474.779	142.802	66.8581	37.8030	23.7743	16.0170
0.90	578.687	169.274	78.0381	43.6647	27.2470	18.2440
0.95	706.188	200.812	91.1335	50.4492	31.2298	20.7790
1.00	862.869	238.441	106.495	58.3134	35.8043	23.6692
1.05	1,055.68	283.395	124.537	67.4412	41.0656	26.9688
1.10	1,293.25	337.169	145.756	78.0489	47.1243	30.7406
1.15	1,586.38	401.569	170.739	90.3903	54.1092	35.0569
1.20	1,948.48	478.785	200.186	104.764	62.1707	40.0018
1.25	2,396.34	571.469	234.931	121.524	71.4840	45.6725
1.30	2,950.93	682.839	275.970	141.082	82.2540	52.1817
1.35	3,638.48	816.803	324.491	163.931	94.7201	59.6603
1.40	4,491.83	978.111	381.913	190.646	109.162	68.2602
1.45	5,552.16	1,172.54	449.933	221.912	125.909	78.1579
1.50	6,871.12	1,407.12	530.581	258.534	145.343	89.5586
1.55	8,513.54	1,690.42	626.289	301.468	167.915	102.701
1.60	10,560.9	2,032.88	739.971	351.845	194.154	117.864
1.65	13,115.8	2,447.25	875.120	411.003	224.679	135.370
1.70	16,307.3	2,949.10	1,035.93	480.531	260.218	155.598
1.75	20,297.9	3,557.43	1,227.43	562.312	301.627	178.988
1.80	25,292.8	4,295.52	1,455.68	658.582	349.911	206.053
1.85	31,550.7	5,191.83	1,727.95	771.997	406.256	237.396
1.90	39,398.8	6,281.25	2,052.99	905.715	472.056	273.716
1.95	49,250.2	7,606.50	2,441.36	1,063.49	548.954	315.836
2.00	61,628.1	9,220.03	2,905.76	1,249.80	638.886	364.714
2.05	77,194.5	11,186.2	3,461.52	1,469.95	744.138	421.476
2.10	96,788.7	13,584.1	4,127.11	1,730.30	867.407	487.436
2.15	121,475	16,510.8	4,924.88	2,038.42	1,011.88	564.138
2.20	152,603	20,085.9	5,881.78	2,403.32	1,181.32	653.392
2.25	191,890	24,456.7	7,030.44	2,835.78	1,380.18	757.319
2.30	241,515	29,804.2	8,410.31	3,348.69	1,613.73	878.414
2.35	304,252	36,352.0	10,069.1	3,957.42	1,888.21	1,019.60
2.40	383,631	44,375.6	12,064.8	4,680.39	2,210.99	1,184.33
2.45	484,148	54,215.1	14,467.4	5,539.63	2,590.84	1,376.63

Table 1.12 (continued)

			t			
v	0.05	0.10	0.15	0.20	0.25	0.30
2.50	611,537	66,290.6	17,362.0	6,561.52	3,038.15	1,601.28
2.55	773,110	81,121.1	20,851.7	7,777.65	3,565.22	1,863.87
2.60	978,200	99,348.7	25,062.0	9,225.94	4,186.68	2,171.01
2.65	1.239E+6	121,768	30,145.0	10,951.8	4,919.91	2,530.48
2.70	1.570	149,361	36,285.8	13,009.9	5,785.55	2,951.44
2.75	1.991	183,348	43,709.6	15,465.6	6,808.16	3,444.72
2.80	2.528	225,239	52,690.4	18,397.7	8,016.95	4,023.09
2.85	3.211	276,908	63,562.0	21,900.9	9,446.68	4,701.63
2.90	4.083	340,681	76,731.0	26,088.9	11,138.8	5,498.16
2.95	5.195	419,445	92,693.2	31,098.8	13,142.6	6,433.75
3.00	6.615	516,791	112,053	37,095.6	15,517.0	7,533.29
3.05	8.430	637,180	135,550	44,278.0	18,332.1	8,826.29
3.10	1.075E+7	786,167	164,084	52,885.7	21,671.7	10,347.6
3.15	1.372	970,661	198,757	63,207.7	25,635.8	12,138.6
3.20	1.752	1.199E+6	240,917	75,592.7	30,344.0	14,248.3
3.25	2.239	1.483	292,210	90,461.7	35,939.1	16,734.7
3.30	2.863	1.834	354,652	108,323	42,591.8	19,666.8
3.35	3.665	2.271	430,713	129,792	50,506.6	23,126.2
3.40	4.693	2.813	523,417	155,612	59,928.1	27,210.1
3.45	6.014	3.487	636,470	186,681	71,149.2	32,033.8
3.50	7.712	4.326	774,421	224,088	84,521.1	37,734.2
3.55	9.896	5.369	942,848	269,151	100,465	44,474.3
3.60	1.271E+8	6.668	1.149E+6	323,467	119,484	52,447.9
3.65	1.633	8.287	1.400	388,971	142,186	61,885.7
3.70	2.099	1.030E+7	1.708	468,009	169,296	73,062.2
3.75	2.700	1.282	2.084	563,431	201,688	86,304.4
3.80	3.476	1.596	2.545	678,691	240,411	102,002
3.85	4.477	1.988	3.109	817,989	286,727	120,620
3.90	5.770	2.478	3.801	986,425	342,150	142,713
3.95	7.441	3.091	4.649	1.190E+6	408,508	168,941
4.00	9.602	3.857	5.690	1.437	487,995	200,094
4.05	1.240E+9	4.816	6.968	1.736	583,257	237,115
4.10	1.6016×10^9	6.016	8.537	2.097	697,480	281,130
4.15	2.070	7.520	1.047E+7	2.536	834,507	333,486
4.20	2.678	9.405	1.284	3.068	998,967	395,793
4.25	3.465	1.177E+8	1.575	3.714	1.196E+6	469,976
4.30	4.486	1.473	1.934	4.498	1.434	558,341
4.35	5.812	1.846	2.376	5.450	1.719	663,647
4.40	7.533	2.313	2.920	6.607	2.062	789,200
4.45	9.770	2.901	3.591	8.013	2.474	938,961
4.50	1.268E+10	3.639	4.418	9.724	2.971	1.118E+6
4.55	1.646	4.568	5.438	1.181E+7	3.569	1.331
4.60	2.138	5.737	6.698	1.434	4.289	1.586

Table 1.12 (continued)

			t			
v	0.05	0.10	0.15	0.20	0.25	0.30
4.65	2.778	7.209	8.253	1.743	5.157	1.890
4.70	3.613	9.063	1.017E+8	2.119	6.203	2.254
4.75	4.700	1.140E+9	1.255	2.578	7.466	2.690
4.80	6.117	1.434	1.548	3.137	8.990	3.211
4.85	7.966	1.806	1.912	3.819	1.083E+7	3.834
4.90	1.038E+11	2.275	2.361	4.653	1.305	4.581
4.95	1.353	2.867	2.917	5.670	1.574	5.476
5.00	1.764	3.614	3.607	6.913	1.898	6.548

			t			
v	0.35	0.40	0.45	0.50	0.55	0.60
0.00	0.000000	0.000000	0.000000	0.000000	0.000000	0.000000
0.05	0.364897	0.278468	0.217199	0.172421	0.138880	0.113240
0.10	0.734729	0.560431	0.436947	0.346744	0.279207	0.227600
0.15	1.11452	0.849440	0.661829	0.524899	0.422449	0.344214
0.20	1.50944	1.14915	0.894507	0.708865	0.570110	0.464243
0.25	1.92494	1.46340	1.137760	0.900706	0.723753	0.588893
0.30	2.36681	1.79624	1.394510	1.10259	0.885020	0.719424
0.35	2.84127	2.15201	1.667890	1.31684	1.05565	0.857174
0.40	3.35511	2.53542	1.961280	1.54592	1.23751	1.003570
0.45	3.91579	2.95163	2.278350	1.79253	1.43262	1.160140
0.50	4.53157	3.40632	2.623130	2.05961	1.64317	1.328560
0.55	5.21166	3.90576	3.000080	2.35040	1.87155	1.510650
0.60	5.96640	4.45698	3.414110	2.66845	2.12041	1.708380
0.65	6.80743	5.06784	3.870740	3.01773	2.39266	1.923970
0.70	7.74791	5.74717	4.376110	3.40265	2.69155	2.159830
0.75	8.80278	6.50496	4.937140	3.82814	3.02068	2.418650
0.80	9.98904	7.35249	5.561600	4.29973	3.38408	2.703440
0.85	11.3261	8.30254	6.258280	4.82363	3.78626	3.017530
0.90	12.8360	9.36965	7.037100	5.40684	4.23227	3.364650
0.95	14.5441	10.5703	7.909290	6.05723	4.72778	3.748990
1.00	16.4794	11.9235	8.887600	6.78371	5.27919	4.175220
1.05	18.6751	13.4506	9.986510	7.59636	5.89370	4.648620
1.10	21.1696	15.1761	11.2225	8.50657	6.57942	5.175090
1.15	24.0066	17.1282	12.6142	9.52727	7.34551	5.761290
1.20	27.2366	19.3391	14.1830	10.6731	8.20235	6.414710
1.25	30.9179	21.8454	15.9533	11.9607	9.16165	7.143810
1.30	35.1175	24.6893	17.9528	13.4091	10.2367	7.958120
1.35	39.9126	27.9193	20.2131	15.0396	11.4424	8.868420
1.40	45.3924	31.5909	22.7704	16.8768	12.7960	9.886850
1.45	51.6600	35.7679	25.6662	18.9485	14.3167	11.0272
1.50	58.8343	40.5236	28.9478	21.2865	16.0265	12.3050

Table 1.12 (continued)

				t		
v	0.35	0.40	0.45	0.50	0.55	0.60
1.55	67.0530	45.9423	32.6694	23.9270	17.9502	13.7378
1.60	76.4753	52.1212	36.8931	26.9112	20.1163	15.3457
1.65	87.2855	59.1720	41.6901	30.2863	22.5569	17.1512
1.70	99.6972	67.2234	47.1420	34.1061	25.3087	19.1800
1.75	113.958	76.4241	53.3424	38.4320	28.4134	21.4612
1.80	130.355	86.9452	60.3990	43.3345	31.9185	24.0278
1.85	149.222	98.9846	68.4351	48.8939	35.8782	26.9172
1.90	170.945	112.770	77.5929	55.2023	40.3543	30.1722
1.95	195.974	128.567	88.0357	62.3651	45.4172	33.8412
2.00	224.831	146.679	99.9512	70.5032	51.1472	37.9793
2.05	258.125	167.459	113.556	79.7549	57.6364	42.6491
2.10	296.564	191.316	129.099	90.2791	64.9895	47.9222
2.15	340.970	218.722	146.867	102.258	73.3265	53.8799
2.20	392.304	250.227	167.192	115.901	82.7846	60.6149
2.25	451.686	286.465	190.456	131.448	93.5207	68.2328
2.30	520.419	328.173	217.098	149.176	105.714	76.8544
2.35	600.028	376.206	247.630	169.402	119.572	86.6172
2.40	692.291	431.557	282.638	192.491	135.328	97.6784
2.45	799.286	495.380	322.802	218.864	153.255	110.218
2.50	923.441	569.015	368.910	249.005	173.660	124.440
2.55	1,067.60	654.021	421.870	283.472	196.902	140.580
2.60	1,235.08	752.212	482.737	322.908	223.387	158.905
2.65	1,429.77	865.698	552.730	368.053	253.585	179.725
2.70	1,656.23	996.939	633.264	419.764	288.036	203.388
2.75	1,919.81	1,148.80	725.977	479.027	327.358	230.300
2.80	2,226.75	1,324.62	832.771	546.982	372.265	260.921
2.85	2,584.41	1,528.29	955.851	624.946	423.576	295.780
2.90	3,001.39	1,764.36	1,097.78	714.441	482.237	335.484
2.95	3,487.83	2,038.14	1,261.53	817.226	549.334	380.730
3.00	4,055.61	2,355.81	1,450.56	935.338	626.120	432.318
3.05	4,718.71	2,724.63	1,668.90	1,071.13	714.039	491.166
3.10	5,493.54	3,153.05	1,921.20	1,227.34	814.756	558.328
3.15	6,399.44	3,650.98	2,212.93	1,407.12	930.194	635.018
3.20	7,459.17	4,230.00	2,550.40	1,614.13	1,062.57	722.631
3.25	8,699.50	4,903.68	2,941.00	1,852.63	1,214.44	822.770
3.30	10,152.0	5,687.90	3,393.32	2,127.53	1,388.77	937.282
3.35	11,853.9	6,601.28	3,917.39	2,444.56	1,588.98	1,068.29
3.40	13,849.0	7,665.64	4,524.88	2,810.35	1,819.01	1,218.25
3.45	16,189.1	8,906.57	5,229.44	3,232.61	2,083.44	1,389.99
3.50	18,935.3	10,354.1	6,046.98	3,720.29	2,387.56	1,586.74
3.55	22,159.7	12,043.5	6,996.08	4,283.80	2,737.49	1,812.27
3.60	25,947.4	14,016.1	8,098.47	4,935.26	3,140.33	2,070.91
3.65	30,399.3	16,320.6	9,379.52	5,688.74	3,604.29	2,367.65

Table 1.12 (continued)

v	0.35	0.40	0.45	0.50	0.55	0.60
3.70	35,634.3	19,014.1	10,868.9	6,560.64	4,138.90	2,708.26
3.75	41,793.4	22,163.9	12,601.3	7,570.05	4,755.19	3,099.40
3.80	49,043.2	25,849.0	14,617.5	8,739.19	5,465.99	3,548.78
3.85	57,580.9	30,162.5	16,964.8	10,094.0	6,286.14	4,065.31
3.90	67,640.5	35,213.9	19,699.1	11,664.6	7,232.92	4,659.27
3.95	79,498.8	41,132.4	22,885.6	13,486.3	8,326.35	5,342.58
4.00	93,484.2	48,069.9	26,600.8	15,600.2	9,589.73	6,129.03
4.05	109,986	56,205.8	30,934.4	18,054.2	11,050.1	7,034.59
4.10	129,466	65,751.5	35,991.8	20,904.4	12,739.0	8,077.74
4.15	152,473	76,956.3	41,896.3	24,216.1	14,692.9	9,279.91
4.20	179,659	90,114.9	48,793.0	28,065.9	16,954.5	10,665.9
4.25	211,795	105,575	56,852.4	32,543.0	19,573.4	12,264.6
4.30	249,802	123,746	66,274.3	37,752.1	22,607.2	14,109.4
4.35	294,773	145,115	77,294.3	43,815.4	26,123.3	16,239.0
4.40	348,007	170,255	90,188.8	50,876.0	30,199.9	18,698.4
4.45	411,051	199,844	105,283	59,101.4	34,928.6	21,540.0
4.50	485,746	234,685	122,961	68,688.0	40,415.9	24,824.4
4.55	574,284	275,727	143,672	79,865.6	46,786.1	28,622.3
4.60	679,277	324,097	167,948	92,903.7	54,184.3	33,015.6
4.65	803,836	381,125	196,415	108,118	62,779.9	38,099.8
4.70	951,671	448,391	229,809	125,880	72,770.9	43,986.0
4.75	1.127E+6	527,765	269,000	146,625	84,388.5	50,803.2
4.80	1.336	621,468	315,012	170,862	97,902.8	58,702.1
4.85	1.584	732,130	369,057	199,191	113,630	67,857.6
4.90	1.878	862,876	432,562	232,317	131,939	78,474.1
4.95	2.229	1.017E+6	507,211	271,068	153,263	90,789.3
5.00	2.645	1.200	594,998	316,415	178,108	10,5081

v	0.65	0.70	0.75	0.80	0.85	0.90
0.00	0.000000	0.000000	0.000000	0.000000	0.000000	0.000000
0.05	0.093304	0.077575	0.065010	0.054860	0.046584	0.039776
0.10	0.187486	0.155848	0.130578	0.110174	0.093537	0.079856
0.15	0.283436	0.235523	0.197273	0.166398	0.141234	0.120548
0.20	0.382062	0.317322	0.265669	0.224000	0.190056	0.162164
0.25	0.484308	0.401990	0.336366	0.283463	0.240395	0.205027
0.30	0.591157	0.490306	0.409983	0.345286	0.292659	0.249472
0.35	0.703649	0.583089	0.487175	0.409998	0.347275	0.295847
0.40	0.822893	0.681210	0.568634	0.478154	0.404696	0.344523
0.45	0.950076	0.785603	0.655101	0.550347	0.465400	0.395891
0.50	1.08649	0.897272	0.747372	0.627216	0.529903	0.450368
0.55	1.23352	1.01731	0.846307	0.709445	0.598757	0.508405

Table 1.12 (continued)

			t			
v	0.65	0.70	0.75	0.80	0.85	0.90
0.60	1.39270	1.14690	0.952841	0.797781	0.672560	0.570486
0.65	1.56571	1.28734	1.06800	0.893032	0.751963	0.637136
0.70	1.75439	1.44007	1.19289	0.996086	0.837673	0.708929
0.75	1.96079	1.60666	1.32876	1.10792	0.930467	0.786487
0.80	2.18718	1.78885	1.47695	1.22959	1.03120	0.870495
0.85	2.43608	1.98857	1.63897	1.36228	1.14079	0.961702
0.90	2.71029	2.20797	1.81647	1.50730	1.26029	1.06093
0.95	3.01295	2.44944	2.01131	1.66608	1.39083	1.16910
1.00	3.34756	2.71563	2.22553	1.84022	1.53367	1.28720
1.05	3.71804	3.00951	2.46141	2.03151	1.69021	1.41634
1.10	4.12878	3.33441	2.72149	2.24190	1.86199	1.55776
1.15	4.58471	3.69402	3.00861	2.47359	2.05073	1.71281
1.20	5.09136	4.09251	3.32593	2.72903	2.25834	1.88299
1.25	5.65495	4.53453	3.67699	3.01095	2.48695	2.06999
1.30	6.28248	5.02531	4.06576	3.32239	2.73893	2.27566
1.35	6.98182	5.57070	4.49667	3.66675	3.01690	2.50207
1.40	7.76184	6.17730	4.97468	4.04781	3.32381	2.75152
1.45	8.63254	6.85250	5.50536	4.46984	3.66294	3.02657
1.50	9.60519	7.60463	6.09495	4.93759	4.03796	3.33008
1.55	10.6925	8.44305	6.75048	5.45637	4.45295	3.66522
1.60	11.9089	9.37832	7.47980	6.03215	4.91248	4.03554
1.65	13.2706	10.4223	8.29177	6.67161	5.42168	4.44502
1.70	14.7959	11.5884	9.19633	7.38225	5.98626	4.89806
1.75	16.5056	12.8918	10.2047	8.17247	6.61263	5.39960
1.80	18.4231	14.3495	11.3294	9.05172	7.30797	5.95517
1.85	20.5752	15.9806	12.5847	10.0306	8.08032	6.57095
1.90	22.9919	17.8071	13.9864	11.1210	8.93870	7.25383
1.95	25.7073	19.8534	15.5527	12.3364	9.89323	8.01155
2.00	28.7602	22.1473	17.3038	13.6918	10.9552	8.85276
2.05	32.1945	24.7202	19.2627	15.2042	12.1375	9.78716
2.10	36.0601	27.6077	21.4551	16.8927	13.4543	10.825600
2.15	40.4135	30.8501	23.9102	18.7787	14.9217	11.9803
2.20	45.3190	34.4930	26.6610	20.8866	16.5578	13.2649
2.25	50.8498	38.5879	29.7448	23.2435	18.3830	14.6948
2.30	57.0890	43.1936	33.2036	25.8804	20.4201	16.2871
2.35	64.1310	48.3765	37.0852	28.8320	22.6948	18.0613
2.40	72.0836	54.2119	41.4434	32.1375	25.2362	20.0389
2.45	81.0692	60.7855	46.3392	35.8412	28.0770	22.2446
2.50	91.2274	68.1946	51.8419	39.9933	31.2538	24.7056
2.55	102.717	76.5496	58.0298	44.6502	34.8084	27.4530
2.60	115.720	85.9760	64.9916	49.8759	38.7874	30.5215
2.65	130.443	96.6168	72.8281	55.7428	43.2438	33.9502
2.70	147.122	108.634	81.6536	62.3327	48.2371	37.7833

Table 1.12 (continued)

			t			
ν	0.65	0.70	0.75	0.80	0.85	0.90
2.75	166.027	122.214	91.5978	69.7385	53.8348	42.0704
2.80	187.465	137.566	102.808	78.0649	60.1129	46.8675
2.85	211.788	154.930	115.451	87.4311	67.1574	52.2378
2.90	239.399	174.580	129.718	97.9717	75.0657	58.2526
2.95	270.758	196.828	145.825	109.840	83.9477	64.9922
3.00	306.390	222.029	164.017	123.209	93.9280	72.5475
3.05	346.899	250.590	184.574	138.275	105.147	81.0209
3.10	392.974	282.972	207.815	155.263	117.766	90.5284
3.15	445.404	319.707	234.103	174.425	131.963	101.201
3.20	505.097	361.397	263.850	196.051	147.946	113.186
3.25	573.090	408.735	297.528	220.468	165.945	126.652
3.30	650.575	462.509	335.673	248.049	186.225	141.788
3.35	738.917	523.624	378.899	279.217	209.085	158.808
3.40	839.687	593.115	427.903	314.455	234.863	177.956
3.45	954.686	672.164	483.483	354.312	263.946	199.507
3.50	1,085.98	762.129	546.550	399.415	296.771	223.772
3.55	1,235.96	864.562	618.146	450.474	333.837	251.105
3.60	1,407.35	981.244	699.458	508.303	375.708	281.908
3.65	1,603.30	1,114.22	791.847	573.827	423.029	316.635
3.70	1,827.43	1,265.82	896.866	648.103	476.530	355.802
3.75	2,083.91	1,438.75	1,016.30	732.336	537.045	399.996
3.80	2,377.54	1,636.08	1,152.17	827.900	605.523	449.883
3.85	2,713.85	1,861.36	1,306.83	936.367	683.043	506.219
3.90	3,099.21	2,118.66	1,482.93	1,059.53	770.835	569.864
3.95	3,540.97	2,412.65	1,683.54	1,199.44	870.304	641.796
4.00	4,047.60	2,748.72	1,912.16	1,358.44	983.046	723.126
4.05	4,628.89	3,133.04	2,172.83	1,539.21	1,110.89	815.120
4.10	5,296.12	3,572.73	2,470.14	1,744.81	1,255.91	919.218
4.15	6,062.31	4,075.99	2,809.39	1,978.75	1,420.48	1,037.06
4.20	6,942.53	4,652.24	3,196.67	2,245.05	1,607.32	1,170.51
4.25	7,954.18	5,312.33	3,638.94	2,548.31	1,819.52	1,321.70
4.30	9,117.34	6,068.80	4,144.23	2,893.78	2,060.63	1,493.05
4.35	10,455.3	6,936.05	4,721.75	3,287.52	2,334.68	1,687.33
4.40	11,994.9	7,930.72	5,382.09	3,736.43	2,646.30	1,907.69
4.45	13,767.3	9,071.99	6,137.42	4,248.45	3,000.77	2,157.73
4.50	15,808.6	10,382.0	7,001.76	4,832.68	3,404.16	2,441.55
4.55	18,160.4	11,886.3	7,991.22	5,499.57	3,863.39	2,763.85
4.60	20,871.1	13,614.4	9,124.38	6,261.12	4,386.38	3,129.97
4.65	23,996.8	15,600.3	10,422.6	7,131.07	4,982.22	3,546.04
4.70	27,602.3	17,883.5	11,910.5	8,125.26	5,661.31	4,019.04
4.75	31,763.1	20,509.4	13,616.4	9,261.86	6,435.57	4,556.96
4.80	36,566.3	23,530.7	15,573.1	10,561.8	7,318.68	5,168.95
4.85	42,113.6	27,008.3	17,818.3	12,049.0	8,326.32	5,865.45

Table 1.12 (continued)

			t			
v	0.65	0.70	0.75	0.80	0.85	0.90
4.90	48,522.5	31,012.5	20,395.5	13,751.1	9,476.46	6,658.44
4.95	55,929.7	35,625.0	23,354.8	15,700.1	10,789.7	7,561.59
5.00	64,494.2	40,940.0	26,754.3	17,932.3	12,289.9	8,590.60

			t			
v	0.95	1.0	2.0	3.0	4.0	5.0
0.00	0.000000	0.000000	0.000000	0.000000	0.000000	0.000000
0.05	0.034134	0.029425	0.002674	0.000411	0.000080	0.000018
0.10	0.068519	0.059060	0.005359	0.000824	0.000160	0.000035
0.15	0.103411	0.089116	0.008068	0.001239	0.000240	0.000053
0.20	0.139068	0.119810	0.010812	0.001658	0.000321	0.000071
0.25	0.175757	0.151363	0.013603	0.002083	0.000403	0.000089
0.30	0.213753	0.184004	0.016454	0.002514	0.000486	0.000108
0.35	0.253346	0.217972	0.019377	0.002953	0.000570	0.000126
0.40	0.294839	0.253519	0.022386	0.003402	0.000656	0.000145
0.45	0.338552	0.290911	0.025493	0.003862	0.000743	0.000164
0.50	0.384831	0.330432	0.028713	0.004334	0.000832	0.000183
0.55	0.434042	0.372384	0.032061	0.004820	0.000923	0.000203
0.60	0.486581	0.417092	0.035553	0.005321	0.001017	0.000223
0.65	0.542877	0.464910	0.039204	0.005840	0.001113	0.000244
0.70	0.603394	0.516217	0.043032	0.006379	0.001212	0.000265
0.75	0.668641	0.571428	0.047056	0.006938	0.001314	0.000287
0.80	0.739169	0.630995	0.051294	0.007520	0.001420	0.000309
0.85	0.815586	0.695411	0.055768	0.008126	0.001529	0.000332
0.90	0.898556	0.765218	0.060499	0.008760	0.001643	0.000356
0.95	0.988811	0.841008	0.065512	0.009424	0.001760	0.000381
1.00	1.087160	0.923433	0.070832	0.010119	0.001883	0.000406
1.05	1.194480	1.01321	0.076487	0.010848	0.002010	0.000432
1.10	1.311770	1.11114	0.082504	0.011614	0.002142	0.000459
1.15	1.440100	1.21808	0.088917	0.012420	0.002281	0.000488
1.20	1.580680	1.33500	0.095758	0.013269	0.002425	0.000517
1.25	1.734830	1.46298	0.103065	0.014164	0.002576	0.000547
1.30	1.904040	1.60319	0.110878	0.015109	0.002734	0.000579
1.35	2.089940	1.75693	0.119238	0.016107	0.002900	0.000612
1.40	2.294350	1.92567	0.128193	0.017161	0.003073	0.000646
1.45	2.519280	2.11100	0.137791	0.018277	0.003255	0.000682
1.50	2.766990	2.31470	0.148088	0.019459	0.003447	0.000719
1.55	3.039970	2.53877	0.159142	0.020711	0.003647	0.000758
1.60	3.341010	2.78540	0.171016	0.022039	0.003859	0.000799
1.65	3.673190	3.05704	0.183780	0.023448	0.004081	0.000842
1.70	4.039990	3.35640	0.197509	0.024944	0.004314	0.000886
1.75	4.445240	3.68653	0.212283	0.026533	0.004561	0.000933

Table 1.12 (continued)

			t			
v	0.95	1.0	2.0	3.0	4.0	5.0
1.80	4.893230	4.05078	0.228191	0.028222	0.004820	0.000982
1.85	5.388770	4.45291	0.245329	0.030018	0.005094	0.001033
1.90	5.937190	4.89711	0.263801	0.031928	0.005382	0.001087
1.95	6.544470	5.38805	0.283721	0.033962	0.005687	0.001143
2.00	7.217300	5.93092	0.305212	0.036128	0.006009	0.001202
2.05	7.963120	6.53152	0.328409	0.038435	0.006348	0.001264
2.10	8.790300	7.19635	0.353458	0.040894	0.006707	0.001329
2.15	9.708170	7.93263	0.380518	0.043516	0.007086	0.001398
2.20	10.7272	8.74846	0.409763	0.046312	0.007488	0.001470
2.25	11.8591	9.65286	0.441381	0.049296	0.007912	0.001545
2.30	13.1169	10.6559	0.475581	0.052481	0.008361	0.001625
2.35	14.5154	11.7690	0.512586	0.055882	0.008837	0.001708
2.40	16.0711	13.0047	0.552642	0.059515	0.009340	0.001796
2.45	17.8025	14.3772	0.596017	0.063397	0.009873	0.001889
2.50	19.7302	15.9023	0.643004	0.067546	0.010439	0.001986
2.55	21.8777	17.5978	0.693923	0.071983	0.011038	0.002089
2.60	24.2710	19.4836	0.749121	0.076728	0.011673	0.002197
2.65	26.9395	21.5821	0.808982	0.081805	0.012346	0.002311
2.70	29.9163	23.9182	0.873921	0.087239	0.013061	0.002431
2.75	33.2386	26.5202	0.944396	0.093057	0.013819	0.002558
2.80	36.9480	29.4194	1.02090	0.099286	0.014624	0.002692
2.85	41.0916	32.6513	1.10399	0.105960	0.015479	0.002833
2.90	45.7223	36.2558	1.19425	0.113111	0.016386	0.002981
2.95	50.8998	40.2775	1.29235	0.120776	0.017350	0.003139
3.00	56.6910	44.7666	1.39898	0.128994	0.018375	0.003304
3.05	63.1718	49.7799	1.51495	0.137807	0.019464	0.003479
3.10	70.4273	55.3807	1.64111	0.147262	0.020621	0.003664
3.15	78.5538	61.6408	1.77839	0.157408	0.021852	0.003860
3.20	87.6598	68.6406	1.92784	0.168299	0.023161	0.004066
3.25	97.8677	76.4711	2.09058	0.179993	0.024555	0.004285
3.30	109.316	85.2345	2.26786	0.192551	0.026037	0.004515
3.35	122.161	95.0461	2.46104	0.206043	0.027615	0.004759
3.40	136.578	106.036	2.67161	0.220541	0.029295	0.005017
3.45	152.769	118.350	2.90122	0.236125	0.031084	0.005290
3.50	170.957	132.155	3.15166	0.252880	0.032990	0.005579
3.55	191.399	147.637	3.42493	0.270900	0.035021	0.005885
3.60	214.383	165.007	3.72319	0.290286	0.037185	0.006208
3.65	240.236	184.503	4.04883	0.311146	0.039492	0.006551
3.70	269.328	206.395	4.40450	0.333598	0.041951	0.006914
3.75	302.078	230.986	4.79308	0.357772	0.044574	0.007298
3.80	338.963	258.621	5.21774	0.383805	0.047373	0.007705
3.85	380.519	289.689	5.68201	0.411849	0.050358	0.008137
3.90	427.359	324.630	6.18971	0.442066	0.053544	0.008594

Table 1.12 (continued)

			t			
v	0.95	1.0	2.0	3.0	4.0	5.0
3.95	480.174	363.944	6.74511	0.474634	0.056945	0.009079
4.00	539.752	408.194	7.35287	0.509745	0.060577	0.009593
4.05	606.985	458.020	8.01814	0.547608	0.064455	0.010138
4.10	682.887	514.146	8.74660	0.588449	0.068597	0.010717
4.15	768.609	577.395	9.54448	0.632514	0.073023	0.011330
4.20	865.459	648.697	10.4187	0.680070	0.077753	0.011981
4.25	974.925	729.108	11.3768	0.731407	0.082809	0.012672
4.30	1,098.70	819.828	12.4273	0.786841	0.088215	0.013406
4.35	1,238.70	922.216	13.5793	0.846714	0.093995	0.014185
4.40	1,397.13	1,037.82	14.8430	0.911398	0.100179	0.015012
4.45	1,576.47	1,168.39	16.2298	0.981297	0.106794	0.015891
4.50	1,779.55	1,315.92	17.7521	1.05685	0.113874	0.016825
4.55	2,009.62	1,482.68	19.4237	1.13855	0.121452	0.017817
4.60	2,270.36	1,671.24	21.2596	1.22689	0.129565	0.018871
4.65	2,565.95	1,884.53	23.2768	1.32246	0.138252	0.019993
4.70	2,901.20	2,125.88	25.4938	1.42588	0.147557	0.021185
4.75	3,281.55	2,399.09	27.9310	1.53780	0.157526	0.022452
4.80	3,713.23	2,708.48	30.6113	1.65897	0.168209	0.023801
4.85	4,203.35	3,058.95	33.5596	1.79018	0.179659	0.025236
4.90	4,760.03	3,456.11	36.8038	1.93230	0.191933	0.026762
4.95	5,392.52	3,906.34	40.3747	2.08628	0.205095	0.028387
5.00	6,111.42	4,416.91	44.3063	2.25315	0.219211	0.030117

2 Numerical Results – Tabulation of First, Second and Third Derivatives with Respect to the Order of the Struve, Modified Struve, Anger and Weber Functions

Derivatives with respect to the order of these Bessel-related functions are not tabulated in literature. The results presented here cover $0 \le v \le 5$ range, with 0.05 intervals. The arguments of functions are in $0.05 \le t \le 5$ range, with 0.05 intervals for $t \le 1.0$ and with 1.0 intervals for $t > 1.0$. Calculations were performed using MATHEMATICA program. Derivatives with respect to the order were calculated by evaluation of integrals demonstrating the Anger and Weber integral representations. In other cases, they were performed directly by using the central-difference formulas of order $O(h^4)$ with $h = 0.001$, when the involved functions were taken from MATHEMATICA program.

2.1 First, Second and Third Derivatives with Respect to the Order of the Struve Functions

Derivatives with respect to the order of the Struve functions were determined using central-difference formulas of order $O(h^4)$ with $h = 10^{-3}$. The Struve functions $H_v(t)$ were taken from MATHEMATICA program:

$$\frac{\partial H_v(t)}{\partial v} = \frac{- H_{v+2h}(t) + 8 H_{v+h}(t) - 8 H_{v-h}(t) + H_{v-2h}(t)}{12h} \tag{2.1.1}$$

$$\frac{\partial^2 H_v(t)}{\partial v^2} = \frac{- H_{v+2h}(t) + 16 H_{v+h}(t) - 30 H_v(t) + 16 H_{v-h}(t) - H_{v-2h}(t)}{12h^2} \tag{2.1.2}$$

$$\frac{\partial^3 H_v(t)}{\partial v^3} = \frac{- H_{v+3h}(t) + 8 H_{v+2h}(t) - 13 H_{v+h}(t)}{8h^3}$$
$$+ \frac{13 H_{v-h}(t) - 8 H_{v-2h}(t) + H_{v-3h}(t)}{8h^3} \tag{2.1.3}$$

Functional behaviour of derivatives with respect to the order of the Struve functions resembles the Bessel functions of the first kind; the oscillatory character is damped and it tends to zero values with an increase in the order v and the argument t (see Figures 2.1–2.18; Tables 2.1–2.12).

https://doi.org/10.1515/9783110682472-002

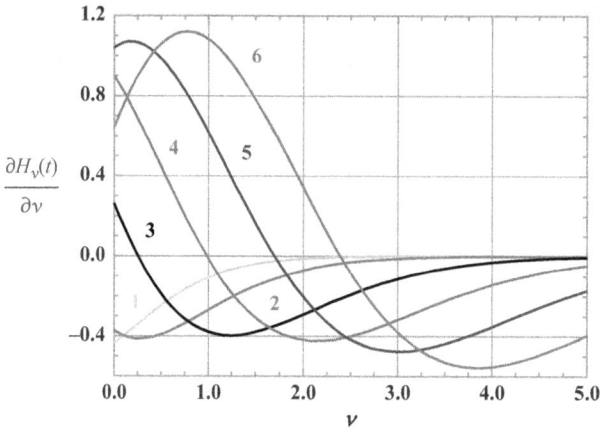

Figure 2.1: First derivatives of the Struve function with respect to the order v as a function of v, at constant values of argument t.

$1 - t = 0.50;\ 2 - t = 1.0;\ 3 - t = 2.0;\ 4 - t = 3.0;\ 5 - t = 4.0;\ 6 - t = 5.0.$

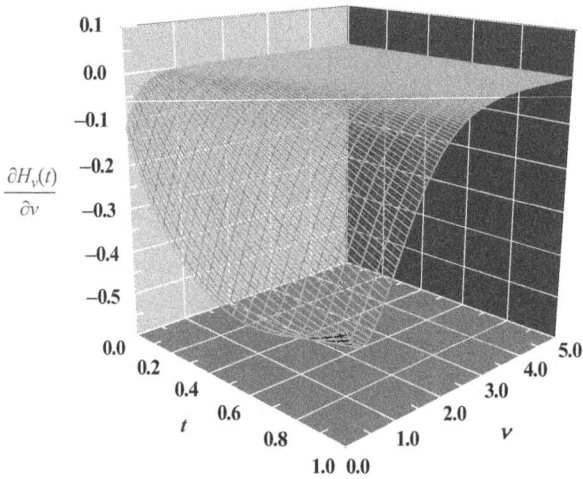

Figure 2.2: First derivatives with respect to the order of the Struve functions $\partial H_v(t)/\partial v$ as a function of v and t.

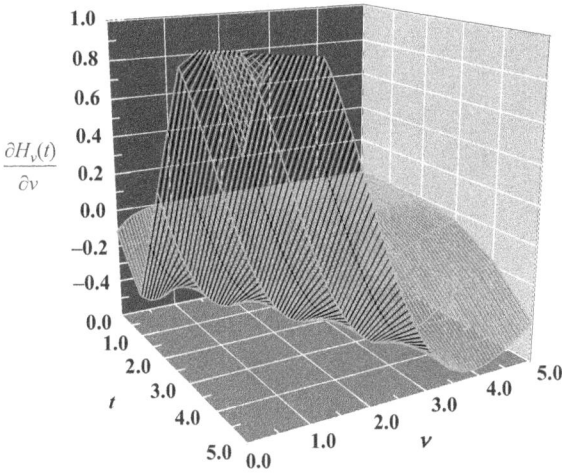

Figure 2.3: First derivatives with respect to the order of the Struve functions $\partial H_\nu(t)/\partial \nu$ as a function of ν and t.

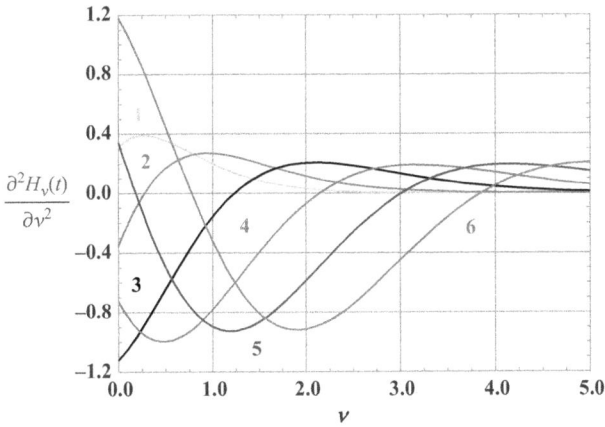

Figure 2.4: Second derivatives of the Struve function with respect to the order ν as a function of ν, at constant values of argument t.

$1 - t = 0.50; 2 - t = 1.0; 3 - t = 2.0; 4 - t = 3.0; 5 - t = 4.0; 6 - t = 5.0.$

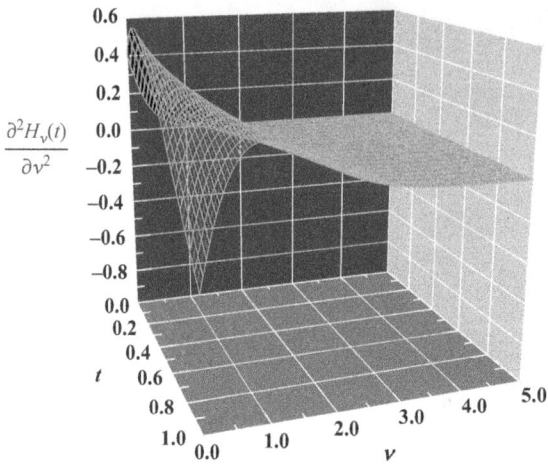

Figure 2.5: Second derivatives with respect to the order of the Struve functions $\partial^2 H_v(t)/\partial v^2$ as a function of v and t.

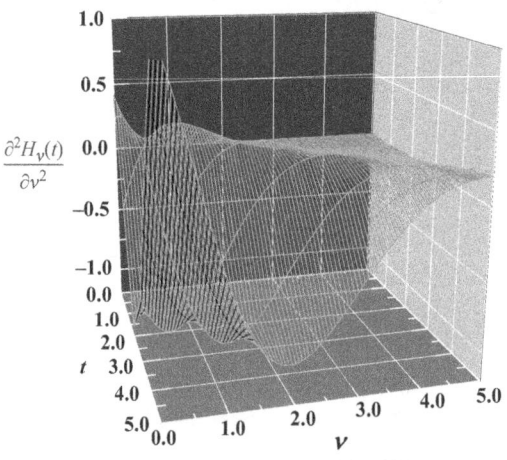

Figure 2.6: Second derivatives with respect to the order of the Struve functions $\partial^2 H_v(t)/\partial v^2$ as a function of v and t.

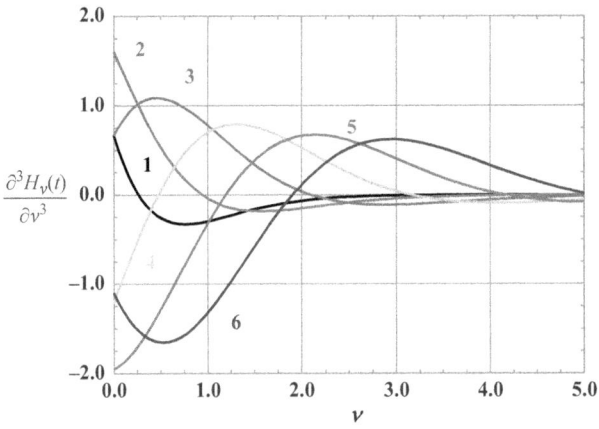

Figure 2.7: Third derivatives of the Struve function with respect to the order v as a function of v, at constant values of argument t.
$1 - t = 0.50$; $2 - t = 1.0$; $3 - t = 2.0$; $4 - t = 3.0$; $5 - t = 4.0$; $6 - t = 5.0$.

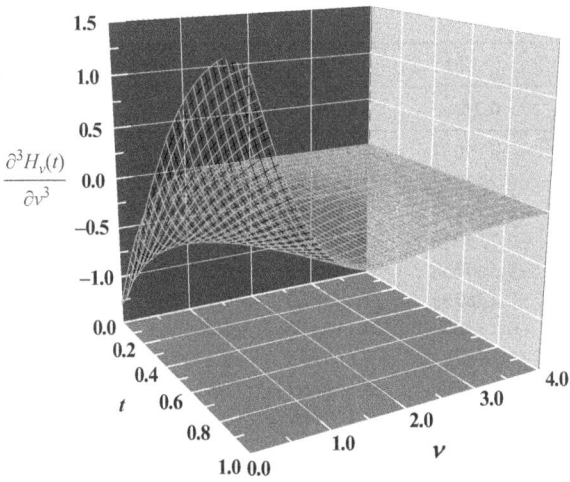

Figure 2.8: Third derivatives with respect to the order of the Struve functions $\partial^3 H_v(t)/\partial v^3$ as a function of v and t.

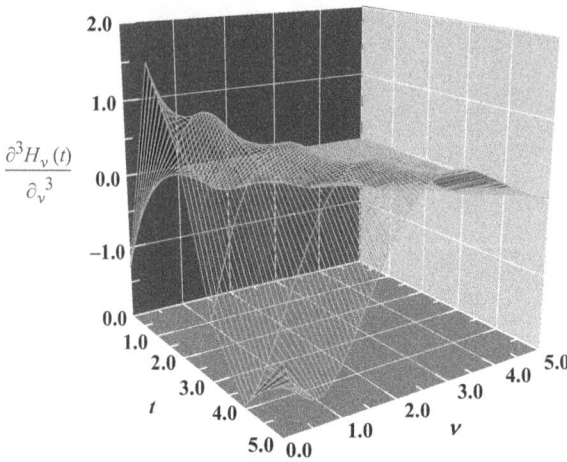

Figure 2.9: Third derivatives with respect to the order of the Struve functions $\partial^3 H_v(t)/\partial v^3$ as a function of v and t.

Table 2.1: First derivatives with respect to the order of the Struve functions $\partial H_v(t)/\partial v$.

				t		
v	0.05	0.10	0.15	0.20	0.25	0.30
0.00	−0.118543	−0.192776	−0.250051	−0.296122	−0.333694	−0.364347
0.05	−0.099492	−0.167971	−0.222866	−0.268346	−0.306454	−0.338414
0.10	−0.083261	−0.145908	−0.197992	−0.242347	−0.280430	−0.313144
0.15	−0.069488	−0.126376	−0.175358	−0.218167	−0.255757	−0.288744
0.20	−0.057843	−0.109158	−0.154864	−0.195808	−0.232520	−0.265370
0.25	−0.048030	−0.094040	−0.136393	−0.175240	−0.210768	−0.243134
0.30	−0.039788	−0.080814	−0.119813	−0.156410	−0.190515	−0.222113
0.35	−0.032885	−0.069284	−0.104989	−0.139246	−0.171751	−0.202350
0.40	−0.027122	−0.059265	−0.091783	−0.123663	−0.154444	−0.183865
0.45	−0.022322	−0.050584	−0.080057	−0.109569	−0.138548	−0.166653
0.50	−0.018335	−0.043086	−0.069679	−0.096865	−0.124003	−0.150695
0.55	−0.015031	−0.036625	−0.060522	−0.085452	−0.110741	−0.135957
0.60	−0.012300	−0.031074	−0.052463	−0.075229	−0.098690	−0.122396
0.65	−0.010047	−0.026315	−0.045391	−0.066100	−0.087773	−0.109959
0.70	−0.008193	−0.022245	−0.039200	−0.057969	−0.077912	−0.098590
0.75	−0.006669	−0.018772	−0.033793	−0.050745	−0.069030	−0.088227
0.80	−0.005421	−0.015815	−0.029082	−0.044344	−0.061051	−0.078809
0.85	−0.004399	−0.013303	−0.024987	−0.038685	−0.053901	−0.070271
0.90	−0.003564	−0.011172	−0.021434	−0.033692	−0.047508	−0.062551
0.95	−0.002884	−0.009368	−0.018357	−0.029297	−0.041805	−0.055587
1.00	−0.002330	−0.007843	−0.015699	−0.025437	−0.036730	−0.049319

Table 2.1 (continued)

			t			
v	0.05	0.10	0.15	0.20	0.25	0.30
1.05	−0.001879	−0.006558	−0.013405	−0.022052	−0.032221	−0.043690
1.10	−0.001514	−0.005475	−0.011431	−0.019090	−0.028225	−0.038646
1.15	−0.001218	−0.004565	−0.009734	−0.016502	−0.024688	−0.034134
1.20	−0.000979	−0.003802	−0.008278	−0.014246	−0.021565	−0.030106
1.25	−0.000786	−0.003162	−0.007030	−0.012282	−0.018811	−0.026517
1.30	−0.000630	−0.002626	−0.005963	−0.010575	−0.016388	−0.023325
1.35	−0.000504	−0.002179	−0.005051	−0.009093	−0.014258	−0.020490
1.40	−0.000403	−0.001805	−0.004274	−0.007810	−0.012389	−0.017977
1.45	−0.000322	−0.001494	−0.003612	−0.006700	−0.010753	−0.015752
1.50	−0.000257	−0.001235	−0.003049	−0.005740	−0.009321	−0.013787
1.55	−0.000205	−0.001020	−0.002571	−0.004913	−0.008071	−0.012052
1.60	−0.000163	−0.000841	−0.002165	−0.004200	−0.006980	−0.010524
1.65	−0.000130	−0.000693	−0.001822	−0.003587	−0.006030	−0.009179
1.70	−0.000103	−0.000571	−0.001531	−0.003059	−0.005204	−0.007997
1.75	−0.000082	−0.000469	−0.001286	−0.002607	−0.004486	−0.006960
1.80	−0.000065	−0.000385	−0.001078	−0.002219	−0.003863	−0.006051
1.85	−0.000051	−0.000316	−0.000903	−0.001887	−0.003324	−0.005255
1.90	−0.000041	−0.000259	−0.000756	−0.001603	−0.002856	−0.004559
1.95	−0.000032	−0.000212	−0.000632	−0.001361	−0.002453	−0.003952
2.00	−0.000025	−0.000174	−0.000528	−0.001154	−0.002104	−0.003422
2.05	−0.000020	−0.000142	−0.000441	−0.000978	−0.001803	−0.002960
2.10	−0.000016	−0.000116	−0.000368	−0.000827	−0.001544	−0.002559
2.15	−0.000012	−0.000095	−0.000307	−0.000700	−0.001320	−0.002209
2.20	−0.000010	−0.000077	−0.000255	−0.000591	−0.001128	−0.001906
2.25	−0.000008	−0.000063	−0.000212	−0.000499	−0.000964	−0.001643
2.30	−0.000006	−0.000051	−0.000176	−0.000421	−0.000822	−0.001415
2.35	−0.000005	−0.000042	−0.000147	−0.000355	−0.000701	−0.001218
2.40	−0.000004	−0.000034	−0.000122	−0.000299	−0.000597	−0.001047
2.45	−0.000003	−0.000028	−0.000101	−0.000251	−0.000508	−0.000899
2.50	−0.000002	−0.000022	−0.000083	−0.000211	−0.000432	−0.000772
2.55	−0.000002	−0.000018	−0.000069	−0.000177	−0.000367	−0.000662
2.60	−0.000001	−0.000015	−0.000057	−0.000149	−0.000312	−0.000567
2.65	−0.000001	−0.000012	−0.000047	−0.000125	−0.000264	−0.000486
2.70	−0.000001	−0.000010	−0.000039	−0.000105	−0.000224	−0.000416
2.75	−0.000001	−0.000008	−0.000032	−0.000088	−0.000190	−0.000356
2.80	−0.000001	−0.000006	−0.000027	−0.000073	−0.000161	−0.000304
2.85		−0.000005	−0.000022	−0.000061	−0.000136	−0.000259
2.90		−0.000004	−0.000018	−0.000051	−0.000115	−0.000221
2.95		−0.000003	−0.000015	−0.000043	−0.000097	−0.000189
3.00		−0.000003	−0.000012	−0.000036	−0.000082	−0.000161
3.05		−0.000002	−0.000010	−0.000030	−0.000069	−0.000137
3.10		−0.000002	−0.000008	−0.000025	−0.000058	−0.000116

Table 2.1 (continued)

				t		
v	0.05	0.10	0.15	0.20	0.25	0.30
3.15		−0.000001	−0.000007	−0.000021	−0.000049	−0.000099
3.20		−0.000001	−0.000006	−0.000017	−0.000041	−0.000084
3.25		−0.000001	−0.000005	−0.000014	−0.000035	−0.000071
3.30		−0.000001	−0.000004	−0.000012	−0.000029	−0.000061
3.35		−0.000001	−0.000003	−0.000010	−0.000025	−0.000051
3.40			−0.000002	−0.000008	−0.000021	−0.000044
3.45			−0.000002	−0.000007	−0.000017	−0.000037
3.50			−0.000002	−0.000006	−0.000015	−0.000031
3.55			−0.000001	−0.000005	−0.000012	−0.000026
3.60			−0.000001	−0.000004	−0.000010	−0.000022
3.65			−0.000001	−0.000003	−0.000009	−0.000019
3.70			−0.000001	−0.000003	−0.000007	−0.000016
3.75			−0.000001	−0.000002	−0.000006	−0.000013
3.80				−0.000002	−0.000005	−0.000011
3.85				−0.000002	−0.000004	−0.000010
3.90				−0.000001	−0.000003	−0.000008
3.95				−0.000001	−0.000003	−0.000007
4.00				−0.000001	−0.000002	−0.000006
4.05				−0.000001	−0.000002	−0.000005
4.10				−0.000001	−0.000002	−0.000004
4.15					−0.000001	−0.000003
4.20					−0.000001	−0.000003
4.25					−0.000001	−0.000002
4.30					−0.000001	−0.000002
4.35					−0.000001	−0.000002
4.40					−0.000001	−0.000001
4.45						−0.000001
4.50						−0.000001
4.55						−0.000001
4.60						−0.000001
4.65						−0.000001

				t		
v	0.35	0.40	0.45	0.50	0.55	0.60
0.00	−0.389117	−0.408741	−0.423769	−0.434630	−0.441673	−0.445187
0.05	−0.365046	−0.386940	−0.404545	−0.418214	−0.428237	−0.434856
0.10	−0.341107	−0.364771	−0.384487	−0.400534	−0.413142	−0.422510
0.15	−0.317564	−0.342542	−0.363935	−0.381952	−0.396767	−0.408531
0.20	−0.294629	−0.320510	−0.343182	−0.362787	−0.379447	−0.393269
0.25	−0.272471	−0.298886	−0.322473	−0.343313	−0.361479	−0.377035
0.30	−0.251215	−0.277842	−0.302018	−0.323769	−0.343124	−0.360108
0.35	−0.230956	−0.257512	−0.281987	−0.304356	−0.324608	−0.342735

Table 2.1 (continued)

v	0.35	0.40	0.45	0.50	0.55	0.60
				t		
0.40	-0.211756	-0.238002	-0.262518	-0.285242	-0.306126	-0.325131
0.45	-0.193654	-0.219385	-0.243723	-0.266568	-0.287843	-0.307484
0.50	-0.176667	-0.201716	-0.225684	-0.248445	-0.269897	-0.289953
0.55	-0.160795	-0.185024	-0.208463	-0.230963	-0.252402	-0.272675
0.60	-0.146024	-0.169326	-0.192103	-0.214191	-0.235450	-0.255764
0.65	-0.132328	-0.154622	-0.176630	-0.198177	-0.219115	-0.239312
0.70	-0.119673	-0.140899	-0.162054	-0.182956	-0.203450	-0.223396
0.75	-0.108015	-0.128136	-0.148375	-0.168548	-0.188495	-0.208074
0.80	-0.097310	-0.116305	-0.135583	-0.154961	-0.174278	-0.193391
0.85	-0.087507	-0.105371	-0.123659	-0.142192	-0.160812	-0.179377
0.90	-0.078554	-0.095294	-0.112578	-0.130232	-0.148104	-0.166054
0.95	-0.070398	-0.086033	-0.102309	-0.119064	-0.136150	-0.153431
1.00	-0.062986	-0.077542	-0.092819	-0.108665	-0.124939	-0.141513
1.05	-0.056265	-0.069777	-0.084072	-0.099009	-0.114456	-0.130292
1.10	-0.050185	-0.062692	-0.076028	-0.090065	-0.104680	-0.119760
1.15	-0.044695	-0.056241	-0.068649	-0.081801	-0.095587	-0.109901
1.20	-0.039749	-0.050381	-0.061893	-0.074182	-0.087149	-0.100696
1.25	-0.035301	-0.045066	-0.055722	-0.067175	-0.079337	-0.092121
1.30	-0.031307	-0.040257	-0.050095	-0.060743	-0.072122	-0.084153
1.35	-0.027729	-0.035913	-0.044976	-0.054851	-0.065470	-0.076764
1.40	-0.024528	-0.031996	-0.040326	-0.049464	-0.059351	-0.069927
1.45	-0.021670	-0.028470	-0.036110	-0.044547	-0.053732	-0.063613
1.50	-0.019121	-0.025301	-0.032295	-0.040069	-0.048583	-0.057794
1.55	-0.016852	-0.022457	-0.028847	-0.035995	-0.043871	-0.052440
1.60	-0.014835	-0.019910	-0.025737	-0.032297	-0.039569	-0.047523
1.65	-0.013044	-0.017631	-0.022935	-0.028945	-0.035645	-0.043015
1.70	-0.011457	-0.015596	-0.020415	-0.025911	-0.032073	-0.038889
1.75	-0.010052	-0.013780	-0.018152	-0.023168	-0.028827	-0.035118
1.80	-0.008810	-0.012163	-0.016122	-0.020694	-0.025880	-0.031677
1.85	-0.007714	-0.010724	-0.014304	-0.018463	-0.023209	-0.028542
1.90	-0.006747	-0.009446	-0.012677	-0.016456	-0.020792	-0.025689
1.95	-0.005895	-0.008312	-0.011225	-0.014652	-0.018607	-0.023098
2.00	-0.005146	-0.007306	-0.009928	-0.013033	-0.016635	-0.020746
2.05	-0.004488	-0.006416	-0.008773	-0.011581	-0.014857	-0.018615
2.10	-0.003910	-0.005629	-0.007745	-0.010281	-0.013256	-0.016687
2.15	-0.003403	-0.004935	-0.006831	-0.009118	-0.011816	-0.014944
2.20	-0.002960	-0.004321	-0.006019	-0.008079	-0.010523	-0.013370
2.25	-0.002572	-0.003781	-0.005299	-0.007152	-0.009363	-0.011951
2.30	-0.002233	-0.003305	-0.004661	-0.006326	-0.008323	-0.010673
2.35	-0.001937	-0.002887	-0.004097	-0.005590	-0.007392	-0.009523
2.40	-0.001678	-0.002520	-0.003597	-0.004936	-0.006560	-0.008490
2.45	-0.001453	-0.002197	-0.003156	-0.004355	-0.005816	-0.007562
2.50	-0.001258	-0.001914	-0.002766	-0.003838	-0.005152	-0.006730

Table 2.1 (continued)

				t		
v	0.35	0.40	0.45	0.50	0.55	0.60
2.55	−0.001087	−0.001666	−0.002423	−0.003381	−0.004561	−0.005984
2.60	−0.000939	−0.001449	−0.002121	−0.002975	−0.004034	−0.005317
2.65	−0.000811	−0.001260	−0.001855	−0.002616	−0.003565	−0.004720
2.70	−0.000699	−0.001094	−0.001621	−0.002299	−0.003148	−0.004187
2.75	−0.000603	−0.000950	−0.001415	−0.002018	−0.002777	−0.003711
2.80	−0.000519	−0.000824	−0.001235	−0.001770	−0.002449	−0.003287
2.85	−0.000447	−0.000714	−0.001077	−0.001552	−0.002157	−0.002909
2.90	−0.000384	−0.000618	−0.000938	−0.001360	−0.001899	−0.002573
2.95	−0.000330	−0.000535	−0.000817	−0.001190	−0.001671	−0.002274
3.00	−0.000284	−0.000462	−0.000710	−0.001041	−0.001469	−0.002008
3.05	−0.000243	−0.000400	−0.000618	−0.000910	−0.001290	−0.001772
3.10	−0.000209	−0.000345	−0.000537	−0.000795	−0.001133	−0.001563
3.15	−0.000179	−0.000298	−0.000466	−0.000694	−0.000994	−0.001377
3.20	−0.000153	−0.000257	−0.000404	−0.000605	−0.000871	−0.001213
3.25	−0.000131	−0.000221	−0.000350	−0.000528	−0.000763	−0.001067
3.30	−0.000112	−0.000191	−0.000304	−0.000460	−0.000668	−0.000939
3.35	−0.000096	−0.000164	−0.000263	−0.000400	−0.000585	−0.000825
3.40	−0.000082	−0.000141	−0.000228	−0.000348	−0.000511	−0.000725
3.45	−0.000070	−0.000121	−0.000197	−0.000303	−0.000447	−0.000636
3.50	−0.000060	−0.000104	−0.000170	−0.000263	−0.000390	−0.000558
3.55	−0.000051	−0.000089	−0.000147	−0.000229	−0.000341	−0.000489
3.60	−0.000043	−0.000077	−0.000127	−0.000198	−0.000297	−0.000429
3.65	−0.000037	−0.000066	−0.000109	−0.000172	−0.000259	−0.000375
3.70	−0.000031	−0.000056	−0.000094	−0.000149	−0.000226	−0.000328
3.75	−0.000027	−0.000048	−0.000081	−0.000129	−0.000196	−0.000287
3.80	−0.000023	−0.000041	−0.000070	−0.000112	−0.000171	−0.000251
3.85	−0.000019	−0.000035	−0.000060	−0.000097	−0.000149	−0.000219
3.90	−0.000016	−0.000030	−0.000052	−0.000084	−0.000129	−0.000192
3.95	−0.000014	−0.000026	−0.000045	−0.000072	−0.000112	−0.000167
4.00	−0.000012	−0.000022	−0.000038	−0.000063	−0.000097	−0.000146
4.05	−0.000010	−0.000019	−0.000033	−0.000054	−0.000085	−0.000127
4.10	−0.000008	−0.000016	−0.000028	−0.000047	−0.000073	−0.000111
4.15	−0.000007	−0.000014	−0.000024	−0.000040	−0.000064	−0.000096
4.20	−0.000006	−0.000012	−0.000021	−0.000035	−0.000055	−0.000084
4.25	−0.000005	−0.000010	−0.000018	−0.000030	−0.000048	−0.000073
4.30	−0.000004	−0.000008	−0.000015	−0.000026	−0.000041	−0.000063
4.35	−0.000004	−0.000007	−0.000013	−0.000022	−0.000036	−0.000055
4.40	−0.000003	−0.000006	−0.000011	−0.000019	−0.000031	−0.000048
4.45	−0.000003	−0.000005	−0.000010	−0.000016	−0.000027	−0.000042
4.50	−0.000002	−0.000004	−0.000008	−0.000014	−0.000023	−0.000036
4.55	−0.000002	−0.000004	−0.000007	−0.000012	−0.000020	−0.000031
4.60	−0.000002	−0.000003	−0.000006	−0.000010	−0.000017	−0.000027
4.65	−0.000001	−0.000003	−0.000005	−0.000009	−0.000015	−0.000023

Table 2.1 (continued)

			t			
v	0.35	0.40	0.45	0.50	0.55	0.60
4.70	−0.000001	−0.000002	−0.000004	−0.000008	−0.000013	−0.000020
4.75	−0.000001	−0.000002	−0.000004	−0.000007	−0.000011	−0.000018
4.80	−0.000001	−0.000002	−0.000003	−0.000006	−0.000009	−0.000015
4.85	−0.000001	−0.000001	−0.000003	−0.000005	−0.000008	−0.000013
4.90	−0.000001	−0.000001	−0.000002	−0.000004	−0.000007	−0.000011
4.95		−0.000001	−0.000002	−0.000004	−0.000006	−0.000010
5.00		−0.000001	−0.000002	−0.000003	−0.000005	−0.000008

			t			
v	0.65	0.70	0.75	0.80	0.85	0.90
0.00	−0.445417	−0.442578	−0.436863	−0.428441	−0.417472	−0.404101
0.05	−0.438282	−0.438696	−0.436265	−0.431138	−0.423453	−0.413340
0.10	−0.428807	−0.432185	−0.432782	−0.430724	−0.426129	−0.419109
0.15	−0.417378	−0.423426	−0.426787	−0.427562	−0.425848	−0.421737
0.20	−0.404349	−0.412776	−0.418633	−0.421997	−0.422945	−0.421549
0.25	−0.390044	−0.400563	−0.408649	−0.414356	−0.417740	−0.418855
0.30	−0.374753	−0.387086	−0.397139	−0.404945	−0.410538	−0.413954
0.35	−0.358738	−0.372618	−0.384384	−0.394048	−0.401623	−0.407128
0.40	−0.342232	−0.357406	−0.370640	−0.381926	−0.391260	−0.398644
0.45	−0.325440	−0.341669	−0.356138	−0.368820	−0.379696	−0.388750
0.50	−0.308542	−0.325604	−0.341087	−0.354950	−0.367157	−0.377679
0.55	−0.291694	−0.309382	−0.325674	−0.340513	−0.353850	−0.365645
0.60	−0.275028	−0.293154	−0.310063	−0.325686	−0.339964	−0.352844
0.65	−0.258657	−0.277049	−0.294399	−0.310629	−0.325669	−0.339455
0.70	−0.242675	−0.261179	−0.278810	−0.295482	−0.311115	−0.325641
0.75	−0.227159	−0.245637	−0.263403	−0.280366	−0.296439	−0.311546
0.80	−0.212170	−0.230500	−0.248272	−0.265388	−0.281759	−0.297303
0.85	−0.197756	−0.215831	−0.233493	−0.250640	−0.267179	−0.283024
0.90	−0.183952	−0.201681	−0.219130	−0.236198	−0.252788	−0.268813
0.95	−0.170783	−0.188087	−0.205236	−0.222127	−0.238664	−0.254756
1.00	−0.158264	−0.175079	−0.191851	−0.208480	−0.224870	−0.240929
1.05	−0.146401	−0.162674	−0.179006	−0.195300	−0.211460	−0.227398
1.10	−0.135196	−0.150883	−0.166723	−0.182618	−0.198478	−0.214215
1.15	−0.124642	−0.139712	−0.155014	−0.170459	−0.185958	−0.201424
1.20	−0.114729	−0.129156	−0.143889	−0.158841	−0.173926	−0.189063
1.25	−0.105441	−0.119211	−0.133348	−0.147772	−0.162401	−0.177158
1.30	−0.096759	−0.109863	−0.123388	−0.137257	−0.151396	−0.165729
1.35	−0.088664	−0.101100	−0.114000	−0.127296	−0.140917	−0.154792
1.40	−0.081132	−0.092902	−0.105174	−0.117883	−0.130966	−0.144355
1.45	−0.074138	−0.085250	−0.096894	−0.109010	−0.121540	−0.134422
1.50	−0.067657	−0.078124	−0.089144	−0.100664	−0.112633	−0.124993
1.55	−0.061663	−0.071499	−0.081904	−0.092832	−0.104235	−0.116063

Table 2.1 (continued)

			t			
v	**0.65**	**0.70**	**0.75**	**0.80**	**0.85**	**0.90**
1.60	−0.056129	−0.065353	−0.075155	−0.085497	−0.096335	−0.107625
1.65	−0.051030	−0.059660	−0.068876	−0.078641	−0.088919	−0.099670
1.70	−0.046338	−0.054398	−0.063044	−0.072245	−0.081970	−0.092184
1.75	−0.042028	−0.049542	−0.057636	−0.066289	−0.075472	−0.085155
1.80	−0.038077	−0.045067	−0.052632	−0.060752	−0.069405	−0.078566
1.85	−0.034458	−0.040950	−0.048007	−0.055613	−0.063752	−0.072402
1.90	−0.031149	−0.037168	−0.043740	−0.050852	−0.058493	−0.066644
1.95	−0.028129	−0.033700	−0.039809	−0.046448	−0.053608	−0.061276
2.00	−0.025374	−0.030523	−0.036193	−0.042380	−0.049078	−0.056278
2.05	−0.022867	−0.027617	−0.032871	−0.038627	−0.044883	−0.051632
2.10	−0.020586	−0.024963	−0.029824	−0.035171	−0.041004	−0.047321
2.15	−0.018515	−0.022542	−0.027032	−0.031991	−0.037422	−0.043324
2.20	−0.016637	−0.020336	−0.024478	−0.029071	−0.034120	−0.039626
2.25	−0.014935	−0.018329	−0.022144	−0.026392	−0.031078	−0.036208
2.30	−0.013395	−0.016504	−0.020014	−0.023937	−0.028281	−0.033053
2.35	−0.012003	−0.014847	−0.018073	−0.021690	−0.025712	−0.030144
2.40	−0.010746	−0.013345	−0.016305	−0.019637	−0.023354	−0.027466
2.45	−0.009612	−0.011985	−0.014697	−0.017762	−0.021194	−0.025004
2.50	−0.008591	−0.010754	−0.013236	−0.016052	−0.019217	−0.022741
2.55	−0.007671	−0.009641	−0.011910	−0.014494	−0.017409	−0.020666
2.60	−0.006845	−0.008636	−0.010708	−0.013077	−0.015757	−0.018764
2.65	−0.006102	−0.007730	−0.009619	−0.011788	−0.014251	−0.017023
2.70	−0.005436	−0.006913	−0.008634	−0.010618	−0.012878	−0.015430
2.75	−0.004839	−0.006177	−0.007744	−0.009556	−0.011628	−0.013975
2.80	−0.004304	−0.005516	−0.006940	−0.008593	−0.010491	−0.012647
2.85	−0.003825	−0.004922	−0.006215	−0.007722	−0.009458	−0.011437
2.90	−0.003397	−0.004388	−0.005562	−0.006934	−0.008520	−0.010334
2.95	−0.003015	−0.003909	−0.004973	−0.006221	−0.007669	−0.009331
3.00	−0.002674	−0.003480	−0.004444	−0.005578	−0.006898	−0.008418
3.05	−0.002369	−0.003096	−0.003968	−0.004997	−0.006200	−0.007590
3.10	−0.002098	−0.002753	−0.003540	−0.004474	−0.005569	−0.006838
3.15	−0.001857	−0.002445	−0.003156	−0.004003	−0.004998	−0.006156
3.20	−0.001642	−0.002171	−0.002812	−0.003579	−0.004483	−0.005538
3.25	−0.001451	−0.001926	−0.002504	−0.003197	−0.004018	−0.004978
3.30	−0.001282	−0.001708	−0.002228	−0.002854	−0.003599	−0.004473
3.35	−0.001131	−0.001513	−0.001981	−0.002547	−0.003221	−0.004016
3.40	−0.000998	−0.001340	−0.001761	−0.002271	−0.002881	−0.003603
3.45	−0.000879	−0.001185	−0.001563	−0.002023	−0.002576	−0.003230
3.50	−0.000775	−0.001048	−0.001388	−0.001802	−0.002301	−0.002894
3.55	−0.000682	−0.000926	−0.001231	−0.001604	−0.002054	−0.002592
3.60	−0.000600	−0.000818	−0.001091	−0.001426	−0.001833	−0.002319
3.65	−0.000528	−0.000722	−0.000966	−0.001268	−0.001634	−0.002074
3.70	−0.000464	−0.000637	−0.000855	−0.001126	−0.001456	−0.001854

Table 2.1 (continued)

			t			
v	**0.65**	**0.70**	**0.75**	**0.80**	**0.85**	**0.90**
3.75	−0.000407	−0.000562	−0.000757	−0.001000	−0.001297	−0.001656
3.80	−0.000357	−0.000495	−0.000669	−0.000887	−0.001154	−0.001478
3.85	−0.000313	−0.000436	−0.000592	−0.000786	−0.001027	−0.001319
3.90	−0.000275	−0.000384	−0.000522	−0.000697	−0.000913	−0.001176
3.95	−0.000241	−0.000337	−0.000461	−0.000617	−0.000811	−0.001048
4.00	−0.000211	−0.000297	−0.000407	−0.000546	−0.000720	−0.000934
4.05	−0.000185	−0.000261	−0.000359	−0.000483	−0.000639	−0.000831
4.10	−0.000161	−0.000229	−0.000316	−0.000427	−0.000567	−0.000739
4.15	−0.000141	−0.000201	−0.000278	−0.000378	−0.000503	−0.000657
4.20	−0.000123	−0.000176	−0.000245	−0.000334	−0.000445	−0.000584
4.25	−0.000108	−0.000154	−0.000216	−0.000295	−0.000394	−0.000519
4.30	−0.000094	−0.000135	−0.000190	−0.000260	−0.000349	−0.000461
4.35	−0.000082	−0.000119	−0.000167	−0.000229	−0.000309	−0.000409
4.40	−0.000072	−0.000104	−0.000146	−0.000202	−0.000273	−0.000362
4.45	−0.000062	−0.000091	−0.000129	−0.000178	−0.000241	−0.000321
4.50	−0.000054	−0.000079	−0.000113	−0.000157	−0.000213	−0.000285
4.55	−0.000047	−0.000069	−0.000099	−0.000138	−0.000188	−0.000252
4.60	−0.000041	−0.000061	−0.000087	−0.000121	−0.000166	−0.000223
4.65	−0.000036	−0.000053	−0.000076	−0.000107	−0.000146	−0.000197
4.70	−0.000031	−0.000046	−0.000067	−0.000094	−0.000129	−0.000174
4.75	−0.000027	−0.000040	−0.000058	−0.000082	−0.000114	−0.000154
4.80	−0.000024	−0.000035	−0.000051	−0.000072	−0.000100	−0.000136
4.85	−0.000020	−0.000031	−0.000045	−0.000063	−0.000088	−0.000120
4.90	−0.000018	−0.000027	−0.000039	−0.000056	−0.000078	−0.000106
4.95	−0.000015	−0.000023	−0.000034	−0.000049	−0.000068	−0.000094
5.00	−0.000013	−0.000020	−0.000030	−0.000043	−0.000060	−0.000083

			t			
v	**0.95**	**1.0**	**2.0**	**3.0**	**4.0**	**5.0**
0.00	−0.388466	−0.370697	0.261262	0.902180	1.04066	0.641559
0.05	−0.400922	−0.386314	0.206022	0.864458	1.05533	0.699135
0.10	−0.409769	−0.398213	0.152672	0.824064	1.06518	0.753704
0.15	−0.415320	−0.406683	0.101416	0.781349	1.07034	0.805038
0.20	−0.417882	−0.412016	0.052428	0.736658	1.07092	0.852933
0.25	−0.417756	−0.414498	0.005851	0.690327	1.06707	0.897211
0.30	−0.415231	−0.414408	−0.038203	0.642684	1.05896	0.937724
0.35	−0.410585	−0.412018	−0.079647	0.594044	1.04676	0.974348
0.40	−0.404083	−0.407588	−0.118421	0.544709	1.03068	1.00698
0.45	−0.395975	−0.401366	−0.154490	0.494968	1.01094	1.03556
0.50	−0.386495	−0.393587	−0.187840	0.445093	0.987744	1.06003
0.55	−0.375862	−0.384474	−0.218475	0.395341	0.961334	1.08037
0.60	−0.364279	−0.374232	−0.246420	0.345952	0.931945	1.09657

Table 2.1 (continued)

v	0.95	1.0	2.0	3.0	4.0	5.0
				t		
0.65	−0.351933	−0.363054	−0.271715	0.297148	0.899820	1.10866
0.70	−0.338994	−0.351118	−0.294414	0.249135	0.865207	1.11669
0.75	−0.325617	−0.338586	−0.314586	0.202101	0.828352	1.12070
0.80	−0.311942	−0.325608	−0.332307	0.156215	0.789503	1.12078
0.85	−0.298095	−0.312317	−0.347666	0.111632	0.748906	1.11704
0.90	−0.284188	−0.298835	−0.360758	0.068486	0.706805	1.10957
0.95	−0.270318	−0.285271	−0.371684	0.026897	0.663438	1.09850
1.00	−0.256573	−0.271719	−0.380552	−0.013033	0.619038	1.08398
1.05	−0.243027	−0.258265	−0.387472	−0.051217	0.573832	1.06616
1.10	−0.229743	−0.244982	−0.392557	−0.087582	0.528041	1.04519
1.15	−0.216776	−0.231933	−0.395921	−0.122072	0.481874	1.02125
1.20	−0.204171	−0.219172	−0.397680	−0.154644	0.435535	0.994516
1.25	−0.191964	−0.206744	−0.397949	−0.185265	0.389216	0.965166
1.30	−0.180184	−0.194687	−0.396841	−0.213918	0.343100	0.933390
1.35	−0.168853	−0.183030	−0.394469	−0.240597	0.297359	0.899382
1.40	−0.157987	−0.171796	−0.390942	−0.265304	0.252155	0.863334
1.45	−0.147598	−0.161004	−0.386368	−0.288053	0.207638	0.825442
1.50	−0.137690	−0.150665	−0.380848	−0.308867	0.163948	0.785903
1.55	−0.128264	−0.140787	−0.374483	−0.327776	0.121212	0.744911
1.60	−0.119320	−0.131372	−0.367370	−0.344817	0.079547	0.702660
1.65	−0.110851	−0.122421	−0.359598	−0.360036	0.039060	0.659340
1.70	−0.102850	−0.113929	−0.351255	−0.373483	−0.000156	0.615139
1.75	−0.095306	−0.105890	−0.342423	−0.385213	−0.038017	0.570240
1.80	−0.088207	−0.098296	−0.333180	−0.395285	−0.074449	0.524820
1.85	−0.081539	−0.091137	−0.323598	−0.403763	−0.109389	0.479053
1.90	−0.075287	−0.084398	−0.313746	−0.410714	−0.142785	0.433106
1.95	−0.069435	−0.078068	−0.303687	−0.416207	−0.174592	0.387138
2.00	−0.063967	−0.072132	−0.293479	−0.420312	−0.204778	0.341302
2.05	−0.058866	−0.066574	−0.283178	−0.423102	−0.233317	0.295745
2.10	−0.054115	−0.061378	−0.272832	−0.424649	−0.260191	0.250605
2.15	−0.049695	−0.056529	−0.262487	−0.425028	−0.285392	0.206012
2.20	−0.045590	−0.052009	−0.252184	−0.424311	−0.308919	0.162088
2.25	−0.041783	−0.047803	−0.241961	−0.422572	−0.330776	0.118948
2.30	−0.038257	−0.043894	−0.231851	−0.419882	−0.350976	0.076698
2.35	−0.034995	−0.040266	−0.221883	−0.416314	−0.369536	0.035434
2.40	−0.031981	−0.036903	−0.212085	−0.411936	−0.386481	−0.004754
2.45	−0.029200	−0.033790	−0.202478	−0.406817	−0.401838	−0.043785
2.50	−0.026637	−0.030911	−0.193084	−0.401024	−0.415641	−0.081587
2.55	−0.024278	−0.028253	−0.183919	−0.394620	−0.427928	−0.118093
2.60	−0.022108	−0.025801	−0.174997	−0.387667	−0.438738	−0.153249
2.65	−0.020115	−0.023541	−0.166330	−0.380227	−0.448117	−0.187004
2.70	−0.018287	−0.021462	−0.157928	−0.372355	−0.456111	−0.219318
2.75	−0.016612	−0.019550	−0.149799	−0.364106	−0.462771	−0.250154
2.80	−0.015077	−0.017794	−0.141947	−0.355534	−0.468147	−0.279488

Table 2.1 (continued)

v	0.95	1.0	2.0	3.0	4.0	5.0
				t		
2.85	-0.013674	-0.016182	-0.134376	-0.346687	-0.472293	-0.307296
2.90	-0.012392	-0.014706	-0.127089	-0.337613	-0.475263	-0.333566
2.95	-0.011221	-0.013353	-0.120086	-0.328356	-0.477113	-0.358289
3.00	-0.010153	-0.012116	-0.113366	-0.318957	-0.477900	-0.381464
3.05	-0.009180	-0.010985	-0.106927	-0.309456	-0.477680	-0.403092
3.10	-0.008294	-0.009953	-0.100766	-0.299889	-0.476509	-0.423182
3.15	-0.007489	-0.009011	-0.094879	-0.290290	-0.474444	-0.441748
3.20	-0.006757	-0.008152	-0.089261	-0.280691	-0.471542	-0.458808
3.25	-0.006092	-0.007370	-0.083908	-0.271119	-0.467858	-0.474382
3.30	-0.005489	-0.006659	-0.078811	-0.261603	-0.463447	-0.488497
3.35	-0.004942	-0.006012	-0.073966	-0.252166	-0.458363	-0.501182
3.40	-0.004446	-0.005424	-0.069365	-0.242829	-0.452658	-0.512470
3.45	-0.003998	-0.004890	-0.065001	-0.233615	-0.446385	-0.522395
3.50	-0.003593	-0.004406	-0.060865	-0.224539	-0.439594	-0.530995
3.55	-0.003226	-0.003968	-0.056951	-0.215618	-0.432332	-0.538311
3.60	-0.002896	-0.003571	-0.053250	-0.206867	-0.424648	-0.544384
3.65	-0.002597	-0.003211	-0.049754	-0.198297	-0.416586	-0.549259
3.70	-0.002328	-0.002886	-0.046455	-0.189920	-0.408191	-0.552980
3.75	-0.002085	-0.002592	-0.043345	-0.181744	-0.399504	-0.555593
3.80	-0.001867	-0.002327	-0.040415	-0.173778	-0.390566	-0.557147
3.85	-0.001670	-0.002088	-0.037658	-0.166027	-0.381414	-0.557689
3.90	-0.001494	-0.001872	-0.035067	-0.158497	-0.372085	-0.557268
3.95	-0.001335	-0.001678	-0.032632	-0.151192	-0.362613	-0.555932
4.00	-0.001192	-0.001502	-0.030347	-0.144114	-0.353032	-0.553729
4.05	-0.001064	-0.001345	-0.028204	-0.137265	-0.343372	-0.550710
4.10	-0.000950	-0.001203	-0.026196	-0.130646	-0.333662	-0.546922
4.15	-0.000847	-0.001076	-0.024316	-0.124258	-0.323929	-0.542412
4.20	-0.000755	-0.000961	-0.022558	-0.118098	-0.314198	-0.537230
4.25	-0.000672	-0.000858	-0.020914	-0.112166	-0.304493	-0.531420
4.30	-0.000598	-0.000766	-0.019378	-0.106459	-0.294836	-0.525029
4.35	-0.000532	-0.000684	-0.017945	-0.100974	-0.285247	-0.518102
4.40	-0.000473	-0.000610	-0.016608	-0.095709	-0.275745	-0.510682
4.45	-0.000421	-0.000543	-0.015363	-0.090660	-0.266347	-0.502813
4.50	-0.000374	-0.000484	-0.014202	-0.085822	-0.257068	-0.494536
4.55	-0.000332	-0.000431	-0.013122	-0.081191	-0.247922	-0.485890
4.60	-0.000295	-0.000383	-0.012118	-0.076763	-0.238922	-0.476915
4.65	-0.000261	-0.000341	-0.011184	-0.072531	-0.230080	-0.467648
4.70	-0.000232	-0.000303	-0.010317	-0.068492	-0.221404	-0.458126
4.75	-0.000205	-0.000269	-0.009512	-0.064639	-0.212906	-0.448383
4.80	-0.000182	-0.000239	-0.008765	-0.060967	-0.204591	-0.438451
4.85	-0.000161	-0.000212	-0.008073	-0.057471	-0.196466	-0.428364
4.90	-0.000143	-0.000188	-0.007431	-0.054144	-0.188538	-0.418150
4.95	-0.000126	-0.000167	-0.006837	-0.050981	-0.180810	-0.407839
5.00	-0.000111	-0.000148	-0.006288	-0.047976	-0.173286	-0.397458

Table 2.2: Second derivatives with respect to the order of the Struve functions $\partial^2 H_\nu(t)/\partial\nu^2$.

				t		
ν	0.05	0.10	0.15	0.20	0.25	0.30
0.00	0.411841	0.524888	0.567151	0.572777	0.555796	0.523616
0.05	0.351555	0.467985	0.520405	0.537971	0.533174	0.512808
0.10	0.298881	0.415229	0.474780	0.501861	0.507338	0.497301
0.15	0.253144	0.366783	0.430909	0.465341	0.479321	0.478187
0.20	0.213653	0.322658	0.389240	0.429125	0.449989	0.456406
0.25	0.179730	0.282761	0.350071	0.393765	0.420053	0.432764
0.30	0.150725	0.246916	0.313578	0.359679	0.390094	0.407941
0.35	0.126032	0.214899	0.279841	0.327171	0.360573	0.382508
0.40	0.105093	0.186449	0.248865	0.296451	0.331852	0.356936
0.45	0.087403	0.161289	0.220597	0.267650	0.304206	0.331606
0.50	0.072509	0.139136	0.194941	0.240835	0.277835	0.306824
0.55	0.060010	0.119710	0.171772	0.216025	0.252878	0.282828
0.60	0.049553	0.102737	0.150943	0.193196	0.229424	0.259797
0.65	0.040829	0.087961	0.132297	0.172297	0.207516	0.237862
0.70	0.033571	0.075138	0.115668	0.153253	0.187167	0.217112
0.75	0.027548	0.064045	0.100893	0.135970	0.168359	0.197601
0.80	0.022562	0.054476	0.087809	0.120349	0.151055	0.179354
0.85	0.018444	0.046243	0.076257	0.106280	0.135201	0.162373
0.90	0.015051	0.039179	0.066090	0.093651	0.120732	0.146639
0.95	0.012261	0.033133	0.057165	0.082351	0.107574	0.132122
1.00	0.009972	0.027970	0.049352	0.072270	0.095649	0.118778
1.05	0.008096	0.023571	0.042530	0.063301	0.084874	0.106555
1.10	0.006564	0.019830	0.036586	0.055344	0.075167	0.095396
1.15	0.005313	0.016656	0.031419	0.048300	0.066446	0.085240
1.20	0.004294	0.013969	0.026938	0.042081	0.058632	0.076022
1.25	0.003466	0.011697	0.023060	0.036602	0.051648	0.067679
1.30	0.002794	0.009781	0.019709	0.031785	0.045419	0.060147
1.35	0.002249	0.008166	0.016821	0.027559	0.039878	0.053364
1.40	0.001808	0.006809	0.014335	0.023860	0.034958	0.047269
1.45	0.001451	0.005670	0.012199	0.020627	0.030599	0.041805
1.50	0.001164	0.004715	0.010367	0.017807	0.026744	0.036917
1.55	0.000932	0.003916	0.008799	0.015351	0.023342	0.032552
1.60	0.000745	0.003248	0.007458	0.013216	0.020345	0.028663
1.65	0.000596	0.002691	0.006314	0.011364	0.017708	0.025203
1.70	0.000475	0.002227	0.005339	0.009759	0.015394	0.022132
1.75	0.000379	0.001841	0.004509	0.008370	0.013364	0.019409
1.80	0.000302	0.001520	0.003803	0.007170	0.011588	0.017000
1.85	0.000240	0.001253	0.003204	0.006135	0.010036	0.014871
1.90	0.000191	0.001033	0.002697	0.005243	0.008682	0.012994
1.95	0.000151	0.000850	0.002267	0.004476	0.007502	0.011340
2.00	0.000120	0.000699	0.001904	0.003817	0.006475	0.009885
2.05	0.000095	0.000574	0.001598	0.003252	0.005582	0.008607
2.10	0.000075	0.000471	0.001339	0.002767	0.004808	0.007486
2.15	0.000060	0.000386	0.001121	0.002353	0.004136	0.006504

Table 2.2 (continued)

				t		
v	**0.05**	**0.10**	**0.15**	**0.20**	**0.25**	**0.30**
2.20	0.000047	0.000316	0.000938	0.001998	0.003555	0.005646
2.25	0.000037	0.000259	0.000784	0.001695	0.003052	0.004895
2.30	0.000029	0.000211	0.000654	0.001437	0.002618	0.004240
2.35	0.000023	0.000173	0.000546	0.001217	0.002243	0.003669
2.40	0.000018	0.000141	0.000455	0.001029	0.001921	0.003172
2.45	0.000014	0.000115	0.000379	0.000870	0.001643	0.002740
2.50	0.000011	0.000093	0.000315	0.000735	0.001404	0.002364
2.55	0.000009	0.000076	0.000262	0.000620	0.001198	0.002038
2.60	0.000007	0.000062	0.000217	0.000523	0.001022	0.001755
2.65	0.000005	0.000050	0.000180	0.000440	0.000871	0.001511
2.70	0.000004	0.000041	0.000150	0.000371	0.000742	0.001299
2.75	0.000003	0.000033	0.000124	0.000312	0.000631	0.001116
2.80	0.000003	0.000027	0.000103	0.000262	0.000537	0.000958
2.85	0.000002	0.000022	0.000085	0.000220	0.000456	0.000822
2.90	0.000002	0.000018	0.000070	0.000184	0.000387	0.000704
2.95	0.000001	0.000014	0.000058	0.000155	0.000328	0.000603
3.00	0.000001	0.000012	0.000048	0.000130	0.000278	0.000516
3.05	0.000001	0.000009	0.000039	0.000108	0.000236	0.000441
3.10	0.000001	0.000007	0.000032	0.000091	0.000199	0.000377
3.15		0.000006	0.000027	0.000076	0.000169	0.000322
3.20		0.000005	0.000022	0.000063	0.000143	0.000275
3.25		0.000004	0.000018	0.000053	0.000120	0.000234
3.30		0.000003	0.000015	0.000044	0.000102	0.000199
3.35		0.000003	0.000012	0.000037	0.000086	0.000170
3.40		0.000002	0.000010	0.000031	0.000072	0.000144
3.45		0.000002	0.000008	0.000025	0.000061	0.000123
3.50		0.000001	0.000007	0.000021	0.000051	0.000104
3.55		0.000001	0.000006	0.000018	0.000043	0.000089
3.60		0.000001	0.000005	0.000015	0.000036	0.000075
3.65		0.000001	0.000004	0.000012	0.000030	0.000064
3.70		0.000001	0.000003	0.000010	0.000025	0.000054
3.75			0.000002	0.000008	0.000021	0.000046
3.80			0.000002	0.000007	0.000018	0.000039
3.85			0.000002	0.000006	0.000015	0.000033
3.90			0.000001	0.000005	0.000013	0.000028
3.95			0.000001	0.000004	0.000011	0.000023
4.00			0.000001	0.000003	0.000009	0.000020
4.05			0.000001	0.000003	0.000070	0.000017
4.10			0.000001	0.000002	0.000006	0.000014
4.15				0.000002	0.000005	0.000012
4.20				0.000002	0.000004	0.000010
4.25				0.000001	0.000003	0.000008
4.30				0.000001	0.000003	0.000007

Table 2.2 (continued)

			t			
v	0.05	0.10	0.15	0.20	0.25	0.30
4.35				0.000001	0.000002	0.000006
4.40				0.000001	0.000002	0.000004
4.45				0.000001	0.000002	0.000004
4.50					0.000001	0.000004
4.55					0.000001	0.000300
4.60					0.000001	0.000002
4.65					0.000001	0.000002
4.70					0.000001	0.000002
4.75					0.000001	0.000001
4.80						0.000001
4.85						0.000001
4.90						0.000001
4.95						0.000001
5.00						0.000001

			t			
v	0.35	0.40	0.45	0.50	0.55	0.60
0.00	0.480710	0.430027	0.373635	0.313052	0.249436	0.183699
0.05	0.481077	0.440812	0.394034	0.342251	0.286633	0.228110
0.10	0.475611	0.444915	0.407141	0.363748	0.315882	0.264469
0.15	0.465402	0.443392	0.413950	0.378456	0.338010	0.293509
0.20	0.451418	0.437197	0.415378	0.387240	0.353812	0.315947
0.25	0.434507	0.427188	0.412267	0.390902	0.364042	0.332478
0.30	0.415409	0.414130	0.405381	0.390189	0.369409	0.343765
0.35	0.394762	0.398698	0.395407	0.385782	0.370576	0.350440
0.40	0.373110	0.381483	0.382961	0.378301	0.368156	0.353093
0.45	0.350914	0.362998	0.368587	0.368308	0.362709	0.352278
0.50	0.328558	0.343683	0.352766	0.356305	0.354749	0.348503
0.55	0.306357	0.323914	0.335914	0.342737	0.344735	0.342238
0.60	0.284566	0.304006	0.318393	0.327998	0.333085	0.333907
0.65	0.263389	0.284220	0.300510	0.312432	0.320166	0.323896
0.70	0.242982	0.264770	0.282526	0.296336	0.306304	0.312551
0.75	0.223462	0.245828	0.264660	0.279966	0.291786	0.300181
0.80	0.204912	0.227529	0.247091	0.263540	0.276859	0.287057
0.85	0.187385	0.209974	0.229962	0.247239	0.261737	0.273420
0.90	0.170913	0.193238	0.213390	0.231214	0.246601	0.259477
0.95	0.155505	0.177371	0.197463	0.215588	0.231605	0.245408
1.00	0.141154	0.162404	0.182245	0.200458	0.216875	0.231364
1.05	0.127842	0.148350	0.167782	0.185900	0.202515	0.217477
1.10	0.115538	0.135209	0.154102	0.171969	0.188608	0.203851
1.15	0.104206	0.122968	0.141220	0.158707	0.175219	0.190576
1.20	0.093801	0.111607	0.129137	0.146138	0.162395	0.177722
1.25	0.084277	0.101097	0.117846	0.134276	0.150171	0.165343

Table 2.2 (continued)

			t			
v	0.35	0.40	0.45	0.50	0.55	0.60
1.30	0.075583	0.091404	0.107332	0.123124	0.138569	0.153480
1.35	0.067668	0.082491	0.097572	0.112677	0.127602	0.142164
1.40	0.060480	0.074318	0.088538	0.102922	0.117273	0.131412
1.45	0.053969	0.066842	0.080202	0.093842	0.107578	0.121236
1.50	0.048083	0.060021	0.072528	0.085414	0.098505	0.111638
1.55	0.042774	0.053812	0.065482	0.077613	0.090042	0.102615
1.60	0.037995	0.048172	0.059029	0.070411	0.082168	0.094158
1.65	0.033703	0.043060	0.053131	0.063777	0.074861	0.086254
1.70	0.029854	0.038435	0.047752	0.057681	0.068098	0.078886
1.75	0.026409	0.034260	0.042857	0.052091	0.061853	0.072035
1.80	0.023331	0.030498	0.038411	0.046976	0.056099	0.065681
1.85	0.020586	0.027113	0.034380	0.042306	0.050808	0.059800
1.90	0.018141	0.024074	0.030732	0.038048	0.045953	0.054369
1.95	0.015968	0.021349	0.027436	0.034175	0.041506	0.049364
2.00	0.014038	0.018909	0.024464	0.030658	0.037442	0.044760
2.05	0.012327	0.016729	0.021787	0.027469	0.033733	0.040533
2.10	0.010813	0.014783	0.019381	0.024582	0.030354	0.036660
2.15	0.009474	0.013049	0.017221	0.021973	0.027281	0.033116
2.20	0.008293	0.011506	0.015284	0.019618	0.024491	0.029879
2.25	0.007251	0.010134	0.013551	0.017497	0.021961	0.026928
2.30	0.006333	0.008917	0.012001	0.015587	0.019671	0.024240
2.35	0.005526	0.007838	0.010618	0.013872	0.017601	0.021797
2.40	0.004818	0.006883	0.009384	0.012333	0.015731	0.019579
2.45	0.004196	0.006038	0.008286	0.010953	0.014046	0.017568
2.50	0.003651	0.005292	0.007309	0.009718	0.012529	0.015748
2.55	0.003173	0.004633	0.006441	0.008614	0.011165	0.014102
2.60	0.002756	0.004053	0.005671	0.007628	0.009939	0.012615
2.65	0.002392	0.003542	0.004988	0.006748	0.008840	0.011275
2.70	0.002073	0.003093	0.004384	0.005965	0.007855	0.010067
2.75	0.001796	0.002699	0.003849	0.005268	0.006973	0.008981
2.80	0.001554	0.002353	0.003377	0.004648	0.006185	0.008004
2.85	0.001344	0.002049	0.002960	0.004098	0.005481	0.007128
2.90	0.001162	0.001783	0.002592	0.003609	0.004854	0.006341
2.95	0.001003	0.001551	0.002269	0.003177	0.004294	0.005637
3.00	0.000865	0.001347	0.001984	0.002794	0.003796	0.005007
3.05	0.000746	0.001170	0.001733	0.002455	0.003352	0.004443
3.10	0.000642	0.001015	0.001513	0.002155	0.002959	0.003940
3.15	0.000553	0.000880	0.001320	0.001891	0.002609	0.003491
3.20	0.000476	0.000762	0.001151	0.001657	0.002299	0.003091
3.25	0.000409	0.000660	0.001002	0.001452	0.002024	0.002734
3.30	0.000351	0.000571	0.000872	0.001271	0.001781	0.002417
3.35	0.000301	0.000493	0.000759	0.001112	0.001566	0.002135
3.40	0.000258	0.000426	0.000659	0.000972	0.001376	0.001885

Table 2.2 (continued)

			t			
v	0.35	0.40	0.45	0.50	0.55	0.60
3.45	0.000222	0.000368	0.000573	0.000849	0.001208	0.001662
3.50	0.000190	0.000317	0.000497	0.000741	0.001060	0.001465
3.55	0.000162	0.000273	0.000431	0.000646	0.000929	0.001291
3.60	0.000139	0.000236	0.000374	0.000563	0.000814	0.001136
3.65	0.000119	0.000203	0.000324	0.000491	0.000713	0.000999
3.70	0.000101	0.000174	0.000280	0.000427	0.000623	0.000878
3.75	0.000087	0.000150	0.000242	0.000372	0.000545	0.000772
3.80	0.000074	0.000129	0.000210	0.000323	0.000476	0.000677
3.85	0.000063	0.000111	0.000181	0.000281	0.000416	0.000594
3.90	0.000054	0.000095	0.000156	0.000244	0.000363	0.000521
3.95	0.000046	0.000081	0.000135	0.000211	0.000317	0.000456
4.00	0.000039	0.000070	0.000116	0.000183	0.000276	0.000400
4.05	0.000033	0.000060	0.000100	0.000159	0.000240	0.000350
4.10	0.000028	0.000051	0.000086	0.000138	0.000209	0.000306
4.15	0.000024	0.000044	0.000074	0.000119	0.000182	0.000267
4.20	0.000020	0.000037	0.000064	0.000103	0.000158	0.000234
4.25	0.000017	0.000032	0.000055	0.000089	0.000138	0.000204
4.30	0.000015	0.000027	0.000047	0.000077	0.000119	0.000178
4.35	0.000012	0.000023	0.000041	0.000067	0.000104	0.000155
4.40	0.000011	0.000020	0.000035	0.000057	0.000090	0.000135
4.45	0.000009	0.000017	0.000030	0.000050	0.000078	0.000118
4.50	0.000007	0.000014	0.000026	0.000043	0.000068	0.000103
4.55	0.000006	0.000012	0.000022	0.000037	0.000059	0.000089
4.60	0.000006	0.000011	0.000019	0.000032	0.000051	0.000078
4.65	0.000005	0.000009	0.000016	0.000027	0.000044	0.000067
4.70	0.000004	0.000008	0.000014	0.000024	0.000038	0.000059
4.75	0.000003	0.000006	0.000012	0.000020	0.000033	0.000051
4.80	0.000002	0.000050	0.000010	0.000017	0.000028	0.000044
4.85	0.000002	0.000005	0.000009	0.000015	0.000024	0.000038
4.90	0.000002	0.000004	0.000007	0.000013	0.000021	0.000033
4.95	0.000002	0.000003	0.000006	0.000011	0.000018	0.000029
5.00	0.000001	0.000020	0.000005	0.000009	0.000016	0.000025

			t			
v	0.65	0.70	0.75	0.80	0.85	0.90
0.00	0.116573	0.048663	−0.019527	−0.087566	−0.155082	−0.221753
0.05	0.167440	0.105256	0.042093	−0.021590	−0.085391	−0.148961
0.10	0.210272	0.153932	0.095999	0.036950	−0.022798	−0.082874
0.15	0.245702	0.195224	0.142627	0.088390	0.032940	−0.023344
0.20	0.274365	0.229682	0.182437	0.133105	0.082111	0.029835
0.25	0.296888	0.257860	0.215910	0.171500	0.125044	0.076916
0.30	0.313883	0.280310	0.243536	0.203998	0.162094	0.118188
0.35	0.325938	0.297576	0.265804	0.231033	0.193638	0.153965

Table 2.2 (continued)

			t			
v	0.65	0.70	0.75	0.80	0.85	0.90
0.40	0.333618	0.310182	0.283197	0.253040	0.220061	0.184583
0.45	0.337452	0.318633	0.296187	0.270455	0.241756	0.210389
0.50	0.337942	0.323410	0.305229	0.283702	0.259113	0.231736
0.55	0.335551	0.324967	0.310760	0.293194	0.272519	0.248978
0.60	0.330709	0.323728	0.313193	0.299328	0.282349	0.262466
0.65	0.323808	0.320087	0.312918	0.302483	0.288965	0.272544
0.70	0.315207	0.314409	0.310297	0.303017	0.292716	0.279545
0.75	0.305231	0.307026	0.305668	0.301265	0.293932	0.283789
0.80	0.294169	0.298243	0.299343	0.297542	0.292925	0.285582
0.85	0.282282	0.288334	0.291605	0.292138	0.289986	0.285214
0.90	0.269799	0.277546	0.282715	0.285320	0.285388	0.282957
0.95	0.256922	0.266098	0.272906	0.277333	0.279381	0.279066
1.00	0.243826	0.254185	0.262388	0.268399	0.272197	0.273778
1.05	0.230663	0.241978	0.251348	0.258718	0.264047	0.267310
1.10	0.217561	0.229625	0.239953	0.248470	0.255121	0.259864
1.15	0.204631	0.217256	0.228347	0.237816	0.245593	0.251621
1.20	0.191962	0.204979	0.216657	0.226897	0.235618	0.242748
1.25	0.179629	0.192886	0.204992	0.215838	0.225332	0.233393
1.30	0.167691	0.181056	0.193446	0.204747	0.214858	0.223691
1.35	0.156196	0.169551	0.182098	0.193717	0.204302	0.213760
1.40	0.145177	0.158422	0.171012	0.182827	0.193758	0.203705
1.45	0.134661	0.147707	0.160243	0.172145	0.183304	0.193617
1.50	0.124662	0.137437	0.149832	0.161726	0.173009	0.183576
1.55	0.115190	0.127632	0.139813	0.151616	0.162929	0.173650
1.60	0.106247	0.118305	0.130211	0.141850	0.153113	0.163898
1.65	0.097828	0.109463	0.121043	0.132456	0.143598	0.154367
1.70	0.089927	0.101108	0.112319	0.123456	0.134415	0.145100
1.75	0.082531	0.093236	0.104046	0.114862	0.125587	0.136126
1.80	0.075627	0.085839	0.096223	0.106684	0.117131	0.127473
1.85	0.069196	0.078907	0.088846	0.098925	0.109058	0.119160
1.90	0.063220	0.072427	0.081909	0.091585	0.101376	0.111200
1.95	0.057680	0.066383	0.075401	0.084660	0.094086	0.103603
2.00	0.052553	0.060759	0.069311	0.078143	0.087187	0.096373
2.05	0.047819	0.055535	0.063624	0.072025	0.080675	0.089512
2.10	0.043456	0.050694	0.058325	0.066293	0.074543	0.083017
2.15	0.039441	0.046216	0.053397	0.060936	0.068783	0.076885
2.20	0.035754	0.042081	0.048823	0.055938	0.063382	0.071107
2.25	0.032373	0.038270	0.044586	0.051286	0.058330	0.065677
2.30	0.029278	0.034763	0.040668	0.046962	0.053613	0.060583
2.35	0.026449	0.031541	0.037050	0.042952	0.049217	0.055814
2.40	0.023868	0.028586	0.033716	0.039238	0.045128	0.051357
2.45	0.021515	0.025879	0.030648	0.035805	0.041330	0.047201
2.50	0.019374	0.023404	0.027828	0.032636	0.037809	0.043331

Table 2.2 (continued)

				t		
v	0.65	0.70	0.75	0.80	0.85	0.90
2.55	0.017428	0.021143	0.025241	0.029715	0.034550	0.039733
2.60	0.015662	0.019081	0.022871	0.027027	0.031538	0.036394
2.65	0.014061	0.017204	0.020703	0.024557	0.028759	0.033300
2.70	0.012612	0.015496	0.018722	0.022290	0.026197	0.030437
2.75	0.011301	0.013944	0.016913	0.020212	0.023840	0.027792
2.80	0.010118	0.012536	0.015265	0.018311	0.021673	0.025352
2.85	0.009050	0.011260	0.013765	0.016572	0.019685	0.023103
2.90	0.008088	0.010104	0.012401	0.014985	0.017862	0.021034
2.95	0.007221	0.009059	0.011162	0.013538	0.016193	0.019132
3.00	0.006442	0.008116	0.010038	0.012219	0.014667	0.017386
3.05	0.005743	0.007264	0.009020	0.011020	0.013273	0.015785
3.10	0.005115	0.006496	0.008098	0.009929	0.012001	0.014319
3.15	0.004552	0.005805	0.007264	0.008939	0.010841	0.012978
3.20	0.004048	0.005183	0.006511	0.008041	0.009786	0.011753
3.25	0.003596	0.004624	0.005831	0.007228	0.008826	0.010635
3.30	0.003193	0.004122	0.005218	0.006491	0.007954	0.009615
3.35	0.002833	0.003672	0.004666	0.005825	0.007162	0.008686
3.40	0.002511	0.003269	0.004169	0.005224	0.006444	0.007840
3.45	0.002225	0.002908	0.003722	0.004681	0.005794	0.007072
3.50	0.001970	0.002584	0.003321	0.004191	0.005205	0.006373
3.55	0.001742	0.002295	0.002961	0.003750	0.004673	0.005740
3.60	0.001540	0.002037	0.002638	0.003353	0.004192	0.005166
3.65	0.001361	0.001807	0.002349	0.002995	0.003758	0.004645
3.70	0.001201	0.001602	0.002089	0.002674	0.003366	0.004175
3.75	0.001060	0.001419	0.001858	0.002386	0.003013	0.003749
3.80	0.000934	0.001256	0.001650	0.002127	0.002696	0.003364
3.85	0.000823	0.001111	0.001465	0.001896	0.002410	0.003017
3.90	0.000725	0.000982	0.001300	0.001688	0.002153	0.002704
3.95	0.000638	0.000868	0.001153	0.001502	0.001922	0.002422
4.00	0.000561	0.000766	0.001022	0.001336	0.001715	0.002167
4.05	0.000493	0.000676	0.000905	0.001187	0.001529	0.001939
4.10	0.000433	0.000596	0.000801	0.001054	0.001363	0.001733
4.15	0.000380	0.000525	0.000708	0.000936	0.001213	0.001548
4.20	0.000333	0.000463	0.000626	0.000830	0.001080	0.001382
4.25	0.000292	0.000407	0.000553	0.000736	0.000961	0.001233
4.30	0.000256	0.000358	0.000489	0.000652	0.000854	0.001099
4.35	0.000224	0.000315	0.000431	0.000577	0.000759	0.000980
4.40	0.000196	0.000277	0.000380	0.000511	0.000674	0.000873
4.45	0.000172	0.000243	0.000335	0.000452	0.000598	0.000777
4.50	0.000150	0.000213	0.000295	0.000400	0.000530	0.000691
4.55	0.000131	0.000187	0.000260	0.000353	0.000470	0.000614
4.60	0.000115	0.000164	0.000229	0.000312	0.000416	0.000546
4.65	0.000100	0.000144	0.000201	0.000275	0.000369	0.000485

Table 2.2 (continued)

v	0.65	0.70	0.75	0.80	0.85	0.90
4.70	0.000087	0.000126	0.000177	0.000243	0.000326	0.000430
4.75	0.000076	0.000110	0.000155	0.000214	0.000288	0.000382
4.80	0.000066	0.000096	0.000136	0.000189	0.000255	0.000339
4.85	0.000058	0.000084	0.000120	0.000166	0.000225	0.000300
4.90	0.000050	0.000074	0.000105	0.000146	0.000199	0.000266
4.95	0.000044	0.000064	0.000092	0.000129	0.000176	0.000235
5.00	0.000038	0.000056	0.000081	0.000113	0.000155	0.000208

v	0.95	1.0	2.0	3.0	4.0	5.0
0.00	-0.287296	-0.351458	-1.12214	-0.725359	0.341809	1.17992
0.05	-0.211985	-0.274188	-1.08664	-0.782337	0.244900	1.12227
0.10	-0.142947	-0.202720	-1.04669	-0.832254	0.149766	1.05976
0.15	-0.080119	-0.137074	-1.00296	-0.875204	0.056904	0.992916
0.20	-0.023375	-0.077204	-0.956075	-0.911329	-0.033232	0.922277
0.25	0.027462	-0.023003	-0.906638	-0.940808	-0.120230	0.848372
0.30	0.072615	0.025685	-0.855218	-0.963859	-0.203724	0.771731
0.35	0.112337	0.069053	-0.802347	-0.980729	-0.283390	0.692876
0.40	0.146911	0.107330	-0.748520	-0.991692	-0.358948	0.612321
0.45	0.176637	0.140768	-0.694196	-0.997044	-0.430160	0.530566
0.50	0.201828	0.169637	-0.639794	-0.997100	-0.496830	0.448097
0.55	0.222804	0.194222	-0.585694	-0.992187	-0.558799	0.365383
0.60	0.239887	0.214814	-0.532241	-0.982647	-0.615952	0.282870
0.65	0.253399	0.231708	-0.479742	-0.968824	-0.668205	0.200986
0.70	0.263654	0.245197	-0.428466	-0.951069	-0.715513	0.120134
0.75	0.270960	0.255572	-0.378650	-0.929734	-0.757862	0.040693
0.80	0.275611	0.263116	-0.330497	-0.905168	-0.795269	-0.036986
0.85	0.277894	0.268105	-0.284178	-0.877717	-0.827782	-0.112574
0.90	0.278076	0.270803	-0.239835	-0.847720	-0.855473	-0.185773
0.95	0.276414	0.271461	-0.197580	-0.815507	-0.878438	-0.256311
1.00	0.273146	0.270319	-0.157503	-0.781400	-0.896798	-0.323943
1.05	0.268495	0.267601	-0.119664	-0.745707	-0.910690	-0.388454
1.10	0.262668	0.263518	-0.084107	-0.708725	-0.920273	-0.449653
1.15	0.255856	0.258265	-0.050851	-0.670735	-0.925717	-0.507378
1.20	0.248232	0.252025	-0.019898	-0.632007	-0.927207	-0.561495
1.25	0.239956	0.244963	0.008765	-0.592791	-0.924941	-0.611893
1.30	0.231171	0.237232	0.035167	-0.553324	-0.919124	-0.658487
1.35	0.222007	0.228971	0.059351	-0.513828	-0.909968	-0.701217
1.40	0.212580	0.220304	0.081368	-0.474504	-0.897693	-0.740046
1.45	0.202991	0.211344	0.101283	-0.435541	-0.882520	-0.774957
1.50	0.193332	0.202192	0.119166	-0.397110	-0.864673	-0.805957
1.55	0.183682	0.192936	0.135096	-0.359366	-0.844377	-0.833070

Table 2.2 (continued)

			t			
v	0.95	1.0	2.0	3.0	4.0	5.0
1.60	0.174108	0.183654	0.149155	−0.322448	−0.821857	−0.856340
1.65	0.164670	0.174415	0.161431	−0.286479	−0.797334	−0.875828
1.70	0.155417	0.165277	0.172016	−0.251570	−0.771027	−0.891610
1.75	0.146390	0.156292	0.181002	−0.217813	−0.743150	−0.903776
1.80	0.137625	0.147500	0.188484	−0.185290	−0.713913	−0.912431
1.85	0.129147	0.138939	0.194557	−0.154069	−0.683519	−0.917690
1.90	0.120979	0.130636	0.199316	−0.124204	−0.652163	−0.919679
1.95	0.113137	0.122615	0.202854	−0.095739	−0.620034	−0.918534
2.00	0.105631	0.114893	0.205263	−0.068706	−0.587313	−0.914397
2.05	0.098470	0.107484	0.206633	−0.043128	−0.554171	−0.907418
2.10	0.091655	0.100396	0.207053	−0.019015	−0.520772	−0.897753
2.15	0.085187	0.093634	0.206607	0.003627	−0.487269	−0.885560
2.20	0.079064	0.087201	0.205376	0.024804	−0.453807	−0.871003
2.25	0.073281	0.081097	0.203439	0.044527	−0.420520	−0.854247
2.30	0.067831	0.075317	0.200870	0.062815	−0.387534	−0.835458
2.35	0.062706	0.069857	0.197742	0.079693	−0.354964	−0.814802
2.40	0.057896	0.064710	0.194121	0.095190	−0.322916	−0.792445
2.45	0.053390	0.059868	0.190071	0.109342	−0.291487	−0.768552
2.50	0.049176	0.055321	0.185651	0.122187	−0.260764	−0.743284
2.55	0.045243	0.051060	0.180917	0.133767	−0.230826	−0.716802
2.60	0.041578	0.047073	0.175921	0.144127	−0.201741	−0.689260
2.65	0.038169	0.043349	0.170712	0.153315	−0.173572	−0.660812
2.70	0.035001	0.039877	0.165334	0.161381	−0.146370	−0.631604
2.75	0.032063	0.036644	0.159829	0.168376	−0.120180	−0.601780
2.80	0.029342	0.033638	0.154232	0.174351	−0.095040	−0.571476
2.85	0.026825	0.030847	0.148580	0.179360	−0.070979	−0.540824
2.90	0.024501	0.028260	0.142903	0.183456	−0.048020	−0.509949
2.95	0.022356	0.025865	0.137229	0.186693	−0.026181	−0.478971
3.00	0.020380	0.023651	0.131583	0.189122	−0.005471	−0.448003
3.05	0.018562	0.021606	0.125988	0.190796	0.014104	−0.417151
3.10	0.016891	0.019720	0.120464	0.191768	0.032544	−0.386515
3.15	0.015357	0.017983	0.115028	0.192087	0.049855	−0.356188
3.20	0.013951	0.016385	0.109695	0.191802	0.066047	−0.326258
3.25	0.012662	0.014915	0.104479	0.190963	0.081132	−0.296804
3.30	0.011483	0.013567	0.099390	0.189616	0.095129	−0.267900
3.35	0.010406	0.012330	0.094437	0.187805	0.108058	−0.239612
3.40	0.009422	0.011197	0.089630	0.185573	0.119943	−0.212002
3.45	0.008524	0.010160	0.084972	0.182963	0.130811	−0.185123
3.50	0.007706	0.009211	0.080470	0.180015	0.140690	−0.159024
3.55	0.006961	0.008345	0.076127	0.176765	0.149611	−0.133747
3.60	0.006284	0.007555	0.071945	0.173249	0.157607	−0.109330
3.65	0.005668	0.006834	0.067925	0.169503	0.164711	−0.085803
3.70	0.005109	0.006178	0.064067	0.165558	0.170959	−0.063192

Table 2.2 (continued)

v	0.95	1.0	2.0	3.0	4.0	5.0
3.75	0.004602	0.005581	0.060371	0.161444	0.176386	−0.041518
3.80	0.004142	0.005038	0.056836	0.157189	0.181030	−0.020798
3.85	0.003726	0.004544	0.053459	0.152821	0.184927	−0.001043
3.90	0.003349	0.004096	0.050239	0.148363	0.188115	0.017739
3.95	0.003008	0.003690	0.047172	0.143839	0.190631	0.035546
4.00	0.002700	0.003322	0.044254	0.139271	0.192513	0.052378
4.05	0.002423	0.002989	0.041483	0.134677	0.193798	0.068238
4.10	0.002172	0.002687	0.038853	0.130075	0.194522	0.083135
4.15	0.001946	0.002414	0.036362	0.125483	0.194721	0.097077
4.20	0.001742	0.002168	0.034003	0.120914	0.194431	0.110080
4.25	0.001559	0.001945	0.031774	0.116383	0.193687	0.122157
4.30	0.001394	0.001745	0.029668	0.111901	0.192521	0.133327
4.35	0.001246	0.001564	0.027681	0.107480	0.190968	0.143611
4.40	0.001113	0.001401	0.025808	0.103130	0.189059	0.153029
4.45	0.000994	0.001254	0.024046	0.098858	0.186824	0.161605
4.50	0.000887	0.001122	0.022387	0.094672	0.184294	0.169365
4.55	0.000791	0.001003	0.020829	0.090578	0.181496	0.176333
4.60	0.000705	0.000897	0.019366	0.086583	0.178459	0.182538
4.65	0.000628	0.000801	0.017994	0.082691	0.175208	0.188008
4.70	0.000559	0.000715	0.016709	0.078905	0.171768	0.192770
4.75	0.000497	0.000638	0.015505	0.075229	0.168163	0.196854
4.80	0.000442	0.000569	0.014378	0.071665	0.164416	0.200291
4.85	0.000393	0.000507	0.013325	0.068215	0.160547	0.203109
4.90	0.000349	0.000452	0.012342	0.064879	0.156577	0.205338
4.95	0.000310	0.000402	0.011424	0.061658	0.152524	0.207009
5.00	0.000275	0.000358	0.010568	0.058553	0.148406	0.208151

Table 2.3: Third derivatives with respect to the order of the Struve functions $\partial^3 H_v(t)/\partial v^3$.

v	0.05	0.10	0.15	0.20	0.25	0.30
0.00	−1.28610	−1.17756	−0.940839	−0.676023	−0.412466	−0.161136
0.05	−1.12741	−1.09741	−0.926163	−0.712486	−0.488331	−0.267035
0.10	−0.981813	−1.01228	−0.896741	−0.728970	−0.541681	−0.349582
0.15	−0.849974	−0.925549	−0.856626	−0.729452	−0.576107	−0.411857
0.20	−0.731912	−0.839755	−0.809153	−0.717366	−0.594823	−0.456705
0.25	−0.627183	−0.756742	−0.757028	−0.695646	−0.600676	−0.486728
0.30	−0.535039	−0.677798	−0.702420	−0.666776	−0.596166	−0.504277
0.35	−0.454552	−0.603770	−0.647034	−0.632833	−0.583468	−0.511462
0.40	−0.384696	−0.535154	−0.592187	−0.595537	−0.564457	−0.510158

Table 2.3 (continued)

				t		
v	0.05	0.10	0.15	0.20	0.25	0.30
0.45	−0.324415	−0.472178	−0.538867	−0.556294	−0.540731	−0.502016
0.50	−0.272669	−0.414869	−0.487795	−0.516236	−0.513644	−0.488476
0.55	−0.228459	−0.363104	−0.439470	−0.476258	−0.484325	−0.470787
0.60	−0.190852	−0.316652	−0.394210	−0.437057	−0.453709	−0.450019
0.65	−0.158992	−0.275214	−0.352193	−0.399158	−0.422557	−0.427081
0.70	−0.132100	−0.238444	−0.313484	−0.362943	−0.391482	−0.402737
0.75	−0.109481	−0.205973	−0.278064	−0.328680	−0.360964	−0.377622
0.80	−0.090519	−0.177426	−0.245847	−0.296537	−0.331376	−0.352257
0.85	−0.074670	−0.152430	−0.216702	−0.266606	−0.302992	−0.327063
0.90	−0.061462	−0.130626	−0.190466	−0.238916	−0.276011	−0.302374
0.95	−0.050484	−0.111673	−0.166954	−0.213450	−0.250565	−0.278450
1.00	−0.041384	−0.095253	−0.145969	−0.190150	−0.226732	−0.255485
1.05	−0.033859	−0.081069	−0.127311	−0.168935	−0.204546	−0.233620
1.10	−0.027651	−0.068854	−0.110781	−0.149703	−0.184007	−0.212950
1.15	−0.022541	−0.058362	−0.096185	−0.132337	−0.165086	−0.193534
1.20	−0.018344	−0.049374	−0.083335	−0.116715	−0.147734	−0.175396
1.25	−0.014903	−0.041693	−0.072056	−0.102710	−0.131888	−0.158540
1.30	−0.012088	−0.035145	−0.062182	−0.090194	−0.117472	−0.142946
1.35	−0.009790	−0.029574	−0.053561	−0.079044	−0.104403	−0.128580
1.40	−0.007917	−0.024845	−0.046053	−0.069138	−0.092595	−0.115398
1.45	−0.006392	−0.020838	−0.039529	−0.060362	−0.081957	−0.103344
1.50	−0.005154	−0.017451	−0.033872	−0.052605	−0.072403	−0.092360
1.55	−0.004150	−0.014592	−0.028978	−0.045766	−0.063845	−0.082381
1.60	−0.003337	−0.012184	−0.024753	−0.039750	−0.056199	−0.073341
1.65	−0.002680	−0.010159	−0.021112	−0.034469	−0.049385	−0.065175
1.70	−0.002149	−0.008459	−0.017980	−0.029843	−0.043325	−0.057818
1.75	−0.001722	−0.007034	−0.015291	−0.025800	−0.037948	−0.051205
1.80	−0.001378	−0.005841	−0.012986	−0.022272	−0.033188	−0.045275
1.85	−0.001101	−0.004845	−0.011014	−0.019199	−0.028982	−0.039969
1.90	−0.000879	−0.004013	−0.009329	−0.016527	−0.025272	−0.035231
1.95	−0.000701	−0.003320	−0.007892	−0.014208	−0.022007	−0.031010
2.00	−0.000558	−0.002744	−0.006668	−0.012199	−0.019137	−0.027255
2.05	−0.000444	−0.002265	−0.005627	−0.010461	−0.016620	−0.023923
2.10	−0.000353	−0.001868	−0.004743	−0.008959	−0.014415	−0.020969
2.15	−0.000280	−0.001538	−0.003993	−0.007664	−0.012487	−0.018357
2.20	−0.000222	−0.001266	−0.003358	−0.006548	−0.010804	−0.016049
2.25	−0.000176	−0.001040	−0.002821	−0.005589	−0.009337	−0.014015
2.30	−0.000139	−0.000854	−0.002368	−0.004765	−0.008059	−0.012224
2.35	−0.000110	−0.000701	−0.001985	−0.004057	−0.006949	−0.010649
2.40	−0.000087	−0.000574	−0.001662	−0.003452	−0.005985	−0.009266
2.45	−0.000069	−0.000470	−0.001391	−0.002933	−0.005149	−0.008054
2.50	−0.000054	−0.000384	−0.001162	−0.002490	−0.004425	−0.006993
2.55	−0.000043	−0.000314	−0.000970	−0.002112	−0.003800	−0.006066
2.60	−0.000034	−0.000256	−0.000810	−0.001789	−0.003259	−0.005256

Table 2.3 (continued)

			t			
v	0.05	0.10	0.15	0.20	0.25	0.30
2.65	−0.000026	−0.000209	−0.000675	−0.001515	−0.002792	−0.004549
2.70	−0.000021	−0.000170	−0.000562	−0.001281	−0.002390	−0.003934
2.75	−0.000016	−0.000139	−0.000467	−0.001082	−0.002044	−0.003398
2.80	−0.000013	−0.000113	−0.000388	−0.000914	−0.001747	−0.002933
2.85	−0.000010	−0.000092	−0.000323	−0.000770	−0.001491	−0.002529
2.90	−0.000008	−0.000075	−0.000268	−0.000649	−0.001272	−0.002178
2.95	−0.000006	−0.000061	−0.000222	−0.000547	−0.001084	−0.001875
3.00	−0.000005	−0.000049	−0.000184	−0.000460	−0.000923	−0.001612
3.05	−0.000004	−0.000040	−0.000152	−0.000387	−0.000785	−0.001385
3.10	−0.000003	−0.000032	−0.000126	−0.000325	−0.000667	−0.001189
3.15	−0.000002	−0.000026	−0.000104	−0.000272	−0.000567	−0.001020
3.20	−0.000002	−0.000021	−0.000086	−0.000228	−0.000481	−0.000874
3.25	−0.000001	−0.000017	−0.000071	−0.000191	−0.000408	−0.000749
3.30	−0.000001	−0.000014	−0.000058	−0.000160	−0.000346	−0.000641
3.35	−0.000001	−0.000011	−0.000048	−0.000134	−0.000293	−0.000548
3.40	−0.000001	−0.000009	−0.000040	−0.000112	−0.000248	−0.000468
3.45	−0.000001	−0.000007	−0.000033	−0.000094	−0.000209	−0.000400
3.50		−0.000006	−0.000027	−0.000078	−0.000177	−0.000341
3.55		−0.000005	−0.000022	−0.000065	−0.000149	−0.000291
3.60		−0.000004	−0.000018	−0.000054	−0.000126	−0.000248
3.65		−0.000003	−0.000015	−0.000045	−0.000106	−0.000211
3.70		−0.000002	−0.000012	−0.000038	−0.000089	−0.000179
3.75		−0.000002	−0.000010	−0.000031	−0.000075	−0.000152
3.80		−0.000002	−0.000008	−0.000026	−0.000063	−0.000129
3.85		−0.000001	−0.000007	−0.000022	−0.000053	−0.000110
3.90		−0.000001	−0.000005	−0.000018	−0.000045	−0.000093
3.95		−0.000001	−0.000004	−0.000015	−0.000038	−0.000079
4.00		−0.000001	−0.000004	−0.000012	−0.000032	−0.000067
4.05		−0.000001	−0.000003	−0.000010	−0.000026	−0.000057
4.10			−0.000002	−0.000009	−0.000022	−0.000048
4.15			−0.000002	−0.000007	−0.000019	−0.000041
4.20			−0.000002	−0.000006	−0.000016	−0.000034
4.25			−0.000001	−0.000005	−0.000013	−0.000029
4.30			−0.000001	−0.000004	−0.000011	−0.000024
4.35			−0.000001	−0.000003	−0.000009	−0.000021
4.40			−0.000001	−0.000003	−0.000008	−0.000017
4.45			−0.000001	−0.000002	−0.000006	−0.000015
4.50				−0.000002	−0.000005	−0.000012
4.55				−0.000002	−0.000004	−0.000010
4.60				−0.000001	−0.000004	−0.000009
4.65				−0.000001	−0.000003	−0.000007
4.70				−0.000001	−0.000003	−0.000006
4.75				−0.000001	−0.000002	−0.000005

Table 2.3 (continued)

			t			
v	0.05	0.10	0.15	0.20	0.25	0.30
4.80				−0.000001	−0.000002	−0.000004
4.85					−0.000001	−0.000004
4.90					−0.000001	−0.000003
4.95					−0.000001	−0.000003
5.00					−0.000001	−0.000002

			t			
v	0.35	0.40	0.45	0.50	0.55	0.60
0.00	0.073574	0.290032	0.487879	0.667370	0.829057	0.973631
0.05	−0.054841	0.145200	0.331626	0.503825	0.661656	0.805249
0.10	−0.160172	0.022450	0.195972	0.359082	0.511066	0.651577
0.15	−0.244953	−0.080206	0.079441	0.232139	0.376714	0.512426
0.20	−0.311597	−0.164738	−0.019498	0.121879	0.257868	0.387416
0.25	−0.362378	−0.233059	−0.102392	0.027115	0.153681	0.276015
0.30	−0.399414	−0.287003	−0.170772	−0.053381	0.063223	0.177579
0.35	−0.424653	−0.328305	−0.226131	−0.120845	−0.014488	0.091375
0.40	−0.439874	−0.358588	−0.269900	−0.176502	−0.080458	0.016614
0.45	−0.446684	−0.379355	−0.303440	−0.221542	−0.135696	−0.047529
0.50	−0.446525	−0.391986	−0.328026	−0.257110	−0.181201	−0.101893
0.55	−0.440677	−0.397735	−0.344845	−0.284295	−0.217943	−0.147319
0.60	−0.430271	−0.397734	−0.354988	−0.304121	−0.246855	−0.184635
0.65	−0.416294	−0.392997	−0.359456	−0.317542	−0.268825	−0.214643
0.70	−0.399601	−0.384423	−0.359154	−0.325439	−0.284686	−0.238117
0.75	−0.380928	−0.372805	−0.354896	−0.328621	−0.295220	−0.255789
0.80	−0.360902	−0.358835	−0.347409	−0.327822	−0.301148	−0.268350
0.85	−0.340049	−0.343115	−0.337338	−0.323709	−0.303134	−0.276447
0.90	−0.318809	−0.326159	−0.325250	−0.316874	−0.301782	−0.280679
0.95	−0.297544	−0.308406	−0.311638	−0.307850	−0.297642	−0.281596
1.00	−0.276545	−0.290227	−0.296932	−0.297103	−0.291205	−0.279702
1.05	−0.256046	−0.271929	−0.281497	−0.285045	−0.282912	−0.275455
1.10	−0.236225	−0.253765	−0.265645	−0.272033	−0.273153	−0.269265
1.15	−0.217220	−0.235938	−0.249638	−0.258375	−0.262273	−0.261503
1.20	−0.199128	−0.218611	−0.233694	−0.244335	−0.250571	−0.252495
1.25	−0.182013	−0.201908	−0.217989	−0.230135	−0.238308	−0.242533
1.30	−0.165916	−0.185923	−0.202666	−0.215961	−0.225708	−0.231870
1.35	−0.150854	−0.170722	−0.187839	−0.201968	−0.212960	−0.220727
1.40	−0.136824	−0.156348	−0.173592	−0.188281	−0.200225	−0.209295
1.45	−0.123811	−0.142826	−0.159989	−0.175000	−0.187635	−0.197736
1.50	−0.111789	−0.130163	−0.147072	−0.162200	−0.175301	−0.186188
1.55	−0.100722	−0.118354	−0.134870	−0.149942	−0.163309	−0.174766
1.60	−0.090567	−0.107386	−0.123395	−0.138265	−0.151728	−0.163565
1.65	−0.081280	−0.097234	−0.112648	−0.127197	−0.140610	−0.152661
1.70	−0.072810	−0.087868	−0.102621	−0.116753	−0.129995	−0.142116

Table 2.3 (continued)

			t			
v	0.35	0.40	0.45	0.50	0.55	0.60
1.75	−0.065107	−0.079256	−0.093300	−0.106939	−0.119907	−0.131976
1.80	−0.058120	−0.071358	−0.084664	−0.097750	−0.110362	−0.122276
1.85	−0.051797	−0.064135	−0.076686	−0.089177	−0.101368	−0.113039
1.90	−0.046088	−0.057548	−0.069337	−0.081205	−0.092923	−0.104281
1.95	−0.040946	−0.051553	−0.062586	−0.073813	−0.085020	−0.096010
2.00	−0.036323	−0.046111	−0.056400	−0.066979	−0.077649	−0.088226
2.05	−0.032176	−0.041182	−0.050745	−0.060677	−0.070794	−0.080925
2.10	−0.028463	−0.036726	−0.045588	−0.054880	−0.064437	−0.074097
2.15	−0.025144	−0.032706	−0.040895	−0.049561	−0.058555	−0.067731
2.20	−0.022183	−0.029086	−0.036633	−0.044691	−0.053128	−0.061812
2.25	−0.019545	−0.025832	−0.032769	−0.040242	−0.048132	−0.056321
2.30	−0.017200	−0.022913	−0.029274	−0.036185	−0.043542	−0.051241
2.35	−0.015118	−0.020298	−0.026117	−0.032493	−0.039335	−0.046550
2.40	−0.013272	−0.017960	−0.023272	−0.029139	−0.035486	−0.042230
2.45	−0.011639	−0.015872	−0.020711	−0.026099	−0.031971	−0.038258
2.50	−0.010195	−0.014011	−0.018410	−0.023347	−0.028768	−0.034613
2.55	−0.008920	−0.012354	−0.016345	−0.020859	−0.025853	−0.031275
2.60	−0.007796	−0.010881	−0.014496	−0.018615	−0.023206	−0.028224
2.65	−0.006807	−0.009573	−0.012841	−0.016594	−0.020805	−0.025440
2.70	−0.005937	−0.008414	−0.011363	−0.014775	−0.018631	−0.022903
2.75	−0.005173	−0.007387	−0.010045	−0.013142	−0.016666	−0.020595
2.80	−0.004504	−0.006479	−0.008870	−0.011677	−0.014891	−0.018500
2.85	−0.003917	−0.005677	−0.007825	−0.010364	−0.013292	−0.016599
2.90	−0.003403	−0.004970	−0.006896	−0.009189	−0.011852	−0.014877
2.95	−0.002954	−0.004347	−0.006072	−0.008140	−0.010557	−0.013321
3.00	−0.002562	−0.003798	−0.005341	−0.007204	−0.009394	−0.011914
3.05	−0.002220	−0.003316	−0.004693	−0.006369	−0.008351	−0.010646
3.10	−0.001922	−0.002892	−0.004121	−0.005625	−0.007417	−0.009504
3.15	−0.001663	−0.002520	−0.003615	−0.004964	−0.006582	−0.008476
3.20	−0.001437	−0.002194	−0.003168	−0.004377	−0.005835	−0.007552
3.25	−0.001241	−0.001909	−0.002775	−0.003856	−0.005168	−0.006723
3.30	−0.001071	−0.001659	−0.002428	−0.003394	−0.004573	−0.005979
3.35	−0.000923	−0.001441	−0.002122	−0.002985	−0.004044	−0.005313
3.40	−0.000795	−0.001251	−0.001854	−0.002623	−0.003572	−0.004717
3.45	−0.000685	−0.001085	−0.001618	−0.002303	−0.003153	−0.004185
3.50	−0.000589	−0.000940	−0.001411	−0.002020	−0.002781	−0.003709
3.55	−0.000506	−0.000814	−0.001230	−0.001771	−0.002451	−0.003285
3.60	−0.000435	−0.000704	−0.001071	−0.001551	−0.002159	−0.002907
3.65	−0.000373	−0.000609	−0.000932	−0.001358	−0.001899	−0.002570
3.70	−0.000320	−0.000526	−0.000810	−0.001187	−0.001670	−0.002271
3.75	−0.000274	−0.000454	−0.000704	−0.001038	−0.001467	−0.002005
3.80	−0.000235	−0.000392	−0.000611	−0.000906	−0.001288	−0.001769

Table 2.3 (continued)

			t			
v	0.35	0.40	0.45	0.50	0.55	0.60
3.85	−0.000201	−0.000338	−0.000531	−0.000791	−0.001130	−0.001560
3.90	−0.000172	−0.000291	−0.000460	−0.000690	−0.000991	−0.001374
3.95	−0.000147	−0.000251	−0.000399	−0.000601	−0.000868	−0.001209
4.00	−0.000126	−0.000216	−0.000345	−0.000524	−0.000760	−0.001064
4.05	−0.000107	−0.000185	−0.000299	−0.000456	−0.000665	−0.000935
4.10	−0.000092	−0.000159	−0.000258	−0.000396	−0.000581	−0.000822
4.15	−0.000078	−0.000137	−0.000223	−0.000344	−0.000508	−0.000721
4.20	−0.000067	−0.000117	−0.000193	−0.000299	−0.000443	−0.000633
4.25	−0.000057	−0.000101	−0.000166	−0.000260	−0.000387	−0.000555
4.30	−0.000048	−0.000086	−0.000144	−0.000225	−0.000337	−0.000486
4.35	−0.000041	−0.000074	−0.000124	−0.000195	−0.000294	−0.000426
4.40	−0.000035	−0.000063	−0.000107	−0.000169	−0.000256	−0.000373
4.45	−0.000030	−0.000054	−0.000092	−0.000147	−0.000223	−0.000326
4.50	−0.000025	−0.000046	−0.000079	−0.000127	−0.000194	−0.000285
4.55	−0.000021	−0.000040	−0.000068	−0.000110	−0.000169	−0.000249
4.60	−0.000018	−0.000034	−0.000058	−0.000095	−0.000146	−0.000217
4.65	−0.000015	−0.000029	−0.000050	−0.000082	−0.000127	−0.000189
4.70	−0.000013	−0.000025	−0.000043	−0.000071	−0.000110	−0.000165
4.75	−0.000011	−0.000021	−0.000037	−0.000061	−0.000096	−0.000144
4.80	−0.000009	−0.000018	−0.000032	−0.000053	−0.000083	−0.000125
4.85	−0.000008	−0.000015	−0.000027	−0.000045	−0.000072	−0.000109
4.90	−0.000007	−0.000013	−0.000023	−0.000039	−0.000062	−0.000095
4.95	−0.000006	−0.000011	−0.000020	−0.000034	−0.000054	−0.000083
5.00	−0.000005	−0.000009	−0.000017	−0.000029	−0.000047	−0.000072

			t			
v	0.65	0.70	0.75	0.80	0.85	0.90
0.00	1.10184	1.21445	1.31222	1.39586	1.46609	1.52358
0.05	0.934890	1.05096	1.15388	1.24412	1.32213	1.38840
0.10	0.780501	0.897876	1.00384	1.09861	1.18242	1.25556
0.15	0.638823	0.755652	0.862798	0.960245	1.048050	1.12632
0.20	0.509793	0.624504	0.731215	0.829718	0.919894	1.001700
0.25	0.393173	0.504457	0.609356	0.707495	0.798609	0.882516
0.30	0.288580	0.395375	0.497307	0.593868	0.684665	0.769400
0.35	0.195522	0.296990	0.395010	0.488966	0.578366	0.662813
0.40	0.113422	0.208927	0.302281	0.392788	0.479875	0.563067
0.45	0.041641	0.130729	0.218838	0.305215	0.389228	0.470343
0.50	−0.020501	0.061878	0.144320	0.226038	0.306361	0.384712
0.55	−0.073706	0.001811	0.078302	0.154969	0.231118	0.306144
0.60	−0.118687	−0.050065	0.020317	0.091663	0.163273	0.234531
0.65	−0.156152	−0.094356	−0.030136	0.035728	0.102544	0.169695

Table 2.3 (continued)

			t			
ν	0.65	0.70	0.75	0.80	0.85	0.90
0.70	−0.186798	−0.131672	−0.073577	−0.013262	0.048601	0.111403
0.75	−0.211297	−0.162618	−0.110532	−0.055751	0.001082	0.059379
0.80	−0.230299	−0.187785	−0.141532	−0.092200	−0.040398	0.013314
0.85	−0.244416	−0.207747	−0.167096	−0.123072	−0.076240	−0.027128
0.90	−0.254227	−0.223050	−0.187733	−0.148828	−0.106852	−0.062296
0.95	−0.260275	−0.234215	−0.203933	−0.169919	−0.132643	−0.092552
1.00	−0.263060	−0.241732	−0.216164	−0.186786	−0.154016	−0.118259
1.05	−0.263045	−0.246059	−0.224870	−0.199851	−0.171368	−0.139780
1.10	−0.260654	−0.247619	−0.230469	−0.209519	−0.185081	−0.157472
1.15	−0.256270	−0.246805	−0.233352	−0.216170	−0.195524	−0.171683
1.20	−0.250242	−0.243974	−0.233881	−0.220165	−0.203045	−0.182747
1.25	−0.242880	−0.239454	−0.232389	−0.221839	−0.207976	−0.190986
1.30	−0.234463	−0.233539	−0.229182	−0.221502	−0.210628	−0.196705
1.35	−0.225237	−0.226494	−0.224540	−0.219442	−0.211290	−0.200193
1.40	−0.215416	−0.218555	−0.218713	−0.215920	−0.210231	−0.201720
1.45	−0.205191	−0.209933	−0.211930	−0.211177	−0.207698	−0.201537
1.50	−0.194724	−0.200812	−0.204391	−0.205429	−0.203919	−0.199878
1.55	−0.184154	−0.191354	−0.196278	−0.198870	−0.199099	−0.196957
1.60	−0.173601	−0.181697	−0.187748	−0.191676	−0.193427	−0.192972
1.65	−0.163163	−0.171964	−0.178941	−0.184000	−0.187070	−0.188101
1.70	−0.152924	−0.162256	−0.169978	−0.175981	−0.180179	−0.182507
1.75	−0.142950	−0.152659	−0.160961	−0.167737	−0.172888	−0.176336
1.80	−0.133294	−0.143245	−0.151980	−0.159373	−0.165315	−0.169718
1.85	−0.123998	−0.134072	−0.143109	−0.150978	−0.157565	−0.162770
1.90	−0.115093	−0.125187	−0.134411	−0.142631	−0.149726	−0.155595
1.95	−0.106601	−0.116626	−0.125936	−0.134394	−0.141879	−0.148282
2.00	−0.098535	−0.108417	−0.117725	−0.126322	−0.134088	−0.140911
2.05	−0.090904	−0.100580	−0.109809	−0.118460	−0.126410	−0.133548
2.10	−0.083709	−0.093127	−0.102215	−0.110843	−0.118892	−0.126252
2.15	−0.076947	−0.086066	−0.094957	−0.103498	−0.111572	−0.119071
2.20	−0.070610	−0.079397	−0.088049	−0.096448	−0.104483	−0.112046
2.25	−0.064689	−0.073119	−0.081496	−0.089707	−0.097646	−0.105211
2.30	−0.059171	−0.067226	−0.075299	−0.083285	−0.091083	−0.098592
2.35	−0.054042	−0.061710	−0.069459	−0.077188	−0.084804	−0.092212
2.40	−0.049284	−0.056560	−0.063968	−0.071418	−0.078819	−0.086084
2.45	−0.044881	−0.051763	−0.058821	−0.065972	−0.073134	−0.080222
2.50	−0.040816	−0.047305	−0.054007	−0.060847	−0.067748	−0.074632
2.55	−0.037069	−0.043171	−0.049516	−0.056037	−0.062661	−0.069318
2.60	−0.033622	−0.039345	−0.045336	−0.051532	−0.057869	−0.064282
2.65	−0.030458	−0.035812	−0.041452	−0.047322	−0.053366	−0.059522
2.70	−0.027558	−0.032555	−0.037852	−0.043398	−0.049144	−0.055033
2.75	−0.024904	−0.029558	−0.034520	−0.039748	−0.045194	−0.050812

Table 2.3 (continued)

			t			
v	0.65	0.70	0.75	0.80	0.85	0.90
2.80	−0.022480	−0.026805	−0.031443	−0.036358	−0.041508	−0.046850
2.85	−0.020269	−0.024280	−0.028606	−0.033216	−0.038073	−0.043141
2.90	−0.018255	−0.021968	−0.025995	−0.030309	−0.034880	−0.039674
2.95	−0.016424	−0.019855	−0.023595	−0.027624	−0.031916	−0.036440
3.00	−0.014761	−0.017925	−0.021394	−0.025149	−0.029169	−0.033430
3.05	−0.013253	−0.016166	−0.019376	−0.022870	−0.026629	−0.030633
3.10	−0.011887	−0.014565	−0.017531	−0.020775	−0.024283	−0.028037
3.15	−0.010651	−0.013109	−0.015845	−0.018852	−0.022120	−0.025633
3.20	−0.009535	−0.011787	−0.014307	−0.017090	−0.020128	−0.023410
3.25	−0.008528	−0.010589	−0.012905	−0.015477	−0.018297	−0.021357
3.30	−0.007620	−0.009503	−0.011630	−0.014002	−0.016616	−0.019464
3.35	−0.006803	−0.008521	−0.010471	−0.012656	−0.015074	−0.017721
3.40	−0.006068	−0.007633	−0.009419	−0.011428	−0.013662	−0.016119
3.45	−0.005408	−0.006832	−0.008465	−0.010311	−0.012371	−0.014647
3.50	−0.004815	−0.006110	−0.007601	−0.009294	−0.011192	−0.013297
3.55	−0.004284	−0.005459	−0.006819	−0.008370	−0.010116	−0.012061
3.60	−0.003809	−0.004874	−0.006112	−0.007531	−0.009135	−0.010929
3.65	−0.003383	−0.004348	−0.005474	−0.006771	−0.008243	−0.009895
3.70	−0.003003	−0.003875	−0.004899	−0.006082	−0.007431	−0.008951
3.75	−0.002663	−0.003452	−0.004381	−0.005459	−0.006694	−0.008091
3.80	−0.002360	−0.003072	−0.003914	−0.004896	−0.006025	−0.007307
3.85	−0.002090	−0.002732	−0.003495	−0.004387	−0.005418	−0.006593
3.90	−0.001849	−0.002428	−0.003118	−0.003929	−0.004869	−0.005945
3.95	−0.001635	−0.002156	−0.002779	−0.003515	−0.004372	−0.005356
4.00	−0.001445	−0.001913	−0.002476	−0.003143	−0.003923	−0.004822
4.05	−0.001276	−0.001696	−0.002204	−0.002808	−0.003517	−0.004337
4.10	−0.001126	−0.001503	−0.001960	−0.002507	−0.003151	−0.003899
4.15	−0.000993	−0.001331	−0.001743	−0.002237	−0.002821	−0.003502
4.20	−0.000875	−0.001177	−0.001548	−0.001994	−0.002524	−0.003144
4.25	−0.000771	−0.001041	−0.001374	−0.001777	−0.002256	−0.002820
4.30	−0.000678	−0.000920	−0.001219	−0.001582	−0.002016	−0.002527
4.35	−0.000596	−0.000813	−0.001081	−0.001408	−0.001800	−0.002264
4.40	−0.000524	−0.000717	−0.000957	−0.001252	−0.001606	−0.002026
4.45	−0.000460	−0.000633	−0.000848	−0.001112	−0.001432	−0.001813
4.50	−0.000404	−0.000558	−0.000750	−0.000988	−0.001276	−0.001620
4.55	−0.000355	−0.000491	−0.000663	−0.000876	−0.001136	−0.001448
4.60	−0.000311	−0.000432	−0.000586	−0.000777	−0.001011	−0.001292
4.65	−0.000272	−0.000380	−0.000518	−0.000689	−0.000899	−0.001153
4.70	−0.000239	−0.000335	−0.000457	−0.000610	−0.000799	−0.001028
4.75	−0.000209	−0.000294	−0.000403	−0.000540	−0.000710	−0.000916
4.80	−0.000183	−0.000258	−0.000355	−0.000478	−0.000630	−0.000816
4.85	−0.000160	−0.000227	−0.000313	−0.000423	−0.000559	−0.000726

Table 2.3 (continued)

			t			
v	0.65	0.70	0.75	0.80	0.85	0.90
4.90	−0.000140	−0.000199	−0.000276	−0.000374	−0.000496	−0.000646
4.95	−0.000122	−0.000174	−0.000243	−0.000330	−0.000439	−0.000574
5.00	−0.000106	−0.000153	−0.000214	−0.000291	−0.000389	−0.000510

			t			
v	0.95	1.0	2.0	3.0	4.0	5.0
0.00	1.56899	1.60292	0.661032	−1.21059	−1.95240	−1.10102
0.05	1.44339	1.48758	0.756728	−1.06871	−1.92216	−1.20338
0.10	1.31834	1.37108	0.838960	−0.928270	−1.88155	−1.29527
0.15	1.19521	1.25489	0.908246	−0.790207	−1.83142	−1.37657
0.20	1.075140	1.14027	0.965179	−0.655375	−1.77263	−1.44721
0.25	0.959099	1.02830	1.01041	−0.524531	−1.70607	−1.50722
0.30	0.847849	0.919850	1.04463	−0.398340	−1.63261	−1.55669
0.35	0.741991	0.815651	1.06856	−0.277377	−1.55311	−1.59579
0.40	0.641973	0.716271	1.08297	−0.162129	−1.46843	−1.62474
0.45	0.548110	0.622145	1.08860	−0.053001	−1.37940	−1.64382
0.50	0.460595	0.533583	1.08623	0.049683	−1.28685	−1.65336
0.55	0.379522	0.450789	1.07661	0.145676	−1.19155	−1.65373
0.60	0.304891	0.373869	1.06049	0.234801	−1.09427	−1.64535
0.65	0.236629	0.302849	1.03860	0.316948	−0.995709	−1.62866
0.70	0.174597	0.237683	1.01165	0.392067	−0.896560	−1.60416
0.75	0.118604	0.178264	0.980311	0.460168	−0.797460	−1.57233
0.80	0.068419	0.124437	0.945239	0.521310	−0.699005	−1.53371
0.85	0.023774	0.076005	0.907046	0.575599	−0.601748	−1.48883
0.90	−0.015621	0.032739	0.866309	0.623185	−0.506198	−1.43823
0.95	−0.050074	−0.005615	0.823571	0.664250	−0.412819	−1.38247
1.00	−0.079905	−0.039329	0.779333	0.699014	−0.322028	−1.32210
1.05	−0.105439	−0.068688	0.734059	0.727719	−0.234201	−1.25768
1.10	−0.127003	−0.093982	0.688174	0.750635	−0.149665	−1.18974
1.15	−0.144919	−0.115507	0.642066	0.768048	−0.068709	−1.11883
1.20	−0.159503	−0.133554	0.596081	0.780259	0.008425	−1.04547
1.25	−0.171063	−0.148412	0.550533	0.787582	0.081532	−0.970162
1.30	−0.179894	−0.160363	0.505698	0.790339	0.150448	−0.893406
1.35	−0.186277	−0.169679	0.461817	0.788856	0.215046	−0.815673
1.40	−0.190480	−0.176620	0.419102	0.783461	0.275233	−0.737414
1.45	−0.192757	−0.181437	0.377731	0.774482	0.330952	−0.659060
1.50	−0.193341	−0.184363	0.337855	0.762242	0.382175	−0.581015
1.55	−0.192455	−0.185622	0.299597	0.747062	0.428904	−0.503660
1.60	−0.190300	−0.185419	0.263054	0.729254	0.471169	−0.427351
1.65	−0.187064	−0.183948	0.228303	0.709120	0.509024	−0.352415
1.70	−0.182919	−0.181387	0.195396	0.686955	0.542542	−0.279154
1.75	−0.178021	−0.177899	0.164369	0.663040	0.571823	−0.207844

Table 2.3 (continued)

			t			
v	0.95	1.0	2.0	3.0	4.0	5.0
1.80	−0.172510	−0.173635	0.135238	0.637644	0.596980	−0.138734
1.85	−0.166514	−0.168730	0.108004	0.611024	0.618145	−0.072044
1.90	−0.160148	−0.163309	0.082655	0.583422	0.635461	−0.007972
1.95	−0.153511	−0.157482	0.059166	0.555067	0.649084	0.053314
2.00	−0.146693	−0.151349	0.037500	0.526172	0.659185	0.111667
2.05	−0.139774	−0.144997	0.017614	0.496936	0.665935	0.166967
2.10	−0.132820	−0.138506	−0.000547	0.467543	0.669517	0.219115
2.15	−0.125892	−0.131943	−0.017041	0.438162	0.670118	0.268038
2.20	−0.119038	−0.125367	−0.031934	0.408948	0.667926	0.313679
2.25	−0.112303	−0.118831	−0.045295	0.380042	0.663134	0.356006
2.30	−0.105721	−0.112377	−0.057196	0.351570	0.655934	0.395006
2.35	−0.099321	−0.106044	−0.067712	0.323646	0.646518	0.430680
2.40	−0.093126	−0.099861	−0.076919	0.296370	0.635075	0.463050
2.45	−0.087156	−0.093854	−0.084895	0.269829	0.621794	0.492149
2.50	−0.081423	−0.088043	−0.091716	0.244100	0.606857	0.518029
2.55	−0.075937	−0.082444	−0.097460	0.219247	0.590444	0.540753
2.60	−0.070705	−0.077069	−0.102200	0.195324	0.572727	0.560395
2.65	−0.065729	−0.071925	−0.106013	0.172374	0.553877	0.577041
2.70	−0.061011	−0.067019	−0.108970	0.150433	0.534054	0.590786
2.75	−0.056549	−0.062352	−0.111141	0.129525	0.513413	0.601732
2.80	−0.052339	−0.057925	−0.112593	0.109668	0.492102	0.609991
2.85	−0.048376	−0.053735	−0.113392	0.090874	0.470261	0.615678
2.90	−0.044654	−0.049781	−0.113600	0.073144	0.448024	0.618916
2.95	−0.041166	−0.046056	−0.113276	0.056477	0.425516	0.619830
3.00	−0.037902	−0.042555	−0.112477	0.040863	0.402852	0.618547
3.05	−0.034856	−0.039271	−0.111254	0.026289	0.380143	0.615200
3.10	−0.032017	−0.036197	−0.109659	0.012737	0.357490	0.609921
3.15	−0.029375	−0.033325	−0.107737	0.000186	0.334985	0.602843
3.20	−0.026921	−0.030645	−0.105534	−0.011390	0.312716	0.594097
3.25	−0.024646	−0.028149	−0.103089	−0.022020	0.290759	0.583818
3.30	−0.022539	−0.025829	−0.100440	−0.031734	0.269186	0.572134
3.35	−0.020591	−0.023675	−0.097622	−0.040565	0.248060	0.559174
3.40	−0.018793	−0.021678	−0.094667	−0.048547	0.227437	0.545065
3.45	−0.017135	−0.019830	−0.091605	−0.055718	0.207368	0.529930
3.50	−0.015608	−0.018121	−0.088462	−0.062112	0.187895	0.513887
3.55	−0.014204	−0.016544	−0.085262	−0.067769	0.169056	0.497054
3.60	−0.012914	−0.015090	−0.082028	−0.072726	0.150883	0.479541
3.65	−0.011731	−0.013751	−0.078779	−0.077021	0.133401	0.461457
3.70	−0.010647	−0.012520	−0.075533	−0.080692	0.116631	0.442904
3.75	−0.009655	−0.011388	−0.072306	−0.083778	0.100589	0.423982
3.80	−0.008747	−0.010351	−0.069111	−0.086314	0.085285	0.404781
3.85	−0.007919	−0.009399	−0.065961	−0.088338	0.070727	0.385392
3.90	−0.007163	−0.008529	−0.062865	−0.089886	0.056919	0.365898
3.95	−0.006474	−0.007732	−0.059834	−0.090992	0.043858	0.346377

Table 2.3 (continued)

			t			
v	0.95	1.0	2.0	3.0	4.0	5.0
4.00	−0.005847	−0.007005	−0.056875	−0.091690	0.031542	0.326901
4.05	−0.005277	−0.006340	−0.053993	−0.092013	0.019964	0.307540
4.10	−0.004758	−0.005735	−0.051196	−0.091992	0.009113	0.288354
4.15	−0.004288	−0.005183	−0.048486	−0.091658	−0.001023	0.269404
4.20	−0.003861	−0.004681	−0.045867	−0.091040	−0.010458	0.250741
4.25	−0.003474	−0.004225	−0.043341	−0.090165	−0.019208	0.232415
4.30	−0.003123	−0.003810	−0.040910	−0.089059	−0.027292	0.214468
4.35	−0.002806	−0.003434	−0.038576	−0.087748	−0.034729	0.196939
4.40	−0.002520	−0.003092	−0.036337	−0.086255	−0.041540	0.179864
4.45	−0.002261	−0.002783	−0.034195	−0.084602	−0.047746	0.163274
4.50	−0.002027	−0.002503	−0.032147	−0.082811	−0.053370	0.147193
4.55	−0.001817	−0.002249	−0.030194	−0.080900	−0.058436	0.131645
4.60	−0.001627	−0.002020	−0.028334	−0.078888	−0.062968	0.116650
4.65	−0.001456	−0.001813	−0.026564	−0.076791	−0.066989	0.102221
4.70	−0.001302	−0.001627	−0.024884	−0.074627	−0.070524	0.088372
4.75	−0.001164	−0.001458	−0.023289	−0.072409	−0.073598	0.075111
4.80	−0.001040	−0.001306	−0.021779	−0.070151	−0.076234	0.062444
4.85	−0.000928	−0.001170	−0.020350	−0.067865	−0.078458	0.050375
4.90	−0.000828	−0.001047	−0.018999	−0.065563	−0.080292	0.038904
4.95	−0.000739	−0.000936	−0.017724	−0.063254	−0.081762	0.028030
5.00	−0.000658	−0.000837	−0.016522	−0.060949	−0.082889	0.017748

2.2 First, Second and Third Derivatives with Respect to the Order of the modified Struve Functions

The modified Struve functions $L_v(t)$ were taken from MATHEMATICA program and their derivatives with respect to the order were evaluated by using the central-difference formulas of order $O(h^4)$ with $h = 10^{-3}$:

$$\frac{\partial L_v(t)}{\partial v} = \frac{-L_{v+2h}(t) + 8L_{v+h}(t) - 8L_{v-h}(t) + L_{v-2h}(t)}{12h} \tag{2.2.1}$$

$$\frac{\partial^2 L_v(t)}{\partial v^2} = \frac{-L_{v+2h}(t) + 16L_{v+h}(t) - 30L_v(t) + 16L_{v-h}(t) - L_{v-2h}(t)}{12h^2} \tag{2.2.2}$$

$$\frac{\partial^3 L_v(t)}{\partial v^3} = \frac{-L_{v+3h}(t) + 8L_{v+2h}(t) - 13L_{v+h}(t)}{8h^3}$$

$$+ \frac{13L_{v-h}(t) - 8L_{v-2h}(t) + L_{v-3h}(t)}{8h^3} \tag{2.2.3}$$

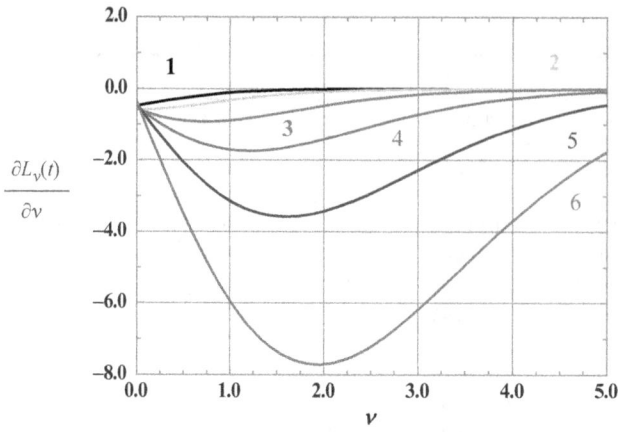

Figure 2.10: First derivatives of the modified Struve function with respect to the order v as a function of v, at constant values of argument t.

$1 - t = 0.50; \ 2 - t = 1.0; \ 3 - t = 2.0; \ 4 - t = 3.0; \ 5 - t = 4.0; \ 6 - t = 5.0.$

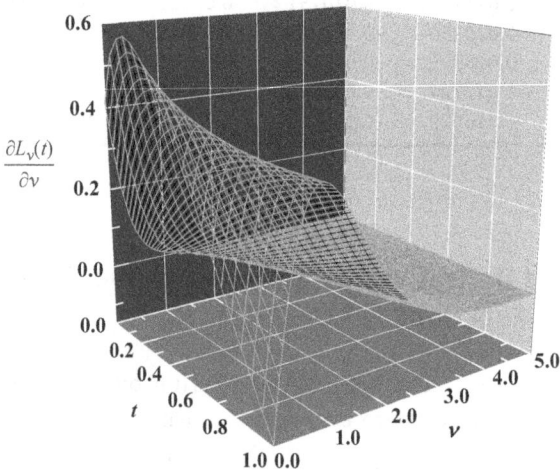

Figure 2.11: First derivatives with respect to the order of the modified Struve functions $\partial L_v(t)/\partial v$ as a function of v and t.

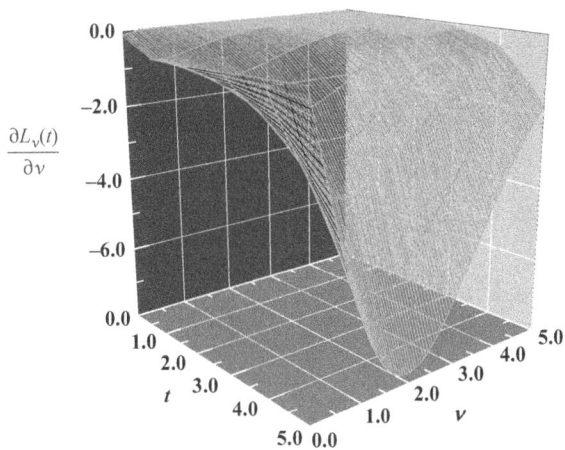

Figure 2.12: First derivatives with respect to the order of the modified Struve functions $\partial L_\nu(t)/\partial \nu$ as a function of ν and t.

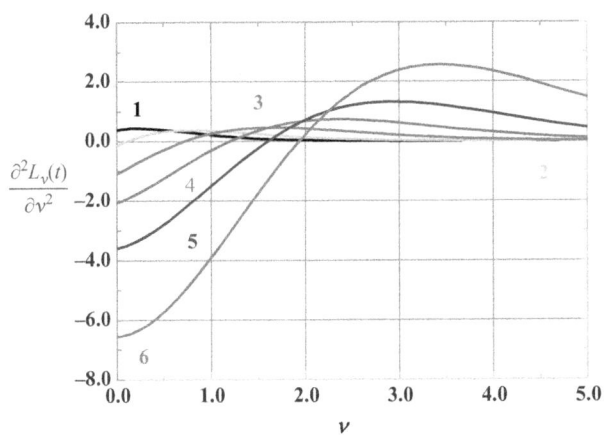

Figure 2.13: Second derivatives of the modified Struve function with respect to the order ν as a function of ν, at constant values of argument t.

$1 - t = 0.50$; $2 - t = 1.0$; $3 - t = 2.0$; $4 - t = 3.0$; $5 - t = 4.0$; $6 - t = 5.0$.

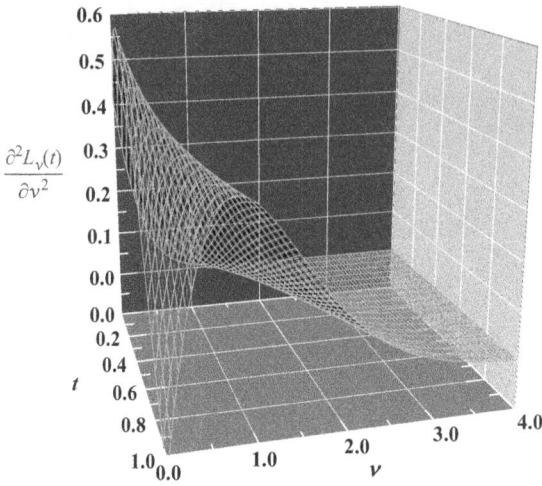

Figure 2.14: Second derivatives with respect to the order of the modified Struve functions $\partial^2 L_\nu(t)/\partial \nu^2$ as a function of ν and t.

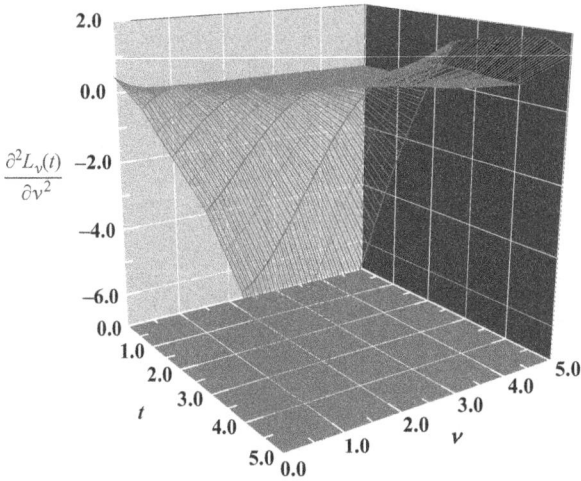

Figure 2.15: Second derivatives with respect to the order of the modified Struve functions $\partial^2 L_\nu(t)/\partial \nu^2$ as a function of ν and t.

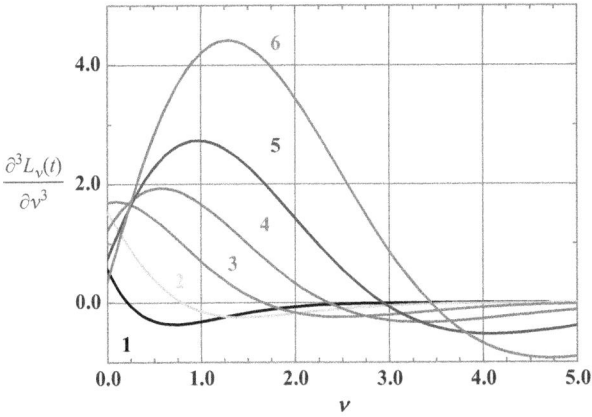

Figure 2.16: Third derivatives of the modified Struve function with respect to the order v as a function of v, at constant values of argument t.

1 – t = 0.50; 2 – t = 1.0; 3 – t = 2.0; 4 – t = 3.0; 5 – t = 4.0; 6 – t = 5.0.

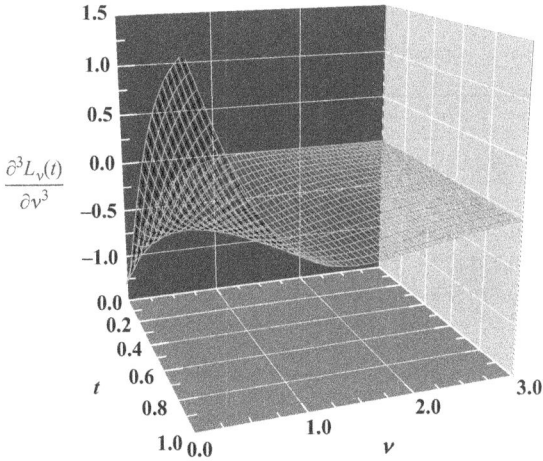

Figure 2.17: Third derivatives with respect to the order of the modified Struve functions $\partial^3 L_v(t)/\partial v^3$ as a function of v and t.

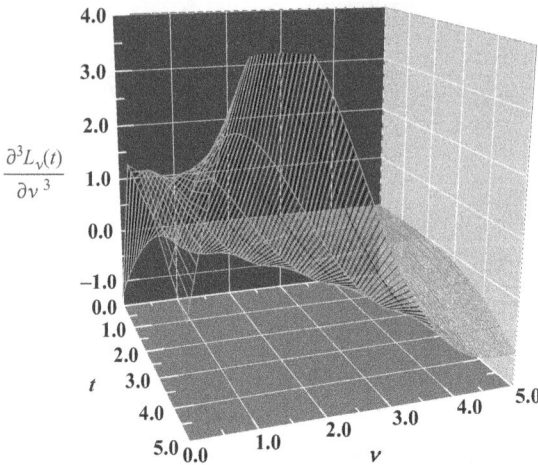

Figure 2.18: Third derivatives with respect to the order of the modified Struve functions $\partial^3 L_v(t)/\partial v^3$ as a function of v and t.

Table 2.4: First derivatives with respect to the order of the modified Struve functions $\partial L_v(t)/\partial v$.

			t			
v	0.05	0.10	0.15	0.20	0.25	0.30
0.00	−0.118621	−0.193299	−0.251624	−0.299524	−0.339845	−0.374279
0.05	−0.099554	−0.168408	−0.224208	−0.271295	−0.311850	−0.347211
0.10	−0.083312	−0.146273	−0.199137	−0.244900	−0.285156	−0.320925
0.15	−0.069529	−0.126680	−0.176333	−0.220373	−0.259890	−0.295615
0.20	−0.057875	−0.109411	−0.155693	−0.197712	−0.236130	−0.271429
0.25	−0.048056	−0.094250	−0.137097	−0.176882	−0.213916	−0.248471
0.30	−0.039809	−0.080989	−0.120410	−0.157824	−0.193258	−0.226806
0.35	−0.032902	−0.069429	−0.105495	−0.140461	−0.174137	−0.206473
0.40	−0.027135	−0.059385	−0.092211	−0.124707	−0.156518	−0.187481
0.45	−0.022333	−0.050684	−0.080419	−0.110464	−0.140347	−0.169821
0.50	−0.018344	−0.043168	−0.069985	−0.097632	−0.125562	−0.153468
0.55	−0.015038	−0.036693	−0.060779	−0.086108	−0.112091	−0.138381
0.60	−0.012305	−0.031130	−0.052680	−0.075791	−0.099857	−0.124512
0.65	−0.010051	−0.026361	−0.045573	−0.066579	−0.088781	−0.111805
0.70	−0.008196	−0.022283	−0.039353	−0.058378	−0.078782	−0.100198
0.75	−0.006672	−0.018804	−0.033922	−0.051094	−0.069781	−0.089626
0.80	−0.005423	−0.015841	−0.029190	−0.044641	−0.061697	−0.080025
0.85	−0.004400	−0.013324	−0.025077	−0.038937	−0.054456	−0.071327
0.90	−0.003565	−0.011189	−0.021509	−0.033906	−0.047986	−0.063467
0.95	−0.002885	−0.009382	−0.018420	−0.029479	−0.042216	−0.056381
1.00	−0.002330	−0.007855	−0.015751	−0.025591	−0.037082	−0.050006

Table 2.4 (continued)

				t		
v	0.05	0.10	0.15	0.20	0.25	0.30
1.05	−0.001880	−0.006567	−0.013450	−0.022182	−0.032523	−0.044285
1.10	−0.001515	−0.005483	−0.011468	−0.019200	−0.028483	−0.039160
1.15	−0.001219	−0.004572	−0.009765	−0.016596	−0.024909	−0.034578
1.20	−0.000979	−0.003807	−0.008303	−0.014325	−0.021754	−0.030489
1.25	−0.000786	−0.003166	−0.007051	−0.012348	−0.018972	−0.026847
1.30	−0.000630	−0.002630	−0.005980	−0.010631	−0.016525	−0.023609
1.35	−0.000505	−0.002181	−0.005066	−0.009141	−0.014375	−0.020734
1.40	−0.000403	−0.001807	−0.004286	−0.007850	−0.012489	−0.018187
1.45	−0.000322	−0.001496	−0.003622	−0.006733	−0.010838	−0.015933
1.50	−0.000257	−0.001237	−0.003057	−0.005769	−0.009393	−0.013942
1.55	−0.000205	−0.001021	−0.002578	−0.004937	−0.008132	−0.012185
1.60	−0.000163	−0.000842	−0.002171	−0.004220	−0.007032	−0.010638
1.65	−0.000130	−0.000694	−0.001827	−0.003603	−0.006074	−0.009276
1.70	−0.000103	−0.000571	−0.001535	−0.003073	−0.005241	−0.008080
1.75	−0.000082	−0.000470	−0.001289	−0.002619	−0.004518	−0.007031
1.80	−0.000065	−0.000386	−0.001081	−0.002229	−0.003890	−0.006112
1.85	−0.000051	−0.000317	−0.000906	−0.001896	−0.003346	−0.005307
1.90	−0.000041	−0.000260	−0.000758	−0.001610	−0.002876	−0.004604
1.95	−0.000032	−0.000213	−0.000634	−0.001367	−0.002469	−0.003990
2.00	−0.000025	−0.000174	−0.000530	−0.001159	−0.002117	−0.003454
2.05	−0.000020	−0.000142	−0.000442	−0.000982	−0.001814	−0.002988
2.10	−0.000016	−0.000116	−0.000369	−0.000831	−0.001553	−0.002582
2.15	−0.000012	−0.000095	−0.000307	−0.000702	−0.001329	−0.002229
2.20	−0.000010	−0.000077	−0.000256	−0.000593	−0.001135	−0.001923
2.25	−0.000008	−0.000063	−0.000213	−0.000501	−0.000969	−0.001657
2.30	−0.000006	−0.000051	−0.000177	−0.000422	−0.000827	−0.001427
2.35	−0.000005	−0.000042	−0.000147	−0.000356	−0.000705	−0.001228
2.40	−0.000004	−0.000034	−0.000122	−0.000300	−0.000600	−0.001056
2.45	−0.000003	−0.000028	−0.000101	−0.000252	−0.000511	−0.000907
2.50	−0.000002	−0.000022	−0.000084	−0.000212	−0.000434	−0.000778
2.55	−0.000002	−0.000018	−0.000069	−0.000178	−0.000369	−0.000667
2.60	−0.000001	−0.000015	−0.000057	−0.000149	−0.000313	−0.000572
2.65	−0.000001	−0.000012	−0.000047	−0.000125	−0.000266	−0.000490
2.70	−0.000001	−0.000010	−0.000039	−0.000105	−0.000225	−0.000419
2.75	−0.000001	−0.000008	−0.000032	−0.000088	−0.000191	−0.000358
2.80	−0.000001	−0.000006	−0.000027	−0.000074	−0.000162	−0.000306
2.85		−0.000005	−0.000022	−0.000062	−0.000137	−0.000261
2.90		−0.000004	−0.000018	−0.000051	−0.000116	−0.000223
2.95		−0.000003	−0.000015	−0.000043	−0.000098	−0.000190
3.00		−0.000003	−0.000012	−0.000036	−0.000082	−0.000162
3.05		−0.000002	−0.000010	−0.000030	−0.000069	−0.000138
3.10		−0.000002	−0.000008	−0.000025	−0.000059	−0.000117
3.15		−0.000001	−0.000007	−0.000021	−0.000049	−0.000100

Table 2.4 (continued)

			t			
v	0.05	0.10	0.15	0.20	0.25	0.30
3.20		−0.000001	−0.000006	−0.000017	−0.000042	−0.000085
3.25		−0.000001	−0.000005	−0.000014	−0.000035	−0.000072
3.30		−0.000001	−0.000004	−0.000012	−0.000029	−0.000061
3.35		−0.000001	−0.000003	−0.000010	−0.000025	−0.000052
3.40			−0.000002	−0.000008	−0.000021	−0.000044
3.45			−0.000002	−0.000007	−0.000017	−0.000037
3.50			−0.000002	−0.000006	−0.000015	−0.000031
3.55			−0.000001	−0.000005	−0.000012	−0.000027
3.60			−0.000001	−0.000004	−0.000010	−0.000023
3.65			−0.000001	−0.000003	−0.000009	−0.000019
3.70			−0.000001	−0.000003	−0.000007	−0.000016
3.75			−0.000001	−0.000002	−0.000006	−0.000014
3.80				−0.000002	−0.000005	−0.000011
3.85				−0.000002	−0.000004	−0.000010
3.90				−0.000001	−0.000003	−0.000008
3.95				−0.000001	−0.000003	−0.000007
4.00				−0.000001	−0.000002	−0.000006
4.05				−0.000001	−0.000002	−0.000005
4.10				−0.000001	−0.000002	−0.000004
4.15					−0.000001	−0.000003
4.20					−0.000001	−0.000003
4.25					−0.000001	−0.000002
4.30					−0.000001	−0.000002
4.35					−0.000001	−0.000002
4.40					−0.000001	−0.000001
4.45						−0.000001
4.50						−0.000001
4.55						−0.000001
4.60						−0.000001
4.65						−0.000001

			t			
v	0.35	0.40	0.45	0.50	0.55	0.60
0.00	−0.403955	−0.429680	−0.452065	−0.471582	−0.488615	−0.503472
0.05	−0.378297	−0.405777	−0.430164	−0.451866	−0.471213	−0.488481
0.10	−0.352923	−0.381690	−0.407646	−0.431131	−0.452425	−0.471765
0.15	−0.328086	−0.357716	−0.384839	−0.409730	−0.432618	−0.453702
0.20	−0.303985	−0.334099	−0.362022	−0.387967	−0.412116	−0.434630
0.25	−0.280778	−0.311039	−0.339430	−0.366106	−0.391205	−0.414851
0.30	−0.258581	−0.288695	−0.317258	−0.344371	−0.370133	−0.394633
0.35	−0.237479	−0.267192	−0.295665	−0.322953	−0.349115	−0.374212
0.40	−0.217525	−0.246624	−0.274779	−0.302007	−0.328333	−0.353789

Table 2.4 (continued)

			t			
v	0.35	0.40	0.45	0.50	0.55	0.60
0.45	-0.198750	-0.227056	-0.254699	-0.281661	-0.307939	-0.333541
0.50	-0.181163	-0.208531	-0.235497	-0.262016	-0.288060	-0.313615
0.55	-0.164757	-0.191073	-0.217227	-0.243151	-0.268797	-0.294135
0.60	-0.149511	-0.174688	-0.199920	-0.225123	-0.250232	-0.275203
0.65	-0.135394	-0.159368	-0.183594	-0.207972	-0.232426	-0.256900
0.70	-0.122365	-0.145097	-0.168252	-0.191722	-0.215423	-0.239290
0.75	-0.110377	-0.131845	-0.153885	-0.176384	-0.199253	-0.222421
0.80	-0.099380	-0.119578	-0.140475	-0.161958	-0.183933	-0.206327
0.85	-0.089319	-0.108256	-0.127999	-0.148433	-0.169469	-0.191029
0.90	-0.080139	-0.097835	-0.116423	-0.135794	-0.155857	-0.176537
0.95	-0.071783	-0.088268	-0.105713	-0.124015	-0.143086	-0.162854
1.00	-0.064194	-0.079507	-0.095830	-0.113068	-0.131138	-0.149973
1.05	-0.057319	-0.071502	-0.086732	-0.102920	-0.119991	-0.137882
1.10	-0.051103	-0.064206	-0.078376	-0.093536	-0.109618	-0.126561
1.15	-0.045494	-0.057568	-0.070719	-0.084879	-0.099987	-0.115990
1.20	-0.040444	-0.051542	-0.063717	-0.076909	-0.091067	-0.106142
1.25	-0.035904	-0.046083	-0.057327	-0.069589	-0.082822	-0.096988
1.30	-0.031831	-0.041145	-0.051507	-0.062877	-0.075219	-0.088498
1.35	-0.028184	-0.036689	-0.046217	-0.056737	-0.068220	-0.080640
1.40	-0.024922	-0.032673	-0.041415	-0.051129	-0.061791	-0.073382
1.45	-0.022011	-0.029060	-0.037066	-0.046016	-0.055895	-0.066690
1.50	-0.019416	-0.025815	-0.033132	-0.041363	-0.050498	-0.060531
1.55	-0.017107	-0.022905	-0.029581	-0.037135	-0.045567	-0.054873
1.60	-0.015055	-0.020299	-0.026379	-0.033300	-0.041068	-0.049684
1.65	-0.013235	-0.017970	-0.023496	-0.029827	-0.036970	-0.044933
1.70	-0.011621	-0.015890	-0.020905	-0.026685	-0.033243	-0.040590
1.75	-0.010194	-0.014035	-0.018580	-0.023848	-0.029858	-0.036626
1.80	-0.008932	-0.012384	-0.016495	-0.021290	-0.026789	-0.033012
1.85	-0.007818	-0.010916	-0.014629	-0.018986	-0.024010	-0.029723
1.90	-0.006837	-0.009612	-0.012961	-0.016914	-0.021497	-0.026734
1.95	-0.005972	-0.008455	-0.011472	-0.015053	-0.019227	-0.024021
2.00	-0.005212	-0.007430	-0.010143	-0.013383	-0.017180	-0.021561
2.05	-0.004545	-0.006524	-0.008960	-0.011887	-0.015336	-0.019335
2.10	-0.003959	-0.005722	-0.007907	-0.010548	-0.013676	-0.017321
2.15	-0.003445	-0.005014	-0.006972	-0.009352	-0.012185	-0.015503
2.20	-0.002996	-0.004390	-0.006142	-0.008283	-0.010846	-0.013862
2.25	-0.002603	-0.003840	-0.005405	-0.007330	-0.009646	-0.012384
2.30	-0.002259	-0.003357	-0.004753	-0.006481	-0.008571	-0.011054
2.35	-0.001959	-0.002931	-0.004176	-0.005725	-0.007609	-0.009858
2.40	-0.001698	-0.002557	-0.003666	-0.005053	-0.006749	-0.008783
2.45	-0.001470	-0.002230	-0.003215	-0.004456	-0.005982	-0.007820
2.50	-0.001272	-0.001942	-0.002818	-0.003927	-0.005297	-0.006956
2.55	-0.001099	-0.001690	-0.002468	-0.003457	-0.004687	-0.006182

Table 2.4 (continued)

				t		
v	0.35	0.40	0.45	0.50	0.55	0.60
2.60	−0.000949	−0.001470	−0.002159	−0.003042	−0.004144	−0.005491
2.65	−0.000819	−0.001278	−0.001888	−0.002674	−0.003660	−0.004872
2.70	−0.000707	−0.001109	−0.001649	−0.002349	−0.003231	−0.004320
2.75	−0.000609	−0.000963	−0.001440	−0.002062	−0.002850	−0.003828
2.80	−0.000524	−0.000835	−0.001256	−0.001808	−0.002512	−0.003389
2.85	−0.000451	−0.000723	−0.001095	−0.001585	−0.002212	−0.002998
2.90	−0.000388	−0.000626	−0.000954	−0.001388	−0.001947	−0.002651
2.95	−0.000333	−0.000542	−0.000830	−0.001215	−0.001712	−0.002342
3.00	−0.000286	−0.000468	−0.000722	−0.001062	−0.001505	−0.002067
3.05	−0.000246	−0.000405	−0.000628	−0.000928	−0.001322	−0.001824
3.10	−0.000211	−0.000349	−0.000545	−0.000811	−0.001160	−0.001608
3.15	−0.000181	−0.000301	−0.000473	−0.000708	−0.001017	−0.001416
3.20	−0.000155	−0.000260	−0.000410	−0.000617	−0.000892	−0.001247
3.25	−0.000132	−0.000224	−0.000356	−0.000538	−0.000781	−0.001097
3.30	−0.000113	−0.000193	−0.000308	−0.000468	−0.000684	−0.000964
3.35	−0.000097	−0.000166	−0.000267	−0.000408	−0.000598	−0.000847
3.40	−0.000083	−0.000143	−0.000231	−0.000355	−0.000523	−0.000744
3.45	−0.000071	−0.000123	−0.000200	−0.000308	−0.000457	−0.000653
3.50	−0.000060	−0.000105	−0.000173	−0.000268	−0.000399	−0.000573
3.55	−0.000051	−0.000090	−0.000149	−0.000233	−0.000348	−0.000502
3.60	−0.000044	−0.000078	−0.000129	−0.000202	−0.000303	−0.000440
3.65	−0.000037	−0.000067	−0.000111	−0.000175	−0.000264	−0.000385
3.70	−0.000032	−0.000057	−0.000096	−0.000152	−0.000230	−0.000337
3.75	−0.000027	−0.000049	−0.000082	−0.000132	−0.000200	−0.000294
3.80	−0.000023	−0.000042	−0.000071	−0.000114	−0.000174	−0.000257
3.85	−0.000019	−0.000036	−0.000061	−0.000099	−0.000152	−0.000225
3.90	−0.000017	−0.000031	−0.000053	−0.000085	−0.000132	−0.000196
3.95	−0.000014	−0.000026	−0.000045	−0.000074	−0.000114	−0.000171
4.00	−0.000012	−0.000022	−0.000039	−0.000064	−0.000099	−0.000149
4.05	−0.000010	−0.000019	−0.000033	−0.000055	−0.000086	−0.000130
4.10	−0.000009	−0.000016	−0.000029	−0.000047	−0.000075	−0.000113
4.15	−0.000007	−0.000014	−0.000025	−0.000041	−0.000065	−0.000099
4.20	−0.000006	−0.000012	−0.000021	−0.000035	−0.000056	−0.000086
4.25	−0.000005	−0.000010	−0.000018	−0.000030	−0.000049	−0.000075
4.30	−0.000004	−0.000009	−0.000015	−0.000026	−0.000042	−0.000065
4.35	−0.000004	−0.000007	−0.000013	−0.000022	−0.000036	−0.000056
4.40	−0.000003	−0.000006	−0.000011	−0.000019	−0.000031	−0.000049
4.45	−0.000003	−0.000005	−0.000010	−0.000017	−0.000027	−0.000042
4.50	−0.000002	−0.000004	−0.000008	−0.000014	−0.000023	−0.000037
4.55	−0.000002	−0.000004	−0.000007	−0.000012	−0.000020	−0.000032
4.60	−0.000002	−0.000003	−0.000006	−0.000011	−0.000017	−0.000028
4.65	−0.000001	−0.000003	−0.000005	−0.000009	−0.000015	−0.000024
4.70	−0.000001	−0.000002	−0.000004	−0.000008	−0.000013	−0.000021

Table 2.4 (continued)

				t		
v	0.35	0.40	0.45	0.50	0.55	0.60
4.75	−0.000001	−0.000002	−0.000004	−0.000007	−0.000011	−0.000018
4.80	−0.000001	−0.000002	−0.000003	−0.000006	−0.000010	−0.000016
4.85	−0.000001	−0.000001	−0.000003	−0.000005	−0.000008	−0.000013
4.90	−0.000001	−0.000001	−0.000002	−0.000004	−0.000007	−0.000012
4.95		−0.000001	−0.000002	−0.000004	−0.000006	−0.000010
5.00		−0.000001	−0.000002	−0.000003	−0.000005	−0.000009

				t		
v	0.65	0.70	0.75	0.80	0.85	0.90
0.00	−0.516416	−0.527664	−0.537405	−0.545799	−0.552987	−0.559092
0.05	−0.503901	−0.517670	−0.529960	−0.540918	−0.550675	−0.559346
0.10	−0.489353	−0.505362	−0.519945	−0.533235	−0.545352	−0.556401
0.15	−0.473153	−0.491121	−0.507739	−0.523124	−0.537384	−0.550612
0.20	−0.455649	−0.475300	−0.493696	−0.510937	−0.527117	−0.542319
0.25	−0.437156	−0.458223	−0.478143	−0.497002	−0.514879	−0.531847
0.30	−0.417956	−0.440181	−0.461381	−0.481626	−0.500979	−0.519502
0.35	−0.398301	−0.421440	−0.443685	−0.465088	−0.485701	−0.505574
0.40	−0.378411	−0.402236	−0.425302	−0.447647	−0.469311	−0.490330
0.45	−0.358481	−0.382778	−0.406455	−0.429536	−0.452049	−0.474020
0.50	−0.338677	−0.363250	−0.387342	−0.410966	−0.434137	−0.456872
0.55	−0.319144	−0.343813	−0.368140	−0.392125	−0.415774	−0.439096
0.60	−0.300001	−0.324606	−0.349002	−0.373181	−0.397140	−0.420880
0.65	−0.281350	−0.305744	−0.330060	−0.354280	−0.378394	−0.402395
0.70	−0.263272	−0.287328	−0.311429	−0.335551	−0.359677	−0.383795
0.75	−0.245831	−0.269438	−0.293205	−0.317104	−0.341113	−0.365214
0.80	−0.229079	−0.252140	−0.275469	−0.299034	−0.322808	−0.346772
0.85	−0.213051	−0.235484	−0.258285	−0.281418	−0.304855	−0.328571
0.90	−0.197773	−0.219510	−0.241706	−0.264323	−0.287330	−0.310702
0.95	−0.183258	−0.204245	−0.225772	−0.247800	−0.270297	−0.293238
1.00	−0.169513	−0.189707	−0.210512	−0.231890	−0.253810	−0.276245
1.05	−0.156535	−0.175903	−0.195945	−0.216624	−0.237908	−0.259773
1.10	−0.144316	−0.162837	−0.182083	−0.202023	−0.222625	−0.243864
1.15	−0.132842	−0.150501	−0.168931	−0.188100	−0.207982	−0.228551
1.20	−0.122094	−0.138886	−0.156485	−0.174863	−0.193994	−0.213856
1.25	−0.112050	−0.127976	−0.144738	−0.162310	−0.180669	−0.199797
1.30	−0.102685	−0.117753	−0.133678	−0.150436	−0.168010	−0.186383
1.35	−0.093973	−0.108195	−0.123288	−0.139232	−0.156013	−0.173616
1.40	−0.085883	−0.099277	−0.113549	−0.128684	−0.144671	−0.161497
1.45	−0.078387	−0.090974	−0.104440	−0.118775	−0.133970	−0.150018
1.50	−0.071453	−0.083258	−0.095937	−0.109486	−0.123898	−0.139169
1.55	−0.065052	−0.076100	−0.088015	−0.100794	−0.114436	−0.128938
1.60	−0.059152	−0.069473	−0.080647	−0.092677	−0.105564	−0.119309
1.65	−0.053724	−0.063347	−0.073808	−0.085111	−0.097261	−0.110263

Table 2.4 (continued)

				t		
v	0.65	0.70	0.75	0.80	0.85	0.90
1.70	−0.048737	−0.057694	−0.067469	−0.078070	−0.089505	−0.101782
1.75	−0.044164	−0.052486	−0.061604	−0.071529	−0.082272	−0.093843
1.80	−0.039975	−0.047695	−0.056187	−0.065463	−0.075538	−0.086426
1.85	−0.036145	−0.043295	−0.051189	−0.059845	−0.069279	−0.079506
1.90	−0.032648	−0.039258	−0.046587	−0.054651	−0.063469	−0.073060
1.95	−0.029458	−0.035562	−0.042354	−0.049855	−0.058086	−0.067066
2.00	−0.026553	−0.032180	−0.038466	−0.045433	−0.053104	−0.061500
2.05	−0.023911	−0.029091	−0.034900	−0.041362	−0.048500	−0.056338
2.10	−0.021511	−0.026273	−0.031634	−0.037618	−0.044252	−0.051558
2.15	−0.019333	−0.023705	−0.028646	−0.034181	−0.040336	−0.047138
2.20	−0.017360	−0.021368	−0.025915	−0.031028	−0.036732	−0.043056
2.25	−0.015574	−0.019244	−0.023424	−0.028139	−0.033419	−0.039290
2.30	−0.013959	−0.017316	−0.021152	−0.025497	−0.030377	−0.035821
2.35	−0.012501	−0.015567	−0.019084	−0.023082	−0.027587	−0.032628
2.40	−0.011185	−0.013982	−0.017203	−0.020877	−0.025031	−0.029694
2.45	−0.009999	−0.012548	−0.015494	−0.018867	−0.022693	−0.027000
2.50	−0.008932	−0.011252	−0.013944	−0.017036	−0.020555	−0.024530
2.55	−0.007971	−0.010081	−0.012538	−0.015369	−0.018603	−0.022267
2.60	−0.007109	−0.009024	−0.011264	−0.013855	−0.016823	−0.020196
2.65	−0.006334	−0.008072	−0.010112	−0.012479	−0.015201	−0.018303
2.70	−0.005640	−0.007215	−0.009071	−0.011232	−0.013724	−0.016574
2.75	−0.005018	−0.006444	−0.008130	−0.010101	−0.012381	−0.014997
2.80	−0.004461	−0.005751	−0.007282	−0.009077	−0.011161	−0.013559
2.85	−0.003963	−0.005128	−0.006517	−0.008151	−0.010054	−0.012250
2.90	−0.003518	−0.004570	−0.005828	−0.007313	−0.009049	−0.011059
2.95	−0.003121	−0.004069	−0.005208	−0.006557	−0.008139	−0.009977
3.00	−0.002766	−0.003621	−0.004651	−0.005875	−0.007316	−0.008993
3.05	−0.002450	−0.003220	−0.004150	−0.005260	−0.006571	−0.008101
3.10	−0.002169	−0.002861	−0.003701	−0.004707	−0.005897	−0.007293
3.15	−0.001919	−0.002541	−0.003298	−0.004208	−0.005289	−0.006560
3.20	−0.001696	−0.002254	−0.002937	−0.003760	−0.004741	−0.005897
3.25	−0.001498	−0.001999	−0.002614	−0.003357	−0.004246	−0.005298
3.30	−0.001323	−0.001772	−0.002325	−0.002996	−0.003801	−0.004756
3.35	−0.001167	−0.001569	−0.002066	−0.002671	−0.003400	−0.004267
3.40	−0.001029	−0.001389	−0.001835	−0.002381	−0.003039	−0.003826
3.45	−0.000907	−0.001228	−0.001629	−0.002120	−0.002715	−0.003428
3.50	−0.000798	−0.001086	−0.001445	−0.001887	−0.002424	−0.003069
3.55	−0.000703	−0.000959	−0.001281	−0.001679	−0.002163	−0.002747
3.60	−0.000618	−0.000847	−0.001135	−0.001492	−0.001929	−0.002456
3.65	−0.000543	−0.000747	−0.001005	−0.001326	−0.001719	−0.002196
3.70	−0.000477	−0.000659	−0.000889	−0.001177	−0.001531	−0.001961
3.75	−0.000419	−0.000581	−0.000787	−0.001044	−0.001363	−0.001751
3.80	−0.000368	−0.000511	−0.000695	−0.000926	−0.001212	−0.001562

Table 2.4 (continued)

			t			
v	0.65	0.70	0.75	0.80	0.85	0.90
3.85	−0.000322	−0.000450	−0.000614	−0.000821	−0.001078	−0.001393
3.90	−0.000283	−0.000396	−0.000542	−0.000727	−0.000958	−0.001241
3.95	−0.000248	−0.000348	−0.000478	−0.000644	−0.000850	−0.001106
4.00	−0.000217	−0.000306	−0.000422	−0.000570	−0.000755	−0.000984
4.05	−0.000190	−0.000269	−0.000372	−0.000504	−0.000670	−0.000876
4.10	−0.000166	−0.000236	−0.000328	−0.000445	−0.000594	−0.000779
4.15	−0.000145	−0.000207	−0.000289	−0.000393	−0.000526	−0.000692
4.20	−0.000127	−0.000182	−0.000254	−0.000347	−0.000466	−0.000615
4.25	−0.000111	−0.000159	−0.000223	−0.000306	−0.000412	−0.000546
4.30	−0.000097	−0.000139	−0.000196	−0.000270	−0.000365	−0.000484
4.35	−0.000084	−0.000122	−0.000172	−0.000238	−0.000323	−0.000429
4.40	−0.000073	−0.000107	−0.000151	−0.000210	−0.000285	−0.000381
4.45	−0.000064	−0.000093	−0.000133	−0.000185	−0.000252	−0.000337
4.50	−0.000056	−0.000082	−0.000117	−0.000163	−0.000222	−0.000299
4.55	−0.000048	−0.000071	−0.000102	−0.000143	−0.000196	−0.000264
4.60	−0.000042	−0.000062	−0.000090	−0.000126	−0.000173	−0.000234
4.65	−0.000037	−0.000054	−0.000079	−0.000111	−0.000153	−0.000207
4.70	−0.000032	−0.000047	−0.000069	−0.000097	−0.000135	−0.000183
4.75	−0.000028	−0.000041	−0.000060	−0.000085	−0.000119	−0.000161
4.80	−0.000024	−0.000036	−0.000053	−0.000075	−0.000104	−0.000143
4.85	−0.000021	−0.000031	−0.000046	−0.000066	−0.000092	−0.000126
4.90	−0.000018	−0.000027	−0.000040	−0.000058	−0.000081	−0.000111
4.95	−0.000016	−0.000024	−0.000035	−0.000051	−0.000071	−0.000098
5.00	−0.000014	−0.000021	−0.000031	−0.000044	−0.000062	−0.000086

			t			
v	0.95	1.0	2.0	3.0	4.0	5.0
0.00	−0.564222	−0.568474	−0.555123	−0.486199	−0.421180	−0.368653
0.05	−0.567034	−0.573830	−0.607642	−0.588360	−0.600152	−0.696047
0.10	−0.566478	−0.575667	−0.655913	−0.687168	−0.776782	−1.02186
0.15	−0.562896	−0.574314	−0.699919	−0.782346	−0.950600	−1.34541
0.20	−0.556621	−0.570093	−0.739676	−0.873650	−1.12116	−1.66603
0.25	−0.547971	−0.563316	−0.775224	−0.960861	−1.28804	−1.98306
0.30	−0.537253	−0.554284	−0.806628	−1.04379	−1.45085	−2.29588
0.35	−0.524753	−0.543284	−0.833975	−1.12229	−1.60919	−2.60386
0.40	−0.510744	−0.530589	−0.857371	−1.19622	−1.76274	−2.90642
0.45	−0.495479	−0.516453	−0.876938	−1.26548	−1.91116	−3.20298
0.50	−0.479193	−0.501118	−0.892815	−1.33000	−2.05417	−3.49300
0.55	−0.462102	−0.484807	−0.905152	−1.38972	−2.19150	−3.77596
0.60	−0.444405	−0.467723	−0.914109	−1.44462	−2.32291	−4.05138
0.65	−0.426283	−0.450058	−0.919854	−1.49470	−2.44819	−4.31880
0.70	−0.407898	−0.431982	−0.922564	−1.53997	−2.56717	−4.57779

Table 2.4 (continued)

				t		
v	0.95	1.0	2.0	3.0	4.0	5.0
0.75	-0.389397	-0.413652	-0.922417	-1.58048	-2.67968	-4.82795
0.80	-0.370909	-0.395208	-0.919596	-1.61629	-2.78561	-5.06893
0.85	-0.352549	-0.376774	-0.914287	-1.64747	-2.88486	-5.30038
0.90	-0.334418	-0.358461	-0.906672	-1.67412	-2.97735	-5.52201
0.95	-0.316601	-0.340367	-0.896936	-1.69635	-3.06305	-5.73356
1.00	-0.299172	-0.322574	-0.885259	-1.71427	-3.14192	-5.93480
1.05	-0.282195	-0.305155	-0.871819	-1.72802	-3.21397	-6.12552
1.10	-0.265719	-0.288172	-0.856790	-1.73774	-3.27922	-6.30557
1.15	-0.249787	-0.271674	-0.840340	-1.74360	-3.33773	-6.47480
1.20	-0.234432	-0.255704	-0.822632	-1.74573	-3.38956	-6.63312
1.25	-0.219677	-0.240293	-0.803823	-1.74433	-3.43480	-6.78047
1.30	-0.205539	-0.225467	-0.784063	-1.73954	-3.47355	-6.91680
1.35	-0.192030	-0.211243	-0.763495	-1.73157	-3.50594	-7.04210
1.40	-0.179154	-0.197634	-0.742256	-1.72057	-3.53210	-7.15640
1.45	-0.166911	-0.184644	-0.720473	-1.70674	-3.55218	-7.25975
1.50	-0.155296	-0.172276	-0.698268	-1.69025	-3.56636	-7.35222
1.55	-0.144301	-0.160525	-0.675754	-1.67129	-3.57480	-7.43391
1.60	-0.133915	-0.149385	-0.653036	-1.65003	-3.57770	-7.50495
1.65	-0.124123	-0.138845	-0.630213	-1.62666	-3.57525	-7.56549
1.70	-0.114909	-0.128893	-0.607375	-1.60136	-3.56766	-7.61571
1.75	-0.106253	-0.119513	-0.584605	-1.57429	-3.55514	-7.65579
1.80	-0.098138	-0.110688	-0.561978	-1.54562	-3.53791	-7.68594
1.85	-0.090541	-0.102399	-0.539565	-1.51553	-3.51619	-7.70640
1.90	-0.083440	-0.094627	-0.517427	-1.48416	-3.49021	-7.71742
1.95	-0.076815	-0.087351	-0.495621	-1.45169	-3.46019	-7.71925
2.00	-0.070642	-0.080550	-0.474195	-1.41825	-3.42636	-7.71216
2.05	-0.064898	-0.074203	-0.453193	-1.38398	-3.38897	-7.69645
2.10	-0.059562	-0.068286	-0.432653	-1.34904	-3.34823	-7.67242
2.15	-0.054611	-0.062780	-0.412609	-1.31354	-3.30437	-7.64037
2.20	-0.050024	-0.057662	-0.393088	-1.27762	-3.25763	-7.60062
2.25	-0.045778	-0.052911	-0.374112	-1.24138	-3.20822	-7.55348
2.30	-0.041855	-0.048507	-0.355702	-1.20496	-3.15638	-7.49930
2.35	-0.038233	-0.044429	-0.337871	-1.16844	-3.10231	-7.43840
2.40	-0.034894	-0.040657	-0.320630	-1.13193	-3.04623	-7.37112
2.45	-0.031818	-0.037173	-0.303987	-1.09552	-2.98834	-7.29780
2.50	-0.028989	-0.033959	-0.287947	-1.05930	-2.92886	-7.21877
2.55	-0.026389	-0.030996	-0.272509	-1.02334	-2.86797	-7.13438
2.60	-0.024002	-0.028268	-0.257675	-0.987724	-2.80587	-7.04496
2.65	-0.021814	-0.025759	-0.243438	-0.952513	-2.74273	-6.95086
2.70	-0.019809	-0.023455	-0.229795	-0.917771	-2.67875	-6.85240
2.75	-0.017974	-0.021339	-0.216737	-0.883551	-2.61408	-6.74991
2.80	-0.016297	-0.019399	-0.204255	-0.849906	-2.54889	-6.64372
2.85	-0.014765	-0.017622	-0.192338	-0.816879	-2.48333	-6.53414

Table 2.4 (continued)

			t			
v	0.95	1.0	2.0	3.0	4.0	5.0
2.90	−0.013367	−0.015996	−0.180975	−0.784511	−2.41756	−6.42150
2.95	−0.012092	−0.014509	−0.170151	−0.752835	−2.35172	−6.30610
3.00	−0.010931	−0.013151	−0.159853	−0.721883	−2.28592	−6.18824
3.05	−0.009874	−0.011911	−0.150066	−0.691679	−2.22032	−6.06821
3.10	−0.008914	−0.010781	−0.140775	−0.662247	−2.15501	−5.94629
3.15	−0.008041	−0.009751	−0.131963	−0.633602	−2.09011	−5.82277
3.20	−0.007248	−0.008813	−0.123615	−0.605760	−2.02572	−5.69790
3.25	−0.006530	−0.007961	−0.115714	−0.578730	−1.96195	−5.57195
3.30	−0.005878	−0.007186	−0.108243	−0.552521	−1.89888	−5.44516
3.35	−0.005288	−0.006482	−0.101185	−0.527135	−1.83658	−5.31778
3.40	−0.004755	−0.005843	−0.094524	−0.502575	−1.77515	−5.19003
3.45	−0.004272	−0.005264	−0.088242	−0.478839	−1.71464	−5.06212
3.50	−0.003836	−0.004739	−0.082325	−0.455923	−1.65511	−4.93428
3.55	−0.003442	−0.004264	−0.076754	−0.433823	−1.59663	−4.80669
3.60	−0.003087	−0.003834	−0.071516	−0.412529	−1.53925	−4.67956
3.65	−0.002767	−0.003445	−0.066592	−0.392033	−1.48300	−4.55304
3.70	−0.002478	−0.003094	−0.061970	−0.372323	−1.42794	−4.42732
3.75	−0.002219	−0.002777	−0.057633	−0.353387	−1.37408	−4.30255
3.80	−0.001985	−0.002491	−0.053567	−0.335210	−1.32147	−4.17888
3.85	−0.001775	−0.002233	−0.049758	−0.317777	−1.27012	−4.05645
3.90	−0.001586	−0.002001	−0.046193	−0.301073	−1.22005	−3.93539
3.95	−0.001417	−0.001792	−0.042859	−0.285081	−1.17128	−3.81581
4.00	−0.001265	−0.001604	−0.039742	−0.269782	−1.12383	−3.69784
4.05	−0.001128	−0.001435	−0.036831	−0.255158	−1.07769	−3.58156
4.10	−0.001006	−0.001283	−0.034115	−0.241192	−1.03287	−3.46708
4.15	−0.000897	−0.001146	−0.031581	−0.227863	−0.989373	−3.35448
4.20	−0.000799	−0.001023	−0.029219	−0.215152	−0.947194	−3.24383
4.25	−0.000711	−0.000914	−0.027019	−0.203040	−0.906327	−3.13520
4.30	−0.000633	−0.000815	−0.024972	−0.191507	−0.866764	−3.02866
4.35	−0.000563	−0.000727	−0.023067	−0.180533	−0.828494	−2.92425
4.40	−0.000500	−0.000648	−0.021296	−0.170098	−0.791504	−2.82202
4.45	−0.000444	−0.000577	−0.019652	−0.160183	−0.755778	−2.72202
4.50	−0.000394	−0.000514	−0.018125	−0.150769	−0.721299	−2.62426
4.55	−0.000350	−0.000457	−0.016708	−0.141836	−0.688048	−2.52879
4.60	−0.000311	−0.000406	−0.015394	−0.133365	−0.656004	−2.43562
4.65	−0.000275	−0.000361	−0.014176	−0.125337	−0.625145	−2.34476
4.70	−0.000244	−0.000321	−0.013049	−0.117735	−0.595448	−2.25623
4.75	−0.000216	−0.000285	−0.012005	−0.110539	−0.566890	−2.17003
4.80	−0.000191	−0.000253	−0.011039	−0.103734	−0.539444	−2.08615
4.85	−0.000169	−0.000225	−0.010146	−0.097301	−0.513085	−2.00460
4.90	−0.000150	−0.000199	−0.009321	−0.091224	−0.487786	−1.92536
4.95	−0.000132	−0.000177	−0.008560	−0.085486	−0.463521	−1.84843
5.00	−0.000117	−0.000156	−0.007856	−0.080072	−0.440260	−1.77377

Table 2.5: Second derivatives with respect to the order of the modified Struve functions $\partial^2 L_\nu(t)/\partial \nu^2$.

ν	0.05	0.10	0.15	0.20	0.25	0.30
0.00	0.412173	0.526754	0.572096	0.582446	0.571828	0.547570
0.05	0.351825	0.469557	0.524666	0.546441	0.547398	0.534291
0.10	0.299099	0.416552	0.478445	0.509264	0.519933	0.516531
0.15	0.253321	0.367894	0.434056	0.471802	0.490453	0.495366
0.20	0.213796	0.323590	0.391938	0.434753	0.459810	0.471724
0.25	0.179846	0.283541	0.352380	0.398660	0.428704	0.446397
0.30	0.150818	0.247569	0.315552	0.363930	0.397700	0.420055
0.35	0.126107	0.215443	0.281526	0.330857	0.367251	0.393255
0.40	0.105153	0.186903	0.250301	0.299642	0.337705	0.366454
0.45	0.087451	0.161667	0.221819	0.270408	0.309328	0.340022
0.50	0.072548	0.139451	0.195979	0.243216	0.282312	0.314255
0.55	0.060041	0.119971	0.172653	0.218077	0.256785	0.289379
0.60	0.049578	0.102954	0.151690	0.194963	0.232829	0.265565
0.65	0.040849	0.088141	0.132929	0.173816	0.210480	0.242933
0.70	0.033587	0.075287	0.116203	0.154557	0.189743	0.221565
0.75	0.027561	0.064168	0.101345	0.137089	0.170595	0.201505
0.80	0.022572	0.054577	0.088189	0.121307	0.152994	0.182773
0.85	0.018452	0.046327	0.076578	0.107100	0.136880	0.165363
0.90	0.015058	0.039248	0.066360	0.094352	0.122185	0.149252
0.95	0.012266	0.033190	0.057392	0.082949	0.108829	0.134402
1.00	0.009976	0.028016	0.049543	0.072780	0.096732	0.120766
1.05	0.008100	0.023609	0.042690	0.063736	0.085807	0.108286
1.10	0.006566	0.019862	0.036720	0.055713	0.075971	0.096901
1.15	0.005315	0.016682	0.031532	0.048614	0.067138	0.086547
1.20	0.004296	0.013990	0.027032	0.042348	0.059227	0.077156
1.25	0.003467	0.011714	0.023138	0.036828	0.052158	0.068662
1.30	0.002795	0.009795	0.019775	0.031977	0.045857	0.060999
1.35	0.002249	0.008178	0.016875	0.027722	0.040253	0.054101
1.40	0.001808	0.006819	0.014380	0.023997	0.035279	0.047906
1.45	0.001452	0.005678	0.012237	0.020743	0.030874	0.042355
1.50	0.001164	0.004721	0.010399	0.017905	0.026979	0.037392
1.55	0.000932	0.003921	0.008825	0.015434	0.023542	0.032962
1.60	0.000746	0.003253	0.007480	0.013286	0.020515	0.029015
1.65	0.000596	0.002695	0.006332	0.011423	0.017854	0.025507
1.70	0.000475	0.002230	0.005354	0.009808	0.015518	0.022393
1.75	0.000379	0.001843	0.004521	0.008411	0.013470	0.019633
1.80	0.000302	0.001522	0.003813	0.007205	0.011678	0.017192
1.85	0.000240	0.001255	0.003213	0.006164	0.010113	0.015036
1.90	0.000191	0.001034	0.002704	0.005268	0.008747	0.013135
1.95	0.000151	0.000851	0.002273	0.004497	0.007557	0.011461
2.00	0.000120	0.000699	0.001909	0.003835	0.006521	0.009989

Table 2.5 (continued)

			t			
v	0.05	0.10	0.15	0.20	0.25	0.30
2.05	0.000095	0.000574	0.001602	0.003266	0.005622	0.008696
2.10	0.000075	0.000471	0.001342	0.002780	0.004841	0.007562
2.15	0.000060	0.000386	0.001124	0.002363	0.004164	0.006569
2.20	0.000047	0.000316	0.000940	0.002006	0.003579	0.005701
2.25	0.000037	0.000259	0.000785	0.001702	0.003072	0.004942
2.30	0.000029	0.000212	0.000656	0.001443	0.002635	0.004280
2.35	0.000023	0.000173	0.000547	0.001222	0.002258	0.003703
2.40	0.000018	0.000141	0.000456	0.001034	0.001933	0.003201
2.45	0.000014	0.000115	0.000379	0.000874	0.001653	0.002764
2.50	0.000011	0.000094	0.000316	0.000738	0.001412	0.002385
2.55	0.000009	0.000076	0.000262	0.000622	0.001206	0.002056
2.60	0.000007	0.000062	0.000218	0.000525	0.001028	0.001771
2.65	0.000005	0.000050	0.000181	0.000442	0.000876	0.001524
2.70	0.000004	0.000041	0.000150	0.000372	0.000746	0.001310
2.75	0.000003	0.000033	0.000124	0.000313	0.000635	0.001125
2.80	0.000003	0.000027	0.000103	0.000263	0.000540	0.000966
2.85	0.000002	0.000022	0.000085	0.000221	0.000459	0.000828
2.90	0.000002	0.000018	0.000070	0.000185	0.000389	0.000710
2.95	0.000001	0.000014	0.000058	0.000155	0.000330	0.000608
3.00	0.000001	0.000012	0.000048	0.000130	0.000280	0.000520
3.05	0.000001	0.000009	0.000039	0.000109	0.000237	0.000445
3.10	0.000001	0.000008	0.000033	0.000091	0.000200	0.000380
3.15		0.000006	0.000027	0.000076	0.000170	0.000324
3.20		0.000005	0.000022	0.000064	0.000143	0.000277
3.25		0.000004	0.000018	0.000053	0.000121	0.000236
3.30		0.000003	0.000015	0.000044	0.000102	0.000201
3.35		0.000003	0.000012	0.000037	0.000086	0.000171
3.40		0.000002	0.000010	0.000031	0.000073	0.000145
3.45		0.000020	0.000008	0.000026	0.000061	0.000124
3.50		0.000001	0.000007	0.000021	0.000051	0.000105
3.55		0.000001	0.000006	0.000018	0.000043	0.000089
3.60		0.000001	0.000005	0.000015	0.000036	0.000076
3.65		0.000001	0.000004	0.000012	0.000031	0.000064
3.70		0.000001	0.000002	0.000010	0.000026	0.000054
3.75			0.000002	0.000008	0.000021	0.000046
3.80			0.000002	0.000070	0.000018	0.000039
3.85			0.000002	0.000006	0.000015	0.000033
3.90			0.000001	0.000005	0.000013	0.000028
3.95			0.000001	0.000004	0.000011	0.000024
4.00			0.000001	0.000003	0.000009	0.000020
4.05			0.000001	0.000003	0.000007	0.000017
4.10			0.000001	0.000002	0.000006	0.000014
4.15				0.000002	0.000005	0.000012

Table 2.5 (continued)

			t			
v	0.05	0.10	0.15	0.20	0.25	0.30
4.20				0.000002	0.000004	0.000010
4.25				0.000001	0.000004	0.000008
4.30				0.000001	0.000003	0.000007
4.35				0.000001	0.000002	0.000006
4.40				0.000001	0.000002	0.000005
4.45				0.000001	0.000002	0.000004
4.50					0.000001	0.000004
4.55					0.000001	0.000003
4.60					0.000001	0.000002
4.65					0.000001	0.000002
4.70					0.000001	0.000002
4.75					0.000001	0.000001
4.80						0.000001
4.85						0.000001
4.90						0.000001
4.95						0.000001
5.00						0.000001

			t			
v	0.35	0.40	0.45	0.50	0.55	0.60
0.00	0.514030	0.474012	0.429418	0.381591	0.331505	0.279879
0.05	0.511244	0.480970	0.445356	0.405759	0.363184	0.318390
0.10	0.502865	0.481500	0.454250	0.422450	0.387103	0.348984
0.15	0.489976	0.476650	0.457096	0.432591	0.404112	0.372426
0.20	0.473531	0.467370	0.454812	0.437052	0.415022	0.389459
0.25	0.454369	0.454510	0.448235	0.436642	0.420597	0.400798
0.30	0.433218	0.438824	0.438124	0.432104	0.421555	0.407122
0.35	0.410702	0.420976	0.425160	0.424118	0.418560	0.409071
0.40	0.387354	0.401548	0.409947	0.413300	0.412225	0.407244
0.45	0.363622	0.381039	0.393023	0.400203	0.403109	0.402195
0.50	0.339877	0.359878	0.374855	0.385321	0.391718	0.394434
0.55	0.316424	0.338430	0.355851	0.369091	0.378509	0.384425
0.60	0.293507	0.316996	0.336358	0.351895	0.363888	0.372591
0.65	0.271318	0.295829	0.316675	0.334068	0.348215	0.359310
0.70	0.250004	0.275130	0.297051	0.315897	0.331806	0.344920
0.75	0.229673	0.255061	0.277692	0.297625	0.314938	0.329722
0.80	0.210398	0.235745	0.258767	0.279459	0.297848	0.313977
0.85	0.192226	0.217277	0.240410	0.261570	0.280738	0.297917
0.90	0.175178	0.199721	0.222727	0.244099	0.263779	0.281738
0.95	0.159258	0.183119	0.205796	0.227158	0.247115	0.265609
1.00	0.144453	0.167495	0.189674	0.210834	0.230861	0.249673
1.05	0.130739	0.152853	0.174396	0.195193	0.215111	0.234049

Table 2.5 (continued)

v	0.35	0.40	0.45	0.50	0.55	0.60
			t			
1.10	0.118079	0.139187	0.159985	0.180284	0.199939	0.218834
1.15	0.106432	0.126479	0.146446	0.166137	0.185399	0.204105
1.20	0.095749	0.114702	0.133774	0.152770	0.171530	0.189923
1.25	0.085980	0.103823	0.121957	0.140190	0.158360	0.176334
1.30	0.077070	0.093802	0.110972	0.128391	0.145903	0.163371
1.35	0.068966	0.084599	0.100791	0.117363	0.134162	0.151055
1.40	0.061611	0.076169	0.091383	0.107088	0.123135	0.139396
1.45	0.054954	0.068466	0.082713	0.097541	0.112810	0.128398
1.50	0.048940	0.061444	0.074743	0.088695	0.103171	0.118056
1.55	0.043519	0.055057	0.067434	0.080521	0.094198	0.108360
1.60	0.038642	0.049261	0.060747	0.072985	0.085867	0.099296
1.65	0.034264	0.044011	0.054642	0.066054	0.078151	0.090844
1.70	0.030340	0.039266	0.049080	0.059693	0.071021	0.082984
1.75	0.026830	0.034985	0.044023	0.053868	0.064447	0.075690
1.80	0.023696	0.031130	0.039433	0.048544	0.058399	0.068937
1.85	0.020901	0.027664	0.035276	0.043687	0.052846	0.062699
1.90	0.018414	0.024553	0.031516	0.039265	0.047757	0.056948
1.95	0.016203	0.021765	0.028123	0.035246	0.043102	0.051656
2.00	0.014241	0.019271	0.025064	0.031599	0.038852	0.046795
2.05	0.012502	0.017043	0.022312	0.028296	0.034978	0.042339
2.10	0.010964	0.015056	0.019839	0.025308	0.031453	0.038260
2.15	0.009604	0.013285	0.017620	0.022610	0.028250	0.034534
2.20	0.008404	0.011710	0.015632	0.020176	0.025345	0.031134
2.25	0.007347	0.010311	0.013854	0.017985	0.022713	0.028038
2.30	0.006416	0.009070	0.012265	0.016015	0.020332	0.025221
2.35	0.005597	0.007970	0.010847	0.014246	0.018182	0.022664
2.40	0.004878	0.006997	0.009584	0.012660	0.016242	0.020344
2.45	0.004248	0.006136	0.008459	0.011239	0.014495	0.018243
2.50	0.003695	0.005377	0.007459	0.009967	0.012922	0.016342
2.55	0.003211	0.004707	0.006571	0.008831	0.011509	0.014625
2.60	0.002789	0.004116	0.005784	0.007817	0.010241	0.013076
2.65	0.002419	0.003597	0.005086	0.006914	0.009104	0.011680
2.70	0.002097	0.003140	0.004468	0.006109	0.008086	0.010423
2.75	0.001816	0.002739	0.003922	0.005393	0.007176	0.009293
2.80	0.001572	0.002387	0.003440	0.004757	0.006362	0.008279
2.85	0.001359	0.002079	0.003015	0.004192	0.005636	0.007368
2.90	0.001174	0.001809	0.002640	0.003691	0.004988	0.006552
2.95	0.001014	0.001573	0.002309	0.003248	0.004411	0.005822
3.00	0.000874	0.001366	0.002019	0.002855	0.003898	0.005169
3.05	0.000754	0.001186	0.001763	0.002508	0.003442	0.004585
3.10	0.000649	0.001029	0.001539	0.002201	0.003036	0.004064
3.15	0.000559	0.000892	0.001342	0.001931	0.002677	0.003599
3.20	0.000480	0.000772	0.001170	0.001692	0.002358	0.003185

Table 2.5 (continued)

			t			
v	0.35	0.40	0.45	0.50	0.55	0.60
3.25	0.000413	0.000668	0.001019	0.001482	0.002075	0.002817
3.30	0.000355	0.000578	0.000887	0.001297	0.001825	0.002489
3.35	0.000304	0.000499	0.000771	0.001134	0.001604	0.002198
3.40	0.000261	0.000431	0.000670	0.000991	0.001409	0.001939
3.45	0.000224	0.000372	0.000582	0.000865	0.001237	0.001710
3.50	0.000192	0.000321	0.000505	0.000755	0.001085	0.001507
3.55	0.000164	0.000277	0.000438	0.000659	0.000951	0.001327
3.60	0.000140	0.000238	0.000379	0.000574	0.000833	0.001167
3.65	0.000120	0.000205	0.000329	0.000500	0.000729	0.001026
3.70	0.000102	0.000176	0.000284	0.000435	0.000637	0.000902
3.75	0.000087	0.000152	0.000246	0.000378	0.000557	0.000792
3.80	0.000075	0.000130	0.000213	0.000329	0.000487	0.000695
3.85	0.000064	0.000112	0.000184	0.000286	0.000425	0.000610
3.90	0.000054	0.000096	0.000159	0.000248	0.000371	0.000534
3.95	0.000046	0.000082	0.000137	0.000215	0.000323	0.000468
4.00	0.000039	0.000071	0.000118	0.000187	0.000282	0.000410
4.05	0.000033	0.000060	0.000102	0.000162	0.000245	0.000358
4.10	0.000028	0.000052	0.000088	0.000140	0.000214	0.000313
4.15	0.000024	0.000044	0.000075	0.000121	0.000186	0.000274
4.20	0.000020	0.000038	0.000065	0.000105	0.000161	0.000239
4.25	0.000017	0.000032	0.000056	0.000091	0.000140	0.000209
4.30	0.000015	0.000028	0.000048	0.000078	0.000122	0.000182
4.35	0.000013	0.000024	0.000041	0.000068	0.000106	0.000159
4.40	0.000011	0.000020	0.000035	0.000058	0.000092	0.000138
4.45	0.000010	0.000017	0.000030	0.000050	0.000080	0.000121
4.50	0.000006	0.000015	0.000026	0.000043	0.000069	0.000105
4.55	0.000004	0.000012	0.000022	0.000037	0.000060	0.000091
4.60	0.000005	0.000011	0.000019	0.000032	0.000052	0.000079
4.65	0.000004	0.000009	0.000016	0.000028	0.000045	0.000069
4.70	0.000004	0.000008	0.000014	0.000024	0.000039	0.000060
4.75	0.000003	0.000007	0.000012	0.000021	0.000033	0.000052
4.80	0.000003	0.000006	0.000010	0.000018	0.000029	0.000045
4.85	0.000002	0.000005	0.000009	0.000015	0.000025	0.000039
4.90	0.000002	0.000004	0.000007	0.000013	0.000021	0.000034
4.95	0.000002	0.000003	0.000001	0.000011	0.000019	0.000029
5.00	0.000001	0.000003	0.000005	0.000010	0.000016	0.000025

			t			
v	0.65	0.70	0.75	0.80	0.85	0.90
0.00	0.227252	0.174035	0.120539	0.067003	0.013607	-0.039511
0.05	0.271961	0.224348	0.175907	0.126915	0.077591	0.028108
0.10	0.308697	0.266720	0.223435	0.179147	0.134104	0.088504

Table 2.5 (continued)

			t			
v	0.65	0.70	0.75	0.80	0.85	0.90
0.15	0.338139	0.301741	0.263626	0.224118	0.183482	0.141936
0.20	0.360960	0.330008	0.297004	0.262276	0.226101	0.188707
0.25	0.377815	0.352119	0.324101	0.294091	0.262365	0.229158
0.30	0.389341	0.368658	0.345451	0.320040	0.292696	0.263654
0.35	0.396145	0.380196	0.361581	0.340604	0.317528	0.292581
0.40	0.398802	0.387279	0.373004	0.356260	0.337298	0.316333
0.45	0.397856	0.390430	0.380216	0.367476	0.352439	0.335310
0.50	0.393808	0.390139	0.383693	0.374703	0.363380	0.349910
0.55	0.387127	0.386869	0.383883	0.378376	0.370534	0.360527
0.60	0.378240	0.381049	0.381213	0.378910	0.374303	0.367542
0.65	0.367539	0.373074	0.376076	0.376694	0.375068	0.371326
0.70	0.355376	0.363306	0.368840	0.372097	0.373192	0.372234
0.75	0.342069	0.352078	0.359844	0.365460	0.369017	0.370602
0.80	0.327903	0.339687	0.349396	0.357098	0.362860	0.366751
0.85	0.313128	0.326402	0.337779	0.347303	0.355019	0.360977
0.90	0.297964	0.312462	0.325247	0.336339	0.345767	0.353561
0.95	0.282603	0.298078	0.312026	0.324448	0.335355	0.344759
1.00	0.267212	0.283436	0.298319	0.311847	0.324010	0.334809
1.05	0.251930	0.268697	0.284306	0.298728	0.311940	0.323928
1.10	0.236878	0.254000	0.270144	0.285265	0.299330	0.312314
1.15	0.222154	0.239464	0.255968	0.271610	0.286347	0.300143
1.20	0.207839	0.225189	0.241895	0.257896	0.273138	0.287577
1.25	0.193998	0.211256	0.228026	0.244237	0.259832	0.274758
1.30	0.180682	0.197734	0.214443	0.230733	0.246541	0.261811
1.35	0.167926	0.184676	0.201215	0.217468	0.233365	0.248847
1.40	0.155758	0.172122	0.188399	0.204510	0.220385	0.235962
1.45	0.144195	0.160105	0.176039	0.191918	0.207672	0.223238
1.50	0.133245	0.148644	0.164167	0.179738	0.195286	0.210746
1.55	0.122908	0.137751	0.152809	0.168006	0.183272	0.198545
1.60	0.113180	0.127434	0.141981	0.156749	0.171670	0.186683
1.65	0.104050	0.117691	0.131692	0.145986	0.160509	0.175201
1.70	0.095506	0.108516	0.121945	0.135730	0.149810	0.164129
1.75	0.087529	0.099899	0.112738	0.125987	0.139588	0.153490
1.80	0.080100	0.091827	0.104064	0.116757	0.129853	0.143302
1.85	0.073196	0.084284	0.095914	0.108038	0.120607	0.133576
1.90	0.066794	0.077250	0.088274	0.099821	0.111849	0.124317
1.95	0.060870	0.070706	0.081128	0.092097	0.103576	0.115526
2.00	0.055398	0.064630	0.074459	0.084853	0.095778	0.107202
2.05	0.050354	0.059000	0.068248	0.078073	0.088446	0.099338
2.10	0.045713	0.053791	0.062475	0.071741	0.081566	0.091926
2.15	0.041449	0.048983	0.057118	0.065838	0.075124	0.084955
2.20	0.037539	0.044551	0.052158	0.060347	0.069104	0.078412

Table 2.5 (continued)

			t			
v	0.65	0.70	0.75	0.80	0.85	0.90
2.25	0.033959	0.040473	0.047571	0.055247	0.063488	0.072283
2.30	0.030686	0.036726	0.043338	0.050518	0.058259	0.066553
2.35	0.027698	0.033289	0.039438	0.046142	0.053399	0.061204
2.40	0.024975	0.030142	0.035849	0.042098	0.048889	0.056221
2.45	0.022495	0.027263	0.032551	0.038366	0.044711	0.051586
2.50	0.020242	0.024633	0.029526	0.034928	0.040845	0.047281
2.55	0.018196	0.022235	0.026755	0.031765	0.037275	0.043289
2.60	0.016340	0.020050	0.024219	0.028860	0.033982	0.039593
2.65	0.014660	0.018063	0.021903	0.026194	0.030948	0.036176
2.70	0.013141	0.016257	0.019789	0.023751	0.028158	0.033021
2.75	0.011768	0.014618	0.017862	0.021516	0.025594	0.030111
2.80	0.010529	0.013132	0.016108	0.019472	0.023242	0.027432
2.85	0.009412	0.011787	0.014513	0.017607	0.021087	0.024968
2.90	0.008406	0.010570	0.013064	0.015906	0.019114	0.022705
2.95	0.007502	0.009471	0.011750	0.014357	0.017311	0.020628
3.00	0.006689	0.008479	0.010559	0.012948	0.015664	0.018724
3.05	0.005959	0.007585	0.009481	0.011667	0.014161	0.016982
3.10	0.005305	0.006779	0.008506	0.010504	0.012792	0.015389
3.15	0.004719	0.006054	0.007625	0.009449	0.011546	0.013933
3.20	0.004194	0.005403	0.006830	0.008494	0.010413	0.012605
3.25	0.003725	0.004817	0.006113	0.007629	0.009384	0.011395
3.30	0.003306	0.004292	0.005467	0.006847	0.008449	0.010292
3.35	0.002931	0.003822	0.004885	0.006140	0.007602	0.009289
3.40	0.002598	0.003400	0.004363	0.005502	0.006835	0.008378
3.45	0.002300	0.003023	0.003893	0.004927	0.006140	0.007550
3.50	0.002036	0.002685	0.003471	0.004409	0.005512	0.006799
3.55	0.001800	0.002384	0.003093	0.003942	0.004945	0.006118
3.60	0.001591	0.002115	0.002754	0.003522	0.004433	0.005502
3.65	0.001405	0.001875	0.002451	0.003145	0.003971	0.004944
3.70	0.001239	0.001661	0.002179	0.002806	0.003555	0.004439
3.75	0.001093	0.001471	0.001937	0.002502	0.003181	0.003984
3.80	0.000963	0.001301	0.001720	0.002230	0.002843	0.003572
3.85	0.000848	0.001151	0.001526	0.001986	0.002540	0.003201
3.90	0.000747	0.001017	0.001354	0.001767	0.002268	0.002867
3.95	0.000657	0.000898	0.001200	0.001572	0.002024	0.002566
4.00	0.000578	0.000793	0.001063	0.001397	0.001805	0.002295
4.05	0.000507	0.000699	0.000941	0.001241	0.001608	0.002052
4.10	0.000446	0.000616	0.000832	0.001102	0.001432	0.001833
4.15	0.000391	0.000543	0.000736	0.000977	0.001275	0.001636
4.20	0.000343	0.000478	0.000650	0.000867	0.001134	0.001460
4.25	0.000301	0.000421	0.000574	0.000768	0.001008	0.001302
4.30	0.000263	0.000370	0.000507	0.000680	0.000896	0.001160

Table 2.5 (continued)

			t			
v	0.65	0.70	0.75	0.80	0.85	0.90
4.35	0.000231	0.000325	0.000447	0.000602	0.000795	0.001033
4.40	0.000202	0.000286	0.000394	0.000533	0.000706	0.000920
4.45	0.000176	0.000251	0.000347	0.000471	0.000626	0.000818
4.50	0.000154	0.000220	0.000306	0.000416	0.000555	0.000728
4.55	0.000135	0.000193	0.000269	0.000367	0.000492	0.000647
4.60	0.000118	0.000169	0.000237	0.000324	0.000435	0.000574
4.65	0.000103	0.000148	0.000208	0.000286	0.000385	0.000510
4.70	0.000090	0.000130	0.000183	0.000252	0.000341	0.000452
4.75	0.000078	0.000114	0.000161	0.000222	0.000301	0.000401
4.80	0.000068	0.000099	0.000141	0.000196	0.000266	0.000355
4.85	0.000059	0.000087	0.000124	0.000172	0.000235	0.000315
4.90	0.000052	0.000076	0.000109	0.000152	0.000208	0.000279
4.95	0.000045	0.000066	0.000095	0.000133	0.000183	0.000247
5.00	0.000039	0.000058	0.000083	0.000117	0.000162	0.000218

			t			
v	0.95	1.0	2.0	3.0	4.0	5.0
0.00	-0.092251	-0.144537	-1.09262	-2.07476	-3.59967	-6.55911
0.05	-0.021400	-0.070827	-1.00800	-2.01065	-3.55761	-6.53435
0.10	0.042510	-0.003750	-0.922771	-1.94073	-3.50601	-6.49587
0.15	0.099661	0.056804	-0.837542	-1.86559	-3.44526	-6.44388
0.20	0.150288	0.111005	-0.752870	-1.78583	-3.37581	-6.37868
0.25	0.194669	0.159069	-0.669264	-1.70202	-3.29813	-6.30059
0.30	0.233115	0.201251	-0.587185	-1.61477	-3.21271	-6.21000
0.35	0.265960	0.237836	-0.507041	-1.52464	-3.12006	-6.10732
0.40	0.293556	0.269135	-0.429196	-1.43219	-3.02073	-5.99302
0.45	0.316268	0.295472	-0.353965	-1.33797	-2.91525	-5.86758
0.50	0.334461	0.317183	-0.281619	-1.24249	-2.80419	-5.73153
0.55	0.348505	0.334607	-0.212385	-1.14626	-2.68811	-5.58544
0.60	0.358761	0.348086	-0.146449	-1.04976	-2.56757	-5.42988
0.65	0.365585	0.357956	-0.083957	-0.953432	-2.44314	-5.26546
0.70	0.369321	0.364549	-0.025020	-0.857708	-2.31537	-5.09279
0.75	0.370299	0.368186	0.030286	-0.762984	-2.18483	-4.91253
0.80	0.368834	0.369175	0.081917	-0.669629	-2.05207	-4.72531
0.85	0.365225	0.367811	0.129856	-0.577982	-1.91760	-4.53180
0.90	0.359752	0.364377	0.174115	-0.488355	-1.78196	-4.33266
0.95	0.352679	0.359135	0.214728	-0.401028	-1.64565	-4.12857
1.00	0.344248	0.352334	0.251752	-0.316253	-1.50916	-3.92018
1.05	0.334685	0.344206	0.285260	-0.234255	-1.37295	-3.70816
1.10	0.324196	0.334963	0.315346	-0.155228	-1.23747	-3.49318
1.15	0.312970	0.324804	0.342114	-0.079341	-1.10314	-3.27586

Table 2.5 (continued)

			t			
v	0.95	1.0	2.0	3.0	4.0	5.0
1.20	0.301177	0.313908	0.365683	-0.006736	-0.970356	-3.05687
1.25	0.288973	0.302438	0.386180	0.062472	-0.839504	-2.83681
1.30	0.276493	0.290545	0.403742	0.128189	-0.710932	-2.61630
1.35	0.263861	0.278359	0.418512	0.190348	-0.584965	-2.39592
1.40	0.251184	0.266001	0.430637	0.248901	-0.461904	-2.17625
1.45	0.238556	0.253576	0.440267	0.303823	-0.342024	-1.95784
1.50	0.226060	0.241175	0.447557	0.355107	-0.225575	-1.74122
1.55	0.213765	0.228880	0.452658	0.402764	-0.112780	-1.52688
1.60	0.201731	0.216759	0.455723	0.446822	-0.003839	-1.31531
1.65	0.190006	0.204873	0.456904	0.487325	0.101075	-1.10696
1.70	0.178632	0.193270	0.456349	0.524329	0.201811	-0.902265
1.75	0.167641	0.181992	0.454205	0.557905	0.298243	-0.701613
1.80	0.157057	0.171073	0.450611	0.588133	0.390268	-0.505381
1.85	0.146900	0.160538	0.445706	0.615105	0.477803	-0.313913
1.90	0.137183	0.150409	0.439622	0.638921	0.560790	-0.127525
1.95	0.127912	0.140698	0.432485	0.659687	0.639187	0.053495
2.00	0.119092	0.131416	0.424416	0.677519	0.712973	0.228888
2.05	0.110722	0.122567	0.415531	0.692534	0.782147	0.398420
2.10	0.102797	0.114153	0.405938	0.704857	0.846723	0.561888
2.15	0.095312	0.106172	0.395741	0.714615	0.906731	0.719112
2.20	0.088256	0.098617	0.385036	0.721936	0.962218	0.869942
2.25	0.081620	0.091482	0.373914	0.726952	1.01324	1.01425
2.30	0.075389	0.084758	0.362458	0.729794	1.05988	1.15194
2.35	0.069551	0.078432	0.350748	0.730594	1.10221	1.28293
2.40	0.064090	0.072493	0.338854	0.729483	1.14033	1.40717
2.45	0.058991	0.066926	0.326844	0.726591	1.17435	1.52463
2.50	0.054238	0.061717	0.314778	0.722045	1.20438	1.63530
2.55	0.049814	0.056851	0.302712	0.715971	1.23054	1.73919
2.60	0.045702	0.052313	0.290695	0.708492	1.25296	1.83634
2.65	0.041887	0.048087	0.278772	0.699728	1.27176	1.92678
2.70	0.038351	0.044158	0.266984	0.689794	1.28710	2.01060
2.75	0.035079	0.040509	0.255366	0.678803	1.29911	2.08787
2.80	0.032056	0.037125	0.243950	0.666864	1.30794	2.15869
2.85	0.029265	0.033991	0.232763	0.654080	1.31374	2.22317
2.90	0.026693	0.031093	0.221829	0.640551	1.31665	2.28144
2.95	0.024324	0.028416	0.211168	0.626373	1.31682	2.33364
3.00	0.022147	0.025946	0.200796	0.611636	1.31441	2.37991
3.05	0.020146	0.023671	0.190727	0.596426	1.30956	2.42041
3.10	0.018311	0.021576	0.180973	0.580824	1.30243	2.45532
3.15	0.016629	0.019649	0.171542	0.564907	1.29315	2.48479
3.20	0.015089	0.017880	0.162440	0.548747	1.28189	2.50902
3.25	0.013680	0.016257	0.153670	0.532411	1.26877	2.52819

Table 2.5 (continued)

v	0.95	1.0	2.0	3.0	4.0	5.0
				t		
3.30	0.012393	0.014770	0.145236	0.515962	1.25394	2.54248
3.35	0.011219	0.013407	0.137136	0.499458	1.23754	2.55211
3.40	0.010148	0.012162	0.129370	0.482954	1.21969	2.55726
3.45	0.009172	0.011023	0.121934	0.466498	1.20054	2.55813
3.50	0.008284	0.009984	0.114826	0.450137	1.18019	2.55493
3.55	0.007476	0.009036	0.108039	0.433913	1.15879	2.54785
3.60	0.006743	0.008172	0.101569	0.417863	1.13643	2.53711
3.65	0.006077	0.007385	0.095407	0.402022	1.11324	2.52290
3.70	0.005473	0.006670	0.089547	0.386420	1.08933	2.50542
3.75	0.004925	0.006019	0.083980	0.371085	1.06479	2.48487
3.80	0.004430	0.005428	0.078698	0.356041	1.03972	2.46145
3.85	0.003981	0.004892	0.073692	0.341310	1.01422	2.43535
3.90	0.003576	0.004406	0.068953	0.326909	0.988373	2.40675
3.95	0.003210	0.003966	0.064470	0.312855	0.962267	2.37585
4.00	0.002879	0.003568	0.060236	0.299160	0.935980	2.34283
4.05	0.002581	0.003207	0.056240	0.285836	0.909586	2.30786
4.10	0.002312	0.002881	0.052471	0.272890	0.883154	2.27112
4.15	0.002070	0.002587	0.048922	0.260331	0.856750	2.23278
4.20	0.001853	0.002321	0.045582	0.248162	0.830433	2.19299
4.25	0.001657	0.002081	0.042441	0.236387	0.804261	2.15191
4.30	0.001481	0.001865	0.039491	0.225007	0.778285	2.10970
4.35	0.001323	0.001670	0.036722	0.214021	0.752553	2.06649
4.40	0.001181	0.001495	0.034125	0.203429	0.727108	2.02244
4.45	0.001053	0.001338	0.031692	0.193228	0.701991	1.97767
4.50	0.000939	0.001196	0.029415	0.183413	0.677239	1.93232
4.55	0.000837	0.001069	0.027284	0.173981	0.652882	1.88650
4.60	0.000746	0.000955	0.025292	0.164925	0.628952	1.84032
4.65	0.000664	0.000852	0.023432	0.156239	0.605473	1.79391
4.70	0.000591	0.000760	0.021695	0.147917	0.582468	1.74736
4.75	0.000525	0.000678	0.020076	0.139950	0.559959	1.70078
4.80	0.000467	0.000604	0.018567	0.132330	0.537960	1.65424
4.85	0.000415	0.000538	0.017162	0.125049	0.516488	1.60785
4.90	0.000368	0.000479	0.015854	0.118098	0.495553	1.56169
4.95	0.000327	0.000426	0.014638	0.111467	0.475166	1.51582
5.00	0.000290	0.000379	0.013508	0.105147	0.455334	1.47032

Table 2.6: Third derivatives with respect to the order of the modified Struve functions $\partial^3 L_\nu(t)/\partial \nu^3$.

ν	0.05	0.10	0.15	0.20	0.25	0.30
0.00	−1.28748	−1.18392	−0.955470	−0.701485	−0.450521	−0.212780
0.05	−1.12854	−1.10282	−0.938923	−0.735104	−0.522653	−0.314239
0.10	−0.982734	−1.01687	−0.907843	−0.749010	−0.572549	−0.392593
0.15	−0.850726	−0.929444	−0.866265	−0.747165	−0.603795	−0.450932
0.20	−0.732524	−0.843049	−0.817503	−0.732986	−0.619595	−0.492107
0.25	−0.627679	−0.759524	−0.764248	−0.709391	−0.622787	−0.518717
0.30	−0.535442	−0.680143	−0.708650	−0.678844	−0.615857	−0.533111
0.35	−0.454878	−0.605743	−0.652400	−0.643408	−0.600966	−0.537392
0.40	−0.384960	−0.536811	−0.596800	−0.604786	−0.579973	−0.533424
0.45	−0.324629	−0.473569	−0.542827	−0.564369	−0.554462	−0.522846
0.50	−0.272841	−0.416034	−0.491188	−0.523272	−0.525772	−0.507088
0.55	−0.228597	−0.364077	−0.442372	−0.482379	−0.495018	−0.487385
0.60	−0.190963	−0.317465	−0.396689	−0.442373	−0.463121	−0.464793
0.65	−0.159081	−0.275892	−0.354307	−0.403767	−0.430827	−0.440208
0.70	−0.132172	−0.239008	−0.315285	−0.366934	−0.398736	−0.414380
0.75	−0.109539	−0.206442	−0.279595	−0.332130	−0.367318	−0.387931
0.80	−0.090565	−0.177816	−0.247148	−0.299515	−0.336931	−0.361371
0.85	−0.074707	−0.152753	−0.217805	−0.269173	−0.307843	−0.335108
0.90	−0.061491	−0.130894	−0.191400	−0.241126	−0.280241	−0.309464
0.95	−0.050508	−0.111895	−0.167744	−0.215349	−0.254248	−0.284690
1.00	−0.041403	−0.095436	−0.146637	−0.191781	−0.229935	−0.260968
1.05	−0.033874	−0.081220	−0.127875	−0.170334	−0.207327	−0.238432
1.10	−0.027663	−0.068979	−0.111256	−0.150900	−0.186419	−0.217168
1.15	−0.022551	−0.058465	−0.096584	−0.133361	−0.167175	−0.197226
1.20	−0.018351	−0.049459	−0.083671	−0.117590	−0.149542	−0.178624
1.25	−0.014909	−0.041763	−0.072338	−0.103456	−0.133450	−0.161358
1.30	−0.012093	−0.035202	−0.062419	−0.090831	−0.118820	−0.145403
1.35	−0.009794	−0.029620	−0.053760	−0.079586	−0.105565	−0.130721
1.40	−0.007919	−0.024883	−0.046220	−0.069600	−0.093596	−0.117260
1.45	−0.006395	−0.020870	−0.039668	−0.060754	−0.082819	−0.104963
1.50	−0.005156	−0.017476	−0.033989	−0.052937	−0.073143	−0.093765
1.55	−0.004151	−0.014613	−0.029076	−0.046048	−0.064481	−0.083600
1.60	−0.003338	−0.012201	−0.024834	−0.039989	−0.056744	−0.074397
1.65	−0.002681	−0.010173	−0.021179	−0.034671	−0.049852	−0.066089
1.70	−0.002150	−0.008470	−0.018036	−0.030014	−0.043725	−0.058608
1.75	−0.001722	−0.007043	−0.015337	−0.025944	−0.038290	−0.051887
1.80	−0.001378	−0.005849	−0.013025	−0.022394	−0.033480	−0.045864
1.85	−0.001101	−0.004851	−0.011046	−0.019302	−0.029231	−0.040477
1.90	−0.000879	−0.004018	−0.009356	−0.016614	−0.025485	−0.035669
1.95	−0.000701	−0.003325	−0.007914	−0.014281	−0.022188	−0.031386
2.00	−0.000558	−0.002747	−0.006686	−0.012261	−0.019291	−0.027579
2.05	−0.000444	−0.002268	−0.005642	−0.010513	−0.016751	−0.024201

Table 2.6 (continued)

v	0.05	0.10	0.15	0.20	0.25	0.30
				t		
2.10	−0.000353	−0.001870	−0.004756	−0.009003	−0.014527	−0.021208
2.15	−0.000280	−0.001540	−0.004004	−0.007700	−0.012582	−0.018562
2.20	−0.000222	−0.001267	−0.003367	−0.006579	−0.010884	−0.016225
2.25	−0.000176	−0.001041	−0.002828	−0.005614	−0.009405	−0.014165
2.30	−0.000139	−0.000855	−0.002373	−0.004786	−0.008117	−0.012352
2.35	−0.000110	−0.000701	−0.001990	−0.004075	−0.006998	−0.010759
2.40	−0.000087	−0.000575	−0.001666	−0.003467	−0.006026	−0.009360
2.45	−0.000069	−0.000470	−0.001394	−0.002946	−0.005184	−0.008135
2.50	−0.000054	−0.000385	−0.001165	−0.002501	−0.004455	−0.007062
2.55	−0.000043	−0.000314	−0.000973	−0.002121	−0.003824	−0.006124
2.60	−0.000034	−0.000257	−0.000811	−0.001797	−0.003280	−0.005305
2.65	−0.000026	−0.000209	−0.000676	−0.001521	−0.002810	−0.004591
2.70	−0.000021	−0.000171	−0.000563	−0.001286	−0.002405	−0.003970
2.75	−0.000016	−0.000139	−0.000468	−0.001086	−0.002057	−0.003429
2.80	−0.000013	−0.000113	−0.000389	−0.000917	−0.001757	−0.002959
2.85	−0.000010	−0.000092	−0.000323	−0.000773	−0.001500	−0.002551
2.90	−0.000008	−0.000075	−0.000268	−0.000652	−0.001279	−0.002197
2.95	−0.000006	−0.000061	−0.000222	−0.000549	−0.001090	−0.001891
3.00	−0.000005	−0.000049	−0.000184	−0.000461	−0.000928	−0.001626
3.05	−0.000004	−0.000040	−0.000153	−0.000388	−0.000789	−0.001397
3.10	−0.000003	−0.000032	−0.000126	−0.000326	−0.000671	−0.001199
3.15	−0.000002	−0.000026	−0.000104	−0.000273	−0.000570	−0.001028
3.20	−0.000002	−0.000021	−0.000086	−0.000229	−0.000484	−0.000881
3.25	−0.000001	−0.000017	−0.000071	−0.000192	−0.000410	−0.000754
3.30	−0.000001	−0.000014	−0.000059	−0.000161	−0.000347	−0.000646
3.35	−0.000001	−0.000011	−0.000048	−0.000135	−0.000294	−0.000552
3.40	−0.000001	−0.000009	−0.000040	−0.000112	−0.000249	−0.000471
3.45	−0.000001	−0.000007	−0.000033	−0.000094	−0.000210	−0.000402
3.50		−0.000006	−0.000027	−0.000078	−0.000178	−0.000343
3.55		−0.000005	−0.000022	−0.000065	−0.000150	−0.000293
3.60		−0.000004	−0.000018	−0.000055	−0.000127	−0.000249
3.65		−0.000003	−0.000015	−0.000045	−0.000107	−0.000212
3.70		−0.000002	−0.000012	−0.000038	−0.000090	−0.000180
3.75		−0.000002	−0.000010	−0.000031	−0.000076	−0.000153
3.80		−0.000002	−0.000008	−0.000026	−0.000064	−0.000130
3.85		−0.000001	−0.000007	−0.000022	−0.000053	−0.000111
3.90		−0.000001	−0.000005	−0.000018	−0.000045	−0.000094
3.95		−0.000001	−0.000004	−0.000015	−0.000038	−0.000080
4.00		−0.000001	−0.000004	−0.000012	−0.000032	−0.000067
4.05		−0.000001	−0.000003	−0.000010	−0.000027	−0.000057
4.10			−0.000002	−0.000009	−0.000022	−0.000048
4.15			−0.000002	−0.000007	−0.000019	−0.000041

Table 2.6 (continued)

			t			
v	0.05	0.10	0.15	0.20	0.25	0.30
4.20			−0.000002	−0.000006	−0.000016	−0.000035
4.25			−0.000001	−0.000005	−0.000013	−0.000029
4.30			−0.000001	−0.000004	−0.000011	−0.000025
4.35			−0.000001	−0.000003	−0.000009	−0.000021
4.40			−0.000001	−0.000003	−0.000008	−0.000018
4.45			−0.000001	−0.000002	−0.000006	−0.000015
4.50				−0.000002	−0.000005	−0.000012
4.55				−0.000002	−0.000004	−0.000010
4.60				−0.000001	−0.000004	−0.000009
4.65				−0.000001	−0.000003	−0.000007
4.70				−0.000001	−0.000003	−0.000006
4.75				−0.000001	−0.000002	−0.000005
4.80				−0.000001	−0.000002	−0.000004
4.85					−0.000001	−0.000004
4.90					−0.000001	−0.000003
4.95					−0.000001	−0.000003
5.00					−0.000001	−0.000002

			t			
v	0.35	0.40	0.45	0.50	0.55	0.60
0.00	0.008052	0.210986	0.396238	0.564571	0.716986	0.854566
0.05	−0.115465	0.071211	0.244873	0.405400	0.553095	0.688480
0.10	−0.216070	−0.046535	0.114208	0.265321	0.406518	0.537842
0.15	−0.296325	−0.144294	0.002693	0.143230	0.276557	0.402316
0.20	−0.358667	−0.224075	−0.091269	0.037925	0.162373	0.281392
0.25	−0.405383	−0.287825	−0.169276	−0.051856	0.063024	0.174425
0.30	−0.438598	−0.337402	−0.232900	−0.127399	−0.022501	0.080668
0.35	−0.460266	−0.374557	−0.283667	−0.189991	−0.095251	−0.000697
0.40	−0.472162	−0.400923	−0.323033	−0.240897	−0.156288	−0.070533
0.45	−0.475891	−0.418010	−0.352376	−0.281339	−0.206670	−0.129728
0.50	−0.472888	−0.427197	−0.372983	−0.312486	−0.247434	−0.179176
0.55	−0.464423	−0.429738	−0.386047	−0.335446	−0.279581	−0.219762
0.60	−0.451617	−0.426759	−0.392664	−0.351253	−0.304066	−0.252349
0.65	−0.435445	−0.419268	−0.393833	−0.360871	−0.321796	−0.277770
0.70	−0.416752	−0.408155	−0.390455	−0.365185	−0.333617	−0.296819
0.75	−0.396261	−0.394204	−0.383341	−0.365003	−0.340318	−0.310248
0.80	−0.374586	−0.378096	−0.373210	−0.361060	−0.342626	−0.318759
0.85	−0.352242	−0.360420	−0.360698	−0.354015	−0.341205	−0.323007
0.90	−0.329656	−0.341682	−0.346363	−0.344457	−0.336659	−0.323596
0.95	−0.307179	−0.322309	−0.330689	−0.332910	−0.329532	−0.321079
1.00	−0.285091	−0.302659	−0.314094	−0.319833	−0.320314	−0.315957
1.05	−0.263614	−0.283029	−0.296933	−0.305628	−0.309436	−0.308686

Table 2.6 (continued)

			t			
v	0.35	0.40	0.45	0.50	0.55	0.60
1.10	-0.242919	-0.263662	-0.279508	-0.290642	-0.297283	-0.299673
1.15	-0.223132	-0.244750	-0.262071	-0.275174	-0.284190	-0.289281
1.20	-0.204343	-0.226445	-0.244827	-0.259477	-0.270447	-0.277831
1.25	-0.186608	-0.208865	-0.227945	-0.243764	-0.256307	-0.265605
1.30	-0.169959	-0.192093	-0.211558	-0.228213	-0.241983	-0.252850
1.35	-0.154406	-0.176187	-0.195770	-0.212966	-0.227657	-0.239777
1.40	-0.139942	-0.161182	-0.180658	-0.198141	-0.213479	-0.226568
1.45	-0.126545	-0.147097	-0.166275	-0.183828	-0.199572	-0.213377
1.50	-0.114184	-0.133932	-0.152659	-0.170095	-0.186038	-0.200333
1.55	-0.102817	-0.121678	-0.139829	-0.156994	-0.172954	-0.187542
1.60	-0.092398	-0.110312	-0.127792	-0.144557	-0.160383	-0.175089
1.65	-0.082879	-0.099808	-0.116542	-0.132804	-0.148367	-0.163045
1.70	-0.074205	-0.090131	-0.106067	-0.121745	-0.136939	-0.151461
1.75	-0.066322	-0.081242	-0.096346	-0.111377	-0.126117	-0.140376
1.80	-0.059177	-0.073100	-0.087352	-0.101693	-0.115909	-0.129818
1.85	-0.052716	-0.065661	-0.079057	-0.092676	-0.106318	-0.119804
1.90	-0.046887	-0.058883	-0.071426	-0.084307	-0.097335	-0.110343
1.95	-0.041639	-0.052721	-0.064425	-0.076560	-0.088950	-0.101436
2.00	-0.036924	-0.047132	-0.058018	-0.069409	-0.081145	-0.093078
2.05	-0.032697	-0.042073	-0.052167	-0.062825	-0.073902	-0.085259
2.10	-0.028913	-0.037503	-0.046836	-0.056778	-0.067196	-0.077965
2.15	-0.025534	-0.033383	-0.041990	-0.051235	-0.061003	-0.071180
2.20	-0.022520	-0.029675	-0.037592	-0.046167	-0.055298	-0.064883
2.25	-0.019836	-0.026345	-0.033609	-0.041541	-0.050053	-0.059055
2.30	-0.017451	-0.023359	-0.030009	-0.037329	-0.045242	-0.053671
2.35	-0.015335	-0.020685	-0.026760	-0.033499	-0.040838	-0.048710
2.40	-0.013459	-0.018296	-0.023833	-0.030023	-0.036813	-0.044147
2.45	-0.011799	-0.016163	-0.021200	-0.026874	-0.033143	-0.039958
2.50	-0.010333	-0.014263	-0.018837	-0.024027	-0.029801	-0.036120
2.55	-0.009039	-0.012572	-0.016717	-0.021456	-0.026763	-0.032610
2.60	-0.007898	-0.011070	-0.014820	-0.019138	-0.024007	-0.029405
2.65	-0.006895	-0.009736	-0.013123	-0.017051	-0.021510	-0.026483
2.70	-0.006012	-0.008555	-0.011608	-0.015175	-0.019251	-0.023825
2.75	-0.005238	-0.007509	-0.010258	-0.013491	-0.017210	-0.021409
2.80	-0.004559	-0.006584	-0.009055	-0.011982	-0.015369	-0.019217
2.85	-0.003964	-0.005768	-0.007985	-0.010630	-0.013711	-0.017231
2.90	-0.003443	-0.005048	-0.007035	-0.009422	-0.012219	-0.015434
2.95	-0.002989	-0.004414	-0.006192	-0.008342	-0.010879	-0.013811
3.00	-0.002591	-0.003856	-0.005445	-0.007380	-0.009676	-0.012346
3.05	-0.002245	-0.003365	-0.004784	-0.006522	-0.008598	-0.011025
3.10	-0.001943	-0.002935	-0.004199	-0.005759	-0.007633	-0.009837
3.15	-0.001681	-0.002557	-0.003682	-0.005080	-0.006770	-0.008768

Table 2.6 (continued)

			t			
v	0.35	0.40	0.45	0.50	0.55	0.60
3.20	−0.001453	−0.002226	−0.003227	−0.004478	−0.005999	−0.007808
3.25	−0.001254	−0.001936	−0.002825	−0.003943	−0.005311	−0.006947
3.30	−0.001082	−0.001683	−0.002471	−0.003470	−0.004698	−0.006176
3.35	−0.000933	−0.001461	−0.002160	−0.003050	−0.004153	−0.005486
3.40	−0.000804	−0.001268	−0.001886	−0.002679	−0.003667	−0.004868
3.45	−0.000692	−0.001099	−0.001646	−0.002352	−0.003236	−0.004317
3.50	−0.000595	−0.000952	−0.001435	−0.002063	−0.002853	−0.003824
3.55	−0.000511	−0.000824	−0.001251	−0.001808	−0.002514	−0.003386
3.60	−0.000439	−0.000713	−0.001089	−0.001583	−0.002213	−0.002995
3.65	−0.000377	−0.000617	−0.000947	−0.001385	−0.001946	−0.002647
3.70	−0.000323	−0.000533	−0.000823	−0.001211	−0.001711	−0.002338
3.75	−0.000277	−0.000460	−0.000715	−0.001058	−0.001503	−0.002063
3.80	−0.000237	−0.000397	−0.000621	−0.000924	−0.001319	−0.001820
3.85	−0.000203	−0.000342	−0.000539	−0.000806	−0.001157	−0.001604
3.90	−0.000174	−0.000295	−0.000467	−0.000703	−0.001014	−0.001412
3.95	−0.000148	−0.000254	−0.000405	−0.000612	−0.000888	−0.001243
4.00	−0.000127	−0.000218	−0.000350	−0.000533	−0.000777	−0.001093
4.05	−0.000108	−0.000188	−0.000303	−0.000464	−0.000680	−0.000961
4.10	−0.000092	−0.000161	−0.000262	−0.000403	−0.000594	−0.000844
4.15	−0.000079	−0.000138	−0.000226	−0.000351	−0.000519	−0.000740
4.20	−0.000067	−0.000119	−0.000196	−0.000305	−0.000453	−0.000649
4.25	−0.000057	−0.000102	−0.000169	−0.000264	−0.000395	−0.000569
4.30	−0.000049	−0.000087	−0.000146	−0.000229	−0.000345	−0.000499
4.35	−0.000041	−0.000075	−0.000125	−0.000199	−0.000300	−0.000436
4.40	−0.000035	−0.000064	−0.000108	−0.000172	−0.000261	−0.000382
4.45	−0.000030	−0.000055	−0.000093	−0.000149	−0.000228	−0.000334
4.50	−0.000025	−0.000047	−0.000080	−0.000129	−0.000198	−0.000292
4.55	−0.000022	−0.000040	−0.000069	−0.000112	−0.000172	−0.000255
4.60	−0.000018	−0.000034	−0.000059	−0.000096	−0.000149	−0.000222
4.65	−0.000015	−0.000029	−0.000051	−0.000083	−0.000130	−0.000194
4.70	−0.000013	−0.000025	−0.000044	−0.000072	−0.000113	−0.000169
4.75	−0.000011	−0.000021	−0.000037	−0.000062	−0.000098	−0.000147
4.80	−0.000009	−0.000018	−0.000032	−0.000053	−0.000085	−0.000128
4.85	−0.000008	−0.000015	−0.000028	−0.000046	−0.000073	−0.000112
4.90	−0.000007	−0.000013	−0.000024	−0.000040	−0.000063	−0.000097
4.95	−0.000006	−0.000011	−0.000020	−0.000034	−0.000055	−0.000084
5.00	−0.000005	−0.000009	−0.000017	−0.000029	−0.000047	−0.000073

			t			
v	0.65	0.70	0.75	0.80	0.85	0.90
0.00	0.978399	1.08954	1.18897	1.27764	1.35640	1.42605
0.05	0.812190	0.924906	1.02732	1.12010	1.20392	1.27938

Table 2.6 (continued)

			t			
v	**0.65**	**0.70**	**0.75**	**0.80**	**0.85**	**0.90**
0.10	0.659531	0.771934	0.875466	0.970572	1.05771	1.13734
0.15	0.520404	0.630884	0.733923	0.829758	0.918666	1.00095
0.20	0.394597	0.501805	0.602971	0.698147	0.787448	0.871036
0.25	0.281736	0.384571	0.482706	0.576031	0.664522	0.748214
0.30	0.181321	0.278910	0.373062	0.463534	0.550181	0.632927
0.35	0.092754	0.184429	0.273840	0.360633	0.444562	0.525459
0.40	0.015366	0.100647	0.184732	0.267184	0.347674	0.425957
0.45	−0.051562	0.027009	0.105345	0.182944	0.259415	0.334452
0.50	−0.108778	−0.037088	0.035215	0.107587	0.179590	0.250871
0.55	−0.157044	−0.092281	−0.026168	0.040724	0.107927	0.175057
0.60	−0.197122	−0.139224	−0.079351	−0.018083	0.044096	0.106783
0.65	−0.229762	−0.178579	−0.124902	−0.069307	−0.012281	0.045763
0.70	−0.255692	−0.211001	−0.163399	−0.113444	−0.061616	−0.008330
0.75	−0.275616	−0.237135	−0.195419	−0.151001	−0.104347	−0.055860
0.80	−0.290202	−0.257603	−0.221534	−0.182493	−0.140924	−0.097213
0.85	−0.300079	−0.273003	−0.242299	−0.208429	−0.171804	−0.132793
0.90	−0.305840	−0.283904	−0.258253	−0.229307	−0.197445	−0.163009
0.95	−0.308032	−0.290839	−0.269908	−0.245613	−0.218297	−0.188274
1.00	−0.307164	−0.294311	−0.277752	−0.257814	−0.234801	−0.208997
1.05	−0.303699	−0.294785	−0.282243	−0.266353	−0.247382	−0.225579
1.10	−0.298059	−0.292690	−0.283809	−0.271653	−0.256448	−0.238410
1.15	−0.290627	−0.288418	−0.282848	−0.274107	−0.262385	−0.247865
1.20	−0.281747	−0.282330	−0.279726	−0.274085	−0.265560	−0.254301
1.25	−0.271723	−0.274747	−0.274778	−0.271929	−0.266315	−0.258058
1.30	−0.260829	−0.265962	−0.268311	−0.267952	−0.264970	−0.259455
1.35	−0.249301	−0.256234	−0.260600	−0.262444	−0.261820	−0.258791
1.40	−0.237349	−0.245793	−0.251895	−0.255667	−0.257137	−0.256345
1.45	−0.225154	−0.234844	−0.242416	−0.247857	−0.251170	−0.252372
1.50	−0.212868	−0.223562	−0.232360	−0.239227	−0.244145	−0.247109
1.55	−0.200624	−0.212103	−0.221902	−0.229969	−0.236266	−0.240771
1.60	−0.188532	−0.200597	−0.211192	−0.220250	−0.227717	−0.233555
1.65	−0.176683	−0.189156	−0.200362	−0.210219	−0.218661	−0.225637
1.70	−0.165151	−0.177876	−0.189525	−0.200007	−0.209244	−0.217175
1.75	−0.153994	−0.166834	−0.178777	−0.189725	−0.199594	−0.208312
1.80	−0.143259	−0.156093	−0.168198	−0.179472	−0.189823	−0.199173
1.85	−0.132979	−0.145704	−0.157856	−0.169328	−0.180026	−0.189868
1.90	−0.123179	−0.135706	−0.147804	−0.159363	−0.170288	−0.180493
1.95	−0.113873	−0.126129	−0.138086	−0.149635	−0.160679	−0.171131
2.00	−0.105069	−0.116993	−0.128735	−0.140189	−0.151258	−0.161855
2.05	−0.096768	−0.108311	−0.119777	−0.131063	−0.142075	−0.152724
2.10	−0.088967	−0.100090	−0.111229	−0.122285	−0.133167	−0.143790
2.15	−0.081657	−0.092330	−0.103101	−0.113876	−0.124568	−0.135094

Table 2.6 (continued)

				t		
v	0.65	0.70	0.75	0.80	0.85	0.90
2.20	−0.074825	−0.085027	−0.095398	−0.105851	−0.116301	−0.126669
2.25	−0.068458	−0.078175	−0.088122	−0.098218	−0.108383	−0.118543
2.30	−0.062538	−0.071762	−0.081269	−0.090982	−0.100827	−0.110735
2.35	−0.057046	−0.065776	−0.074831	−0.084141	−0.093640	−0.103260
2.40	−0.051963	−0.060201	−0.068799	−0.077694	−0.086824	−0.096127
2.45	−0.047268	−0.055021	−0.063161	−0.071632	−0.080378	−0.089343
2.50	−0.042941	−0.050218	−0.057903	−0.065946	−0.074298	−0.082907
2.55	−0.038959	−0.045773	−0.053010	−0.060627	−0.068579	−0.076820
2.60	−0.035302	−0.041668	−0.048466	−0.055660	−0.063211	−0.071077
2.65	−0.031950	−0.037883	−0.044255	−0.051033	−0.058183	−0.065670
2.70	−0.028881	−0.034401	−0.040359	−0.046730	−0.053485	−0.060593
2.75	−0.026078	−0.031201	−0.036762	−0.042737	−0.049103	−0.055834
2.80	−0.023520	−0.028267	−0.033445	−0.039038	−0.045024	−0.051384
2.85	−0.021190	−0.025580	−0.030393	−0.035617	−0.041235	−0.047230
2.90	−0.019070	−0.023123	−0.027589	−0.032458	−0.037720	−0.043359
2.95	−0.017144	−0.020880	−0.025016	−0.029547	−0.034465	−0.039759
3.00	−0.015397	−0.018834	−0.022658	−0.026867	−0.031456	−0.036416
3.05	−0.013815	−0.016972	−0.020502	−0.024405	−0.028678	−0.033318
3.10	−0.012383	−0.015279	−0.018532	−0.022145	−0.026118	−0.030450
3.15	−0.011088	−0.013741	−0.016735	−0.020074	−0.023762	−0.027800
3.20	−0.009920	−0.012346	−0.015097	−0.018179	−0.021597	−0.025354
3.25	−0.008867	−0.011083	−0.013607	−0.016447	−0.019610	−0.023100
3.30	−0.007918	−0.009940	−0.012252	−0.014865	−0.017788	−0.021026
3.35	−0.007065	−0.008906	−0.011022	−0.013424	−0.016121	−0.019120
3.40	−0.006298	−0.007974	−0.009907	−0.012111	−0.014596	−0.017370
3.45	−0.005610	−0.007132	−0.008897	−0.010917	−0.013203	−0.015766
3.50	−0.004993	−0.006374	−0.007983	−0.009832	−0.011933	−0.014297
3.55	−0.004440	−0.005692	−0.007157	−0.008847	−0.010776	−0.012954
3.60	−0.003945	−0.005079	−0.006411	−0.007955	−0.009722	−0.011726
3.65	−0.003503	−0.004528	−0.005738	−0.007146	−0.008765	−0.010606
3.70	−0.003108	−0.004034	−0.005132	−0.006414	−0.007895	−0.009585
3.75	−0.002755	−0.003591	−0.004586	−0.005753	−0.007106	−0.008655
3.80	−0.002440	−0.003194	−0.004095	−0.005156	−0.006390	−0.007809
3.85	−0.002160	−0.002839	−0.003654	−0.004618	−0.005742	−0.007041
3.90	−0.001911	−0.002522	−0.003258	−0.004132	−0.005156	−0.006343
3.95	−0.001689	−0.002238	−0.002903	−0.003695	−0.004627	−0.005710
4.00	−0.001492	−0.001985	−0.002584	−0.003302	−0.004148	−0.005136
4.05	−0.001317	−0.001759	−0.002299	−0.002948	−0.003717	−0.004616
4.10	−0.001161	−0.001558	−0.002044	−0.002630	−0.003327	−0.004146
4.15	−0.001024	−0.001379	−0.001816	−0.002345	−0.002977	−0.003722
4.20	−0.000902	−0.001220	−0.001613	−0.002090	−0.002662	−0.003338
4.25	−0.000794	−0.001078	−0.001431	−0.001861	−0.002378	−0.002992

Table 2.6 (continued)

			t			
v	0.65	0.70	0.75	0.80	0.85	0.90
4.30	-0.000699	-0.000952	-0.001269	-0.001656	-0.002123	-0.002680
4.35	-0.000614	-0.000841	-0.001124	-0.001473	-0.001895	-0.002399
4.40	-0.000540	-0.000742	-0.000996	-0.001309	-0.001689	-0.002146
4.45	-0.000474	-0.000654	-0.000881	-0.001162	-0.001505	-0.001918
4.50	-0.000416	-0.000576	-0.000779	-0.001032	-0.001341	-0.001714
4.55	-0.000365	-0.000507	-0.000689	-0.000915	-0.001193	-0.001530
4.60	-0.000320	-0.000447	-0.000609	-0.000811	-0.001061	-0.001365
4.65	-0.000280	-0.000393	-0.000537	-0.000719	-0.000943	-0.001217
4.70	-0.000245	-0.000345	-0.000474	-0.000636	-0.000838	-0.001085
4.75	-0.000215	-0.000303	-0.000418	-0.000563	-0.000744	-0.000966
4.80	-0.000188	-0.000266	-0.000368	-0.000498	-0.000660	-0.000860
4.85	-0.000164	-0.000234	-0.000324	-0.000440	-0.000586	-0.000765
4.90	-0.000143	-0.000205	-0.000286	-0.000389	-0.000519	-0.000680
4.95	-0.000125	-0.000180	-0.000251	-0.000343	-0.000460	-0.000604
5.00	-0.000109	-0.000157	-0.000221	-0.000303	-0.000407	-0.000537

			t			
v	0.95	1.0	2.0	3.0	4.0	5.0
0.00	1.48734	1.54094	1.68209	1.22021	0.743116	0.356483
0.05	1.34710	1.40760	1.70062	1.34229	0.937920	0.633133
0.10	1.20990	1.27584	1.70653	1.45259	1.12494	0.905633
0.15	1.07692	1.14690	1.70079	1.55103	1.30349	1.17295
0.20	0.949100	1.021850	1.68439	1.63762	1.47293	1.43407
0.25	0.827185	0.901547	1.65830	1.71246	1.63272	1.68810
0.30	0.711753	0.786681	1.62352	1.77571	1.78239	1.93415
0.35	0.603219	0.677787	1.58101	1.82762	1.92155	2.17142
0.40	0.501858	0.575254	1.53173	1.86851	2.04991	2.39912
0.45	0.407821	0.479344	1.47660	1.89873	2.16723	2.61662
0.50	0.321152	0.390208	1.41649	1.91871	2.27333	2.82326
0.55	0.241800	0.307897	1.35226	1.92889	2.36813	3.01851
0.60	0.169638	0.232379	1.28471	1.92977	2.45162	3.20190
0.65	0.104473	0.163550	1.21460	1.92188	2.52383	3.37299
0.70	0.046059	0.101243	1.14263	1.90575	2.58486	3.53146
0.75	-0.005894	0.045243	1.06946	1.88197	2.63488	3.67703
0.80	-0.051706	-0.004705	0.995703	1.85110	2.67408	3.80948
0.85	-0.091723	-0.048890	0.921904	1.81373	2.70273	3.92868
0.90	-0.126307	-0.087619	0.848568	1.77043	2.72113	4.03454
0.95	-0.155828	-0.121221	0.776148	1.72181	2.72961	4.12705
1.00	-0.180661	-0.150033	0.705045	1.66843	2.72857	4.20624
1.05	-0.201177	-0.174394	0.635613	1.61086	2.71839	4.27223
1.10	-0.217746	-0.194648	0.568163	1.54965	2.69952	4.32513
1.15	-0.230724	-0.211131	0.502956	1.48535	2.67240	4.36518

Table 2.6 (continued)

				t		
v	0.95	1.0	2.0	3.0	4.0	5.0
1.20	−0.240458	−0.224174	0.440216	1.41847	2.63752	4.39261
1.25	−0.247279	−0.234097	0.380127	1.34951	2.59537	4.40771
1.30	−0.251503	−0.241210	0.322833	1.27895	2.54644	4.41082
1.35	−0.253428	−0.245806	0.268447	1.20724	2.49124	4.40231
1.40	−0.253336	−0.248165	0.217049	1.13481	2.43028	4.38258
1.45	−0.251487	−0.248550	0.168689	1.062050	2.36408	4.35208
1.50	−0.248125	−0.247210	0.123391	0.989343	2.29314	4.31125
1.55	−0.243474	−0.244372	0.081157	0.917033	2.21798	4.26060
1.60	−0.237738	−0.240252	0.041964	0.845440	2.13909	4.20061
1.65	−0.231107	−0.235044	0.005773	0.774857	2.05696	4.13183
1.70	−0.223750	−0.228928	−0.027474	0.705548	1.97206	4.05478
1.75	−0.215820	−0.222070	−0.057848	0.637752	1.88487	3.97001
1.80	−0.207456	−0.214616	−0.085434	0.571683	1.79584	3.87809
1.85	−0.198780	−0.206700	−0.110326	0.507527	1.70539	3.77957
1.90	−0.189900	−0.198443	−0.132629	0.445449	1.61393	3.67500
1.95	−0.180912	−0.189951	−0.152453	0.385587	1.52187	3.56495
2.00	−0.171899	−0.181317	−0.169913	0.328061	1.42959	3.44998
2.05	−0.162932	−0.172624	−0.185131	0.272967	1.33742	3.33064
2.10	−0.154074	−0.163945	−0.198229	0.220379	1.24571	3.20747
2.15	−0.145375	−0.155340	−0.209332	0.170355	1.15478	3.08101
2.20	−0.136880	−0.146864	−0.218563	0.122936	1.064900	2.95177
2.25	−0.128624	−0.138559	−0.226049	0.078144	0.976353	2.82028
2.30	−0.120636	−0.130464	−0.231911	0.035986	0.889381	2.68702
2.35	−0.112937	−0.122609	−0.236270	−0.003543	0.804211	2.55246
2.40	−0.105545	−0.115017	−0.239246	−0.040463	0.721046	2.41707
2.45	−0.098471	−0.107707	−0.240951	−0.074806	0.640069	2.28129
2.50	−0.091723	−0.100693	−0.241498	−0.106612	0.561444	2.14554
2.55	−0.085304	−0.093985	−0.240993	−0.135934	0.485313	2.01022
2.60	−0.079217	−0.087588	−0.239539	−0.162829	0.411798	1.87571
2.65	−0.073457	−0.081505	−0.237234	−0.187365	0.341002	1.74237
2.70	−0.068021	−0.075735	−0.234171	−0.209615	0.273017	1.61054
2.75	−0.062901	−0.070275	−0.230437	−0.229657	0.207911	1.48053
2.80	−0.058091	−0.065122	−0.226116	−0.247573	0.145735	1.35262
2.85	−0.053581	−0.060267	−0.221285	−0.263450	0.086532	1.22710
2.90	−0.049360	−0.055704	−0.216019	−0.277378	0.030322	1.10421
2.95	−0.045417	−0.051423	−0.210384	−0.289447	−0.022883	0.98418
3.00	−0.041740	−0.047415	−0.204443	−0.299750	−0.073088	0.867210
3.05	−0.038318	−0.043668	−0.198256	−0.308382	−0.120300	0.753487
3.10	−0.035137	−0.040172	−0.191875	−0.315434	−0.164555	0.643175
3.15	−0.032185	−0.036915	−0.185351	−0.321002	−0.205879	0.536414
3.20	−0.029451	−0.033886	−0.178728	−0.325177	−0.244322	0.433321
3.25	−0.026921	−0.031073	−0.172046	−0.328052	−0.279934	0.334015

Table 2.6 (continued)

			t			
v	0.95	1.0	2.0	3.0	4.0	5.0
3.30	−0.024584	−0.028464	−0.165343	−0.329715	−0.312777	0.238571
3.35	−0.022428	−0.026049	−0.158652	−0.330256	−0.342918	0.147060
3.40	−0.020441	−0.023815	−0.152003	−0.329758	−0.370429	0.059538
3.45	−0.018613	−0.021752	−0.145421	−0.328307	−0.395387	−0.023966
3.50	−0.016933	−0.019849	−0.138931	−0.325982	−0.417877	−0.103427
3.55	−0.015391	−0.018097	−0.132551	−0.322861	−0.437984	−0.178840
3.60	−0.013977	−0.016484	−0.126301	−0.319019	−0.455799	−0.250217
3.65	−0.012682	−0.015002	−0.120194	−0.314527	−0.471410	−0.317569
3.70	−0.011497	−0.013641	−0.114243	−0.309454	−0.484916	−0.380937
3.75	−0.010414	−0.012394	−0.108459	−0.303864	−0.496408	−0.440358
3.80	−0.009426	−0.011251	−0.102850	−0.297821	−0.505985	−0.495888
3.85	−0.008524	−0.010205	−0.097423	−0.291381	−0.513742	−0.547590
3.90	−0.007703	−0.009249	−0.092183	−0.284601	−0.519774	−0.595537
3.95	−0.006956	−0.008377	−0.087134	−0.277532	−0.524179	−0.639803
4.00	−0.006276	−0.007580	−0.082277	−0.270222	−0.527050	−0.680475
4.05	−0.005659	−0.006855	−0.077613	−0.262718	−0.528484	−0.717649
4.10	−0.005098	−0.006194	−0.073143	−0.255062	−0.528571	−0.751423
4.15	−0.004590	−0.005592	−0.068865	−0.247293	−0.527402	−0.781898
4.20	−0.004129	−0.005046	−0.064777	−0.239447	−0.525066	−0.809180
4.25	−0.003713	−0.004550	−0.060877	−0.231558	−0.521649	−0.833380
4.30	−0.003335	−0.004099	−0.057162	−0.223656	−0.517235	−0.854611
4.35	−0.002995	−0.003691	−0.053627	−0.215771	−0.511905	−0.872987
4.40	−0.002687	−0.003321	−0.050268	−0.207926	−0.505738	−0.888631
4.45	−0.002409	−0.002987	−0.047081	−0.200146	−0.498811	−0.901653
4.50	−0.002159	−0.002684	−0.044060	−0.192451	−0.491194	−0.912177
4.55	−0.001933	−0.002410	−0.041200	−0.184861	−0.482960	−0.920322
4.60	−0.001730	−0.002163	−0.038496	−0.177390	−0.474174	−0.926207
4.65	−0.001547	−0.001940	−0.035941	−0.170055	−0.464902	−0.929948
4.70	−0.001383	−0.001739	−0.033532	−0.162867	−0.455202	−0.931666
4.75	−0.001235	−0.001558	−0.031260	−0.155838	−0.445133	−0.931474
4.80	−0.001103	−0.001395	−0.029122	−0.148978	−0.434750	−0.929488
4.85	−0.000984	−0.001248	−0.027111	−0.142293	−0.424104	−0.925820
4.90	−0.000877	−0.001116	−0.025221	−0.135791	−0.413244	−0.920581
4.95	−0.000782	−0.000997	−0.023446	−0.129478	−0.402216	−0.913878
5.00	−0.000697	−0.000891	−0.021782	−0.123356	−0.391061	−0.905818

2.3 First, Second and Third Derivatives with Respect to the Order of the Anger Functions

Derivatives with respect to the order of the Anger functions were evaluated using MATHEMATICA program by applying the following integral representations:

$$\frac{\partial J_v(z)}{\partial v} = -\frac{1}{\pi} \int_0^\pi t \sin(vt - z \sin t) \, dt \tag{2.3.1}$$

$$\frac{\partial^2 J_v(z)}{\partial v^2} = -\frac{1}{\pi} \int_0^\pi t^2 \cos(vt - z \sin t) \, dt \tag{2.3.2}$$

$$\frac{\partial^3 J_v(z)}{\partial v^3} = \frac{1}{\pi} \int_0^\pi t^3 \sin(vt - z \sin t) \, dt \tag{2.3.3}$$

In the case of these functions, it is evident that in their oscillatory form, their values are smaller as v and t increase and they tend to near zero values (Figures 2.19–2.36).

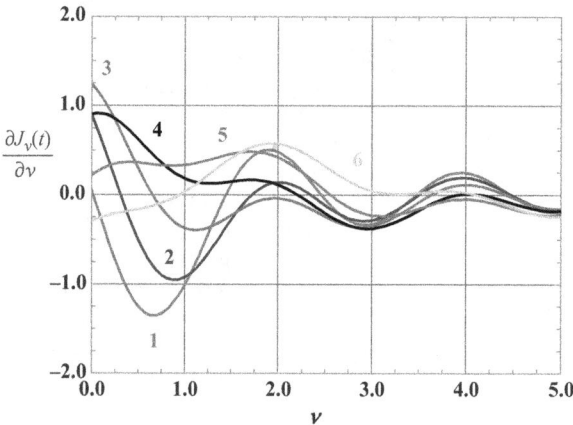

Figure 2.19: First derivatives of the Anger function with respect to the order v as a function of v, at constant values of argument t.

$1 - t = 0.05; 2 - t = 1.0; 3 - t = 2.0; 4 - t = 3.0; 5 - t = 4.0; 6 - t = 5.0.$

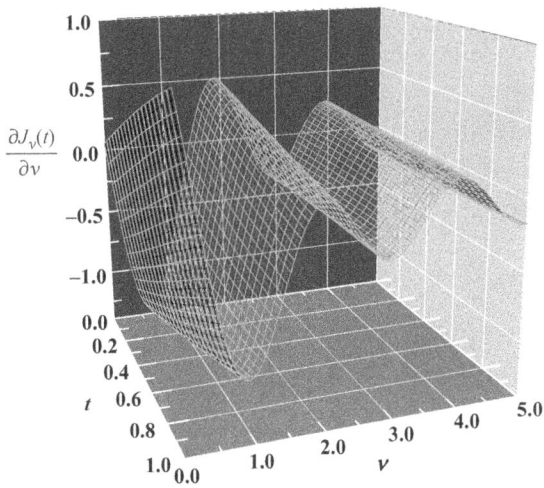

Figure 2.20: First derivatives with respect to the order of the Anger functions $\partial \mathbf{J}_\nu(t)/\partial \nu$ as a function of ν and t.

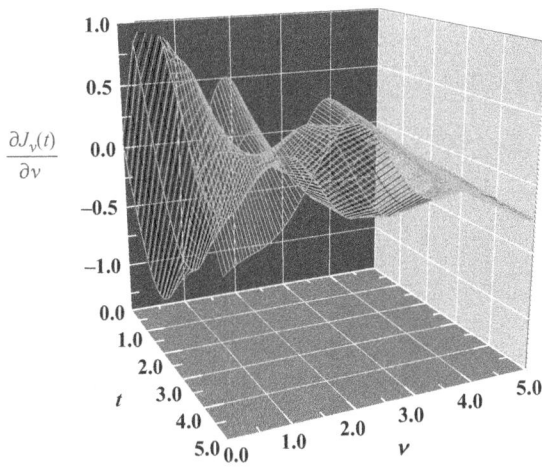

Figure 2.21: First derivatives with respect to the order of the Anger functions $\partial \mathbf{J}_\nu(t)/\partial \nu$ as a function of ν and t.

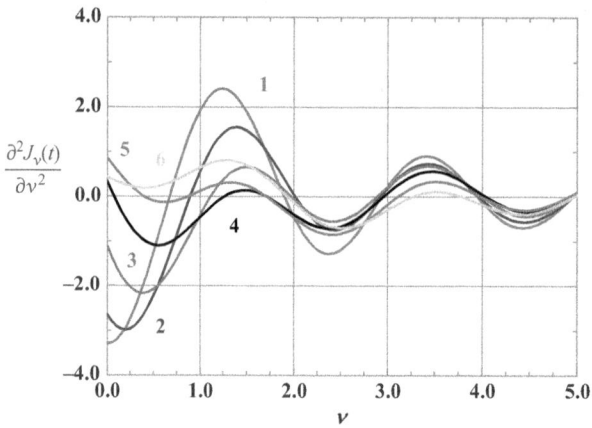

Figure 2.22: Second derivatives of the Anger function with respect to the order v as a function of v, at constant values of argument t.

$1 - t = 0.05; 2 - t = 1.0; 3 - t = 2.0; 4 - t = 3.0; 5 - t = 4.0; 6 - t = 5.0.$

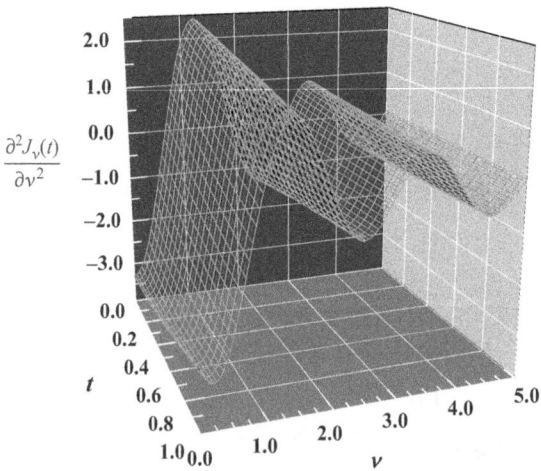

Figure 2.23: Second derivatives with respect to the order of the Anger functions $\partial^2 J_v(t)/\partial v^2$ as a function of v and t.

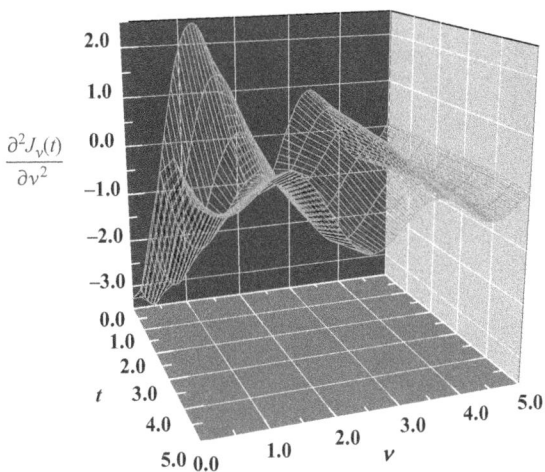

Figure 2.24: Second derivatives with respect to the order of the Anger functions $\partial^2 J_v(t)/\partial v^2$ as a function of v and t.

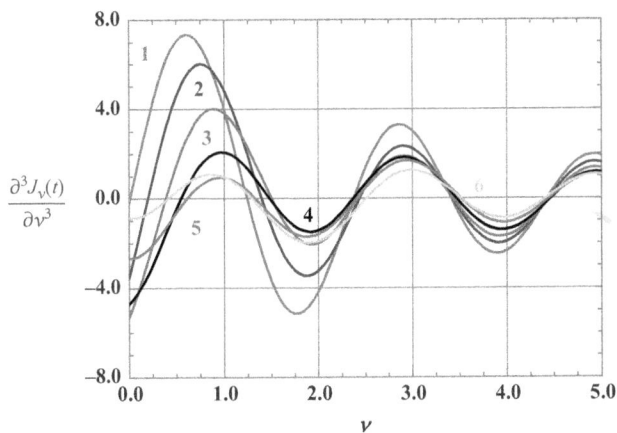

Figure 2.25: Third derivatives of the Anger function with respect to the order v as a function of v, at constant values of argument t.

$1 - t = 0.05$; $2 - t = 1.0$; $3 - t = 2.0$; $4 - t = 3.0$; $5 - t = 4.0$; $5 - t = 5.0$.

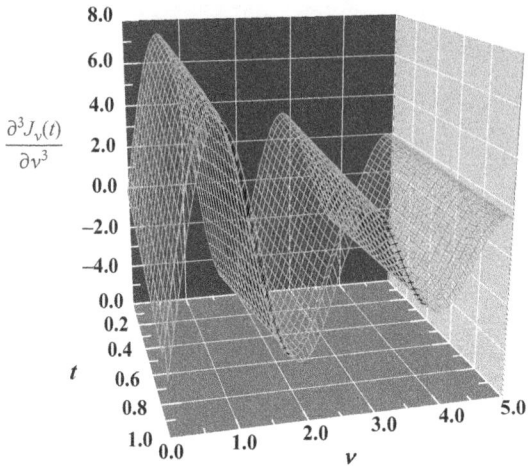

Figure 2.26: Third derivatives with respect to the order of the Anger functions $\partial^3 J_v(t)/\partial v^3$ as a function of v and t.

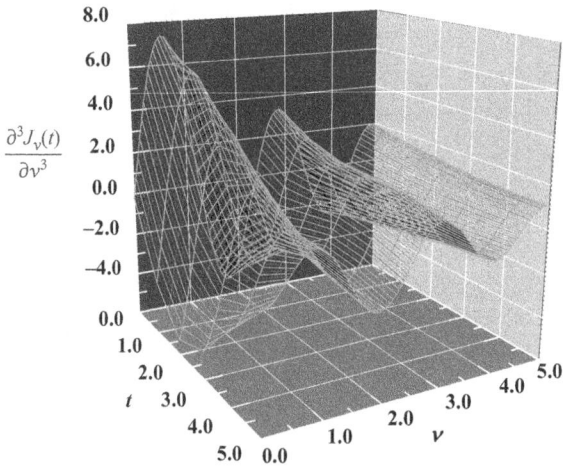

Figure 2.27: Third derivatives with respect to the order of the Anger functions $\partial^3 J_v(t)/\partial v^3$ as a function of v and t.

Table 2.7: First derivatives with respect to the order of the Anger functions $\partial J_\nu(t)/\partial \nu$.

			t			
ν	0.05	0.10	0.15	0.20	0.25	0.30
0.00	0.049986	0.099889	0.149625	0.199113	0.248268	0.297011
0.05	-0.114256	-0.064334	-0.014404	0.035451	0.085147	0.134602
0.10	-0.276555	-0.227095	-0.177453	-0.127713	-0.077957	-0.028270
0.15	-0.434520	-0.385997	-0.337121	-0.287974	-0.238639	-0.189198
0.20	-0.585838	-0.538715	-0.491071	-0.442986	-0.394543	-0.345824
0.25	-0.728318	-0.683038	-0.637074	-0.590507	-0.543414	-0.495876
0.30	-0.859922	-0.816904	-0.773048	-0.728430	-0.683125	-0.637213
0.35	-0.978808	-0.938442	-0.897091	-0.854827	-0.811725	-0.767859
0.40	-1.08335	-1.04599	-1.00751	-0.967979	-0.927462	-0.886034
0.45	-1.17219	-1.13815	-1.10287	-1.06640	-1.02882	-0.990188
0.50	-1.24422	-1.21378	-1.18198	-1.14888	-1.11454	-1.07903
0.55	-1.29863	-1.27202	-1.24395	-1.21447	-1.18364	-1.15152
0.60	-1.33493	-1.31233	-1.28818	-1.26253	-1.23543	-1.20694
0.65	-1.35292	-1.33445	-1.31437	-1.29271	-1.26952	-1.24484
0.70	-1.35270	-1.33845	-1.32253	-1.30497	-1.28581	-1.26510
0.75	-1.33469	-1.32469	-1.31298	-1.29958	-1.28453	-1.26787
0.80	-1.29960	-1.29383	-1.28632	-1.27709	-1.26618	-1.25361
0.85	-1.24840	-1.24679	-1.24343	-1.23834	-1.23154	-1.22305
0.90	-1.18234	-1.18477	-1.18546	-1.18442	-1.18166	-1.17720
0.95	-1.10288	-1.10919	-1.11378	-1.11665	-1.11782	-1.11729
1.00	-1.01166	-1.02165	-1.02996	-1.03658	-1.04151	-1.04476
1.05	-0.910536	-0.923961	-0.935753	-0.945902	-0.954401	-0.961248
1.10	-0.801444	-0.818032	-0.833048	-0.846478	-0.858309	-0.868532
1.15	-0.686435	-0.705881	-0.723830	-0.740262	-0.755157	-0.768501
1.20	-0.567608	-0.589583	-0.610147	-0.629273	-0.646938	-0.663120
1.25	-0.447073	-0.471230	-0.494070	-0.515563	-0.535679	-0.554392
1.30	-0.326915	-0.352891	-0.377655	-0.401170	-0.423402	-0.444319
1.35	-0.209154	-0.236580	-0.262904	-0.288085	-0.312084	-0.334866
1.40	-0.095712	-0.124213	-0.151728	-0.178212	-0.203623	-0.227922
1.45	0.011627	-0.017577	-0.045915	-0.073340	-0.099806	-0.125268
1.50	0.111240	0.081699	0.052899	0.024893	-0.002272	-0.028548
1.55	0.201697	0.172171	0.143263	0.115025	0.087509	0.060764
1.60	0.281781	0.252608	0.223928	0.195795	0.168263	0.141383
1.65	0.350506	0.322004	0.293870	0.266160	0.238931	0.212234
1.70	0.407133	0.379592	0.352300	0.325311	0.298683	0.272468
1.75	0.451172	0.424859	0.398676	0.372681	0.346928	0.321474
1.80	0.482391	0.457541	0.432709	0.407950	0.383322	0.358878
1.85	0.500811	0.477627	0.454354	0.431046	0.407761	0.384552
1.90	0.506700	0.485351	0.463812	0.442138	0.420383	0.398602
1.95	0.500559	0.481179	0.461517	0.441625	0.421557	0.401365
2.00	0.483107	0.465796	0.448118	0.430123	0.411863	0.393390
2.05	0.455263	0.440084	0.424462	0.408445	0.392082	0.375425
2.10	0.418118	0.405101	0.391573	0.377581	0.363172	0.348393

Table 2.7 (continued)

			t			
v	0.05	0.10	0.15	0.20	0.25	0.30
2.15	0.372910	0.362050	0.350622	0.338670	0.326238	0.313370
2.20	0.320996	0.312258	0.302905	0.292976	0.282512	0.271557
2.25	0.263819	0.257138	0.249804	0.241852	0.233321	0.224250
2.30	0.202878	0.198163	0.192765	0.186717	0.180054	0.172812
2.35	0.139696	0.136830	0.133261	0.129020	0.124137	0.118644
2.40	0.075784	0.074630	0.072763	0.070208	0.066994	0.063150
2.45	0.012614	0.013018	0.012706	0.011701	0.010027	0.007709
2.50	-0.048411	-0.046619	-0.045536	-0.045145	-0.045423	-0.046350
2.55	-0.105986	-0.102984	-0.100680	-0.099056	-0.098096	-0.097783
2.60	-0.158926	-0.154903	-0.151555	-0.148873	-0.146841	-0.145445
2.65	-0.206192	-0.201337	-0.197131	-0.193567	-0.190634	-0.188322
2.70	-0.246904	-0.241407	-0.236527	-0.232261	-0.228601	-0.225540
2.75	-0.280358	-0.274405	-0.269034	-0.264244	-0.260030	-0.256387
2.80	-0.306034	-0.299807	-0.294121	-0.288979	-0.284379	-0.280319
2.85	-0.323603	-0.317273	-0.311442	-0.306114	-0.301290	-0.296972
2.90	-0.332929	-0.326657	-0.320838	-0.315479	-0.310584	-0.306157
2.95	-0.334067	-0.327998	-0.322336	-0.317089	-0.312265	-0.307868
3.00	-0.327252	-0.321518	-0.316143	-0.311137	-0.306511	-0.302270
3.05	-0.312895	-0.307609	-0.302634	-0.297983	-0.293667	-0.289696
3.10	-0.291565	-0.286822	-0.282343	-0.278144	-0.274235	-0.270628
3.15	-0.263970	-0.259847	-0.255943	-0.252273	-0.248851	-0.245690
3.20	-0.230943	-0.227498	-0.224227	-0.221148	-0.218276	-0.215623
3.25	-0.193417	-0.190687	-0.188091	-0.185646	-0.183367	-0.181269
3.30	-0.152399	-0.150405	-0.148506	-0.146719	-0.145061	-0.143548
3.35	-0.108950	-0.107694	-0.106497	-0.105377	-0.104350	-0.103433
3.40	-0.064158	-0.063624	-0.063118	-0.062656	-0.062255	-0.061932
3.45	-0.019109	-0.019268	-0.019425	-0.019598	-0.019803	-0.020057
3.50	0.025133	0.024325	0.023545	0.022774	0.021997	0.021197
3.55	0.067548	0.066149	0.064797	0.063478	0.062174	0.060870
3.60	0.107190	0.105265	0.103406	0.101597	0.099822	0.098066
3.65	0.143198	0.140823	0.138527	0.136296	0.134114	0.131967
3.70	0.174821	0.172075	0.169420	0.166840	0.164322	0.161851
3.75	0.201426	0.198395	0.195461	0.192611	0.189831	0.187108
3.80	0.222513	0.219284	0.216155	0.213115	0.210151	0.207250
3.85	0.237725	0.234384	0.231144	0.227995	0.224924	0.221921
3.90	0.246849	0.243481	0.240212	0.237033	0.233932	0.230900
3.95	0.249820	0.246505	0.243286	0.240152	0.237096	0.234107
4.00	0.246718	0.243532	0.240435	0.237420	0.234477	0.231599
4.05	0.237763	0.234775	0.231868	0.229036	0.226270	0.223564
4.10	0.223307	0.220578	0.217921	0.215330	0.212798	0.210319
4.15	0.203826	0.201407	0.199050	0.196749	0.194499	0.192295
4.20	0.179897	0.177832	0.175816	0.173847	0.171919	0.170029
4.25	0.152195	0.150514	0.148871	0.147265	0.145690	0.144144

Table 2.7 (continued)

			t			
v	**0.05**	**0.10**	**0.15**	**0.20**	**0.25**	**0.30**
4.30	0.121463	0.120189	0.118941	0.117718	0.116517	0.115336
4.35	0.088503	0.087646	0.086805	0.085977	0.085162	0.084357
4.40	0.054150	0.053712	0.053277	0.052847	0.052420	0.051993
4.45	0.019254	0.019225	0.019190	0.019150	0.019103	0.019048
4.50	−0.015341	−0.014978	−0.014631	−0.014298	−0.013980	−0.013677
4.55	−0.048815	−0.048087	−0.047382	−0.046700	−0.046040	−0.045403
4.60	−0.080392	−0.079332	−0.078302	−0.077303	−0.076332	−0.075389
4.65	−0.109358	−0.108006	−0.106690	−0.105410	−0.104164	−0.102952
4.70	−0.135075	−0.133475	−0.131917	−0.130399	−0.128920	−0.127478
4.75	−0.156995	−0.155197	−0.153443	−0.151733	−0.150065	−0.148438
4.80	−0.174671	−0.172725	−0.170827	−0.168974	−0.167165	−0.165399
4.85	−0.187766	−0.185726	−0.183733	−0.181788	−0.179887	−0.178030
4.90	−0.196058	−0.193976	−0.191941	−0.189953	−0.188009	−0.186109
4.95	−0.199444	−0.197371	−0.195345	−0.193364	−0.191426	−0.189530
5.00	−0.197940	−0.195926	−0.193956	−0.192029	−0.190142	−0.188296

			t			
v	**0.35**	**0.40**	**0.45**	**0.50**	**0.55**	**0.60**
0.00	0.345259	0.392934	0.439957	0.486249	0.531736	0.576343
0.05	0.183733	0.232459	0.280700	0.328375	0.375407	0.421718
0.10	0.021267	0.070570	0.119558	0.168149	0.216264	0.263822
0.15	−0.139735	−0.090333	−0.041074	0.007961	0.056688	0.105027
0.20	−0.296909	−0.247883	−0.198828	−0.149826	−0.100959	−0.052310
0.25	−0.447975	−0.399792	−0.351408	−0.302907	−0.254371	−0.205881
0.30	−0.590773	−0.543885	−0.496629	−0.449087	−0.401341	−0.353471
0.35	−0.723306	−0.678145	−0.632453	−0.586311	−0.539799	−0.492997
0.40	−0.843768	−0.800740	−0.757025	−0.712701	−0.667847	−0.622541
0.45	−0.950576	−0.910057	−0.868702	−0.826587	−0.783788	−0.740382
0.50	−1.04240	−1.00473	−0.966083	−0.926536	−0.886159	−0.845027
0.55	−1.11817	−1.08365	−1.04803	−1.01137	−0.973747	−0.935229
0.60	−1.17711	−1.14600	−1.11367	−1.08019	−1.04561	−1.01001
0.65	−1.21874	−1.19126	−1.16245	−1.13238	−1.10110	−1.06868
0.70	−1.24287	−1.21918	−1.19408	−1.16761	−1.13984	−1.11082
0.75	−1.24963	−1.22985	−1.20858	−1.18587	−1.16176	−1.13632
0.80	−1.23941	−1.22361	−1.20627	−1.18741	−1.16708	−1.14534
0.85	−1.21290	−1.20111	−1.18773	−1.17278	−1.15630	−1.13834
0.90	−1.17106	−1.16325	−1.15382	−1.14278	−1.13017	−1.11602
0.95	−1.11507	−1.11117	−1.10563	−1.09846	−1.08969	−1.07935
1.00	−1.04633	−1.04624	−1.04449	−1.04111	−1.03610	−1.02951
1.05	−0.966443	−0.969990	−0.97189	−0.972167	−0.970823	−0.967876
1.10	−0.877141	−0.884133	−0.889509	−0.893273	−0.895432	−0.895996

Table 2.7 (continued)

				t		
v	0.35	0.40	0.45	0.50	0.55	0.60
1.15	−0.780282	−0.790491	−0.799123	−0.806175	−0.811649	−0.815548
1.20	−0.677801	−0.690968	−0.702609	−0.712716	−0.721284	−0.728310
1.25	−0.571678	−0.587518	−0.601895	−0.614795	−0.626206	−0.636122
1.30	−0.463894	−0.482102	−0.498919	−0.514326	−0.528309	−0.540852
1.35	−0.356399	−0.376651	−0.395596	−0.413209	−0.429468	−0.444356
1.40	−0.251071	−0.273035	−0.293782	−0.313284	−0.331513	−0.348448
1.45	−0.149687	−0.173024	−0.195241	−0.216307	−0.236189	−0.254860
1.50	−0.053891	−0.078259	−0.101612	−0.123912	−0.145125	−0.165218
1.55	0.034838	0.009775	−0.014380	−0.037587	−0.059808	−0.081006
1.60	0.115205	0.089777	0.065144	0.041352	0.018442	−0.003547
1.65	0.186122	0.160644	0.135850	0.111785	0.088495	0.066021
1.70	0.246722	0.221495	0.196838	0.172799	0.149425	0.126762
1.75	0.296372	0.271675	0.247436	0.223703	0.200527	0.177953
1.80	0.334675	0.310766	0.287202	0.264036	0.241317	0.219094
1.85	0.361475	0.338583	0.315929	0.293565	0.271542	0.249910
1.90	0.376849	0.355177	0.333640	0.312288	0.291174	0.270348
1.95	0.381103	0.360824	0.340581	0.320424	0.300405	0.280575
2.00	0.374756	0.356012	0.337209	0.318400	0.299635	0.280963
2.05	0.358522	0.341423	0.324180	0.306841	0.289458	0.272078
2.10	0.333293	0.317919	0.302320	0.286545	0.270643	0.254661
2.15	0.300113	0.286512	0.272613	0.258465	0.244113	0.229606
2.20	0.260152	0.248341	0.236170	0.223681	0.210921	0.197935
2.25	0.214678	0.204648	0.194200	0.183377	0.172222	0.160778
2.30	0.165028	0.156740	0.147987	0.138808	0.129244	0.119337
2.35	0.112576	0.105968	0.098855	0.091275	0.083265	0.074864
2.40	0.058705	0.053692	0.048144	0.042094	0.035577	0.028629
2.45	0.004775	0.001253	−0.002827	−0.007435	−0.012539	−0.018105
2.50	−0.047902	−0.050054	−0.052780	−0.056051	−0.059839	−0.064113
2.55	−0.098095	−0.099011	−0.100508	−0.102562	−0.105146	−0.108234
2.60	−0.144669	−0.144494	−0.144901	−0.145868	−0.147373	−0.149391
2.65	−0.186616	−0.185502	−0.184964	−0.184983	−0.185540	−0.186614
2.70	−0.223067	−0.221170	−0.219837	−0.219052	−0.218800	−0.219062
2.75	−0.253309	−0.250787	−0.248810	−0.247368	−0.246447	−0.246031
2.80	−0.276796	−0.273804	−0.271335	−0.269381	−0.267931	−0.266974
2.85	−0.293158	−0.289845	−0.287030	−0.284707	−0.282867	−0.281502
2.90	−0.302199	−0.298711	−0.295689	−0.293133	−0.291035	−0.289391
2.95	−0.303903	−0.300372	−0.297276	−0.294615	−0.292386	−0.290585
3.00	−0.298423	−0.294973	−0.291923	−0.289275	−0.287030	−0.285186
3.05	−0.286077	−0.282817	−0.279922	−0.277395	−0.275239	−0.273454
3.10	−0.267334	−0.264361	−0.261715	−0.259403	−0.257428	−0.255795
3.15	−0.242801	−0.240194	−0.237878	−0.235860	−0.234147	−0.232743
3.20	−0.213203	−0.211028	−0.209106	−0.207448	−0.206061	−0.204952

Table 2.7 (continued)

			t			
v	0.35	0.40	0.45	0.50	0.55	0.60
3.25	−0.179366	−0.177670	−0.176193	−0.174945	−0.173935	−0.173170
3.30	−0.142192	−0.141009	−0.140010	−0.139207	−0.138609	−0.138227
3.35	−0.102641	−0.101987	−0.101485	−0.101147	−0.100984	−0.101005
3.40	−0.061702	−0.061580	−0.061579	−0.061712	−0.061991	−0.062426
3.45	−0.020375	−0.020772	−0.021262	−0.021859	−0.022574	−0.023420
3.50	0.020358	0.019466	0.018507	0.017466	0.016332	0.015093
3.55	0.059550	0.058200	0.056805	0.055352	0.053827	0.052219
3.60	0.096314	0.094552	0.092765	0.090939	0.089063	0.087123
3.65	0.129841	0.127720	0.125592	0.123444	0.121261	0.119033
3.70	0.159414	0.156997	0.154587	0.152170	0.149734	0.147267
3.75	0.184429	0.181781	0.179151	0.176526	0.173895	0.171245
3.80	0.204401	0.201590	0.198806	0.196037	0.193271	0.190496
3.85	0.218973	0.216070	0.213199	0.210350	0.207511	0.204672
3.90	0.227926	0.224999	0.222109	0.219244	0.216395	0.213551
3.95	0.231176	0.228293	0.225447	0.222630	0.219831	0.217040
4.00	0.228775	0.225999	0.223260	0.220549	0.217858	0.215177
4.05	0.220909	0.218297	0.215721	0.213172	0.210642	0.208122
4.10	0.207886	0.205492	0.203130	0.200792	0.198470	0.196158
4.15	0.190131	0.187999	0.185895	0.183810	0.181739	0.179675
4.20	0.168171	0.166339	0.164529	0.162734	0.160948	0.159164
4.25	0.142621	0.141119	0.139631	0.138152	0.136678	0.135202
4.30	0.114170	0.113016	0.111870	0.110728	0.109584	0.108434
4.35	0.083559	0.082765	0.081973	0.081177	0.080375	0.079561
4.40	0.051566	0.051135	0.050698	0.050252	0.049793	0.049319
4.45	0.018985	0.018911	0.018824	0.018723	0.018602	0.018461
4.50	−0.013390	−0.013121	−0.012871	−0.012642	−0.012436	−0.012257
4.55	−0.044788	−0.044197	−0.043631	−0.043092	−0.042580	−0.042098
4.60	−0.074476	−0.073591	−0.072736	−0.071912	−0.071120	−0.070362
4.65	−0.101773	−0.100628	−0.099517	−0.098441	−0.097400	−0.096396
4.70	−0.126074	−0.124707	−0.123377	−0.122085	−0.120831	−0.119617
4.75	−0.146851	−0.145304	−0.143797	−0.142329	−0.140901	−0.139515
4.80	−0.163675	−0.161992	−0.160350	−0.158750	−0.157190	−0.155673
4.85	−0.176215	−0.174442	−0.172711	−0.171022	−0.169374	−0.167768
4.90	−0.184251	−0.182435	−0.180660	−0.178926	−0.177233	−0.175581
4.95	−0.187675	−0.185861	−0.184086	−0.182352	−0.180657	−0.179001
5.00	−0.186489	−0.184720	−0.182989	−0.181296	−0.179640	−0.178022

			t			
v	0.65	0.70	0.75	0.80	0.85	0.90
0.00	0.619997	0.662628	0.704168	0.744549	0.783707	0.821581
0.05	0.467232	0.511875	0.555575	0.598263	0.639868	0.680326

Table 2.7 (continued)

				t		
v	0.65	0.70	0.75	0.80	0.85	0.90
0.10	0.310746	0.356958	0.402383	0.446948	0.490580	0.533210
0.15	0.152897	0.200221	0.246920	0.292917	0.338138	0.382509
0.20	-0.003960	0.044011	0.091522	0.138496	0.184854	0.230521
0.25	-0.157519	-0.109368	-0.061508	-0.014019	0.033020	0.079530
0.30	-0.305560	-0.257690	-0.209943	-0.162398	-0.115136	-0.068238
0.35	-0.445987	-0.398849	-0.351666	-0.304519	-0.257487	-0.210653
0.40	-0.576863	-0.530894	-0.484713	-0.438403	-0.392043	-0.345714
0.45	-0.696446	-0.652059	-0.607299	-0.562248	-0.516984	-0.471588
0.50	-0.803213	-0.760796	-0.717851	-0.674457	-0.630693	-0.586637
0.55	-0.895888	-0.855797	-0.815030	-0.773665	-0.731776	-0.689442
0.60	-0.973456	-0.936013	-0.897755	-0.858756	-0.819088	-0.778828
0.65	-1.03518	-1.00067	-0.965213	-0.928882	-0.891747	-0.853881
0.70	-1.08061	-1.04927	-1.01687	-0.983473	-0.949144	-0.913954
0.75	-1.10958	-1.08162	-1.05248	-1.02224	-0.990949	-0.958677
0.80	-1.12222	-1.09778	-1.07208	-1.04517	-1.01711	-0.987959
0.85	-1.11893	-1.09813	-1.07598	-1.05254	-1.02785	-1.00198
0.90	-1.10037	-1.08327	-1.06475	-1.04486	-1.02366	-1.00118
0.95	-1.06747	-1.05408	-1.03922	-1.02293	-1.00526	-0.986251
1.00	-1.02134	-1.01164	-1.00043	-0.987745	-0.973625	-0.958109
1.05	-0.963348	-0.957260	-0.949639	-0.940514	-0.929916	-0.917880
1.10	-0.894979	-0.892397	-0.888271	-0.882622	-0.875478	-0.866865
1.15	-0.817881	-0.818657	-0.817890	-0.815598	-0.811799	-0.806517
1.20	-0.733797	-0.737749	-0.740173	-0.741080	-0.740485	-0.738404
1.25	-0.644538	-0.651453	-0.656868	-0.660788	-0.663221	-0.664179
1.30	-0.551946	-0.561584	-0.569763	-0.576480	-0.581739	-0.585545
1.35	-0.457857	-0.469957	-0.480649	-0.489924	-0.497781	-0.504218
1.40	-0.364066	-0.378350	-0.391285	-0.402861	-0.413067	-0.421898
1.45	-0.272295	-0.288470	-0.303367	-0.316969	-0.329261	-0.340232
1.50	-0.184161	-0.201928	-0.218493	-0.233836	-0.247938	-0.260783
1.55	-0.101147	-0.120200	-0.138137	-0.154932	-0.170561	-0.185004
1.60	-0.024577	-0.044614	-0.063624	-0.081578	-0.098449	-0.114212
1.65	0.044405	0.023684	0.003894	-0.014931	-0.032759	-0.049563
1.70	0.104851	0.083736	0.063453	0.044041	0.025533	0.007962
1.75	0.156027	0.134793	0.114293	0.094565	0.075647	0.057575
1.80	0.197414	0.176323	0.155863	0.136078	0.117007	0.098688
1.85	0.228716	0.208009	0.187832	0.168231	0.149247	0.130921
1.90	0.249859	0.229755	0.210082	0.190887	0.172212	0.154100
1.95	0.260983	0.241678	0.222707	0.204117	0.185952	0.168257
2.00	0.262435	0.244099	0.226003	0.208194	0.190718	0.173619
2.05	0.254753	0.237529	0.220455	0.203579	0.186945	0.170599
2.10	0.238649	0.222653	0.206722	0.190903	0.175240	0.159781
2.15	0.214989	0.200310	0.185615	0.170951	0.156364	0.141898

Table 2.7 (continued)

				t		
v	0.65	0.70	0.75	0.80	0.85	0.90
2.20	0.184769	0.171467	0.158075	0.144640	0.131205	0.117815
2.25	0.149089	0.137198	0.125151	0.112990	0.100760	0.088505
2.30	0.109127	0.098657	0.087969	0.077105	0.066108	0.055021
2.35	0.066111	0.057045	0.047708	0.038140	0.028381	0.018473
2.40	0.021288	0.013590	0.005574	-0.002723	-0.011261	-0.020000
2.45	-0.024099	-0.030487	-0.037234	-0.044302	-0.051655	-0.059255
2.50	-0.068844	-0.073997	-0.079542	-0.085442	-0.091665	-0.098175
2.55	-0.111798	-0.115808	-0.120233	-0.125043	-0.130206	-0.135688
2.60	-0.151898	-0.154866	-0.158269	-0.162078	-0.166263	-0.170795
2.65	-0.188183	-0.190223	-0.192710	-0.195618	-0.198919	-0.202587
2.70	-0.219819	-0.221050	-0.222734	-0.224848	-0.227367	-0.230267
2.75	-0.246106	-0.246652	-0.247652	-0.249084	-0.250928	-0.253161
2.80	-0.266496	-0.266483	-0.266919	-0.267785	-0.269063	-0.270734
2.85	-0.280601	-0.280153	-0.280144	-0.280558	-0.281381	-0.282595
2.90	-0.288193	-0.287430	-0.287093	-0.287169	-0.287645	-0.288506
2.95	-0.289207	-0.288245	-0.287692	-0.287538	-0.287771	-0.288380
3.00	-0.283740	-0.282687	-0.282023	-0.281740	-0.281829	-0.282281
3.05	-0.272041	-0.270996	-0.270317	-0.269998	-0.270035	-0.270418
3.10	-0.254503	-0.253555	-0.252947	-0.252677	-0.252742	-0.253136
3.15	-0.231652	-0.230875	-0.230413	-0.230266	-0.230431	-0.230906
3.20	-0.204124	-0.203583	-0.203331	-0.203368	-0.203695	-0.204310
3.25	-0.172658	-0.172403	-0.172410	-0.172682	-0.173219	-0.174024
3.30	-0.138066	-0.138136	-0.138440	-0.138984	-0.139769	-0.140799
3.35	-0.101221	-0.101639	-0.102265	-0.103106	-0.104165	-0.105446
3.40	-0.063028	-0.063806	-0.064766	-0.065917	-0.067263	-0.068809
3.45	-0.024406	-0.025542	-0.026838	-0.028300	-0.029936	-0.031750
3.50	0.013736	0.012252	0.010632	0.008868	0.006951	0.004875
3.55	0.050517	0.048709	0.046785	0.044737	0.042556	0.040236
3.60	0.085107	0.083006	0.080809	0.078505	0.076087	0.073547
3.65	0.116746	0.114391	0.111957	0.109434	0.106813	0.104085
3.70	0.144758	0.142195	0.139568	0.136866	0.134081	0.131205
3.75	0.168565	0.165845	0.163073	0.160240	0.157336	0.154352
3.80	0.187702	0.184878	0.182013	0.179097	0.176122	0.173079
3.85	0.201821	0.198949	0.196044	0.193099	0.190102	0.187047
3.90	0.210701	0.207836	0.204947	0.202023	0.199055	0.196036
3.95	0.214248	0.211446	0.208624	0.205772	0.202883	0.199948
4.00	0.212497	0.209809	0.207104	0.204374	0.201611	0.198806
4.05	0.205604	0.203079	0.200539	0.197977	0.195383	0.192751
4.10	0.193846	0.191528	0.189196	0.186841	0.184457	0.182037
4.15	0.177610	0.175536	0.173448	0.171338	0.169198	0.167022
4.20	0.157378	0.155581	0.153767	0.151931	0.150064	0.148162
4.25	0.133719	0.132224	0.130709	0.129169	0.127598	0.125990

Table 2.7 (continued)

				t		
v	0.65	0.70	0.75	0.80	0.85	0.90
4.30	0.107273	0.106096	0.104897	0.103670	0.102411	0.101112
4.35	0.078732	0.077883	0.077008	0.076104	0.075164	0.074184
4.40	0.048824	0.048305	0.047757	0.047177	0.046559	0.045899
4.45	0.018296	0.018102	0.017876	0.017615	0.017313	0.016968
4.50	-0.012105	-0.011986	-0.011902	-0.011856	-0.011852	-0.011895
4.55	-0.041649	-0.041235	-0.040859	-0.040524	-0.040234	-0.039991
4.60	-0.069639	-0.068955	-0.068312	-0.067711	-0.067157	-0.066652
4.65	-0.095431	-0.094506	-0.093624	-0.092787	-0.091997	-0.091258
4.70	-0.118443	-0.117311	-0.116224	-0.115182	-0.114189	-0.113247
4.75	-0.138170	-0.136869	-0.135613	-0.134403	-0.133242	-0.132132
4.80	-0.154198	-0.152766	-0.151380	-0.150040	-0.148749	-0.147508
4.85	-0.166204	-0.164684	-0.163208	-0.161778	-0.160395	-0.159061
4.90	-0.173971	-0.172403	-0.170878	-0.169398	-0.167964	-0.166577
4.95	-0.177386	-0.175811	-0.174277	-0.172787	-0.171340	-0.169938
5.00	-0.176441	-0.174899	-0.173396	-0.171933	-0.170511	-0.169132

				t		
v	0.95	1.0	2.0	3.0	4.0	5.0
0.00	0.858112	0.893244	1.242280	0.902118	0.212080	-0.290938
0.05	0.719572	0.757545	1.181000	0.913094	0.252053	-0.270826
0.10	0.574769	0.615192	1.107920	0.912911	0.285386	-0.252896
0.15	0.425958	0.468417	1.024590	0.902301	0.312272	-0.237063
0.20	0.275423	0.319486	0.932675	0.882151	0.333036	-0.223156
0.25	0.125435	0.170659	0.833943	0.853477	0.348120	-0.210919
0.30	-0.021782	0.024156	0.730210	0.817399	0.358063	-0.200032
0.35	-0.164095	-0.117892	0.623311	0.775113	0.363488	-0.190116
0.40	-0.299497	-0.253470	0.515071	0.727865	0.365077	-0.180753
0.45	-0.426140	-0.380720	0.407263	0.676921	0.363552	-0.171498
0.50	-0.542368	-0.497966	0.301581	0.623541	0.359657	-0.161894
0.55	-0.646739	-0.603748	0.199608	0.568955	0.354133	-0.151490
0.60	-0.738051	-0.696833	0.102793	0.514332	0.347705	-0.139853
0.65	-0.815355	-0.776246	0.012423	0.460765	0.341055	-0.126583
0.70	-0.877971	-0.841269	-0.070394	0.409243	0.334813	-0.111328
0.75	-0.925491	-0.891459	-0.144744	0.360641	0.329540	-0.093791
0.80	-0.957783	-0.926643	-0.209918	0.315702	0.325712	-0.073743
0.85	-0.974982	-0.946918	-0.265421	0.275029	0.323715	-0.051032
0.90	-0.977489	-0.952639	-0.310969	0.239073	0.323833	-0.025584
0.95	-0.965951	-0.944411	-0.346491	0.208136	0.326248	0.002588
1.00	-0.941241	-0.923066	-0.372121	0.182367	0.331035	0.033389
1.05	-0.904443	-0.889645	-0.388186	0.161766	0.338164	0.066636
1.10	-0.856817	-0.845367	-0.395195	0.146191	0.347502	0.102067
1.15	-0.799778	-0.791609	-0.393817	0.135370	0.358823	0.139344

Table 2.7 (continued)

			t			
v	0.95	1.0	2.0	3.0	4.0	5.0
1.20	−0.734857	−0.729867	−0.384861	0.128912	0.371814	0.178060
1.25	−0.663675	−0.661727	−0.369257	0.126321	0.386085	0.217750
1.30	−0.587905	−0.588832	−0.348026	0.127014	0.401183	0.257900
1.35	−0.509239	−0.512848	−0.322259	0.130341	0.416606	0.297958
1.40	−0.429352	−0.435429	−0.293085	0.135601	0.431817	0.337350
1.45	−0.349875	−0.358184	−0.261649	0.142068	0.446257	0.375490
1.50	−0.272357	−0.282652	−0.229083	0.149004	0.459368	0.411796
1.55	−0.198243	−0.210264	−0.196481	0.155686	0.470601	0.445703
1.60	−0.128844	−0.142326	−0.164879	0.161418	0.479437	0.476676
1.65	−0.065316	−0.079994	−0.135228	0.165554	0.485396	0.504222
1.70	−0.008642	−0.024252	−0.108378	0.167514	0.488054	0.527902
1.75	0.040382	0.024098	−0.085065	0.166795	0.487056	0.547341
1.80	0.081158	0.064449	−0.065890	0.162983	0.482119	0.562239
1.85	0.113291	0.096393	−0.051320	0.155766	0.473047	0.572373
1.90	0.136592	0.119725	−0.041671	0.144937	0.459733	0.577606
1.95	0.151073	0.134440	−0.037115	0.130399	0.442162	0.577889
2.00	0.156941	0.140725	−0.037674	0.112166	0.420416	0.573262
2.05	0.154586	0.138948	−0.043228	0.090362	0.394665	0.563853
2.10	0.144569	0.129647	−0.053520	0.065213	0.365173	0.549876
2.15	0.127598	0.113508	−0.068172	0.037047	0.332285	0.531624
2.20	0.104515	0.091347	−0.086689	0.006276	0.296421	0.509463
2.25	0.076267	0.064090	−0.108484	−0.026609	0.258068	0.483826
2.30	0.043885	0.032743	−0.132888	−0.061053	0.217770	0.455199
2.35	0.008456	−0.001628	−0.159174	−0.096453	0.176110	0.424114
2.40	−0.028902	−0.037927	−0.186572	−0.132170	0.133703	0.391134
2.45	−0.067066	−0.075049	−0.214294	−0.167551	0.091180	0.356845
2.50	−0.104936	−0.111912	−0.241550	−0.201939	0.049171	0.321837
2.55	−0.141456	−0.147477	−0.267571	−0.234692	0.008296	0.286697
2.60	−0.175641	−0.180770	−0.291627	−0.265199	−0.030852	0.251993
2.65	−0.206592	−0.210905	−0.313042	−0.292894	−0.067720	0.218263
2.70	−0.233520	−0.237101	−0.331214	−0.317271	−0.101806	0.186004
2.75	−0.255759	−0.258698	−0.345625	−0.337892	−0.132669	0.155664
2.80	−0.272775	−0.275166	−0.355856	−0.354403	−0.159940	0.127630
2.85	−0.284181	−0.286120	−0.361593	−0.366535	−0.183328	0.102221
2.90	−0.289736	−0.291318	−0.362633	−0.374116	−0.202625	0.079685
2.95	−0.289351	−0.290669	−0.358890	−0.377069	−0.217711	0.060196
3.00	−0.283084	−0.284227	−0.350393	−0.375418	−0.228556	0.043847
3.05	−0.271139	−0.272188	−0.337287	−0.369284	−0.235215	0.030654
3.10	−0.253852	−0.254883	−0.319822	−0.358882	−0.237832	0.020556
3.15	−0.231686	−0.232764	−0.298352	−0.344514	−0.236629	0.013422
3.20	−0.205211	−0.206394	−0.273322	−0.326563	−0.231904	0.009049
3.25	−0.175093	−0.176426	−0.245256	−0.305484	−0.224020	0.007177

Table 2.7 (continued)

				t		
v	0.95	1.0	2.0	3.0	4.0	5.0
3.30	−0.142074	−0.143593	−0.214743	−0.281788	−0.213398	0.007491
3.35	−0.106951	−0.108681	−0.182427	−0.256036	−0.200504	0.009631
3.40	−0.070559	−0.072514	−0.148982	−0.228822	−0.185839	0.013206
3.45	−0.033748	−0.035933	−0.115104	−0.200759	−0.169927	0.017800
3.50	0.002635	0.000226	−0.081491	−0.172465	−0.153299	0.022984
3.55	0.037770	0.035152	−0.048824	−0.144550	−0.136486	0.028330
3.60	0.070878	0.068073	−0.017757	−0.117600	−0.120001	0.033417
3.65	0.101244	0.098282	0.011105	−0.092167	−0.104333	0.037847
3.70	0.128228	0.125144	0.037213	−0.068751	−0.089928	0.041251
3.75	0.151282	0.148116	0.060091	−0.047796	−0.077189	0.043299
3.80	0.169959	0.166754	0.079343	−0.029677	−0.066458	0.043709
3.85	0.183923	0.180724	0.094665	−0.014693	−0.058016	0.042251
3.90	0.192956	0.189808	0.105846	−0.003061	−0.052074	0.038756
3.95	0.196958	0.193907	0.112774	0.005087	−0.048769	0.033115
4.00	0.195951	0.193039	0.115437	0.009706	−0.048163	0.025287
4.05	0.190072	0.187340	0.113924	0.010837	−0.050244	0.015292
4.10	0.179571	0.177055	0.108417	0.008605	−0.054926	0.003215
4.15	0.164803	0.162535	0.099186	0.003214	−0.062052	−0.010796
4.20	0.146216	0.144221	0.086588	−0.005058	−0.071402	−0.026539
4.25	0.124339	0.122639	0.071046	−0.015868	−0.082697	−0.043760
4.30	0.099770	0.098378	0.053047	−0.028817	−0.095606	−0.062161
4.35	0.073159	0.072083	0.033126	−0.043463	−0.109759	−0.081407
4.40	0.045192	0.044433	0.011854	−0.059328	−0.124754	−0.101136
4.45	0.016574	0.016126	−0.010181	−0.075912	−0.140167	−0.120968
4.50	−0.011988	−0.012135	−0.032380	−0.092707	−0.155567	−0.140512
4.55	−0.039800	−0.039664	−0.054154	−0.109209	−0.170521	−0.159383
4.60	−0.066200	−0.065803	−0.074936	−0.124929	−0.184609	−0.177201
4.65	−0.090571	−0.089941	−0.094194	−0.139405	−0.197434	−0.193610
4.70	−0.112358	−0.111525	−0.111441	−0.152214	−0.208630	−0.208282
4.75	−0.131074	−0.130073	−0.126251	−0.162979	−0.217875	−0.220925
4.80	−0.146319	−0.145184	−0.138264	−0.171384	−0.224892	−0.231293
4.85	−0.157779	−0.156549	−0.147193	−0.177172	−0.229461	−0.239188
4.90	−0.165239	−0.163952	−0.152835	−0.180157	−0.231423	−0.244466
4.95	−0.168583	−0.167277	−0.155069	−0.180227	−0.230681	−0.247044
5.00	−0.167797	−0.166507	−0.153861	−0.177343	−0.227206	−0.246895

Table 2.8: Second derivatives with respect to the order of the Anger functions $\partial^2 J_\nu(t)/\partial\nu^2$.

			t			
ν	0.05	0.10	0.15	0.20	0.25	0.30
0.00	−3.28812	−3.28290	−3.27420	−3.26204	−3.24644	−3.22743
0.05	−3.27347	−3.27790	−3.27886	−3.27633	−3.27033	−3.26087
0.10	−3.21050	−3.22453	−3.23509	−3.24219	−3.24582	−3.24597
0.15	−3.10037	−3.12379	−3.14381	−3.16041	−3.17357	−3.18328
0.20	−2.94506	−2.97756	−3.00675	−3.03260	−3.05508	−3.07416
0.25	−2.74733	−2.78848	−2.82644	−2.86118	−2.89264	−2.92079
0.30	−2.51069	−2.55994	−2.60617	−2.64930	−2.68930	−2.72611
0.35	−2.23931	−2.29602	−2.34989	−2.40085	−2.44883	−2.49377
0.40	−1.93794	−2.00138	−2.06219	−2.12029	−2.17561	−2.22806
0.45	−1.61184	−1.68119	−1.74816	−1.81264	−1.87454	−1.93379
0.50	−1.26665	−1.34104	−1.41330	−1.48332	−1.55100	−1.61626
0.55	−0.908320	−0.986813	−1.06345	−1.13811	−1.21069	−1.28110
0.60	−0.542965	−0.624582	−0.704628	−0.782986	−0.859542	−0.934189
0.65	−0.176766	−0.260503	−0.342971	−0.424046	−0.503608	−0.581543
0.70	0.184151	0.099308	0.015426	−0.067367	−0.148948	−0.229195
0.75	0.533824	0.448888	0.364601	0.281092	0.198489	0.116918
0.80	0.866568	0.782535	0.698837	0.615606	0.532971	0.451062
0.85	1.17707	1.09491	1.01277	0.930786	0.849090	0.767812
0.90	1.46051	1.38114	1.30148	1.22168	1.14186	1.06215
0.95	1.71261	1.63689	1.56059	1.48385	1.40679	1.32955
1.00	1.92972	1.85845	1.78633	1.71347	1.64001	1.56607
1.05	2.10890	2.04280	1.97558	1.90736	1.83826	1.76840
1.10	2.24794	2.18765	2.12598	2.06306	1.99900	1.93393
1.15	2.34541	2.29146	2.23590	2.17886	2.12044	2.06077
1.20	2.40065	2.35348	2.30450	2.25382	2.20155	2.14779
1.25	2.41380	2.37374	2.33170	2.28777	2.24205	2.19464
1.30	2.38577	2.35306	2.31821	2.28130	2.24244	2.20170
1.35	2.31823	2.29298	2.26547	2.23576	2.20395	2.17011
1.40	2.21351	2.19575	2.17561	2.15317	2.12851	2.10170
1.45	2.07462	2.06425	2.05143	2.03623	2.01872	1.99896
1.50	1.90513	1.90196	1.89630	1.88821	1.87773	1.86494
1.55	1.70909	1.71284	1.71409	1.71288	1.70925	1.70325
1.60	1.49098	1.50128	1.50909	1.51445	1.51737	1.51791
1.65	1.25557	1.27198	1.28593	1.29745	1.30657	1.31330
1.70	1.00785	1.02985	1.04946	1.06668	1.08154	1.09404
1.75	0.752929	0.779941	0.804645	0.827041	0.847135	0.864934
1.80	0.495908	0.527307	0.556505	0.583492	0.608263	0.630816
1.85	0.241800	0.276924	0.309973	0.340926	0.369771	0.396495
1.90	−0.004571	0.033589	0.069816	0.104082	0.136364	0.166642
1.95	−0.238668	−0.198174	−0.159457	−0.122552	−0.087492	−0.054302
2.00	−0.456319	−0.414198	−0.373685	−0.334823	−0.297650	−0.262203
2.05	−0.653796	−0.610746	−0.569125	−0.528984	−0.490367	−0.453318
2.10	−0.827879	−0.784579	−0.742523	−0.701767	−0.662362	−0.624358

Table 2.8 (continued)

			t			
v	0.05	0.10	0.15	0.20	0.25	0.30
2.15	−0.975911	−0.933011	−0.891166	−0.850435	−0.810877	−0.772545
2.20	−1.09584	−1.05395	−1.01292	−0.972826	−0.933719	−0.895660
2.25	−1.18623	−1.14592	−1.10628	−1.06738	−1.029290	−0.992069
2.30	−1.24631	−1.20809	−1.17035	−1.13317	−1.096610	−1.06075
2.35	−1.27594	−1.24026	−1.20489	−1.16989	−1.135330	−1.10129
2.40	−1.27563	−1.24288	−1.21026	−1.17784	−1.145690	−1.11388
2.45	−1.24648	−1.21698	−1.18744	−1.15793	−1.128530	−1.09931
2.50	−1.19016	−1.16415	−1.13796	−1.11164	−1.085270	−1.05892
2.55	−1.10887	−1.08654	−1.06387	−1.04094	−1.017810	−0.994559
2.60	−1.00528	−0.986728	−0.967709	−0.948293	−0.928550	−0.908547
2.65	−0.882430	−0.867693	−0.852374	−0.836540	−0.820259	−0.803596
2.70	−0.743701	−0.732746	−0.721110	−0.708856	−0.696048	−0.682751
2.75	−0.592707	−0.585439	−0.577406	−0.568666	−0.559283	−0.549315
2.80	−0.433221	−0.429486	−0.424917	−0.419569	−0.413501	−0.406770
2.85	−0.269090	−0.268680	−0.267383	−0.265252	−0.262338	−0.258697
2.90	−0.104151	−0.106811	−0.108547	−0.109407	−0.109439	−0.108693
2.95	0.057854	0.052419	0.047930	0.044343	0.041616	0.039703
3.00	0.213353	0.205471	0.198543	0.192529	0.187391	0.183089
3.05	0.359020	0.349046	0.340017	0.331902	0.324667	0.318275
3.10	0.491842	0.480145	0.469375	0.459504	0.450503	0.442343
3.15	0.609171	0.596133	0.583991	0.572721	0.562302	0.552707
3.20	0.708783	0.694786	0.681644	0.669339	0.657852	0.647162
3.25	0.788906	0.774329	0.760556	0.747575	0.735371	0.723928
3.30	0.848258	0.833466	0.819420	0.806113	0.793534	0.781672
3.35	0.886055	0.871396	0.857419	0.844120	0.831494	0.819534
3.40	0.902022	0.887819	0.874229	0.861252	0.848888	0.837132
3.45	0.896385	0.882933	0.870020	0.857653	0.845832	0.834559
3.50	0.869854	0.857413	0.845438	0.833935	0.822912	0.812372
3.55	0.823598	0.812392	0.801575	0.791160	0.781155	0.771567
3.60	0.759202	0.749415	0.739943	0.730800	0.721997	0.713547
3.65	0.678631	0.670405	0.662422	0.654697	0.647245	0.640078
3.70	0.584167	0.577604	0.571214	0.565013	0.559020	0.553247
3.75	0.478357	0.473517	0.468783	0.464173	0.459707	0.455401
3.80	0.363950	0.360850	0.357794	0.354802	0.351895	0.349090
3.85	0.243822	0.242441	0.241048	0.239664	0.238309	0.237002
3.90	0.120918	0.121198	0.121415	0.121592	0.121748	0.121904
3.95	−0.001824	0.000026	0.001768	0.003425	0.005019	0.006568
4.00	−0.121536	−0.118239	−0.115088	−0.112058	−0.109131	−0.106285
4.05	−0.235486	−0.230891	−0.226471	−0.222205	−0.218072	−0.214052
4.10	−0.341136	−0.335412	−0.329887	−0.324541	−0.319353	−0.314305
4.15	−0.436199	−0.429533	−0.423084	−0.416831	−0.410756	−0.404840
4.20	−0.518682	−0.511274	−0.504093	−0.497120	−0.490338	−0.483729
4.25	−0.586930	−0.578985	−0.571273	−0.563774	−0.556474	−0.549355

Table 2.8 (continued)

			t			
ν	0.05	0.10	0.15	0.20	0.25	0.30
4.30	-0.639653	-0.631381	-0.623340	-0.615514	-0.607887	-0.600444
4.35	-0.675952	-0.667560	-0.659392	-0.651435	-0.643672	-0.636092
4.40	-0.695331	-0.687018	-0.678919	-0.671020	-0.663309	-0.655772
4.45	-0.697696	-0.689653	-0.681807	-0.674148	-0.666665	-0.659346
4.50	-0.683358	-0.675758	-0.668338	-0.661086	-0.653995	-0.647054
4.55	-0.653008	-0.646009	-0.639167	-0.632475	-0.625925	-0.619508
4.60	-0.607700	-0.601438	-0.595310	-0.589309	-0.583430	-0.577664
4.65	-0.548817	-0.543406	-0.538104	-0.532906	-0.527806	-0.522800
4.70	-0.478033	-0.473563	-0.469176	-0.464868	-0.460636	-0.456475
4.75	-0.397269	-0.393804	-0.390396	-0.387042	-0.383740	-0.380488
4.80	-0.308642	-0.306220	-0.303830	-0.301470	-0.299139	-0.296834
4.85	-0.214412	-0.213048	-0.211689	-0.210338	-0.208993	-0.207653
4.90	-0.116929	-0.116610	-0.116274	-0.115922	-0.115555	-0.115174
4.95	-0.018575	-0.019266	-0.019918	-0.020535	-0.021116	-0.021664
5.00	0.078295	0.076652	0.075065	0.073533	0.072054	0.070626

			t			
ν	0.35	0.40	0.45	0.50	0.55	0.60
0.00	-3.20505	-3.17933	-3.15033	-3.11809	-3.08266	-3.04412
0.05	-3.24797	-3.23166	-3.21198	-3.18896	-3.16264	-3.13309
0.10	-3.24266	-3.23590	-3.22570	-3.21210	-3.19513	-3.17482
0.15	-3.18953	-3.19232	-3.19166	-3.18756	-3.18004	-3.16913
0.20	-3.08982	-3.10205	-3.11084	-3.11620	-3.11813	-3.11663
0.25	-2.94559	-2.96703	-2.98508	-2.99973	-3.01097	-3.01880
0.30	-2.75968	-2.78997	-2.81696	-2.84062	-2.86092	-2.87785
0.35	-2.53561	-2.57430	-2.60980	-2.64207	-2.67107	-2.69678
0.40	-2.27757	-2.32410	-2.36758	-2.40795	-2.44518	-2.47923
0.45	-1.99030	-2.04400	-2.09483	-2.14271	-2.18760	-2.22944
0.50	-1.67900	-1.73914	-1.79659	-1.85130	-1.90319	-1.95219
0.55	-1.34923	-1.41499	-1.47829	-1.53906	-1.59720	-1.65265
0.60	-1.00682	-1.07733	-1.14563	-1.21162	-1.27522	-1.33633
0.65	-0.657738	-0.732084	-0.804476	-0.874813	-0.942999	-1.00894
0.70	-0.307990	-0.385219	-0.460768	-0.534531	-0.606404	-0.676286
0.75	0.036502	-0.042640	-0.120390	-0.196633	-0.271260	-0.344163
0.80	0.370005	0.289923	0.210939	0.133171	0.056736	-0.018254
0.85	0.687080	0.607022	0.527763	0.449425	0.372129	0.295991
0.90	0.982684	0.903590	0.824995	0.747023	0.669799	0.593444
0.95	1.25225	1.17503	1.09800	1.02131	0.945062	0.869394
1.00	1.49179	1.41729	1.34270	1.26814	1.19374	1.11963
1.05	1.69792	1.62694	1.55558	1.48398	1.41225	1.34053
1.10	1.86796	1.80123	1.73384	1.66594	1.59764	1.52906

Table 2.8 (continued)

			t			
v	0.35	0.40	0.45	0.50	0.55	0.60
1.15	1.99995	1.93810	1.87536	1.81183	1.74764	1.68292
1.20	2.09266	2.03627	1.97874	1.92019	1.86072	1.80046
1.25	2.14565	2.09518	2.04336	1.99028	1.93606	1.88081
1.30	2.15920	2.11503	2.06930	2.02212	1.97358	1.92380
1.35	2.13435	2.09676	2.05742	2.01644	1.97393	1.92997
1.40	2.07283	2.04198	2.00923	1.97469	1.93844	1.90058
1.45	1.97702	1.95298	1.92693	1.89894	1.86910	1.83750
1.50	1.84989	1.83265	1.81329	1.79188	1.76850	1.74324
1.55	1.69494	1.68437	1.67161	1.65672	1.63976	1.62082
1.60	1.51610	1.51200	1.50566	1.49712	1.48646	1.47374
1.65	1.31768	1.31975	1.31956	1.31715	1.31256	1.30586
1.70	1.10422	1.11210	1.11772	1.12111	1.12231	1.12136
1.75	0.880452	0.893707	0.904722	0.913524	0.920143	0.924617
1.80	0.651156	0.669292	0.685235	0.699004	0.710621	0.720111
1.85	0.421095	0.443569	0.463920	0.482158	0.498294	0.512345
1.90	0.194902	0.221134	0.245334	0.267499	0.287634	0.305746
1.95	−0.023006	0.006378	0.033835	0.059356	0.082937	0.104575
2.00	−0.228511	−0.196600	−0.166494	−0.138210	−0.111760	−0.087156
2.05	−0.417873	−0.384067	−0.351929	−0.321485	−0.292756	−0.265759
2.10	−0.587797	−0.552722	−0.519168	−0.487168	−0.456751	−0.427942
2.15	−0.735490	−0.699758	−0.665392	−0.632432	−0.600911	−0.570863
2.20	−0.858705	−0.822906	−0.788311	−0.754965	−0.722910	−0.692182
2.25	−0.955775	−0.920467	−0.886196	−0.853014	−0.820966	−0.790097
2.30	−1.02563	−0.991333	−0.957903	−0.925398	−0.893869	−0.863363
2.35	−1.06782	−1.03500	−1.00288	−0.971525	−0.940985	−0.911312
2.40	−1.08248	−1.05156	−1.02117	−0.991388	−0.962260	−0.933846
2.45	−1.07033	−1.04167	−1.01339	−0.985546	−0.958205	−0.931424
2.50	−1.03266	−1.00656	−0.980687	−0.955102	−0.929867	−0.905044
2.55	−0.971251	−0.947955	−0.924738	−0.901664	−0.878796	−0.856197
2.60	−0.888354	−0.868037	−0.847662	−0.827294	−0.806997	−0.786831
2.65	−0.786620	−0.769394	−0.751986	−0.734457	−0.716872	−0.699292
2.70	−0.669030	−0.654947	−0.640566	−0.625951	−0.611162	−0.596261
2.75	−0.538826	−0.527877	−0.516528	−0.504840	−0.492874	−0.480690
2.80	−0.399434	−0.391552	−0.383182	−0.374382	−0.365211	−0.355726
2.85	−0.254382	−0.249448	−0.243952	−0.237948	−0.231492	−0.224640
2.90	−0.107219	−0.105069	−0.102294	−0.098947	−0.095081	−0.090747
2.95	0.038557	0.038132	0.038379	0.039250	0.040696	0.042666
3.00	0.179581	0.176824	0.174774	0.173387	0.172616	0.172416
3.05	0.312691	0.307875	0.303788	0.300389	0.297637	0.295488
3.10	0.434990	0.428410	0.422569	0.417430	0.412955	0.409106
3.15	0.543907	0.535875	0.528580	0.521988	0.516067	0.510783
3.20	0.637247	0.628082	0.619641	0.611896	0.604817	0.598374

Table 2.8 (continued)

				t		
v	0.35	0.40	0.45	0.50	0.55	0.60
3.25	0.713228	0.703252	0.693977	0.685380	0.677436	0.670119
3.30	0.770514	0.760043	0.750244	0.741095	0.732577	0.724667
3.35	0.808231	0.797573	0.787547	0.778139	0.769331	0.761105
3.40	0.825980	0.815425	0.805458	0.796067	0.787240	0.778963
3.45	0.823833	0.813650	0.804005	0.794891	0.786299	0.778218
3.50	0.802319	0.792752	0.783669	0.775068	0.766942	0.759284
3.55	0.762403	0.753666	0.745358	0.737478	0.730024	0.722992
3.60	0.705456	0.697731	0.690377	0.683398	0.676793	0.670562
3.65	0.633208	0.626644	0.620393	0.614461	0.608851	0.603565
3.70	0.547710	0.542418	0.537383	0.532611	0.528110	0.523883
3.75	0.451270	0.447328	0.443586	0.440054	0.436742	0.433655
3.80	0.346403	0.343850	0.341445	0.339198	0.337121	0.335222
3.85	0.235762	0.234604	0.233544	0.232594	0.231766	0.231071
3.90	0.122078	0.122287	0.122548	0.122874	0.123279	0.123775
3.95	0.008093	0.009611	0.011138	0.012690	0.014281	0.015924
4.00	-0.103501	-0.100762	-0.098051	-0.095350	-0.092647	-0.089926
4.05	-0.210127	-0.206279	-0.202489	-0.198743	-0.195024	-0.191318
4.10	-0.309377	-0.304552	-0.299812	-0.295142	-0.290526	-0.285949
4.15	-0.399065	-0.393414	-0.387869	-0.382414	-0.377036	-0.371718
4.20	-0.477275	-0.470961	-0.464768	-0.458683	-0.452690	-0.446775
4.25	-0.542400	-0.535594	-0.528921	-0.522366	-0.515916	-0.509555
4.30	-0.593170	-0.586049	-0.579067	-0.572210	-0.565464	-0.558816
4.35	-0.628679	-0.621420	-0.614301	-0.607309	-0.600431	-0.593654
4.40	-0.648398	-0.641173	-0.634086	-0.627124	-0.620274	-0.613525
4.45	-0.652179	-0.645155	-0.638260	-0.631485	-0.624818	-0.618248
4.50	-0.640253	-0.633582	-0.627032	-0.620591	-0.614251	-0.608001
4.55	-0.613215	-0.607038	-0.600969	-0.594998	-0.589116	-0.583316
4.60	-0.572006	-0.566446	-0.560979	-0.555596	-0.550290	-0.545052
4.65	-0.517881	-0.513044	-0.508282	-0.503588	-0.498956	-0.494379
4.70	-0.452381	-0.448349	-0.444374	-0.440451	-0.436574	-0.432737
4.75	-0.377281	-0.374117	-0.370991	-0.367899	-0.364837	-0.361799
4.80	-0.294555	-0.292298	-0.290061	-0.287839	-0.285630	-0.283431
4.85	-0.206318	-0.204985	-0.203654	-0.202321	-0.200984	-0.199640
4.90	-0.114777	-0.114364	-0.113935	-0.113487	-0.113020	-0.112530
4.95	-0.022179	-0.022660	-0.023108	-0.023523	-0.023902	-0.024244
5.00	0.069247	0.067917	0.066635	0.065402	0.064217	0.063082

				t		
v	0.65	0.70	0.75	0.80	0.85	0.90
0.00	-3.00254	-2.95797	-2.91052	-2.86025	-2.80726	-2.75164
0.05	-3.10034	-3.06448	-3.02555	-2.98365	-2.93884	-2.89121

Table 2.8 (continued)

				t		
v	0.65	0.70	0.75	0.80	0.85	0.90
0.10	−3.15122	−3.12438	−3.09436	−3.06121	−3.02501	−2.98582
0.15	−3.15485	−3.13725	−3.11637	−3.09226	−3.06496	−3.03455
0.20	−3.11174	−3.10347	−3.09186	−3.07693	−3.05874	−3.03733
0.25	−3.02322	−3.02425	−3.02191	−3.01621	−3.00719	−2.99488
0.30	−2.89141	−2.90160	−2.90841	−2.91185	−2.91196	−2.90874
0.35	−2.71918	−2.73825	−2.75398	−2.76637	−2.77543	−2.78117
0.40	−2.51005	−2.53763	−2.56195	−2.58298	−2.60072	−2.61516
0.45	−2.26819	−2.30380	−2.33625	−2.36550	−2.39153	−2.41433
0.50	−1.99825	−2.04132	−2.08136	−2.11831	−2.15216	−2.18287
0.55	−1.70535	−1.75522	−1.80220	−1.84626	−1.88735	−1.92541
0.60	−1.39490	−1.45083	−1.50407	−1.55454	−1.60221	−1.64700
0.65	−1.07256	−1.13375	−1.19246	−1.24861	−1.30212	−1.35295
0.70	−0.744081	−0.809699	−0.873053	−0.93406	−0.99265	−1.04873
0.75	−0.415240	−0.484391	−0.551522	−0.616541	−0.679364	−0.739909
0.80	−0.091689	−0.163463	−0.233476	−0.301628	−0.367826	−0.431980
0.85	0.221127	0.147649	0.075663	0.005275	−0.063415	−0.130310
0.90	0.518075	0.443809	0.370759	0.299034	0.228742	0.159986
0.95	0.794425	0.720272	0.647053	0.574883	0.503872	0.434129
1.00	1.04594	0.972768	0.900251	0.828503	0.757637	0.687768
1.05	1.26893	1.19757	1.12659	1.05610	0.986214	0.917049
1.10	1.46034	1.39158	1.32292	1.25447	1.18635	1.11869
1.15	1.61777	1.55233	1.48671	1.42104	1.35543	1.29001
1.20	1.73954	1.67806	1.61615	1.55392	1.49150	1.42900
1.25	1.82466	1.76771	1.71009	1.65190	1.59327	1.53432
1.30	1.87289	1.82096	1.76813	1.71450	1.66020	1.60532
1.35	1.88469	1.83818	1.79055	1.74191	1.69238	1.64206
1.40	1.86120	1.82042	1.77832	1.73502	1.69063	1.64523
1.45	1.80423	1.76939	1.73306	1.69536	1.65637	1.61620
1.50	1.71617	1.68739	1.65698	1.62503	1.59164	1.55691
1.55	1.59996	1.57728	1.55283	1.52672	1.49903	1.46985
1.60	1.45901	1.44236	1.42385	1.40357	1.38158	1.35798
1.65	1.29710	1.28634	1.27365	1.25910	1.24276	1.22470
1.70	1.11832	1.11324	1.10617	1.09718	1.08633	1.07368
1.75	0.926984	0.927287	0.925576	0.921899	0.916314	0.908877
1.80	0.727505	0.732838	0.736148	0.737477	0.736872	0.734382
1.85	0.524333	0.534283	0.542225	0.548191	0.552220	0.554352
1.90	0.321848	0.335956	0.348091	0.358277	0.366543	0.372921
1.95	0.124275	0.142044	0.157894	0.171841	0.183906	0.194112
2.00	−0.064400	−0.043495	−0.024437	−0.007218	0.008174	0.021754
2.05	−0.240508	−0.217010	−0.195271	−0.175291	−0.157067	−0.140591
2.10	−0.400760	−0.375222	−0.351340	−0.329123	−0.308575	−0.289697
2.15	−0.542313	−0.515286	−0.489801	−0.465874	−0.443516	−0.422735

Table 2.8 (continued)

| v | t | | | | | |
	0.65	0.70	0.75	0.80	0.85	0.90
2.20	-0.662816	-0.634841	-0.608285	-0.583168	-0.559510	-0.537325
2.25	-0.760444	-0.732045	-0.704931	-0.679131	-0.654670	-0.631569
2.30	-0.833926	-0.805599	-0.778419	-0.752420	-0.727634	-0.704087
2.35	-0.882557	-0.854765	-0.827978	-0.802235	-0.777573	-0.754024
2.40	-0.906197	-0.879363	-0.853390	-0.828323	-0.804202	-0.781064
2.45	-0.905259	-0.879762	-0.854983	-0.830970	-0.807766	-0.785414
2.50	-0.880688	-0.856857	-0.833603	-0.810976	-0.789024	-0.767791
2.55	-0.833924	-0.812036	-0.790587	-0.769630	-0.749215	-0.729390
2.60	-0.766858	-0.747135	-0.727719	-0.708664	-0.690023	-0.671845
2.65	-0.681777	-0.664387	-0.647177	-0.630205	-0.613522	-0.597182
2.70	-0.581308	-0.566361	-0.551478	-0.536714	-0.522125	-0.507761
2.75	-0.468346	-0.455900	-0.443409	-0.430929	-0.418514	-0.406218
2.80	-0.345986	-0.336045	-0.325961	-0.315789	-0.305582	-0.295393
2.85	-0.217447	-0.209968	-0.202257	-0.194368	-0.186354	-0.178267
2.90	-0.086000	-0.080891	-0.075473	-0.069799	-0.063919	-0.057885
2.95	0.045112	0.047982	0.051227	0.054796	0.058641	0.062712
3.00	0.172739	0.173539	0.174768	0.176378	0.178323	0.180555
3.05	0.293900	0.292829	0.292230	0.292058	0.292270	0.292819
3.10	0.405842	0.403124	0.400911	0.399160	0.397830	0.396878
3.15	0.506098	0.501977	0.498382	0.495274	0.492615	0.490366
3.20	0.592536	0.587269	0.582539	0.578311	0.574550	0.571220
3.25	0.663400	0.657250	0.651640	0.646537	0.641911	0.637726
3.30	0.717341	0.710575	0.704341	0.698611	0.693358	0.688551
3.35	0.753441	0.746318	0.739713	0.733601	0.727959	0.722758
3.40	0.771218	0.763990	0.757258	0.751002	0.745201	0.739830
3.45	0.770636	0.763538	0.756910	0.750734	0.744992	0.739664
3.50	0.752086	0.745337	0.739026	0.733137	0.727657	0.722569
3.55	0.716377	0.710172	0.704368	0.698955	0.693920	0.689251
3.60	0.664703	0.659211	0.654081	0.649306	0.644877	0.640783
3.65	0.598605	0.593968	0.589652	0.585653	0.581964	0.578579
3.70	0.519934	0.516265	0.512874	0.509760	0.506921	0.504350
3.75	0.430799	0.428177	0.425793	0.423646	0.421736	0.420060
3.80	0.333508	0.331986	0.330659	0.329531	0.328602	0.327872
3.85	0.230519	0.230116	0.229868	0.229781	0.229857	0.230099
3.90	0.124371	0.125078	0.125903	0.126852	0.127929	0.129140
3.95	0.017630	0.019410	0.021273	0.023226	0.025275	0.027427
4.00	-0.087176	-0.084385	-0.081544	-0.078643	-0.075676	-0.072635
4.05	-0.187614	-0.183898	-0.180160	-0.176390	-0.172580	-0.168722
4.10	-0.281399	-0.276863	-0.272329	-0.267788	-0.263230	-0.258647
4.15	-0.366448	-0.361214	-0.356003	-0.350804	-0.345609	-0.340407
4.20	-0.440924	-0.435126	-0.429367	-0.423638	-0.417928	-0.412229
4.25	-0.503272	-0.497054	-0.490888	-0.484764	-0.478672	-0.472603
4.30	-0.552253	-0.545764	-0.539336	-0.532960	-0.526624	-0.520320

Table 2.8 (continued)

			t			
v	0.65	0.70	0.75	0.80	0.85	0.90
4.35	−0.586967	−0.580358	−0.573816	−0.567330	−0.560890	−0.554488
4.40	−0.606867	−0.600287	−0.593776	−0.587323	−0.580918	−0.574554
4.45	−0.611765	−0.605358	−0.599018	−0.592736	−0.586501	−0.580306
4.50	−0.601832	−0.595734	−0.589698	−0.583715	−0.577777	−0.571875
4.55	−0.577586	−0.571921	−0.566310	−0.560746	−0.555221	−0.549727
4.60	−0.539876	−0.534753	−0.529675	−0.524636	−0.519628	−0.514644
4.65	−0.489851	−0.485364	−0.480912	−0.476489	−0.472087	−0.467700
4.70	−0.428934	−0.425161	−0.421409	−0.417675	−0.413952	−0.410234
4.75	−0.358781	−0.355778	−0.352784	−0.349795	−0.346806	−0.343812
4.80	−0.281235	−0.279041	−0.276843	−0.274637	−0.272419	−0.270184
4.85	−0.198286	−0.196919	−0.195535	−0.194130	−0.192701	−0.191245
4.90	−0.112016	−0.111475	−0.110903	−0.110300	−0.109660	−0.108981
4.95	−0.024549	−0.024814	−0.025036	−0.025215	−0.025346	−0.025429
5.00	0.061997	0.060964	0.059984	0.059060	0.058192	0.057383

			t			
v	0.95	1.0	2.0	3.0	4.0	5.0
0.00	−2.69350	−2.63292	−1.09804	0.335139	0.866446	0.424125
0.05	−2.84085	−2.78785	−1.34850	0.105762	0.732643	0.380302
0.10	−2.94372	−2.89880	−1.56949	−0.110643	0.601313	0.337194
0.15	−3.00109	−2.96464	−1.75819	−0.310781	0.475214	0.296681
0.20	−3.01275	−2.98507	−1.91243	−0.491813	0.356830	0.260473
0.25	−2.97933	−2.96057	−2.03077	−0.651400	0.248322	0.230062
0.30	−2.90222	−2.89245	−2.11244	−0.787733	0.151499	0.206693
0.35	−2.78359	−2.78273	−2.15740	−0.899551	0.067779	0.191338
0.40	−2.62632	−2.63419	−2.16629	−0.986154	−0.001819	0.184672
0.45	−2.43390	−2.45021	−2.14038	−1.04740	−0.056691	0.187070
0.50	−2.21041	−2.23478	−2.08161	−1.08367	−0.096642	0.198597
0.55	−1.96042	−1.99236	−1.99244	−1.09587	−0.121878	0.219017
0.60	−1.68888	−1.72781	−1.87586	−1.08539	−0.132986	0.247804
0.65	−1.40102	−1.44629	−1.73525	−1.05402	−0.130903	0.284160
0.70	−1.10226	−1.15316	−1.57438	−1.00394	−0.116883	0.327045
0.75	−0.798099	−0.853863	−1.39726	−0.937650	−0.092455	0.375203
0.80	−0.494005	−0.553822	−1.20807	−0.857881	−0.059374	0.427204
0.85	−0.195316	−0.258347	−1.01109	−0.767548	−0.019570	0.481484
0.90	0.092864	0.027473	−0.810587	−0.669673	0.024909	0.536383
0.95	0.365759	0.298863	−0.610742	−0.567307	0.071948	0.590197
1.00	0.619003	0.551451	−0.415555	−0.463467	0.119426	0.641219
1.05	0.848720	0.781336	−0.228773	−0.361063	0.165268	0.687788
1.10	1.05159	0.985162	−0.053822	−0.262835	0.207502	0.728329
1.15	1.22488	1.16017	0.106258	−0.171298	0.244300	0.761395

Table 2.8 (continued)

v			t			
	0.95	1.0	2.0	3.0	4.0	5.0
1.20	1.36653	1.30423	0.248868	−0.088687	0.274032	0.785706
1.25	1.47515	1.41588	0.371884	−0.016916	0.295297	0.800179
1.30	1.55000	1.49434	0.473696	0.042460	0.306957	0.803959
1.35	1.59106	1.53950	0.553220	0.088266	0.308166	0.796438
1.40	1.59896	1.55190	0.609911	0.119729	0.298380	0.777271
1.45	1.57496	1.53274	0.643750	0.136480	0.277372	0.746383
1.50	1.52092	1.48379	0.655238	0.138554	0.245233	0.703977
1.55	1.43926	1.40737	0.645360	0.126377	0.202362	0.650523
1.60	1.33285	1.30628	0.615553	0.100742	0.149455	0.586747
1.65	1.20499	1.18373	0.567659	0.062783	0.087485	0.513621
1.70	1.05930	1.04327	0.503868	0.013937	0.017670	0.432330
1.75	0.899650	0.888699	0.426664	−0.044099	−0.058555	0.344250
1.80	0.730061	0.723965	0.338755	−0.109418	−0.139579	0.250915
1.85	0.554633	0.553111	0.243005	−0.179953	−0.223658	0.153977
1.90	0.377449	0.380166	0.142367	−0.253533	−0.308956	0.055172
1.95	0.202487	0.209063	0.039808	−0.327939	−0.393595	−0.043721
2.00	0.033542	0.043561	−0.061757	−0.400962	−0.475698	−0.140926
2.05	−0.125851	−0.112832	−0.159533	−0.470454	−0.553440	−0.234704
2.10	−0.272483	−0.256927	−0.250907	−0.534385	−0.625088	−0.323398
2.15	−0.403535	−0.385915	−0.333502	−0.590889	−0.689045	−0.405466
2.20	−0.516624	−0.497414	−0.405229	−0.638308	−0.743887	−0.479519
2.25	−0.609846	−0.589514	−0.464330	−0.675229	−0.788398	−0.544349
2.30	−0.681802	−0.660801	−0.509405	−0.700518	−0.821594	−0.598955
2.35	−0.731617	−0.710378	−0.539446	−0.713342	−0.842751	−0.642566
2.40	−0.758943	−0.737869	−0.553842	−0.713186	−0.851415	−0.674654
2.45	−0.763952	−0.743414	−0.552390	−0.699859	−0.847414	−0.694942
2.50	−0.747320	−0.727650	−0.535289	−0.673500	−0.830862	−0.703411
2.55	−0.710200	−0.691687	−0.503124	−0.634566	−0.802147	−0.700297
2.60	−0.654178	−0.637068	−0.456848	−0.583819	−0.761930	−0.686078
2.65	−0.581233	−0.565722	−0.397746	−0.522301	−0.711117	−0.661465
2.70	−0.493675	−0.479914	−0.327403	−0.451312	−0.650844	−0.627380
2.75	−0.394090	−0.382182	−0.247653	−0.372367	−0.582445	−0.584931
2.80	−0.285273	−0.275273	−0.160537	−0.287165	−0.507419	−0.535387
2.85	−0.170158	−0.162077	−0.068245	−0.197541	−0.427394	−0.480144
2.90	−0.051746	−0.045553	0.026937	−0.105423	−0.344087	−0.420690
2.95	0.066959	0.071336	0.122685	−0.012784	−0.259267	−0.358570
3.00	0.183027	0.185693	0.216693	0.078407	−0.174708	−0.295351
3.05	0.293661	0.294752	0.306729	0.166227	−0.092153	−0.232582
3.10	0.396263	0.395942	0.390683	0.248851	−0.013268	−0.171760
3.15	0.488486	0.486935	0.466622	0.324589	0.060388	−0.114295
3.20	0.568283	0.565703	0.532831	0.391925	0.127404	−0.061479
3.25	0.633951	0.630550	0.587848	0.449555	0.186545	−0.014456

Table 2.8 (continued)

			t			
v	0.95	1.0	2.0	3.0	4.0	5.0
3.30	0.684159	0.680152	0.630497	0.496410	0.236780	0.025802
3.35	0.717973	0.713574	0.659913	0.531681	0.277302	0.058512
3.40	0.734867	0.730285	0.675557	0.554835	0.307542	0.083102
3.45	0.734729	0.730166	0.677221	0.565625	0.327182	0.099216
3.50	0.717856	0.713497	0.665033	0.564088	0.336157	0.106718
3.55	0.684932	0.680948	0.639446	0.550548	0.334654	0.105698
3.60	0.637012	0.633553	0.601221	0.525596	0.323101	0.096459
3.65	0.575487	0.572680	0.551410	0.490079	0.302155	0.079508
3.70	0.502042	0.499989	0.491319	0.445075	0.272685	0.055544
3.75	0.418614	0.417393	0.422481	0.391864	0.235744	0.025434
3.80	0.327341	0.327004	0.346611	0.331894	0.192543	−0.009812
3.85	0.230506	0.231078	0.265564	0.266746	0.144421	−0.049063
3.90	0.130485	0.131965	0.181290	0.198097	0.092808	−0.091102
3.95	0.029683	0.032048	0.095787	0.127675	0.039191	−0.134653
4.00	−0.069516	−0.066314	0.011051	0.057217	−0.014920	−0.178418
4.05	−0.164810	−0.160840	−0.070972	−0.011566	−0.068034	−0.221104
4.10	−0.254032	−0.249379	−0.148426	−0.077037	−0.118712	−0.261462
4.15	−0.335193	−0.329958	−0.219595	−0.137663	−0.165600	−0.298310
4.20	−0.406531	−0.400828	−0.282937	−0.192058	−0.207462	−0.330568
4.25	−0.466548	−0.460499	−0.337118	−0.239008	−0.243208	−0.357279
4.30	−0.514039	−0.507773	−0.381043	−0.277498	−0.271918	−0.377635
4.35	−0.548115	−0.541762	−0.413877	−0.306738	−0.292863	−0.390994
4.40	−0.568221	−0.561911	−0.435060	−0.326171	−0.305520	−0.396895
4.45	−0.574142	−0.568002	−0.444320	−0.335492	−0.309581	−0.395066
4.50	−0.566002	−0.560151	−0.441669	−0.334642	−0.304959	−0.385433
4.55	−0.544257	−0.538805	−0.427407	−0.323813	−0.291785	−0.368114
4.60	−0.509677	−0.504721	−0.402102	−0.303440	−0.270407	−0.343421
4.65	−0.463322	−0.458948	−0.366579	−0.274182	−0.241372	−0.311848
4.70	−0.406516	−0.402793	−0.321893	−0.236905	−0.205417	−0.274055
4.75	−0.340807	−0.337787	−0.269303	−0.192662	−0.163444	−0.230856
4.80	−0.267928	−0.265647	−0.210238	−0.142662	−0.116497	−0.183193
4.85	−0.189756	−0.188233	−0.146258	−0.088236	−0.065739	−0.132114
4.90	−0.108261	−0.107495	−0.079020	−0.030808	−0.012415	−0.078747
4.95	−0.025459	−0.025435	−0.010232	0.028141	0.042174	−0.024271
5.00	0.056635	0.055950	0.058385	0.087114	0.096705	0.030112

Table 2.9: Third derivatives with respect to the order of the Anger functions $\partial^3 J_\nu(t)/\partial \nu^3$.

ν	0.05	0.10	0.15	0.20	0.25	0.30
0.00	−0.193436	−0.386605	−0.579241	−0.771078	−0.961853	−1.15130
0.05	0.778699	0.585739	0.392619	0.199605	0.006963	−0.185042
0.10	1.73623	1.54597	1.35486	1.16316	0.971143	0.779071
0.15	2.66235	2.47724	2.29059	2.10268	1.91375	1.72409
0.20	3.54087	3.36326	3.18346	3.00173	2.81832	2.63350
0.25	4.35649	4.18864	4.01798	3.84473	3.66916	3.49152
0.30	5.09516	4.93917	4.77977	4.61718	4.45165	4.28341
0.35	5.74428	5.60208	5.45590	5.30597	5.15252	4.99577
0.40	6.29297	6.16627	6.03509	5.89963	5.76010	5.61673
0.45	6.73229	6.62256	6.50790	6.38850	6.26454	6.13624
0.50	7.05533	6.96381	6.86696	6.76496	6.65798	6.54619
0.55	7.25741	7.18506	7.10705	7.02353	6.93466	6.84060
0.60	7.33613	7.28362	7.22519	7.16097	7.09110	7.01570
0.65	7.29139	7.25911	7.22072	7.17633	7.12603	7.06996
0.70	7.12538	7.11343	7.09525	7.07092	7.04052	7.00414
0.75	6.84255	6.85074	6.85264	6.84832	6.83783	6.82124
0.80	6.44946	6.47731	6.49889	6.51424	6.52340	6.52641
0.85	5.95467	6.00143	6.04201	6.07643	6.10470	6.12683
0.90	5.36856	5.43322	5.49186	5.54448	5.59105	5.63157
0.95	4.70308	4.78440	4.85992	4.92961	4.99344	5.05136
1.00	3.97154	4.06805	4.15905	4.24449	4.32429	4.39840
1.05	3.18831	3.29836	3.40325	3.50289	3.59718	3.68607
1.10	2.36855	2.49034	2.60735	2.71947	2.82661	2.92867
1.15	1.52790	1.65947	1.78670	1.90946	2.02764	2.14111
1.20	0.682158	0.821470	0.956917	1.08835	1.21564	1.33865
1.25	−0.153020	−0.008075	0.133513	0.271581	0.405975	0.536551
1.30	−0.962422	−0.813983	−0.668370	−0.525759	−0.386315	−0.250201
1.35	−1.73158	−1.58178	−1.43427	−1.28921	−1.14680	−1.00720
1.40	−2.44704	−2.29799	−2.15066	−2.00523	−1.86191	−1.72088
1.45	−3.09661	−2.95034	−2.80523	−2.66146	−2.51926	−2.37879
1.50	−3.66960	−3.52805	−3.38709	−3.24693	−3.10777	−2.96982
1.55	−4.15698	−4.02196	−3.88698	−3.75225	−3.61799	−3.48438
1.60	−4.55158	−4.42474	−4.29742	−4.16982	−4.04214	−3.91460
1.65	−4.84814	−4.73098	−4.61282	−4.49388	−4.37435	−4.25443
1.70	−5.04346	−4.93727	−4.82959	−4.72065	−4.61064	−4.49976
1.75	−5.13637	−5.04222	−4.94616	−4.84837	−4.74906	−4.64842
1.80	−5.12774	−5.04651	−4.96296	−4.87727	−4.78963	−4.70024
1.85	−5.02045	−4.95277	−4.88240	−4.80953	−4.73433	−4.65699
1.90	−4.81925	−4.76553	−4.70880	−4.64925	−4.58703	−4.52232
1.95	−4.53069	−4.49110	−4.44824	−4.40226	−4.35334	−4.30162
2.00	−4.16290	−4.13738	−4.10838	−4.07605	−4.04051	−4.00192
2.05	−3.72543	−3.71372	−3.69836	−3.67949	−3.65722	−3.63169

Table 2.9 (continued)

			t			
v	0.05	0.10	0.15	0.20	0.25	0.30
2.10	−3.22903	−3.23064	−3.22850	−3.22270	−3.21338	−3.20063
2.15	−2.68537	−2.69964	−2.71008	−2.71680	−2.71990	−2.71947
2.20	−2.10684	−2.13291	−2.15515	−2.17363	−2.18845	−2.19968
2.25	−1.50625	−1.54311	−1.57618	−1.60552	−1.63119	−1.65327
2.30	−0.896528	−0.943052	−0.985865	−1.02501	−1.06054	−1.09251
2.35	−0.290522	−0.345455	−0.396804	−0.444595	−0.488863	−0.529646
2.40	0.299325	0.237317	0.178729	0.123548	0.071757	0.023333
2.45	0.861203	0.793514	0.729043	0.667795	0.609767	0.554951
2.50	1.38417	1.31223	1.24327	1.17732	1.11439	1.054480
2.55	1.85836	1.78360	1.71157	1.64229	1.57581	1.512130
2.60	2.27519	2.19903	2.12531	2.05409	1.98540	1.919280
2.65	2.62748	2.55130	2.47726	2.40543	2.33586	2.268590
2.70	2.90961	2.83471	2.76165	2.69050	2.62132	2.554170
2.75	3.11756	3.04517	2.97432	2.90506	2.83748	2.771630
2.80	3.24902	3.18028	3.11275	3.04650	2.98163	2.918210
2.85	3.30338	3.23929	3.17609	3.11387	3.05272	2.992720
2.90	3.28168	3.22311	3.16512	3.10782	3.05128	2.995600
2.95	3.18658	3.13427	3.08224	3.03060	2.97944	2.928860
3.00	3.02228	2.97681	2.93134	2.88598	2.84081	2.795950
3.05	2.79436	2.75616	2.71769	2.67907	2.64039	2.601740
3.10	2.50966	2.47900	2.44784	2.41627	2.38440	2.352330
3.15	2.17609	2.15310	2.12938	2.10503	2.08016	2.054860
3.20	1.80245	1.78709	1.77081	1.75371	1.73588	1.717420
3.25	1.39820	1.39030	1.38131	1.37133	1.36045	1.348760
3.30	0.973245	0.972501	0.970527	0.967415	0.963255	0.958136
3.35	0.537712	0.543695	0.548337	0.551724	0.553942	0.555074
3.40	0.101726	0.113898	0.124648	0.134054	0.142196	0.149154
3.45	−0.324824	−0.307091	−0.290836	−0.275984	−0.262464	−0.250202
3.50	−0.732489	−0.709904	−0.688822	−0.669179	−0.650908	−0.633941
3.55	−1.11247	−1.08580	−1.06063	−1.03691	−1.01458	−0.993572
3.60	−1.45679	−1.42684	−1.39837	−1.37134	−1.34568	−1.321350
3.65	−1.75845	−1.72606	−1.69511	−1.66554	−1.63733	−1.610420
3.70	−2.01160	−1.97761	−1.94499	−1.91370	−1.88371	−1.854970
3.75	−2.21159	−2.17683	−2.14336	−2.11114	−2.08015	−2.050350
3.80	−2.35510	−2.32038	−2.28684	−2.25447	−2.22324	−2.193120
3.85	−2.44017	−2.40625	−2.37339	−2.34160	−2.31085	−2.281130
3.90	−2.46618	−2.43377	−2.40231	−2.37179	−2.34220	−2.313540
3.95	−2.43391	−2.40366	−2.37422	−2.34559	−2.31779	−2.290800
4.00	−2.34544	−2.31790	−2.29104	−2.26488	−2.23941	−2.214630
4.05	−2.20408	−2.17974	−2.15595	−2.13271	−2.11004	−2.087950
4.10	−2.01430	−1.99354	−1.97320	−1.95328	−1.93381	−1.914790
4.15	−1.78156	−1.76469	−1.74809	−1.73179	−1.71582	−1.700170
4.20	−1.51223	−1.49943	−1.48678	−1.47431	−1.46203	−1.449980

Table 2.9 (continued)

			t			
v	0.05	0.10	0.15	0.20	0.25	0.30
4.25	−1.21337	−1.20473	−1.19613	−1.18759	−1.17914	−1.170790
4.30	−0.892557	−0.888098	−0.883560	−0.878974	−0.874367	−0.869764
4.35	−0.557757	−0.557379	−0.556823	−0.556121	−0.555300	−0.554389
4.40	−0.217074	−0.220602	−0.223864	−0.226892	−0.229715	−0.232363
4.45	0.121408	0.114227	0.107389	0.100862	0.094613	0.088614
4.50	0.449816	0.439308	0.429205	0.419475	0.410087	0.401012
4.55	0.760667	0.747216	0.734219	0.721644	0.709463	0.697644
4.60	1.04703	1.03107	1.01560	1.00059	0.986006	0.971826
4.65	1.30269	1.28469	1.26720	1.25019	1.23365	1.21752
4.70	1.52224	1.50270	1.48368	1.46515	1.44710	1.42948
4.75	1.70125	1.68068	1.66062	1.64106	1.62197	1.60333
4.80	1.83629	1.81520	1.79462	1.77451	1.75487	1.73567
4.85	1.92504	1.90394	1.88332	1.86316	1.84344	1.82414
4.90	1.96630	1.94568	1.92549	1.90574	1.88640	1.86746
4.95	1.96001	1.94031	1.92101	1.90211	1.88358	1.86541
5.00	1.90720	1.88884	1.87084	1.85319	1.83587	1.81888

			t			
v	0.35	0.40	0.45	0.50	0.55	0.60
0.00	−1.33917	−1.52520	−1.70913	−1.89073	−2.06973	−2.24591
0.05	−0.376145	−0.566084	−0.754599	−0.941435	−1.12634	−1.30906
0.10	0.587212	0.395829	0.205187	0.015548	−0.172830	−0.359689
0.15	1.53395	1.34360	1.15330	0.963320	0.773919	0.585360
0.20	2.44752	2.26064	2.07313	1.88526	1.69728	1.50947
0.25	3.31205	3.13103	2.94870	2.76534	2.58119	2.39653
0.30	4.11272	3.93982	3.76497	3.58842	3.41043	3.23126
0.35	4.83596	4.67332	4.50811	4.34057	4.17095	3.99951
0.40	5.46973	5.31934	5.16577	5.00928	4.85010	4.68848
0.45	6.00379	5.86740	5.72731	5.58373	5.43688	5.28701
0.50	6.42979	6.30898	6.18396	6.05494	5.92212	5.78574
0.55	6.74152	6.63760	6.52902	6.41598	6.29867	6.17730
0.60	6.93493	6.84896	6.75794	6.66205	6.56147	6.45639
0.65	7.00823	6.94099	6.86838	6.79056	6.70768	6.61991
0.70	6.96190	6.91390	6.86028	6.80115	6.73668	6.66699
0.75	6.79864	6.77011	6.73577	6.69571	6.65006	6.59894
0.80	6.52332	6.51421	6.49916	6.47825	6.45157	6.41924
0.85	6.14286	6.15284	6.15681	6.15485	6.14703	6.13342
0.90	5.66605	5.69452	5.71699	5.73352	5.74416	5.74895
0.95	5.10338	5.14948	5.18968	5.22398	5.25242	5.27503
1.00	4.46680	4.52946	4.58635	4.63747	4.68282	4.72242
1.05	3.76950	3.84740	3.91975	3.98651	4.04767	4.10320
1.10	3.02557	3.11724	3.20361	3.28464	3.36028	3.43050

Table 2.9 (continued)

				t		
v	0.35	0.40	0.45	0.50	0.55	0.60
1.15	2.24978	2.35356	2.45236	2.54611	2.63474	2.71821
1.20	1.45726	1.57136	1.68085	1.78563	1.88562	1.98075
1.25	0.663170	0.785704	0.904031	1.01804	1.12763	1.23270
1.30	−0.117569	0.011437	0.136679	0.258029	0.375367	0.488581
1.35	−0.870579	−0.737101	−0.606915	−0.480163	−0.356982	−0.237499
1.40	−1.58230	−1.44636	−1.31321	−1.18302	−1.05593	−0.932079
1.45	−2.24025	−2.10383	−1.96968	−1.83799	−1.70892	−1.58262
1.50	−2.83325	−2.69827	−2.56505	−2.43377	−2.30461	−2.17772
1.55	−3.35163	−3.21993	−3.08947	−2.96044	−2.83302	−2.70737
1.60	−3.78739	−3.66071	−3.53475	−3.40970	−3.28575	−3.16308
1.65	−4.13434	−4.01427	−3.89441	−3.77497	−3.65611	−3.53804
1.70	−4.38821	−4.27618	−4.16388	−4.05150	−3.93923	−3.82725
1.75	−4.54665	−4.44393	−4.34047	−4.23646	−4.13209	−4.02755
1.80	−4.60928	−4.51695	−4.42344	−4.32892	−4.23360	−4.13766
1.85	−4.57769	−4.49661	−4.41393	−4.32985	−4.24454	−4.15818
1.90	−4.45529	−4.38612	−4.31498	−4.24205	−4.16752	−4.09155
1.95	−4.24727	−4.19045	−4.13134	−4.07009	−4.00689	−3.94190
2.00	−3.96044	−3.91621	−3.86939	−3.82015	−3.76863	−3.71502
2.05	−3.60305	−3.57143	−3.53697	−3.49984	−3.46017	−3.41812
2.10	−3.18460	−3.16541	−3.14318	−3.11806	−3.09019	−3.05970
2.15	−2.71562	−2.70848	−2.69815	−2.68477	−2.66845	−2.64933
2.20	−2.20742	−2.21177	−2.21283	−2.21072	−2.20555	−2.19743
2.25	−1.67183	−1.68696	−1.69874	−1.70728	−1.71266	−1.71500
2.30	−1.12098	−1.14602	−1.16770	−1.18611	−1.20132	−1.21343
2.35	−0.566993	−0.600954	−0.631589	−0.658963	−0.683144	−0.704208
2.40	−0.021756	−0.063546	−0.102081	−0.137410	−0.169588	−0.198675
2.45	0.503332	0.454886	0.409588	0.367401	0.328285	0.292196
2.50	0.997590	0.943721	0.892854	0.844970	0.800043	0.758042
2.55	1.45127	1.39324	1.33804	1.28566	1.23609	1.18932
2.60	1.85575	1.79483	1.73654	1.68088	1.62786	1.57747
2.65	2.20367	2.14112	2.08098	2.02326	1.96797	1.91514
2.70	2.48909	2.42614	2.36535	2.30675	2.25038	2.19623
2.75	2.70759	2.64541	2.58512	2.52679	2.47043	2.41609
2.80	2.85630	2.79597	2.73728	2.68028	2.62501	2.57152
2.85	2.93395	2.87648	2.82038	2.76570	2.71251	2.66085
2.90	2.94087	2.88715	2.83453	2.78308	2.73284	2.68390
2.95	2.87893	2.82975	2.78139	2.73394	2.68745	2.64200
3.00	2.75148	2.70749	2.66405	2.62127	2.57919	2.53791
3.05	2.56323	2.52494	2.48697	2.44938	2.41227	2.37571
3.10	2.32014	2.28794	2.25581	2.22383	2.19210	2.16069
3.15	2.02923	2.00336	1.97733	1.95125	1.92518	1.89922
3.20	1.69842	1.67899	1.65920	1.63914	1.61890	1.59856

Table 2.9 (continued)

			t			
v	0.35	0.40	0.45	0.50	0.55	0.60
3.25	1.33636	1.32333	1.30976	1.29575	1.28137	1.26671
3.30	0.952147	0.945376	0.937908	0.929829	0.921222	0.912168
3.35	0.555207	0.554423	0.552806	0.550439	0.547401	0.543773
3.40	0.155008	0.159836	0.163719	0.166734	0.168959	0.170472
3.45	−0.239123	−0.229152	−0.220216	−0.212241	−0.205151	−0.198875
3.50	−0.618211	−0.603650	−0.590187	−0.577753	−0.566280	−0.555698
3.55	−0.973833	−0.955296	−0.937899	−0.921577	−0.906267	−0.891904
3.60	−1.29829	−1.27644	−1.25575	−1.23617	−1.21762	−1.20006
3.65	−1.58476	−1.56031	−1.53701	−1.51481	−1.49367	−1.47352
3.70	−1.82746	−1.80112	−1.77591	−1.75179	−1.72870	−1.70661
3.75	−2.02172	−1.99422	−1.96780	−1.94244	−1.91809	−1.89470
3.80	−2.16410	−2.13614	−2.10922	−2.08329	−2.05833	−2.03430
3.85	−2.25242	−2.22469	−2.19792	−2.17209	−2.14717	−2.12312
3.90	−2.28579	−2.25893	−2.23296	−2.20784	−2.18356	−2.16009
3.95	−2.26462	−2.23923	−2.21463	−2.19080	−2.16773	−2.14540
4.00	−2.19055	−2.16717	−2.14447	−2.12244	−2.10109	−2.08039
4.05	−2.06644	−2.04552	−2.02518	−2.00541	−1.98622	−1.96760
4.10	−1.89624	−1.87816	−1.86056	−1.84343	−1.82678	−1.81061
4.15	−1.68488	−1.66995	−1.65539	−1.64120	−1.62739	−1.61397
4.20	−1.43816	−1.42659	−1.41529	−1.40426	−1.39352	−1.38307
4.25	−1.16258	−1.15451	−1.14660	−1.13888	−1.13134	−1.12400
4.30	−0.865190	−0.860665	−0.856209	−0.851837	−0.847566	−0.843407
4.35	−0.553413	−0.552396	−0.551359	−0.550320	−0.549298	−0.548307
4.40	−0.234862	−0.237238	−0.239513	−0.241708	−0.243843	−0.245935
4.45	0.082837	0.077256	0.071846	0.066587	0.061456	0.056435
4.50	0.392220	0.383687	0.375386	0.367296	0.359393	0.351659
4.55	0.686160	0.674984	0.664093	0.653461	0.643067	0.632890
4.60	0.958019	0.944561	0.931426	0.918590	0.906032	0.893730
4.65	1.20180	1.18646	1.17147	1.15680	1.14245	1.12837
4.70	1.41228	1.39548	1.37904	1.36295	1.34719	1.33174
4.75	1.58510	1.56728	1.54983	1.53274	1.51599	1.49955
4.80	1.71688	1.69849	1.68047	1.66281	1.64548	1.62847
4.85	1.80524	1.78672	1.76857	1.75076	1.73328	1.71612
4.90	1.84889	1.83068	1.81282	1.79528	1.77806	1.76113
4.95	1.84759	1.83010	1.81293	1.79606	1.77948	1.76317
5.00	1.80219	1.78580	1.76969	1.75385	1.73827	1.72294

			t			
v	0.65	0.70	0.75	0.80	0.85	0.90
0.00	−2.41902	−2.58885	−2.75516	−2.91773	−3.07637	−3.23086
0.05	−1.48935	−1.66697	−1.84168	−2.01326	−2.18148	−2.34612
0.10	−0.544776	−0.727840	−0.908637	−1.08693	−1.26247	−1.43504

Table 2.9 (continued)

				t		
v	0.65	0.70	0.75	0.80	0.85	0.90
0.15	0.397902	0.211801	0.027309	−0.15532	−0.33585	−0.51403
0.20	1.32207	1.13536	0.949591	0.76502	0.58189	0.40046
0.25	2.21161	2.02669	1.84205	1.65793	1.47459	1.29229
0.30	3.05116	2.87041	2.68925	2.50794	2.32676	2.14595
0.35	3.82649	3.65215	3.47676	3.30056	3.12382	2.94680
0.40	4.52466	4.35889	4.19143	4.02252	3.85243	3.68140
0.45	5.13435	4.97915	4.82163	4.66206	4.50068	4.33774
0.50	5.64602	5.50319	5.35748	5.20912	5.05836	4.90545
0.55	6.05208	5.92322	5.79096	5.65550	5.51709	5.37596
0.60	6.34700	6.23351	6.11612	5.99504	5.87049	5.74269
0.65	6.52744	6.43045	6.32912	6.22366	6.11426	6.00113
0.70	6.59226	6.51264	6.42832	6.33946	6.24625	6.14889
0.75	6.54250	6.48088	6.41422	6.34270	6.26648	6.18572
0.80	6.38137	6.33808	6.28951	6.23579	6.17707	6.11352
0.85	6.11413	6.08925	6.05889	6.02318	5.98224	5.93621
0.90	5.74797	5.74131	5.72904	5.71127	5.68811	5.65966
0.95	5.29186	5.30295	5.30839	5.30825	5.30260	5.29154
1.00	4.75629	4.78446	4.80697	4.82388	4.83525	4.84114
1.05	4.15312	4.19743	4.23616	4.26932	4.29697	4.31915
1.10	3.49527	3.55459	3.60845	3.65687	3.69984	3.73741
1.15	2.79647	2.86947	2.93721	2.99966	3.05682	3.10870
1.20	2.07095	2.15616	2.23634	2.31146	2.38148	2.44640
1.25	1.33318	1.42898	1.52004	1.60631	1.68773	1.76427
1.30	0.597569	0.702236	0.802499	0.89828	0.98952	1.07615
1.35	−0.121835	−0.010099	0.097606	0.20119	0.30056	0.39564
1.40	−0.811611	−0.694650	−0.581312	−0.47171	−0.36594	−0.26411
1.45	−1.45923	−1.33890	−1.22176	−1.10793	−0.99754	−0.89069
1.50	−2.05327	−1.93142	−1.81229	−1.69605	−1.58280	−1.47269
1.55	−2.58367	−2.46208	−2.34276	−2.22586	−2.11151	−1.99986
1.60	−3.04186	−2.92227	−2.80447	−2.68861	−2.57486	−2.46336
1.65	−3.42093	−3.30496	−3.19031	−3.07713	−2.96559	−2.85584
1.70	−3.71575	−3.60491	−3.49491	−3.38592	−3.27811	−3.17163
1.75	−3.92301	−3.81867	−3.71470	−3.61129	−3.50859	−3.40679
1.80	−4.04128	−3.94464	−3.84793	−3.75133	−3.65500	−3.55912
1.85	−4.07097	−3.98307	−3.89468	−3.80596	−3.71710	−3.62826
1.90	−4.01432	−3.93602	−3.85682	−3.77689	−3.69641	−3.61555
1.95	−3.87530	−3.80725	−3.73793	−3.66751	−3.59616	−3.52405
2.00	−3.65946	−3.60213	−3.54319	−3.48281	−3.42115	−3.35838
2.05	−3.37384	−3.32750	−3.27925	−3.22925	−3.17765	−3.12463
2.10	−3.02675	−2.99147	−2.95402	−2.91455	−2.87321	−2.83016
2.15	−2.62755	−2.60323	−2.57651	−2.54755	−2.51647	−2.48343
2.20	−2.18648	−2.17283	−2.15661	−2.13795	−2.11697	−2.09382
2.25	−1.71440	−1.71097	−1.70483	−1.69610	−1.68490	−1.67135

Table 2.9 (continued)

			t			
v	0.65	0.70	0.75	0.80	0.85	0.90
2.30	−1.22253	−1.22872	−1.23211	−1.23279	−1.23088	−1.22650
2.35	−0.722235	−0.737310	−0.749523	−0.758967	−0.765741	−0.769946
2.40	−0.224738	−0.247848	−0.268080	−0.285515	−0.300238	−0.312339
2.45	0.259080	0.228882	0.201537	0.176981	0.155139	0.135937
2.50	0.718929	0.682660	0.649188	0.618457	0.590411	0.564984
2.55	1.14531	1.10404	1.06548	1.02958	0.996304	0.965592
2.60	1.52971	1.48454	1.44197	1.40194	1.36444	1.32943
2.65	1.86475	1.81680	1.77130	1.72821	1.68752	1.64921
2.70	2.14435	2.09472	2.04736	2.00226	1.95942	1.91882
2.75	2.36380	2.31356	2.26540	2.21933	2.17534	2.13343
2.80	2.51985	2.47003	2.42207	2.37601	2.33185	2.28960
2.85	2.61077	2.56231	2.51550	2.47038	2.42696	2.38527
2.90	2.63629	2.59007	2.54528	2.50195	2.46013	2.41984
2.95	2.59765	2.55444	2.51244	2.47169	2.43223	2.39410
3.00	2.49748	2.45797	2.41944	2.38194	2.34551	2.31020
3.05	2.33977	2.30452	2.27001	2.23632	2.20350	2.17159
3.10	2.12967	2.09912	2.06911	2.03969	2.01092	1.98288
3.15	1.87344	1.84791	1.82271	1.79791	1.77356	1.74973
3.20	1.57821	1.55791	1.53774	1.51778	1.49808	1.47871
3.25	1.25185	1.23686	1.22182	1.20680	1.19188	1.17710
3.30	0.902747	0.893035	0.883109	0.873039	0.862896	0.852746
3.35	0.539632	0.535054	0.530114	0.524884	0.519433	0.513829
3.40	0.171347	0.171659	0.171481	0.170884	0.169936	0.168704
3.45	−0.193339	−0.188472	−0.184205	−0.180466	−0.177190	−0.174310
3.50	−0.545938	−0.536932	−0.528614	−0.520916	−0.513772	−0.507120
3.55	−0.878425	−0.865763	−0.853857	−0.842642	−0.832057	−0.822039
3.60	−1.18341	−1.16763	−1.15266	−1.13842	−1.12488	−1.11195
3.65	−1.45431	−1.43599	−1.41851	−1.40179	−1.38581	−1.37048
3.70	−1.68546	−1.66521	−1.64579	−1.62717	−1.60929	−1.59211
3.75	−1.87224	−1.85067	−1.82993	−1.80998	−1.79078	−1.77227
3.80	−2.01117	−1.98888	−1.96742	−1.94673	−1.92677	−1.90750
3.85	−2.09992	−2.07753	−2.05592	−2.03506	−2.01490	−1.99542
3.90	−2.13741	−2.11549	−2.09431	−2.07382	−2.05400	−2.03482
3.95	−2.12378	−2.10286	−2.08262	−2.06302	−2.04404	−2.02566
4.00	−2.06034	−2.04090	−2.02208	−2.00384	−1.98617	−1.96904
4.05	−1.94954	−1.93202	−1.91504	−1.89858	−1.88261	−1.86713
4.10	−1.79491	−1.77968	−1.76490	−1.75056	−1.73666	−1.72318
4.15	−1.60093	−1.58827	−1.57598	−1.56407	−1.55251	−1.54131
4.20	−1.37291	−1.36305	−1.35348	−1.34420	−1.33521	−1.32651
4.25	−1.11687	−1.10995	−1.10324	−1.09676	−1.09048	−1.08443
4.30	−0.839371	−0.835466	−0.831699	−0.828075	−0.824597	−0.821265
4.35	−0.547360	−0.546468	−0.545642	−0.544888	−0.544212	−0.543618
4.40	−0.247998	−0.250048	−0.252094	−0.254147	−0.256216	−0.258305

Table 2.9 (continued)

			t			
v	**0.65**	**0.70**	**0.75**	**0.80**	**0.85**	**0.90**
4.45	0.051507	0.046656	0.041871	0.037137	0.032446	0.027789
4.50	0.344074	0.336623	0.329290	0.322062	0.314928	0.307877
4.55	0.622911	0.613113	0.603479	0.593995	0.584649	0.575428
4.60	0.881666	0.869820	0.858177	0.846721	0.835439	0.824317
4.65	1.11457	1.10102	1.08769	1.07458	1.06168	1.04896
4.70	1.31657	1.30167	1.28702	1.27261	1.25842	1.24444
4.75	1.48341	1.46755	1.45196	1.43661	1.42150	1.40661
4.80	1.61176	1.59534	1.57918	1.56328	1.54761	1.53217
4.85	1.69924	1.68265	1.66631	1.65023	1.63438	1.61876
4.90	1.74448	1.72810	1.71197	1.69608	1.68042	1.66497
4.95	1.74712	1.73132	1.71576	1.70042	1.68529	1.67036
5.00	1.70784	1.69296	1.67830	1.66384	1.64957	1.63548

			t			
v	**0.95**	**1.0**	**2.0**	**3.0**	**4.0**	**5.0**
0.00	−3.38101	−3.52664	−5.28211	−4.69326	−2.68126	−0.870330
0.05	−3.50697	−2.66384	−4.72502	−4.46951	−2.66093	−0.875832
0.10	−1.60440	−1.77035	−4.10514	−4.17568	−2.58305	−0.842224
0.15	−0.68963	−0.86241	−3.43543	−3.82036	−2.45261	−0.772566
0.20	0.22097	0.04367	−2.72952	−3.41313	−2.27556	−0.670776
0.25	1.11128	0.93181	−2.00145	−2.96432	−2.05870	−0.541504
0.30	1.96576	1.78646	−1.26535	−2.48477	−1.80950	−0.390001
0.35	2.76974	2.59291	−0.53517	−1.98562	−1.53588	−0.221969
0.40	3.50970	3.33757	0.17557	−1.47803	−1.24603	−0.043408
0.45	4.17350	4.00819	0.854082	−0.972960	−0.948199	0.139542
0.50	4.75062	4.59412	1.48852	−0.480941	−0.650539	0.320757
0.55	5.23234	5.08646	2.06823	−0.011862	−0.360877	0.494290
0.60	5.61187	5.47826	2.58390	0.425228	−0.086563	0.654513
0.65	5.88449	5.76455	3.02777	0.822289	0.165697	0.796254
0.70	6.04757	5.94250	3.39369	1.17244	0.389980	0.914923
0.75	6.10063	6.01137	3.67730	1.47008	0.581264	1.00662
0.80	6.04529	5.97255	3.87598	1.71095	0.735525	1.06821
0.85	5.88522	5.82944	3.98893	1.89225	0.849808	1.09742
0.90	5.62605	5.58741	4.01714	2.01257	0.922271	1.09285
0.95	5.27518	5.25362	3.96329	2.07198	0.952195	1.05401
1.00	4.84165	4.83686	3.83169	2.07191	0.939985	0.981339
1.05	4.33592	4.34734	3.62814	2.01512	0.887121	0.876157
1.10	3.76961	3.79649	3.35975	1.90560	0.796108	0.740634
1.15	3.15530	3.19665	3.03480	1.74842	0.670382	0.577725
1.20	2.50620	2.56088	2.66250	1.54963	0.514213	0.391090
1.25	1.83590	1.90261	2.25278	1.31603	0.332576	0.184991

Table 2.9 (continued)

			t			
v	0.95	1.0	2.0	3.0	4.0	5.0
1.30	1.15812	1.23541	1.81608	1.05507	0.131018	-0.035812
1.35	0.486372	0.572693	1.36308	0.774594	-0.084489	-0.266187
1.40	-0.166283	-0.072549	0.904480	0.482691	-0.307708	-0.500764
1.45	-0.787482	-0.688010	0.450781	0.187478	-0.532292	-0.734058
1.50	-1.36582	-1.26230	0.012035	-0.103080	-0.751945	-0.960615
1.55	-1.89104	-1.78516	-0.402344	-0.381361	-0.960575	-1.17514
1.60	-2.35424	-2.24765	-0.783777	-0.640250	-1.15244	-1.37262
1.65	-2.74804	-2.64233	-1.12469	-0.873299	-1.32230	-1.54847
1.70	-3.06666	-2.96333	-1.41864	-1.07486	-1.46549	-1.69858
1.75	-3.30603	-3.20649	-1.66047	-1.24021	-1.57811	-1.81949
1.80	-3.46386	-3.36938	-1.84633	-1.36564	-1.65703	-1.90841
1.85	-3.53961	-3.45132	-1.97379	-1.44850	-1.69999	-1.96330
1.90	-3.53448	-3.45336	-2.04182	-1.48728	-1.70566	-1.98293
1.95	-3.45134	-3.37821	-2.05080	-1.48158	-1.67362	-1.96687
2.00	-3.29466	-3.23015	-2.00247	-1.43214	-1.60442	-1.91551
2.05	-3.07033	-3.01492	-1.89985	-1.34076	-1.49948	-1.83007
2.10	-2.78554	-2.73951	-1.74716	-1.21025	-1.36111	-1.71252
2.15	-2.44857	-2.41204	-1.54966	-1.04437	-1.19241	-1.56556
2.20	-2.06863	-2.04154	-1.31356	-0.847663	-0.997207	-1.39253
2.25	-1.65558	-1.63773	-1.04579	-0.625400	-0.779925	-1.19733
2.30	-1.21975	-1.21077	-0.753872	-0.383389	-0.545502	-0.984353
2.35	-0.771686	-0.771071	-0.445722	-0.127850	-0.299252	-0.758329
2.40	-0.321911	-0.329049	-0.129442	0.134749	-0.046736	-0.524245
2.45	0.119293	0.105122	0.186855	0.397851	0.206373	-0.287214
2.50	0.542109	0.521713	0.495243	0.654969	0.454429	-0.052354
2.55	0.937393	0.911646	0.788163	0.899839	0.691947	0.175328
2.60	1.29686	1.26668	1.05859	1.12657	0.913729	0.391055
2.65	1.61324	1.57958	1.30018	1.32979	1.114990	0.590386
2.70	1.88044	1.84425	1.50742	1.50473	1.29146	0.769321
2.75	2.09361	2.05585	1.67574	1.64739	1.43949	0.924384
2.80	2.24927	2.21085	1.80158	1.75456	1.55612	1.05270
2.85	2.34532	2.30712	1.88246	1.82395	1.63916	1.15206
2.90	2.38110	2.34393	1.91706	1.85418	1.68720	1.22095
2.95	2.35732	2.32193	1.90517	1.84481	1.69968	1.25861
3.00	2.27605	2.24310	1.84770	1.79638	1.67685	1.26498
3.05	2.14064	2.11069	1.74666	1.71033	1.61980	1.24078
3.10	1.95559	1.92912	1.60505	1.58898	1.53037	1.18739
3.15	1.72647	1.70384	1.42680	1.43546	1.41115	1.10689
3.20	1.45973	1.44120	1.21667	1.25364	1.26537	1.00194
3.25	1.16255	1.14827	0.980074	1.04799	1.09684	0.875738
3.30	0.842655	0.832682	0.723007	0.823504	0.909868	0.731944
3.35	0.508138	0.502421	0.451845	0.585560	0.709095	0.574561

Table 2.9 (continued)

				t		
v	0.95	1.0	2.0	3.0	4.0	5.0
3.40	0.167255	0.165650	0.173207	0.339787	0.499440	0.407854
3.45	−0.171762	−0.169483	−0.106211	0.091925	0.285950	0.236237
3.50	−0.500896	−0.495040	−0.379792	−0.152305	0.073689	0.064168
3.55	−0.812529	−0.803466	−0.641158	−0.387342	−0.132387	−0.103960
3.60	−1.09960	−1.08776	−0.884311	−0.607913	−0.327542	−0.263923
3.65	−1.35578	−1.34162	−1.10378	−0.809154	−0.507373	−0.411765
3.70	−1.57556	−1.55960	−1.29473	−0.986723	−0.667907	−0.543892
3.75	−1.75442	−1.73717	−1.45307	−1.13689	−0.805692	−0.657150
3.80	−1.88888	−1.87087	−1.57555	−1.25663	−0.917871	−0.748900
3.85	−1.97657	−1.95831	−1.65982	−1.34366	−1.00225	−0.817073
3.90	−2.01625	−1.99824	−1.70445	−1.39653	−1.05731	−0.860211
3.95	−2.00785	−1.99057	−1.70900	−1.41458	−1.08230	−0.877498
4.00	−1.95242	−1.93630	−1.67395	−1.39801	−1.07717	−0.868771
4.05	−1.85212	−1.83755	−1.60075	−1.34784	−1.04260	−0.834516
4.10	−1.71010	−1.69741	−1.49169	−1.26585	−0.979994	−0.775845
4.15	−1.53045	−1.51992	−1.34992	−1.15457	−0.891390	−0.694469
4.20	−1.31808	−1.30992	−1.17929	−1.01718	−0.779439	−0.592641
4.25	−1.07858	−1.07293	−0.984294	−0.857425	−0.647317	−0.473102
4.30	−0.818078	−0.815036	−0.769921	−0.679555	−0.498650	−0.339003
4.35	−0.543108	−0.542682	−0.541563	−0.488174	−0.337414	−0.193827
4.40	−0.260421	−0.262566	−0.304859	−0.288145	−0.167841	−0.041301
4.45	0.023159	0.018550	−0.065566	−0.084471	0.005691	0.114701
4.50	0.300900	0.293991	0.170583	0.117830	0.178763	0.270251
4.55	0.566323	0.557325	0.398011	0.313845	0.347021	0.421465
4.60	0.813345	0.802514	0.611430	0.498879	0.506285	0.564597
4.65	1.03642	1.02404	0.805964	0.668568	0.652643	0.696129
4.70	1.23065	1.21706	0.977255	0.818977	0.782546	0.812856
4.75	1.39192	1.37744	1.12156	0.946691	0.892889	0.911957
4.80	1.51695	1.50192	1.23584	1.04889	0.981079	0.991067
4.85	1.60335	1.58814	1.31782	1.12340	1.04509	1.04833
4.90	1.64972	1.63467	1.36600	1.16875	1.08353	1.08242
4.95	1.65562	1.64106	1.37975	1.18421	1.09561	1.09260
5.00	1.62156	1.60780	1.35925	1.16975	1.08123	1.07873

2.4 First, Second and Third Derivatives with Respect to the Order of the Weber Functions

Derivatives with respect to the order of the Weber functions were determined using MATHEMATICA program by applying the following integral representations:

$$\frac{\partial E_v(z)}{\partial v} = -\frac{1}{\pi} \int_0^{\pi} t \cos(vt - z \sin t)\, dt \qquad (2.4.1)$$

$$\frac{\partial^2 E_v(z)}{\partial v^2} = -\frac{1}{\pi} \int_0^{\pi} t^2 \sin(vt - z \sin t)\, dt \qquad (2.4.2)$$

$$\frac{\partial^3 E_v(z)}{\partial v^3} = \frac{1}{\pi} \int_0^{\pi} t^3 \cos(vt - z \sin t)\, dt \qquad (2.4.3)$$

Oscillatory form of these derivatives is similar to that observed in the case of Anger functions.

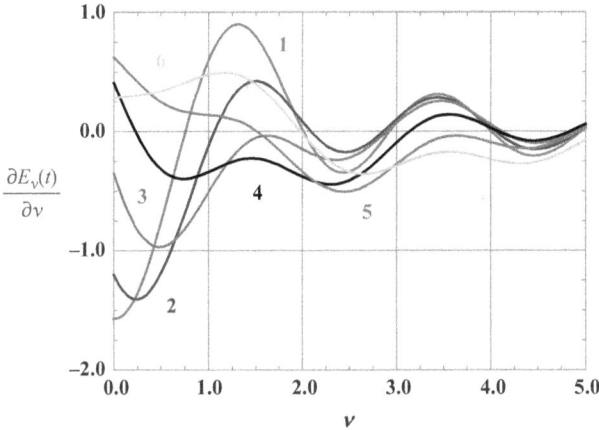

Figure 2.28: First derivatives of the Weber function with respect to the order v as a function of v, at constant values of argument t.

$1 - t = 0.05$; $2 - t = 1.0$; $3 - t = 2.0$; $4 - t = 3.0$; $5 - t = 4.0$; $6 - t = 5.0$.

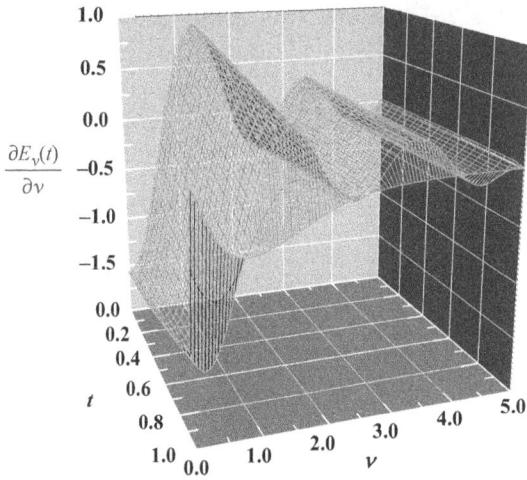

Figure 2.29: First derivatives with respect to the order of the Weber functions $\partial E_\nu(t)/\partial \nu$ as a function of ν and t.

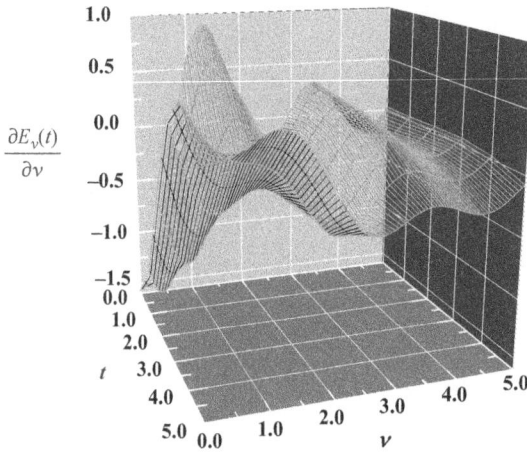

Figure 2.30: First derivatives with respect to the order of the Weber functions $\partial E_\nu(t)/\partial \nu$ as a function of ν and t.

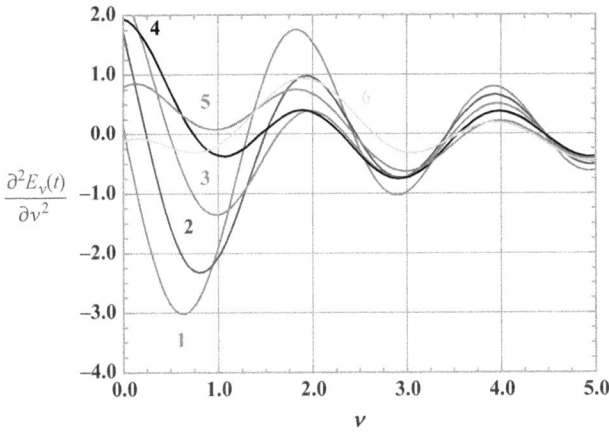

Figure 2.31: Second derivatives of the Weber function with respect to the order v as a function of v, at constant values of argument t.

$1 - t = 0.05$; $2 - t = 1.0$; $3 - t = 2.0$; $4 - t = 3.0$; $5 - t = 4.0$; $6 - t = 5.0$.

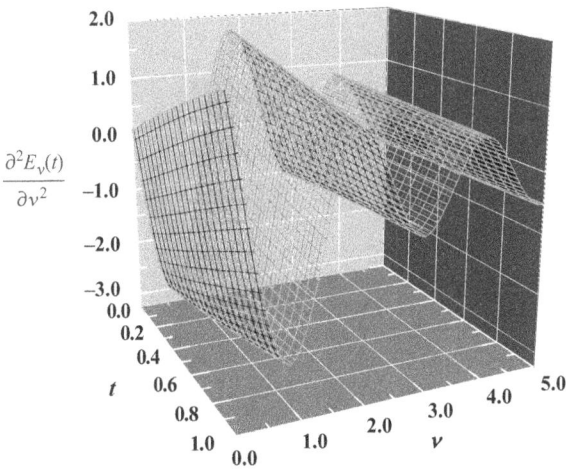

Figure 2.32: Second derivatives with respect to the order of the Weber functions $\partial^2 E_v(t)/\partial v^2$ as a function of v and t.

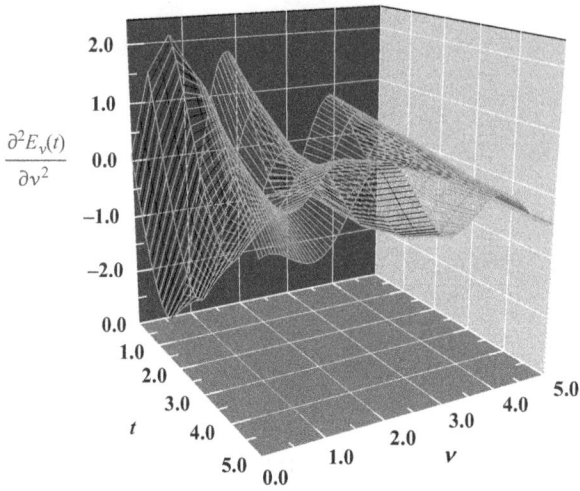

Figure 2.33: Second derivatives with respect to the order of the Weber functions $\partial^2 E_v(t)/\partial v^2$ as a function of v and t.

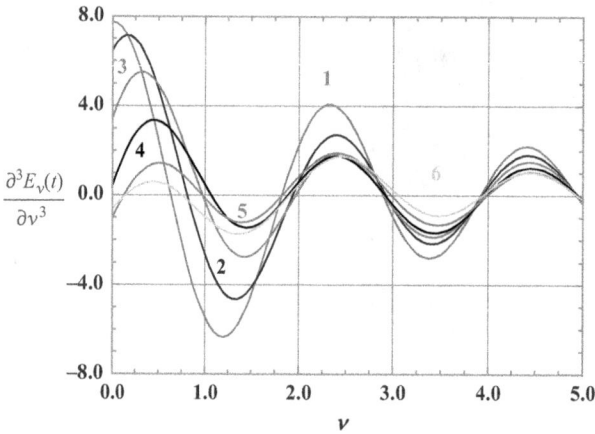

Figure 2.34: Third derivatives of the Weber function with respect to the order v as a function of v, at constant values of argument t.

$1 - t = 0.05;\ 2 - t = 1.0;\ 3 - t = 2.0;\ 4 - t = 3.0;\ 5 - t = 4.0;\ 6 - t = 5.0.$

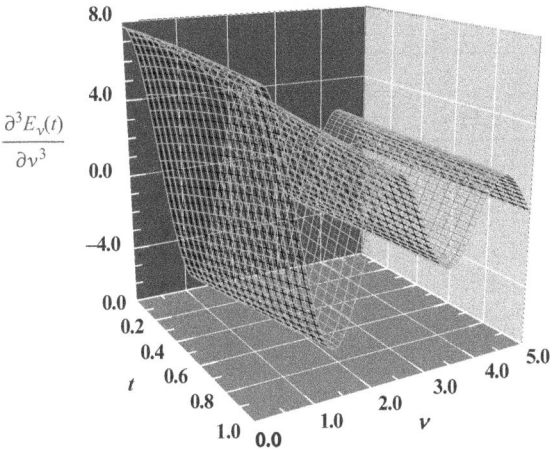

$$\frac{\partial^3 E_\nu(t)}{\partial \nu^3}$$

Figure 2.35: Third derivatives with respect to the order of the Weber functions $\partial^3 E_\nu(t)/\partial \nu^3$ as a function of v and t.

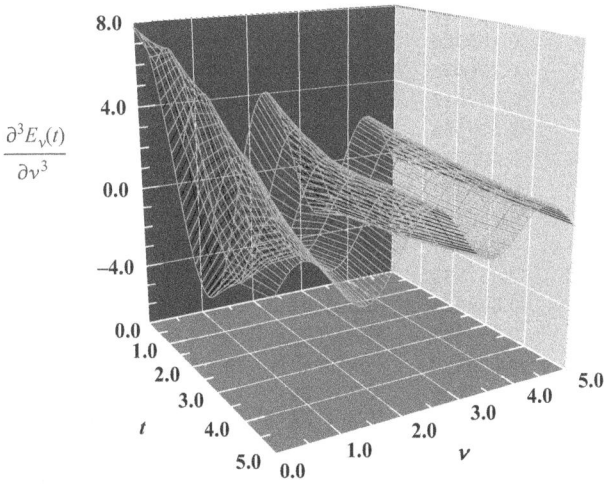

$$\frac{\partial^3 E_\nu(t)}{\partial \nu^3}$$

Figure 2.36: Third derivatives with respect to the order of the Weber functions $\partial^3 E_\nu(t)/\partial \nu^3$ as a function of v and t.

Table 2.10: First derivatives with respect to the order of the Weber functions $\partial E_\nu(t)/\partial \nu$.

			t			
ν	0.05	0.10	0.15	0.20	0.25	0.30
0.00	−1.56981	−1.56687	−1.56197	−1.55513	−1.54635	−1.53565
0.05	−1.56480	−1.56653	−1.56629	−1.56409	−1.55994	−1.55384
0.10	−1.54055	−1.54691	−1.55133	−1.55378	−1.55428	−1.55282
0.15	−1.49749	−1.50840	−1.51740	−1.52446	−1.52957	−1.53274
0.20	−1.43633	−1.45167	−1.46513	−1.47669	−1.48634	−1.49406
0.25	−1.35811	−1.37769	−1.39545	−1.41136	−1.42541	−1.43757
0.30	−1.26411	−1.28770	−1.30955	−1.32962	−1.34788	−1.36431
0.35	−1.15590	−1.18322	−1.20889	−1.23287	−1.25512	−1.27560
0.40	−1.03523	−1.06598	−1.09517	−1.12277	−1.14872	−1.17300
0.45	−0.904066	−0.937885	−0.970265	−1.00116	−1.03051	−1.05828
0.50	−0.764525	−0.801034	−0.836230	−0.870056	−0.902462	−0.933396
0.55	−0.618832	−0.657625	−0.695239	−0.731615	−0.766695	−0.800425
0.60	−0.469289	−0.509940	−0.549554	−0.588070	−0.625424	−0.661558
0.65	−0.318233	−0.360301	−0.401482	−0.441710	−0.480919	−0.519047
0.70	−0.167991	−0.211027	−0.253332	−0.294834	−0.335466	−0.375161
0.75	−0.020840	−0.064394	−0.107374	−0.149708	−0.191324	−0.232153
0.80	0.121034	0.077408	0.034197	−0.008524	−0.050684	−0.092210
0.85	0.255569	0.212309	0.169306	0.126633	0.084365	0.042576
0.90	0.380870	0.338399	0.296026	0.253827	0.211876	0.170248
0.95	0.495236	0.453955	0.412617	0.371299	0.330075	0.289019
1.00	0.597185	0.557471	0.517551	0.477499	0.437390	0.397298
1.05	0.685485	0.647685	0.609535	0.571108	0.532478	0.493717
1.10	0.759164	0.723591	0.687532	0.651058	0.614240	0.577150
1.15	0.817529	0.784460	0.750777	0.716549	0.681844	0.646732
1.20	0.860170	0.829841	0.798781	0.767054	0.734727	0.701867
1.25	0.886963	0.859571	0.831339	0.802329	0.772605	0.742231
1.30	0.898067	0.873764	0.848524	0.822406	0.795470	0.767776
1.35	0.893918	0.872810	0.850683	0.827589	0.803583	0.778724
1.40	0.875207	0.857361	0.838422	0.818440	0.797466	0.775554
1.45	0.842872	0.828307	0.812590	0.795766	0.777882	0.758987
1.50	0.798066	0.786760	0.774257	0.760596	0.745818	0.729968
1.55	0.742137	0.734028	0.724689	0.714153	0.702458	0.689642
1.60	0.676592	0.671580	0.665317	0.657832	0.649157	0.639326
1.65	0.603071	0.601019	0.597709	0.593164	0.587411	0.580478
1.70	0.523305	0.524046	0.523533	0.521784	0.518821	0.514666
1.75	0.439089	0.442426	0.444524	0.445398	0.445063	0.443537
1.80	0.352238	0.357949	0.362449	0.365746	0.367851	0.368776
1.85	0.264556	0.272400	0.279070	0.284569	0.288903	0.292080
1.90	0.177799	0.187518	0.196110	0.203573	0.209907	0.215116
1.95	0.093643	0.104969	0.115222	0.124396	0.132487	0.139494
2.00	0.013654	0.026310	0.037955	0.048580	0.058175	0.066734
2.05	−0.060744	−0.047036	−0.034270	−0.022461	−0.011622	−0.001761
2.10	−0.128288	−0.113805	−0.100190	−0.087463	−0.075638	−0.064731

Table 2.10 (continued)

			t			
v	0.05	0.10	0.15	0.20	0.25	0.30
2.15	-0.187899	-0.172912	-0.158716	-0.145332	-0.132781	-0.121079
2.20	-0.238702	-0.223472	-0.208952	-0.195167	-0.182139	-0.169890
2.25	-0.280031	-0.264806	-0.250209	-0.236266	-0.223003	-0.210445
2.30	-0.311442	-0.296454	-0.282009	-0.268138	-0.254868	-0.242226
2.35	-0.332714	-0.318174	-0.304094	-0.290506	-0.277441	-0.264925
2.40	-0.343844	-0.329943	-0.316420	-0.303309	-0.290639	-0.278443
2.45	-0.345047	-0.331952	-0.319155	-0.306689	-0.294589	-0.282885
2.50	-0.336739	-0.324592	-0.312665	-0.300993	-0.289611	-0.278550
2.55	-0.319525	-0.308442	-0.297505	-0.286750	-0.276211	-0.265922
2.60	-0.294182	-0.284251	-0.274398	-0.264657	-0.255064	-0.245651
2.65	-0.261633	-0.252919	-0.244217	-0.235563	-0.226992	-0.218536
2.70	-0.222928	-0.215467	-0.207960	-0.200440	-0.192943	-0.185501
2.75	-0.179215	-0.173020	-0.166725	-0.160363	-0.153967	-0.147572
2.80	-0.131714	-0.126773	-0.121683	-0.116479	-0.111191	-0.105853
2.85	-0.081687	-0.077965	-0.074053	-0.069985	-0.065789	-0.061497
2.90	-0.030408	-0.027851	-0.025071	-0.022095	-0.018955	-0.015680
2.95	0.020859	0.022323	0.024039	0.025981	0.028119	0.030426
3.00	0.070894	0.071353	0.072087	0.073071	0.074278	0.075681
3.05	0.118544	0.118099	0.117946	0.118061	0.118419	0.118996
3.10	0.162746	0.161509	0.160574	0.159920	0.159524	0.159363
3.15	0.202549	0.200638	0.199035	0.197720	0.196672	0.195872
3.20	0.237131	0.234670	0.232517	0.230654	0.229065	0.227728
3.25	0.265814	0.262929	0.260348	0.258055	0.256034	0.254269
3.30	0.288077	0.284893	0.282005	0.279399	0.277060	0.274975
3.35	0.303561	0.300202	0.297125	0.294320	0.291775	0.289477
3.40	0.312074	0.308656	0.305506	0.302613	0.299969	0.297563
3.45	0.313592	0.310225	0.307108	0.304233	0.301592	0.299176
3.50	0.308252	0.305038	0.302052	0.299291	0.296747	0.294414
3.55	0.296348	0.293376	0.290612	0.288051	0.285690	0.283523
3.60	0.278318	0.275668	0.273202	0.270917	0.268813	0.266885
3.65	0.254733	0.252470	0.250366	0.248423	0.246640	0.245013
3.70	0.226278	0.224453	0.222764	0.221214	0.219802	0.218528
3.75	0.193733	0.192385	0.191150	0.190032	0.189030	0.188147
3.80	0.157957	0.157110	0.156354	0.155693	0.155128	0.154662
3.85	0.119861	0.119527	0.119261	0.119070	0.118956	0.118922
3.90	0.080391	0.080567	0.080791	0.081070	0.081407	0.081807
3.95	0.040501	0.041172	0.041873	0.042611	0.043390	0.044214
4.00	0.001133	0.002273	0.003428	0.004602	0.005801	0.007031
4.05	-0.036805	-0.035231	-0.033659	-0.032080	-0.030490	-0.028884
4.10	-0.072460	-0.070497	-0.068548	-0.066606	-0.064666	-0.062720
4.15	-0.105050	-0.102751	-0.100476	-0.098219	-0.095973	-0.093732
4.20	-0.133885	-0.131307	-0.128762	-0.126242	-0.123743	-0.121258

Table 2.10 (continued)

			t			
v	0.05	0.10	0.15	0.20	0.25	0.30
4.25	−0.158376	−0.155582	−0.152826	−0.150102	−0.147405	−0.144729
4.30	−0.178051	−0.175104	−0.172200	−0.169334	−0.166498	−0.163688
4.35	−0.192560	−0.189526	−0.186538	−0.183590	−0.180676	−0.177790
4.40	−0.201682	−0.198627	−0.195618	−0.192650	−0.189716	−0.186812
4.45	−0.205330	−0.202316	−0.199348	−0.196418	−0.193524	−0.190659
4.50	−0.203544	−0.200632	−0.197761	−0.194928	−0.192128	−0.189355
4.55	−0.196495	−0.193738	−0.191020	−0.188335	−0.185679	−0.183050
4.60	−0.184472	−0.181920	−0.179401	−0.176912	−0.174448	−0.172006
4.65	−0.167876	−0.165572	−0.163295	−0.161041	−0.158809	−0.156593
4.70	−0.147208	−0.145187	−0.143186	−0.141203	−0.139235	−0.137279
4.75	−0.123055	−0.121344	−0.119647	−0.117961	−0.116283	−0.114612
4.80	−0.096074	−0.094693	−0.093318	−0.091948	−0.090580	−0.089213
4.85	−0.066977	−0.065937	−0.064896	−0.063852	−0.062805	−0.061752
4.90	−0.036510	−0.035814	−0.035109	−0.034396	−0.033673	−0.032938
4.95	−0.005438	−0.005081	−0.004709	−0.004321	−0.003918	−0.003497
5.00	0.025475	0.025506	0.025558	0.025631	0.025725	0.025843

			t			
v	0.35	0.40	0.45	0.50	0.55	0.60
0.00	−1.52306	−1.50859	−1.49228	−1.47414	−1.45423	−1.43257
0.05	−1.54581	−1.53587	−1.52403	−1.51032	−1.49476	−1.47740
0.10	−1.54941	−1.54406	−1.53678	−1.52760	−1.51652	−1.50358
0.15	−1.53395	−1.53322	−1.53054	−1.52593	−1.51940	−1.51097
0.20	−1.49985	−1.50370	−1.50560	−1.50557	−1.50361	−1.49972
0.25	−1.44783	−1.45617	−1.46259	−1.46708	−1.46964	−1.47028
0.30	−1.37888	−1.39158	−1.40240	−1.41131	−1.41831	−1.42340
0.35	−1.29429	−1.31116	−1.32620	−1.33938	−1.35068	−1.36010
0.40	−1.19556	−1.21638	−1.23543	−1.25267	−1.26810	−1.28168
0.45	−1.08444	−1.10893	−1.13172	−1.15279	−1.17211	−1.18964
0.50	−0.962813	−0.990667	−1.01692	−1.04153	−1.06447	−1.08570
0.55	−0.832752	−0.863627	−0.893005	−0.920842	−0.947097	−0.971734
0.60	−0.696415	−0.729941	−0.762085	−0.792799	−0.822035	−0.849753
0.65	−0.556031	−0.591815	−0.626342	−0.659558	−0.691413	−0.721859
0.70	−0.413855	−0.451486	−0.487993	−0.523318	−0.557407	−0.590207
0.75	−0.272127	−0.311182	−0.349254	−0.386281	−0.422203	−0.456965
0.80	−0.133033	−0.173085	−0.212298	−0.250609	−0.287954	−0.324274
0.85	0.001335	−0.039285	−0.079216	−0.118390	−0.156743	−0.194210
0.90	0.129015	0.088249	0.048021	0.008400	−0.030545	−0.068749
0.95	0.248204	0.207704	0.167590	0.127933	0.088804	0.050269
1.00	0.357295	0.317456	0.277853	0.238557	0.199640	0.161170
1.05	0.454898	0.416095	0.377379	0.338823	0.300498	0.262475
1.10	0.539860	0.502441	0.464966	0.427507	0.390133	0.352918

Table 2.10 (continued)

			t			
v	0.35	0.40	0.45	0.50	0.55	0.60
1.15	0.611285	0.575571	0.539661	0.503628	0.467540	0.431469
1.20	0.668541	0.634819	0.600769	0.566461	0.531965	0.497349
1.25	0.711271	0.679792	0.647861	0.615545	0.582912	0.550031
1.30	0.739387	0.710365	0.680775	0.650682	0.620151	0.589249
1.35	0.753069	0.726679	0.699614	0.671937	0.643710	0.614998
1.40	0.752757	0.729132	0.704736	0.679629	0.653870	0.627520
1.45	0.739129	0.718363	0.696741	0.674318	0.651151	0.627298
1.50	0.713091	0.695236	0.676451	0.656788	0.636299	0.615039
1.55	0.675747	0.660815	0.644892	0.628025	0.610261	0.591652
1.60	0.628374	0.616341	0.603266	0.589192	0.574164	0.558227
1.65	0.572396	0.563199	0.552923	0.541604	0.529284	0.516004
1.70	0.509347	0.502891	0.495329	0.486694	0.477021	0.466349
1.75	0.440841	0.436999	0.432036	0.425981	0.418864	0.410718
1.80	0.368539	0.367156	0.364649	0.361042	0.356359	0.350629
1.85	0.294111	0.295009	0.294790	0.293472	0.291077	0.287627
1.90	0.219206	0.222185	0.224064	0.224858	0.224581	0.223253
1.95	0.145418	0.150262	0.154034	0.156741	0.158396	0.159011
2.00	0.074255	0.080736	0.086180	0.090591	0.093974	0.096341
2.05	0.007112	0.014994	0.021881	0.027774	0.032674	0.036587
2.10	-0.054752	-0.045711	-0.037615	-0.030467	-0.024270	-0.019024
2.15	-0.110242	-0.100284	-0.091213	-0.083040	-0.075769	-0.069405
2.20	-0.158438	-0.147800	-0.137989	-0.129017	-0.120895	-0.113629
2.25	-0.198612	-0.187525	-0.177200	-0.167653	-0.158897	-0.150943
2.30	-0.230236	-0.218920	-0.208299	-0.198391	-0.189212	-0.180777
2.35	-0.252987	-0.241650	-0.230939	-0.220873	-0.211472	-0.202753
2.40	-0.266749	-0.255584	-0.244972	-0.234937	-0.225501	-0.216683
2.45	-0.271608	-0.260787	-0.250448	-0.240617	-0.231318	-0.222573
2.50	-0.267843	-0.257520	-0.247608	-0.238136	-0.229128	-0.220609
2.55	-0.255916	-0.246222	-0.236872	-0.227893	-0.219313	-0.211156
2.60	-0.236453	-0.227499	-0.218823	-0.210451	-0.202414	-0.194737
2.65	-0.210229	-0.202104	-0.194190	-0.186520	-0.179120	-0.172020
2.70	-0.178147	-0.170913	-0.163832	-0.156933	-0.150245	-0.143798
2.75	-0.141209	-0.134910	-0.128706	-0.122628	-0.116705	-0.110967
2.80	-0.100496	-0.095152	-0.089850	-0.084622	-0.079497	-0.074504
2.85	-0.057141	-0.052750	-0.048356	-0.043986	-0.039672	-0.035441
2.90	-0.012300	-0.008843	-0.005340	-0.001820	0.001690	0.005160
2.95	0.032874	0.035433	0.038077	0.040776	0.043504	0.046232
3.00	0.077254	0.078970	0.080802	0.082723	0.084706	0.086726
3.05	0.119767	0.120706	0.121788	0.122988	0.124279	0.125637
3.10	0.159415	0.159656	0.160062	0.160608	0.161271	0.162025
3.15	0.195298	0.194927	0.194739	0.194709	0.194816	0.195036
3.20	0.226626	0.225739	0.225045	0.224524	0.224155	0.223916

Table 2.10 (continued)

			t			
v	0.35	0.40	0.45	0.50	0.55	0.60
3.25	0.252742	0.251436	0.250331	0.249408	0.248648	0.248032
3.30	0.273128	0.271503	0.270083	0.268850	0.267787	0.266876
3.35	0.287414	0.285570	0.283930	0.282480	0.281203	0.280082
3.40	0.295384	0.293418	0.291654	0.290077	0.288672	0.287426
3.45	0.296977	0.294983	0.293184	0.291567	0.290119	0.288828
3.50	0.292285	0.290349	0.288599	0.287024	0.285612	0.284352
3.55	0.281544	0.279746	0.278123	0.276664	0.275360	0.274202
3.60	0.265129	0.263540	0.262111	0.260835	0.259705	0.258711
3.65	0.243541	0.242220	0.241045	0.240009	0.239108	0.238334
3.70	0.217391	0.216387	0.215513	0.214766	0.214141	0.213631
3.75	0.187382	0.186734	0.186200	0.185778	0.185464	0.185255
3.80	0.154296	0.154030	0.153862	0.153791	0.153815	0.153931
3.85	0.118969	0.119099	0.119313	0.119608	0.119985	0.120441
3.90	0.082271	0.082802	0.083401	0.084068	0.084802	0.085604
3.95	0.045087	0.046012	0.046989	0.048022	0.049109	0.050252
4.00	0.008294	0.009595	0.010936	0.012319	0.013745	0.015215
4.05	−0.027258	−0.025606	−0.023927	−0.022218	−0.020476	−0.018701
4.10	−0.060766	−0.058799	−0.056814	−0.054810	−0.052783	−0.050732
4.15	−0.091493	−0.089250	−0.087000	−0.084739	−0.082464	−0.080173
4.20	−0.118782	−0.116310	−0.113839	−0.111364	−0.108883	−0.106393
4.25	−0.142069	−0.139419	−0.136777	−0.134137	−0.131496	−0.128852
4.30	−0.160898	−0.158124	−0.155362	−0.152607	−0.149856	−0.147106
4.35	−0.174927	−0.172084	−0.169255	−0.166437	−0.163626	−0.160820
4.40	−0.183934	−0.181076	−0.178234	−0.175406	−0.172586	−0.169772
4.45	−0.187819	−0.185000	−0.182199	−0.179410	−0.176631	−0.173859
4.50	−0.186607	−0.183879	−0.181167	−0.178467	−0.175777	−0.173093
4.55	−0.180442	−0.177851	−0.175275	−0.172711	−0.170154	−0.167602
4.60	−0.169581	−0.167172	−0.164775	−0.162386	−0.160002	−0.157621
4.65	−0.154391	−0.152201	−0.150019	−0.147841	−0.145666	−0.143492
4.70	−0.135332	−0.133391	−0.131455	−0.129520	−0.127583	−0.125644
4.75	−0.112945	−0.111280	−0.109614	−0.107945	−0.106270	−0.104589
4.80	−0.087844	−0.086472	−0.085093	−0.083708	−0.082312	−0.080906
4.85	−0.060692	−0.059623	−0.058544	−0.057452	−0.056346	−0.055225
4.90	−0.032191	−0.031429	−0.030652	−0.029857	−0.029044	−0.028212
4.95	−0.003059	−0.002601	−0.002122	−0.001622	−0.001099	−0.000553
5.00	0.025983	0.026147	0.026336	0.026552	0.026793	0.027063

			t			
v	0.65	0.70	0.75	0.80	0.85	0.90
0.00	−1.40921	−1.38419	−1.35755	−1.32935	−1.29963	−1.26846
0.05	−1.45825	−1.43736	−1.41477	−1.39052	−1.36465	−1.33723
0.10	−1.48880	−1.47221	−1.45385	−1.43375	−1.41195	−1.38851

Table 2.10 (continued)

			t			
v	0.65	0.70	0.75	0.80	0.85	0.90
0.15	−1.50066	−1.48849	−1.47449	−1.45869	−1.44112	−1.42182
0.20	−1.49392	−1.48624	−1.47668	−1.46527	−1.45204	−1.43703
0.25	−1.46900	−1.46580	−1.46072	−1.45375	−1.44492	−1.43427
0.30	−1.42657	−1.42784	−1.42720	−1.42467	−1.42026	−1.41398
0.35	−1.36764	−1.37327	−1.37701	−1.37885	−1.37881	−1.37690
0.40	−1.29341	−1.30328	−1.31128	−1.31740	−1.32164	−1.32401
0.45	−1.20538	−1.21930	−1.23138	−1.24162	−1.25001	−1.25655
0.50	−1.10520	−1.12294	−1.13890	−1.15307	−1.16543	−1.17597
0.55	−0.994719	−1.01602	−1.03561	−1.05347	−1.06958	−1.08391
0.60	−0.875912	−0.900476	−0.923412	−0.944689	−0.964282	−0.982166
0.65	−0.750852	−0.778348	−0.804310	−0.828702	−0.851491	−0.872647
0.70	−0.621669	−0.651744	−0.680389	−0.707563	−0.733227	−0.757347
0.75	−0.490512	−0.522792	−0.553755	−0.583357	−0.611553	−0.638304
0.80	−0.359509	−0.393605	−0.426507	−0.458165	−0.488531	−0.517558
0.85	−0.230730	−0.266244	−0.300693	−0.334024	−0.366183	−0.397121
0.90	−0.106148	−0.142677	−0.178278	−0.212891	−0.246460	−0.278932
0.95	0.012396	−0.024751	−0.061107	−0.096611	−0.131204	−0.164829
1.00	0.123217	0.085847	0.049126	0.013117	−0.022116	−0.056514
1.05	0.224822	0.187609	0.150902	0.114768	0.079271	0.044473
1.10	0.315929	0.279237	0.242909	0.207014	0.171616	0.136780
1.15	0.395485	0.359657	0.324053	0.288742	0.253791	0.219265
1.20	0.462683	0.428037	0.393478	0.359076	0.324897	0.291008
1.25	0.516968	0.483793	0.450574	0.417378	0.384273	0.351326
1.30	0.558042	0.526598	0.494982	0.463262	0.431505	0.399776
1.35	0.585864	0.556374	0.526593	0.496587	0.466421	0.436160
1.40	0.600640	0.573295	0.545546	0.517458	0.489094	0.460518
1.45	0.602817	0.577769	0.552214	0.526213	0.499829	0.473123
1.50	0.593063	0.570428	0.547192	0.523412	0.499150	0.474464
1.55	0.572250	0.552107	0.531278	0.509818	0.487784	0.465234
1.60	0.541429	0.523820	0.505451	0.486373	0.466641	0.446307
1.65	0.501807	0.486739	0.470847	0.454179	0.436785	0.418716
1.70	0.454714	0.442160	0.428728	0.414464	0.399413	0.383623
1.75	0.401576	0.391477	0.380457	0.368559	0.355824	0.342295
1.80	0.343881	0.336148	0.327464	0.317864	0.307388	0.296075
1.85	0.283147	0.277666	0.271212	0.263818	0.255518	0.246346
1.90	0.220893	0.217525	0.213174	0.207868	0.201635	0.194507
1.95	0.158603	0.157191	0.154794	0.151436	0.147141	0.141938
2.00	0.097701	0.098069	0.097461	0.095895	0.093392	0.089975
2.05	0.039519	0.041480	0.042481	0.042537	0.041664	0.039881
2.10	−0.014724	−0.011367	−0.008945	−0.007448	−0.006864	−0.007179
2.15	−0.063947	−0.059397	−0.055749	−0.052999	−0.051139	−0.050157
2.20	−0.107224	−0.101684	−0.097010	−0.093198	−0.090245	−0.088146

Table 2.10 (continued)

			t			
v	0.65	0.70	0.75	0.80	0.85	0.90
2.25	−0.143799	−0.137473	−0.131968	−0.127286	−0.123428	−0.120390
2.30	−0.173097	−0.166183	−0.160041	−0.154678	−0.150097	−0.146299
2.35	−0.194731	−0.187419	−0.180827	−0.174966	−0.169841	−0.165456
2.40	−0.208502	−0.200972	−0.194108	−0.187922	−0.182424	−0.177620
2.45	−0.214401	−0.206822	−0.199851	−0.193503	−0.187791	−0.182725
2.50	−0.212601	−0.205124	−0.198198	−0.191838	−0.186061	−0.180879
2.55	−0.203448	−0.196210	−0.189463	−0.183226	−0.177518	−0.172352
2.60	−0.187446	−0.180566	−0.174117	−0.168123	−0.162600	−0.157569
2.65	−0.165245	−0.158822	−0.152773	−0.147121	−0.141888	−0.137092
2.70	−0.137618	−0.131732	−0.126165	−0.120940	−0.116079	−0.111604
2.75	−0.105441	−0.100154	−0.095132	−0.090399	−0.085979	−0.081893
2.80	−0.069672	−0.065026	−0.060594	−0.056400	−0.052469	−0.048824
2.85	−0.031322	−0.027342	−0.023526	−0.019902	−0.016493	−0.013323
2.90	0.008563	0.011872	0.015061	0.018103	0.020974	0.023651
2.95	0.048933	0.051580	0.054149	0.056612	0.058945	0.061125
3.00	0.088754	0.090766	0.092735	0.094636	0.096446	0.098138
3.05	0.127035	0.128449	0.129853	0.131224	0.132536	0.133766
3.10	0.162848	0.163714	0.164599	0.165480	0.166333	0.167134
3.15	0.195347	0.195724	0.196146	0.196589	0.197030	0.197446
3.20	0.223786	0.223742	0.223764	0.223827	0.223912	0.223996
3.25	0.247537	0.247145	0.246834	0.246583	0.246373	0.246181
3.30	0.266097	0.265432	0.264862	0.264366	0.263925	0.263519
3.35	0.279101	0.278240	0.277484	0.276813	0.276209	0.275653
3.40	0.286322	0.285343	0.284475	0.283699	0.283000	0.282358
3.45	0.287679	0.286657	0.285749	0.284938	0.284208	0.283543
3.50	0.283232	0.282238	0.281358	0.280576	0.279880	0.279253
3.55	0.273179	0.272278	0.271489	0.270799	0.270193	0.269660
3.60	0.257846	0.257098	0.256457	0.255912	0.255451	0.255062
3.65	0.237679	0.237135	0.236692	0.236340	0.236069	0.235869
3.70	0.213231	0.212933	0.212729	0.212611	0.212569	0.212595
3.75	0.185144	0.185125	0.185193	0.185340	0.185558	0.185839
3.80	0.154134	0.154420	0.154783	0.155218	0.155717	0.156274
3.85	0.120973	0.121578	0.122251	0.122988	0.123782	0.124627
3.90	0.086470	0.087399	0.088387	0.089430	0.090523	0.091662
3.95	0.051448	0.052697	0.053995	0.055341	0.056730	0.058158
4.00	0.016728	0.018284	0.019881	0.021518	0.023190	0.024896
4.05	−0.016892	−0.015049	−0.013174	−0.011266	−0.009329	−0.007365
4.10	−0.048656	−0.046554	−0.044427	−0.042274	−0.040098	−0.037901
4.15	−0.077864	−0.075537	−0.073191	−0.070826	−0.068443	−0.066043
4.20	−0.103892	−0.101379	−0.098852	−0.096312	−0.093758	−0.091191
4.25	−0.126203	−0.123546	−0.120880	−0.118205	−0.115520	−0.112826
4.30	−0.144354	−0.141599	−0.138839	−0.136073	−0.133300	−0.130520

Table 2.10 (continued)

			t			
v	0.65	0.70	0.75	0.80	0.85	0.90
4.35	−0.158015	−0.155209	−0.152400	−0.149588	−0.146771	−0.143949
4.40	−0.166962	−0.164152	−0.161342	−0.158529	−0.155712	−0.152891
4.45	−0.171091	−0.168325	−0.165558	−0.162789	−0.160016	−0.157240
4.50	−0.170413	−0.167734	−0.165055	−0.162372	−0.159686	−0.156995
4.55	−0.165052	−0.162503	−0.159951	−0.157396	−0.154836	−0.152269
4.60	−0.155241	−0.152859	−0.150473	−0.148081	−0.145682	−0.143276
4.65	−0.141315	−0.139133	−0.136945	−0.134749	−0.132544	−0.130328
4.70	−0.123698	−0.121745	−0.119783	−0.117810	−0.115824	−0.113826
4.75	−0.102898	−0.101196	−0.099482	−0.097753	−0.096009	−0.094248
4.80	−0.079486	−0.078051	−0.076600	−0.075132	−0.073645	−0.072138
4.85	−0.054086	−0.052929	−0.051751	−0.050553	−0.049332	−0.048088
4.90	−0.027358	−0.026481	−0.025581	−0.024657	−0.023706	−0.022729
4.95	0.000019	0.000617	0.001242	0.001895	0.002577	0.003289
5.00	0.027361	0.027689	0.028048	0.028437	0.028859	0.029314

			t			
v	0.95	1.0	2.0	3.0	4.0	5.0
0.00	−1.23588	−1.20197	−0.351687	0.408489	0.623841	0.278968
0.05	−1.30830	−1.27791	−0.468230	0.312247	0.583204	0.284284
0.10	−1.36345	−1.33684	−0.574757	0.218523	0.541071	0.288600
0.15	−1.40084	−1.37821	−0.670045	0.128633	0.498408	0.292447
0.20	−1.42026	−1.40177	−0.753105	0.043751	0.456112	0.296315
0.25	−1.42180	−1.40756	−0.823190	−0.035108	0.414992	0.300637
0.30	−1.40586	−1.39593	−0.879805	−0.107099	0.375753	0.305783
0.35	−1.37312	−1.36750	−0.922706	−0.171562	0.338988	0.312044
0.40	−1.32451	−1.32316	−0.951899	−0.228022	0.305171	0.319632
0.45	−1.26123	−1.26407	−0.967632	−0.276194	0.274646	0.328669
0.50	−1.18468	−1.19157	−0.970382	−0.315977	0.247631	0.339190
0.55	−1.09646	−1.10722	−0.960840	−0.347452	0.224215	0.351142
0.60	−0.998323	−1.01273	−0.939888	−0.370869	0.204368	0.364383
0.65	−0.892145	−0.909962	−0.908579	−0.386634	0.187940	0.378694
0.70	−0.779890	−0.800828	−0.868109	−0.395293	0.174678	0.393780
0.75	−0.663570	−0.687319	−0.819788	−0.397513	0.164235	0.409284
0.80	−0.545206	−0.571434	−0.765012	−0.394063	0.156186	0.424794
0.85	−0.426791	−0.455148	−0.705228	−0.385788	0.150038	0.439858
0.90	−0.310255	−0.340380	−0.641909	−0.373589	0.145254	0.453996
0.95	−0.197430	−0.228955	−0.576520	−0.358399	0.141266	0.466714
1.00	−0.090019	−0.122573	−0.510488	−0.341157	0.137494	0.477517
1.05	0.010435	−0.022783	−0.445175	−0.322789	0.133363	0.485925
1.10	0.102570	0.069047	−0.381855	−0.304184	0.128320	0.491486
1.15	0.185229	0.151746	−0.321684	−0.286171	0.121849	0.493788

Table 2.10 (continued)

				t		
v	0.95	1.0	2.0	3.0	4.0	5.0
1.20	0.257475	0.224361	−0.265688	−0.269508	0.113488	0.492473
1.25	0.318602	0.286166	−0.214743	−0.254860	0.102841	0.487245
1.30	0.368142	0.336669	−0.169561	−0.242788	0.089586	0.477879
1.35	0.405870	0.375617	−0.130683	−0.233739	0.073486	0.464231
1.40	0.431796	0.402990	−0.098474	−0.228039	0.054398	0.446239
1.45	0.446158	0.418998	−0.073121	−0.225889	0.032272	0.423929
1.50	0.449416	0.424065	−0.054637	−0.227363	0.007154	0.397415
1.55	0.442226	0.418818	−0.042869	−0.232412	−0.020814	0.366898
1.60	0.425429	0.404062	−0.037504	−0.240868	−0.051398	0.332662
1.65	0.400024	0.380764	−0.038087	−0.252452	−0.084280	0.295066
1.70	0.367142	0.350022	−0.044034	−0.266786	−0.119063	0.254545
1.75	0.328019	0.313041	−0.054654	−0.283403	−0.155283	0.211590
1.80	0.283966	0.271104	−0.069166	−0.301765	−0.192420	0.166748
1.85	0.236341	0.225541	−0.086722	−0.321277	−0.229911	0.120601
1.90	0.186517	0.177700	−0.106430	−0.341303	−0.267166	0.073763
1.95	0.135854	0.128921	−0.127378	−0.361187	−0.303578	0.026859
2.00	0.085668	0.080498	−0.148657	−0.380267	−0.338541	−0.019485
2.05	0.037207	0.033666	−0.169380	−0.397897	−0.371465	−0.064653
2.10	−0.008376	−0.010437	−0.188707	−0.413459	−0.401789	−0.108058
2.15	−0.050043	−0.050782	−0.205862	−0.426385	−0.428996	−0.149147
2.20	−0.086892	−0.086471	−0.220151	−0.436164	−0.452621	−0.187420
2.25	−0.118169	−0.116757	−0.230979	−0.442363	−0.472270	−0.222431
2.30	−0.143284	−0.141047	−0.237857	−0.444631	−0.487619	−0.253808
2.35	−0.161816	−0.158919	−0.240417	−0.442710	−0.498433	−0.281249
2.40	−0.173517	−0.170119	−0.238416	−0.436441	−0.504560	−0.304536
2.45	−0.178314	−0.174565	−0.231738	−0.425769	−0.505943	−0.323532
2.50	−0.176303	−0.172345	−0.220396	−0.410742	−0.502617	−0.338187
2.55	−0.167743	−0.163703	−0.204527	−0.391508	−0.494707	−0.348537
2.60	−0.153043	−0.149038	−0.184390	−0.368315	−0.482428	−0.354697
2.65	−0.132751	−0.128882	−0.160352	−0.341502	−0.466079	−0.356865
2.70	−0.107534	−0.103887	−0.132879	−0.311490	−0.446033	−0.355308
2.75	−0.078163	−0.074807	−0.102526	−0.278774	−0.422731	−0.350360
2.80	−0.045486	−0.042476	−0.069916	−0.243907	−0.396671	−0.342411
2.85	−0.010414	−0.007789	−0.035728	−0.207490	−0.368395	−0.331895
2.90	0.026109	0.028328	−0.000676	−0.170158	−0.338479	−0.319285
2.95	0.063127	0.064929	0.034504	−0.132562	−0.307517	−0.305073
3.00	0.099692	0.101083	0.069079	−0.095354	−0.276111	−0.289768
3.05	0.134890	0.135887	0.102331	−0.059176	−0.244853	−0.273874
3.10	0.167861	0.168492	0.133577	−0.024640	−0.214317	−0.257885
3.15	0.197817	0.198118	0.162187	0.007683	−0.185043	−0.242273
3.20	0.224057	0.224075	0.187593	0.037274	−0.157526	−0.227473
3.25	0.245988	0.245772	0.209306	0.063684	−0.132208	−0.213880

Table 2.10 (continued)

v	0.95	1.0	2.0	3.0	4.0	5.0
3.30	0.263129	0.262735	0.226927	0.086535	−0.109464	−0.201834
3.35	0.275127	0.274612	0.240154	0.105534	−0.089602	−0.191618
3.40	0.281757	0.281180	0.248787	0.120475	−0.072850	−0.183451
3.45	0.282928	0.282345	0.252733	0.131247	−0.059357	−0.177483
3.50	0.278680	0.278147	0.252007	0.137828	−0.049190	−0.173795
3.55	0.269184	0.268753	0.246730	0.140293	−0.042334	−0.172397
3.60	0.254732	0.254448	0.237123	0.138804	−0.038694	−0.173230
3.65	0.235727	0.235632	0.223504	0.133608	−0.038100	−0.176169
3.70	0.212677	0.212805	0.206276	0.125028	−0.040311	−0.181026
3.75	0.186173	0.186551	0.185917	0.113458	−0.045023	−0.187562
3.80	0.156880	0.157527	0.162970	0.099349	−0.051878	−0.195484
3.85	0.125518	0.126445	0.138026	0.083199	−0.060475	−0.204463
3.90	0.092840	0.094051	0.111711	0.065539	−0.070377	−0.214138
3.95	0.059620	0.061111	0.084672	0.046924	−0.081126	−0.224129
4.00	0.026630	0.028389	0.057559	0.027914	−0.092254	−0.234044
4.05	−0.005377	−0.003369	0.031012	0.009066	−0.103293	−0.243491
4.10	−0.035684	−0.033452	0.005645	−0.009083	−0.113788	−0.252089
4.15	−0.063629	−0.061201	−0.017968	−0.026026	−0.123309	−0.259475
4.20	−0.088614	−0.086026	−0.039307	−0.041300	−0.131458	−0.265319
4.25	−0.110123	−0.107414	−0.057914	−0.054493	−0.137883	−0.269324
4.30	−0.127734	−0.124943	−0.073407	−0.065257	−0.142281	−0.271243
4.35	−0.141121	−0.138290	−0.085484	−0.073311	−0.144412	−0.270876
4.40	−0.150066	−0.147237	−0.093934	−0.078449	−0.144095	−0.268082
4.45	−0.154459	−0.151673	−0.098633	−0.080545	−0.141219	−0.262778
4.50	−0.154299	−0.151597	−0.099554	−0.079552	−0.135743	−0.254942
4.55	−0.149696	−0.147115	−0.096757	−0.075503	−0.127695	−0.244617
4.60	−0.140860	−0.138435	−0.090396	−0.068510	−0.117171	−0.231904
4.65	−0.128100	−0.125861	−0.080706	−0.058758	−0.104332	−0.216962
4.70	−0.111813	−0.109785	−0.067999	−0.046499	−0.089400	−0.200005
4.75	−0.092470	−0.090674	−0.052656	−0.032050	−0.072651	−0.181295
4.80	−0.070610	−0.069061	−0.035117	−0.015778	−0.054409	−0.161137
4.85	−0.046821	−0.045528	−0.015866	0.001908	−0.035034	−0.139870
4.90	−0.021725	−0.020694	0.004577	0.020566	−0.014918	−0.117860
4.95	0.004032	0.004805	0.025672	0.039731	0.005529	−0.095489
5.00	0.029802	0.030324	0.046869	0.058932	0.025889	−0.073150

Table 2.11: Second derivatives with respect to the order of the Weber functions $\partial^2 E_\nu(t)/\partial \nu^2$.

ν	t					
	0.05	0.10	0.15	0.20	0.25	0.30
0.00	0.093394	0.186644	0.279606	0.372138	0.464097	0.555343
0.05	−0.293491	−0.200272	−0.107003	−0.013829	0.079107	0.171662
0.10	−0.675095	−0.582976	−0.490471	−0.397724	−0.304879	−0.212080
0.15	−1.045170	−0.955202	−0.864517	−0.773258	−0.681567	−0.589586
0.20	−1.397680	−1.310890	−1.223050	−1.134310	−1.044820	−0.954703
0.25	−1.726930	−1.644280	−1.560280	−1.475060	−1.388770	−1.301530
0.30	−2.027650	−1.950050	−1.870810	−1.790060	−1.707930	−1.624540
0.35	−2.295080	−2.223390	−2.149780	−2.074370	−1.997280	−1.918660
0.40	−2.525110	−2.460080	−2.392870	−2.323610	−2.252400	−2.179370
0.45	−2.714290	−2.656580	−2.596470	−2.534060	−2.469470	−2.402790
0.50	−2.859930	−2.810110	−2.757690	−2.702750	−2.645400	−2.585750
0.55	−2.960120	−2.918650	−2.874390	−2.827440	−2.777880	−2.725810
0.60	−3.013780	−2.981000	−2.945290	−2.906720	−2.865370	−2.821340
0.65	−3.020680	−2.996810	−2.969890	−2.939990	−2.907180	−2.871530
0.70	−2.981400	−2.966540	−2.948550	−2.927480	−2.903400	−2.876370
0.75	−2.897350	−2.891480	−2.882430	−2.870250	−2.854980	−2.836670
0.80	−2.770700	−2.773690	−2.773490	−2.770130	−2.763640	−2.754050
0.85	−2.604390	−2.615990	−2.624420	−2.629690	−2.631820	−2.630840
0.90	−2.401990	−2.421850	−2.438580	−2.452200	−2.462690	−2.470080
0.95	−2.167690	−2.195350	−2.219960	−2.241520	−2.260020	−2.275450
1.00	−1.906180	−1.941090	−1.973060	−2.002070	−2.028110	−2.051140
1.05	−1.622570	−1.664100	−1.702820	−1.738710	−1.771730	−1.801850
1.10	−1.322280	−1.369720	−1.414520	−1.456620	−1.495990	−1.532600
1.15	−1.010940	−1.063530	−1.113650	−1.161250	−1.206280	−1.248700
1.20	−0.694271	−0.751184	−0.805835	−0.858160	−0.908098	−0.955596
1.25	−0.377982	−0.438372	−0.496720	−0.552952	−0.607000	−0.658801
1.30	−0.067660	−0.130655	−0.191841	−0.251137	−0.308468	−0.363760
1.35	0.231331	0.166614	0.103461	0.041958	−0.017809	−0.075762
1.40	0.513956	0.448394	0.384143	0.321296	0.259942	0.200167
1.45	0.775596	0.710051	0.645557	0.582214	0.520116	0.459355
1.50	1.012130	0.947433	0.883526	0.820514	0.758494	0.697563
1.55	1.220010	1.156950	1.094420	1.032530	0.971379	0.911067
1.60	1.396310	1.335620	1.275210	1.215190	1.155660	1.096710
1.65	1.538770	1.481140	1.423540	1.366070	1.308850	1.251980
1.70	1.645860	1.591880	1.537710	1.483430	1.429150	1.374990
1.75	1.716730	1.666950	1.616740	1.566200	1.515440	1.464560
1.80	1.751290	1.706150	1.660380	1.614060	1.567290	1.520190
1.85	1.750130	1.710010	1.669040	1.627330	1.584980	1.542090
1.90	1.714550	1.679700	1.643830	1.607050	1.569440	1.531100
1.95	1.646450	1.617070	1.586510	1.554870	1.522240	1.488710
2.00	1.548370	1.524540	1.499410	1.473050	1.445560	1.417000
2.05	1.423320	1.405070	1.385390	1.364360	1.342070	1.318590
2.10	1.274820	1.262060	1.247770	1.232050	1.214940	1.196540

Table 2.11 (continued)

			t			
v	**0.05**	**0.10**	**0.15**	**0.20**	**0.25**	**0.30**
2.15	1.106720	1.099290	1.090270	1.079720	1.067720	1.054330
2.20	0.923170	0.920842	0.916875	0.911326	0.904255	0.895722
2.25	0.728514	0.730989	0.731797	0.730988	0.728613	0.724728
2.30	0.527194	0.534112	0.539357	0.542970	0.544996	0.545482
2.35	0.323660	0.334611	0.343902	0.351566	0.357641	0.362166
2.40	0.122276	0.136806	0.149708	0.161007	0.170733	0.178919
2.45	-0.072766	-0.055145	-0.039102	-0.024620	-0.011676	-0.000246
2.50	-0.257533	-0.237331	-0.218643	-0.201457	-0.185758	-0.171531
2.55	-0.428423	-0.406166	-0.385342	-0.365949	-0.347978	-0.331419
2.60	-0.582233	-0.558450	-0.536010	-0.514914	-0.495162	-0.476751
2.65	-0.716218	-0.691436	-0.667893	-0.645599	-0.624559	-0.604776
2.70	-0.828138	-0.802868	-0.778727	-0.755728	-0.733885	-0.713206
2.75	-0.916291	-0.891024	-0.866768	-0.843542	-0.821365	-0.800251
2.80	-0.979537	-0.954735	-0.930820	-0.907819	-0.885754	-0.864645
2.85	-1.017310	-0.993396	-0.970247	-0.947890	-0.926353	-0.905660
2.90	-1.029610	-1.006970	-0.984970	-0.963639	-0.943010	-0.923110
2.95	-1.017000	-0.995971	-0.975457	-0.955491	-0.936108	-0.917340
3.00	-0.980565	-0.961437	-0.942699	-0.924390	-0.906547	-0.889205
3.05	-0.921894	-0.904899	-0.888175	-0.871765	-0.855707	-0.840038
3.10	-0.843015	-0.828332	-0.813808	-0.799487	-0.785409	-0.771613
3.15	-0.746350	-0.734104	-0.721912	-0.709817	-0.697862	-0.686086
3.20	-0.634652	-0.624911	-0.615128	-0.605345	-0.595604	-0.585948
3.25	-0.510931	-0.503713	-0.496363	-0.488925	-0.481441	-0.473953
3.30	-0.378388	-0.373657	-0.368716	-0.363608	-0.358373	-0.353054
3.35	-0.240334	-0.238010	-0.235405	-0.232563	-0.229525	-0.226330
3.40	-0.100119	-0.100074	-0.099692	-0.099012	-0.098074	-0.096918
3.45	0.038947	0.036877	0.035192	0.033854	0.032825	0.032067
3.50	0.173668	0.169681	0.166116	0.162937	0.160109	0.157595
3.55	0.301026	0.295348	0.290119	0.285305	0.280873	0.276789
3.60	0.418248	0.411128	0.404473	0.398253	0.392436	0.386991
3.65	0.522864	0.514565	0.506739	0.499358	0.492392	0.485813
3.70	0.612750	0.603548	0.594816	0.586529	0.578661	0.571187
3.75	0.686176	0.676347	0.666977	0.658046	0.649531	0.641407
3.80	0.741828	0.731647	0.721909	0.712595	0.703684	0.695156
3.85	0.778835	0.768571	0.758725	0.749281	0.740223	0.731533
3.90	0.796776	0.786683	0.776978	0.767648	0.758681	0.750059
3.95	0.795679	0.785994	0.776662	0.767675	0.759020	0.750686
4.00	0.776013	0.766951	0.758204	0.749766	0.741628	0.733781
4.05	0.738666	0.730416	0.722441	0.714737	0.707297	0.700116
4.10	0.684916	0.677640	0.670597	0.663784	0.657199	0.650836
4.15	0.616392	0.610221	0.604240	0.598449	0.592847	0.587430
4.20	0.535033	0.530067	0.525248	0.520578	0.516058	0.511687
4.25	0.443032	0.439339	0.435750	0.432271	0.428902	0.425646

Table 2.11 (continued)

			t			
v	0.05	0.10	0.15	0.20	0.25	0.30
4.30	0.342785	0.340399	0.338078	0.335826	0.333649	0.331548
4.35	0.236828	0.235753	0.234704	0.233689	0.232712	0.231779
4.40	0.127777	0.127986	0.128186	0.128386	0.128591	0.128809
4.45	0.018266	0.019703	0.021100	0.022466	0.023809	0.025134
4.50	−0.089117	−0.086532	−0.084015	−0.081556	−0.079147	−0.076781
4.55	−0.191883	−0.188255	−0.184717	−0.181261	−0.177877	−0.174559
4.60	−0.287705	−0.283156	−0.278717	−0.274379	−0.270132	−0.265970
4.65	−0.374464	−0.369134	−0.363929	−0.358839	−0.353857	−0.348972
4.70	−0.450296	−0.444336	−0.438512	−0.432815	−0.427235	−0.421765
4.75	−0.513631	−0.507202	−0.500915	−0.494761	−0.488732	−0.482819
4.80	−0.563222	−0.556488	−0.549899	−0.543445	−0.537118	−0.530912
4.85	−0.598173	−0.591299	−0.584567	−0.577970	−0.571501	−0.565150
4.90	−0.617946	−0.611093	−0.604378	−0.597793	−0.591331	−0.584985
4.95	−0.622374	−0.615697	−0.609149	−0.602724	−0.596416	−0.590217
5.00	−0.611652	−0.605296	−0.599058	−0.592932	−0.586914	−0.580997

			t			
v	0.35	0.40	0.45	0.50	0.55	0.60
0.00	0.645736	0.735137	0.823411	0.910424	0.996043	1.080140
0.05	0.263695	0.355063	0.445628	0.535251	0.623797	0.711132
0.10	−0.119469	−0.027189	0.064618	0.155811	0.246250	0.335799
0.15	−0.497460	−0.405330	−0.313339	−0.221631	−0.130345	−0.039624
0.20	−0.864110	−0.773182	−0.682060	−0.590888	−0.499807	−0.408960
0.25	−1.213500	−1.124800	−1.035570	−0.945974	−0.856135	−0.766199
0.30	−1.540040	−1.454560	−1.368230	−1.281210	−1.193620	−1.105610
0.35	−1.838610	−1.757290	−1.674810	−1.591320	−1.506950	−1.421840
0.40	−2.104640	−2.028340	−1.950600	−1.871540	−1.791300	−1.710010
0.45	−2.334150	−2.263680	−2.191480	−2.117690	−2.042420	−1.965820
0.50	−2.523890	−2.459940	−2.394020	−2.326240	−2.256730	−2.185600
0.55	−2.671320	−2.614520	−2.555510	−2.494400	−2.431310	−2.366360
0.60	−2.774700	−2.725560	−2.674000	−2.620130	−2.564050	−2.505870
0.65	−2.833120	−2.792030	−2.748350	−2.702170	−2.653580	−2.602690
0.70	−2.846440	−2.813700	−2.778220	−2.740070	−2.699350	−2.656150
0.75	−2.815380	−2.791160	−2.764080	−2.734200	−2.701610	−2.666390
0.80	−2.741410	−2.725770	−2.707180	−2.685700	−2.661390	−2.634330
0.85	−2.626770	−2.619660	−2.609530	−2.596450	−2.580460	−2.561630
0.90	−2.474390	−2.475630	−2.473840	−2.469060	−2.461320	−2.450660
0.95	−2.287820	−2.297150	−2.303460	−2.306760	−2.307080	−2.304470
1.00	−2.071190	−2.088230	−2.102290	−2.113360	−2.121480	−2.126650
1.05	−1.829070	−1.853360	−1.874730	−1.893180	−1.908700	−1.921310
1.10	−1.566420	−1.597410	−1.625580	−1.650900	−1.673360	−1.692980
1.15	−1.288460	−1.325540	−1.359900	−1.391530	−1.420400	−1.446490

Table 2.11 (continued)

			t			
v	0.35	0.40	0.45	0.50	0.55	0.60
1.20	−1.000600	−1.043080	−1.082980	−1.120270	−1.154930	−1.186930
1.25	−0.708295	−0.755428	−0.800152	−0.842423	−0.882202	−0.919456
1.30	−0.416947	−0.467966	−0.516758	−0.563271	−0.607457	−0.649272
1.35	−0.131826	−0.185929	−0.238006	−0.287994	−0.335837	−0.381482
1.40	0.142054	0.085681	0.031121	−0.021557	−0.072286	−0.121006
1.45	0.400018	0.342190	0.285951	0.231378	0.178543	0.127514
1.50	0.637814	0.579336	0.522215	0.466534	0.412371	0.359802
1.55	0.851691	0.793345	0.736119	0.680101	0.625375	0.572022
1.60	1.038460	0.980995	0.924411	0.868799	0.814249	0.760847
1.65	1.195550	1.139670	1.084430	1.029930	0.976258	0.923508
1.70	1.321030	1.267380	1.214150	1.161410	1.109280	1.057840
1.75	1.413650	1.362830	1.312190	1.261830	1.211840	1.162330
1.80	1.472860	1.425380	1.377870	1.330420	1.283120	1.236070
1.85	1.498750	1.455070	1.411150	1.367070	1.322940	1.278860
1.90	1.492120	1.452610	1.412650	1.372350	1.331790	1.291080
1.95	1.454370	1.419310	1.383630	1.347420	1.310770	1.273780
2.00	1.387490	1.357090	1.325910	1.294030	1.261550	1.228540
2.05	1.294010	1.268400	1.241870	1.214480	1.186330	1.157510
2.10	1.176920	1.156150	1.134330	1.111520	1.087810	1.063280
2.15	1.039610	1.023660	1.006520	0.988296	0.969049	0.948862
2.20	0.885794	0.874535	0.862015	0.848304	0.833472	0.817594
2.25	0.719390	0.712659	0.704596	0.695266	0.684734	0.673068
2.30	0.544478	0.542037	0.538214	0.533068	0.526656	0.519042
2.35	0.365184	0.366741	0.366885	0.365668	0.363142	0.359362
2.40	0.185600	0.190814	0.194603	0.197011	0.198085	0.197872
2.45	0.009698	0.018188	0.025259	0.030946	0.035290	0.038334
2.50	−0.158753	−0.147400	−0.137445	−0.128858	−0.121605	−0.115652
2.55	−0.316259	−0.302480	−0.290061	−0.278980	−0.269210	−0.260722
2.60	−0.459673	−0.443918	−0.429473	−0.416320	−0.404441	−0.393812
2.65	−0.586250	−0.568976	−0.552949	−0.538156	−0.524586	−0.512223
2.70	−0.693697	−0.675359	−0.658192	−0.642193	−0.627353	−0.613664
2.75	−0.780211	−0.761253	−0.743383	−0.726601	−0.710907	−0.696298
2.80	−0.844507	−0.825356	−0.807199	−0.790047	−0.773902	−0.758766
2.85	−0.885832	−0.866888	−0.848842	−0.831708	−0.815494	−0.800209
2.90	−0.903966	−0.885600	−0.868031	−0.851278	−0.835354	−0.820270
2.95	−0.899216	−0.881763	−0.865004	−0.848962	−0.833653	−0.819095
3.00	−0.872396	−0.856150	−0.840496	−0.825456	−0.811055	−0.797310
3.05	−0.824794	−0.810008	−0.795708	−0.781925	−0.768682	−0.756003
3.10	−0.758135	−0.745012	−0.732276	−0.719957	−0.708083	−0.696681
3.15	−0.674529	−0.663227	−0.652215	−0.641524	−0.631187	−0.621230
3.20	−0.576415	−0.567044	−0.557870	−0.548927	−0.540247	−0.531862
3.25	−0.466500	−0.459121	−0.451853	−0.444730	−0.437787	−0.431056

Table 2.11 (continued)

				t		
v	0.35	0.40	0.45	0.50	0.55	0.60
3.30	−0.347689	−0.342319	−0.336979	−0.331706	−0.326534	−0.321496
3.35	−0.223019	−0.219628	−0.216197	−0.212759	−0.209351	−0.206005
3.40	−0.095581	−0.094102	−0.092517	−0.090862	−0.089172	−0.087479
3.45	0.031544	0.031218	0.031054	0.031016	0.031072	0.031187
3.50	0.155360	0.153368	0.151586	0.149979	0.148514	0.147158
3.55	0.273019	0.269530	0.266289	0.263263	0.260420	0.257728
3.60	0.381887	0.377091	0.372574	0.368303	0.364248	0.360380
3.65	0.479592	0.473700	0.468107	0.462784	0.457703	0.452835
3.70	0.564080	0.557313	0.550859	0.544691	0.538782	0.533105
3.75	0.633651	0.626239	0.619146	0.612347	0.605817	0.599531
3.80	0.686990	0.679164	0.671656	0.664444	0.657504	0.650813
3.85	0.723191	0.715180	0.707480	0.700069	0.692928	0.686036
3.90	0.741769	0.733793	0.726114	0.718715	0.711576	0.704680
3.95	0.742659	0.734926	0.727473	0.720282	0.713340	0.706628
4.00	0.726214	0.718916	0.711875	0.705077	0.698509	0.692157
4.05	0.693183	0.686492	0.680032	0.673791	0.667760	0.661925
4.10	0.644690	0.638754	0.633022	0.627483	0.622129	0.616951
4.15	0.582196	0.577141	0.572258	0.567541	0.562984	0.558578
4.20	0.507463	0.503385	0.499449	0.495650	0.491983	0.488442
4.25	0.422503	0.419472	0.416552	0.413740	0.411032	0.408425
4.30	0.329527	0.327585	0.325724	0.323942	0.322236	0.320606
4.35	0.230892	0.230054	0.229267	0.228531	0.227845	0.227209
4.40	0.129042	0.129295	0.129571	0.129871	0.130197	0.130548
4.45	0.026447	0.027754	0.029057	0.030360	0.031665	0.032973
4.50	−0.074451	−0.072153	−0.069881	−0.067631	−0.065401	−0.063188
4.55	−0.171299	−0.168091	−0.164930	−0.161811	−0.158730	−0.155684
4.60	−0.261883	−0.257866	−0.253913	−0.250019	−0.246179	−0.242390
4.65	−0.344179	−0.339469	−0.334838	−0.330278	−0.325786	−0.321358
4.70	−0.416396	−0.411122	−0.405936	−0.400832	−0.395807	−0.390854
4.75	−0.477014	−0.471312	−0.465706	−0.460189	−0.454756	−0.449404
4.80	−0.524818	−0.518829	−0.512940	−0.507145	−0.501438	−0.495816
4.85	−0.558913	−0.552782	−0.546751	−0.540815	−0.534969	−0.529207
4.90	−0.578750	−0.572618	−0.566584	−0.560643	−0.554790	−0.549020
4.95	−0.584123	−0.578127	−0.572225	−0.566411	−0.560680	−0.555029
5.00	−0.575175	−0.569445	−0.563801	−0.558238	−0.552752	−0.547339

				t		
v	0.65	0.70	0.75	0.80	0.85	0.90
0.00	1.162590	1.243270	1.322070	1.398850	1.473520	1.545970
0.05	0.797123	0.881643	0.964566	1.045770	1.125130	1.202540
0.10	0.424322	0.511683	0.597754	0.682403	0.765507	0.846943
0.15	0.050395	0.139574	0.227778	0.314873	0.400728	0.485216

Table 2.11 (continued)

			t			
v	0.65	0.70	0.75	0.80	0.85	0.90
0.20	−0.318487	−0.228527	−0.139220	−0.050702	0.036892	0.123429
0.25	−0.676308	−0.586603	−0.497224	−0.408310	−0.319999	−0.232426
0.30	−1.017320	−0.928885	−0.840452	−0.752159	−0.664146	−0.576549
0.35	−1.336130	−1.249950	−1.163460	−1.076780	−0.990062	−0.903438
0.40	−1.627820	−1.544850	−1.461240	−1.377140	−1.292670	−1.207980
0.45	−1.888020	−1.809140	−1.729310	−1.648690	−1.567390	−1.485560
0.50	−2.112980	−2.039010	−1.963800	−1.887500	−1.810230	−1.732120
0.55	−2.299650	−2.231320	−2.161490	−2.090270	−2.017820	−1.944240
0.60	−2.445710	−2.383670	−2.319880	−2.254450	−2.187510	−2.119180
0.65	−2.549590	−2.494410	−2.437230	−2.378180	−2.317380	−2.254930
0.70	−2.610550	−2.562670	−2.512580	−2.460420	−2.406270	−2.350260
0.75	−2.628610	−2.588370	−2.545760	−2.500870	−2.453800	−2.404660
0.80	−2.604580	−2.572220	−2.537340	−2.500020	−2.460350	−2.418430
0.85	−2.540000	−2.515660	−2.488660	−2.459100	−2.427040	−2.392580
0.90	−2.437150	−2.420840	−2.401780	−2.380050	−2.355710	−2.328840
0.95	−2.298950	−2.290580	−2.279400	−2.265470	−2.248850	−2.229600
1.00	−2.128910	−2.128280	−2.124820	−2.118560	−2.109550	−2.097850
1.05	−1.931020	−1.937860	−1.941860	−1.943040	−1.941440	−1.937110
1.10	−1.709740	−1.723660	−1.734770	−1.743070	−1.748590	−1.751370
1.15	−1.469820	−1.490370	−1.508140	−1.523150	−1.535420	−1.544960
1.20	−1.216250	−1.242890	−1.266830	−1.288070	−1.306620	−1.322490
1.25	−0.954157	−0.986281	−1.015810	−1.042740	−1.067050	−1.088750
1.30	−0.688678	−0.725642	−0.760137	−0.792139	−0.821632	−0.848604
1.35	−0.424882	−0.465994	−0.504782	−0.541212	−0.575257	−0.606896
1.40	−0.167662	−0.212202	−0.254580	−0.294754	−0.332686	−0.368347
1.45	0.078355	0.031126	−0.014119	−0.057330	−0.098461	−0.137471
1.50	0.308897	0.259722	0.212341	0.166811	0.123185	0.081512
1.55	0.520119	0.469738	0.420948	0.373815	0.328400	0.284758
1.60	0.708673	0.657807	0.608324	0.560295	0.513786	0.468862
1.65	0.871763	0.821108	0.771621	0.723379	0.676456	0.630919
1.70	1.007190	0.957401	0.908566	0.860765	0.814076	0.768572
1.75	1.113370	1.065060	1.017490	0.970747	0.924903	0.880041
1.80	1.189370	1.143110	1.097370	1.052240	1.007800	0.964150
1.85	1.234900	1.191180	1.147770	1.104770	1.062260	1.020330
1.90	1.250310	1.209560	1.168930	1.128500	1.088380	1.048630
1.95	1.236530	1.199130	1.161650	1.124190	1.086830	1.049670
2.00	1.195110	1.161340	1.127320	1.093140	1.058880	1.024640
2.05	1.128100	1.098180	1.067850	1.037200	1.006300	0.975237
2.10	1.038020	1.012110	0.985634	0.958676	0.931319	0.903648
2.15	0.927813	0.905982	0.883449	0.860297	0.836604	0.812453
2.20	0.800744	0.782997	0.764429	0.745118	0.725142	0.704577
2.25	0.660337	0.646612	0.631964	0.616466	0.600192	0.583216

Table 2.11 (continued)

v	0.65	0.70	0.75	0.80	0.85	0.90
2.30	0.510289	0.500462	0.489626	0.477852	0.465206	0.451761
2.35	0.354387	0.348276	0.341090	0.332891	0.323744	0.313713
2.40	0.196426	0.193798	0.190045	0.185224	0.179393	0.172612
2.45	0.040122	0.040701	0.040122	0.038434	0.035692	0.031951
2.50	-0.110960	-0.107487	-0.105191	-0.104027	-0.103945	-0.104898
2.55	-0.253484	-0.247462	-0.242618	-0.238914	-0.236307	-0.234754
2.60	-0.384409	-0.376202	-0.369162	-0.363254	-0.358444	-0.354692
2.65	-0.501046	-0.491034	-0.482163	-0.474405	-0.467730	-0.462106
2.70	-0.601112	-0.589680	-0.579352	-0.570104	-0.561914	-0.554755
2.75	-0.682765	-0.670299	-0.658886	-0.648513	-0.639159	-0.630805
2.80	-0.744638	-0.731513	-0.719384	-0.708241	-0.698071	-0.688860
2.85	-0.785854	-0.772432	-0.759940	-0.748373	-0.737725	-0.727984
2.90	-0.806036	-0.792657	-0.780136	-0.768473	-0.757666	-0.747710
2.95	-0.805300	-0.792278	-0.780037	-0.768582	-0.757915	-0.748036
3.00	-0.784241	-0.771860	-0.760179	-0.749208	-0.738954	-0.729420
3.05	-0.743909	-0.732418	-0.721545	-0.711303	-0.701702	-0.692752
3.10	-0.685774	-0.675384	-0.665529	-0.656225	-0.647487	-0.639327
3.15	-0.611681	-0.602563	-0.593898	-0.585705	-0.578002	-0.570804
3.20	-0.523799	-0.516084	-0.508742	-0.501795	-0.495262	-0.489161
3.25	-0.424565	-0.418343	-0.412417	-0.406810	-0.401545	-0.396642
3.30	-0.316622	-0.311943	-0.307486	-0.303276	-0.299338	-0.295694
3.35	-0.202754	-0.199627	-0.196653	-0.193860	-0.191272	-0.188913
3.40	-0.085816	-0.084214	-0.082702	-0.081308	-0.080058	-0.078977
3.45	0.031329	0.031469	0.031575	0.031620	0.031576	0.031419
3.50	0.145880	0.144649	0.143436	0.142213	0.140951	0.139625
3.55	0.255158	0.252678	0.250260	0.247875	0.245496	0.243098
3.60	0.356668	0.353082	0.349596	0.346180	0.342809	0.339456
3.65	0.448151	0.443623	0.439224	0.434927	0.430706	0.426536
3.70	0.527633	0.522339	0.517198	0.512183	0.507270	0.502433
3.75	0.593464	0.587591	0.581887	0.576328	0.570889	0.565548
3.80	0.644348	0.638087	0.632005	0.626081	0.620290	0.614611
3.85	0.679371	0.672912	0.666638	0.660528	0.654559	0.648712
3.90	0.698007	0.691538	0.685254	0.679134	0.673159	0.667309
3.95	0.700131	0.693830	0.687708	0.681748	0.675931	0.670238
4.00	0.686005	0.680039	0.674243	0.668600	0.663095	0.657710
4.05	0.656274	0.650794	0.645471	0.640290	0.635237	0.630297
4.10	0.611936	0.607074	0.602354	0.597761	0.593285	0.588910
4.15	0.554314	0.550183	0.546175	0.542279	0.538484	0.534779
4.20	0.485021	0.481710	0.478504	0.475391	0.472363	0.469410
4.25	0.405913	0.403490	0.401150	0.398885	0.396688	0.394549
4.30	0.319046	0.317553	0.316122	0.314747	0.313420	0.312137
4.35	0.226620	0.226075	0.225571	0.225104	0.224668	0.224258

Table 2.11 (continued)

			t			
v	0.65	0.70	0.75	0.80	0.85	0.90
4.40	0.130924	0.131322	0.131742	0.132179	0.132630	0.133090
4.45	0.034285	0.035599	0.036915	0.038231	0.039545	0.040852
4.50	-0.060991	-0.058810	-0.056644	-0.054495	-0.052363	-0.050252
4.55	-0.152672	-0.149691	-0.146742	-0.143823	-0.140936	-0.138082
4.60	-0.238649	-0.234954	-0.231303	-0.227695	-0.224131	-0.220611
4.65	-0.316989	-0.312678	-0.308422	-0.304219	-0.300070	-0.295974
4.70	-0.385971	-0.381154	-0.376400	-0.371709	-0.367078	-0.362507
4.75	-0.444127	-0.438923	-0.433789	-0.428722	-0.423720	-0.418784
4.80	-0.490272	-0.484805	-0.479411	-0.474088	-0.468833	-0.463646
4.85	-0.523526	-0.517923	-0.512393	-0.506935	-0.501546	-0.496225
4.90	-0.543330	-0.537715	-0.532174	-0.526702	-0.521298	-0.515960
4.95	-0.549454	-0.543951	-0.538516	-0.533149	-0.527846	-0.522605
5.00	-0.541997	-0.536720	-0.531508	-0.526356	-0.521264	-0.516229

			t			
v	0.95	1.0	2.0	3.0	4.0	5.0
0.00	1.616090	1.683780	2.421810	1.940640	0.791070	-0.119980
0.05	1.277880	1.351050	2.235140	1.904260	0.831029	-0.094492
0.10	0.926591	1.004340	2.021840	1.840300	0.851074	-0.079928
0.15	0.568210	0.649590	1.786380	1.751380	0.852434	-0.075605
0.20	0.208780	0.292815	1.533530	1.640510	0.836698	-0.080558
0.25	-0.145726	-0.060031	1.268230	1.511020	0.805772	-0.093572
0.30	-0.489507	-0.403154	0.995550	1.366440	0.761813	-0.113225
0.35	-0.817048	-0.731028	0.720510	1.210500	0.707180	-0.137929
0.40	-1.123210	-1.038500	0.448041	1.046950	0.644360	-0.165984
0.45	-1.403340	-1.320860	0.182871	0.879572	0.575912	-0.195620
0.50	-1.653320	-1.573960	-0.070561	0.712026	0.504399	-0.225053
0.55	-1.869680	-1.794260	-0.308185	0.547819	0.432326	-0.252531
0.60	-2.049590	-1.978860	-0.526369	0.390221	0.362082	-0.276383
0.65	-2.190970	-2.125620	-0.721981	0.242204	0.295888	-0.295063
0.70	-2.292490	-2.233080	-0.892437	0.106388	0.235743	-0.307191
0.75	-2.353560	-2.300600	-1.035740	-0.015004	0.183383	-0.311588
0.80	-2.374360	-2.328230	-1.150500	-0.120182	0.140251	-0.307308
0.85	-2.355800	-2.316800	-1.235940	-0.207811	0.107461	-0.293657
0.90	-2.299520	-2.267830	-1.291920	-0.277028	0.085790	-0.270213
0.95	-2.207790	-2.183500	-1.318880	-0.327442	0.075660	-0.236837
1.00	-2.083510	-2.066590	-1.317850	-0.359130	0.077140	-0.193672
1.05	-1.930100	-1.920440	-1.290400	-0.372614	0.089956	-0.141140
1.10	-1.751440	-1.748840	-1.238570	-0.368839	0.113503	-0.079928
1.15	-1.551790	-1.555950	-1.164840	-0.349133	0.146872	-0.010974
1.20	-1.335680	-1.346230	-1.072060	-0.315164	0.188876	0.064562

Table 2.11 (continued)

			t			
v	0.95	1.0	2.0	3.0	4.0	5.0
1.25	−1.107830	−1.124320	−0.963366	−0.268891	0.238095	0.145324
1.30	−0.873046	−0.894958	−0.842113	−0.212503	0.292909	0.229797
1.35	−0.636110	−0.662888	−0.711793	−0.148366	0.351547	0.316345
1.40	−0.401708	−0.432748	−0.575954	−0.078957	0.412139	0.403251
1.45	−0.174326	−0.208992	−0.438125	−0.006799	0.472760	0.488758
1.50	0.041837	0.004200	−0.301736	0.065597	0.531488	0.571115
1.55	0.242942	0.203000	−0.170047	0.135798	0.586447	0.648617
1.60	0.425581	0.383998	−0.046079	0.201502	0.635859	0.719647
1.65	0.586835	0.544265	0.067444	0.260595	0.678085	0.782717
1.70	0.724324	0.681399	0.168146	0.311196	0.711672	0.836500
1.75	0.836235	0.793557	0.254046	0.351698	0.735381	0.879862
1.80	0.921352	0.879486	0.323587	0.380802	0.748220	0.911892
1.85	0.979061	0.938526	0.375661	0.397543	0.749470	0.931917
1.90	1.009350	0.970611	0.409620	0.401305	0.738693	0.939522
1.95	1.012780	0.976253	0.425280	0.391830	0.715752	0.934555
2.00	0.990487	0.956518	0.422907	0.369216	0.680802	0.917135
2.05	0.944105	0.912983	0.403204	0.333909	0.634292	0.887639
2.10	0.875745	0.847691	0.367277	0.286688	0.576949	0.846704
2.15	0.787923	0.763096	0.316604	0.228634	0.509758	0.795198
2.20	0.683501	0.661994	0.252983	0.161102	0.433940	0.734209
2.25	0.565612	0.547455	0.178487	0.085683	0.350918	0.665014
2.30	0.437585	0.422750	0.095407	0.004160	0.262283	0.589048
2.35	0.302867	0.291271	0.006190	−0.081539	0.169755	0.507873
2.40	0.164945	0.156453	−0.086623	−0.169390	0.075141	0.423136
2.45	0.027267	0.021698	−0.180457	−0.257324	−0.019706	0.336539
2.50	−0.106834	−0.109699	−0.272768	−0.343278	−0.112936	0.249790
2.55	−0.234208	−0.234623	−0.361103	−0.425247	−0.202741	0.164573
2.60	−0.351960	−0.350204	−0.443155	−0.501332	−0.287400	0.082505
2.65	−0.457499	−0.453871	−0.516817	−0.569783	−0.365314	0.005101
2.70	−0.548597	−0.543410	−0.580228	−0.629040	−0.435050	−0.066258
2.75	−0.623426	−0.616998	−0.631814	−0.677772	−0.495363	−0.130357
2.80	−0.680588	−0.673236	−0.670318	−0.714903	−0.545234	−0.186169
2.85	−0.719139	−0.711174	−0.694824	−0.739636	−0.583881	−0.232877
2.90	−0.738597	−0.730317	−0.704777	−0.751469	−0.610782	−0.269893
2.95	−0.738942	−0.730628	−0.699982	−0.750206	−0.625681	−0.296862
3.00	−0.720608	−0.712516	−0.680612	−0.735955	−0.628594	−0.313669
3.05	−0.684457	−0.676820	−0.647189	−0.709124	−0.619799	−0.320439
3.10	−0.631753	−0.624773	−0.600570	−0.670407	−0.599834	−0.317528
3.15	−0.564123	−0.557968	−0.541920	−0.620765	−0.569476	−0.305510
3.20	−0.483508	−0.478316	−0.472679	−0.561401	−0.529722	−0.285163
3.25	−0.392117	−0.387989	−0.394522	−0.493728	−0.481764	−0.257442
3.30	−0.292363	−0.289365	−0.309319	−0.419332	−0.426955	−0.223459
3.35	−0.186806	−0.184970	−0.219081	−0.339934	−0.366779	−0.184447

Table 2.11 (continued)

v	0.95	1.0	2.0	3.0	4.0	5.0
3.40	−0.078089	−0.077415	−0.125916	−0.257349	−0.302812	−0.141734
3.45	0.031123	0.030667	−0.031973	−0.173439	−0.236684	−0.096706
3.50	0.138211	0.136684	0.060606	−0.090071	−0.170042	−0.050775
3.55	0.240655	0.238143	0.149742	−0.009072	−0.104509	−0.005341
3.60	0.336097	0.332707	0.233462	0.067813	−0.041647	0.038238
3.65	0.422391	0.418250	0.309948	0.138960	0.017079	0.078684
3.70	0.497650	0.492896	0.377578	0.202907	0.070335	0.114821
3.75	0.560281	0.555065	0.434958	0.258379	0.116947	0.145610
3.80	0.609022	0.603501	0.480957	0.304322	0.155925	0.170169
3.85	0.642964	0.637297	0.514725	0.339922	0.186486	0.187791
3.90	0.661565	0.655907	0.535713	0.364620	0.208068	0.197962
3.95	0.664653	0.659157	0.543680	0.378123	0.220342	0.200372
4.00	0.652430	0.647237	0.538693	0.380407	0.223214	0.194914
4.05	0.625455	0.620695	0.521124	0.371716	0.216825	0.181690
4.10	0.584625	0.580415	0.491634	0.352546	0.201542	0.161003
4.15	0.531151	0.527588	0.451154	0.323636	0.177951	0.133350
4.20	0.466521	0.463685	0.400862	0.285944	0.146832	0.099402
4.25	0.392460	0.390412	0.342144	0.240622	0.109144	0.059991
4.30	0.310888	0.309666	0.276567	0.188984	0.065997	0.016086
4.35	0.223867	0.223489	0.205834	0.132476	0.018619	−0.031236
4.40	0.133555	0.134019	0.131741	0.072635	−0.031669	−0.080812
4.45	0.042150	0.043433	0.056137	0.011053	−0.083491	−0.131426
4.50	−0.048164	−0.046102	−0.019124	−0.050660	−0.135448	−0.181839
4.55	−0.135262	−0.132480	−0.092226	−0.110917	−0.186155	−0.230822
4.60	−0.217136	−0.213707	−0.161434	−0.168194	−0.234274	−0.277180
4.65	−0.291931	−0.287943	−0.225133	−0.221060	−0.278543	−0.319790
4.70	−0.357996	−0.353545	−0.281867	−0.268215	−0.317811	−0.357619
4.75	−0.413912	−0.409103	−0.330368	−0.308519	−0.351060	−0.389756
4.80	−0.458525	−0.453470	−0.369588	−0.341014	−0.377432	−0.415429
4.85	−0.490971	−0.485782	−0.398715	−0.364950	−0.396250	−0.434027
4.90	−0.510687	−0.505479	−0.417196	−0.379793	−0.407025	−0.445107
4.95	−0.517426	−0.512307	−0.424741	−0.385245	−0.409473	−0.448412
5.00	−0.511250	−0.506327	−0.421327	−0.381238	−0.403520	−0.443870

Table 2.12: Third derivatives with respect to the order of the Weber functions $\partial^3 E_\nu(t)/\partial \nu^3$.

				t		
ν	0.05	0.10	0.15	0.20	0.25	0.30
0.00	7.74820	7.73809	7.72126	7.69774	7.66756	7.63078
0.05	7.70597	7.71730	7.72190	7.71976	7.71090	7.69535
0.10	7.53734	7.56993	7.59583	7.61504	7.62753	7.63331
0.15	7.24552	7.29887	7.34567	7.38588	7.41947	7.44639
0.20	6.83599	6.90932	6.97631	7.03689	7.09100	7.13858
0.25	6.31637	6.40863	6.49482	6.57486	6.64865	6.71611
0.30	5.69631	5.80618	5.91033	6.00863	6.10099	6.18728
0.35	4.98728	5.11320	5.23380	5.34895	5.45849	5.56232
0.40	4.20233	4.34253	4.47787	4.60819	4.73333	4.85314
0.45	3.35585	3.50837	3.65654	3.80017	3.93910	4.07315
0.50	2.46327	2.62600	2.78492	2.93985	3.09059	3.23695
0.55	1.54081	1.71150	1.87898	2.04303	2.20346	2.36005
0.60	0.605082	0.781413	0.955148	1.12606	1.29394	1.45856
0.65	−0.327158	−0.147567	0.030066	0.205507	0.378524	0.548893
0.70	−1.23937	−1.05892	−0.879768	−0.702168	−0.526356	−0.352567
0.75	−2.11555	−1.93660	−1.758310	−1.58091	−1.40466	−1.22979
0.80	−2.94051	−2.76538	−2.590260	−2.41539	−2.24101	−2.06738
0.85	−3.70017	−3.53110	−3.361390	−3.19129	−3.02104	−2.85089
0.90	−4.38185	−4.22094	−4.058760	−3.89557	−3.73160	−3.56711
0.95	−4.97448	−4.82367	−4.671000	−4.51671	−4.36105	−4.20425
1.00	−5.46877	−5.32984	−5.188480	−5.04493	−4.89943	−4.75221
1.05	−5.85744	−5.73194	−5.603490	−5.47233	−5.33866	−5.20272
1.10	−6.13528	−6.02455	−5.910400	−5.79303	−5.67267	−5.54952
1.15	−6.29924	−6.20438	−6.105670	−6.00330	−5.89748	−5.78840
1.20	−6.34850	−6.27035	−6.187970	−6.10155	−6.01126	−5.91731
1.25	−6.28443	−6.22356	−6.158150	−6.08836	−6.01436	−5.93632
1.30	−6.11051	−6.06724	−6.019170	−5.96644	−5.90922	−5.84764
1.35	−5.83231	−5.80668	−5.776050	−5.74056	−5.70033	−5.65550
1.40	−5.45727	−5.44906	−5.435730	−5.41738	−5.39412	−5.36607
1.45	−4.99458	−5.00333	−5.006880	−5.00533	−4.99876	−4.98727
1.50	−4.45495	−4.47995	−4.499750	−4.51441	−4.52401	−4.52862
1.55	−3.85040	−3.89072	−3.925900	−3.95597	−3.98099	−4.00101
1.60	−3.19397	−3.24848	−3.297960	−3.34242	−3.38191	−3.41645
1.65	−2.49946	−2.56685	−2.629380	−2.68704	−2.73985	−2.78782
1.70	−1.78115	−1.85996	−1.934130	−2.00363	−2.06845	−2.12860
1.75	−1.05351	−1.14215	−1.226410	−1.30625	−1.38165	−1.45257
1.80	−0.33087	−0.427652	−0.520362	−0.608946	−0.69336	−0.77355
1.85	0.37280	0.269639	0.170197	0.074551	−0.01723	−0.10509
1.90	1.04425	0.936490	0.832076	0.731098	0.63364	0.53977
1.95	1.67112	1.56056	1.45295	1.34839	1.24697	1.14879

Table 2.12 (continued)

			t			
v	0.05	0.10	0.15	0.20	0.25	0.30
2.00	2.24222	2.13065	2.02160	1.91519	1.81153	1.71072
2.05	2.74776	2.63690	2.52812	2.42157	2.31734	2.21558
2.10	3.17945	3.07095	2.96410	2.85902	2.75587	2.65475
2.15	3.53072	3.42613	3.32274	3.22070	3.12014	3.02122
2.20	3.79677	3.69750	3.59901	3.50143	3.40492	3.30961
2.25	3.97465	3.88199	3.78969	3.69787	3.60671	3.51633
2.30	4.06329	3.97837	3.89338	3.80848	3.72383	3.63956
2.35	4.06349	3.98725	3.91055	3.83356	3.75641	3.67927
2.40	3.97786	3.91106	3.84344	3.77515	3.70635	3.63719
2.45	3.81070	3.75392	3.69598	3.63704	3.57724	3.51674
2.50	3.56795	3.52156	3.47371	3.42456	3.37424	3.32290
2.55	3.25694	3.22114	3.18361	3.14450	3.10395	3.06209
2.60	2.88629	2.86107	2.83390	2.80491	2.77423	2.74199
2.65	2.46564	2.45083	2.43387	2.41489	2.39402	2.37137
2.70	2.00547	2.00071	1.99366	1.98442	1.97311	1.95986
2.75	1.51683	1.52162	1.52399	1.52406	1.52194	1.51771
2.80	1.01111	1.02479	1.03598	1.04478	1.05129	1.05561
2.85	0.499771	0.521560	0.540833	0.557669	0.572155	0.584378
2.90	−0.005874	0.023143	0.049646	0.073704	0.095390	0.114780
2.95	−0.494923	−0.459649	−0.426852	−0.396474	−0.368455	−0.342729
3.00	−0.957098	−0.916604	−0.878517	−0.842792	−0.809378	−0.778222
3.05	−1.38296	−1.33833	−1.29600	−1.25595	−1.21813	−1.18251
3.10	−1.76410	−1.71643	−1.67094	−1.62761	−1.58641	−1.54731
3.15	−2.09328	−2.04368	−1.99610	−1.95055	−1.90699	−1.86543
3.20	−2.36460	−2.31414	−2.26554	−2.21881	−2.17393	−2.13089
3.25	−2.57355	−2.52328	−2.47469	−2.42779	−2.38257	−2.33905
3.30	−2.71712	−2.66803	−2.62043	−2.57431	−2.52971	−2.48663
3.35	−2.79383	−2.74684	−2.70112	−2.65669	−2.61359	−2.57183
3.40	−2.80371	−2.75964	−2.71662	−2.67471	−2.63391	−2.59427
3.45	−2.74827	−2.70785	−2.66827	−2.62958	−2.59182	−2.55501
3.50	−2.63048	−2.59432	−2.55879	−2.52395	−2.48983	−2.45648
3.55	−2.45458	−2.42319	−2.39222	−2.36174	−2.33179	−2.30241
3.60	−2.22607	−2.19982	−2.17380	−2.14808	−2.12269	−2.09769
3.65	−1.95147	−1.93062	−1.90982	−1.88913	−1.86860	−1.84828
3.70	−1.63821	−1.62290	−1.60746	−1.59196	−1.57645	−1.56099
3.75	−1.29443	−1.28466	−1.27462	−1.26436	−1.25393	−1.24340
3.80	−0.928750	−0.924436	−0.919700	−0.914604	−0.909206	−0.903563
3.85	−0.550129	−0.551053	−0.551434	−0.551331	−0.550804	−0.549910
3.90	−0.167602	−0.173455	−0.178662	−0.183283	−0.187374	−0.190992
3.95	0.209902	0.199517	0.189860	0.180875	0.172506	0.164698
4.00	0.573760	0.559317	0.545665	0.532749	0.520518	0.508919
4.05	0.915851	0.897886	0.880756	0.864409	0.848795	0.833866

Table 2.12 (continued)

			t			
v	0.05	0.10	0.15	0.20	0.25	0.30
4.10	1.22872	1.20782	1.18778	1.16855	1.15009	1.13234
4.15	1.50575	1.48254	1.46019	1.43867	1.41792	1.39791
4.20	1.74128	1.71640	1.69237	1.66916	1.64672	1.62502
4.25	1.93073	1.90483	1.87975	1.85546	1.83194	1.80913
4.30	2.07067	2.04438	2.01888	1.99414	1.97012	1.94680
4.35	2.15887	2.13283	2.10752	2.08291	2.05898	2.03572
4.40	2.19438	2.16916	2.14461	2.12071	2.09743	2.07476
4.45	2.17744	2.15360	2.13034	2.10767	2.08556	2.06399
4.50	2.10955	2.08756	2.06609	2.04512	2.02464	2.00464
4.55	1.99333	1.97362	1.95435	1.93550	1.91707	1.89904
4.60	1.83247	1.81541	1.79869	1.78231	1.76627	1.75056
4.65	1.63165	1.61751	1.60363	1.59001	1.57664	1.56353
4.70	1.39638	1.38537	1.37453	1.36386	1.35337	1.34306
4.75	1.13285	1.12510	1.11743	1.10986	1.10238	1.09501
4.80	0.847823	0.843374	0.838931	0.834504	0.830101	0.825732
4.85	0.548416	0.547242	0.546001	0.544704	0.543362	0.541986
4.90	0.241956	0.243957	0.245823	0.247568	0.249205	0.250745
4.95	−0.064200	−0.059193	−0.054381	−0.049747	−0.045278	−0.040962
5.00	−0.362836	−0.355054	−0.347517	−0.340209	−0.333114	−0.326220

			t			
v	0.35	0.40	0.45	0.50	0.55	0.60
0.00	7.58746	7.53767	7.48149	7.41902	7.35036	7.27563
0.05	7.67313	7.64430	7.60892	7.56705	7.51877	7.46417
0.10	7.63239	7.62478	7.61052	7.58965	7.56223	7.52831
0.15	7.46664	7.48021	7.48711	7.48735	7.48096	7.46798
0.20	7.17959	7.21400	7.24179	7.26295	7.27748	7.28540
0.25	6.77718	6.83180	6.87992	6.92152	6.95655	6.98503
0.30	6.26743	6.34135	6.40897	6.47023	6.52509	6.57349
0.35	5.66031	5.75236	5.83837	5.91826	5.99196	6.05941
0.40	4.96749	5.07626	5.17931	5.27655	5.36788	5.45322
0.45	4.20218	4.32602	4.44454	4.55762	4.66512	4.76695
0.50	3.37876	3.51586	3.64808	3.77527	3.89729	4.01402
0.55	2.51261	2.66097	2.80494	2.94436	3.07906	3.20889
0.60	1.61972	1.77721	1.93086	2.08046	2.22585	2.36685
0.65	0.716394	0.880814	1.04195	1.19959	1.35355	1.50365
0.70	−0.181030	−0.011968	0.154399	0.317859	0.478206	0.635241
0.75	−1.05654	−0.885140	−0.715826	−0.548815	−0.384325	−0.222568
0.80	−1.89472	−1.72329	−1.55332	−1.38503	−1.21865	−1.05441
0.85	−2.68110	−2.51188	−2.34350	−2.17618	−2.01015	−1.84565
0.90	−3.40234	−3.23752	−3.07291	−2.90874	−2.74525	−2.58267

Table 2.12 (continued)

			t			
v	0.35	0.40	0.45	0.50	0.55	0.60
0.95	-4.04656	-3.88821	-3.72946	-3.57053	-3.41166	-3.25310
1.00	-4.60351	-4.45356	-4.30260	-4.15088	-3.99862	-3.84607
1.05	-5.06475	-4.92496	-4.78360	-4.64090	-4.49708	-4.35239
1.10	-5.42381	-5.29577	-5.16560	-5.03356	-4.89985	-4.76471
1.15	-5.67629	-5.56134	-5.44378	-5.32383	-5.20170	-5.07762
1.20	-5.81988	-5.71918	-5.61540	-5.50877	-5.39948	-5.28775
1.25	-5.85443	-5.76887	-5.67983	-5.58750	-5.49209	-5.39380
1.30	-5.78188	-5.71212	-5.63851	-5.56126	-5.48053	-5.39651
1.35	-5.60623	-5.55265	-5.49494	-5.43326	-5.36777	-5.29865
1.40	-5.33336	-5.29613	-5.25452	-5.20867	-5.15873	-5.10487
1.45	-4.97098	-4.94999	-4.92443	-4.89443	-4.86012	-4.82165
1.50	-4.52834	-4.52325	-4.51347	-4.49910	-4.48025	-4.45707
1.55	-4.01611	-4.02636	-4.03184	-4.03265	-4.02888	-4.02064
1.60	-3.44610	-3.47090	-3.49093	-3.50625	-3.51694	-3.52309
1.65	-2.83099	-2.86937	-2.90302	-2.93200	-2.95635	-2.97614
1.70	-2.18408	-2.23490	-2.28108	-2.32267	-2.35969	-2.39220
1.75	-1.51902	-1.58097	-1.63845	-1.69145	-1.74001	-1.78414
1.80	-0.84950	-0.921171	-0.988559	-1.051660	-1.11046	-1.16498
1.85	-0.18898	-0.268854	-0.344683	-0.416441	-0.484111	-0.547686
1.90	0.44957	0.363076	0.280354	0.201443	0.126376	0.055179
1.95	1.05392	0.962445	0.874425	0.789920	0.708981	0.631649
2.00	1.61286	1.51804	1.42633	1.33782	1.25256	1.17062
2.05	2.11637	2.01982	1.92603	1.83509	1.74706	1.66203
2.10	2.55579	2.45911	2.36480	2.27297	2.18371	2.09711
2.15	2.92404	2.82873	2.73542	2.64420	2.55518	2.46846
2.20	3.21564	3.12314	3.03224	2.94305	2.85568	2.77025
2.25	3.42689	3.33851	3.25134	3.16549	3.08109	2.99825
2.30	3.55582	3.47276	3.39051	3.30921	3.22897	3.14992
2.35	3.60227	3.52557	3.44931	3.37361	3.29861	3.22443
2.40	3.56781	3.49836	3.42898	3.35980	3.29097	3.22260
2.45	3.45568	3.39420	3.33244	3.27055	3.20866	3.14690
2.50	3.27068	3.21772	3.16417	3.11016	3.05583	3.00131
2.55	3.01906	2.97501	2.93007	2.88437	2.83805	2.79124
2.60	2.70833	2.67339	2.63728	2.60016	2.56214	2.52336
2.65	2.34708	2.32128	2.29408	2.26562	2.23603	2.20542
2.70	1.94477	1.92798	1.90959	1.88973	1.86853	1.84609
2.75	1.51151	1.50344	1.49361	1.48214	1.46914	1.45473
2.80	1.05783	1.05806	1.05639	1.05295	1.04783	1.04114
2.85	0.594427	0.602395	0.608378	0.612471	0.614775	0.615388
2.90	0.131955	0.146996	0.159988	0.171020	0.180180	0.187560

Table 2.12 (continued)

			t			
v	0.35	0.40	0.45	0.50	0.55	0.60
2.95	−0.319227	−0.297878	−0.278608	−0.261337	−0.245987	−0.232475
3.00	−0.749268	−0.722453	−0.697715	−0.674984	−0.654192	−0.635265
3.05	−1.14903	−1.11765	−1.08831	−1.06096	−1.03554	−1.01198
3.10	−1.51027	−1.47527	−1.44224	−1.41116	−1.38196	−1.35460
3.15	−1.82582	−1.78815	−1.75237	−1.71846	−1.68636	−1.65605
3.20	−2.08969	−2.05030	−2.01271	−1.97688	−1.94279	−1.91040
3.25	−2.29721	−2.25705	−2.21856	−2.18171	−2.14650	−2.11289
3.30	−2.44508	−2.40505	−2.36654	−2.32955	−2.29407	−2.26008
3.35	−2.53142	−2.49237	−2.45469	−2.41838	−2.38344	−2.34987
3.40	−2.55580	−2.51852	−2.48244	−2.44758	−2.41394	−2.38151
3.45	−2.51919	−2.48438	−2.45061	−2.41788	−2.38622	−2.35564
3.50	−2.42393	−2.39221	−2.36136	−2.33138	−2.30231	−2.27416
3.55	−2.27365	−2.24554	−2.21812	−2.19141	−2.16545	−2.14025
3.60	−2.07313	−2.04905	−2.02548	−2.00246	−1.98003	−1.95820
3.65	−1.82822	−1.80847	−1.78907	−1.77006	−1.75148	−1.73336
3.70	−1.54563	−1.53042	−1.51540	−1.50061	−1.48610	−1.47191
3.75	−1.23281	−1.22222	−1.21168	−1.20124	−1.19092	−1.18078
3.80	−0.897730	−0.891758	−0.885699	−0.879597	−0.873499	−0.867446
3.85	−0.548703	−0.547235	−0.545557	−0.543718	−0.541762	−0.539734
3.90	−0.194190	−0.197021	−0.199536	−0.201781	−0.203805	−0.205652
3.95	0.157400	0.150559	0.144127	0.138054	0.132295	0.126806
4.00	0.497901	0.487415	0.477412	0.467845	0.458669	0.449839
4.05	0.819573	0.805869	0.792706	0.780040	0.767827	0.756023
4.10	1.11527	1.09883	1.08297	1.06765	1.05283	1.03847
4.15	1.37859	1.35993	1.34187	1.32439	1.30744	1.29098
4.20	1.60402	1.58367	1.56396	1.54482	1.52624	1.50816
4.25	1.78702	1.76556	1.74472	1.72446	1.70476	1.68557
4.30	1.92414	1.90212	1.88071	1.85986	1.83956	1.81977
4.35	2.01308	1.99104	1.96958	1.94867	1.92828	1.90839
4.40	2.05267	2.03115	2.01016	1.98969	1.96970	1.95018
4.45	2.04296	2.02243	2.00240	1.98283	1.96372	1.94503
4.50	1.98511	1.96603	1.94739	1.92916	1.91134	1.89390
4.55	1.88141	1.86418	1.84731	1.83081	1.81466	1.79884
4.60	1.73518	1.72012	1.70537	1.69093	1.67677	1.66290
4.65	1.55067	1.53806	1.52569	1.51357	1.50168	1.49001
4.70	1.33293	1.32299	1.31321	1.30362	1.29419	1.28493
4.75	1.08775	1.08059	1.07355	1.06662	1.05980	1.05308
4.80	0.821400	0.817112	0.812871	0.808678	0.804535	0.800442
4.85	0.540582	0.539159	0.537721	0.536272	0.534818	0.533358
4.90	0.252198	0.253573	0.254877	0.256116	0.257296	0.258420

Table 2.12 (continued)

v	0.35	0.40	0.45	0.50	0.55	0.60
			t			
4.95	−0.036786	−0.032742	−0.028819	−0.025010	−0.021309	−0.017709
5.00	−0.319514	−0.312985	−0.306622	−0.300417	−0.294361	−0.288448

v	0.65	0.70	0.75	0.80	0.85	0.90
			t			
0.00	7.19494	7.10844	7.01626	6.91856	6.81550	6.70724
0.05	7.40335	7.33643	7.26352	7.18475	7.10026	7.01020
0.10	7.48799	7.44133	7.38844	7.32942	7.26440	7.19350
0.15	7.44847	7.42247	7.39007	7.35135	7.30640	7.25532
0.20	7.28672	7.28150	7.26976	7.25158	7.22701	7.19614
0.25	7.00693	7.02228	7.03110	7.03341	7.02926	7.01870
0.30	6.61543	6.65088	6.67983	6.70231	6.71831	6.72787
0.35	6.12055	6.17534	6.22376	6.26578	6.30140	6.33062
0.40	5.53248	5.60560	5.67253	5.73322	5.78764	5.83578
0.45	4.86301	4.95320	5.03745	5.11568	5.18785	5.25391
0.50	4.12533	4.23110	4.33125	4.42567	4.51430	4.59705
0.55	3.33372	3.45341	3.56783	3.67688	3.78046	3.87846
0.60	2.50330	2.63505	2.76197	2.88391	3.00075	3.11238
0.65	1.64970	1.79154	1.92901	2.06195	2.19022	2.31368
0.70	0.788770	0.938609	1.08458	1.22652	1.36426	1.49764
0.75	−0.063750	0.091933	0.244288	0.393131	0.538284	0.679581
0.80	−0.892518	−0.733186	−0.576616	−0.423006	−0.272548	−0.125423
0.85	−1.68290	−1.52211	−1.36350	−1.20728	−1.05365	−0.902812
0.90	−2.42122	−2.26115	−2.10266	−1.94597	−1.79129	−1.63884
0.95	−3.09507	−2.93780	−2.78152	−2.62646	−2.47282	−2.32084
1.00	−3.69345	−3.54100	−3.38895	−3.23752	−3.08694	−2.93742
1.05	−4.20705	−4.06130	−3.91537	−3.76948	−3.62386	−3.47873
1.10	−4.62836	−4.49104	−4.35298	−4.21440	−4.07552	−3.93657
1.15	−4.95180	−4.82447	−4.69586	−4.56619	−4.43568	−4.30456
1.20	−5.17379	−5.05783	−4.94007	−4.82074	−4.70005	−4.57823
1.25	−5.29282	−5.18938	−5.08366	−4.97590	−4.86629	−4.75506
1.30	−5.30941	−5.21942	−5.12673	−5.03155	−4.93408	−4.83452
1.35	−5.22608	−5.15024	−5.07133	−4.98952	−4.90502	−4.81802
1.40	−5.04726	−4.98605	−4.92143	−4.85356	−4.78264	−4.70885
1.45	−4.77917	−4.73282	−4.68278	−4.62919	−4.57223	−4.51208
1.50	−4.42966	−4.39818	−4.36277	−4.32357	−4.28073	−4.23442
1.55	−4.00804	−3.99120	−3.97024	−3.94531	−3.91653	−3.88405
1.60	−3.52480	−3.52216	−3.51529	−3.50429	−3.48929	−3.47041
1.65	−2.99146	−3.00238	−3.00900	−3.01141	−3.00971	−3.00401
1.70	−2.42026	−2.44391	−2.46325	−2.47834	−2.48928	−2.49615
1.75	−1.82388	−1.85929	−1.89041	−1.91730	−1.94004	−1.95869

Table 2.12 (continued)

			t			
v	0.65	0.70	0.75	0.80	0.85	0.90
1.80	−1.21524	−1.26127	−1.30309	−1.34075	−1.37430	−1.40380
1.85	−0.607165	−0.662555	−0.713874	−0.761145	−0.804399	−0.843675
1.90	−0.012131	−0.075543	−0.135056	−0.190677	−0.242419	−0.290303
1.95	0.557960	0.487939	0.421606	0.358971	0.300038	0.244801
2.00	1.09203	1.01686	0.945130	0.876868	0.812093	0.750818
2.05	1.58006	1.50121	1.42552	1.35304	1.28381	1.21784
2.10	2.01323	1.93216	1.85396	1.77868	1.70637	1.63707
2.15	2.38413	2.30227	2.22296	2.14627	2.07225	2.00097
2.20	2.68685	2.60557	2.52651	2.44975	2.37535	2.30339
2.25	2.91708	2.83768	2.76016	2.68460	2.61109	2.53970
2.30	3.07219	2.99587	2.92107	2.84791	2.77646	2.70681
2.35	3.15121	3.07905	3.00807	2.93838	2.87007	2.80325
2.40	3.15484	3.08780	3.02159	2.95634	2.89215	2.82912
2.45	3.08540	3.02429	2.96369	2.90371	2.84447	2.78608
2.50	2.94673	2.89222	2.83790	2.78389	2.73030	2.67726
2.55	2.74407	2.69667	2.64916	2.60167	2.55431	2.50719
2.60	2.48394	2.44401	2.40369	2.36311	2.32239	2.28164
2.65	2.17393	2.14167	2.10878	2.07536	2.04155	2.00744
2.70	1.82255	1.79802	1.77263	1.74648	1.71969	1.69238
2.75	1.43901	1.42211	1.40414	1.38522	1.36545	1.34495
2.80	1.03299	1.02348	1.01274	1.00086	0.987959	0.974139
2.85	0.614411	0.611947	0.608099	0.602968	0.596659	0.589275
2.90	0.193253	0.197353	0.199956	0.201159	0.201058	0.199751
2.95	−0.220717	−0.210625	−0.202113	−0.195092	−0.189471	−0.185159
3.00	−0.618128	−0.602703	−0.588911	−0.576670	−0.565898	−0.556511
3.05	−0.990217	−0.970183	−0.951807	−0.935016	−0.919735	−0.905887
3.10	−1.32901	−1.30514	−1.28293	−1.26231	−1.24321	−1.22557
3.15	−1.62747	−1.60058	−1.57531	−1.55162	−1.52944	−1.50872
3.20	−1.87967	−1.85057	−1.82305	−1.79706	−1.77256	−1.74948
3.25	−2.08085	−2.05037	−2.02139	−1.99388	−1.96781	−1.94313
3.30	−2.22756	−2.19648	−2.16683	−2.13858	−2.11168	−2.08611
3.35	−2.31764	−2.28675	−2.25718	−2.22891	−2.20191	−2.17616
3.40	−2.35031	−2.32033	−2.29154	−2.26395	−2.23753	−2.21227
3.45	−2.32613	−2.29771	−2.27036	−2.24409	−2.21888	−2.19472
3.50	−2.24694	−2.22067	−2.19534	−2.17096	−2.14753	−2.12503
3.55	−2.11583	−2.09222	−2.06941	−2.04742	−2.02626	−2.00592
3.60	−1.93701	−1.91648	−1.89662	−1.87744	−1.85896	−1.84119
3.65	−1.71572	−1.69860	−1.68201	−1.66597	−1.65051	−1.63563
3.70	−1.45806	−1.44458	−1.43151	−1.41886	−1.40666	−1.39492
3.75	−1.17085	−1.16116	−1.15175	−1.14264	−1.13386	−1.12543
3.80	−0.861477	−0.855629	−0.849934	−0.844424	−0.839126	−0.834066
3.85	−0.537674	−0.535621	−0.533610	−0.531674	−0.529845	−0.528150

Table 2.12 (continued)

			t			
v	0.65	0.70	0.75	0.80	0.85	0.90
3.90	-0.207362	-0.208976	-0.210532	-0.212064	-0.213605	-0.215184
3.95	0.121543	0.116467	0.111539	0.106724	0.101987	0.097296
4.00	0.441314	0.433053	0.425018	0.417171	0.409479	0.401908
4.05	0.744587	0.733479	0.722661	0.712097	0.701751	0.691591
4.10	1.024520	1.010960	0.997736	0.984818	0.972170	0.959760
4.15	1.27497	1.25937	1.24416	1.22929	1.21474	1.20046
4.20	1.49056	1.47340	1.45664	1.44026	1.42421	1.40848
4.25	1.66687	1.64862	1.63078	1.61333	1.59624	1.57947
4.30	1.80046	1.78159	1.76315	1.74510	1.72741	1.71005
4.35	1.88896	1.86997	1.85138	1.83318	1.81534	1.79783
4.40	1.93111	1.91245	1.89417	1.87627	1.85871	1.84147
4.45	1.92675	1.90886	1.89133	1.87414	1.85727	1.84070
4.50	1.87683	1.86011	1.84371	1.82763	1.81183	1.79631
4.55	1.78335	1.76815	1.75325	1.73862	1.72424	1.71010
4.60	1.64929	1.63594	1.62283	1.60995	1.59729	1.58483
4.65	1.47855	1.46730	1.45624	1.44537	1.43467	1.42413
4.70	1.27582	1.26687	1.25806	1.24939	1.24085	1.23242
4.75	1.04647	1.03996	1.03354	1.02721	1.02095	1.01478
4.80	0.796397	0.792399	0.788444	0.784530	0.780650	0.776802
4.85	0.531894	0.530427	0.528956	0.527478	0.525992	0.524493
4.90	0.259491	0.260512	0.261482	0.262404	0.263275	0.264094
4.95	-0.014207	-0.010799	-0.007481	-0.004254	-0.001114	0.001936
5.00	-0.282672	-0.277028	-0.271511	-0.266120	-0.260852	-0.255705

			t			
v	0.95	1.0	2.0	3.0	4.0	5.0
0.00	6.59398	6.47588	3.43888	0.436503	-1.00450	-0.621986
0.05	6.91474	6.81402	4.01403	1.01143	-0.596603	-0.398755
0.10	7.11684	7.03459	4.50311	1.53821	-0.209321	-0.186108
0.15	7.19822	7.13523	4.89942	2.00844	0.149619	0.009959
0.20	7.15906	7.11586	5.19819	2.41514	0.473434	0.184096
0.25	7.00179	6.97862	5.39662	2.75284	0.756418	0.331717
0.30	6.73103	6.72785	5.49391	3.01771	0.994035	0.449090
0.35	6.35345	6.36991	5.49124	3.20751	1.18299	0.533415
0.40	5.87761	5.91313	5.39172	3.32169	1.32126	0.582865
0.45	5.31381	5.36754	5.20026	3.36128	1.40810	0.596616
0.50	4.67386	4.74470	4.92344	3.32887	1.44406	0.574840
0.55	3.97081	4.05744	4.56939	3.22849	1.43088	0.518685
0.60	3.21869	3.31959	4.14751	3.06550	1.37143	0.430219
0.65	2.43222	2.54572	3.66833	2.84643	1.26965	0.312368
0.70	1.62654	1.75081	3.14323	2.57879	1.13037	0.168819

Table 2.12 (continued)

				t		
v	0.95	1.0	2.0	3.0	4.0	5.0
0.75	0.816858	0.949966	2.58419	2.27092	0.959199	0.003919
0.80	0.018192	0.158129	2.00355	1.93174	0.762381	−0.177451
0.85	−0.754943	−0.610232	1.41371	1.57055	0.546602	−0.370007
0.90	−1.48881	−1.34139	0.826923	1.19683	0.318833	−0.568198
0.95	−2.17070	−2.02262	0.255007	0.820004	0.086149	−0.766357
1.00	−2.78919	−2.64244	−0.290881	0.449250	−0.144441	−0.958840
1.05	−3.33431	−3.19082	−0.800459	0.093280	−0.366170	−1.14017
1.10	−3.79777	−3.65934	−1.26452	−0.239833	−0.572666	−1.30517
1.15	−4.17304	−4.04134	−1.67510	−0.542825	−0.758105	−1.44911
1.20	−4.45549	−4.33204	−2.02560	−0.809378	−0.917337	−1.56778
1.25	−4.64242	−4.52857	−2.31092	−1.03425	−1.04600	−1.65763
1.30	−4.73309	−4.62999	−2.52752	−1.21335	−1.14062	−1.71584
1.35	−4.72872	−4.63732	−2.67344	−1.34384	−1.19866	−1.74036
1.40	−4.63237	−4.55340	−2.74832	−1.42410	−1.21858	−1.73000
1.45	−4.44890	−4.38288	−2.75337	−1.45382	−1.19986	−1.68443
1.50	−4.18480	−4.13203	−2.69130	−1.43390	−1.14299	−1.60414
1.55	−3.84801	−3.80858	−2.56619	−1.36644	−1.04942	−1.49052
1.60	−3.44780	−3.42157	−2.38344	−1.25466	−0.921560	−1.34572
1.65	−2.99443	−2.98110	−2.14953	−1.10277	−0.762645	−1.17265
1.70	−2.49905	−2.49809	−1.87194	−0.915889	−0.576687	−0.974879
1.75	−1.97334	−1.98407	−1.55886	−0.699853	−0.368350	−0.756576
1.80	−1.42930	−1.45089	−1.21908	−0.461100	−0.142823	−0.522367
1.85	−0.879020	−0.910487	−0.861719	−0.206490	0.094308	−0.277245
1.90	−0.334359	−0.374623	−0.496044	0.056874	0.337206	−0.026439
1.95	0.193249	0.145361	−0.131238	0.321815	0.579927	0.224704
2.00	0.693048	0.638778	0.223795	0.581256	0.816566	0.470857
2.05	1.15517	1.09580	0.560628	0.828391	1.04140	0.706830
2.10	1.57083	1.50766	0.871492	1.05684	1.24904	0.927695
2.15	1.93247	1.86680	1.14944	1.26080	1.43452	1.12890
2.20	2.23393	2.16703	1.38848	1.43516	1.59345	1.30638
2.25	2.47051	2.40359	1.58369	1.57564	1.72212	1.45663
2.30	2.63906	2.57328	1.73133	1.67884	1.81754	1.57679
2.35	2.73801	2.67442	1.82887	1.74235	1.87754	1.66473
2.40	2.76735	2.70694	1.87504	1.76476	1.90081	1.71905
2.45	2.72863	2.67224	1.86982	1.74569	1.88690	1.73914
2.50	2.62487	2.57323	1.81443	1.68578	1.83626	1.72518
2.55	2.46044	2.41415	1.71125	1.58666	1.75019	1.67809
2.60	2.24097	2.20049	1.56378	1.45089	1.63081	1.59957
2.65	1.97317	1.93882	1.37649	1.28192	1.48098	1.49198

Table 2.12 (continued)

v	0.95	1.0	2.0	3.0	4.0	5.0
2.70	1.66467	1.63665	1.15472	1.08392	1.30428	1.35833
2.75	1.32382	1.30218	0.904530	0.861762	1.10486	1.20219
2.80	0.959508	0.944172	0.632560	0.620808	0.887352	1.02756
2.85	0.580917	0.571688	0.345835	0.366833	0.656783	0.838828
2.90	0.197335	0.193907	0.051610	0.105854	0.418411	0.640629
2.95	−0.182065	−0.180097	−0.242813	−0.156006	0.177625	0.437739
3.00	−0.548424	−0.541551	−0.530250	−0.412673	−0.060196	0.234964
3.05	−0.893396	−0.882181	−0.803811	−0.658262	−0.289804	0.037024
3.10	−1.20932	−1.19439	−1.05705	−0.887216	−0.506206	−0.151557
3.15	−1.48940	−1.47140	−1.28413	−1.09444	−0.704780	−0.326555
3.20	−1.72779	−1.70742	−1.47991	−1.27539	−0.881379	−0.484141
3.25	−1.91978	−1.89773	−1.64010	−1.42623	−1.03242	−0.620969
3.30	−2.06183	−2.03879	−1.76131	−1.54386	−1.15497	−0.734246
3.35	−2.15163	−2.12828	−1.84116	−1.62598	−1.24681	−0.821794
3.40	−2.18814	−2.16512	−1.87826	−1.67120	−1.30644	−0.882090
3.45	−2.17160	−2.14950	−1.87229	−1.67897	−1.33317	−0.914296
3.50	−2.10348	−2.08284	−1.82394	−1.64965	−1.32708	−0.918269
3.55	−1.98640	−1.96771	−1.73490	−1.58448	−1.28901	−0.894553
3.60	−1.82412	−1.80777	−1.60783	−1.48549	−1.22057	−0.844361
3.65	−1.62135	−1.60766	−1.44621	−1.35553	−1.12403	−0.769532
3.70	−1.38367	−1.37290	−1.25432	−1.19808	−1.00230	−0.672482
3.75	−1.11736	−1.10967	−1.03710	−1.01728	−0.858861	−0.556134
3.80	−0.829266	−0.824745	−0.800006	−0.817723	−0.697638	−0.423847
3.85	−0.526614	−0.525260	−0.548883	−0.604394	−0.522926	−0.279321
3.90	−0.216831	−0.218569	−0.289834	−0.382535	−0.339275	−0.126509
3.95	0.092622	0.087939	−0.029059	−0.157512	−0.151382	0.030484
4.00	0.394429	0.387012	0.227292	0.065308	0.036030	0.187503
4.05	0.681585	0.671704	0.473261	0.280688	0.218318	0.340445
4.10	0.947557	0.935531	0.703224	0.483643	0.391032	0.485358
4.15	1.18643	1.17261	0.912017	0.669553	0.550022	0.618534
4.20	1.39302	1.37780	1.09505	0.834273	0.691528	0.736600
4.25	1.56300	1.54680	1.24842	0.974220	0.812272	0.836592
4.30	1.69300	1.67622	1.36896	1.08645	0.909526	0.916027
4.35	1.78062	1.76369	1.45437	1.16874	0.981170	0.972956
4.40	1.82452	1.80784	1.50317	1.21959	1.02574	1.00601
4.45	1.82441	1.80837	1.51480	1.23829	1.04245	1.01441
4.50	1.78104	1.76600	1.48958	1.22491	1.03121	0.998019
4.55	1.69619	1.68247	1.42871	1.18029	0.992617	0.957292
4.60	1.57256	1.56045	1.33422	1.10600	0.927918	0.893290
4.65	1.41374	1.40349	1.20889	1.00430	0.838996	0.807637
4.70	1.22411	1.21589	1.05623	0.878104	0.728302	0.702479
4.75	1.00867	1.00261	0.880320	0.730851	0.598792	0.580424

Table 2.12 (continued)

			t			
v	0.95	1.0	2.0	3.0	4.0	5.0
4.80	0.772978	0.769172	0.685753	0.566453	0.453853	0.444474
4.85	0.522980	0.521446	0.477496	0.389183	0.297210	0.297951
4.90	0.264859	0.265568	0.260773	0.203567	0.132837	0.144412
4.95	0.004898	0.007769	0.040936	0.014278	−0.035144	−0.012443
5.00	−0.250679	−0.245774	−0.176662	−0.173986	−0.202565	−0.168856

3 Numerical Results – Tabulation of the Kelvin Functions, Their First and Second Derivatives with Respect to the Order

The Kelvin functions ber(t), bei(t), ker(t) and kei(t) and their first derivatives with regard to argument t were tabulated by McLachlan in his book *Bessel Functions for Engineers* [22] in $0 \le v \le 10$ range with 0.10 intervals. For the same range of t, he also gave values of ber$_n$(t), bei$_n$(t), ker$_n$(t) and kei$_n$(t) with $n = 1, 2, 3, 4$ and 5. In MATHEMATICA program, the Kelvin functions are available for any argument t and order v. Similarly as in cases of other functions, the Kelvin functions are presented here in $0 \le v \le 5$ range with 0.05 intervals, and arguments t are in $0.05 \le t \le 5$ range with 0.05 intervals for $t \le 1.0$ and with 1.0 intervals for $t > 1.0$. The first and second derivatives with respect to the order of the Bessel functions were always calculated by using the central-difference formulas of order $O(h^4)$ with $h = 0.001$.

3.1 The Kelvin Functions ber$_v$(t), and Their First and Second Derivatives with Respect to the Order

Applying the central-difference formulas of order $O(h^4)$ with $h = 10^{-3}$, derivatives with respect to the order of the Kelvin function (Figures 3.1–3.18) of the first kind were evaluated using MATHEMATICA program (Tables 3.1–3.6):

$$\frac{\partial \mathrm{ber}_v(t)}{\partial v} = \frac{-\mathrm{ber}_{v+2h}(t) + 8\,\mathrm{ber}_{v+h}(t) - 8\,\mathrm{ber}_{v-h}(t) + \mathrm{ber}_{v-2h}(t)}{12h} \tag{3.1.1}$$

$$\frac{\partial^2 \mathrm{ber}_v(t)}{\partial v^2} = \frac{-\mathrm{ber}_{v+2h}(t) + 16\,\mathrm{ber}_{v+h}(t) - 30\,\mathrm{ber}_v(t) + 16\,\mathrm{ber}_{v-h}(t) - \mathrm{ber}_{v-2h}(t)}{12h^2} \tag{3.1.2}$$

https://doi.org/10.1515/9783110682472-003

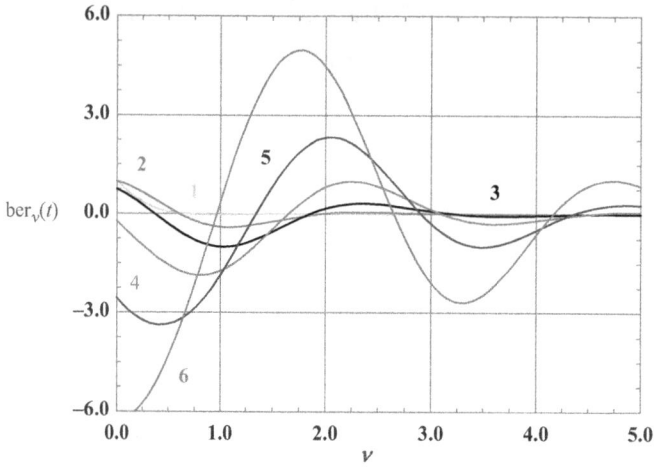

Figure 3.1: The Kelvin functions ber$_v$(t) plotted as a function of the order v, at constant values of argument t.

$1 - t = 0.05; 2 - t = 1.0; 3 - t = 2.0; 4 - t = 3.0; 5 - t = 4.0; 6 - t = 5.0.$

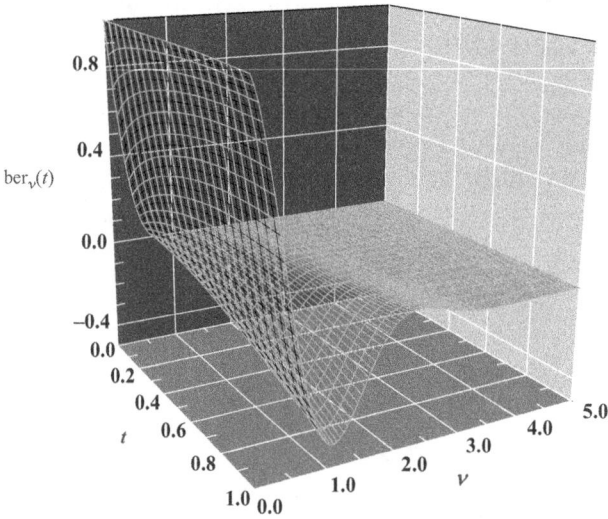

Figure 3.2: The Kelvin functions ber$_v$(t) as a function of v and t.

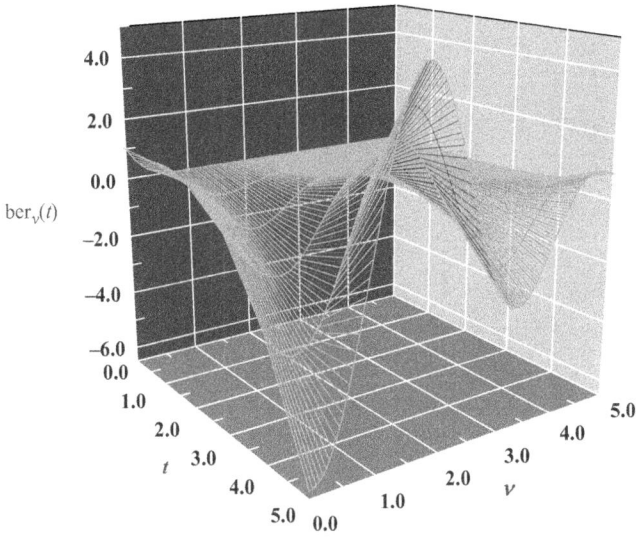

Figure 3.3: The Kelvin functions ber$_v$(t) as a function of v and t.

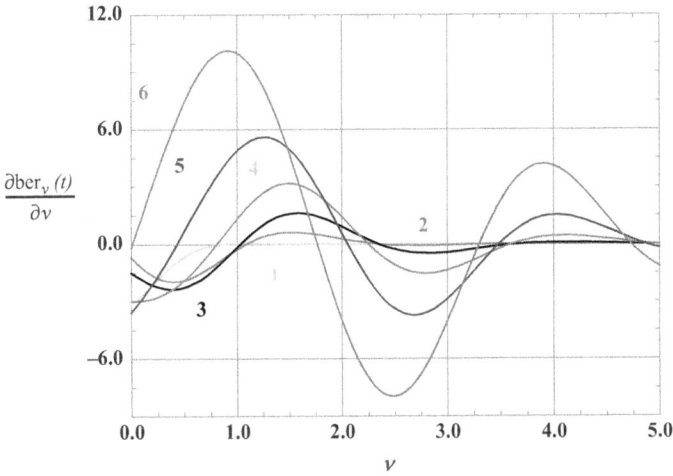

Figure 3.4: First derivatives of the Kelvin functions ber$_v$(t) with respect to the order v as a function of v at constant values of argument t.

$1 - t = 0.05$; $2 - t = 1.0$; $3 - t = 2.0$; $4 - t = 3.0$; $5 - t = 4.0$; $6 - t = 5.0$.

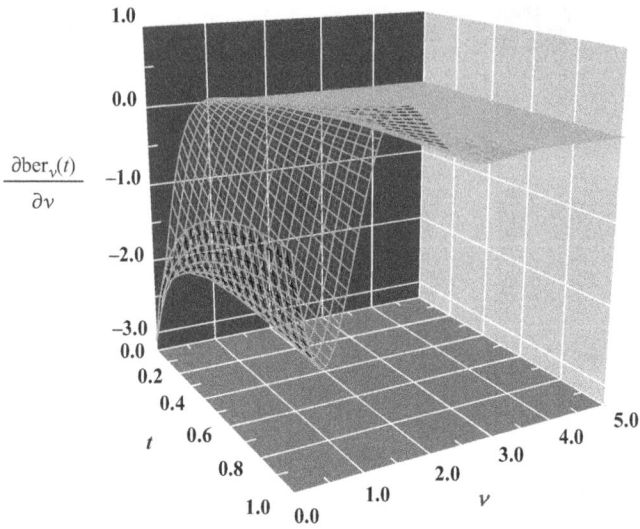

Figure 3.5: First derivatives with respect to the order of the Kelvin functions $\partial \text{ber}_\nu(t)/\partial \nu$ as a function of ν and t.

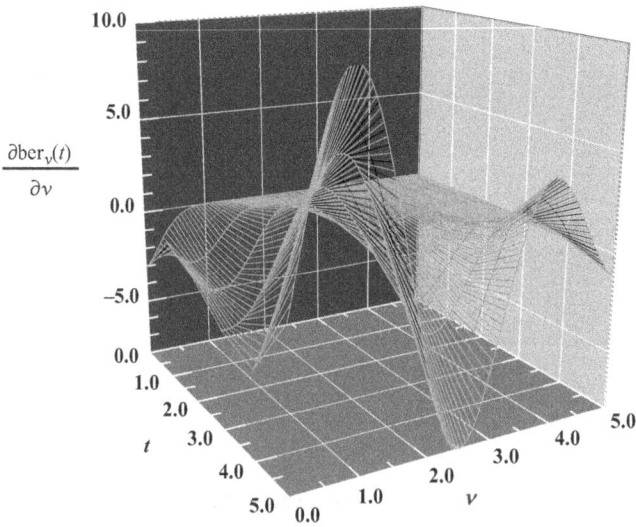

Figure 3.6: First derivatives with respect to the order of the Kelvin functions $\partial \text{ber}_\nu(t)/\partial \nu$ as a function of ν and t.

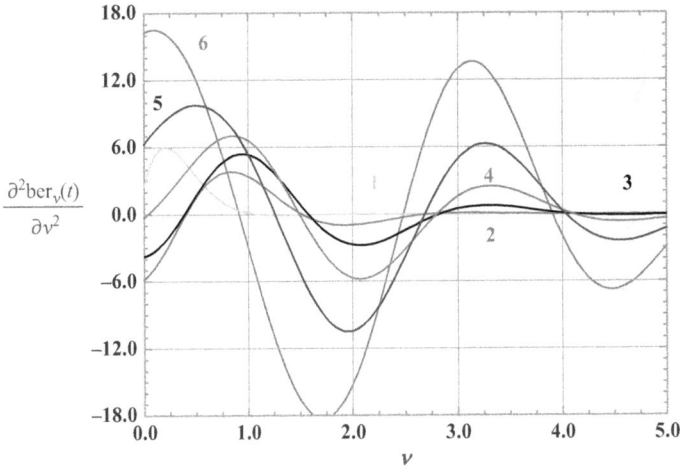

Figure 3.7: Second derivatives of the Kelvin functions $ber_v(t)$ with respect to the order v as a function of v, at constant values of argument t.

$1 - t = 0.05; 2 - t = 1.0; 3 - t = 2.0; 4 - t = 3.0; 5 - t = 4.0; 6 - t = 5.0.$

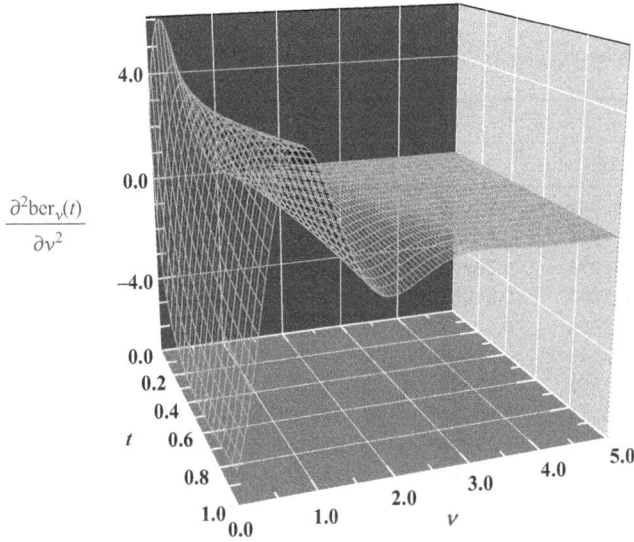

Figure 3.8: Second derivatives with respect to the order of the Kelvin functions $\partial^2 ber_v(t)/\partial v^2$ as a function of v and t.

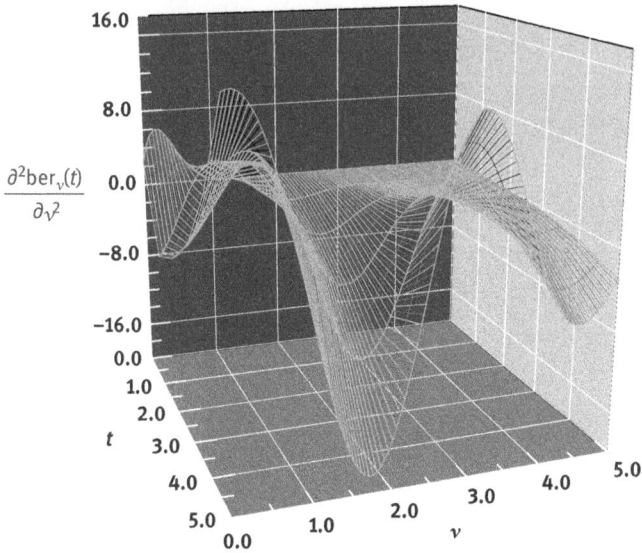

Figure 3.9: Second derivatives with respect to the order of the Kelvin functions $\partial^2 \mathrm{ber}_\nu(t)/\partial\nu^2$ as a function of ν and t.

Table 3.1: The Kelvin functions $\mathrm{ber}_\nu(t)$.

			t			
ν	0.05	0.10	0.15	0.20	0.25	0.30
0.00	1.00000	0.999998	0.999992	0.999975	0.999939	0.999873
0.05	0.848218	0.877944	0.895603	0.908116	0.917691	0.925320
0.10	0.706684	0.757095	0.787881	0.810089	0.827323	0.841229
0.15	0.578089	0.641043	0.680554	0.709556	0.732358	0.750954
0.20	0.463907	0.532464	0.576673	0.609680	0.635959	0.657606
0.25	0.364641	0.433198	0.478608	0.513086	0.540868	0.563967
0.30	0.280046	0.344351	0.388091	0.421847	0.449365	0.472441
0.35	0.209340	0.266415	0.306265	0.337512	0.363262	0.385022
0.40	0.151382	0.199381	0.233765	0.261137	0.283917	0.303287
0.45	0.104820	0.142857	0.170790	0.193338	0.212256	0.228408
0.50	0.068207	0.096168	0.117185	0.134351	0.148824	0.161173
0.55	0.040094	0.058449	0.072526	0.084097	0.093825	0.102026
0.60	0.019095	0.028728	0.036184	0.042242	0.047176	0.051107
0.65	0.003927	0.005983	0.007396	0.008258	0.008568	0.008297
0.70	−0.006559	−0.010804	−0.014678	−0.018516	−0.022493	−0.026730
0.75	−0.013370	−0.022606	−0.030913	−0.038832	−0.046626	−0.054468
0.80	−0.017365	−0.030331	−0.042175	−0.053483	−0.064539	−0.075527

Table 3.1 (continued)

			t			
v	0.05	0.10	0.15	0.20	0.25	0.30
0.85	-0.019263	-0.034797	-0.049295	-0.063271	-0.076983	-0.090597
0.90	-0.019652	-0.036730	-0.053047	-0.068980	-0.084726	-0.100413
0.95	-0.019002	-0.036755	-0.054135	-0.071350	-0.088518	-0.105721
1.00	-0.017683	-0.035400	-0.053182	-0.071064	-0.089077	-0.107255
1.05	-0.015976	-0.033102	-0.050731	-0.068736	-0.087070	-0.105715
1.10	-0.014088	-0.030215	-0.047240	-0.064908	-0.083100	-0.101756
1.15	-0.012169	-0.027015	-0.043092	-0.060046	-0.077705	-0.095970
1.20	-0.010320	-0.023715	-0.038595	-0.054543	-0.071349	-0.088887
1.25	-0.008606	-0.020472	-0.033993	-0.048722	-0.064428	-0.080966
1.30	-0.007064	-0.017394	-0.029469	-0.042839	-0.057267	-0.072600
1.35	-0.005709	-0.014554	-0.025158	-0.037094	-0.050129	-0.064110
1.40	-0.004545	-0.011993	-0.021153	-0.031634	-0.043217	-0.055758
1.45	-0.003562	-0.009730	-0.017511	-0.026561	-0.036684	-0.047746
1.50	-0.002747	-0.007767	-0.014262	-0.021941	-0.030635	-0.040224
1.55	-0.002081	-0.006092	-0.011413	-0.017809	-0.025137	-0.033295
1.60	-0.001547	-0.004686	-0.008957	-0.014175	-0.020225	-0.027023
1.65	-0.001124	-0.003524	-0.006872	-0.011030	-0.015908	-0.021440
1.70	-0.000794	-0.002579	-0.005132	-0.008352	-0.012176	-0.016550
1.75	-0.000543	-0.001824	-0.003702	-0.006109	-0.009000	-0.012335
1.80	-0.000354	-0.001230	-0.002547	-0.004262	-0.006343	-0.008764
1.85	-0.000215	-0.000773	-0.001633	-0.002768	-0.004160	-0.005790
1.90	-0.000115	-0.000429	-0.000924	-0.001585	-0.002402	-0.003362
1.95	-0.000046	-0.000177	-0.000388	-0.000670	-0.001019	-0.001424
2.00	0.000000	0.000001	0.000005	0.000017	0.000041	0.000084
2.05	0.000029	0.000122	0.000282	0.000514	0.000824	0.001219
2.10	0.000046	0.000197	0.000465	0.000855	0.001376	0.002037
2.15	0.000054	0.000240	0.000575	0.001071	0.001739	0.002589
2.20	0.000056	0.000258	0.000630	0.001189	0.001949	0.002923
2.25	0.000054	0.000258	0.000643	0.001231	0.002039	0.003083
2.30	0.000050	0.000247	0.000627	0.001217	0.002038	0.003107
2.35	0.000045	0.000228	0.000591	0.001163	0.001969	0.003027
2.40	0.000039	0.000205	0.000542	0.001083	0.001852	0.002873
2.45	0.000033	0.000180	0.000487	0.000986	0.001704	0.002667
2.50	0.000027	0.000155	0.000429	0.000880	0.001539	0.002429
2.55	0.000023	0.000132	0.000371	0.000773	0.001366	0.002175
2.60	0.000018	0.000110	0.000316	0.000668	0.001193	0.001917
2.65	0.000014	0.000091	0.000265	0.000568	0.001027	0.001665
2.70	0.000011	0.000073	0.000219	0.000477	0.000871	0.001425
2.75	0.000009	0.000059	0.000179	0.000394	0.000728	0.001201
2.80	0.000007	0.000046	0.000143	0.000321	0.000599	0.000997
2.85	0.000005	0.000036	0.000113	0.000257	0.000485	0.000815
2.90	0.000004	0.000027	0.000088	0.000202	0.000386	0.000654
2.95	0.000003	0.000020	0.000067	0.000156	0.000301	0.000515

Table 3.1 (continued)

				t		
v	0.05	0.10	0.15	0.20	0.25	0.30
3.00	0.000002	0.000015	0.000050	0.000118	0.000229	0.000396
3.05	0.000001	0.000010	0.000036	0.000086	0.000170	0.000295
3.10	0.000001	0.000007	0.000025	0.000061	0.000121	0.000212
3.15	0.000001	0.000005	0.000016	0.000041	0.000082	0.000145
3.20		0.000003	0.000010	0.000025	0.000051	0.000090
3.25		0.000001	0.000005	0.000013	0.000027	0.000048
3.30			0.000002	0.000004	0.000009	0.000016
3.35			−0.000001	−0.000002	−0.000004	−0.000008
3.40			−0.000002	−0.000006	−0.000013	−0.000025
3.45			−0.000003	−0.000009	−0.000019	−0.000037
3.50			−0.000004	−0.000010	−0.000023	−0.000044
3.55			−0.000004	−0.000011	−0.000025	−0.000047
3.60			−0.000004	−0.000011	−0.000025	−0.000048
3.65			−0.000004	−0.000011	−0.000024	−0.000047
3.70			−0.000003	−0.000010	−0.000023	−0.000044
3.75			−0.000003	−0.000009	−0.000021	−0.000041
3.80			−0.000003	−0.000008	−0.000019	−0.000037
3.85			−0.000002	−0.000007	−0.000016	−0.000033
3.90			−0.000002	−0.000006	−0.000014	−0.000029
3.95			−0.000002	−0.000005	−0.000012	−0.000025
4.00			−0.000001	−0.000004	−0.000010	−0.000021
4.05			−0.000001	−0.000003	−0.000008	−0.000018
4.10			−0.000001	−0.000003	−0.000007	−0.000015
4.15			−0.000001	−0.000002	−0.000006	−0.000012
4.20			−0.000001	−0.000002	−0.000004	−0.000009
4.25				−0.000001	−0.000003	−0.000007
4.30				−0.000001	−0.000003	−0.000006
4.35				−0.000001	−0.000002	−0.000004
4.40				−0.000001	−0.000001	−0.000003
4.45					−0.000001	−0.000002
4.50					−0.000001	−0.000001
4.55						−0.000001

				t		
v	0.35	0.40	0.45	0.50	0.55	0.60
0.00	0.999766	0.999600	0.999359	0.999023	0.998570	0.997975
0.05	0.931529	0.936630	0.940817	0.944216	0.946909	0.948951
0.10	0.852694	0.862252	0.870243	0.876896	0.882369	0.886772
0.15	0.766435	0.779461	0.790458	0.799717	0.807441	0.813776
0.20	0.675777	0.691180	0.704276	0.715379	0.724713	0.732438
0.25	0.583503	0.600164	0.614404	0.626531	0.636765	0.645267
0.30	0.492085	0.508920	0.523357	0.535678	0.546083	0.554716
0.35	0.403643	0.419653	0.433398	0.445119	0.454982	0.463109

Table 3.1 (continued)

			t			
v	0.35	0.40	0.45	0.50	0.55	0.60
0.40	0.319921	0.334231	0.346493	0.356890	0.365553	0.372574
0.45	0.242280	0.254172	0.264280	0.272734	0.279620	0.285000
0.50	0.171714	0.180639	0.188068	0.194077	0.198712	0.202003
0.55	0.108870	0.114454	0.118830	0.122026	0.124053	0.124909
0.60	0.054083	0.056123	0.057226	0.057378	0.056559	0.054745
0.65	0.007411	0.005866	0.003620	0.000630	−0.003149	−0.007758
0.70	−0.031323	−0.036351	−0.041885	−0.047990	−0.054724	−0.062144
0.75	−0.062481	−0.070768	−0.079414	−0.088493	−0.098074	−0.108218
0.80	−0.086579	−0.097801	−0.109282	−0.121100	−0.133324	−0.146017
0.85	−0.104235	−0.117996	−0.131961	−0.146204	−0.160790	−0.175778
0.90	−0.116142	−0.131993	−0.148036	−0.164332	−0.180939	−0.197907
0.95	−0.123027	−0.140494	−0.158174	−0.176113	−0.194356	−0.212945
1.00	−0.125629	−0.144231	−0.163092	−0.182243	−0.201714	−0.221534
1.05	−0.124670	−0.143938	−0.163529	−0.183458	−0.203738	−0.224387
1.10	−0.120839	−0.140331	−0.160220	−0.180504	−0.201183	−0.222259
1.15	−0.114779	−0.134091	−0.153877	−0.174118	−0.194802	−0.215919
1.20	−0.107073	−0.125848	−0.145170	−0.165007	−0.185332	−0.206127
1.25	−0.098240	−0.116176	−0.134721	−0.153832	−0.173474	−0.193619
1.30	−0.088731	−0.105580	−0.123087	−0.141200	−0.159877	−0.179084
1.35	−0.078928	−0.094501	−0.110761	−0.127655	−0.145135	−0.163161
1.40	−0.069150	−0.083309	−0.098169	−0.113673	−0.129771	−0.146422
1.45	−0.059648	−0.072309	−0.085665	−0.099660	−0.114244	−0.129371
1.50	−0.050618	−0.061744	−0.073541	−0.085955	−0.098937	−0.112443
1.55	−0.042204	−0.051799	−0.062024	−0.072829	−0.084168	−0.095999
1.60	−0.034502	−0.042605	−0.051284	−0.060492	−0.070187	−0.080330
1.65	−0.027570	−0.034250	−0.041439	−0.049096	−0.057182	−0.065662
1.70	−0.021431	−0.026781	−0.032563	−0.038742	−0.045286	−0.052160
1.75	−0.016083	−0.020210	−0.024688	−0.029487	−0.034579	−0.039933
1.80	−0.011499	−0.014525	−0.017817	−0.021350	−0.025100	−0.029040
1.85	−0.007640	−0.009691	−0.011923	−0.014316	−0.016849	−0.019497
1.90	−0.004452	−0.005657	−0.006961	−0.008347	−0.009796	−0.011288
1.95	−0.001875	−0.002361	−0.002868	−0.003382	−0.003886	−0.004362
2.00	0.000156	0.000267	0.000427	0.000651	0.000953	0.001350
2.05	0.001708	0.002299	0.003005	0.003837	0.004809	0.005937
2.10	0.002845	0.003812	0.004947	0.006264	0.007776	0.009498
2.15	0.003631	0.004876	0.006337	0.008025	0.009955	0.012141
2.20	0.004124	0.005563	0.007254	0.009210	0.011446	0.013976
2.25	0.004378	0.005937	0.007775	0.009907	0.012348	0.015113
2.30	0.004441	0.006058	0.007972	0.010200	0.012757	0.015659
2.35	0.004358	0.005980	0.007910	0.010164	0.012760	0.015715
2.40	0.004166	0.005751	0.007646	0.009870	0.012441	0.015375
2.45	0.003895	0.005411	0.007233	0.009381	0.011873	0.014726
2.50	0.003574	0.004996	0.006715	0.008751	0.011122	0.013846

Table 3.1 (continued)

v	0.35	0.40	0.45	0.50	0.55	0.60
2.55	0.003224	0.004536	0.006130	0.008026	0.010245	0.012803
2.60	0.002863	0.004054	0.005508	0.007248	0.009291	0.011656
2.65	0.002505	0.003569	0.004877	0.006448	0.008301	0.010455
2.70	0.002160	0.003096	0.004255	0.005653	0.007309	0.009242
2.75	0.001834	0.002647	0.003657	0.004883	0.006342	0.008049
2.80	0.001535	0.002229	0.003096	0.004154	0.005419	0.006905
2.85	0.001263	0.001846	0.002579	0.003477	0.004555	0.005827
2.90	0.001022	0.001502	0.002110	0.002859	0.003761	0.004830
2.95	0.000810	0.001198	0.001692	0.002304	0.003044	0.003922
3.00	0.000627	0.000933	0.001325	0.001812	0.002404	0.003110
3.05	0.000471	0.000706	0.001008	0.001384	0.001844	0.002393
3.10	0.000341	0.000514	0.000737	0.001017	0.001359	0.001770
3.15	0.000234	0.000354	0.000511	0.000707	0.000948	0.001237
3.20	0.000147	0.000224	0.000324	0.000450	0.000604	0.000790
3.25	0.000079	0.000120	0.000174	0.000241	0.000323	0.000420
3.30	0.000026	0.000039	0.000055	0.000074	0.000097	0.000122
3.35	−0.000014	−0.000023	−0.000036	−0.000054	−0.000079	−0.000112
3.40	−0.000043	−0.000069	−0.000104	−0.000151	−0.000212	−0.000291
3.45	−0.000063	−0.000100	−0.000151	−0.000220	−0.000308	−0.000421
3.50	−0.000075	−0.000120	−0.000183	−0.000266	−0.000373	−0.000510
3.55	−0.000081	−0.000131	−0.000200	−0.000293	−0.000412	−0.000565
3.60	−0.000083	−0.000135	−0.000208	−0.000304	−0.000431	−0.000591
3.65	−0.000082	−0.000134	−0.000207	−0.000304	−0.000432	−0.000595
3.70	−0.000078	−0.000129	−0.000199	−0.000295	−0.000421	−0.000582
3.75	−0.000073	−0.000121	−0.000188	−0.000279	−0.000400	−0.000556
3.80	−0.000067	−0.000111	−0.000173	−0.000259	−0.000373	−0.000520
3.85	−0.000060	−0.000100	−0.000157	−0.000236	−0.000341	−0.000477
3.90	−0.000053	−0.000089	−0.000140	−0.000212	−0.000307	−0.000432
3.95	−0.000046	−0.000077	−0.000123	−0.000187	−0.000273	−0.000384
4.00	−0.000039	−0.000067	−0.000107	−0.000163	−0.000238	−0.000337
4.05	−0.000033	−0.000057	−0.000091	−0.000140	−0.000205	−0.000292
4.10	−0.000027	−0.000047	−0.000077	−0.000118	−0.000174	−0.000249
4.15	−0.000022	−0.000039	−0.000064	−0.000098	−0.000146	−0.000209
4.20	−0.000018	−0.000032	−0.000052	−0.000080	−0.000120	−0.000173
4.25	−0.000014	−0.000025	−0.000041	−0.000065	−0.000097	−0.000140
4.30	−0.000011	−0.000020	−0.000032	−0.000051	−0.000077	−0.000111
4.35	−0.000008	−0.000015	−0.000025	−0.000039	−0.000059	−0.000086
4.40	−0.000006	−0.000011	−0.000018	−0.000029	−0.000044	−0.000064
4.45	−0.000004	−0.000008	−0.000013	−0.000021	−0.000032	−0.000046
4.50	−0.000003	−0.000005	−0.000009	−0.000014	−0.000021	−0.000031
4.55	−0.000002	−0.000003	−0.000005	−0.000008	−0.000013	−0.000019
4.60	−0.000001	−0.000001	−0.000003	−0.000004	−0.000006	−0.000009
4.65		0.000000	0.000000	−0.000001	−0.000001	−0.000001

Table 3.1 (continued)

			t			
v	0.35	0.40	0.45	0.50	0.55	0.60
4.70		0.000001	0.000001	0.000002	0.000003	0.000004
4.75		0.000001	0.000002	0.000004	0.000006	0.000009
4.80		0.000002	0.000002	0.000004	0.000008	0.000012
4.85		0.000002	0.000003	0.000006	0.000006	0.000014
4.90		0.000002	0.000004	0.000006	0.000006	0.000014
4.95		0.000002	0.000004	0.000006	0.000010	0.000015
5.00		0.000002	0.000003	0.000006	0.000006	0.000015

			t			
v	0.65	0.70	0.75	0.80	0.85	0.90
0.00	0.997211	0.996249	0.995057	0.993601	0.991846	0.989751
0.05	0.950374	0.951198	0.951427	0.951060	0.950086	0.948488
0.10	0.890181	0.892650	0.894213	0.894891	0.894692	0.893617
0.15	0.818825	0.822664	0.825345	0.826903	0.827362	0.826732
0.20	0.738671	0.743496	0.746975	0.749153	0.750057	0.749705
0.25	0.652152	0.657505	0.661389	0.663848	0.664912	0.664599
0.30	0.561685	0.567064	0.570911	0.573265	0.574153	0.573589
0.35	0.469587	0.474480	0.477830	0.479670	0.480018	0.478884
0.40	0.378015	0.381923	0.384327	0.385246	0.384688	0.382656
0.45	0.288911	0.291379	0.292418	0.292031	0.290219	0.286973
0.50	0.203964	0.204599	0.203906	0.201876	0.198496	0.193750
0.55	0.124588	0.123075	0.120352	0.116398	0.111190	0.104703
0.60	0.051908	0.048020	0.043048	0.036962	0.029728	0.021315
0.65	−0.013240	−0.019638	−0.026991	−0.035340	−0.044726	−0.055185
0.70	−0.070301	−0.079247	−0.089027	−0.099690	−0.111277	−0.123832
0.75	−0.118984	−0.130425	−0.142594	−0.155537	−0.169302	−0.183932
0.80	−0.159237	−0.173039	−0.187475	−0.202591	−0.218435	−0.235050
0.85	−0.191223	−0.207176	−0.223687	−0.240800	−0.258558	−0.277002
0.90	−0.215285	−0.233119	−0.251451	−0.270323	−0.289771	−0.309833
0.95	−0.231918	−0.251313	−0.271165	−0.291508	−0.312374	−0.333794
1.00	−0.241731	−0.262333	−0.283367	−0.304859	−0.326832	−0.349310
1.05	−0.245421	−0.266857	−0.288712	−0.311004	−0.333747	−0.356959
1.10	−0.243739	−0.265628	−0.287935	−0.310666	−0.333829	−0.357432
1.15	−0.237463	−0.259432	−0.281822	−0.304632	−0.327862	−0.351511
1.20	−0.227376	−0.249066	−0.271186	−0.293726	−0.316678	−0.340032
1.25	−0.214242	−0.235323	−0.256842	−0.278783	−0.301129	−0.323864
1.30	−0.198789	−0.218964	−0.239584	−0.260624	−0.282064	−0.303880
1.35	−0.181696	−0.200709	−0.220167	−0.240043	−0.260308	−0.280937
1.40	−0.163583	−0.181220	−0.199298	−0.217784	−0.236645	−0.255852
1.45	−0.145002	−0.161096	−0.177618	−0.194532	−0.211802	−0.229395
1.50	−0.126432	−0.140863	−0.155699	−0.170903	−0.186439	−0.202269
1.55	−0.108280	−0.120973	−0.134039	−0.147441	−0.161142	−0.175104
1.60	−0.090881	−0.101803	−0.113058	−0.124609	−0.136419	−0.148450

Table 3.1 (continued)

				t		
v	0.65	0.70	0.75	0.80	0.85	0.90
1.65	-0.074499	-0.083657	-0.093100	-0.102792	-0.112696	-0.122775
1.70	-0.059332	-0.066769	-0.074435	-0.082298	-0.090320	-0.098466
1.75	-0.045519	-0.051307	-0.057265	-0.063360	-0.069559	-0.075826
1.80	-0.033143	-0.037381	-0.041725	-0.046144	-0.050607	-0.055080
1.85	-0.022238	-0.025046	-0.027893	-0.030752	-0.033593	-0.036384
1.90	-0.012801	-0.014312	-0.015797	-0.017229	-0.018581	-0.019822
1.95	-0.004791	-0.005150	-0.005419	-0.005572	-0.005583	-0.005425
2.00	0.001859	0.002501	0.003295	0.004266	0.005436	0.006832
2.05	0.007237	0.008728	0.010430	0.012363	0.014550	0.017016
2.10	0.011446	0.013637	0.016090	0.018826	0.021865	0.025230
2.15	0.014600	0.017347	0.020402	0.023782	0.027509	0.031605
2.20	0.016817	0.019985	0.023497	0.027373	0.031631	0.036293
2.25	0.018219	0.021681	0.025518	0.029748	0.034389	0.039461
2.30	0.018923	0.022567	0.026606	0.031060	0.035946	0.041284
2.35	0.019045	0.022767	0.026900	0.031461	0.036467	0.041938
2.40	0.018691	0.022405	0.026535	0.031099	0.036115	0.041600
2.45	0.017960	0.021591	0.025637	0.030115	0.035044	0.040440
2.50	0.016943	0.020429	0.024323	0.028641	0.033401	0.038619
2.55	0.015720	0.019013	0.022699	0.026796	0.031320	0.036287
2.60	0.014361	0.017424	0.020861	0.024688	0.028923	0.033581
2.65	0.012926	0.015733	0.018890	0.022414	0.026321	0.030625
2.70	0.011466	0.014000	0.016857	0.020055	0.023606	0.027526
2.75	0.010022	0.012275	0.014824	0.017681	0.020862	0.024378
2.80	0.008627	0.010600	0.012836	0.015351	0.018154	0.021260
2.85	0.007305	0.009004	0.010935	0.013110	0.015540	0.018236
2.90	0.006076	0.007512	0.009148	0.010994	0.013061	0.015358
2.95	0.004951	0.006139	0.007496	0.009030	0.010751	0.012665
3.00	0.003938	0.004896	0.005993	0.007236	0.008631	0.010185
3.05	0.003039	0.003788	0.004647	0.005621	0.006716	0.007935
3.10	0.002253	0.002815	0.003460	0.004191	0.005012	0.005926
3.15	0.001578	0.001974	0.002429	0.002943	0.003520	0.004160
3.20	0.001007	0.001260	0.001548	0.001873	0.002234	0.002632
3.25	0.000534	0.000664	0.000810	0.000971	0.001147	0.001335
3.30	0.000149	0.000177	0.000203	0.000227	0.000245	0.000255
3.35	-0.000155	-0.000211	-0.000283	-0.000373	-0.000485	-0.000624
3.40	-0.000389	-0.000511	-0.000660	-0.000841	-0.001059	-0.001318
3.45	-0.000561	-0.000734	-0.000943	-0.001194	-0.001493	-0.001846
3.50	-0.000680	-0.000889	-0.001142	-0.001445	-0.001805	-0.002228
3.55	-0.000755	-0.000988	-0.001271	-0.001609	-0.002010	-0.002482
3.60	-0.000792	-0.001040	-0.001340	-0.001699	-0.002126	-0.002627
3.65	-0.000800	-0.001053	-0.001360	-0.001728	-0.002166	-0.002681
3.70	-0.000785	-0.001035	-0.001341	-0.001708	-0.002145	-0.002660
3.75	-0.000752	-0.000995	-0.001291	-0.001649	-0.002076	-0.002579

Table 3.1 (continued)

			t			
v	0.65	0.70	0.75	0.80	0.85	0.90
3.80	-0.000705	-0.000936	-0.001219	-0.001561	-0.001970	-0.002453
3.85	-0.000650	-0.000866	-0.001131	-0.001452	-0.001837	-0.002292
3.90	-0.000590	-0.000789	-0.001033	-0.001330	-0.001686	-0.002109
3.95	-0.000528	-0.000707	-0.000929	-0.001199	-0.001524	-0.001911
4.00	-0.000465	-0.000625	-0.000824	-0.001066	-0.001359	-0.001707
4.05	-0.000404	-0.000545	-0.000720	-0.000934	-0.001194	-0.001503
4.10	-0.000345	-0.000468	-0.000620	-0.000807	-0.001033	-0.001305
4.15	-0.000291	-0.000395	-0.000526	-0.000686	-0.000881	-0.001115
4.20	-0.000241	-0.000329	-0.000438	-0.000574	-0.000738	-0.000936
4.25	-0.000196	-0.000268	-0.000359	-0.000471	-0.000607	-0.000772
4.30	-0.000156	-0.000214	-0.000287	-0.000378	-0.000489	-0.000623
4.35	-0.000121	-0.000167	-0.000225	-0.000296	-0.000384	-0.000490
4.40	-0.000091	-0.000126	-0.000170	-0.000224	-0.000291	-0.000372
4.45	-0.000066	-0.000091	-0.000123	-0.000162	-0.000211	-0.000270
4.50	-0.000044	-0.000061	-0.000083	-0.000110	-0.000143	-0.000183
4.55	-0.000027	-0.000037	-0.000050	-0.000066	-0.000086	-0.000110
4.60	-0.000013	-0.000018	-0.000023	-0.000031	-0.000039	-0.000050
4.65	-0.000002	-0.000002	-0.000002	-0.000002	-0.000002	-0.000001
4.70	0.000007	0.000010	0.000014	0.000020	0.000027	0.000037
4.75	0.000013	0.000019	0.000026	0.000036	0.000049	0.000066
4.80	0.000017	0.000025	0.000035	0.000048	0.000065	0.000086
4.85	0.000020	0.000029	0.000041	0.000056	0.000076	0.000101
4.90	0.000022	0.000031	0.000044	0.000060	0.000082	0.000109
4.95	0.000022	0.000032	0.000045	0.000062	0.000084	0.000112
5.00	0.000022	0.000032	0.000045	0.000062	0.000084	0.000112

			t			
v	0.95	1.0	2.0	3.0	4.0	5.0
0.00	0.987278	0.984382	0.751734	-0.221380	-2.56342	-6.23008
0.05	0.946245	0.943328	0.672808	-0.370423	-2.73358	-6.21819
0.10	0.891658	0.888801	0.584932	-0.519205	-2.88659	-6.16518
0.15	0.825016	0.822209	0.488823	-0.666613	-3.02107	-6.07094
0.20	0.748105	0.745257	0.385453	-0.811446	-3.13578	-5.93567
0.25	0.662919	0.659874	0.276031	-0.952421	-3.22956	-5.75995
0.30	0.571582	0.568133	0.161971	-1.08819	-3.30143	-5.54469
0.35	0.476272	0.472177	0.044861	-1.21735	-3.35054	-5.29113
0.40	0.379146	0.374151	-0.073583	-1.33849	-3.37618	-5.00087
0.45	0.282283	0.276133	-0.191561	-1.45016	-3.37784	-4.67583
0.50	0.187619	0.180082	-0.307238	-1.55098	-3.35518	-4.31827
0.55	0.096910	0.087783	-0.418791	-1.63960	-3.30806	-3.93074
0.60	0.011688	0.000816	-0.524454	-1.71476	-3.23655	-3.51608
0.65	-0.066757	-0.079477	-0.622570	-1.77533	-3.14094	-3.07739
0.70	-0.137396	-0.152008	-0.711625	-1.82031	-3.02173	-2.61805

Table 3.1 (continued)

				t		
v	0.95	1.0	2.0	3.0	4.0	5.0
0.75	−0.199469	−0.215954	−0.790287	−1.84890	−2.87968	−2.14163
0.80	−0.252476	−0.270754	−0.857438	−1.86048	−2.71576	−1.65189
0.85	−0.296171	−0.316102	−0.912193	−1.85466	−2.53119	−1.15278
0.90	−0.330544	−0.351934	−0.953920	−1.83131	−2.32743	−0.648350
0.95	−0.355795	−0.378405	−0.982251	−1.79051	−2.10614	−0.142762
1.00	−0.372315	−0.395868	−0.997078	−1.73264	−1.86925	0.359777
1.05	−0.380652	−0.404842	−0.998554	−1.65832	−1.61885	0.855037
1.10	−0.381482	−0.405984	−0.987080	−1.56841	−1.35723	1.33882
1.15	−0.375577	−0.400061	−0.963288	−1.46404	−1.08686	1.80697
1.20	−0.363781	−0.387916	−0.928016	−1.34654	−0.810330	2.25547
1.25	−0.346975	−0.370444	−0.882288	−1.21746	−0.530360	2.68044
1.30	−0.326053	−0.348560	−0.827279	−1.07852	−0.249734	3.07817
1.35	−0.301901	−0.323175	−0.764286	−0.931596	0.028717	3.44522
1.40	−0.275373	−0.295178	−0.694691	−0.778662	0.302155	3.77839
1.45	−0.247277	−0.265413	−0.619935	−0.621779	0.567776	4.07482
1.50	−0.218358	−0.234668	−0.541478	−0.463045	0.822847	4.33196
1.55	−0.189288	−0.203658	−0.460769	−0.304558	1.06475	4.54769
1.60	−0.160663	−0.173020	−0.379219	−0.148384	1.29101	4.72025
1.65	−0.132992	−0.143306	−0.298173	0.003487	1.49935	4.84835
1.70	−0.106699	−0.114980	−0.218885	0.149169	1.68771	4.93113
1.75	−0.082126	−0.088421	−0.142498	0.286917	1.85427	4.96822
1.80	−0.059530	−0.063919	−0.070032	0.415151	1.99753	4.95971
1.85	−0.039092	−0.041684	−0.002364	0.532479	2.11624	4.90618
1.90	−0.020923	−0.021851	0.059772	0.637723	2.20952	4.80869
1.95	−0.005069	−0.004485	0.115796	0.729926	2.27679	4.66878
2.00	0.008481	0.010411	0.165279	0.808368	2.31785	4.48843
2.05	0.019786	0.022887	0.207943	0.872569	2.33280	4.27006
2.10	0.028945	0.033036	0.243653	0.922288	2.32212	4.01650
2.15	0.036091	0.040992	0.272409	0.957518	2.28660	3.73095
2.20	0.041380	0.046914	0.294337	0.978480	2.22734	3.41693
2.25	0.044986	0.050983	0.309674	0.985608	2.14575	3.07828
2.30	0.047093	0.053395	0.318757	0.979531	2.04350	2.71905
2.35	0.047893	0.054350	0.322005	0.961056	1.92250	2.34350
2.40	0.047574	0.054054	0.319907	0.931142	1.78485	1.95600
2.45	0.046322	0.052706	0.313003	0.890882	1.63285	1.56101
2.50	0.044313	0.050499	0.301870	0.841469	1.46890	1.16303
2.55	0.041714	0.047617	0.287110	0.784178	1.29551	0.766478
2.60	0.038677	0.044227	0.269330	0.720333	1.11523	0.375708
2.65	0.035341	0.040483	0.249137	0.651285	0.930629	−0.005091
2.70	0.031827	0.036524	0.227121	0.578387	0.744238	−0.371936
2.75	0.028243	0.032468	0.203846	0.502968	0.558522	−0.721096
2.80	0.024678	0.028420	0.179841	0.426312	0.375843	−1.04915
2.85	0.021208	0.024465	0.155598	0.349642	0.198427	−1.35300

Table 3.1 (continued)

v	0.95	1.0	2.0	3.0	4.0	5.0
2.90	0.017893	0.020674	0.131559	0.274095	0.028333	-1.62995
2.95	0.014780	0.017101	0.108117	0.200717	-0.132570	-1.87769
3.00	0.011902	0.013788	0.085611	0.130443	-0.282627	-2.09435
3.05	0.009283	0.010763	0.064329	0.064094	-0.420411	-2.27851
3.10	0.006936	0.008042	0.044503	0.002365	-0.544741	-2.42919
3.15	0.004864	0.005633	0.026312	-0.054172	-0.654682	-2.54587
3.20	0.003067	0.003535	0.009886	-0.105073	-0.749555	-2.62848
3.25	0.001534	0.001740	-0.004692	-0.150023	-0.828932	-2.67740
3.30	0.000252	0.000234	-0.017384	-0.188823	-0.892635	-2.69342
3.35	-0.000794	-0.001000	-0.028191	-0.221394	-0.940720	-2.67774
3.40	-0.001624	-0.001984	-0.037149	-0.247765	-0.973469	-2.63190
3.45	-0.002260	-0.002741	-0.044323	-0.268062	-0.991372	-2.55782
3.50	-0.002722	-0.003294	-0.049805	-0.282503	-0.995112	-2.45766
3.55	-0.003032	-0.003669	-0.053707	-0.291381	-0.985536	-2.33389
3.60	-0.003212	-0.003889	-0.056159	-0.295059	-0.963640	-2.18916
3.65	-0.003282	-0.003978	-0.057298	-0.293951	-0.930540	-2.02629
3.70	-0.003262	-0.003959	-0.057274	-0.288518	-0.887450	-1.84822
3.75	-0.003169	-0.003853	-0.056235	-0.279247	-0.835656	-1.65797
3.80	-0.003019	-0.003677	-0.054335	-0.266649	-0.776491	-1.45856
3.85	-0.002827	-0.003450	-0.051719	-0.251241	-0.711310	-1.25303
3.90	-0.002606	-0.003187	-0.048532	-0.233541	-0.641468	-1.04432
3.95	-0.002367	-0.002900	-0.044908	-0.214055	-0.568300	-0.835300
4.00	-0.002119	-0.002601	-0.040973	-0.193273	-0.493097	-0.628675
4.05	-0.001870	-0.002300	-0.036843	-0.171659	-0.417090	-0.426994
4.10	-0.001626	-0.002004	-0.032621	-0.149645	-0.341438	-0.232602
4.15	-0.001392	-0.001719	-0.028399	-0.127629	-0.267207	-0.047618
4.20	-0.001172	-0.001450	-0.024256	-0.105969	-0.195365	0.126085
4.25	-0.000969	-0.001200	-0.020260	-0.084981	-0.126772	0.286890
4.30	-0.000783	-0.000972	-0.016465	-0.064938	-0.062172	0.433455
4.35	-0.000616	-0.000766	-0.012914	-0.046069	-0.002193	0.564710
4.40	-0.000469	-0.000584	-0.009642	-0.028558	0.052660	0.679864
4.45	-0.000341	-0.000425	-0.006669	-0.012550	0.101998	0.778396
4.50	-0.000231	-0.000287	-0.004011	0.001855	0.145549	0.860053
4.55	-0.000138	-0.000171	-0.001672	0.014593	0.183154	0.924835
4.60	-0.000062	-0.000075	0.000349	0.025636	0.214757	0.972982
4.65	0.000000	0.000003	0.002059	0.034989	0.240403	1.00496
4.70	0.000049	0.000065	0.003471	0.042683	0.260223	1.02142
4.75	0.000086	0.000112	0.004603	0.048778	0.274432	1.02321
4.80	0.000113	0.000147	0.005472	0.053354	0.283313	1.01133
4.85	0.000132	0.000170	0.006103	0.056507	0.287209	0.986899
4.90	0.000143	0.000185	0.006519	0.058349	0.286511	0.951158
4.95	0.000148	0.000191	0.006743	0.059002	0.281649	0.905416
5.00	0.000148	0.000192	0.006802	0.058594	0.273082	0.851038

Table 3.2: First derivatives with respect to the order of the Kelvin functions $\partial \mathrm{ber}_\nu(t)/\partial \nu$.

			t			
ν	0.05	0.10	0.15	0.20	0.25	0.30
0.00	−3.11314	−2.42440	−2.02628	−1.74885	−1.53886	−1.37256
0.05	−2.94439	−2.44260	−2.13466	−1.91184	−1.73841	−1.59793
0.10	−2.70808	−2.37941	−2.16180	−1.99702	−1.86453	−1.75449
0.15	−2.43091	−2.25376	−2.12110	−2.01374	−1.92352	−1.84610
0.20	−2.13476	−2.08329	−2.02626	−1.97253	−1.92328	−1.87855
0.25	−1.83678	−1.88371	−1.89057	−1.88435	−1.87276	−1.85899
0.30	−1.54969	−1.66856	−1.72634	−1.76004	−1.78132	−1.79540
0.35	−1.28239	−1.44896	−1.54459	−1.60985	−1.65827	−1.69610
0.40	−1.04047	−1.23374	−1.35478	−1.44313	−1.51248	−1.56939
0.45	−0.826879	−1.02952	−1.16478	−1.26807	−1.35213	−1.42318
0.50	−0.642492	−0.840966	−0.980831	−1.09168	−1.18449	−1.26476
0.55	−0.486646	−0.671031	−0.807656	−0.919625	−1.01579	−1.10066
0.60	−0.357604	−0.521240	−0.648589	−0.756372	−0.851197	−0.936507
0.65	−0.252943	−0.391976	−0.505738	−0.605191	−0.694809	−0.776970
0.70	−0.169864	−0.282739	−0.380176	−0.468299	−0.549699	−0.625784
0.75	−0.105433	−0.192385	−0.272123	−0.346981	−0.418003	−0.485770
0.80	−0.056759	−0.119333	−0.181130	−0.241738	−0.301018	−0.358899
0.85	−0.021111	−0.061743	−0.106249	−0.152428	−0.199321	−0.246380
0.90	0.003994	−0.017652	−0.046180	−0.078409	−0.112884	−0.148757
0.95	0.020755	0.014909	0.000596	−0.018664	−0.041201	−0.066013
1.00	0.031071	0.037836	0.035710	0.028077	0.016606	0.002324
1.05	0.036540	0.052894	0.060818	0.063242	0.061676	0.057077
1.10	0.038480	0.061680	0.077529	0.088327	0.095320	0.099329
1.15	0.037952	0.065601	0.087361	0.104831	0.118942	0.130336
1.20	0.035793	0.065871	0.091703	0.114202	0.133969	0.151452
1.25	0.032646	0.063511	0.091796	0.117796	0.141799	0.164062
1.30	0.028990	0.059364	0.088722	0.116850	0.143761	0.169535
1.35	0.025172	0.054105	0.083404	0.112467	0.141083	0.169182
1.40	0.021433	0.048263	0.076612	0.105608	0.134874	0.164221
1.45	0.017927	0.042240	0.068969	0.097090	0.126112	0.155763
1.50	0.014747	0.036330	0.060971	0.087591	0.115640	0.144792
1.55	0.011940	0.030737	0.052997	0.077661	0.104167	0.132163
1.60	0.009515	0.025593	0.045325	0.067732	0.092274	0.118602
1.65	0.007463	0.020973	0.038148	0.058130	0.080425	0.104706
1.70	0.005757	0.016909	0.031593	0.049090	0.068975	0.090957
1.75	0.004361	0.013400	0.025726	0.040771	0.058186	0.077726
1.80	0.003239	0.010421	0.020572	0.033267	0.048238	0.065287
1.85	0.002352	0.007936	0.016123	0.026622	0.039239	0.053830
1.90	0.001661	0.005895	0.012347	0.020838	0.031246	0.043473
1.95	0.001132	0.004247	0.009194	0.015888	0.024265	0.034274
2.00	0.000736	0.002941	0.006607	0.011722	0.018272	0.026241
2.05	0.000445	0.001924	0.004521	0.008276	0.013213	0.019344
2.10	0.000237	0.001150	0.002873	0.005480	0.009020	0.013528

Table 3.2 (continued)

			t			
v	0.05	0.10	0.15	0.20	0.25	0.30
2.15	0.000092	0.000574	0.001598	0.003256	0.005611	0.008713
2.20	−0.000004	0.000160	0.000639	0.001528	0.002900	0.004808
2.25	−0.000064	−0.000127	−0.000061	0.000224	0.000798	0.001717
2.30	−0.000098	−0.000315	−0.000550	−0.000726	−0.000781	−0.000663
2.35	−0.000114	−0.000427	−0.000870	−0.001385	−0.001920	−0.002429
2.40	−0.000118	−0.000483	−0.001060	−0.001810	−0.002694	−0.003677
2.45	−0.000114	−0.000498	−0.001150	−0.002049	−0.003171	−0.004494
2.50	−0.000105	−0.000486	−0.001166	−0.002146	−0.003415	−0.004962
2.55	−0.000093	−0.000455	−0.001131	−0.002137	−0.003477	−0.005152
2.60	−0.000081	−0.000414	−0.001061	−0.002051	−0.003403	−0.005129
2.65	−0.000069	−0.000367	−0.000969	−0.001915	−0.003233	−0.004946
2.70	−0.000057	−0.000319	−0.000866	−0.001746	−0.002998	−0.004650
2.75	−0.000047	−0.000272	−0.000759	−0.001561	−0.002722	−0.004279
2.80	−0.000037	−0.000229	−0.000654	−0.001371	−0.002428	−0.003865
2.85	−0.000030	−0.000189	−0.000554	−0.001184	−0.002129	−0.003432
2.90	−0.000023	−0.000154	−0.000463	−0.001007	−0.001838	−0.003000
2.95	−0.000018	−0.000123	−0.000381	−0.000844	−0.001563	−0.002582
3.00	−0.000013	−0.000097	−0.000308	−0.000697	−0.001309	−0.002189
3.05	−0.000010	−0.000076	−0.000246	−0.000566	−0.001079	−0.001827
3.10	−0.000007	−0.000058	−0.000193	−0.000452	−0.000875	−0.001501
3.15	−0.000005	−0.000043	−0.000148	−0.000355	−0.000697	−0.001211
3.20	−0.000004	−0.000032	−0.000112	−0.000273	−0.000545	−0.000958
3.25	−0.000002	−0.000022	−0.000082	−0.000204	−0.000415	−0.000741
3.30	−0.000002	−0.000015	−0.000058	−0.000149	−0.000308	−0.000558
3.35	−0.000001	−0.000010	−0.000039	−0.000104	−0.000220	−0.000405
3.40	−0.000001	−0.000006	−0.000025	−0.000069	−0.000149	−0.000280
3.45		−0.000003	−0.000014	−0.000041	−0.000093	−0.000180
3.50		−0.000001	−0.000006	−0.000021	−0.000050	−0.000101
3.55		0.000000	−0.000001	−0.000005	−0.000017	−0.000040
3.60		0.000001	0.000003	0.000005	0.000007	0.000005
3.65		0.000002	0.000005	0.000013	0.000023	0.000037
3.70		0.000002	0.000007	0.000017	0.000034	0.000059
3.75		0.000002	0.000007	0.000020	0.000041	0.000073
3.80		0.000002	0.000007	0.000020	0.000044	0.000080
3.85		0.000002	0.000007	0.000020	0.000044	0.000083
3.90		0.000002	0.000007	0.000019	0.000043	0.000081
3.95		0.000001	0.000006	0.000018	0.000040	0.000078
4.00		0.000001	0.000005	0.000016	0.000037	0.000072
4.05		0.000001	0.000005	0.000014	0.000033	0.000065
4.10		0.000001	0.000004	0.000012	0.000029	0.000058
4.15		0.000001	0.000003	0.000010	0.000025	0.000051
4.20		0.000001	0.000002	0.000008	0.000021	0.000044
4.25			0.000002	0.000007	0.000018	0.000037

Table 3.2 (continued)

			t			
v	0.05	0.10	0.15	0.20	0.25	0.30
4.30			0.000002	0.000006	0.000015	0.000031
4.35			0.000001	0.000005	0.000012	0.000026
4.40			0.000001	0.000004	0.000010	0.000021
4.45			0.000001	0.000003	0.000008	0.000017
4.50			0.000001	0.000002	0.000006	0.000013
4.55				0.000002	0.000005	0.000010
4.60				0.000001	0.000003	0.000008
4.65				0.000001	0.000002	0.000005
4.70				0.000001	0.000002	0.000003
4.75					0.000001	0.000002
4.80					0.000001	0.000001
4.85						0.000001

			t			
v	0.35	0.40	0.45	0.50	0.55	0.60
0.00	−1.23729	−1.12545	−1.03217	−0.954070	−0.888762	−0.834461
0.05	−1.48143	−1.38347	−1.30046	−1.22992	−1.17005	−1.11950
0.10	−1.66134	−1.58163	−1.51306	−1.45398	−1.40321	−1.35984
0.15	−1.77890	−1.72024	−1.66893	−1.62414	−1.58523	−1.55170
0.20	−1.83812	−1.80177	−1.76929	−1.74051	−1.71526	−1.69342
0.25	−1.84460	−1.83048	−1.81719	−1.80506	−1.79435	−1.78522
0.30	−1.80504	−1.81192	−1.81712	−1.82137	−1.82521	−1.82904
0.35	−1.72682	−1.75254	−1.77466	−1.79419	−1.81184	−1.82815
0.40	−1.61755	−1.65927	−1.69609	−1.72914	−1.75921	−1.78695
0.45	−1.48478	−1.53922	−1.58804	−1.63236	−1.67300	−1.71061
0.50	−1.33570	−1.39938	−1.45721	−1.51024	−1.55924	−1.60482
0.55	−1.17693	−1.24634	−1.31011	−1.36914	−1.42409	−1.47552
0.60	−1.01438	−1.08617	−1.15286	−1.21516	−1.27361	−1.32864
0.65	−0.853128	−0.924256	−0.991045	−1.05401	−1.11354	−1.16995
0.70	−0.697427	−0.765216	−0.829572	−0.890810	−0.949172	−1.00485
0.75	−0.550646	−0.612878	−0.672641	−0.730064	−0.785243	−0.838254
0.80	−0.415331	−0.470274	−0.523694	−0.575559	−0.625839	−0.674504
0.85	−0.293247	−0.339661	−0.385422	−0.430369	−0.474367	−0.517299
0.90	−0.185459	−0.222579	−0.259797	−0.296857	−0.333543	−0.369673
0.95	−0.092419	−0.119919	−0.148119	−0.176703	−0.205403	−0.233990
1.00	−0.014057	−0.032000	−0.051085	−0.070963	−0.091341	−0.111963
1.05	0.050125	0.041341	0.031141	0.019875	0.007840	−0.004701
1.10	0.100953	0.100662	0.098840	0.095812	0.091864	0.087248
1.15	0.139502	0.146833	0.152658	0.157266	0.160911	0.163823
1.20	0.167011	0.180948	0.193529	0.204989	0.215542	0.225388
1.25	0.184813	0.204259	0.222592	0.239991	0.256623	0.272651
1.30	0.194277	0.218101	0.241125	0.263470	0.285255	0.306600
1.35	0.196758	0.223841	0.250481	0.276742	0.302697	0.328426

Table 3.2 (continued)

			t			
v	0.35	0.40	0.45	0.50	0.55	0.60
1.40	0.193553	0.222827	0.252034	0.281187	0.310315	0.339459
1.45	0.185876	0.216353	0.247134	0.278191	0.309513	0.341109
1.50	0.174834	0.205627	0.237073	0.269110	0.301698	0.334815
1.55	0.161413	0.191749	0.223051	0.255232	0.288227	0.321994
1.60	0.146471	0.175703	0.206165	0.237752	0.270388	0.304012
1.65	0.130739	0.158345	0.187386	0.217755	0.249365	0.282146
1.70	0.114821	0.140400	0.167563	0.196202	0.226228	0.257568
1.75	0.099204	0.122472	0.147412	0.173923	0.201919	0.231329
1.80	0.084264	0.105044	0.127525	0.151619	0.177248	0.204347
1.85	0.070279	0.088488	0.108375	0.129865	0.152895	0.177404
1.90	0.057441	0.073077	0.090321	0.109116	0.129408	0.151150
1.95	0.045866	0.058997	0.073626	0.089713	0.107219	0.126107
2.00	0.035609	0.046357	0.058461	0.071899	0.086645	0.102672
2.05	0.026674	0.035202	0.044923	0.055828	0.067904	0.081134
2.10	0.019025	0.025528	0.033044	0.041577	0.051125	0.061681
2.15	0.012595	0.017286	0.022803	0.029160	0.036363	0.044414
2.20	0.007296	0.010397	0.014138	0.018537	0.023607	0.029358
2.25	0.003025	0.004762	0.006955	0.009628	0.012798	0.016477
2.30	−0.000327	0.000264	0.001138	0.002322	0.003833	0.005685
2.35	−0.002874	−0.003221	−0.003442	−0.003514	−0.003419	−0.003142
2.40	−0.004727	−0.005817	−0.006923	−0.008026	−0.009107	−0.010157
2.45	−0.005994	−0.007651	−0.009446	−0.011366	−0.013397	−0.015530
2.50	−0.006774	−0.008840	−0.011148	−0.013690	−0.016457	−0.019444
2.55	−0.007160	−0.009498	−0.012161	−0.015147	−0.018455	−0.022084
2.60	−0.007234	−0.009725	−0.012607	−0.015882	−0.019556	−0.023634
2.65	−0.007069	−0.009616	−0.012599	−0.016028	−0.019914	−0.024268
2.70	−0.006727	−0.009250	−0.012238	−0.015707	−0.019674	−0.024155
2.75	−0.006262	−0.008699	−0.011613	−0.015028	−0.018966	−0.023447
2.80	−0.005719	−0.008021	−0.010802	−0.014088	−0.017907	−0.022283
2.85	−0.005133	−0.007268	−0.009870	−0.012970	−0.016599	−0.020786
2.90	−0.004534	−0.006479	−0.008871	−0.011744	−0.015131	−0.019064
2.95	−0.003943	−0.005687	−0.007849	−0.010467	−0.013576	−0.017209
3.00	−0.003378	−0.004916	−0.006840	−0.009188	−0.011995	−0.015297
3.05	−0.002849	−0.004185	−0.005870	−0.007943	−0.010438	−0.013390
3.10	−0.002365	−0.003506	−0.004958	−0.006758	−0.008941	−0.011540
3.15	−0.001930	−0.002887	−0.004118	−0.005655	−0.007532	−0.009782
3.20	−0.001544	−0.002333	−0.003357	−0.004646	−0.006232	−0.008145
3.25	−0.001209	−0.001846	−0.002680	−0.003739	−0.005052	−0.006647
3.30	−0.000921	−0.001423	−0.002086	−0.002936	−0.003999	−0.005299
3.35	−0.000679	−0.001062	−0.001574	−0.002237	−0.003073	−0.004104
3.40	−0.000478	−0.000759	−0.001139	−0.001638	−0.002273	−0.003063
3.45	−0.000314	−0.000509	−0.000777	−0.001133	−0.001592	−0.002170
3.50	−0.000183	−0.000306	−0.000479	−0.000714	−0.001023	−0.001417

Table 3.2 (continued)

			t			
v	0.35	0.40	0.45	0.50	0.55	0.60
3.55	−0.000081	−0.000146	−0.000241	−0.000375	−0.000557	−0.000794
3.60	−0.000003	−0.000022	−0.000054	−0.000106	−0.000183	−0.000290
3.65	0.000054	0.000071	0.000088	0.000102	0.000109	0.000108
3.70	0.000094	0.000138	0.000192	0.000256	0.000330	0.000413
3.75	0.000120	0.000183	0.000264	0.000365	0.000489	0.000636
3.80	0.000135	0.000210	0.000309	0.000437	0.000596	0.000790
3.85	0.000141	0.000223	0.000334	0.000478	0.000661	0.000887
3.90	0.000141	0.000226	0.000342	0.000495	0.000691	0.000936
3.95	0.000136	0.000220	0.000337	0.000492	0.000693	0.000947
4.00	0.000128	0.000209	0.000323	0.000476	0.000675	0.000929
4.05	0.000117	0.000194	0.000302	0.000448	0.000641	0.000888
4.10	0.000105	0.000176	0.000277	0.000414	0.000596	0.000832
4.15	0.000093	0.000157	0.000249	0.000376	0.000545	0.000764
4.20	0.000081	0.000138	0.000221	0.000336	0.000490	0.000691
4.25	0.000070	0.000120	0.000193	0.000295	0.000433	0.000615
4.30	0.000059	0.000102	0.000166	0.000256	0.000378	0.000539
4.35	0.000049	0.000086	0.000141	0.000218	0.000324	0.000466
4.40	0.000040	0.000071	0.000117	0.000183	0.000274	0.000396
4.45	0.000033	0.000058	0.000096	0.000152	0.000228	0.000332
4.50	0.000026	0.000046	0.000078	0.000123	0.000187	0.000273
4.55	0.000020	0.000036	0.000061	0.000098	0.000150	0.000221
4.60	0.000015	0.000028	0.000047	0.000076	0.000118	0.000174
4.65	0.000011	0.000021	0.000036	0.000058	0.000090	0.000134
4.70	0.000008	0.000015	0.000026	0.000042	0.000066	0.000100
4.75	0.000005	0.000010	0.000018	0.000029	0.000047	0.000071
4.80	0.000003	0.000006	0.000011	0.000019	0.000030	0.000047
4.85	0.000002	0.000003	0.000006	0.000011	0.000017	0.000027
4.90	0.000000	0.000001	0.000002	0.000004	0.000007	0.000012
4.95	0.000000	−0.000001	−0.000001	−0.000001	−0.000001	0.000000
5.00	−0.000001	−0.000002	−0.000003	−0.000005	−0.000007	−0.000009

			t			
v	0.65	0.70	0.75	0.80	0.85	0.90
0.00	−0.789802	−0.753721	−0.725362	−0.704031	−0.689151	−0.680239
0.05	−1.07723	−1.04242	−1.01441	−0.992646	−0.976680	−0.966122
0.10	−1.32316	−1.29260	−1.26770	−1.24808	−1.23340	−1.22339
0.15	−1.52317	−1.49930	−1.47983	−1.46454	−1.45323	−1.44573
0.20	−1.67487	−1.65953	−1.64729	−1.63808	−1.63183	−1.62845
0.25	−1.77780	−1.77216	−1.76838	−1.76649	−1.76653	−1.76852
0.30	−1.83314	−1.83776	−1.84308	−1.84923	−1.85634	−1.86449
0.35	−1.84355	−1.85837	−1.87287	−1.88727	−1.90175	−1.91644
0.40	−1.81283	−1.83725	−1.86052	−1.88290	−1.90460	−1.92579
0.45	−1.74570	−1.77866	−1.80984	−1.83950	−1.86786	−1.89513

Table 3.2 (continued)

			t			
v	**0.65**	**0.70**	**0.75**	**0.80**	**0.85**	**0.90**
0.50	−1.64746	−1.68756	−1.72542	−1.76131	−1.79546	−1.82805
0.55	−1.52383	−1.56937	−1.61243	−1.65324	−1.69201	−1.72889
0.60	−1.38059	−1.42975	−1.47635	−1.52057	−1.56259	−1.60254
0.65	−1.22349	−1.27437	−1.32276	−1.36879	−1.41258	−1.45423
0.70	−1.05799	−1.10873	−1.15715	−1.20335	−1.24740	−1.28934
0.75	−0.889153	−0.937987	−0.984787	−1.02958	−1.07238	−1.11319
0.80	−0.721525	−0.766873	−0.810518	−0.852430	−0.892576	−0.930924
0.85	−0.559062	−0.599564	−0.638719	−0.676448	−0.712675	−0.747328
0.90	−0.405086	−0.439637	−0.473197	−0.505645	−0.536870	−0.566766
0.95	−0.262260	−0.290035	−0.317149	−0.343452	−0.368804	−0.393074
1.00	−0.132604	−0.153058	−0.173142	−0.192684	−0.211525	−0.229515
1.05	−0.017512	−0.030383	−0.043117	−0.055533	−0.067462	−0.078744
1.10	0.082193	0.076910	0.071591	0.066420	0.061567	0.057196
1.15	0.166215	0.168283	0.170212	0.172175	0.174339	0.176861
1.20	0.234712	0.243690	0.252489	0.261269	0.270183	0.279380
1.25	0.288228	0.303503	0.318621	0.333720	0.348939	0.364410
1.30	0.327626	0.348450	0.369191	0.389966	0.410891	0.432081
1.35	0.354011	0.379542	0.405106	0.430798	0.456710	0.482938
1.40	0.368669	0.398002	0.427520	0.457291	0.487384	0.517873
1.45	0.372997	0.405206	0.437771	0.470736	0.504149	0.538060
1.50	0.368453	0.402617	0.437319	0.472580	0.508425	0.544888
1.55	0.356503	0.391737	0.427688	0.464358	0.501755	0.539892
1.60	0.338578	0.374054	0.410415	0.447649	0.485745	0.524702
1.65	0.316041	0.351007	0.387008	0.424017	0.462013	0.500980
1.70	0.290161	0.323955	0.358907	0.394980	0.432145	0.470377
1.75	0.262091	0.294149	0.327457	0.361973	0.397660	0.434488
1.80	0.232853	0.262716	0.293885	0.326319	0.359976	0.394820
1.85	0.203338	0.230649	0.259289	0.289215	0.320388	0.352768
1.90	0.174295	0.198800	0.224621	0.251719	0.280053	0.309587
1.95	0.146339	0.167879	0.190690	0.214736	0.239981	0.266387
2.00	0.119954	0.138460	0.158162	0.179028	0.201025	0.224121
2.05	0.095502	0.110986	0.127563	0.145206	0.163890	0.183583
2.10	0.073237	0.085777	0.099286	0.113744	0.129128	0.145412
2.15	0.053310	0.063045	0.073608	0.084984	0.097154	0.110098
2.20	0.035791	0.042905	0.050694	0.059148	0.068254	0.077992
2.25	0.020672	0.025386	0.030615	0.036355	0.042593	0.049316
2.30	0.007888	0.010447	0.013361	0.016627	0.020237	0.024178
2.35	−0.002674	−0.002009	−0.001146	−0.000087	0.001160	0.002586
2.40	−0.011164	−0.012126	−0.013039	−0.013907	−0.014736	−0.015536
2.45	−0.017759	−0.020080	−0.022493	−0.025001	−0.027612	−0.030335
2.50	−0.022649	−0.026071	−0.029711	−0.033576	−0.037673	−0.042013
2.55	−0.026036	−0.030312	−0.034920	−0.039865	−0.045159	−0.050813
2.60	−0.028120	−0.033024	−0.038353	−0.044118	−0.050332	−0.057008

Table 3.2 (continued)

				t		
v	0.65	0.70	0.75	0.80	0.85	0.90
2.65	−0.029101	−0.034423	−0.040249	−0.046591	−0.053465	−0.060889
2.70	−0.029166	−0.034722	−0.040841	−0.047540	−0.054838	−0.062753
2.75	−0.028492	−0.034121	−0.040355	−0.047214	−0.054721	−0.062898
2.80	−0.027241	−0.032805	−0.039000	−0.045850	−0.053378	−0.061611
2.85	−0.025559	−0.030945	−0.036972	−0.043666	−0.051056	−0.059167
2.90	−0.023574	−0.028691	−0.034445	−0.040866	−0.047982	−0.055823
2.95	−0.021398	−0.026177	−0.031577	−0.037628	−0.044363	−0.051812
3.00	−0.019126	−0.023517	−0.028501	−0.034113	−0.040383	−0.047343
3.05	−0.016835	−0.020806	−0.025335	−0.030456	−0.036201	−0.042603
3.10	−0.014589	−0.018123	−0.022173	−0.026773	−0.031954	−0.037749
3.15	−0.012437	−0.015531	−0.019093	−0.023158	−0.027755	−0.032916
3.20	−0.010417	−0.013077	−0.016157	−0.019687	−0.023696	−0.028214
3.25	−0.008553	−0.010798	−0.013410	−0.016417	−0.019848	−0.023730
3.30	−0.006862	−0.008715	−0.010883	−0.013391	−0.016265	−0.019531
3.35	−0.005353	−0.006843	−0.008596	−0.010636	−0.012984	−0.015665
3.40	−0.004028	−0.005187	−0.006560	−0.008167	−0.010027	−0.012161
3.45	−0.002882	−0.003745	−0.004776	−0.005990	−0.007405	−0.009037
3.50	−0.001909	−0.002512	−0.003239	−0.004103	−0.005118	−0.006296
3.55	−0.001097	−0.001475	−0.001937	−0.002495	−0.003157	−0.003933
3.60	−0.000434	−0.000621	−0.000857	−0.001150	−0.001507	−0.001933
3.65	0.000095	0.000067	0.000019	−0.000052	−0.000149	−0.000275
3.70	0.000504	0.000604	0.000711	0.000823	0.000942	0.001065
3.75	0.000809	0.001009	0.001238	0.001497	0.001789	0.002116
3.80	0.001024	0.001299	0.001621	0.001994	0.002421	0.002908
3.85	0.001162	0.001492	0.001881	0.002337	0.002865	0.003472
3.90	0.001237	0.001602	0.002036	0.002549	0.003147	0.003839
3.95	0.001262	0.001645	0.002105	0.002651	0.003292	0.004038
4.00	0.001245	0.001634	0.002103	0.002663	0.003324	0.004098
4.05	0.001198	0.001581	0.002046	0.002604	0.003266	0.004044
4.10	0.001129	0.001497	0.001948	0.002490	0.003137	0.003901
4.15	0.001043	0.001391	0.001819	0.002336	0.002956	0.003689
4.20	0.000948	0.001271	0.001669	0.002154	0.002737	0.003429
4.25	0.000849	0.001143	0.001508	0.001955	0.002494	0.003137
4.30	0.000748	0.001013	0.001343	0.001747	0.002238	0.002826
4.35	0.000650	0.000884	0.001177	0.001539	0.001979	0.002508
4.40	0.000556	0.000760	0.001017	0.001335	0.001724	0.002193
4.45	0.000468	0.000643	0.000865	0.001140	0.001478	0.001888
4.50	0.000387	0.000535	0.000723	0.000958	0.001247	0.001599
4.55	0.000315	0.000437	0.000594	0.000790	0.001033	0.001330
4.60	0.000250	0.000350	0.000477	0.000638	0.000838	0.001083
4.65	0.000194	0.000272	0.000374	0.000503	0.000663	0.000862
4.70	0.000145	0.000206	0.000284	0.000384	0.000509	0.000665
4.75	0.000104	0.000148	0.000206	0.000281	0.000375	0.000493

Table 3.2 (continued)

			t			
v	0.65	0.70	0.75	0.80	0.85	0.90
4.80	0.000070	0.000100	0.000141	0.000194	0.000261	0.000345
4.85	0.000041	0.000061	0.000087	0.000121	0.000165	0.000220
4.90	0.000019	0.000029	0.000043	0.000061	0.000085	0.000116
4.95	0.000000	0.000004	0.000007	0.000013	0.000021	0.000032
5.00	-0.000012	-0.000016	-0.000020	-0.000024	-0.000029	-0.000035

			t			
v	0.95	1.00	2.00	3.00	4.00	5.00
0.00	-0.676881	-0.678724	-1.48561	-2.97652	-3.56517	-0.170755
0.05	-0.960639	-0.959939	-1.66996	-2.98180	-3.23645	0.648041
0.10	-1.21781	-1.21643	-1.84267	-2.96576	-2.87931	1.47266
0.15	-1.44188	-1.44154	-1.99844	-2.92656	-2.49587	2.29637
0.20	-1.62789	-1.63006	-2.13232	-2.86248	-2.08847	3.11228
0.25	-1.77245	-1.77832	-2.23985	-2.77203	-1.65968	3.91348
0.30	-1.87375	-1.88417	-2.31723	-2.65404	-1.21233	4.69303
0.35	-1.93146	-1.94690	-2.36143	-2.50773	-0.749473	5.44408
0.40	-1.94661	-1.96720	-2.37032	-2.33282	-0.274440	6.15989
0.45	-1.92145	-1.94696	-2.34269	-2.12954	0.209204	6.83394
0.50	-1.85923	-1.88913	-2.27830	-1.89872	0.697646	7.45993
0.55	-1.76403	-1.79754	-2.17789	-1.64180	1.18684	8.03191
0.60	-1.64053	-1.67666	-2.04311	-1.36084	1.67252	8.54427
0.65	-1.49383	-1.53143	-1.87647	-1.05852	2.15026	8.99184
0.70	-1.32922	-1.36708	-1.68128	-0.738096	2.61545	9.36995
0.75	-1.15203	-1.18889	-1.46148	-0.403361	3.06339	9.67446
0.80	-0.967438	-1.00208	-1.22159	-0.058576	3.48935	9.90182
0.85	-0.780337	-0.811633	-0.966497	0.291614	3.88859	10.0491
0.90	-0.595232	-0.622173	-0.701373	0.642273	4.25646	10.1141
0.95	-0.416137	-0.437873	-0.431491	0.988285	4.58846	10.0954
1.00	-0.246511	-0.262380	-0.162101	1.32446	4.88034	9.99207
1.05	-0.089225	-0.098762	0.101708	1.64567	5.12814	9.80432
1.10	0.053462	0.050514	0.355132	1.94693	5.32834	9.53299
1.15	0.179895	0.183588	0.593763	2.22356	5.47785	9.17980
1.20	0.289004	0.299195	0.813675	2.47124	5.57415	8.74731
1.25	0.380264	0.396631	1.01151	2.68616	5.61536	8.23893
1.30	0.453650	0.475712	1.18452	2.86506	5.60025	7.65891
1.35	0.509576	0.536720	1.33061	3.00535	5.52832	7.01232
1.40	0.548834	0.580343	1.44836	3.10512	5.39984	6.30499
1.45	0.572527	0.607607	1.53701	3.16320	5.21586	5.54354
1.50	0.582004	0.619813	1.59645	3.17917	4.97823	4.73523
1.55	0.578789	0.618467	1.62719	3.15339	4.68958	3.88797
1.60	0.564524	0.605218	1.63030	3.08693	4.35328	3.01021
1.65	0.540910	0.581796	1.60735	2.98161	3.97344	2.11086
1.70	0.509654	0.549961	1.56036	2.83987	3.55482	1.19915

Table 3.2 (continued)

			t			
v	0.95	1.00	2.00	3.00	4.00	5.00
1.75	0.472426	0.511450	1.49170	2.66478	3.10279	0.284598
1.80	0.430818	0.467938	1.40403	2.45991	2.62321	-0.623164
1.85	0.386318	0.421004	1.30023	2.22929	2.12236	-1.51449
1.90	0.340282	0.372101	1.18328	1.97727	1.60684	-2.37986
1.95	0.293919	0.322538	1.05625	1.70846	1.08346	-3.20999
2.00	0.248280	0.273466	0.922159	1.42764	0.559109	-3.99595
2.05	0.204254	0.225869	0.783954	1.13960	0.040644	-4.72935
2.10	0.162568	0.180564	0.644437	0.849120	-0.465211	-5.40236
2.15	0.123791	0.138204	0.506221	0.560824	-0.951992	-6.00793
2.20	0.088343	0.099280	0.371682	0.279115	-1.41360	-6.53981
2.25	0.056506	0.064139	0.242932	0.008100	-1.84440	-6.99270
2.30	0.028435	0.032987	0.121792	-0.248484	-2.23931	-7.36229
2.35	0.004177	0.005913	0.009780	-0.487321	-2.59388	-7.64535
2.40	-0.016319	-0.017103	-0.091897	-0.705562	-2.90436	-7.83977
2.45	-0.033182	-0.036172	-0.182342	-0.900861	-3.16775	-7.94460
2.50	-0.046608	-0.051478	-0.260958	-1.07139	-3.38184	-7.96004
2.55	-0.056842	-0.063264	-0.327432	-1.21584	-3.54522	-7.88745
2.60	-0.064164	-0.071819	-0.381721	-1.33345	-3.65729	-7.72935
2.65	-0.068880	-0.077460	-0.424022	-1.42394	-3.71827	-7.48932
2.70	-0.071308	-0.080523	-0.454749	-1.48755	-3.72914	-7.17200
2.75	-0.071767	-0.081354	-0.474501	-1.52494	-3.69162	-6.78298
2.80	-0.070573	-0.080292	-0.484031	-1.53721	-3.60811	-6.32870
2.85	-0.068029	-0.077668	-0.484219	-1.52582	-3.48167	-5.81638
2.90	-0.064418	-0.073798	-0.476037	-1.49255	-3.31586	-5.25387
2.95	-0.060005	-0.068973	-0.460520	-1.43945	-3.11474	-4.64957
3.00	-0.055025	-0.063461	-0.438738	-1.36876	-2.88277	-4.01222
3.05	-0.049692	-0.057501	-0.411773	-1.28289	-2.62467	-3.35083
3.10	-0.044188	-0.051304	-0.380689	-1.18434	-2.34539	-2.67454
3.15	-0.038671	-0.045051	-0.346518	-1.07565	-2.04998	-1.99242
3.20	-0.033271	-0.038895	-0.310239	-0.959365	-1.74352	-1.31340
3.25	-0.028091	-0.032958	-0.272764	-0.837952	-1.43100	-0.646108
3.30	-0.023214	-0.027338	-0.234925	-0.713800	-1.11730	0.001243
3.35	-0.018699	-0.022111	-0.197469	-0.589154	-0.807036	0.620978
3.40	-0.014587	-0.017326	-0.161048	-0.466097	-0.504563	1.20605
3.45	-0.010902	-0.013016	-0.126220	-0.346516	-0.213868	1.75015
3.50	-0.007652	-0.009196	-0.093445	-0.232087	0.061456	2.24774
3.55	-0.004834	-0.005868	-0.063090	-0.124256	0.318259	2.69417
3.60	-0.002435	-0.003020	-0.035429	-0.024237	0.553862	3.08565
3.65	-0.000436	-0.000632	-0.010651	0.066998	0.766077	3.41935
3.70	0.001192	0.001325	0.011135	0.148713	0.953215	3.69337
3.75	0.002479	0.002882	0.029893	0.220406	1.11409	3.90673
3.80	0.003459	0.004079	0.045650	0.281791	1.24800	4.05937
3.85	0.004165	0.004952	0.058491	0.332792	1.35472	4.15212

Table 3.2 (continued)

v	0.95	1.00	2.00	3.00	4.00	5.00
			t			
3.90	0.004634	0.005542	0.068547	0.373523	1.43449	4.18661
3.95	0.004900	0.005888	0.075988	0.404268	1.48793	4.16527
4.00	0.004994	0.006027	0.081014	0.425463	1.51609	4.09122
4.05	0.004949	0.005996	0.083847	0.437671	1.52030	3.96818
4.10	0.004793	0.005828	0.084724	0.441561	1.50223	3.80043
4.15	0.004551	0.005553	0.083892	0.437886	1.46378	3.59269
4.20	0.004245	0.005198	0.081596	0.427463	1.40703	3.35002
4.25	0.003897	0.004788	0.078081	0.411146	1.33422	3.07773
4.30	0.003523	0.004343	0.073583	0.389813	1.24768	2.78130
4.35	0.003138	0.003881	0.068323	0.364341	1.14982	2.46626
4.40	0.002753	0.003417	0.062512	0.335596	1.04301	2.13813
4.45	0.002379	0.002962	0.056339	0.304411	0.929609	1.80230
4.50	0.002022	0.002527	0.049978	0.271581	0.811912	1.46398
4.55	0.001688	0.002117	0.043579	0.237847	0.692090	1.12812
4.60	0.001381	0.001738	0.037275	0.203886	0.572185	0.799319
4.65	0.001103	0.001394	0.031176	0.170313	0.454077	0.481827
4.70	0.000855	0.001085	0.025374	0.137666	0.339464	0.179452
4.75	0.000637	0.000813	0.019941	0.106412	0.229850	−0.104459
4.80	0.000449	0.000576	0.014931	0.076939	0.126533	−0.367044
4.85	0.000289	0.000374	0.010383	0.049563	0.030597	−0.605947
4.90	0.000156	0.000205	0.006318	0.024527	−0.057084	−0.819319
4.95	0.000047	0.000066	0.002746	0.002004	−0.135850	−1.00582
5.00	−0.000040	−0.000046	−0.000333	−0.017899	−0.205243	−1.16461

Table 3.3: Second derivatives with respect to the order of the Kelvin functions $\partial^2 \text{ber}_v(t)/\partial v^2$.

v	0.05	0.10	0.15	0.20	0.25	0.30
			t			
0.00	2.49797	−1.30711	−3.06439	−4.09137	−4.75585	−5.20872
0.05	4.14751	0.51436	−1.31191	−2.45324	−3.23932	−3.81033
0.10	5.21584	1.94977	0.181345	−0.985308	−1.82694	−2.46528
0.15	5.79799	3.01716	1.40078	0.281191	−0.55895	−1.21862
0.20	5.99083	3.74938	2.34828	1.33061	0.53835	−0.103319
0.25	5.88638	4.18842	3.03826	2.16027	1.45099	0.858983
0.30	5.56760	4.38078	3.49403	2.77779	2.17551	1.65676
0.35	5.10607	4.37381	3.74451	3.19851	2.71701	2.28722
0.40	4.56109	4.21303	3.82155	3.44314	3.08717	2.75476
0.45	3.97979	3.94020	3.75772	3.53561	3.30247	3.06942
0.50	3.39805	3.59217	3.58456	3.50137	3.38246	3.24542
0.55	2.84172	3.20027	3.33137	3.36585	3.34834	3.29975

Table 3.3 (continued)

				t		
v	0.05	0.10	0.15	0.20	0.25	0.30
0.60	2.32814	2.79018	3.02436	3.15341	3.22171	3.25096
0.65	1.86770	2.38212	2.68616	2.88647	3.02360	3.11812
0.70	1.46528	1.99124	2.33566	2.58498	2.77370	2.91993
0.75	1.12162	1.62825	1.98802	2.26611	2.48985	2.67409
0.80	0.834494	1.30000	1.65489	1.94413	2.18766	2.39675
0.85	0.599696	1.01020	1.34472	1.63045	1.88038	2.10223
0.90	0.411835	0.760026	1.06314	1.33376	1.57882	1.80279
0.95	0.264955	0.548746	0.813379	1.06030	1.29147	1.50858
1.00	0.153000	0.374231	0.596738	0.814134	1.02462	1.22765
1.05	0.070137	0.233421	0.412947	0.597461	0.782569	0.966104
1.10	0.010975	0.122698	0.260572	0.410952	0.567907	0.728206
1.15	−0.029309	0.038191	0.137350	0.254059	0.381746	0.516622
1.20	−0.054910	−0.023991	0.040475	0.125309	0.224001	0.332629
1.25	−0.069388	−0.067585	−0.033152	0.022568	0.093642	0.176346
1.30	−0.075695	−0.096063	−0.086734	−0.056729	−0.011066	0.046966
1.35	−0.076221	−0.112551	−0.123422	−0.115380	−0.092349	−0.057033
1.40	−0.072860	−0.119795	−0.146209	−0.156264	−0.152741	−0.137685
1.45	−0.067072	−0.120147	−0.157848	−0.182211	−0.194924	−0.197364
1.50	−0.059957	−0.115579	−0.160807	−0.195911	−0.221601	−0.238630
1.55	−0.052313	−0.107702	−0.157245	−0.199844	−0.235395	−0.264102
1.60	−0.044699	−0.097804	−0.149007	−0.196243	−0.238777	−0.276359
1.65	−0.037483	−0.086883	−0.137632	−0.187067	−0.234008	−0.277857
1.70	−0.030892	−0.075691	−0.124374	−0.173999	−0.223110	−0.270873
1.75	−0.025041	−0.064771	−0.110226	−0.158446	−0.207851	−0.257468
1.80	−0.019971	−0.054495	−0.095952	−0.141557	−0.189734	−0.239460
1.85	−0.015668	−0.045097	−0.082115	−0.124246	−0.170014	−0.218419
1.90	−0.012084	−0.036705	−0.069110	−0.107212	−0.149705	−0.195667
1.95	−0.009150	−0.029366	−0.057194	−0.090972	−0.129604	−0.172288
2.00	−0.006790	−0.023069	−0.046513	−0.075882	−0.110312	−0.149145
2.05	−0.004922	−0.017761	−0.037125	−0.062168	−0.092261	−0.126899
2.10	−0.003469	−0.013362	−0.029022	−0.049950	−0.075738	−0.106033
2.15	−0.002359	−0.009779	−0.022149	−0.039264	−0.060908	−0.086876
2.20	−0.001527	−0.006911	−0.016421	−0.030080	−0.047842	−0.069626
2.25	−0.000918	−0.004658	−0.011729	−0.022326	−0.036532	−0.054376
2.30	−0.000482	−0.002924	−0.007958	−0.015893	−0.026913	−0.041131
2.35	−0.000181	−0.001621	−0.004987	−0.010657	−0.018880	−0.029831
2.40	0.000018	−0.000668	−0.002701	−0.006482	−0.012299	−0.020367
2.45	0.000142	0.000004	−0.000988	−0.003230	−0.007020	−0.012596
2.50	0.000212	0.000456	0.000251	−0.000766	−0.002888	−0.006356
2.55	0.000244	0.000739	0.001108	0.001036	0.000253	−0.001472
2.60	0.000251	0.000895	0.001662	0.002293	0.002553	0.002232
2.65	0.000241	0.000958	0.001981	0.003111	0.004154	0.004928
2.70	0.000221	0.000956	0.002123	0.003582	0.005184	0.006781
2.75	0.000196	0.000910	0.002134	0.003785	0.005758	0.007939

Table 3.3 (continued)

			t			
v	0.05	0.10	0.15	0.20	0.25	0.30
2.80	0.000169	0.000837	0.002055	0.003789	0.005977	0.008541
2.85	0.000143	0.000750	0.001915	0.003649	0.005927	0.008707
2.90	0.000118	0.000657	0.001739	0.003412	0.005683	0.008542
2.95	0.000096	0.000564	0.001545	0.003114	0.005306	0.008137
3.00	0.000077	0.000476	0.001347	0.002784	0.004845	0.007567
3.05	0.000061	0.000395	0.001154	0.002444	0.004338	0.006893
3.10	0.000047	0.000323	0.000973	0.002110	0.003817	0.006165
3.15	0.000036	0.000260	0.000808	0.001792	0.003304	0.005421
3.20	0.000027	0.000206	0.000660	0.001499	0.002815	0.004691
3.25	0.000020	0.000160	0.000531	0.001234	0.002361	0.003997
3.30	0.000015	0.000123	0.000420	0.001000	0.001949	0.003352
3.35	0.000010	0.000092	0.000327	0.000796	0.001582	0.002765
3.40	0.000007	0.000068	0.000249	0.000622	0.001262	0.002241
3.45	0.000005	0.000048	0.000185	0.000476	0.000986	0.001782
3.50	0.000003	0.000033	0.000134	0.000355	0.000752	0.001386
3.55	0.000002	0.000022	0.000093	0.000256	0.000557	0.001049
3.60	0.000001	0.000014	0.000061	0.000177	0.000398	0.000768
3.65	0.000001	0.000007	0.000037	0.000115	0.000270	0.000538
3.70		0.000003	0.000019	0.000067	0.000168	0.000352
3.75		0.000000	0.000006	0.000031	0.000090	0.000205
3.80		−0.000002	−0.000002	0.000005	0.000031	0.000091
3.85		−0.000003	−0.000008	−0.000013	−0.000012	0.000006
3.90		−0.000004	−0.000012	−0.000026	−0.000042	−0.000055
3.95		−0.000004	−0.000014	−0.000033	−0.000061	−0.000097
4.00		−0.000004	−0.000015	−0.000037	−0.000073	−0.000124
4.05		−0.000004	−0.000014	−0.000038	−0.000078	−0.000139
4.10		−0.000003	−0.000013	−0.000037	−0.000079	−0.000144
4.15		−0.000003	−0.000012	−0.000035	−0.000076	−0.000143
4.20		−0.000002	−0.000011	−0.000032	−0.000071	−0.000137
4.25		−0.000002	−0.000010	−0.000028	−0.000065	−0.000127
4.30		−0.000002	−0.000008	−0.000025	−0.000058	−0.000116
4.35		−0.000001	−0.000007	−0.000021	−0.000051	−0.000104
4.40		−0.000001	−0.000006	−0.000018	−0.000044	−0.000091
4.45		−0.000001	−0.000005	−0.000015	−0.000038	−0.000078
4.50		−0.000001	−0.000004	−0.000013	−0.000031	−0.000067
4.55		−0.000001	−0.000003	−0.000010	−0.000026	−0.000056
4.60			−0.000002	−0.000008	−0.000021	−0.000046
4.65			−0.000002	−0.000006	−0.000017	−0.000037
4.70			−0.000001	−0.000005	−0.000013	−0.000030
4.75			−0.000001	−0.000004	−0.000010	−0.000023
4.80			−0.000001	−0.000003	−0.000008	−0.000018
4.85				−0.000002	−0.000006	−0.000013
4.90				−0.000001	−0.000004	−0.000010

Table 3.3 (continued)

			t			
v	0.05	0.10	0.15	0.20	0.25	0.30
4.95				−0.000001	−0.000003	−0.000007
5.00				−0.000001	−0.000002	−0.000004

			t			
v	0.35	0.40	0.45	0.50	0.55	0.60
0.00	−5.525330	−5.748250	−5.90364	−6.00864	−6.07503	−6.11118
0.05	−4.238300	−4.564910	−4.81636	−5.01012	−5.15842	−5.27012
0.10	−2.964540	−3.362730	−3.68417	−3.94541	−4.15816	−4.33105
0.15	−1.751450	−2.189870	−2.55503	−2.86153	−3.11990	−3.33798
0.20	−0.636106	−1.08584	−1.46969	−1.79971	−2.08474	−2.33143
0.25	0.354545	−0.081229	−0.461203	−0.794581	−1.08821	−1.34734
0.30	1.20302	0.802042	0.445219	0.126247	−0.159670	−0.416284
0.35	1.90038	1.55020	1.23199	0.942182	0.677942	0.436995
0.40	2.44506	2.15678	1.88863	1.63941	1.40807	1.193690
0.45	2.84162	2.62167	2.41100	2.21038	2.02024	1.84082
0.50	3.09948	2.95003	2.80049	2.65314	2.50956	2.37091
0.55	3.23166	3.15119	3.06305	2.97055	2.87608	2.78146
0.60	3.25361	3.23756	3.20821	3.16941	3.12405	3.07436
0.65	3.18217	3.22367	3.24814	3.25960	3.26115	3.25518
0.70	3.03465	3.12519	3.19677	3.25327	3.29769	3.33243
0.75	2.82810	2.95820	3.06899	3.16396	3.24585	3.31688
0.80	2.57866	2.73847	2.87993	3.00598	3.11896	3.22078
0.85	2.30118	2.48095	2.64438	2.79373	2.93085	3.05729
0.90	2.00884	2.19938	2.37631	2.54118	2.69533	2.83987
0.95	1.71302	1.90599	2.08851	2.26146	2.42564	2.58177
1.00	1.42318	1.61133	1.79236	1.96654	2.13418	2.29561
1.05	1.14687	1.32419	1.49764	1.66700	1.83216	1.99309
1.10	0.889868	1.05159	1.21249	1.37191	1.52942	1.68469
1.15	0.656240	0.798907	0.943394	1.08878	1.23435	1.37956
1.20	0.448562	0.569913	0.695266	0.823521	0.953803	1.08540
1.25	0.268097	0.366986	0.471537	0.580578	0.693148	0.808452
1.30	0.115007	0.191263	0.274310	0.362987	0.456326	0.553501
1.35	−0.011444	0.042836	0.104520	0.172531	0.245950	0.323980
1.40	−0.112691	−0.079061	−0.037888	0.009888	0.063446	0.122056
1.45	−0.190682	−0.175867	−0.153782	−0.125197	−0.090802	−0.051229
1.50	−0.247713	−0.249520	−0.244671	−0.233746	−0.217289	−0.195811
1.55	−0.286282	−0.302298	−0.312536	−0.317383	−0.317228	−0.312452
1.60	−0.308964	−0.336684	−0.359686	−0.378176	−0.392389	−0.402582
1.65	−0.318313	−0.355243	−0.388621	−0.418494	−0.444956	−0.468142
1.70	−0.316781	−0.360524	−0.401921	−0.440882	−0.477383	−0.511448
1.75	−0.306656	−0.354981	−0.402144	−0.447951	−0.492276	−0.535055
1.80	−0.290023	−0.340914	−0.391758	−0.442282	−0.492288	−0.541641
1.85	−0.268737	−0.320426	−0.373073	−0.426356	−0.480030	−0.533908

Table 3.3 (continued)

			t			
v	0.35	0.40	0.45	0.50	0.55	0.60
1.90	-0.244408	-0.295395	-0.348203	-0.402494	-0.457998	-0.514497
1.95	-0.218404	-0.267456	-0.319038	-0.372819	-0.428522	-0.485919
2.00	-0.191854	-0.238003	-0.287229	-0.339224	-0.393725	-0.450509
2.05	-0.165662	-0.208192	-0.254181	-0.303360	-0.355494	-0.410380
2.10	-0.140528	-0.178952	-0.221059	-0.266631	-0.315473	-0.367407
2.15	-0.116969	-0.151002	-0.188799	-0.230200	-0.275051	-0.323212
2.20	-0.095338	-0.124874	-0.158129	-0.194997	-0.235374	-0.279160
2.25	-0.075852	-0.100933	-0.129579	-0.161738	-0.197355	-0.236369
2.30	-0.058615	-0.079404	-0.103514	-0.130945	-0.161687	-0.195718
2.35	-0.043636	-0.060388	-0.080150	-0.102968	-0.128869	-0.157871
2.40	-0.030854	-0.043891	-0.059581	-0.078005	-0.099226	-0.123288
2.45	-0.020151	-0.029842	-0.041800	-0.056133	-0.072929	-0.092261
2.50	-0.011373	-0.018112	-0.026719	-0.037321	-0.050026	-0.064925
2.55	-0.004341	-0.008528	-0.014189	-0.021460	-0.030459	-0.041291
2.60	0.001140	-0.000893	-0.004020	-0.008378	-0.014091	-0.021268
2.65	0.005265	0.005006	0.004009	0.002140	-0.000720	-0.004681
2.70	0.008228	0.009389	0.010134	0.010341	0.009899	0.008702
2.75	0.010215	0.012470	0.014596	0.016485	0.018038	0.019160
2.80	0.011398	0.014458	0.017633	0.020837	0.023984	0.026994
2.85	0.011933	0.015544	0.019475	0.023659	0.028029	0.032520
2.90	0.011962	0.015906	0.020334	0.025200	0.030456	0.036054
2.95	0.011606	0.015702	0.020406	0.025694	0.031537	0.037906
3.00	0.010972	0.015072	0.019868	0.025356	0.031527	0.038369
3.05	0.010148	0.014135	0.018873	0.024377	0.030656	0.037714
3.10	0.009208	0.012993	0.017556	0.022928	0.029133	0.036191
3.15	0.008211	0.011730	0.016030	0.021153	0.027139	0.034020
3.20	0.007203	0.010415	0.014389	0.019178	0.024832	0.031397
3.25	0.006219	0.009100	0.012708	0.017103	0.022345	0.028488
3.30	0.005286	0.007827	0.011047	0.015012	0.019787	0.025434
3.35	0.004421	0.006625	0.009451	0.012968	0.017245	0.022348
3.40	0.003635	0.005515	0.007954	0.011022	0.014789	0.019323
3.45	0.002933	0.004508	0.006577	0.009207	0.012469	0.016430
3.50	0.002318	0.003612	0.005333	0.007547	0.010321	0.013721
3.55	0.001787	0.002827	0.004230	0.006056	0.008368	0.011230
3.60	0.001336	0.002150	0.003266	0.004738	0.006623	0.008981
3.65	0.000959	0.001577	0.002438	0.003591	0.005089	0.006983
3.70	0.000650	0.001099	0.001739	0.002611	0.003762	0.005238
3.75	0.000401	0.000708	0.001158	0.001787	0.002634	0.003738
3.80	0.000205	0.000395	0.000686	0.001108	0.001691	0.002472
3.85	0.000055	0.000149	0.000310	0.000558	0.000920	0.001423
3.90	-0.000057	-0.000038	0.000018	0.000125	0.000303	0.000573
3.95	-0.000137	-0.000175	-0.000202	-0.000207	-0.000178	-0.000099
4.00	-0.000191	-0.000271	-0.000359	-0.000452	-0.000540	-0.000614

Table 3.3 (continued)

			t			
v	0.35	0.40	0.45	0.50	0.55	0.60
4.05	−0.000223	−0.000332	−0.000466	−0.000623	−0.000800	−0.000992
4.10	−0.000239	−0.000367	−0.000531	−0.000733	−0.000974	−0.001254
4.15	−0.000242	−0.000380	−0.000562	−0.000793	−0.001078	−0.001419
4.20	−0.000236	−0.000377	−0.000568	−0.000814	−0.001124	−0.001505
4.25	−0.000224	−0.000363	−0.000553	−0.000805	−0.001126	−0.001526
4.30	−0.000207	−0.000340	−0.000525	−0.000773	−0.001094	−0.001499
4.35	−0.000187	−0.000312	−0.000487	−0.000725	−0.001036	−0.001433
4.40	−0.000166	−0.000280	−0.000443	−0.000666	−0.000961	−0.001342
4.45	−0.000145	−0.000248	−0.000396	−0.000601	−0.000875	−0.001232
4.50	−0.000125	−0.000216	−0.000348	−0.000534	−0.000784	−0.001112
4.55	−0.000106	−0.000185	−0.000302	−0.000467	−0.000691	−0.000988
4.60	−0.000089	−0.000156	−0.000257	−0.000402	−0.000600	−0.000865
4.65	−0.000073	−0.000130	−0.000216	−0.000341	−0.000513	−0.000745
4.70	−0.000059	−0.000106	−0.000179	−0.000284	−0.000432	−0.000632
4.75	−0.000047	−0.000086	−0.000145	−0.000233	−0.000357	−0.000527
4.80	−0.000036	−0.000067	−0.000116	−0.000188	−0.000291	−0.000432
4.85	−0.000028	−0.000052	−0.000090	−0.000148	−0.000231	−0.000348
4.90	−0.000020	−0.000039	−0.000069	−0.000114	−0.000180	−0.000273
4.95	−0.000014	−0.000028	−0.000050	−0.000085	−0.000136	−0.000208
5.00	−0.000010	−0.000019	−0.000035	−0.000061	−0.000099	−0.000153

			t			
v	0.65	0.70	0.75	0.80	0.85	0.90
0.00	−6.12328	−6.11599	−6.09287	−6.05673	−6.00979	−5.95383
0.05	−5.35185	−5.40868	−5.44458	−5.46270	−5.46556	−5.45525
0.10	−4.47064	−4.58198	−4.66909	−4.73520	−4.78293	−4.81445
0.15	−3.52179	−3.67605	−3.80453	−3.91031	−3.99591	−4.06345
0.20	−2.54496	−2.72945	−2.88821	−3.02398	−3.13903	−3.23526
0.25	−1.57612	−1.77786	−1.95530	−2.11068	−2.24590	−2.36260
0.30	−0.646597	−0.853060	−1.03771	−1.20225	−1.34815	−1.47665
0.35	0.217470	0.017804	−0.163327	−0.327056	−0.474370	−0.606127
0.40	0.995441	0.812606	0.644542	0.490677	0.350496	0.223534
0.45	1.67221	1.51444	1.36747	1.23125	1.10569	0.990711
0.50	2.23802	2.11154	1.99195	1.87964	1.77489	1.67795
0.55	2.68809	2.59707	2.50930	2.42550	2.34627	2.27211
0.60	3.02211	2.96871	2.91533	2.86295	2.81240	2.76435
0.65	3.24364	3.22812	3.20995	3.19024	3.16996	3.14993
0.70	3.35946	3.38040	3.39661	3.40928	3.41942	3.42793
0.75	3.37888	3.43341	3.48179	3.52517	3.56455	3.60081
0.80	3.31306	3.39718	3.47435	3.54560	3.61187	3.67399
0.85	3.17437	3.28324	3.38491	3.48028	3.57016	3.65528
0.90	2.97581	3.10403	3.22532	3.34038	3.44988	3.55440
0.95	2.73049	2.87240	3.00805	3.13795	3.26258	3.38239

Table 3.3 (continued)

v	0.65	0.70	0.75	0.80	0.85	0.90
1.00	2.45114	2.60109	2.74576	2.88546	3.02049	3.15114
1.05	2.14979	2.30234	2.45081	2.59532	2.73597	2.87290
1.10	1.83747	1.98760	2.13496	2.27949	2.42113	2.55988
1.15	1.52396	1.66722	1.80903	1.94919	2.08751	2.22384
1.20	1.21774	1.35032	1.48274	1.61465	1.74575	1.87580
1.25	0.925814	1.044660	1.16449	1.28488	1.40544	1.52584
1.30	0.653803	0.756612	0.861381	0.967624	1.07491	1.18284
1.35	0.405918	0.491140	0.579087	0.669253	0.761176	0.854432
1.40	0.185062	0.251872	0.321943	0.394776	0.469910	0.546914
1.45	−0.007056	0.041181	0.092982	0.147880	0.205434	0.265229
1.50	−0.169795	−0.139700	−0.105963	−0.069002	−0.029217	0.013006
1.55	−0.303434	−0.290542	−0.274138	−0.254577	−0.232208	−0.207372
1.60	−0.409019	−0.411977	−0.411735	−0.408579	−0.402795	−0.394672
1.65	−0.488212	−0.505347	−0.519747	−0.531621	−0.541192	−0.548690
1.70	−0.543142	−0.572558	−0.599816	−0.625055	−0.648433	−0.670121
1.75	−0.576268	−0.615932	−0.654093	−0.690828	−0.726229	−0.760414
1.80	−0.590253	−0.638076	−0.685098	−0.731332	−0.776821	−0.821625
1.85	−0.587853	−0.641767	−0.695586	−0.749279	−0.802839	−0.856283
1.90	−0.571820	−0.629834	−0.688441	−0.747570	−0.807178	−0.867244
1.95	−0.544823	−0.605080	−0.666567	−0.729189	−0.792875	−0.857575
2.00	−0.509386	−0.570196	−0.632806	−0.697104	−0.763003	−0.830432
2.05	−0.467837	−0.527711	−0.589866	−0.654190	−0.720583	−0.788967
2.10	−0.422276	−0.479939	−0.540270	−0.603159	−0.668509	−0.736237
2.15	−0.374553	−0.428955	−0.486309	−0.546516	−0.609487	−0.675142
2.20	−0.326257	−0.376573	−0.430021	−0.486518	−0.545989	−0.608362
2.25	−0.278718	−0.324340	−0.373171	−0.425152	−0.480222	−0.538325
2.30	−0.233012	−0.273535	−0.317252	−0.364125	−0.414112	−0.467174
2.35	−0.189977	−0.225184	−0.263484	−0.304860	−0.349292	−0.396757
2.40	−0.150229	−0.180070	−0.212827	−0.248505	−0.287105	−0.328621
2.45	−0.114187	−0.138755	−0.165999	−0.195947	−0.228616	−0.264016
2.50	−0.082095	−0.101602	−0.123498	−0.147827	−0.174623	−0.203912
2.55	−0.054045	−0.068800	−0.085623	−0.104569	−0.125686	−0.149010
2.60	−0.030006	−0.040393	−0.052502	−0.066400	−0.082142	−0.099775
2.65	−0.009844	−0.016296	−0.024117	−0.033378	−0.044140	−0.056454
2.70	0.006655	0.003670	−0.000331	−0.005421	−0.011661	−0.019107
2.75	0.019762	0.019763	0.019089	0.017670	0.015447	0.012365
2.80	0.029788	0.032295	0.034447	0.036184	0.037447	0.038189
2.85	0.037068	0.041614	0.046101	0.050475	0.054688	0.058697
2.90	0.041948	0.048090	0.054436	0.060945	0.067577	0.074299
2.95	0.044770	0.052097	0.059858	0.068022	0.076563	0.085456
3.00	0.045867	0.054007	0.062772	0.072145	0.082114	0.092664
3.05	0.045554	0.054174	0.063574	0.073752	0.084706	0.096434
3.10	0.044120	0.052933	0.062644	0.073264	0.084803	0.097273

Table 3.3 (continued)

				t		
v	0.65	0.70	0.75	0.80	0.85	0.90
3.15	0.041829	0.050591	0.060334	0.071080	0.082853	0.095675
3.20	0.038916	0.047426	0.056967	0.067571	0.079275	0.092110
3.25	0.035585	0.043684	0.052832	0.063074	0.074454	0.087015
3.30	0.032011	0.039576	0.048184	0.057887	0.068737	0.080786
3.35	0.028340	0.035285	0.043242	0.052272	0.062433	0.073781
3.40	0.024692	0.030960	0.038193	0.046453	0.055805	0.066310
3.45	0.021159	0.026722	0.033186	0.040617	0.049081	0.058643
3.50	0.017813	0.022665	0.028344	0.034915	0.042446	0.051003
3.55	0.014706	0.018861	0.023759	0.029467	0.036050	0.043574
3.60	0.011872	0.015358	0.019500	0.024362	0.030007	0.036498
3.65	0.009331	0.012188	0.015612	0.019664	0.024401	0.029886
3.70	0.007089	0.009366	0.012123	0.015412	0.019291	0.023814
3.75	0.005144	0.006896	0.009042	0.011629	0.014708	0.018331
3.80	0.003485	0.004770	0.006367	0.008318	0.010668	0.013461
3.85	0.002097	0.002973	0.004086	0.005471	0.007165	0.009208
3.90	0.000958	0.001484	0.002177	0.003067	0.004185	0.005562
3.95	0.000047	0.000278	0.000615	0.001081	0.001699	0.002496
4.00	−0.000663	−0.000674	−0.000632	−0.000522	−0.000326	−0.000026
4.05	−0.001194	−0.001398	−0.001596	−0.001778	−0.001932	−0.002046
4.10	−0.001572	−0.001926	−0.002311	−0.002725	−0.003160	−0.003611
4.15	−0.001821	−0.002284	−0.002811	−0.003401	−0.004055	−0.004771
4.20	−0.001962	−0.002501	−0.003127	−0.003846	−0.004661	−0.005577
4.25	−0.002015	−0.002600	−0.003291	−0.004095	−0.005022	−0.006078
4.30	−0.001999	−0.002605	−0.003330	−0.004183	−0.005177	−0.006324
4.35	−0.001929	−0.002537	−0.003270	−0.004142	−0.005166	−0.006358
4.40	−0.001821	−0.002414	−0.003134	−0.003999	−0.005023	−0.006224
4.45	−0.001686	−0.002251	−0.002943	−0.003780	−0.004779	−0.005957
4.50	−0.001533	−0.002062	−0.002714	−0.003508	−0.004461	−0.005593
4.55	−0.001372	−0.001858	−0.002462	−0.003201	−0.004094	−0.005160
4.60	−0.001209	−0.001648	−0.002197	−0.002874	−0.003697	−0.004685
4.65	−0.001049	−0.001440	−0.001932	−0.002542	−0.003287	−0.004187
4.70	−0.000897	−0.001239	−0.001672	−0.002213	−0.002878	−0.003686
4.75	−0.000754	−0.001049	−0.001425	−0.001897	−0.002481	−0.003194
4.80	−0.000623	−0.000873	−0.001194	−0.001599	−0.002103	−0.002723
4.85	−0.000505	−0.000712	−0.000981	−0.001323	−0.001752	−0.002281
4.90	−0.000400	−0.000569	−0.000790	−0.001073	−0.001430	−0.001873
4.95	−0.000308	−0.000443	−0.000620	−0.000849	−0.001139	−0.001503
5.00	−0.000230	−0.000334	−0.000472	−0.000652	−0.000882	−0.001173

				t		
v	0.95	1.00	2.00	3.00	4.00	5.00
0.00	−5.89030	−5.82035	−3.76727	−0.305815	6.27746	16.2735
0.05	−5.43345	−5.40158	−3.58860	0.101345	6.86526	16.4564

Table 3.3 (continued)

			t			
v	0.95	1.00	2.00	3.00	4.00	5.00
0.10	−4.83158	−4.83582	−3.30206	0.546356	7.41317	16.5060
0.15	−4.11469	−4.15114	−2.91237	1.02739	7.91661	16.4192
0.20	−3.31431	−3.37757	−2.42794	1.54079	8.37089	16.1942
0.25	−2.46214	−2.54574	−1.86047	2.08111	8.77123	15.8305
0.30	−1.58885	−1.68570	−1.22443	2.64114	9.11271	15.3287
0.35	−0.723086	−0.825922	−0.536516	3.21205	9.39034	14.6908
0.40	0.109369	0.007616	0.185008	3.78369	9.59908	13.9200
0.45	0.886201	0.792054	0.920971	4.34476	9.73397	13.0209
0.50	1.58899	1.50816	1.65196	4.88324	9.79021	11.9990
0.55	2.20344	2.14061	2.35896	5.38670	9.76331	10.8613
0.60	2.71943	2.67813	3.02396	5.84271	9.64928	9.61578
0.65	3.13088	3.11343	3.63050	6.23925	9.44474	8.27162
0.70	3.43557	3.44305	4.16409	6.56507	9.14713	6.83909
0.75	3.63474	3.66704	4.61265	6.81011	8.75491	5.32950
0.80	3.73269	3.78865	4.96671	6.96578	8.26768	3.75515
0.85	3.73630	3.81381	5.21964	7.02531	7.68637	2.12933
0.90	3.65450	3.75068	5.36770	6.98396	7.01333	0.46615
0.95	3.49778	3.60915	5.41003	6.83919	6.25249	−1.21948
1.00	3.27767	3.40035	5.34855	6.59078	5.40936	−2.91194
1.05	3.00625	3.13614	5.18776	6.24087	4.49107	−4.59509
1.10	2.69573	2.82869	4.93449	5.79393	3.50638	−6.25232
1.15	2.35805	2.49007	4.59762	5.25667	2.46557	−7.86677
1.20	2.00457	2.13187	4.18767	4.63783	1.38039	−9.42147
1.25	1.64577	1.76497	3.71649	3.94805	0.263821	−10.8995
1.30	1.29106	1.39924	3.19686	3.19953	−0.870046	−12.2844
1.35	0.948631	1.043410	2.64203	2.40573	−2.00628	−13.5600
1.40	0.625387	0.704949	2.06543	1.58109	−3.12937	−14.7110
1.45	0.326870	0.389980	1.48027	0.740592	−4.22351	−15.7230
1.50	0.057297	0.103295	0.899207	−0.100540	−5.27289	−16.5829
1.55	−0.180403	−0.151629	0.334056	−0.927236	−6.26206	−17.2791
1.60	−0.384498	−0.372562	−0.204435	−1.72492	−7.17621	−17.8015
1.65	−0.554354	−0.558425	−0.706804	−2.47985	−8.00147	−18.1420
1.70	−0.690303	−0.709173	−1.16502	−3.17939	−8.72525	−18.2945
1.75	−0.793511	−0.825666	−1.57257	−3.81230	−9.33648	−18.2553
1.80	−0.865827	−0.909525	−1.92454	−4.36895	−9.82585	−18.0230
1.85	−0.909647	−0.962988	−2.21759	−4.84148	−10.1860	−17.5984
1.90	−0.927768	−0.988768	−2.44993	−5.22393	−10.4119	−16.9854
1.95	−0.923260	−0.989917	−2.62124	−5.51232	−10.5003	−16.1899
2.00	−0.899342	−0.969696	−2.73260	−5.70465	−10.4509	−15.2207
2.05	−0.859276	−0.931460	−2.78628	−5.80088	−10.2652	−14.0889
2.10	−0.806272	−0.878554	−2.78567	−5.80288	−9.94737	−12.8081
2.15	−0.743409	−0.814228	−2.73505	−5.71428	−9.50352	−11.3938
2.20	−0.673573	−0.741562	−2.63945	−5.54035	−8.94199	−9.86376

Table 3.3 (continued)

				t		
v	0.95	1.00	2.00	3.00	4.00	5.00
2.25	−0.599404	−0.663409	−2.50447	−5.28778	−8.27301	−8.23737
2.30	−0.523268	−0.582352	−2.33606	−4.96450	−7.50848	−6.53544
2.35	−0.447228	−0.500676	−2.14040	−4.57947	−6.66178	−4.77998
2.40	−0.373041	−0.420350	−1.92371	−4.14238	−5.74744	−2.99376
2.45	−0.302153	−0.343023	−1.69211	−3.66347	−4.78085	−1.20002
2.50	−0.235709	−0.270027	−1.45148	−3.15326	−3.77802	0.577933
2.55	−0.174572	−0.202391	−1.20736	−2.62233	−2.75519	2.31708
2.60	−0.119337	−0.140857	−0.964865	−2.08105	−1.72854	3.99506
2.65	−0.070365	−0.085909	−0.728573	−1.53945	−0.713912	5.59055
2.70	−0.027805	−0.037794	−0.502506	−1.00698	0.273488	7.08360
2.75	0.008377	0.003444	−0.290078	−0.492343	1.21937	8.45591
2.80	0.038364	0.037935	−0.094084	−0.003403	2.11060	9.69120
2.85	0.062464	0.065955	0.083304	0.452960	2.93538	10.7754
2.90	0.081078	0.087890	0.240523	0.870912	3.68346	11.6967
2.95	0.094679	0.104216	0.376590	1.24572	4.34624	12.4463
3.00	0.103785	0.115467	0.491065	1.57379	4.91692	13.0177
3.05	0.108936	0.122215	0.584004	1.85263	5.39049	13.4072
3.10	0.110684	0.125049	0.655907	2.08087	5.76382	13.6143
3.15	0.109569	0.124556	0.707655	2.25816	6.03560	13.6407
3.20	0.106110	0.121307	0.740456	2.38515	6.20629	13.4912
3.25	0.100798	0.115845	0.755776	2.46339	6.27806	13.1730
3.30	0.094083	0.108677	0.755284	2.49521	6.25466	12.6955
3.35	0.086373	0.100265	0.740790	2.48366	6.14127	12.0704
3.40	0.078030	0.091024	0.714187	2.43233	5.94435	11.3115
3.45	0.069367	0.081318	0.677406	2.34532	5.67144	10.4339
3.50	0.060652	0.071458	0.632365	2.22704	5.33101	9.45423
3.55	0.052104	0.061708	0.580930	2.08214	4.93219	8.39007
3.60	0.043901	0.052280	0.524880	1.91539	4.48465	7.25976
3.65	0.036180	0.043344	0.465877	1.73157	3.99831	6.08198
3.70	0.029040	0.035026	0.405447	1.53538	3.48323	4.87548
3.75	0.022548	0.027413	0.344961	1.33136	2.94934	3.65876
3.80	0.016743	0.020563	0.285622	1.12379	2.40633	2.44976
3.85	0.011640	0.014500	0.228463	0.916647	1.86343	1.26560
3.90	0.007232	0.009229	0.174344	0.713551	1.32933	0.122290
3.95	0.003496	0.004729	0.123953	0.517723	0.811997	−0.965432
4.00	0.000399	0.000969	0.077814	0.331952	0.318629	−1.98434
4.05	−0.002106	−0.002099	0.036296	0.158588	−0.144470	−2.92288
4.10	−0.004070	−0.004530	−0.000374	−0.000471	−0.571925	−3.77124
4.15	−0.005547	−0.006383	−0.032101	−0.143774	−0.959334	−4.52151
4.20	−0.006596	−0.007723	−0.058905	−0.270299	−1.30328	−5.16766
4.25	−0.007273	−0.008614	−0.080904	−0.379438	−1.60132	−5.70560
4.30	−0.007635	−0.009121	−0.098303	−0.470968	−1.85199	−6.13313
4.35	−0.007733	−0.009305	−0.111374	−0.545014	−2.05473	−6.44990

Table 3.3 (continued)

ν	0.95	1.00	2.00	3.00	4.00	5.00
4.40	−0.007618	−0.009223	−0.120445	−0.602021	−2.20983	−6.65732
4.45	−0.007334	−0.008930	−0.125881	−0.642705	−2.31839	−6.75845
4.50	−0.006924	−0.008474	−0.128074	−0.668019	−2.38224	−6.75785
4.55	−0.006422	−0.007899	−0.127433	−0.679105	−2.40381	−6.66144
4.60	−0.005859	−0.007242	−0.124366	−0.677258	−2.38611	−6.47633
4.65	−0.005263	−0.006536	−0.119278	−0.663878	−2.33256	−6.21062
4.70	−0.004656	−0.005809	−0.112560	−0.640439	−2.24694	−5.87323
4.75	−0.004055	−0.005083	−0.104581	−0.608447	−2.13329	−5.47367
4.80	−0.003474	−0.004377	−0.095686	−0.569411	−1.99579	−5.02189
4.85	−0.002926	−0.003704	−0.086189	−0.524807	−1.83872	−4.52805
4.90	−0.002416	−0.003076	−0.076373	−0.476062	−1.66633	−4.00237
4.95	−0.001951	−0.002498	−0.066487	−0.424520	−1.48278	−3.45493
5.00	−0.001533	−0.001977	−0.056744	−0.371434	−1.29208	−2.89549

3.2 The Kelvin Functions bei$_\nu$(t), and Their First and Second Derivatives with Respect to the Order

These derivatives with respect to the order were evaluated directly, by using the central-difference formulas of order $O(h^4)$ with $h = 10^{-3}$. The Kelvin functions bei$_\nu$(t) were taken from MATHEMATICA program:

$$\frac{\partial \text{bei}_\nu(t)}{\partial \nu} = \frac{-\text{bei}_{\nu+2h}(t) + 8\,\text{bei}_{\nu+h}(t) - 8\,\text{bei}_{\nu-h}(t) + \text{bei}_{\nu-2h}(t)}{12h} \tag{3.2.1}$$

$$\frac{\partial^2 \text{bei}_\nu(t)}{\partial \nu^2} = \frac{-\text{bei}_{\nu+2h}(t) + 16\,\text{bei}_{\nu+h}(t) - 30\,\text{bei}_\nu(t) + 16\,\text{bei}_{\nu-h}(t) - \text{bei}_{\nu-2h}(t)}{12h^2} \tag{3.2.2}$$

Functional form of ber$_\nu$(t) and bei$_\nu$(t) functions and their derivatives with respect to the order is similar.

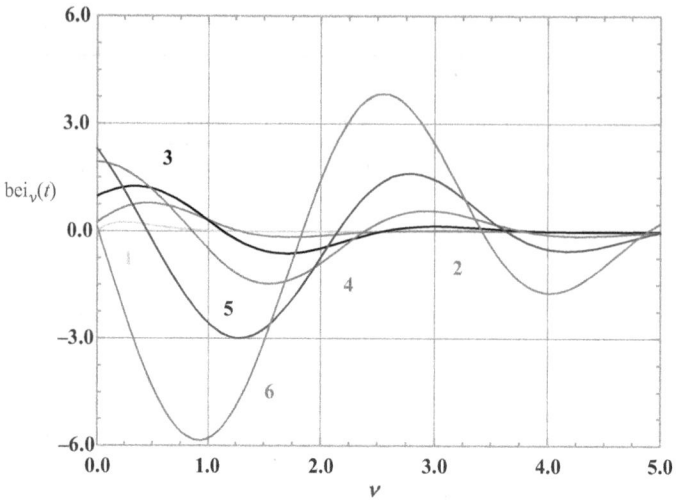

Figure 3.10: The Kelvin functions $bei_\nu(t)$ plotted as a function of the order ν, at constant values of argument t.

$1 - t = 0.05; 2 - t = 1.0; 3 - t = 2.0; 4 - t = 3.0; 5 - t = 4.0; 6 - t = 5.0.$

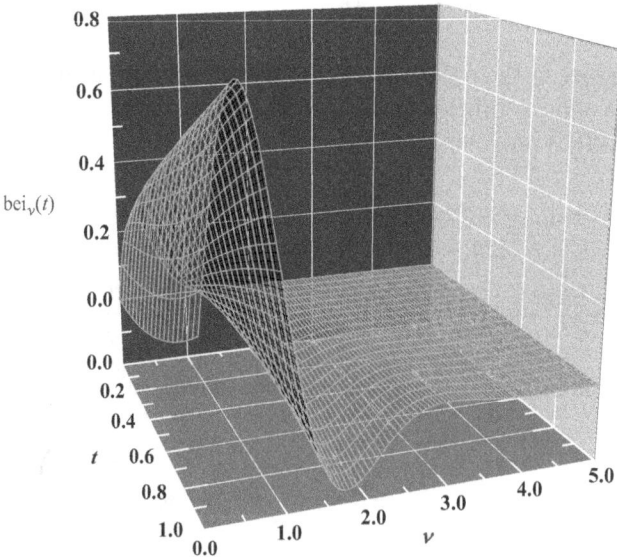

Figure 3.11: The Kelvin functions $bei_\nu(t)$ as a function of ν and t.

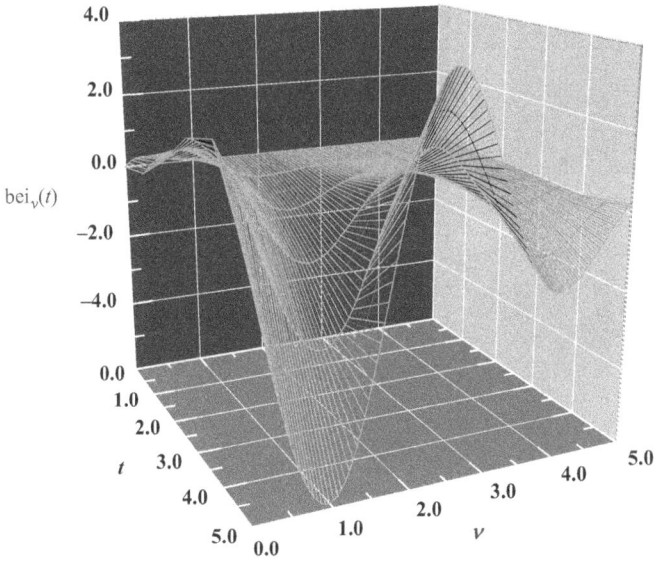

bei$_v$(t)

Figure 3.12: The Kelvin functions bei$_v$(t) as a function of v and t.

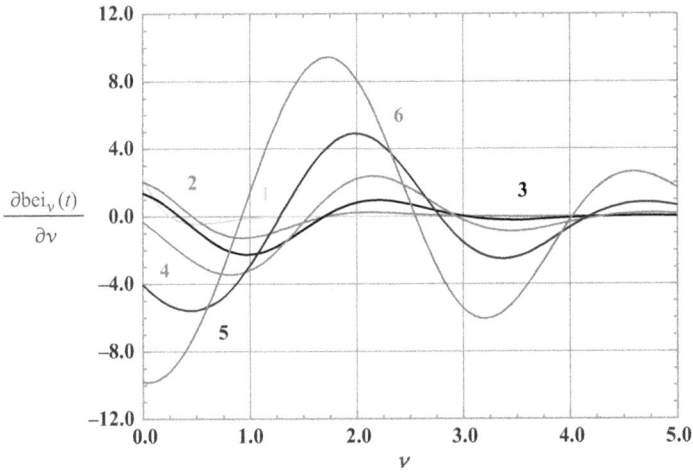

$\dfrac{\partial bei_v(t)}{\partial v}$

Figure 3.13: First derivatives of the Kelvin functions bei$_v$(t) with respect to the order v as a function of v, at constant values of argument t.

$1 - t = 0.05;\ 2 - t = 1.0;\ 3 - t = 2.0;\ 4 - t = 3.0;\ 5 - t = 4.0;\ 6 - t = 5.0.$

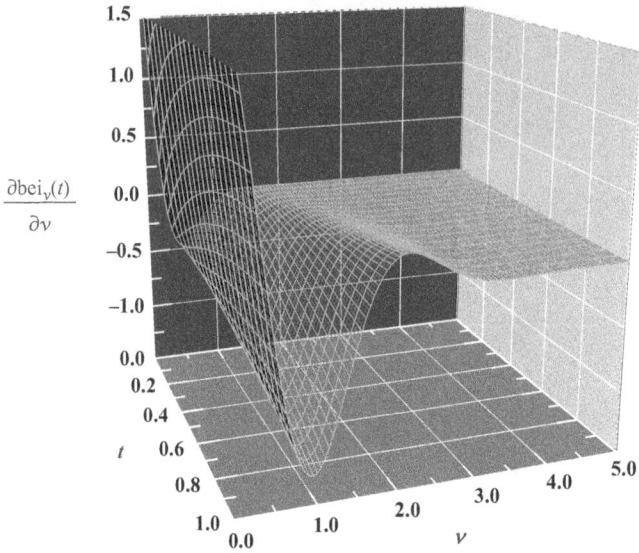

Figure 3.14: First derivatives with respect to the order of the Kelvin functions $\partial\mathrm{bei}_\nu(t)/\partial\nu$ as a function of ν and t.

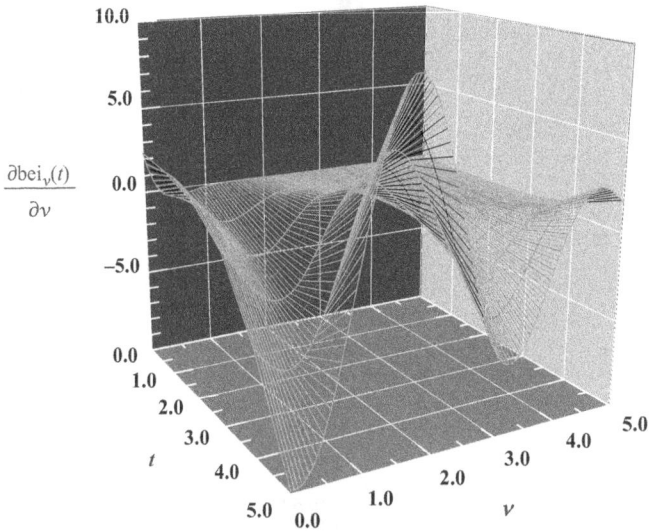

Figure 3.15: First derivatives with respect to the order of the Kelvin functions $\partial\mathrm{bei}_\nu(t)/\partial\nu$ as a function of ν and t.

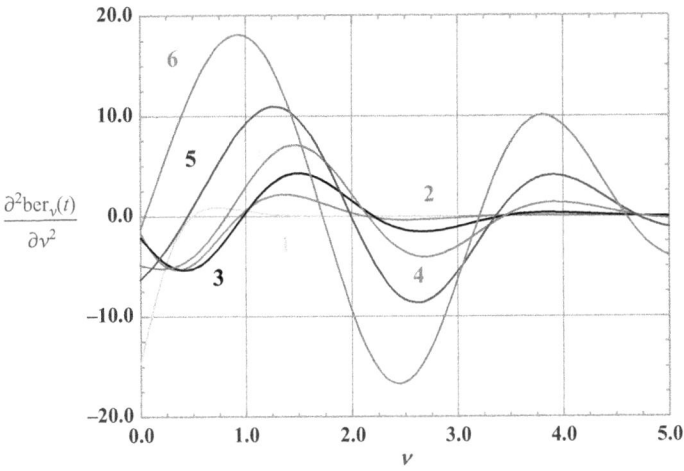

$$\frac{\partial^2 \text{ber}_v(t)}{\partial v^2}$$

Figure 3.16: Second derivatives of the Kelvin functions bei$_v(t)$ with respect to the order v as a function of v, at constant values of argument t.

$1 - t = 0.05; 2 - t = 1.0; 3 - t = 2.0; 4 - t = 3.0; 5 - t = 4.0; 6 - t = 5.0.$

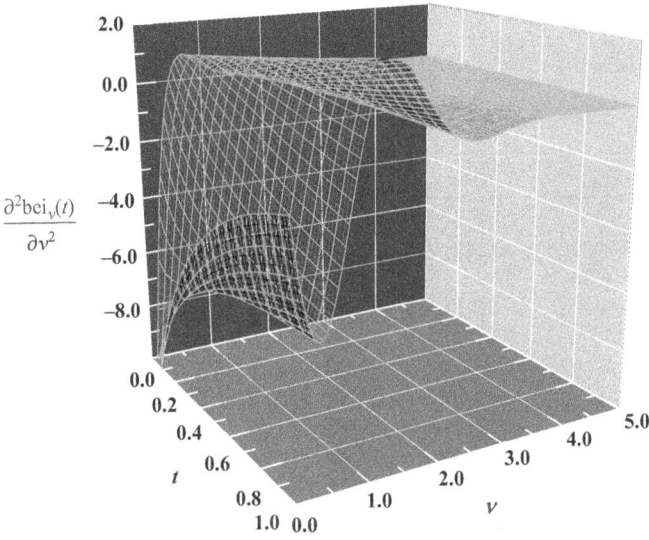

$$\frac{\partial^2 \text{bei}_v(t)}{\partial v^2}$$

Figure 3.17: Second derivatives with respect to the order of the Kelvin functions $\partial^2 \text{bei}_v(t)/\partial v^2$ as a function of v and t.

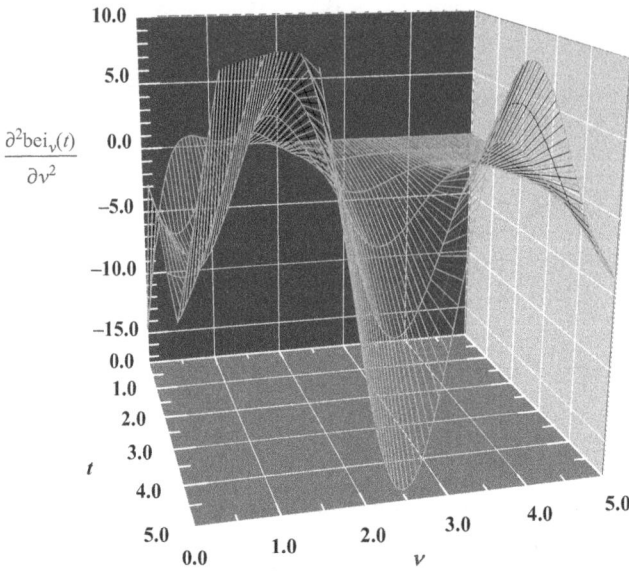

Figure 3.18: Second derivatives with respect to the order of the Kelvin functions $\partial^2 \mathrm{bei}_v(t)/\partial v^2$ as a function of v and t.

Table 3.4: The Kelvin functions $\mathrm{bei}_v(t)$.

				t		
v	0.05	0.10	0.15	0.20	0.25	0.30
0.00	0.000625	0.002500	0.005625	0.010000	0.015625	0.022500
0.05	0.100905	0.106032	0.110870	0.116263	0.122488	0.129678
0.10	0.170085	0.183583	0.193420	0.202291	0.211095	0.220252
0.15	0.213625	0.238078	0.254858	0.268801	0.281543	0.293857
0.20	0.236677	0.272703	0.297243	0.317075	0.334538	0.350749
0.25	0.243909	0.290709	0.322921	0.348802	0.371258	0.391694
0.30	0.239415	0.295251	0.334376	0.365941	0.393232	0.417858
0.35	0.226673	0.289279	0.334098	0.370589	0.402216	0.430698
0.40	0.208556	0.275458	0.324484	0.364877	0.400093	0.431868
0.45	0.187359	0.256124	0.307761	0.350884	0.388780	0.423120
0.50	0.164860	0.233268	0.285939	0.330569	0.370153	0.406235
0.55	0.142383	0.208533	0.260777	0.305722	0.345990	0.382948
0.60	0.120867	0.183234	0.233773	0.277934	0.317923	0.354905
0.65	0.100933	0.158388	0.206164	0.248579	0.287413	0.323618
0.70	0.082949	0.134740	0.178935	0.218807	0.255728	0.290439
0.75	0.067089	0.112806	0.152843	0.189553	0.223938	0.256544
0.80	0.053380	0.092909	0.128435	0.161543	0.192916	0.222926
0.85	0.041747	0.075213	0.106080	0.135318	0.163347	0.190395
0.90	0.032045	0.059761	0.085993	0.111247	0.135741	0.159586

Table 3.4 (continued)

v	0.05	0.10	0.15	0.20	0.25	0.30
0.95	0.024088	0.046500	0.068264	0.089559	0.110452	0.130970
1.00	0.017672	0.035311	0.052884	0.070357	0.087696	0.104868
1.05	0.012586	0.026030	0.039768	0.053645	0.067571	0.081474
1.10	0.008627	0.018466	0.028775	0.039351	0.050077	0.060866
1.15	0.005606	0.012416	0.019729	0.027344	0.035140	0.043031
1.20	0.003350	0.007676	0.012431	0.017449	0.022624	0.027879
1.25	0.001709	0.004048	0.006673	0.009466	0.012351	0.015265
1.30	0.000554	0.001350	0.002247	0.003184	0.004116	0.005000
1.35	−0.000226	−0.000587	−0.001049	−0.001616	−0.002303	−0.003134
1.40	−0.000721	−0.001912	−0.003401	−0.005145	−0.007134	−0.009368
1.45	−0.001006	−0.002755	−0.004982	−0.007608	−0.010599	−0.013940
1.50	−0.001139	−0.003226	−0.005945	−0.009191	−0.012914	−0.017087
1.55	−0.001166	−0.003420	−0.006425	−0.010065	−0.014280	−0.019034
1.60	−0.001124	−0.003411	−0.006537	−0.010382	−0.014881	−0.019993
1.65	−0.001039	−0.003264	−0.006380	−0.010274	−0.014881	−0.020159
1.70	−0.000931	−0.003026	−0.006034	−0.009853	−0.014424	−0.019707
1.75	−0.000813	−0.002735	−0.005564	−0.009215	−0.013636	−0.018792
1.80	−0.000695	−0.002420	−0.005024	−0.008439	−0.012623	−0.017547
1.85	−0.000583	−0.002102	−0.004452	−0.007585	−0.011470	−0.016085
1.90	−0.000481	−0.001795	−0.003880	−0.006704	−0.010249	−0.014500
1.95	−0.000391	−0.001509	−0.003328	−0.005834	−0.009016	−0.012869
2.00	−0.000313	−0.001250	−0.002813	−0.005000	−0.007812	−0.011250
2.05	−0.000246	−0.001020	−0.002341	−0.004222	−0.006670	−0.009689
2.10	−0.000191	−0.000820	−0.001920	−0.003512	−0.005609	−0.008221
2.15	−0.000146	−0.000649	−0.001550	−0.002876	−0.004643	−0.006866
2.20	−0.000110	−0.000505	−0.001231	−0.002316	−0.003780	−0.005639
2.25	−0.000081	−0.000385	−0.000959	−0.001830	−0.003021	−0.004546
2.30	−0.000059	−0.000288	−0.000732	−0.001417	−0.002363	−0.003588
2.35	−0.000041	−0.000210	−0.000544	−0.001069	−0.001803	−0.002761
2.40	−0.000028	−0.000149	−0.000393	−0.000782	−0.001332	−0.002058
2.45	−0.000018	−0.000101	−0.000272	−0.000548	−0.000944	−0.001471
2.50	−0.000011	−0.000064	−0.000177	−0.000362	−0.000629	−0.000988
2.55	−0.000006	−0.000037	−0.000104	−0.000216	−0.000379	−0.000599
2.60	−0.000003	−0.000017	−0.000050	−0.000104	−0.000184	−0.000291
2.65	−0.000001	−0.000003	−0.000010	−0.000021	−0.000036	−0.000055
2.70	0.000001	0.000006	0.000018	0.000039	0.000072	0.000121
2.75	0.000002	0.000012	0.000036	0.000079	0.000148	0.000246
2.80	0.000002	0.000015	0.000047	0.000105	0.000197	0.000331
2.85	0.000002	0.000016	0.000052	0.000119	0.000226	0.000381
2.90	0.000002	0.000017	0.000054	0.000125	0.000239	0.000406
2.95	0.000002	0.000016	0.000053	0.000124	0.000239	0.000411
3.00	0.000002	0.000015	0.000050	0.000118	0.000231	0.000400
3.05	0.000002	0.000013	0.000046	0.000110	0.000217	0.000379
3.10	0.000001	0.000012	0.000041	0.000100	0.000199	0.000351

Table 3.4 (continued)

			t			
v	0.05	0.10	0.15	0.20	0.25	0.30
3.15	0.000001	0.000010	0.000036	0.000089	0.000179	0.000318
3.20	0.000001	0.000008	0.000031	0.000077	0.000158	0.000284
3.25	0.000001	0.000007	0.000026	0.000067	0.000138	0.000249
3.30	0.000001	0.000006	0.000022	0.000056	0.000118	0.000215
3.35		0.000005	0.000018	0.000047	0.000100	0.000183
3.40		0.000004	0.000015	0.000039	0.000083	0.000154
3.45		0.000003	0.000012	0.000031	0.000068	0.000127
3.50		0.000002	0.000009	0.000025	0.000055	0.000104
3.55		0.000002	0.000007	0.000020	0.000043	0.000083
3.60		0.000001	0.000005	0.000015	0.000034	0.000065
3.65		0.000001	0.000004	0.000011	0.000026	0.000050
3.70		0.000001	0.000003	0.000008	0.000019	0.000037
3.75			0.000002	0.000006	0.000014	0.000027
3.80			0.000001	0.000004	0.000009	0.000019
3.85			0.000001	0.000003	0.000006	0.000012
3.90				0.000001	0.000003	0.000007
3.95				0.000001	0.000001	0.000003
4.00				0.000000	0.000000	0.000000
4.05				0.000000	−0.000001	−0.000002
4.10				−0.000001	−0.000002	−0.000004
4.15				−0.000001	−0.000002	−0.000004
4.20				−0.000001	−0.000002	−0.000005
4.25				−0.000001	−0.000002	−0.000005
4.30				−0.000001	−0.000002	−0.000005
4.35				−0.000001	−0.000002	−0.000005
4.40				−0.000001	−0.000002	−0.000004
4.45				−0.000001	−0.000002	−0.000004
4.50				−0.000001	−0.000002	−0.000003
4.55					−0.000001	−0.000003
4.60					−0.000001	−0.000003
4.65					−0.000001	−0.000002
4.70					−0.000001	−0.000002
4.75					−0.000001	−0.000002
4.80					−0.000001	−0.000001
4.85						−0.000001
4.90						−0.000001
4.95						−0.000001

			t			
v	0.35	0.40	0.45	0.50	0.55	0.60
0.00	0.030624	0.039998	0.050621	0.062493	0.075613	0.089980
0.05	0.137905	0.147214	0.157638	0.169197	0.181907	0.195780
0.10	0.229995	0.240473	0.251781	0.263989	0.277147	0.291292

Table 3.4 (continued)

			t			
v	0.35	0.40	0.45	0.50	0.55	0.60
0.15	0.306175	0.318769	0.331820	0.345455	0.359770	0.374834
0.20	0.366339	0.381704	0.397110	0.412748	0.428756	0.445240
0.25	0.410910	0.429411	0.447540	0.465543	0.483598	0.501845
0.30	0.440743	0.462478	0.483463	0.503987	0.524262	0.544452
0.35	0.457034	0.481863	0.505625	0.528635	0.551129	0.573287
0.40	0.461220	0.488808	0.515085	0.540378	0.564935	0.588945
0.45	0.454897	0.484753	0.513137	0.540373	0.566706	0.592324
0.50	0.439737	0.471264	0.501238	0.529967	0.557683	0.584565
0.55	0.417423	0.449958	0.480934	0.510633	0.539263	0.566990
0.60	0.389589	0.422444	0.453801	0.483902	0.512932	0.541032
0.65	0.357776	0.390274	0.421387	0.451314	0.480209	0.508186
0.70	0.323395	0.354902	0.385172	0.414364	0.442595	0.469953
0.75	0.287707	0.317651	0.346532	0.374464	0.401531	0.427795
0.80	0.251802	0.279695	0.306708	0.332914	0.358365	0.383096
0.85	0.216600	0.242047	0.266794	0.290878	0.314320	0.337135
0.90	0.182845	0.205551	0.227724	0.249369	0.270484	0.291062
0.95	0.151116	0.170887	0.190268	0.209242	0.227788	0.245881
1.00	0.121839	0.138574	0.155038	0.171195	0.187009	0.202444
1.05	0.095297	0.108986	0.122492	0.135768	0.148768	0.161446
1.10	0.071648	0.082360	0.092947	0.103354	0.113531	0.123427
1.15	0.050944	0.058817	0.066591	0.074212	0.081626	0.088783
1.20	0.033150	0.038376	0.043502	0.048478	0.053250	0.057770
1.25	0.018155	0.020969	0.023660	0.026180	0.028483	0.030522
1.30	0.005796	0.006464	0.006965	0.007259	0.007305	0.007062
1.35	−0.004132	−0.005325	−0.006744	−0.008420	−0.010388	−0.012681
1.40	−0.011859	−0.014622	−0.017679	−0.021052	−0.024768	−0.028855
1.45	−0.017630	−0.021674	−0.026083	−0.030872	−0.036061	−0.041673
1.50	−0.021697	−0.026741	−0.032222	−0.038148	−0.044533	−0.051392
1.55	−0.024306	−0.030086	−0.036371	−0.043164	−0.050472	−0.058304
1.60	−0.025693	−0.031968	−0.038808	−0.046211	−0.054181	−0.062723
1.65	−0.026082	−0.032633	−0.039800	−0.047579	−0.055968	−0.064969
1.70	−0.025677	−0.032313	−0.039606	−0.047547	−0.056133	−0.065363
1.75	−0.024659	−0.031220	−0.038463	−0.046379	−0.054965	−0.064217
1.80	−0.023192	−0.029543	−0.036588	−0.044321	−0.052735	−0.061829
1.85	−0.021416	−0.027449	−0.034177	−0.041593	−0.049693	−0.058473
1.90	−0.019447	−0.025083	−0.031400	−0.038394	−0.046062	−0.054401
1.95	−0.017387	−0.022566	−0.028403	−0.034896	−0.042042	−0.049839
2.00	−0.015312	−0.019999	−0.025310	−0.031245	−0.037804	−0.044985
2.05	−0.013286	−0.017463	−0.022222	−0.027565	−0.033493	−0.040007
2.10	−0.011356	−0.015020	−0.019218	−0.023954	−0.029230	−0.035048
2.15	−0.009555	−0.012718	−0.016361	−0.020491	−0.025111	−0.030222
2.20	−0.007905	−0.010588	−0.013696	−0.017235	−0.021210	−0.025622
2.25	−0.006420	−0.008652	−0.011253	−0.014227	−0.017581	−0.021317
2.30	−0.005104	−0.006921	−0.009049	−0.011494	−0.014261	−0.017353

Table 3.4 (continued)

				t		
v	0.35	0.40	0.45	0.50	0.55	0.60
2.35	−0.003955	−0.005397	−0.007093	−0.009050	−0.011272	−0.013762
2.40	−0.002970	−0.004076	−0.005383	−0.006898	−0.008623	−0.010560
2.45	−0.002136	−0.002948	−0.003912	−0.005033	−0.006311	−0.007750
2.50	−0.001444	−0.002003	−0.002668	−0.003443	−0.004328	−0.005322
2.55	−0.000879	−0.001224	−0.001635	−0.002112	−0.002655	−0.003263
2.60	−0.000429	−0.000596	−0.000793	−0.001019	−0.001272	−0.001549
2.65	−0.000077	−0.000101	−0.000124	−0.000143	−0.000154	−0.000153
2.70	0.000188	0.000277	0.000393	0.000541	0.000726	0.000954
2.75	0.000380	0.000556	0.000779	0.001056	0.001395	0.001803
2.80	0.000512	0.000750	0.001052	0.001426	0.001881	0.002426
2.85	0.000595	0.000874	0.001230	0.001672	0.002209	0.002853
2.90	0.000637	0.000942	0.001331	0.001816	0.002407	0.003115
2.95	0.000649	0.000964	0.001370	0.001876	0.002495	0.003240
3.00	0.000636	0.000952	0.001359	0.001870	0.002497	0.003253
3.05	0.000607	0.000914	0.001312	0.001813	0.002431	0.003178
3.10	0.000566	0.000857	0.001237	0.001718	0.002313	0.003036
3.15	0.000517	0.000789	0.001144	0.001597	0.002159	0.002844
3.20	0.000465	0.000713	0.001040	0.001459	0.001981	0.002620
3.25	0.000411	0.000634	0.000931	0.001312	0.001789	0.002375
3.30	0.000358	0.000556	0.000820	0.001162	0.001591	0.002121
3.35	0.000307	0.000480	0.000713	0.001014	0.001395	0.001867
3.40	0.000260	0.000409	0.000610	0.000872	0.001206	0.001620
3.45	0.000216	0.000343	0.000514	0.000739	0.001026	0.001384
3.50	0.000178	0.000283	0.000427	0.000617	0.000860	0.001165
3.55	0.000143	0.000230	0.000349	0.000506	0.000709	0.000963
3.60	0.000113	0.000183	0.000279	0.000407	0.000573	0.000781
3.65	0.000088	0.000143	0.000219	0.000320	0.000452	0.000620
3.70	0.000066	0.000108	0.000167	0.000245	0.000348	0.000478
3.75	0.000048	0.000079	0.000123	0.000181	0.000258	0.000356
3.80	0.000033	0.000055	0.000086	0.000128	0.000183	0.000253
3.85	0.000022	0.000036	0.000056	0.000084	0.000120	0.000166
3.90	0.000012	0.000021	0.000032	0.000048	0.000069	0.000095
3.95	0.000005	0.000009	0.000013	0.000020	0.000028	0.000038
4.00	0.000000	−0.000001	−0.000001	−0.000002	−0.000004	−0.000006
4.05	−0.000004	−0.000007	−0.000012	−0.000018	−0.000027	−0.000040
4.10	−0.000007	−0.000012	−0.000019	−0.000030	−0.000045	−0.000064
4.15	−0.000008	−0.000015	−0.000024	−0.000038	−0.000056	−0.000081
4.20	−0.000009	−0.000016	−0.000027	−0.000042	−0.000063	−0.000092
4.25	−0.000010	−0.000017	−0.000028	−0.000044	−0.000067	−0.000097
4.30	−0.000010	−0.000017	−0.000028	−0.000045	−0.000067	−0.000098
4.35	−0.000009	−0.000016	−0.000027	−0.000043	−0.000066	−0.000096
4.40	−0.000009	−0.000015	−0.000026	−0.000041	−0.000063	−0.000092
4.45	−0.000008	−0.000014	−0.000024	−0.000038	−0.000058	−0.000086

Table 3.4 (continued)

			t			
v	0.35	0.40	0.45	0.50	0.55	0.60
4.50	−0.000007	−0.000013	−0.000022	−0.000035	−0.000053	−0.000079
4.55	−0.000006	−0.000011	−0.000019	−0.000031	−0.000048	−0.000071
4.60	−0.000005	−0.000010	−0.000017	−0.000027	−0.000042	−0.000063
4.65	−0.000005	−0.000008	−0.000015	−0.000024	−0.000037	−0.000055
4.70	−0.000004	−0.000007	−0.000012	−0.000020	−0.000032	−0.000048
4.75	−0.000003	−0.000006	−0.000010	−0.000017	−0.000027	−0.000041
4.80	−0.000003	−0.000005	−0.000009	−0.000014	−0.000023	−0.000034
4.85	−0.000002	−0.000004	−0.000007	−0.000012	−0.000019	−0.000028
4.90	−0.000002	−0.000003	−0.000006	−0.000009	−0.000015	−0.000023
4.95	−0.000001	−0.000002	−0.000004	−0.000007	−0.000012	−0.000018
5.00	−0.000001	−0.000002	−0.000003	−0.000006	−0.000009	−0.000014

			t			
v	0.65	0.70	0.75	0.80	0.85	0.90
0.00	0.105592	0.122449	0.140548	0.159886	0.180461	0.202269
0.05	0.210823	0.227043	0.244443	0.263025	0.282789	0.303733
0.10	0.306453	0.322652	0.339904	0.358220	0.377609	0.398076
0.15	0.390702	0.407416	0.425008	0.443502	0.462918	0.483268
0.20	0.462282	0.479944	0.498277	0.517320	0.537103	0.557649
0.25	0.520389	0.539314	0.558687	0.578560	0.598974	0.619962
0.30	0.564684	0.585059	0.605657	0.626541	0.647762	0.669360
0.35	0.595253	0.617136	0.639028	0.660999	0.683108	0.705400
0.40	0.612555	0.635885	0.659028	0.682061	0.705045	0.728028
0.45	0.617375	0.641978	0.666228	0.690200	0.713957	0.737546
0.50	0.610756	0.636366	0.661488	0.686193	0.710539	0.734573
0.55	0.593941	0.620218	0.645905	0.671065	0.695751	0.720004
0.60	0.568313	0.594864	0.620754	0.646038	0.670760	0.694953
0.65	0.535338	0.561735	0.587433	0.612474	0.636892	0.660710
0.70	0.496509	0.522314	0.547408	0.571822	0.595575	0.618681
0.75	0.453301	0.478085	0.502168	0.525567	0.548291	0.570341
0.80	0.407132	0.430489	0.453174	0.475188	0.496529	0.517188
0.85	0.359326	0.380891	0.401824	0.422114	0.441745	0.460699
0.90	0.311090	0.330551	0.349426	0.367692	0.385323	0.402291
0.95	0.263494	0.280599	0.297165	0.313158	0.328546	0.343292
1.00	0.217461	0.232022	0.246089	0.259620	0.272575	0.284912
1.05	0.173757	0.185656	0.197099	0.208039	0.218430	0.228225
1.10	0.132993	0.142180	0.150939	0.159222	0.166978	0.174159
1.15	0.095631	0.102120	0.108200	0.113820	0.118932	0.123483
1.20	0.061988	0.065853	0.069318	0.072332	0.074845	0.076807
1.25	0.032250	0.033621	0.034588	0.035103	0.035118	0.034585
1.30	0.006487	0.005538	0.004171	0.002343	0.000008	−0.002878
1.35	−0.015337	−0.018394	−0.021890	−0.025866	−0.030361	−0.035418
1.40	−0.033345	−0.038269	−0.043661	−0.049555	−0.055989	−0.062999

Table 3.4 (continued)

				t		
v	0.65	0.70	0.75	0.80	0.85	0.90
1.45	−0.047730	−0.054260	−0.061291	−0.068853	−0.076977	−0.085696
1.50	−0.058743	−0.066607	−0.075008	−0.083970	−0.093519	−0.103683
1.55	−0.066674	−0.075599	−0.085094	−0.095181	−0.105881	−0.117216
1.60	−0.071844	−0.081557	−0.091875	−0.102812	−0.114386	−0.126615
1.65	−0.074586	−0.084828	−0.095704	−0.107224	−0.119401	−0.132248
1.70	−0.075239	−0.085766	−0.096949	−0.108796	−0.121316	−0.134519
1.75	−0.074137	−0.084725	−0.095985	−0.107922	−0.120541	−0.133848
1.80	−0.071600	−0.082050	−0.093178	−0.104988	−0.117483	−0.130667
1.85	−0.067931	−0.078068	−0.088882	−0.100374	−0.112546	−0.125398
1.90	−0.063410	−0.073086	−0.083429	−0.094439	−0.106114	−0.118455
1.95	−0.058287	−0.067384	−0.077127	−0.087517	−0.098551	−0.110228
2.00	−0.052788	−0.061212	−0.070255	−0.079915	−0.090190	−0.101077
2.05	−0.047106	−0.054790	−0.063058	−0.071906	−0.081332	−0.091334
2.10	−0.041407	−0.048309	−0.055750	−0.063731	−0.072246	−0.081293
2.15	−0.035827	−0.041925	−0.048515	−0.055595	−0.063162	−0.071210
2.20	−0.030475	−0.035769	−0.041501	−0.047671	−0.054275	−0.061307
2.25	−0.025437	−0.029941	−0.034829	−0.040099	−0.045745	−0.051764
2.30	−0.020772	−0.024518	−0.028590	−0.032985	−0.037700	−0.042728
2.35	−0.016522	−0.019552	−0.022849	−0.026412	−0.030234	−0.034310
2.40	−0.012712	−0.015076	−0.017650	−0.020431	−0.023414	−0.026592
2.45	−0.009348	−0.011104	−0.013015	−0.015076	−0.017282	−0.019626
2.50	−0.006426	−0.007636	−0.008948	−0.010357	−0.011856	−0.013438
2.55	−0.003933	−0.004661	−0.005441	−0.006269	−0.007136	−0.008033
2.60	−0.001845	−0.002155	−0.002473	−0.002793	−0.003104	−0.003398
2.65	−0.000134	−0.000090	−0.000014	0.000102	0.000269	0.000496
2.70	0.001233	0.001570	0.001974	0.002454	0.003023	0.003690
2.75	0.002289	0.002861	0.003531	0.004308	0.005205	0.006234
2.80	0.003071	0.003826	0.004703	0.005713	0.006868	0.008184
2.85	0.003614	0.004503	0.005534	0.006718	0.008068	0.009600
2.90	0.003954	0.004934	0.006070	0.007374	0.008861	0.010545
2.95	0.004122	0.005156	0.006355	0.007732	0.009303	0.011082
3.00	0.004151	0.005206	0.006430	0.007839	0.009448	0.011272
3.05	0.004068	0.005116	0.006336	0.007741	0.009348	0.011172
3.10	0.003899	0.004918	0.006106	0.007479	0.009051	0.010838
3.15	0.003666	0.004638	0.005775	0.007091	0.008601	0.010320
3.20	0.003388	0.004300	0.005370	0.006610	0.008037	0.009664
3.25	0.003083	0.003925	0.004915	0.006067	0.007394	0.008911
3.30	0.002763	0.003529	0.004433	0.005486	0.006703	0.008096
3.35	0.002441	0.003128	0.003940	0.004890	0.005989	0.007250
3.40	0.002125	0.002732	0.003452	0.004295	0.005274	0.006399
3.45	0.001823	0.002351	0.002979	0.003717	0.004575	0.005564
3.50	0.001539	0.001991	0.002531	0.003166	0.003906	0.004761
3.55	0.001277	0.001658	0.002113	0.002650	0.003278	0.004003

Table 3.4 (continued)

			t			
ν	0.65	0.70	0.75	0.80	0.85	0.90
3.60	0.001040	0.001354	0.001730	0.002175	0.002697	0.003300
3.65	0.000827	0.001080	0.001385	0.001745	0.002168	0.002658
3.70	0.000640	0.000839	0.001077	0.001361	0.001694	0.002081
3.75	0.000478	0.000628	0.000809	0.001023	0.001276	0.001569
3.80	0.000340	0.000447	0.000577	0.000731	0.000912	0.001122
3.85	0.000224	0.000295	0.000380	0.000482	0.000601	0.000739
3.90	0.000128	0.000169	0.000217	0.000274	0.000340	0.000415
3.95	0.000051	0.000066	0.000083	0.000103	0.000124	0.000147
4.00	−0.000010	−0.000015	−0.000023	−0.000034	−0.000049	−0.000069
4.05	−0.000056	−0.000078	−0.000106	−0.000141	−0.000185	−0.000239
4.10	−0.000091	−0.000124	−0.000167	−0.000221	−0.000287	−0.000369
4.15	−0.000114	−0.000157	−0.000210	−0.000278	−0.000360	−0.000462
4.20	−0.000129	−0.000177	−0.000238	−0.000315	−0.000409	−0.000524
4.25	−0.000137	−0.000188	−0.000254	−0.000336	−0.000437	−0.000560
4.30	−0.000139	−0.000192	−0.000259	−0.000343	−0.000447	−0.000575
4.35	−0.000137	−0.000189	−0.000256	−0.000340	−0.000444	−0.000572
4.40	−0.000131	−0.000182	−0.000247	−0.000329	−0.000430	−0.000555
4.45	−0.000123	−0.000171	−0.000233	−0.000311	−0.000408	−0.000528
4.50	−0.000113	−0.000158	−0.000216	−0.000289	−0.000380	−0.000493
4.55	−0.000103	−0.000144	−0.000197	−0.000264	−0.000349	−0.000453
4.60	−0.000091	−0.000129	−0.000177	−0.000238	−0.000315	−0.000410
4.65	−0.000080	−0.000114	−0.000156	−0.000211	−0.000280	−0.000365
4.70	−0.000070	−0.000099	−0.000137	−0.000185	−0.000246	−0.000321
4.75	−0.000060	−0.000085	−0.000117	−0.000159	−0.000212	−0.000278
4.80	−0.000050	−0.000071	−0.000099	−0.000135	−0.000181	−0.000238
4.85	−0.000042	−0.000059	−0.000083	−0.000113	−0.000152	−0.000200
4.90	−0.000034	−0.000048	−0.000068	−0.000093	−0.000125	−0.000165
4.95	−0.000027	−0.000039	−0.000054	−0.000075	−0.000101	−0.000133
5.00	−0.000021	−0.000030	−0.000043	−0.000059	−0.000079	−0.000105

			t			
ν	0.95	1.0	2.0	3.0	4.0	5.0
0.00	0.225306	0.249566	0.972292	1.93759	2.29269	0.116034
0.05	0.325854	0.349147	1.03857	1.91662	2.08351	−0.374801
0.10	0.419622	0.442246	1.09806	1.88315	1.85972	−0.865163
0.15	0.504561	0.526803	1.14932	1.83688	1.62275	−1.35140
0.20	0.578975	0.601091	1.19097	1.77761	1.37414	−1.82982
0.25	0.641548	0.663750	1.22180	1.70527	1.11557	−2.29674
0.30	0.691367	0.713805	1.24081	1.61991	0.848823	−2.74851
0.35	0.727911	0.750669	1.24720	1.52176	0.575805	−3.18156
0.40	0.751050	0.774138	1.24043	1.41122	0.298511	−3.59240
0.45	0.761006	0.784366	1.22023	1.28888	0.019029	−3.97768
0.50	0.758330	0.781837	1.18663	1.15550	−0.260480	−4.33423

Table 3.4 (continued)

v	0.95	1.0	2.0	3.0	4.0	5.0
0.55	0.743853	0.767323	1.13992	1.01207	−0.537795	−4.65907
0.60	0.718641	0.741841	1.08068	0.859732	−0.810651	−4.94945
0.65	0.683944	0.706604	1.00977	0.699848	−1.07676	−5.20287
0.70	0.641146	0.662973	0.928264	0.533933	−1.33382	−5.41711
0.75	0.591715	0.612404	0.837462	0.363655	−1.57958	−5.59028
0.80	0.537152	0.556405	0.738841	0.190812	−1.81179	−5.72082
0.85	0.478954	0.496484	0.634019	0.017302	−2.02831	−5.80749
0.90	0.418567	0.434118	0.524713	−0.154912	−2.22708	−5.84947
0.95	0.357359	0.370708	0.412701	−0.323830	−2.40618	−5.84630
1.00	0.296587	0.307557	0.299775	−0.487454	−2.56382	−5.79791
1.05	0.237379	0.245841	0.187703	−0.643829	−2.69843	−5.70466
1.10	0.180715	0.186595	0.078190	−0.791075	−2.80864	−5.56731
1.15	0.127423	0.130700	−0.027161	−0.927424	−2.89329	−5.38704
1.20	0.078167	0.078875	−0.126876	−1.05126	−2.95152	−5.16543
1.25	0.033454	0.031678	−0.219639	−1.16114	−2.98274	−4.90447
1.30	−0.006363	−0.010493	−0.304316	−1.25583	−2.98663	−4.60654
1.35	−0.041080	−0.047389	−0.379969	−1.33433	−2.96322	−4.27437
1.40	−0.070623	−0.078902	−0.445866	−1.39588	−2.91281	−3.91107
1.45	−0.095042	−0.105052	−0.501491	−1.43999	−2.83607	−3.52005
1.50	−0.114490	−0.125971	−0.546539	−1.46644	−2.73396	−3.10502
1.55	−0.129210	−0.141889	−0.580914	−1.47528	−2.60774	−2.66996
1.60	−0.139517	−0.153113	−0.604719	−1.46682	−2.45899	−2.21905
1.65	−0.145782	−0.160016	−0.618242	−1.44164	−2.28956	−1.75664
1.70	−0.148415	−0.163016	−0.621939	−1.40055	−2.10155	−1.28722
1.75	−0.147852	−0.162560	−0.616415	−1.34460	−1.89729	−0.815354
1.80	−0.144543	−0.159116	−0.602402	−1.27503	−1.67930	−0.345630
1.85	−0.138934	−0.153153	−0.580735	−1.19327	−1.45028	0.117388
1.90	−0.131462	−0.145133	−0.552330	−1.10088	−1.21302	0.569218
1.95	−0.122545	−0.135500	−0.518159	−0.999546	−0.970406	1.00551
2.00	−0.112573	−0.124675	−0.479225	−0.891022	−0.725363	1.42210
2.05	−0.101906	−0.113043	−0.436541	−0.777119	−0.480801	1.81507
2.10	−0.090865	−0.100958	−0.391112	−0.659656	−0.239582	2.18076
2.15	−0.079735	−0.088730	−0.343909	−0.540437	−0.004476	2.51588
2.20	−0.068761	−0.076631	−0.295860	−0.421215	0.221879	2.81749
2.25	−0.058148	−0.064889	−0.247828	−0.303665	0.437016	3.08305
2.30	−0.048062	−0.053694	−0.200606	−0.189358	0.638673	3.31049
2.35	−0.038633	−0.043193	−0.154902	−0.079737	0.824822	3.49816
2.40	−0.029957	−0.033499	−0.111334	0.023900	0.993699	3.64493
2.45	−0.022098	−0.024688	−0.070427	0.120417	1.14382	3.75016
2.50	−0.015091	−0.016807	−0.032610	0.208852	1.27400	3.81369
2.55	−0.008949	−0.009874	0.001784	0.288422	1.38337	3.83587
2.60	−0.003662	−0.003886	0.032516	0.358531	1.47136	3.81753
2.65	0.000797	0.001183	0.059435	0.418769	1.53774	3.75999

Table 3.4 (continued)

			t			
v	0.95	1.0	2.0	3.0	4.0	5.0
2.70	0.004469	0.005374	0.082474	0.468905	1.58256	3.66500
2.75	0.007408	0.008742	0.101645	0.508885	1.60619	3.53472
2.80	0.009672	0.011349	0.117027	0.538819	1.60928	3.37171
2.85	0.011328	0.013267	0.128761	0.558973	1.59272	3.17887
2.90	0.012443	0.014569	0.137040	0.569750	1.55768	2.95938
2.95	0.013086	0.015331	0.142100	0.571675	1.50551	2.71668
3.00	0.013327	0.015629	0.144210	0.565381	1.43777	2.45441
3.05	0.013229	0.015535	0.143665	0.551587	1.35613	2.17634
3.10	0.012856	0.015120	0.140775	0.531082	1.26243	1.88633
3.15	0.012264	0.014449	0.135862	0.504705	1.15856	1.58827
3.20	0.011507	0.013581	0.129246	0.473325	1.04648	1.28604
3.25	0.010632	0.012571	0.121244	0.437829	0.928158	0.983424
3.30	0.009679	0.011467	0.112163	0.399100	0.805561	0.684092
3.35	0.008686	0.010310	0.102293	0.358003	0.680605	0.391535
3.40	0.007682	0.009136	0.091905	0.315373	0.555138	0.109029
3.45	0.006693	0.007974	0.081249	0.272003	0.430911	−0.160407
3.50	0.005739	0.006850	0.070549	0.228629	0.309557	−0.414045
3.55	0.004835	0.005782	0.060002	0.185929	0.192572	−0.649474
3.60	0.003994	0.004783	0.049781	0.144508	0.081297	−0.864625
3.65	0.003222	0.003865	0.040029	0.104900	−0.023094	−1.05779
3.70	0.002526	0.003033	0.030863	0.067561	−0.119600	−1.22761
3.75	0.001906	0.002291	0.022375	0.032868	−0.207403	−1.37313
3.80	0.001364	0.001638	0.014632	0.001121	−0.285866	−1.49375
3.85	0.000896	0.001074	0.007679	−0.027456	−0.354535	−1.58923
3.90	0.000500	0.000595	0.001538	−0.052715	−0.413132	−1.65968
3.95	0.000171	0.000195	−0.003785	−0.074575	−0.461551	−1.70555
4.00	−0.000096	−0.000130	−0.008300	−0.093022	−0.499846	−1.72762
4.05	−0.000307	−0.000388	−0.012034	−0.108098	−0.528220	−1.72692
4.10	−0.000467	−0.000586	−0.015023	−0.119900	−0.547014	−1.70478
4.15	−0.000584	−0.000731	−0.017314	−0.128571	−0.556688	−1.66271
4.20	−0.000663	−0.000829	−0.018960	−0.134290	−0.557809	−1.60245
4.25	−0.000709	−0.000888	−0.020019	−0.137273	−0.551031	−1.52588
4.30	−0.000729	−0.000913	−0.020555	−0.137756	−0.537081	−1.43501
4.35	−0.000726	−0.000911	−0.020631	−0.135995	−0.516741	−1.33193
4.40	−0.000706	−0.000887	−0.020310	−0.132261	−0.490831	−1.21877
4.45	−0.000673	−0.000847	−0.019656	−0.126826	−0.460191	−1.09770
4.50	−0.000629	−0.000794	−0.018729	−0.119965	−0.425670	−0.970840
4.55	−0.000579	−0.000733	−0.017586	−0.111949	−0.388108	−0.840292
4.60	−0.000526	−0.000666	−0.016281	−0.103039	−0.348323	−0.708063
4.65	−0.000470	−0.000596	−0.014864	−0.093482	−0.307103	−0.576065
4.70	−0.000414	−0.000527	−0.013376	−0.083512	−0.265190	−0.446081
4.75	−0.000360	−0.000458	−0.011860	−0.073343	−0.223277	−0.319753
4.80	−0.000308	−0.000393	−0.010347	−0.063169	−0.181996	−0.198559

Table 3.4 (continued)

				t		
v	0.95	1.0	2.0	3.0	4.0	5.0
4.85	−0.000259	−0.000331	−0.008867	−0.053163	−0.141916	−0.083810
4.90	−0.000214	−0.000274	−0.007444	−0.043474	−0.103538	0.023368
4.95	−0.000173	−0.000223	−0.006098	−0.034231	−0.067291	0.122034
5.00	−0.000137	−0.000176	−0.004843	−0.025538	−0.033532	0.211434

Table 3.5: First derivatives with respect to the order of the Kelvin functions $\partial \text{bei}_v(t)/\partial v$.

				t		
v	0.05	0.10	0.15	0.20	0.25	0.30
0.00	2.35362	2.347640	2.339230	2.328880	2.31695	2.30370
0.05	1.67609	1.801760	1.873660	1.921630	1.95551	1.97986
0.10	1.10937	1.310130	1.433690	1.521950	1.58912	1.64190
0.15	0.649472	0.880350	1.030770	1.142780	1.23120	1.30320
0.20	0.288191	0.515542	0.672475	0.793837	0.892559	0.975081
0.25	0.014746	0.215243	0.362967	0.481825	0.581389	0.666639
0.30	−0.182926	−0.023690	0.103574	0.210753	0.303441	0.384792
0.35	−0.317172	−0.206286	−0.106667	−0.017707	0.062220	0.134368
0.40	−0.399896	−0.338763	−0.270458	−0.203812	−0.140730	−0.081715
0.45	−0.442078	−0.427953	−0.391746	−0.349354	−0.305521	−0.262230
0.50	−0.453475	−0.480837	−0.475275	−0.457270	−0.433607	−0.407389
0.55	−0.442483	−0.504197	−0.526209	−0.531294	−0.527466	−0.518575
0.60	−0.416113	−0.504357	−0.549825	−0.575640	−0.590317	−0.598087
0.65	−0.380045	−0.487024	−0.551267	−0.594739	−0.625847	−0.648883
0.70	−0.338742	−0.457191	−0.535364	−0.593020	−0.637987	−0.674356
0.75	−0.295586	−0.419099	−0.506510	−0.574741	−0.630713	−0.678131
0.80	−0.253038	−0.376250	−0.468585	−0.543857	−0.607896	−0.663890
0.85	−0.212791	−0.331441	−0.424917	−0.503940	−0.573177	−0.635234
0.90	−0.175922	−0.286832	−0.378281	−0.458122	−0.529880	−0.595569
0.95	−0.143023	−0.244014	−0.330918	−0.409075	−0.480961	−0.548024
1.00	−0.114324	−0.204098	−0.284574	−0.359010	−0.428971	−0.495397
1.05	−0.089790	−0.167786	−0.240547	−0.309697	−0.376052	−0.440120
1.10	−0.069204	−0.135456	−0.199747	−0.262496	−0.323943	−0.384249
1.15	−0.052237	−0.107230	−0.162750	−0.218399	−0.274002	−0.329463
1.20	−0.038495	−0.083038	−0.129860	−0.178075	−0.227241	−0.277087
1.25	−0.027560	−0.062671	−0.101162	−0.141919	−0.184360	−0.228113
1.30	−0.019018	−0.045829	−0.076570	−0.110099	−0.145791	−0.183235
1.35	−0.012477	−0.032157	−0.055877	−0.082601	−0.111738	−0.142885
1.40	−0.007578	−0.021275	−0.038789	−0.059273	−0.082220	−0.107271
1.45	−0.004001	−0.012801	−0.024958	−0.039859	−0.057109	−0.076417

Table 3.5 (continued)

v			t			
	0.05	0.10	0.15	0.20	0.25	0.30
1.50	-0.001471	-0.006364	-0.014011	-0.024036	-0.036170	-0.050199
1.55	0.000247	-0.001620	-0.005566	-0.011436	-0.019084	-0.028379
1.60	0.001348	0.001743	0.000746	-0.001674	-0.005487	-0.010637
1.65	0.001990	0.004002	0.005277	0.005636	0.005015	0.003403
1.70	0.002303	0.005398	0.008349	0.010870	0.012825	0.014147
1.75	0.002385	0.006136	0.010251	0.014380	0.018336	0.022013
1.80	0.002313	0.006385	0.011236	0.016487	0.021925	0.027414
1.85	0.002146	0.006287	0.011524	0.017483	0.023941	0.030750
1.90	0.001927	0.005951	0.011299	0.017621	0.024702	0.032392
1.95	0.001685	0.005467	0.010714	0.017121	0.024490	0.032681
2.00	0.001441	0.004901	0.009893	0.016166	0.023551	0.031922
2.05	0.001209	0.004303	0.008934	0.014910	0.022094	0.030383
2.10	0.000998	0.003709	0.007914	0.013477	0.020297	0.028294
2.15	0.000810	0.003144	0.006889	0.011964	0.018302	0.025850
2.20	0.000647	0.002622	0.005900	0.010445	0.016225	0.023211
2.25	0.000509	0.002154	0.004974	0.008976	0.014155	0.020505
2.30	0.000394	0.001742	0.004130	0.007595	0.012158	0.017833
2.35	0.000301	0.001387	0.003376	0.006327	0.010281	0.015269
2.40	0.000225	0.001086	0.002714	0.005186	0.008557	0.012868
2.45	0.000166	0.000836	0.002145	0.004179	0.007002	0.010666
2.50	0.000119	0.000630	0.001662	0.003305	0.005626	0.008682
2.55	0.000084	0.000463	0.001260	0.002558	0.004427	0.006927
2.60	0.000057	0.000331	0.000929	0.001930	0.003400	0.005398
2.65	0.000037	0.000228	0.000662	0.001411	0.002534	0.004088
2.70	0.000022	0.000149	0.000451	0.000988	0.001816	0.002984
2.75	0.000012	0.000089	0.000286	0.000651	0.001231	0.002069
2.80	0.000006	0.000046	0.000160	0.000388	0.000764	0.001325
2.85	0.000005	0.000015	0.000067	0.000186	0.000398	0.000733
2.90	-0.000002	-0.000006	0.000007	0.000036	0.000120	0.000272
2.95	-0.000004	-0.000020	-0.000045	-0.000071	-0.000086	-0.000077
3.00	-0.000005	-0.000028	-0.000075	-0.000144	-0.000231	-0.000330
3.05	-0.000005	-0.000032	-0.000092	-0.000189	-0.000327	-0.000505
3.10	-0.000005	-0.000033	-0.000099	-0.000214	-0.000385	-0.000616
3.15	-0.000004	-0.000032	-0.000100	-0.000223	-0.000412	-0.000676
3.20	-0.000004	-0.000030	-0.000097	-0.000221	-0.000417	-0.000697
3.25	-0.000003	-0.000027	-0.000090	-0.000211	-0.000405	-0.000689
3.30	-0.000003	-0.000024	-0.000082	-0.000195	-0.000382	-0.000659
3.35	-0.000002	-0.000021	-0.000073	-0.000177	-0.000351	-0.000614
3.40	-0.000002	-0.000017	-0.000063	-0.000157	-0.000317	-0.000561
3.45	-0.000002	-0.000015	-0.000054	-0.000137	-0.000280	-0.000503
3.50	-0.000001	-0.000012	-0.000045	-0.000117	-0.000244	-0.000443
3.55	-0.000001	-0.000010	-0.000038	-0.000099	-0.000209	-0.000384
3.60	-0.000001	-0.000008	-0.000031	-0.000082	-0.000176	-0.000328

Table 3.5 (continued)

				t		
v	0.05	0.10	0.15	0.20	0.25	0.30
3.65	−0.000001	−0.000006	−0.000025	−0.000068	−0.000147	−0.000276
3.70		−0.000005	−0.000020	−0.000055	−0.000120	−0.000229
3.75		−0.000004	−0.000015	−0.000043	−0.000097	−0.000187
3.80		−0.000003	−0.000012	−0.000034	−0.000077	−0.000150
3.85		−0.000002	−0.000009	−0.000026	−0.000060	−0.000118
3.90		−0.000001	−0.000006	−0.000019	−0.000045	−0.000091
3.95		−0.000001	−0.000005	−0.000014	−0.000033	−0.000068
4.00		−0.000001	−0.000003	−0.000010	−0.000024	−0.000049
4.05			−0.000002	−0.000006	−0.000016	−0.000034
4.10			−0.000001	−0.000004	−0.000010	−0.000022
4.15			−0.000001	−0.000002	−0.000006	−0.000013
4.20				−0.000001	−0.000002	−0.000005
4.25				0.000000	0.000000	0.000000
4.30				0.000001	0.000002	0.000004
4.35				0.000001	0.000003	0.000006
4.40				0.000001	0.000004	0.000009
4.45				0.000001	0.000004	0.000009
4.50				0.000001	0.000004	0.000009
4.55				0.000001	0.000004	0.000009
4.60				0.000001	0.000004	0.000008
4.65				0.000001	0.000003	0.000008
4.70				0.000001	0.000003	0.000007
4.75				0.000001	0.000003	0.000006
4.80				0.000001	0.000002	0.000005
4.85				0.000001	0.000002	0.000005
4.90				0.000001	0.000002	0.000004
4.95					0.000001	0.000003
5.00					0.000001	0.000003

				t		
v	0.35	0.40	0.45	0.50	0.55	0.60
0.00	2.28932	2.27397	2.25778	2.24084	2.22325	2.20507
0.05	1.99722	2.00915	2.01671	2.02063	2.02146	2.01960
0.10	1.68403	1.71785	1.74491	1.76634	1.78292	1.79529
0.15	1.36285	1.41271	1.45452	1.48952	1.51862	1.54251
0.20	1.04515	1.10516	1.15676	1.20113	1.23915	1.27152
0.25	0.740560	0.805107	0.861645	0.911171	0.954440	0.992036
0.30	0.456795	0.520803	0.577784	0.628463	0.673396	0.713022
0.35	0.199658	0.258773	0.312236	0.360456	0.403756	0.442398
0.40	−0.026869	0.023841	0.070490	0.113163	0.151946	0.186915
0.45	−0.220468	−0.180784	−0.143509	−0.108856	−0.076966	−0.047940
0.50	−0.380303	−0.353373	−0.327268	−0.302453	−0.279269	−0.257972
0.55	−0.506798	−0.493503	−0.479617	−0.465802	−0.452553	−0.440253

Table 3.5 (continued)

			t			
v	0.35	0.40	0.45	0.50	0.55	0.60
0.60	−0.601409	−0.601867	−0.600553	−0.598265	−0.595608	−0.593055
0.65	−0.666398	−0.680066	−0.691061	−0.700251	−0.708305	−0.715750
0.70	−0.704612	−0.730408	−0.752924	−0.773041	−0.791445	−0.808685
0.75	−0.719282	−0.755713	−0.788544	−0.818619	−0.846603	−0.873035
0.80	−0.713843	−0.759130	−0.800759	−0.839500	−0.875966	−0.910657
0.85	−0.691780	−0.743976	−0.792684	−0.838572	−0.882175	−0.923936
0.90	−0.656499	−0.713597	−0.767562	−0.818939	−0.868173	−0.915635
0.95	−0.611224	−0.671252	−0.728635	−0.783793	−0.837069	−0.888755
1.00	−0.558924	−0.620020	−0.679046	−0.736296	−0.792015	−0.846413
1.05	−0.502256	−0.562727	−0.621746	−0.679487	−0.736102	−0.791722
1.10	−0.443537	−0.501905	−0.559437	−0.616205	−0.672275	−0.727705
1.15	−0.384726	−0.439755	−0.494528	−0.549032	−0.603261	−0.657212
1.20	−0.327430	−0.378136	−0.429104	−0.480255	−0.531524	−0.582860
1.25	−0.272916	−0.318571	−0.364923	−0.411845	−0.459232	−0.506994
1.30	−0.222132	−0.262252	−0.303409	−0.345446	−0.388233	−0.431653
1.35	−0.175741	−0.210069	−0.245672	−0.282383	−0.320056	−0.358561
1.40	−0.134150	−0.162632	−0.192528	−0.223671	−0.255914	−0.289123
1.45	−0.097548	−0.120307	−0.144521	−0.170038	−0.196719	−0.224435
1.50	−0.065944	−0.083245	−0.101960	−0.121955	−0.143107	−0.165298
1.55	−0.039197	−0.051424	−0.064947	−0.079660	−0.095461	−0.112246
1.60	−0.017057	−0.024675	−0.033415	−0.043198	−0.053943	−0.065568
1.65	0.000814	−0.002720	−0.007157	−0.012445	−0.018528	−0.025344
1.70	0.014807	0.014801	0.014140	0.012850	0.010966	0.008529
1.75	0.025346	0.028302	0.030867	0.033044	0.034846	0.036298
1.80	0.032868	0.038231	0.043467	0.048563	0.053514	0.058329
1.85	0.037809	0.045046	0.052414	0.059883	0.067437	0.075071
1.90	0.040585	0.049202	0.058189	0.067505	0.077126	0.087038
1.95	0.041590	0.051139	0.061270	0.071937	0.083113	0.094775
2.00	0.041186	0.051269	0.062114	0.073678	0.085929	0.098844
2.05	0.039696	0.049969	0.061151	0.073204	0.086094	0.099802
2.10	0.037406	0.047581	0.058777	0.070960	0.084104	0.098189
2.15	0.034564	0.044405	0.055344	0.067355	0.080417	0.094515
2.20	0.031378	0.040702	0.051166	0.062753	0.075452	0.089252
2.25	0.028020	0.036691	0.046512	0.057477	0.069583	0.082826
2.30	0.024629	0.032553	0.041609	0.051803	0.063137	0.075617
2.35	0.021314	0.028434	0.036645	0.045961	0.056393	0.067954
2.40	0.018156	0.024448	0.031770	0.040143	0.049586	0.060117
2.45	0.015213	0.020680	0.027099	0.034499	0.042904	0.052338
2.50	0.012523	0.017189	0.022718	0.029144	0.036497	0.044803
2.55	0.010107	0.014014	0.018687	0.024164	0.030477	0.037657
2.60	0.007974	0.011175	0.015042	0.019613	0.024924	0.031006
2.65	0.006121	0.008677	0.011799	0.015525	0.019889	0.024923
2.70	0.004537	0.006516	0.008963	0.011912	0.015398	0.019452

Table 3.5 (continued)

				t		
v	0.35	0.40	0.45	0.50	0.55	0.60
2.75	0.003205	0.004677	0.006521	0.008771	0.011458	0.014612
2.80	0.002106	0.003139	0.004456	0.006086	0.008059	0.010400
2.85	0.001216	0.001877	0.002741	0.003833	0.005178	0.006798
2.90	0.000513	0.000864	0.001346	0.001979	0.002783	0.003777
2.95	−0.000030	0.000070	0.000238	0.000489	0.000836	0.001295
3.00	−0.000433	−0.000532	−0.000617	−0.000678	−0.000706	−0.000693
3.05	−0.000720	−0.000970	−0.001252	−0.001560	−0.001890	−0.002238
3.10	−0.000910	−0.001271	−0.001699	−0.002195	−0.002760	−0.003393
3.15	−0.001023	−0.001459	−0.001990	−0.002623	−0.003360	−0.004207
3.20	−0.001073	−0.001555	−0.002153	−0.002876	−0.003734	−0.004733
3.25	−0.001075	−0.001579	−0.002213	−0.002989	−0.003920	−0.005017
3.30	−0.001042	−0.001548	−0.002192	−0.002990	−0.003955	−0.005104
3.35	−0.000984	−0.001477	−0.002111	−0.002904	−0.003872	−0.005034
3.40	−0.000908	−0.001377	−0.001986	−0.002754	−0.003700	−0.004843
3.45	−0.000823	−0.001259	−0.001832	−0.002560	−0.003463	−0.004562
3.50	−0.000733	−0.001132	−0.001660	−0.002337	−0.003183	−0.004219
3.55	−0.000642	−0.001001	−0.001480	−0.002099	−0.002878	−0.003837
3.60	−0.000554	−0.000872	−0.001299	−0.001856	−0.002561	−0.003436
3.65	−0.000471	−0.000748	−0.001124	−0.001617	−0.002245	−0.003030
3.70	−0.000395	−0.000632	−0.000957	−0.001387	−0.001939	−0.002632
3.75	−0.000325	−0.000526	−0.000803	−0.001172	−0.001649	−0.002252
3.80	−0.000264	−0.000430	−0.000662	−0.000974	−0.001380	−0.001895
3.85	−0.000210	−0.000346	−0.000537	−0.000795	−0.001134	−0.001568
3.90	−0.000164	−0.000272	−0.000426	−0.000636	−0.000914	−0.001272
3.95	−0.000124	−0.000209	−0.000330	−0.000497	−0.000719	−0.001008
4.00	−0.000091	−0.000155	−0.000248	−0.000377	−0.000551	−0.000777
4.05	−0.000064	−0.000111	−0.000179	−0.000276	−0.000406	−0.000578
4.10	−0.000042	−0.000075	−0.000122	−0.000191	−0.000285	−0.000410
4.15	−0.000025	−0.000045	−0.000076	−0.000121	−0.000184	−0.000269
4.20	−0.000012	−0.000023	−0.000040	−0.000065	−0.000102	−0.000153
4.25	−0.000002	−0.000005	−0.000011	−0.000022	−0.000038	−0.000061
4.30	0.000006	0.000008	0.000010	0.000012	0.000012	0.000011
4.35	0.000011	0.000017	0.000026	0.000036	0.000050	0.000066
4.40	0.000014	0.000023	0.000036	0.000054	0.000076	0.000105
4.45	0.000016	0.000027	0.000043	0.000065	0.000094	0.000132
4.50	0.000017	0.000029	0.000046	0.000071	0.000105	0.000149
4.55	0.000017	0.000029	0.000047	0.000074	0.000109	0.000157
4.60	0.000016	0.000028	0.000047	0.000073	0.000109	0.000158
4.65	0.000015	0.000027	0.000045	0.000070	0.000106	0.000155
4.70	0.000014	0.000025	0.000041	0.000066	0.000100	0.000147
4.75	0.000012	0.000022	0.000038	0.000061	0.000093	0.000137
4.80	0.000011	0.000020	0.000034	0.000055	0.000085	0.000126
4.85	0.000009	0.000017	0.000030	0.000049	0.000076	0.000113

Table 3.5 (continued)

			t			
v	0.35	0.40	0.45	0.50	0.55	0.60
4.90	0.000008	0.000015	0.000026	0.000043	0.000067	0.000100
4.95	0.000007	0.000013	0.000022	0.000037	0.000058	0.000087
5.00	0.000006	0.000011	0.000019	0.000031	0.000050	0.000075

			t			
v	0.65	0.70	0.75	0.80	0.85	0.90
0.00	2.18633	2.16708	2.14734	2.12711	2.10641	2.08521
0.05	2.01536	2.00899	2.00068	1.99058	1.97880	1.96543
0.10	1.80391	1.80915	1.81132	1.81065	1.80733	1.80153
0.15	1.56176	1.57677	1.58791	1.59544	1.59961	1.60060
0.20	1.29875	1.32128	1.33946	1.35357	1.36383	1.37046
0.25	1.02442	1.05198	1.07501	1.09376	1.10845	1.11925
0.30	0.747692	0.777694	0.803265	0.824600	0.841864	0.855193
0.35	0.476598	0.506533	0.532350	0.554176	0.572112	0.586248
0.40	0.218139	0.245676	0.269579	0.289892	0.306652	0.319889
0.45	-0.021852	0.001242	0.021298	0.038279	0.052153	0.062894
0.50	-0.238762	-0.221801	-0.207221	-0.195133	-0.185634	-0.178805
0.55	-0.429207	-0.419665	-0.411833	-0.405888	-0.401979	-0.400239
0.60	-0.590989	-0.589727	-0.589531	-0.590628	-0.593215	-0.597460
0.65	-0.723020	-0.730470	-0.738406	-0.747088	-0.756741	-0.767564
0.70	-0.825212	-0.841406	-0.857588	-0.874036	-0.890995	-0.908678
0.75	-0.898360	-0.922954	-0.947142	-0.971203	-0.995386	-1.01991
0.80	-0.943991	-0.976324	-1.00796	-1.03918	-1.07021	-1.10126
0.85	-0.964228	-1.00337	-1.04164	-1.07929	-1.11653	-1.15356
0.90	-0.961639	-1.00646	-1.05033	-1.09348	-1.13608	-1.17832
0.95	-0.939102	-0.988327	-1.03663	-1.08417	-1.13112	-1.17762
1.00	-0.899671	-0.951952	-1.00340	-1.05415	-1.10432	-1.15401
1.05	-0.846463	-0.900430	-0.953718	-1.00642	-1.05860	-1.11035
1.10	-0.782552	-0.836867	-0.890698	-0.944091	-0.997087	-1.04973
1.15	-0.710887	-0.764290	-0.817427	-0.870302	-0.922924	-0.975299
1.20	-0.634220	-0.685566	-0.736867	-0.788096	-0.839228	-0.890240
1.25	-0.555053	-0.603340	-0.651794	-0.700360	-0.748985	-0.797623
1.30	-0.475604	-0.519994	-0.564739	-0.609762	-0.654991	-0.700357
1.35	-0.397781	-0.437610	-0.477949	-0.518706	-0.559793	-0.601127
1.40	-0.323176	-0.357960	-0.393368	-0.429298	-0.465655	-0.502344
1.45	-0.253065	-0.282496	-0.312619	-0.343331	-0.374530	-0.406118
1.50	-0.188417	-0.212355	-0.237007	-0.262272	-0.288048	-0.314237
1.55	-0.129918	-0.148377	-0.167527	-0.187272	-0.207515	-0.228161
1.60	-0.077990	-0.091123	-0.104881	-0.119176	-0.133921	-0.149026
1.65	-0.032825	-0.040901	-0.049500	-0.058545	-0.067956	-0.077653
1.70	0.005590	0.002202	-0.001575	-0.005675	-0.010032	-0.014573
1.75	0.037434	0.038291	0.038918	0.039365	0.039689	0.039952
1.80	0.063026	0.067631	0.072179	0.076708	0.081264	0.085899

Table 3.5 (continued)

				t		
v	0.65	0.70	0.75	0.80	0.85	0.90
1.85	0.082792	0.090614	0.098559	0.106655	0.114939	0.123451
1.90	0.097237	0.107727	0.118523	0.129643	0.141113	0.152966
1.95	0.106914	0.119528	0.132623	0.146210	0.160308	0.174941
2.00	0.112408	0.126616	0.141468	0.156970	0.173135	0.189980
2.05	0.114310	0.129611	0.145702	0.162585	0.180269	0.198766
2.10	0.113201	0.129131	0.145976	0.163737	0.182418	0.202030
2.15	0.109639	0.125781	0.142938	0.161111	0.180305	0.200527
2.20	0.104147	0.120133	0.137209	0.155377	0.174642	0.195011
2.25	0.097205	0.112721	0.129375	0.147172	0.166118	0.186219
2.30	0.089247	0.104032	0.119978	0.137092	0.155381	0.174855
2.35	0.080653	0.094500	0.109505	0.125679	0.143032	0.161573
2.40	0.071752	0.084505	0.098391	0.113423	0.129614	0.146975
2.45	0.062821	0.074373	0.087010	0.100750	0.115608	0.131598
2.50	0.054087	0.064373	0.075680	0.088027	0.101433	0.115913
2.55	0.045731	0.054724	0.064661	0.075560	0.087443	0.100325
2.60	0.037889	0.045599	0.054161	0.063597	0.073928	0.085172
2.65	0.030659	0.037122	0.044338	0.052331	0.061120	0.070724
2.70	0.024104	0.029381	0.035307	0.041904	0.049194	0.057195
2.75	0.018260	0.022428	0.027139	0.032416	0.038276	0.044738
2.80	0.013134	0.016286	0.019875	0.023922	0.028444	0.033457
2.85	0.008716	0.010951	0.013522	0.016446	0.019737	0.023410
2.90	0.004978	0.006402	0.008065	0.009981	0.012161	0.014617
2.95	0.001879	0.002600	0.003469	0.004497	0.005693	0.007063
3.00	−0.000629	−0.000505	−0.000316	−0.000054	0.000286	0.000707
3.05	−0.002600	−0.002973	−0.003352	−0.003736	−0.004124	−0.004515
3.10	−0.004094	−0.004866	−0.005708	−0.006623	−0.007613	−0.008681
3.15	−0.005169	−0.006251	−0.007457	−0.008793	−0.010266	−0.011883
3.20	−0.005884	−0.007194	−0.008673	−0.010330	−0.012175	−0.014219
3.25	−0.006294	−0.007761	−0.009431	−0.011317	−0.013432	−0.015790
3.30	−0.006451	−0.008012	−0.009801	−0.011835	−0.014129	−0.016700
3.35	−0.006406	−0.008006	−0.009852	−0.011961	−0.014353	−0.017047
3.40	−0.006201	−0.007795	−0.009645	−0.011770	−0.014190	−0.016928
3.45	−0.005877	−0.007428	−0.009237	−0.011326	−0.013717	−0.016432
3.50	−0.005466	−0.006946	−0.008680	−0.010692	−0.013005	−0.015641
3.55	−0.004999	−0.006385	−0.008018	−0.009921	−0.012117	−0.014631
3.60	−0.004501	−0.005778	−0.007290	−0.009059	−0.011110	−0.013466
3.65	−0.003991	−0.005149	−0.006527	−0.008146	−0.010031	−0.012204
3.70	−0.003486	−0.004520	−0.005756	−0.007216	−0.008921	−0.010896
3.75	−0.002998	−0.003908	−0.005000	−0.006295	−0.007815	−0.009581
3.80	−0.002538	−0.003325	−0.004274	−0.005406	−0.006738	−0.008293
3.85	−0.002111	−0.002781	−0.003592	−0.004564	−0.005713	−0.007059
3.90	−0.001723	−0.002281	−0.002962	−0.003781	−0.004754	−0.005898
3.95	−0.001375	−0.001831	−0.002390	−0.003066	−0.003873	−0.004825

Table 3.5 (continued)

			t			
v	0.65	0.70	0.75	0.80	0.85	0.90
4.00	−0.001067	−0.001431	−0.001879	−0.002423	−0.003075	−0.003850
4.05	−0.000800	−0.001081	−0.001428	−0.001853	−0.002365	−0.002976
4.10	−0.000572	−0.000779	−0.001038	−0.001356	−0.001743	−0.002206
4.15	−0.000380	−0.000524	−0.000705	−0.000931	−0.001206	−0.001538
4.20	−0.000222	−0.000312	−0.000427	−0.000572	−0.000751	−0.000969
4.25	−0.000094	−0.000139	−0.000198	−0.000275	−0.000372	−0.000492
4.30	0.000007	−0.000002	−0.000015	−0.000035	−0.000064	−0.000102
4.35	0.000084	0.000105	0.000128	0.000153	0.000181	0.000210
4.40	0.000141	0.000184	0.000236	0.000297	0.000369	0.000452
4.45	0.000180	0.000240	0.000313	0.000401	0.000507	0.000631
4.50	0.000205	0.000276	0.000364	0.000472	0.000602	0.000757
4.55	0.000218	0.000297	0.000394	0.000515	0.000661	0.000837
4.60	0.000222	0.000304	0.000407	0.000534	0.000690	0.000878
4.65	0.000218	0.000301	0.000405	0.000535	0.000694	0.000888
4.70	0.000209	0.000290	0.000392	0.000521	0.000679	0.000873
4.75	0.000196	0.000273	0.000372	0.000496	0.000650	0.000839
4.80	0.000181	0.000253	0.000346	0.000463	0.000610	0.000790
4.85	0.000164	0.000230	0.000316	0.000425	0.000562	0.000731
4.90	0.000146	0.000206	0.000284	0.000384	0.000510	0.000665
4.95	0.000128	0.000182	0.000252	0.000342	0.000455	0.000597
5.00	0.000111	0.000158	0.000220	0.000300	0.000401	0.000527

			t			
v	0.95	1.00	2.00	3.00	4.00	5.00
0.00	2.06350	2.04126	1.38322	−0.29662	−4.02880	−9.79738
0.05	1.95054	1.93417	1.26271	−0.54318	−4.33418	−9.82407
0.10	1.79335	1.78290	1.11222	−0.79655	−4.61254	−9.77821
0.15	1.59856	1.59362	0.933342	−1.05494	−4.86119	−9.65884
0.20	1.37359	1.37338	0.728628	−1.31614	−5.07759	−9.46566
0.25	1.12630	1.12973	0.501512	−1.57752	−5.25933	−9.19903
0.30	0.864703	0.870490	0.256194	−1.83607	−5.40414	−8.85999
0.35	0.596656	0.603398	−0.002502	−2.08847	−5.50994	−8.45024
0.40	0.329628	0.335890	−0.269284	−2.33108	−5.57485	−7.97212
0.45	0.070476	0.074876	−0.538562	−2.56013	−5.59723	−7.42866
0.50	−0.174721	−0.173445	−0.804626	−2.77170	−5.57573	−6.82351
0.55	−0.400784	−0.403716	−1.06181	−2.96190	−5.50931	−6.16093
0.60	−0.603515	−0.611515	−1.30469	−3.12697	−5.39728	−5.44578
0.65	−0.779735	−0.793409	−1.52818	−3.26332	−5.23936	−4.68348
0.70	−0.927279	−0.946967	−1.72773	−3.36773	−5.03572	−3.87998
0.75	−1.04497	−1.07074	−1.89941	−3.43738	−4.78700	−3.04170
0.80	−1.13254	−1.16420	−2.04002	−3.47000	−4.49438	−2.17553
0.85	−1.19056	−1.22769	−2.14713	−3.46388	−4.15956	−1.28872
0.90	−1.22034	−1.26229	−2.21915	−3.41800	−3.78482	−0.388881

Table 3.5 (continued)

				t		
v	0.95	1.00	2.00	3.00	4.00	5.00
0.95	−1.22379	−1.26976	−2.25534	−3.33203	−3.37300	0.516123
1.00	−1.20333	−1.25237	−2.25579	−3.20642	−2.92752	1.41822
1.05	−1.16173	−1.21281	−2.22139	−3.04231	−2.45234	2.30919
1.10	−1.10204	−1.15407	−2.15380	−2.84162	−1.95196	3.18077
1.15	−1.02743	−1.07933	−2.05532	−2.60696	−1.43137	4.02473
1.20	−0.941111	−0.991821	−1.92888	−2.34161	−0.895973	4.83296
1.25	−0.846228	−0.894757	−1.77787	−2.04944	−0.351564	5.59752
1.30	−0.745798	−0.791248	−1.60608	−1.73484	0.195788	6.31081
1.35	−0.642629	−0.684221	−1.41758	−1.40260	0.739804	6.96559
1.40	−0.539276	−0.576363	−1.21659	−1.05785	1.27410	7.55513
1.45	−0.437999	−0.470080	−1.00739	−0.705932	1.79229	8.07325
1.50	−0.340741	−0.367465	−0.794211	−0.352264	2.28807	8.51446
1.55	−0.249114	−0.270279	−0.581140	−0.002261	2.75537	8.87402
1.60	−0.164399	−0.179948	−0.372026	0.338796	3.18842	9.14800
1.65	−0.087552	−0.097566	−0.170416	0.665863	3.58186	9.33340
1.70	−0.019224	−0.023906	0.020512	0.974232	3.93085	9.42814
1.75	0.040219	0.040560	0.197998	1.25962	4.23118	9.43119
1.80	0.090667	0.095630	0.359731	1.51825	4.47927	9.34253
1.85	0.132238	0.141349	0.503874	1.74691	4.67235	9.16324
1.90	0.165239	0.177973	0.629070	1.94300	4.80840	8.89544
1.95	0.190138	0.205934	0.734436	2.10457	4.88629	8.54235
2.00	0.207528	0.225807	0.819554	2.23035	4.90570	8.10820
2.05	0.218095	0.238276	0.884435	2.31973	4.86723	7.59823
2.10	0.222587	0.244105	0.929495	2.37276	4.77229	7.01861
2.15	0.221788	0.244102	0.955511	2.39014	4.62314	6.37641
2.20	0.216493	0.239101	0.963574	2.37314	4.42283	5.67942
2.25	0.207486	0.229929	0.955041	2.32361	4.17511	4.93616
2.30	0.195522	0.217394	0.931483	2.24387	3.88440	4.15567
2.35	0.181314	0.202265	0.894634	2.13665	3.55568	3.34746
2.40	0.165519	0.185258	0.846340	2.00505	3.19442	2.52132
2.45	0.148733	0.167025	0.788509	1.85242	2.80645	1.68721
2.50	0.131482	0.148153	0.723064	1.68234	2.39788	0.855112
2.55	0.114224	0.129153	0.651903	1.49848	1.97495	0.034893
2.60	0.097345	0.110463	0.576859	1.30453	1.54400	−0.763848
2.65	0.081161	0.092444	0.499669	1.10420	1.11126	−1.53191
2.70	0.065922	0.075390	0.421946	0.901056	0.682835	−2.26062
2.75	0.051816	0.059524	0.345156	0.698511	0.264550	−2.94197
2.80	0.038974	0.045007	0.270602	0.499763	−0.138120	−3.56871
2.85	0.027476	0.031944	0.199414	0.307743	−0.520131	−4.13443
2.90	0.017357	0.020388	0.132540	0.125075	−0.876948	−4.63370
2.95	0.008614	0.010350	0.070750	−0.045951	−1.20460	−5.06208
3.00	0.001212	0.001803	0.014629	−0.203401	−1.49972	−5.41620

Table 3.5 (continued)

v	0.95	1.00	2.00	3.00	4.00	5.00
3.05	−0.004909	−0.005311	−0.035406	−0.345710	−1.75959	−5.69377
3.10	−0.009832	−0.011072	−0.079103	−0.471687	−1.98217	−5.89363
3.15	−0.013651	−0.015580	−0.116361	−0.580509	−2.16607	−6.01568
3.20	−0.016472	−0.018947	−0.147217	−0.671715	−2.31059	−6.06094
3.25	−0.018406	−0.021294	−0.171827	−0.745182	−2.41565	−6.03144
3.30	−0.019565	−0.022743	−0.190454	−0.801108	−2.48182	−5.93020
3.35	−0.020062	−0.023419	−0.203448	−0.839983	−2.51024	−5.76114
3.40	−0.020005	−0.023443	−0.211228	−0.862555	−2.50259	−5.52902
3.45	−0.019495	−0.022928	−0.214268	−0.869799	−2.46103	−5.23933
3.50	−0.018626	−0.021984	−0.213081	−0.862880	−2.38812	−4.89815
3.55	−0.017486	−0.020710	−0.208199	−0.843116	−2.28680	−4.51211
3.60	−0.016152	−0.019194	−0.200168	−0.811940	−2.16029	−4.08821
3.65	−0.014691	−0.017517	−0.189527	−0.770865	−2.01200	−3.63373
3.70	−0.013163	−0.015747	−0.176801	−0.721450	−1.84552	−3.15610
3.75	−0.011616	−0.013943	−0.162495	−0.665265	−1.66448	−2.66276
3.80	−0.010091	−0.012154	−0.147080	−0.603860	−1.47254	−2.16111
3.85	−0.008621	−0.010420	−0.130990	−0.538741	−1.27330	−1.65831
3.90	−0.007231	−0.008771	−0.114620	−0.471344	−1.07024	−1.16124
3.95	−0.005940	−0.007231	−0.098319	−0.403014	−0.866701	−0.676388
4.00	−0.004759	−0.005817	−0.082390	−0.334988	−0.665779	−0.209759
4.05	−0.003697	−0.004539	−0.067090	−0.268382	−0.470338	0.233186
4.10	−0.002755	−0.003400	−0.052629	−0.204184	−0.282955	0.647601
4.15	−0.001935	−0.002403	−0.039175	−0.143241	−0.105899	1.02929
4.20	−0.001232	−0.001544	−0.026851	−0.086265	0.058889	1.37477
4.25	−0.000640	−0.000818	−0.015742	−0.033823	0.209801	1.68123
4.30	−0.000152	−0.000215	−0.005899	0.013651	0.345574	1.94662
4.35	0.000241	0.000273	0.002663	0.055856	0.465281	2.16960
4.40	0.000547	0.000656	0.009953	0.092615	0.568326	2.34951
4.45	0.000777	0.000946	0.016007	0.123867	0.654430	2.48638
4.50	0.000940	0.001155	0.020881	0.149656	0.723608	2.58090
4.55	0.001046	0.001293	0.024644	0.170119	0.776153	2.63432
4.60	0.001104	0.001371	0.027383	0.185475	0.812607	2.64844
4.65	0.001121	0.001399	0.029189	0.196010	0.833729	2.62554
4.70	0.001107	0.001386	0.030161	0.202070	0.840472	2.56834
4.75	0.001067	0.001341	0.030402	0.204040	0.833946	2.47988
4.80	0.001009	0.001272	0.030014	0.202342	0.815388	2.36349
4.85	0.000937	0.001186	0.029096	0.197415	0.786129	2.22272
4.90	0.000856	0.001087	0.027745	0.189710	0.747566	2.06126
4.95	0.000770	0.000982	0.026053	0.179677	0.701130	1.88286
5.00	0.000683	0.000873	0.024105	0.167760	0.648261	1.69129

Table 3.6: Second derivatives with respect to the order of the Kelvin functions $\partial^2 bei_v(t)/\partial v^2$.

			t			
v	0.05	0.10	0.15	0.20	0.25	0.30
0.00	−14.6567	−11.3832	−9.46994	−8.11792	−7.07720	−6.23655
0.05	−12.4381	−10.4090	−9.10048	−8.11793	−7.32735	−6.66597
0.10	−10.2447	−9.23134	−8.46026	−7.82617	−7.28346	−6.80789
0.15	−8.17922	−7.94965	−7.63134	−7.30834	−6.99725	−6.70221
0.20	−6.30842	−6.64450	−6.68683	−6.62718	−6.52128	−6.39229
0.25	−4.66991	−5.37820	−5.68925	−5.84005	−5.90643	−5.92237
0.30	−3.27836	−4.19612	−4.68989	−4.99741	−5.19991	−5.33548
0.35	−2.13131	−3.12863	−3.72892	−4.14197	−4.44388	−4.67181
0.40	−1.21437	−2.19338	−2.83603	−3.30834	−3.67459	−3.96750
0.45	−0.505479	−1.39761	−2.03142	−2.52324	−2.92201	−3.25390
0.50	0.021618	−0.740456	−1.32705	−1.80596	−2.20977	−2.55714
0.55	0.394715	−0.215054	−0.728002	−1.16912	−1.55543	−1.89806
0.60	0.641193	0.189599	−0.233852	−0.619554	−0.970970	−1.29235
0.65	0.786680	0.487059	0.160051	−0.159232	−0.463397	−0.750871
0.70	0.854227	0.692133	0.461388	0.213740	−0.035486	−0.280193
0.75	0.863864	0.819870	0.679820	0.504226	0.313491	0.116886
0.80	0.832472	0.884795	0.826106	0.719207	0.587023	0.440710
0.85	0.773857	0.900383	0.911383	0.867031	0.790653	0.694150
0.90	0.698974	0.878717	0.946601	0.956780	0.931343	0.882006
0.95	0.616237	0.830308	0.942104	0.997736	1.01691	1.01045
1.00	0.531867	0.764041	0.907345	0.998970	1.05555	1.08652
1.05	0.450253	0.687206	0.850703	0.969029	1.05543	1.11770
1.10	0.374295	0.605603	0.779399	0.915714	1.02443	1.11157
1.15	0.305720	0.523683	0.699479	0.845950	0.969842	1.07546
1.20	0.245364	0.444717	0.615857	0.765707	0.898275	1.01629
1.25	0.193407	0.370975	0.532391	0.679993	0.815530	0.940372
1.30	0.149570	0.303895	0.451988	0.592878	0.726569	0.853325
1.35	0.113275	0.244252	0.376725	0.507559	0.635508	0.760003
1.40	0.083769	0.192302	0.307970	0.426434	0.545654	0.664484
1.45	0.060216	0.147919	0.246508	0.351203	0.459559	0.570091
1.50	0.041765	0.110704	0.192654	0.282964	0.379098	0.479431
1.55	0.027599	0.080078	0.146361	0.222314	0.305543	0.394448
1.60	0.016962	0.055359	0.107315	0.169443	0.239658	0.316505
1.65	0.009175	0.035815	0.075012	0.124227	0.181783	0.246448
1.70	0.003650	0.020716	0.048830	0.086306	0.131915	0.184697
1.75	−0.000118	0.009358	0.028081	0.055152	0.089787	0.131315
1.80	−0.002548	0.001091	0.012058	0.030131	0.054938	0.086088
1.85	−0.003983	−0.004673	0.000064	0.010551	0.026771	0.048586
1.90	−0.004698	−0.008454	−0.008561	−0.004298	0.004606	0.018230
1.95	−0.004912	−0.010698	−0.014427	−0.015114	−0.012275	−0.005658
2.00	−0.004789	−0.011781	−0.018081	−0.022561	−0.024598	−0.023812
2.05	−0.004457	−0.012016	−0.020005	−0.027255	−0.033068	−0.036984
2.10	−0.004008	−0.011657	−0.020613	−0.029750	−0.038350	−0.045914

Table 3.6 (continued)

				t		
v	0.05	0.10	0.15	0.20	0.25	0.30
2.15	−0.003507	−0.010904	−0.020255	−0.030535	−0.041054	−0.051310
2.20	−0.003001	−0.009914	−0.019220	−0.030035	−0.041731	−0.053831
2.25	−0.002517	−0.008805	−0.017743	−0.028609	−0.040862	−0.054074
2.30	−0.002074	−0.007662	−0.016008	−0.026555	−0.038865	−0.052573
2.35	−0.001681	−0.006548	−0.014157	−0.024116	−0.036092	−0.049792
2.40	−0.001342	−0.005503	−0.012297	−0.021483	−0.032831	−0.046126
2.45	−0.001054	−0.004551	−0.010504	−0.018804	−0.029319	−0.041906
2.50	−0.000815	−0.003704	−0.008828	−0.016189	−0.025737	−0.037402
2.55	−0.000620	−0.002967	−0.007303	−0.013715	−0.022227	−0.032829
2.60	−0.000463	−0.002338	−0.005944	−0.011434	−0.018888	−0.028351
2.65	−0.000339	−0.001809	−0.004757	−0.009375	−0.015791	−0.024089
2.70	−0.000243	−0.001373	−0.003737	−0.007554	−0.012979	−0.020128
2.75	−0.000169	−0.001018	−0.002876	−0.005970	−0.010473	−0.016523
2.80	−0.000114	−0.000735	−0.002162	−0.004617	−0.008280	−0.013301
2.85	−0.000073	−0.000512	−0.001578	−0.003479	−0.006393	−0.010473
2.90	−0.000043	−0.000340	−0.001109	−0.002538	−0.004795	−0.008031
2.95	−0.000023	−0.000210	−0.000739	−0.001773	−0.003466	−0.005958
3.00	−0.000009	−0.000114	−0.000454	−0.001164	−0.002379	−0.004228
3.05	0.000001	−0.000046	−0.000238	−0.000687	−0.001507	−0.002810
3.10	0.000006	0.000002	−0.000080	−0.000324	−0.000823	−0.001672
3.15	0.000009	0.000033	0.000031	−0.000056	−0.000300	−0.000779
3.20	0.000011	0.000052	0.000107	0.000136	0.000088	−0.000097
3.25	0.000011	0.000062	0.000153	0.000265	0.000364	0.000407
3.30	0.000010	0.000065	0.000179	0.000346	0.000549	0.000761
3.35	0.000009	0.000064	0.000188	0.000389	0.000662	0.000994
3.40	0.000008	0.000060	0.000187	0.000404	0.000719	0.001130
3.45	0.000007	0.000055	0.000178	0.000398	0.000733	0.001190
3.50	0.000006	0.000049	0.000164	0.000378	0.000716	0.001192
3.55	0.000005	0.000042	0.000147	0.000349	0.000677	0.001152
3.60	0.000004	0.000036	0.000129	0.000315	0.000624	0.001082
3.65	0.000003	0.000030	0.000111	0.000279	0.000563	0.000994
3.70	0.000002	0.000025	0.000095	0.000242	0.000499	0.000895
3.75	0.000002	0.000020	0.000079	0.000207	0.000434	0.000791
3.80	0.000001	0.000016	0.000065	0.000174	0.000372	0.000688
3.85	0.000001	0.000013	0.000053	0.000144	0.000314	0.000590
3.90	0.000001	0.000010	0.000042	0.000118	0.000261	0.000497
3.95	0.000001	0.000007	0.000033	0.000095	0.000213	0.000413
4.00		0.000005	0.000025	0.000075	0.000172	0.000338
4.05		0.000004	0.000019	0.000058	0.000136	0.000271
4.10		0.000003	0.000014	0.000044	0.000105	0.000214
4.15		0.000002	0.000010	0.000032	0.000080	0.000165
4.20		0.000001	0.000007	0.000023	0.000058	0.000124
4.25		0.000001	0.000005	0.000016	0.000041	0.000090

Table 3.6 (continued)

			t			
v	0.05	0.10	0.15	0.20	0.25	0.30
4.30			0.000003	0.000010	0.000028	0.000062
4.35			0.000001	0.000006	0.000017	0.000040
4.40			0.000000	0.000003	0.000009	0.000023
4.45			0.000000	0.000000	0.000003	0.000010
4.50			−0.000001	−0.000001	−0.000001	0.000007
4.55			−0.000001	−0.000002	−0.000004	−0.000006
4.60			−0.000001	−0.000003	−0.000006	−0.000011
4.65			−0.000001	−0.000003	−0.000007	−0.000014
4.70			−0.000001	−0.000003	−0.000008	−0.000015
4.75			−0.000001	−0.000003	−0.000008	−0.000016
4.80			−0.000001	−0.000003	−0.000007	−0.000015
4.85			−0.000001	−0.000003	−0.000007	−0.000015
4.90			−0.000001	−0.000002	−0.000006	−0.000014
4.95			−0.000001	−0.000002	−0.000005	−0.000012
5.00				−0.000002	−0.000005	−0.000011

			t			
v	0.35	0.40	0.45	0.50	0.55	0.60
0.00	−5.53667	−4.94213	−4.43007	−3.98481	−3.59515	−3.25276
0.05	−6.09904	−5.60517	−5.17021	−4.78424	−4.44004	−4.13215
0.10	−6.38486	−6.00481	−5.66112	−5.34896	−5.06469	−4.80547
0.15	−6.42406	−6.16257	−5.91717	−5.68724	−5.47213	−5.27127
0.20	−6.25181	−6.10635	−5.95987	−5.81495	−5.67329	−5.53611
0.25	−5.90601	−5.86818	−5.81587	−5.75382	−5.68542	−5.61318
0.30	−5.42525	−5.48222	−5.51500	−5.52963	−5.53050	−5.52091
0.35	−4.84702	−4.98293	−5.08863	−5.17061	−5.23367	−5.28149
0.40	−4.20637	−4.40370	−4.56812	−4.70592	−4.82182	−4.91952
0.45	−3.53489	−3.77560	−3.98358	−4.16439	−4.32231	−4.46071
0.50	−2.86001	−3.12661	−3.36291	−3.57347	−3.76183	−3.93083
0.55	−2.20471	−2.48098	−2.73110	−2.95839	−3.16552	−3.35468
0.60	−1.58735	−1.85898	−2.10970	−2.34157	−2.55630	−2.75536
0.65	−1.02177	−1.27674	−1.51661	−1.74220	−1.95430	−2.15366
0.70	−0.517610	−0.746413	−0.965980	−1.17606	−1.37661	−1.56769
0.75	−0.080678	−0.276338	−0.468322	−0.655485	−0.837068	−1.01256
0.80	0.286581	0.128567	−0.030719	−0.189454	−0.346321	−0.500341
0.85	0.584551	0.466408	0.342863	0.216192	0.088105	−0.040078
0.90	0.815919	0.737853	0.651193	0.558454	0.461573	0.362085
0.95	0.985147	0.945659	0.895363	0.836822	0.772052	0.702677
1.00	1.09798	1.09420	1.07836	1.05289	1.01978	0.980620
1.05	1.16100	1.18904	1.20463	1.20998	1.20690	1.19691
1.10	1.18125	1.23650	1.27968	1.31269	1.33710	1.35425
1.15	1.16585	1.24332	1.30973	1.36661	1.41523	1.45674
1.20	1.12178	1.21635	1.30135	1.37790	1.44699	1.50950

Table 3.6 (continued)

v	0.35	0.40	0.45	0.50	0.55	0.60
1.25	1.05564	1.16229	1.26117	1.35303	1.43854	1.51835
1.30	0.973504	1.087490	1.19569	1.29850	1.39632	1.48954
1.35	0.880793	0.997800	1.11105	1.22065	1.32672	1.42946
1.40	0.782231	0.898464	1.01291	1.12543	1.23593	1.34441
1.45	0.681819	0.794064	0.906345	1.01832	1.12974	1.24045
1.50	0.582836	0.688491	0.795779	0.904228	1.01347	1.12324
1.55	0.487874	0.584943	0.684973	0.787420	0.891845	0.997894
1.60	0.398889	0.485956	0.577017	0.671507	0.768955	0.868967
1.65	0.317261	0.393445	0.474356	0.559447	0.648253	0.740372
1.70	0.243867	0.308766	0.378830	0.453566	0.532543	0.615377
1.75	0.179150	0.232776	0.291731	0.355601	0.424009	0.496614
1.80	0.123197	0.165904	0.213862	0.266749	0.324260	0.386108
1.85	0.075807	0.108222	0.145606	0.187733	0.234377	0.285314
1.90	0.036558	0.059515	0.086992	0.118860	0.154973	0.195176
1.95	0.004865	0.019341	0.037764	0.060095	0.086263	0.116182
2.00	-0.019964	-0.012907	-0.002559	0.011117	0.028123	0.048432
2.05	-0.038684	-0.037948	-0.034623	-0.028608	-0.019844	-0.008302
2.10	-0.052077	-0.056569	-0.059186	-0.059781	-0.058245	-0.054507
2.15	-0.060920	-0.069586	-0.077076	-0.083209	-0.087843	-0.090873
2.20	-0.065959	-0.077814	-0.089149	-0.099767	-0.109506	-0.118237
2.25	-0.067896	-0.082038	-0.096258	-0.110355	-0.124159	-0.137531
2.30	-0.067372	-0.082998	-0.099228	-0.115867	-0.132750	-0.149737
2.35	-0.064961	-0.081375	-0.098835	-0.117166	-0.136216	-0.155852
2.40	-0.061169	-0.077779	-0.095794	-0.115065	-0.135459	-0.156857
2.45	-0.056428	-0.072752	-0.090752	-0.110311	-0.131323	-0.153689
2.50	-0.051104	-0.066756	-0.084275	-0.103578	-0.124585	-0.147224
2.55	-0.045495	-0.060185	-0.076855	-0.095458	-0.115943	-0.138263
2.60	-0.039841	-0.053363	-0.068908	-0.086464	-0.106010	-0.127525
2.65	-0.034328	-0.046548	-0.060776	-0.077028	-0.095314	-0.115639
2.70	-0.029093	-0.039942	-0.052732	-0.067507	-0.084301	-0.103143
2.75	-0.024232	-0.033695	-0.044990	-0.058184	-0.073334	-0.090490
2.80	-0.019808	-0.027911	-0.037707	-0.049280	-0.062705	-0.078048
2.85	-0.015855	-0.022659	-0.030993	-0.040955	-0.052633	-0.066106
2.90	-0.012381	-0.017972	-0.024916	-0.033320	-0.043279	-0.054884
2.95	-0.009381	-0.013861	-0.019511	-0.026441	-0.034751	-0.044537
3.00	-0.006834	-0.010315	-0.014785	-0.020349	-0.027110	-0.035165
3.05	-0.004708	-0.007310	-0.010722	-0.015046	-0.020381	-0.026822
3.10	-0.002968	-0.004809	-0.007291	-0.010509	-0.014555	-0.019520
3.15	-0.001574	-0.002769	-0.004449	-0.006698	-0.009602	-0.013243
3.20	-0.000484	-0.001142	-0.002144	-0.003563	-0.005472	-0.007946
3.25	0.000344	0.000121	-0.000321	-0.001041	-0.002103	-0.003570
3.30	0.000948	0.001068	0.001079	0.000932	0.000576	-0.000040
3.35	0.001365	0.001748	0.002112	0.002423	0.002641	0.002728

Table 3.6 (continued)

				t		
v	0.35	0.40	0.45	0.50	0.55	0.60
3.40	0.001630	0.002204	0.002834	0.003498	0.004169	0.004818
3.45	0.001772	0.002477	0.003297	0.004221	0.005234	0.006320
3.50	0.001819	0.002603	0.003547	0.004651	0.005910	0.007317
3.55	0.001793	0.002615	0.003629	0.004843	0.006263	0.007892
3.60	0.001714	0.002540	0.003580	0.004847	0.006357	0.008121
3.65	0.001599	0.002404	0.003433	0.004707	0.006247	0.008072
3.70	0.001460	0.002225	0.003216	0.004460	0.005983	0.007809
3.75	0.001310	0.002020	0.002953	0.004139	0.005607	0.007386
3.80	0.001155	0.001803	0.002664	0.003771	0.005156	0.006851
3.85	0.001002	0.001583	0.002364	0.003379	0.004661	0.006244
3.90	0.000856	0.001368	0.002065	0.002979	0.004145	0.005598
3.95	0.000721	0.001165	0.001776	0.002586	0.003629	0.004940
4.00	0.000597	0.000976	0.001504	0.002211	0.003129	0.004293
4.05	0.000486	0.000804	0.001253	0.001859	0.002655	0.003672
4.10	0.000389	0.000651	0.001026	0.001537	0.002215	0.003089
4.15	0.000304	0.000517	0.000823	0.001248	0.001815	0.002553
4.20	0.000232	0.000401	0.000647	0.000991	0.001456	0.002068
4.25	0.000172	0.000302	0.000494	0.000767	0.001141	0.001636
4.30	0.000123	0.000219	0.000365	0.000576	0.000867	0.001258
4.35	0.000082	0.000151	0.000258	0.000414	0.000634	0.000933
4.40	0.000050	0.000096	0.000170	0.000280	0.000438	0.000657
4.45	0.000025	0.000053	0.000099	0.000171	0.000277	0.000428
4.50	0.000007	0.000019	0.000043	0.000084	0.000147	0.000240
4.55	−0.000007	−0.000006	0.000001	0.000016	0.000044	0.000091
4.60	−0.000017	−0.000024	−0.000031	−0.000035	−0.000034	−0.000026
4.65	−0.000023	−0.000036	−0.000053	−0.000072	−0.000093	−0.000113
4.70	−0.000027	−0.000044	−0.000067	−0.000097	−0.000133	−0.000176
4.75	−0.000029	−0.000048	−0.000076	−0.000113	−0.000160	−0.000219
4.80	−0.000029	−0.000050	−0.000080	−0.000121	−0.000175	−0.000244
4.85	−0.000028	−0.000049	−0.000080	−0.000123	−0.000181	−0.000256
4.90	−0.000026	−0.000047	−0.000077	−0.000120	−0.000179	−0.000258
4.95	−0.000024	−0.000043	−0.000072	−0.000114	−0.000173	−0.000251
5.00	−0.000022	−0.000040	−0.000067	−0.000107	−0.000162	−0.000238

				t		
v	0.65	0.70	0.75	0.80	0.85	0.90
0.00	−2.95125	−2.68564	−2.45192	−2.24683	−2.06764	−1.91209
0.05	−3.85632	−3.60918	−3.38797	−3.19043	−3.01464	−2.85896
0.10	−4.56904	−4.35355	−4.15745	−3.97943	−3.81838	−3.67331
0.15	−5.08411	−4.91015	−4.74893	−4.60001	−4.46301	−4.33753
0.20	−5.40424	−5.27831	−5.15874	−5.04587	−4.93991	−4.84101
0.25	−5.53898	−5.46426	−5.39016	−5.31759	−5.24726	−5.17975
0.30	−5.50343	−5.48007	−5.45243	−5.42184	−5.38937	−5.35590

Table 3.6 (continued)

			t			
v	0.65	0.70	0.75	0.80	0.85	0.90
0.35	−5.31698	−5.34242	−5.35971	−5.37040	−5.37577	−5.37692
0.40	−5.00192	−5.07141	−5.12994	−5.17913	−5.22037	−5.25481
0.45	−4.58232	−4.68937	−4.78372	−4.86694	−4.94037	−5.00516
0.50	−4.08282	−4.21974	−4.34323	−4.45469	−4.55534	−4.64623
0.55	−3.52769	−3.68611	−3.83126	−3.96430	−4.08624	−4.19796
0.60	−2.94000	−3.11133	−3.27031	−3.41778	−3.55452	−3.68119
0.65	−2.34096	−2.51683	−2.68184	−2.83651	−2.98134	−3.11676
0.70	−1.74944	−1.92204	−2.08567	−2.24055	−2.38688	−2.52485
0.75	−1.18160	−1.34398	−1.49952	−1.64814	−1.78977	−1.92439
0.80	−0.650767	−0.797025	−0.938660	−1.07531	−1.20669	−1.33256
0.85	−0.167313	−0.292759	−0.415724	−0.535636	−0.652013	−0.764445
0.90	0.261229	0.160032	0.059352	−0.040081	−0.137639	−0.232776
0.95	0.630039	0.555264	0.479311	0.403010	0.327085	0.252175
1.00	0.936771	0.889383	0.839451	0.787849	0.735349	0.682642
1.05	1.18128	1.16113	1.13744	1.11106	1.08279	1.05331
1.10	1.36530	1.37126	1.37305	1.37148	1.36729	1.36116
1.15	1.49211	1.52224	1.54792	1.56988	1.58881	1.60532
1.20	1.56623	1.61791	1.66520	1.70872	1.74907	1.78678
1.25	1.59305	1.66318	1.72927	1.79182	1.85131	1.90818
1.30	1.57855	1.66373	1.74545	1.82407	1.89996	1.97348
1.35	1.52904	1.62570	1.71965	1.81114	1.90040	1.98769
1.40	1.45090	1.55545	1.65818	1.75918	1.85861	1.95660
1.45	1.35034	1.45934	1.56743	1.67463	1.78096	1.88649
1.50	1.23330	1.34351	1.45374	1.56391	1.67396	1.78389
1.55	1.10528	1.21375	1.32314	1.43328	1.54405	1.65536
1.60	0.971208	1.07540	1.18129	1.28868	1.39741	1.50732
1.65	0.835453	0.933191	1.03332	1.13560	1.23983	1.34582
1.70	0.701726	0.791286	0.883779	0.978960	1.07660	1.17650
1.75	0.573101	0.653182	0.736593	0.823086	0.912437	1.00443
1.80	0.452020	0.521742	0.595031	0.671659	0.751409	0.834073
1.85	0.340326	0.399200	0.461726	0.527704	0.596935	0.669228
1.90	0.239306	0.287200	0.338689	0.393604	0.451775	0.513035
1.95	0.149748	0.186845	0.227351	0.271132	0.318053	0.367971
2.00	0.071996	0.098750	0.128612	0.161492	0.197285	0.235880
2.05	0.006020	0.023102	0.042903	0.065370	0.090434	0.118013
2.10	−0.048522	−0.040271	−0.029757	−0.017001	−0.002038	0.015080
2.15	−0.092222	−0.091839	−0.089694	−0.085779	−0.080103	−0.072690
2.20	−0.125862	−0.132304	−0.137511	−0.141451	−0.144107	−0.145482
2.25	−0.150357	−0.162547	−0.174032	−0.184760	−0.194699	−0.203833
2.30	−0.166709	−0.183568	−0.200236	−0.216652	−0.232770	−0.248560
2.35	−0.175959	−0.196440	−0.217213	−0.238212	−0.259386	−0.280699
2.40	−0.179156	−0.202265	−0.226106	−0.250615	−0.275740	−0.301441
2.45	−0.177321	−0.202140	−0.228079	−0.255079	−0.283092	−0.312079

Table 3.6 (continued)

				t		
v	0.65	0.70	0.75	0.80	0.85	0.90
2.50	−0.171425	−0.197128	−0.224276	−0.252823	−0.282728	−0.313958
2.55	−0.162374	−0.188231	−0.215797	−0.245037	−0.275922	−0.308429
2.60	−0.150989	−0.176378	−0.203672	−0.232852	−0.263904	−0.296814
2.65	−0.138003	−0.162406	−0.188846	−0.217322	−0.247831	−0.280375
2.70	−0.124056	−0.147060	−0.172171	−0.199404	−0.228774	−0.260293
2.75	−0.109694	−0.130984	−0.154394	−0.179956	−0.207698	−0.237646
2.80	−0.095370	−0.114724	−0.136160	−0.159724	−0.185457	−0.213400
2.85	−0.081449	−0.098729	−0.118009	−0.139345	−0.162792	−0.188401
2.90	−0.068219	−0.083360	−0.100380	−0.119348	−0.140326	−0.163374
2.95	−0.055889	−0.068891	−0.083623	−0.100159	−0.118570	−0.138924
3.00	−0.044607	−0.055523	−0.067996	−0.082106	−0.097929	−0.115536
3.05	−0.034461	−0.043387	−0.053683	−0.065432	−0.078709	−0.093590
3.10	−0.025493	−0.032559	−0.040800	−0.050297	−0.061127	−0.073364
3.15	−0.017703	−0.023063	−0.029402	−0.036796	−0.045319	−0.055042
3.20	−0.011060	−0.014886	−0.019496	−0.024961	−0.031351	−0.038733
3.25	−0.005507	−0.007978	−0.011047	−0.014778	−0.019233	−0.024474
3.30	−0.000970	−0.002269	−0.003991	−0.006191	−0.008924	−0.012243
3.35	0.002639	0.002333	0.001763	0.000885	−0.000345	−0.001973
3.40	0.005416	0.005928	0.006322	0.006562	0.006613	0.006440
3.45	0.007458	0.008626	0.009802	0.010962	0.012080	0.013130
3.50	0.008864	0.010539	0.012329	0.014221	0.016200	0.018250
3.55	0.009730	0.011776	0.014026	0.016478	0.019125	0.021963
3.60	0.010147	0.012443	0.015016	0.017870	0.021010	0.024439
3.65	0.010198	0.012641	0.015414	0.018532	0.022006	0.025849
3.70	0.009961	0.012461	0.015331	0.018591	0.022260	0.026359
3.75	0.009504	0.011989	0.014867	0.018165	0.021909	0.026125
3.80	0.008887	0.011297	0.014111	0.017361	0.021079	0.025295
3.85	0.008162	0.010451	0.013144	0.016277	0.019886	0.024005
3.90	0.007373	0.009506	0.012036	0.014998	0.018431	0.022374
3.95	0.006555	0.008510	0.010844	0.013595	0.016803	0.020509
4.00	0.005738	0.007500	0.009618	0.012131	0.015079	0.018504
4.05	0.004944	0.006508	0.008399	0.010657	0.013322	0.016435
4.10	0.004191	0.005555	0.007217	0.009214	0.011585	0.014369
4.15	0.003491	0.004661	0.006097	0.007833	0.009907	0.012356
4.20	0.002852	0.003837	0.005055	0.006538	0.008320	0.010438
4.25	0.002277	0.003091	0.004103	0.005345	0.006848	0.008645
4.30	0.001770	0.002425	0.003248	0.004265	0.005505	0.006997
4.35	0.001329	0.001842	0.002492	0.003303	0.004300	0.005508
4.40	0.000952	0.001338	0.001834	0.002460	0.003235	0.004184
4.45	0.000635	0.000911	0.001272	0.001732	0.002311	0.003026
4.50	0.000373	0.000555	0.000799	0.001116	0.001521	0.002030
4.55	0.000162	0.000265	0.000409	0.000603	0.000859	0.001189
4.60	−0.000005	0.000033	0.000095	0.000186	0.000316	0.000492

Table 3.6 (continued)

			t			
v	0.65	0.70	0.75	0.80	0.85	0.90
4.65	−0.000132	−0.000146	−0.000152	−0.000145	−0.000120	−0.000071
4.70	−0.000226	−0.000280	−0.000339	−0.000399	−0.000458	−0.000514
4.75	−0.000291	−0.000375	−0.000474	−0.000586	−0.000712	−0.000850
4.80	−0.000331	−0.000438	−0.000566	−0.000716	−0.000892	−0.001093
4.85	−0.000353	−0.000473	−0.000621	−0.000798	−0.001008	−0.001255
4.90	−0.000359	−0.000487	−0.000646	−0.000839	−0.001073	−0.001350
4.95	−0.000353	−0.000483	−0.000647	−0.000848	−0.001094	−0.001389
5.00	−0.000338	−0.000466	−0.000629	−0.000832	−0.001081	−0.001382

			t			
v	0.95	1.00	2.00	3.00	4.00	5.00
0.00	−1.77821	−1.66434	−2.10488	−4.85329	−6.35943	−1.25003
0.05	−2.72198	−2.60248	−2.71363	−5.00458	−5.84643	0.187456
0.10	−3.54335	−3.42771	−3.30063	−5.12426	−5.27880	1.65029
0.15	−4.22323	−4.11975	−3.84598	−5.20393	−4.65887	3.12563
0.20	−4.74926	−4.66470	−4.33111	−5.23517	−3.98921	4.60037
0.25	−5.11553	−5.05494	−4.73948	−5.20998	−3.27271	6.06119
0.30	−5.32217	−5.28880	−5.05711	−5.12112	−2.51268	7.49473
0.35	−5.37475	−5.37004	−5.27303	−4.96250	−1.71295	8.88767
0.40	−5.28346	−5.30717	−5.37955	−4.72949	−0.87798	10.2268
0.45	−5.06232	−5.11270	−5.37248	−4.41919	−0.01290	11.4993
0.50	−4.72827	−4.80227	−5.25113	−4.03064	0.87633	12.6926
0.55	−4.30026	−4.39381	−5.01824	−3.56498	1.78294	13.7945
0.60	−3.79840	−3.90670	−4.67977	−3.02553	2.69931	14.7935
0.65	−3.24318	−3.36096	−4.24464	−2.41775	3.61695	15.6787
0.70	−2.65467	−2.77652	−3.72430	−1.74920	4.52657	16.4400
0.75	−2.05199	−2.17258	−3.13231	−1.02938	5.41814	17.0681
0.80	−1.45273	−1.56704	−2.48385	−0.26952	6.28103	17.5546
0.85	−0.872579	−0.976108	−1.79520	0.51774	7.10412	17.8923
0.90	−0.325014	−0.413929	−1.08319	1.31858	7.87605	18.0751
0.95	0.178848	0.107616	−0.36472	2.11845	8.58536	18.0982
1.00	0.630349	0.579033	0.34373	2.90242	9.22076	17.9581
1.05	1.02326	0.993241	1.02658	3.65555	9.77140	17.6531
1.10	1.35371	1.34552	1.66951	4.36332	10.2271	17.1828
1.15	1.61999	1.63336	2.25980	5.01199	10.5786	16.5487
1.20	1.82237	1.85634	2.78661	5.58897	10.8178	15.7538
1.25	1.96288	2.01582	3.24113	6.08312	10.9383	14.8033
1.30	2.04496	2.11475	3.61678	6.48506	10.9348	13.7040
1.35	2.07326	2.15737	3.90917	6.78735	10.8045	12.4648
1.40	2.05331	2.14892	4.11614	6.98470	10.5460	11.0961
1.45	1.99128	2.09542	4.23768	7.07404	10.1604	9.61051
1.50	1.89367	2.00333	4.27578	7.05460	9.65067	8.02220
1.55	1.76715	1.87935	4.23423	6.92787	9.02206	6.34700

Table 3.6 (continued)

v	0.95	1.00	2.00	3.00	4.00	5.00
				t		
1.60	1.61829	1.73022	4.11848	6.69755	8.28190	4.60225
1.65	1.45342	1.56249	3.93531	6.36940	7.43954	2.80655
1.70	1.27848	1.38236	3.69261	5.95107	6.50620	0.979662
1.75	1.09888	1.19560	3.39911	5.45188	5.49481	-0.857852
1.80	0.919458	1.00738	3.06409	4.88258	4.41975	-2.68482
1.85	0.744398	0.822264	2.69711	4.25504	3.29659	-4.47978
1.90	0.577213	0.644143	2.30778	3.58198	2.14182	-6.22130
1.95	0.420740	0.476213	1.90550	2.87664	0.972527	-7.88829
2.00	0.277157	0.320991	1.49924	2.15248	-0.193946	-9.46030
2.05	0.148015	0.180341	1.09740	1.42286	-1.34032	-10.9179
2.10	0.034287	0.055508	0.707622	0.700788	-2.44972	-12.2431
2.15	-0.063579	-0.052824	0.336675	-0.001387	-3.50600	-13.4192
2.20	-0.145594	-0.144473	-0.009620	-0.672192	-4.49403	-14.4318
2.25	-0.212161	-0.219696	-0.326457	-1.30125	-5.40002	-15.2685
2.30	-0.264009	-0.279116	-0.610066	-1.87943	-6.21171	-15.9192
2.35	-0.302126	-0.323656	-0.857712	-2.39903	-6.91864	-16.3766
2.40	-0.327689	-0.354468	-1.06767	-2.85382	-7.51227	-16.6359
2.45	-0.342013	-0.372875	-1.23916	-3.23913	-7.98617	-16.6953
2.50	-0.346489	-0.380304	-1.37231	-3.55186	-8.33605	-16.5557
2.55	-0.342539	-0.378241	-1.46802	-3.79045	-8.55981	-16.2210
2.60	-0.331573	-0.368179	-1.52793	-3.95487	-8.65756	-15.6979
2.65	-0.314956	-0.351579	-1.55427	-4.04648	-8.63154	-14.9956
2.70	-0.293977	-0.329839	-1.54976	-4.06796	-8.48600	-14.1261
2.75	-0.269827	-0.304267	-1.51750	-4.02317	-8.22713	-13.1036
2.80	-0.243589	-0.276061	-1.46089	-3.91696	-7.86282	-11.9443
2.85	-0.216220	-0.246293	-1.38347	-3.75510	-7.40250	-10.6665
2.90	-0.188550	-0.215907	-1.28890	-3.54399	-6.85691	-9.28956
2.95	-0.161283	-0.185709	-1.18078	-3.29059	-6.23785	-7.83444
3.00	-0.134996	-0.156373	-1.06266	-3.00218	-5.55792	-6.32269
3.05	-0.110146	-0.128442	-0.937897	-2.68621	-4.83025	-4.77633
3.10	-0.087078	-0.102337	-0.809651	-2.35012	-4.06828	-3.21750
3.15	-0.066036	-0.078365	-0.680806	-2.00124	-3.28543	-1.66802
3.20	-0.047172	-0.056730	-0.553951	-1.64658	-2.49492	-0.149097
3.25	-0.030558	-0.037546	-0.431345	-1.29277	-1.70949	1.31905
3.30	-0.016200	-0.020846	-0.314913	-0.945926	-0.941212	2.71749
3.35	-0.004043	-0.006598	-0.206235	-0.611571	-0.201294	4.02883
3.40	0.006008	0.005283	-0.106552	-0.294571	0.500093	5.23752
3.45	0.014088	0.014928	-0.016775	0.000910	1.15394	6.33000
3.50	0.020355	0.022500	0.062494	0.271443	1.75252	7.29494
3.55	0.024984	0.028183	0.130940	0.514340	2.28945	8.12328
3.60	0.028160	0.032177	0.188511	0.727648	2.75974	8.80840
3.65	0.030073	0.034687	0.235386	0.910130	3.15980	9.34606
3.70	0.030906	0.035920	0.271944	1.06123	3.48748	9.73446
3.75	0.030839	0.036078	0.298732	1.18102	3.74193	9.97418

Table 3.6 (continued)

			t			
ν	0.95	1.00	2.00	3.00	4.00	5.00
3.80	0.030042	0.035351	0.316434	1.27015	3.92364	10.0680
3.85	0.028670	0.033919	0.325836	1.32980	4.03430	10.0210
3.90	0.026866	0.031946	0.327800	1.36159	4.07670	9.84009
3.95	0.024754	0.029581	0.323236	1.36749	4.05462	9.53396
4.00	0.022447	0.026953	0.313071	1.34982	3.97271	9.11295
4.05	0.020039	0.024177	0.298235	1.31110	3.83633	8.58870
4.10	0.017609	0.021348	0.279633	1.25401	3.65142	7.97391
4.15	0.015222	0.018545	0.258132	1.18132	3.42435	7.28213
4.20	0.012929	0.015833	0.234547	1.09585	3.16178	6.52744
4.25	0.010771	0.013262	0.209629	1.00037	2.87050	5.72424
4.30	0.008774	0.010868	0.184058	0.897575	2.55733	4.88697
4.35	0.006957	0.008676	0.158438	0.790028	2.22898	4.02985
4.40	0.005331	0.006702	0.133291	0.680136	1.89191	3.16670
4.45	0.003900	0.004953	0.109060	0.570104	1.55228	2.31069
4.50	0.002660	0.003428	0.086109	0.461920	1.21583	1.47417
4.55	0.001605	0.002123	0.064725	0.357335	0.887796	0.668489
4.60	0.000725	0.001026	0.045123	0.257852	0.572884	-0.096103
4.65	0.000006	0.000124	0.027450	0.164722	0.275201	-0.810609
4.70	-0.000563	-0.000599	0.011792	0.078947	-0.001761	-1.46734
4.75	-0.001000	-0.001161	-0.001818	0.001283	-0.255149	-2.05999
4.80	-0.001321	-0.001578	-0.013397	-0.067745	-0.482742	-2.58364
4.85	-0.001542	-0.001871	-0.023001	-0.127832	-0.682950	-3.03476
4.90	-0.001676	-0.002056	-0.030719	-0.178875	-0.854785	-3.41121
4.95	-0.001739	-0.002150	-0.036668	-0.220953	-0.997833	-3.71218
5.00	-0.001743	-0.002170	-0.040987	-0.254305	-1.112210	-3.93810

3.3 The Kelvin Functions ker$_\nu$(t), and Their First and Second Derivatives with Respect to the Order

The first and second derivatives with respect to the order were evaluated using the central-difference formulas of order $O(h^4)$ with $h = 10^{-3}$, and the Kelvin functions ker$_\nu$(t) were taken from MATHEMATICA program:

$$\frac{\partial \ker_\nu(t)}{\partial \nu} = \frac{-\ker_{\nu+2h}(t) + 8\ker_{\nu+h}(t) - 8\ker_{\nu-h}(t) + \ker_{\nu-2h}(t)}{12h} \tag{3.3.1}$$

$$\frac{\partial^2 \ker_\nu(t)}{\partial \nu^2} = \frac{-\ker_{\nu+2h}(t) + 16\ker_{\nu+h}(t) - 30\ker_\nu(t) + 16\ker_{\nu-h}(t) - \ker_{\nu-2h}(t)}{12h^2} \tag{3.3.2}$$

The behaviour of the Kelvin functions $\ker_v(t)$ and its derivatives with respect to the order is complicated (Tables 3.7–3.9), and the functions have large positive and negative values which depend strongly on v and t (Figures 3.19–3.24).

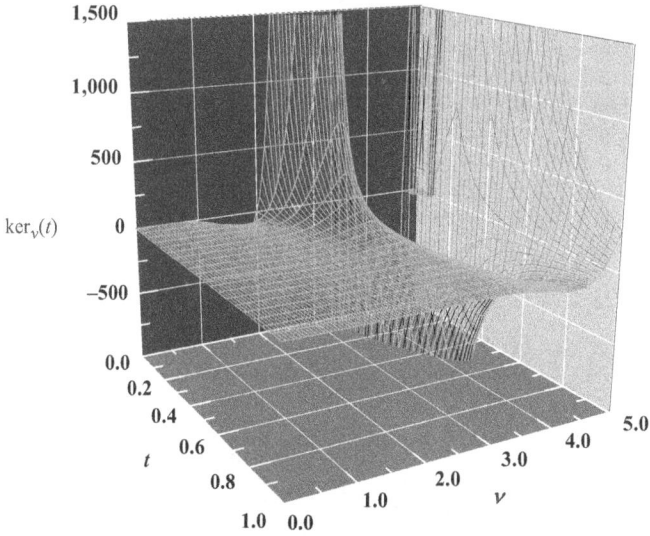

Figure 3.19: The Kelvin functions $\ker_v(t)$ as a function of v and t.

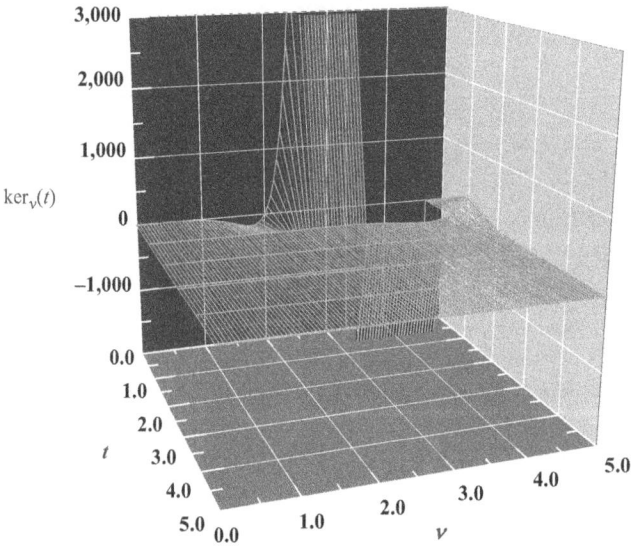

Figure 3.20: The Kelvin functions $\ker_v(t)$ as a function of v and t.

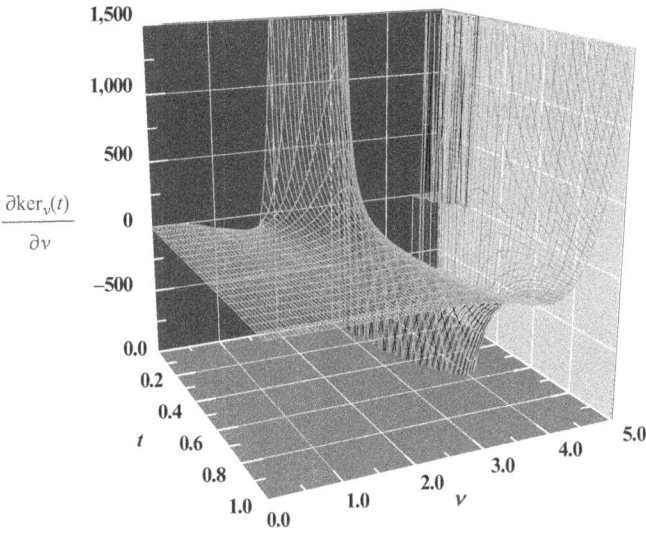

Figure 3.21: First derivatives with respect to the order of the Kelvin functions ∂ker$_v$(t)/∂v as a function of v and t.

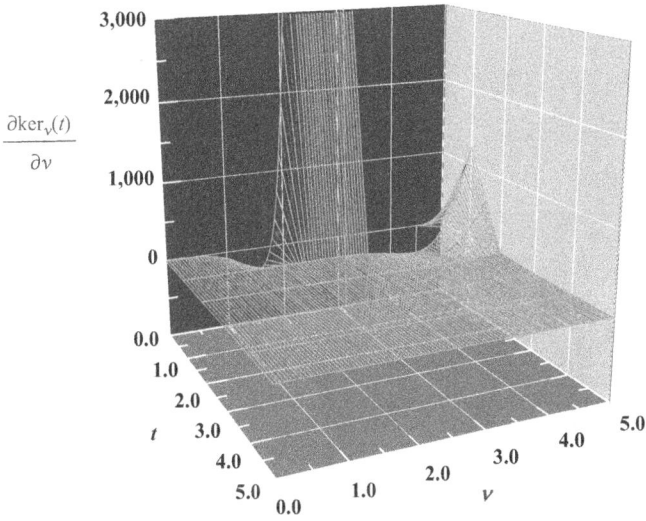

Figure 3.22: First derivatives with respect to the order of the Kelvin functions ∂ker$_v$(t)/∂v as a function of v and t.

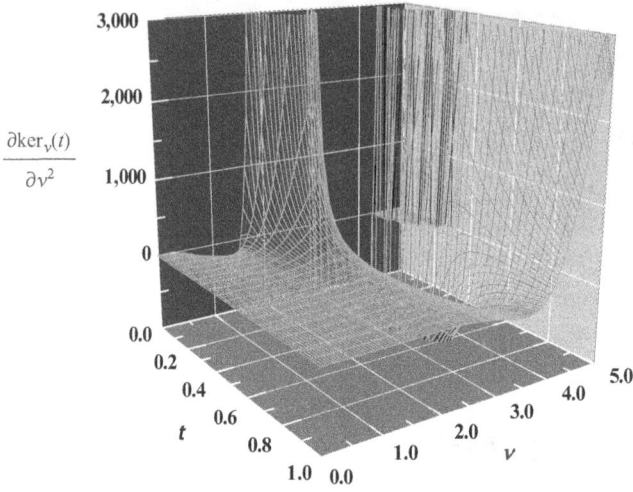

Figure 3.23: Second derivatives with respect to the order of the Kelvin functions $\partial^2 \mathrm{ker}_v(t)/\partial v^2$ as a function of v and t.

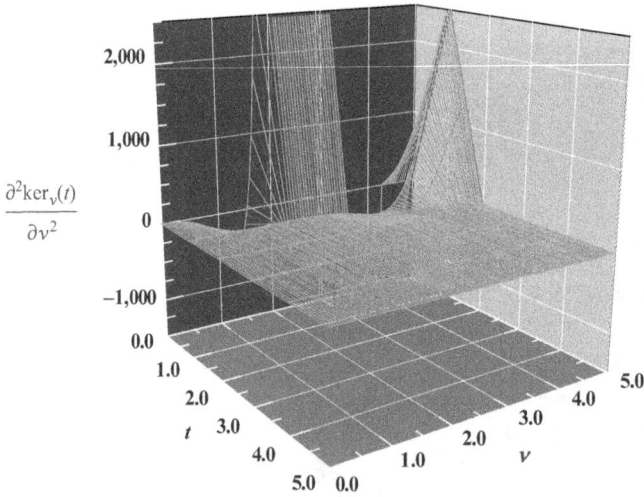

Figure 3.24: Second derivatives with respect to the order of the Kelvin functions $\partial^2 \mathrm{ker}_v(t)/\partial v^2$ as a function of v and t.

Table 3.7: The Kelvin functions ker$_\nu(t)$.

			t		
ν	0.05	0.10	0.15	0.20	0.25
0.00	3.11215	2.42047	2.01744	1.73314	1.51431
0.05	3.05582	2.35952	1.95555	1.67143	1.45329
0.10	3.00634	2.29654	1.88906	1.60424	1.38656
0.15	2.95799	2.22799	1.81548	1.52971	1.31269
0.20	2.90412	2.14985	1.73203	1.44581	1.23014
0.25	2.83684	2.05752	1.63562	1.35031	1.13724
0.30	2.74674	1.94573	1.52278	1.24076	1.03219
0.35	2.62246	1.80842	1.38963	1.11448	0.913061
0.40	2.45028	1.63863	1.23187	0.968552	0.777795
0.45	2.21361	1.42839	1.04470	0.799831	0.624212
0.50	1.89245	1.16859	0.822836	0.604924	0.450023
0.55	1.46275	0.848844	0.560466	0.380207	0.252847
0.60	0.895740	0.457434	0.251253	0.121851	0.030240
0.65	0.157167	−0.018835	−0.111655	−0.174149	−0.220271
0.70	−0.793470	−0.594674	−0.535593	−0.511923	−0.501137
0.75	−2.00372	−1.28634	−1.02831	−0.895653	−0.814724
0.80	−3.52894	−2.11159	−1.59787	−1.32948	−1.16323
0.85	−5.43298	−3.08960	−2.25249	−1.81736	−1.54860
0.90	−7.78868	−4.24068	−3.00035	−2.36294	−1.97237
0.95	−10.6781	−5.58593	−3.84925	−2.96928	−2.43555
1.00	−14.1921	−7.14668	−4.80627	−3.63868	−2.93842
1.05	−18.4298	−8.94365	−5.87727	−4.37228	−3.48033
1.10	−23.4961	−10.9958	−7.06620	−5.16970	−4.05944
1.15	−29.4989	−13.3189	−8.37435	−6.02857	−4.67238
1.20	−36.5434	−15.9234	−9.79930	−6.94395	−5.31399
1.25	−44.7251	−18.8117	−11.3337	−7.90766	−5.97680
1.30	−54.1179	−21.9750	−12.9639	−8.90750	−6.65067
1.35	−64.7596	−25.3890	−14.6678	−9.92628	−7.32221
1.40	−76.6302	−29.0084	−16.4131	−10.9408	−7.97421
1.45	−89.6238	−32.7602	−18.1544	−11.9207	−8.58498
1.50	−103.511	−36.5359	−19.8305	−12.8267	−9.12763
1.55	−117.892	−40.1813	−21.3605	−13.6098	−9.56929
1.60	−132.128	−43.4848	−22.6403	−14.2087	−9.87028
1.65	−145.267	−46.1629	−23.5377	−14.5484	−9.98324
1.70	−155.938	−47.8439	−23.8871	−14.5381	−9.85217
1.75	−162.224	−48.0484	−23.4843	−14.0692	−9.41158
1.80	−161.505	−46.1663	−22.0799	−13.0126	−8.58560
1.85	−150.262	−41.4320	−19.3728	−11.2170	−7.28715
1.90	−123.846	−32.8952	−15.0032	−8.50657	−5.41731
1.95	−76.1932	−19.3900	−8.54565	−4.67896	−2.86484
2.00	0.499755	0.499022	0.497809	0.496125	0.493983
2.05	116.165	28.4718	12.7037	7.27787	4.79506
2.10	283.296	66.5478	28.7319	15.9544	10.1857

Table 3.7 (continued)

			t		
v	0.05	0.10	0.15	0.20	0.25
2.15	517.439	117.098	49.3284	26.8415	16.8229
2.20	837.714	182.875	75.3248	40.2800	24.8711
2.25	1,267.36	267.025	107.634	56.6302	34.4982
2.30	1,834.26	373.096	147.237	76.2644	45.8706
2.35	2,571.39	505.013	195.166	99.5551	59.1459
2.40	3,517.17	667.024	252.473	126.859	74.4638
2.45	4,715.42	863.594	320.185	158.494	91.9342
2.50	6,215.07	1,099.24	399.240	194.712	111.622
2.55	8,069.04	1,378.29	490.399	235.659	133.528
2.60	10,332.2	1,704.50	594.138	281.332	157.567
2.65	13,058.2	2,080.52	710.489	331.514	183.536
2.70	16,293.7	2,507.26	838.852	385.701	211.084
2.75	20,070.6	2,982.88	977.745	443.014	239.670
2.80	24,394.8	3,501.58	1,124.50	502.086	268.514
2.85	29,228.9	4,052.03	1,274.88	560.934	296.543
2.90	34,470.3	4,615.24	1,422.62	616.798	322.326
2.95	39,919.9	5,162.04	1,558.83	665.966	344.000
3.00	45,240.7	5,649.77	1,671.38	703.554	359.189
3.05	49,902.5	6,018.33	1,744.03	723.273	364.912
3.10	53,110.8	6,185.23	1,755.55	717.151	357.486
3.15	53,713.5	6,039.70	1,678.61	675.236	332.414
3.20	50,083.1	5,435.67	1,478.59	585.265	284.282
3.25	39,967.5	4,183.44	1,112.18	432.313	206.638
3.30	20,304.9	2,040.16	525.885	198.435	91.8925
3.35	−13,004.8	−1,301.14	−345.539	−137.697	−68.7864
3.40	−65,375.0	−6,223.62	−1,580.75	−601.126	−285.548
3.45	−143,862	−13,199.8	−3,273.44	−1,220.90	−569.906
3.50	−257,548	−22,805.4	−5,533.79	−2,030.27	−934.740
3.55	−417,979	−35,732.8	−8,489.65	−3,066.74	−1,394.23
3.60	−639,652	−52,803.9	−12,287.2	−4,371.92	−1,963.69
3.65	−940,526	−74,979.6	−17,090.8	−5,991.08	−2,659.31
3.70	−1.343E+06	−103365	−23,081.5	−7,972.35	−3,497.70
3.75	−1.872	−1,39,206	−30,454.0	−10,365.3	−4,495.25
3.80	−2.560	−1,83,874	−39,411.3	−13,218.9	−5,667.28
3.85	−3.443	−2,38,834	−50,155.6	−16,578.6	−7,026.75
3.90	−4.561	−3,05,583	−62,875.9	−20,482.3	−8,582.73
3.95	−5.958	−3,85,559	−77,729.1	−24,954.5	−10,338.3
4.00	−7.680	−4,80,000	−94,814.6	−29,999.8	−12,287.8
4.05	−9.769	−589738	−114140	−35,592.5	−14,413.6
4.10	−1.226E+07	−714915	−135573	−41,665.4	−16,682.1
4.15	−1.517	−854587	−158788	−48,094.7	−19,038.3
4.20	−1.850	−1.006E+06	−183181	−54,680.5	−21,400.1
4.25	−2.217	−1.165	−2,07,778	−61,124.7	−23,650.5

Table 3.7 (continued)

			t		
v	0.05	0.10	0.15	0.20	0.25
4.30	−2.606	−1.322	−2,31,113	−67,002.7	−25,629.6
4.35	−2.992	−1.466	−2,51,075	−71,730.3	−27,123.6
4.40	−3.334	−1.358	−2,64,728	−74,523.9	−27,854.4
4.45	−3.569	−1.363	−2,68,086	−74,354.5	−27,465.1
4.50	−3.602	−1.359	−2,55,852	−69,894.9	−25,506.3
4.55	−3.294	−1.340	−2,21,109	−59,459.4	−21,419.3
4.60	−2.448	−1.006	−1,54,957	−40,936.8	−14,519.2
4.65	−7.905E+06	−3,10,736	−46,113.3	−11,717.8	−3,976.52
4.70	2.054E+7	7,95,376	1,19,547	31,382.0	11,200.5
4.75	6.593	2.456E+06	3,59,479	92,200.7	32,183.1
4.80	1.35E+8	4.854	694984	1,75,306	60,337.0
4.85	2.367	8.215	1.152E+6	2,86,061	9,72,32.0
4.90	3.824	1.282E+07	1.760	4,30,677	1,44,646
4.95	5.870	1.900	2.556	6,16,232	2,04,559
5.00	8.690	2.717	3.581	8,50,646	2,79,129

			t		
v	0.30	0.35	0.40	0.45	0.50
0.00	1.33722	1.18918	1.06262	0.952650	0.855906
0.05	1.27717	1.13028	1.00499	0.896375	0.801044
0.10	1.21147	1.06594	0.942204	0.835248	0.741654
0.15	1.13901	0.995279	0.873543	0.768702	0.677278
0.20	1.058590	0.917350	0.798270	0.696149	0.607453
0.25	0.968936	0.831170	0.715618	0.616991	0.531713
0.30	0.868691	0.735707	0.624795	0.530621	0.449596
0.35	0.756432	0.629896	0.524999	0.436434	0.360651
0.40	0.630666	0.512639	0.415420	0.333838	0.264448
0.45	0.489848	0.382825	0.295259	0.222259	0.160586
0.50	0.332390	0.239341	0.163736	0.101163	0.048708
0.55	0.156685	0.081096	0.020115	−0.029935	−0.071484
0.60	−0.038865	−0.092961	−0.136279	−0.171447	−0.200214
0.65	−0.255817	−0.283802	−0.306032	−0.323692	−0.337616
0.70	−0.495645	−0.492292	−0.489610	−0.486870	−0.483706
0.75	−0.759683	−0.719139	−0.687323	−0.661030	−0.638363
0.80	−1.04906	−0.964843	−0.899283	−0.846041	−0.801297
0.85	−1.36464	−1.22964	−1.12536	−1.04155	−0.972022
0.90	−1.70689	−1.51342	−1.36511	−1.24693	−1.14982
0.95	−2.07580	−1.81565	−1.61774	−1.46126	−1.33372
1.00	−2.47074	−2.13529	−1.88202	−1.68325	−1.52240
1.05	−2.89034	−2.47066	−2.15620	−1.91116	−1.71424
1.10	−3.33225	−2.81934	−2.43792	−2.14278	−1.90718
1.15	−3.79300	−3.17802	−2.72415	−2.37536	−2.09873
1.20	−4.26775	−3.54234	−3.01101	−2.60548	−2.28588

Table 3.7 (continued)

			t		
v	0.30	0.35	0.40	0.45	0.50
1.25	−4.75004	−3.90676	−3.29374	−2.82903	−2.46508
1.30	−5.23148	−4.26432	−3.56652	−3.04111	−2.63216
1.35	−5.70147	−4.60647	−3.82237	−3.23594	−2.78228
1.40	−6.14683	−4.92288	−4.05300	−3.40676	−2.90987
1.45	−6.55142	−5.20115	−4.24867	−3.54578	−3.00860
1.50	−6.89578	−5.42664	−4.39807	−3.64407	−3.07130
1.55	−7.15665	−5.58222	−4.48818	−3.69151	−3.08996
1.60	−7.30657	−5.64802	−4.50414	−3.67670	−3.05567
1.65	−7.31345	−5.60121	−4.42914	−3.58693	−2.95861
1.70	−7.14010	−5.41581	−4.24433	−3.40815	−2.78807
1.75	−6.74389	−5.06252	−3.92878	−3.12495	−2.53248
1.80	−6.07634	−4.50859	−3.45947	−2.72064	−2.17944
1.85	−5.08291	−3.71781	−2.81129	−2.17729	−1.71584
1.90	−3.70280	−2.65049	−1.95723	−1.47590	−1.12800
1.95	−1.86903	−1.26372	−0.868533	−0.596591	−0.401823
2.00	0.491395	0.488378	0.484946	0.481117	0.476909
2.05	3.45692	2.6541	2.1343	1.77789	1.52224
2.10	7.11056	5.2834	4.1109	3.31408	2.74763
2.15	11.5386	8.4267	6.4458	5.10910	4.16545
2.20	16.8285	12.1339	9.1681	7.18062	5.78636
2.25	23.0667	16.4517	12.3043	9.54354	7.61860
2.30	30.3346	21.4224	15.8762	12.2089	9.66711
2.35	38.7046	27.0797	19.8992	15.1823	11.9325
2.40	48.2342	33.4460	24.3794	18.4625	14.4098
2.45	58.9589	40.5278	29.3113	22.0391	17.0873
2.50	70.8840	48.3100	34.6740	25.8906	19.9447
2.55	83.9733	56.7501	40.4272	29.9816	22.9516
2.60	98.1371	65.7705	46.5071	34.2599	26.0654
2.65	113.217	75.2500	52.8206	38.6533	29.2290
2.70	128.968	85.0138	59.2397	43.0655	32.3690
2.75	145.039	94.8225	65.5947	47.3727	35.3922
2.80	160.946	104.359	71.6672	51.4187	38.1838
2.85	176.051	113.214	77.1817	55.0105	40.6038
2.90	189.526	120.871	81.7976	57.9130	42.4846
2.95	200.322	126.691	85.0994	59.8443	43.6277
3.00	207.131	129.889	86.5878	60.4703	43.8010
3.05	208.349	129.522	85.6700	59.4000	42.7363
3.10	202.031	124.464	81.6506	56.1807	40.1267
3.15	185.851	113.394	73.7227	50.2939	35.6251
3.20	157.057	94.7693	60.9610	41.1529	28.8434
3.25	112.431	66.8183	42.3155	28.1008	19.3522
3.30	48.2533	27.5250	16.6099	10.4112	6.68320
3.35	−39.7225	−25.3758	−17.4583	−12.7086	−9.66829

Table 3.7 (continued)

			t		
v	0.30	0.35	0.40	0.45	0.50
3.40	−156.288	−94.3959	−61.3106	−42.1112	−30.2370
3.45	−306.773	−182.281	−116.475	−78.6979	−55.5793
3.50	−497.018	−291.982	−184.564	−123.403	−86.2615
3.55	−733.312	−426.609	−267.241	−177.170	−122.843
3.60	−1,022.28	−589.362	−366.178	−240.923	−165.856
3.65	−1,370.74	−783.432	−482.988	−315.523	−215.778
3.70	−1,785.41	−1,011.86	−619.146	−401.719	−272.998
3.75	−2,272.66	−1,277.37	−775.886	−500.082	−337.774
3.80	−2,838.02	−1,582.12	−954.063	−610.926	−410.187
3.85	−3,485.63	−1,927.41	−1,153.99	−734.208	−490.072
3.90	−4,217.50	−2,313.33	−1,375.24	−869.413	−576.958
3.95	−5,032.59	−2,738.27	−1,616.39	−1,015.41	−669.972
4.00	−5,925.68	−3,198.42	−1,874.75	−1,170.30	−767.750
4.05	−6,885.94	−3,687.03	−2,145.99	−1,331.21	−868.326
4.10	−7,895.28	−4,193.70	−2,423.77	−1,494.09	−969.001
4.15	−8,926.23	−4,703.38	−2,699.25	−1,653.43	−1,066.21
4.20	−9,939.62	−5,195.36	−2,960.58	−1,802.05	−1,155.35
4.25	−10,881.7	−5,641.97	−3,192.29	−1,930.71	−1,230.65
4.30	−11,680.9	−6,007.23	−3,374.63	−2,027.84	−1,284.94
4.35	−12,244.3	−6,245.29	−3,482.89	−2,079.16	−1,309.51
4.40	−12,452.9	−6,298.68	−3,486.54	−2,067.26	−1,293.87
4.45	−12,157.9	−6,096.50	−3,348.52	−1,971.30	−1,225.60
4.50	−11,174.7	−5,552.48	−3,024.32	−1,766.57	−1,090.15
4.55	−9,278.62	−4,562.94	−2,461.22	−1,424.15	−870.695
4.60	−6,198.46	−3,004.88	−1,597.48	−910.583	−547.976
4.65	−1,611.71	−734.110	−361.737	−187.682	−100.261
4.70	4,860.74	2,416.29	1,327.53	787.628	496.666
4.75	13,658.9	6,637.08	3,562.16	2,063.25	1,269.41
4.80	25,286.6	12,143.2	6,444.15	3,691.64	2,246.73
4.85	40,312.5	19,173.3	10,084.9	5,729.28	3,459.12
4.90	59,368.1	27,987.6	14,604.0	8,235.71	4,938.23
4.95	83,141.7	38,864.1	20,126.5	11,272.1	6,715.97
5.00	1,12,367	52,092.1	26,779.9	14,899.4	8,823.26

			t		
v	0.55	0.60	0.65	0.70	0.75
0.00	0.769990	0.693121	0.623938	0.561378	0.504590
0.05	0.716586	0.641205	0.573531	0.512492	0.457231
0.10	0.658974	0.585395	0.519532	0.460304	0.406844
0.15	0.596790	0.525402	0.461717	0.404640	0.353301
0.20	0.529669	0.460944	0.399867	0.345339	0.296485
0.25	0.457253	0.391746	0.333780	0.282256	0.236298
0.30	0.379192	0.317552	0.263272	0.215264	0.172663

Table 3.7 (continued)

			t		
v	0.55	0.60	0.65	0.70	0.75
0.35	0.295156	0.238124	0.188183	0.144266	0.105531
0.40	0.204840	0.153259	0.108384	0.069195	0.034884
0.45	0.107976	0.062788	0.023787	-0.009977	-0.039259
0.50	0.004342	-0.033409	-0.065649	-0.093231	-0.116830
0.55	-0.106225	-0.135391	-0.159908	-0.180493	-0.197714
0.60	-0.223813	-0.243148	-0.258906	-0.271624	-0.281734
0.65	-0.348422	-0.356584	-0.362479	-0.366415	-0.368646
0.70	-0.479942	-0.475503	-0.470373	-0.464570	-0.458129
0.75	-0.618138	-0.599594	-0.582227	-0.565699	-0.549779
0.80	-0.762621	-0.728408	-0.697560	-0.669306	-0.643093
0.85	-0.912831	-0.861346	-0.815754	-0.774771	-0.737465
0.90	-1.06801	-0.997630	-0.936040	-0.881346	-0.832172
0.95	-1.22715	-1.13629	-1.05748	-0.988133	-0.926363
1.00	-1.38901	-1.27612	-1.17894	-1.09407	-1.01905
1.05	-1.55204	-1.41568	-1.29908	-1.19793	-1.10910
1.10	-1.71435	-1.55326	-1.41634	-1.29829	-1.19522
1.15	-1.87371	-1.68683	-1.52892	-1.39351	-1.27595
1.20	-2.02746	-1.81406	-1.63473	-1.48176	-1.34964
1.25	-2.17252	-1.93225	-1.73141	-1.56097	-1.41448
1.30	-2.30531	-2.03833	-1.81631	-1.62883	-1.46845
1.35	-2.42174	-2.12883	-1.88645	-1.68277	-1.50935
1.40	-2.51719	-2.19984	-1.93854	-1.72000	-1.53476
1.45	-2.58644	-2.24704	-1.96895	-1.73745	-1.54211
1.50	-2.62367	-2.26563	-1.97371	-1.73182	-1.52861
1.55	-2.62243	-2.25040	-1.94853	-1.69955	-1.49131
1.60	-2.57565	-2.19565	-1.88881	-1.63688	-1.42710
1.65	-2.47564	-2.09526	-1.78962	-1.53983	-1.33276
1.70	-2.31409	-1.94273	-1.64579	-1.40426	-1.20495
1.75	-2.08214	-1.73117	-1.45194	-1.22592	-1.04028
1.80	-1.77043	-1.45341	-1.20252	-1.00047	-0.835362
1.85	-1.36921	-1.10209	-0.891898	-0.723595	-0.586860
1.90	-0.868422	-0.669744	-0.514489	-0.391076	-0.291582
1.95	-0.257922	-0.148962	-0.064841	0.001109	0.053435
2.00	0.472340	0.467427	0.462191	0.456649	0.450820
2.05	1.33204	1.18614	1.07127	0.978754	0.902738
2.10	2.33017	2.01318	1.76636	1.56998	1.41075
2.15	3.47461	2.95352	2.55049	2.23201	1.97564
2.20	4.77171	4.01072	3.42542	2.96547	2.59726
2.25	6.22566	5.18653	4.39138	3.76961	3.27427
2.30	7.83792	6.48037	5.44661	4.64204	4.00396
2.35	9.60639	7.88877	6.58700	5.57840	4.78197
2.40	11.5246	9.40474	7.80558	6.57200	5.60201
2.45	13.5808	11.0171	9.09199	7.61343	6.45560
2.50	15.7566	12.7096	10.4320	8.69014	7.33170

Table 3.7 (continued)

			t		
v	0.55	0.60	0.65	0.70	0.75
2.55	18.0261	14.4601	11.8066	9.78597	8.21642
2.60	20.3544	16.2397	13.1918	10.8807	9.09266
2.65	22.6958	18.0116	14.5575	11.9494	9.93974
2.70	24.9930	19.7302	15.8670	12.9624	10.7331
2.75	27.1746	21.3399	17.0762	13.8840	11.4438
2.80	29.1540	22.7738	18.1328	14.6728	12.0386
2.85	30.8270	23.9530	18.9755	15.2807	12.4789
2.90	32.0707	24.7850	19.5337	15.6525	12.7214
2.95	32.7411	25.1630	19.7264	15.7259	12.7174
3.00	32.6722	24.9649	19.4621	15.4311	12.4126
3.05	31.6739	24.0528	18.6383	14.6904	11.7475
3.10	29.5314	22.2721	17.1414	13.4190	10.6574
3.15	26.0045	19.4519	14.8471	11.5246	9.07292
3.20	20.8274	15.4053	11.6207	8.90827	6.92017
3.25	13.7099	9.93037	7.31840	5.46561	4.12206
3.30	4.33885	2.81175	1.78869	1.08785	0.599245
3.35	-7.61841	-6.17635	-5.12537	-4.33620	-3.72831
3.40	-22.5095	-17.2665	-13.5831	-10.918	-8.93937
3.45	-40.6905	-30.6930	-23.7421	-18.765	-15.1090
3.50	-62.5167	-46.6850	-35.7533	-27.978	-22.3055
3.55	-88.3310	-65.4582	-49.7549	-38.647	-30.5868
3.60	-118.450	-87.2040	-65.8640	-50.844	-39.9963
3.65	-153.143	-112.077	-84.1677	-64.616	-50.5578
3.70	-192.613	-140.178	-104.712	-79.978	-62.2691
3.75	-236.965	-171.537	-127.489	-96.904	-75.0962
3.80	-286.178	-206.090	-152.420	-115.314	-88.9646
3.85	-340.060	-243.653	-179.341	-135.066	-103.752
3.90	-398.208	-283.893	-207.979	-155.937	-119.276
3.95	-459.955	-326.295	-237.934	-177.613	-135.287
4.00	-524.305	-370.121	-268.649	-199.668	-151.455
4.05	-589.872	-414.370	-299.386	-221.547	-167.356
4.10	-654.802	-457.732	-329.197	-242.548	-182.463
4.15	-716.690	-498.533	-356.886	-261.800	-196.125
4.20	-772.489	-534.687	-380.984	-278.243	-207.563
4.25	-818.416	-563.632	-399.707	-290.603	-215.849
4.30	-849.849	-582.279	-410.927	-297.376	-219.896
4.35	-861.226	-586.948	-412.132	-296.803	-218.445
4.40	-845.940	-573.312	-400.401	-286.859	-210.056
4.45	-796.240	-536.348	-372.371	-265.230	-193.103
4.50	-703.147	-470.295	-324.223	-229.312	-165.767
4.55	-556.378	-368.616	-251.663	-176.206	-126.042
4.60	-344.299	-223.995	-149.930	-102.724	-71.7410
4.65	-53.9172	-28.340	-13.8126	-5.41023	-0.515933

Table 3.7 (continued)

			t		
v	0.55	0.60	0.65	0.70	0.75
4.70	329.084	227.168	162.312	119.425	90.1155
4.75	820.246	551.994	384.411	275.654	202.729
4.80	1,436.15	956.109	658.691	467.254	339.937
4.85	2,194.15	1,449.80	991.461	698.209	504.315
4.90	3,111.94	2,043.37	1,388.94	972.372	698.305
4.95	4,207.00	2,746.81	1,857.00	1,293.30	924.099
5.00	5,495.84	3,569.26	2,400.87	1,664.02	1,183.48

			t		
v	0.80	0.85	0.90	0.95	1.0
0.00	0.452882	0.405683	0.362515	0.322972	0.286706
0.05	0.407050	0.361373	0.319718	0.281675	0.246894
0.10	0.358450	0.314540	0.274629	0.238303	0.205209
0.15	0.306990	0.265124	0.227212	0.192841	0.161657
0.20	0.252593	0.213078	0.177449	0.145294	0.116257
0.25	0.195198	0.158375	0.125339	0.095682	0.069050
0.30	0.134770	0.101010	0.070904	0.044048	0.020096
0.35	0.071297	0.041005	0.014189	−0.009541	−0.030522
0.40	0.004800	−0.021589	−0.044728	−0.064990	−0.082693
0.45	−0.064662	−0.086684	−0.105738	−0.122171	−0.136280
0.50	−0.136989	−0.154153	−0.168694	−0.180923	−0.191109
0.55	−0.212031	−0.223824	−0.233406	−0.241047	−0.246975
0.60	−0.289586	−0.295473	−0.299641	−0.302299	−0.303630
0.65	−0.369387	−0.368821	−0.367110	−0.364390	−0.360786
0.70	−0.451098	−0.443527	−0.435470	−0.426980	−0.418110
0.75	−0.534308	−0.519178	−0.504316	−0.489673	−0.475218
0.80	−0.618520	−0.595289	−0.573177	−0.552015	−0.531676
0.85	−0.703145	−0.671292	−0.641510	−0.613490	−0.586993
0.90	−0.787492	−0.746532	−0.708694	−0.673514	−0.640623
0.95	−0.870760	−0.820257	−0.774029	−0.731432	−0.691956
1.00	−0.952033	−0.891617	−0.836728	−0.786519	−0.740322
1.05	−1.03027	−0.959657	−0.895913	−0.837972	−0.784987
1.10	−1.10429	−1.02331	−0.950616	−0.884909	−0.825152
1.15	−1.17278	−1.08138	−0.999770	−0.926372	−0.859952
1.20	−1.23428	−1.13259	−1.04221	−0.961322	−0.888458
1.25	−1.28718	−1.17549	−1.07669	−0.988643	−0.909682
1.30	−1.32973	−1.20856	−1.10184	−1.007140	−0.922573
1.35	−1.36000	−1.23013	−1.11622	−1.015560	−0.926028
1.40	−1.37594	−1.23841	−1.11829	−1.012560	−0.918897
1.45	−1.37534	−1.23154	−1.10644	−0.996774	−0.899992
1.50	−1.35586	−1.20752	−1.07899	−0.966756	−0.868093
1.55	−1.31504	−1.16428	−1.03419	−0.921050	−0.821968
1.60	−1.25029	−1.09971	−0.970288	−0.858177	−0.760386

Table 3.7 (continued)

			t		
ν	0.80	0.85	0.90	0.95	1.0
1.65	−1.15899	−1.011620	−0.885481	−0.776664	−0.682134
1.70	−1.03843	−0.897822	−0.777997	−0.675069	−0.586042
1.75	−0.885909	−0.756160	−0.646102	−0.552008	−0.471009
1.80	−0.698757	−0.584530	−0.488147	−0.406187	−0.336028
1.85	−0.474402	−0.380948	−0.302605	−0.236445	−0.180228
1.90	−0.210430	−0.143599	−0.088127	−0.041794	−0.002902
1.95	0.095331	0.129091	0.156403	0.178532	0.196440
2.00	0.444724	0.438379	0.431803	0.425016	0.418034
2.05	0.839158	0.785119	0.738519	0.697803	0.661800
2.10	1.27949	1.16966	1.07654	0.996629	0.927294
2.15	1.76588	1.59175	1.44532	1.32075	1.21364
2.20	2.29767	2.05039	1.84366	1.66884	1.51947
2.25	2.87318	2.54372	2.26962	2.03896	1.84285
2.30	3.48957	3.06888	2.72039	2.42840	2.18120
2.35	4.14261	3.62185	3.19219	2.83361	2.53124
2.40	4.82647	4.19726	3.68011	3.25013	2.88888
2.45	5.53351	4.78826	4.17800	3.67242	3.24917
2.50	6.25405	5.38631	4.67832	4.09383	3.60619
2.55	6.97608	5.98102	5.17202	4.50645	3.95302
2.60	7.68506	6.55994	5.64840	4.90102	4.28161
2.65	8.36363	7.10837	6.09496	5.26684	4.58277
2.70	8.99138	7.60925	6.49729	5.59169	4.84606
2.75	9.54459	8.04292	6.83898	5.86178	5.05981
2.80	9.99608	8.38703	7.10153	6.06165	5.21105
2.85	10.3149	8.61643	7.26427	6.17421	5.28552
2.90	10.4664	8.70309	7.30437	6.18070	5.26771
2.95	10.4120	8.61610	7.19684	6.06075	5.14090
3.00	10.1090	8.32170	6.91461	5.79249	4.88727
3.05	9.51124	7.78341	6.42870	5.35267	4.48805
3.10	8.56877	6.96232	5.70842	4.71688	3.92367
3.15	7.22849	5.81735	4.72169	3.85986	3.17410
3.20	5.43458	4.30578	3.43549	2.75586	2.21911
3.25	3.12933	2.38391	1.81638	1.37911	1.03870
3.30	0.254039	0.00783	−0.168750	−0.295583	−0.386380
3.35	−3.24965	−2.86546	−2.55188	−2.29211	−2.07407
3.40	−7.43822	−6.27684	−5.36277	−4.63223	−4.04029
3.45	−12.3643	−10.2635	−8.62773	−7.33454	−6.29808
3.50	−18.0744	−14.8570	−12.3682	−10.4133	−8.85674
3.55	−24.6061	−20.0814	−16.5989	−13.8772	−11.7207
3.60	−31.9846	−25.9503	−21.3262	−17.7276	−14.8882
3.65	−40.2191	−32.4643	−26.5453	−21.9569	−18.3502
3.70	−49.2986	−39.6076	−32.2386	−26.5469	−22.0888
3.75	−59.1865	−47.3442	−38.3719	−31.4660	−26.0752

Table 3.7 (continued)

			t		
v	0.80	0.85	0.90	0.95	1.0
3.80	−69.8153	−55.6139	−44.8921	−36.6677	−30.2686
3.85	−81.0803	−64.3277	−51.7235	−42.0873	−34.6139
3.90	−92.8327	−73.3628	−58.7642	−47.6399	−39.0396
3.95	−104.872	−82.5574	−65.8823	−53.2170	−43.4560
4.00	−116.939	−91.7048	−72.9118	−58.6840	−47.7531
4.05	−128.705	−100.548	−79.6484	−63.8772	−51.7981
4.10	−139.767	−108.772	−85.8455	−68.6004	−55.4338
4.15	−149.632	−116.002	−91.2097	−72.6226	−58.4763
4.20	−157.717	−121.791	−95.3972	−75.6753	−60.7134
4.25	−163.331	−125.619	−98.0101	−77.4500	−61.9031
4.30	−165.676	−126.887	−98.5933	−77.5963	−61.7729
4.35	−163.832	−124.914	−96.6324	−75.7213	−60.0188
4.40	−156.756	−118.930	−91.5526	−71.3891	−56.3062
4.45	−143.280	−108.081	−82.7190	−64.1216	−50.2707
4.50	−122.108	−91.4282	−69.4391	−53.4011	−41.5206
4.55	−91.8219	−67.9496	−50.9666	−38.6744	−29.6404
4.60	−50.8903	−36.5530	−26.5092	−19.3588	−14.1965
4.65	2.31670	3.91390	4.76080	5.14878	5.25618
4.70	69.505	54.645	43.6925	35.4605	29.1643
4.75	152.424	116.851	91.1366	72.1824	57.9643
4.80	252.827	191.727	147.921	115.895	92.068
4.85	372.411	280.413	214.821	167.131	131.844
4.90	512.753	383.941	292.519	226.345	177.596
4.95	675.224	503.176	381.562	293.882	229.538
5.00	860.880	638.736	482.305	369.935	287.765

			t		
v	1.5	2.0	3.0	4.0	5.0
0.00	0.052935	−0.041665	−0.067029	−0.036179	−0.011512
0.05	0.026647	−0.057506	−0.070865	−0.035902	−0.010598
0.10	−0.000112	−0.073193	−0.074324	−0.035414	−0.009619
0.15	−0.027263	−0.088654	−0.077383	−0.034716	−0.008576
0.20	−0.054718	−0.103814	−0.080017	−0.033806	−0.007477
0.25	−0.082377	−0.118595	−0.082203	−0.032685	−0.006324
0.30	−0.110129	−0.132913	−0.083918	−0.031355	−0.005125
0.35	−0.137853	−0.146681	−0.085140	−0.029820	−0.003884
0.40	−0.165415	−0.159808	−0.085851	−0.028084	−0.002609
0.45	−0.192667	−0.172202	−0.086030	−0.026150	−0.001305
0.50	−0.219451	−0.183763	−0.085662	−0.024027	0.000020
0.55	−0.245594	−0.194392	−0.084732	−0.021721	0.001360
0.60	−0.270909	−0.203986	−0.083227	−0.019242	0.002706
0.65	−0.295198	−0.212441	−0.081138	−0.016599	0.004051

Table 3.7 (continued)

v	1.5	2.0	3.0	4.0	5.0
0.70	−0.318249	−0.219650	−0.078455	−0.013804	0.005387
0.75	−0.339834	−0.225506	−0.075176	−0.010871	0.006705
0.80	−0.359717	−0.229903	−0.071298	−0.007811	0.007998
0.85	−0.377646	−0.232735	−0.066824	−0.004642	0.009256
0.90	−0.393359	−0.233896	−0.061758	−0.001380	0.010471
0.95	−0.406586	−0.233286	−0.056112	0.001957	0.011635
1.00	−0.417044	−0.230806	−0.049898	0.005351	0.012737
1.05	−0.424448	−0.226364	−0.043135	0.008781	0.013771
1.10	−0.428503	−0.219875	−0.035845	0.012225	0.014728
1.15	−0.428915	−0.211259	−0.028057	0.015662	0.015598
1.20	−0.425387	−0.200449	−0.019801	0.019067	0.016374
1.25	−0.417629	−0.187386	−0.011117	0.022416	0.017048
1.30	−0.405352	−0.172027	−0.002047	0.025684	0.017613
1.35	−0.388284	−0.154340	0.007360	0.028845	0.018061
1.40	−0.366163	−0.134314	0.017051	0.031872	0.018386
1.45	−0.338751	−0.111953	0.026965	0.034739	0.018581
1.50	−0.305834	−0.087284	0.037037	0.037419	0.018642
1.55	−0.267229	−0.060356	0.047197	0.039884	0.018563
1.60	−0.222793	−0.031243	0.057369	0.042108	0.018340
1.65	−0.172427	−0.000047	0.067471	0.044062	0.017970
1.70	−0.116086	0.033102	0.077416	0.045722	0.017450
1.75	−0.053785	0.068041	0.087114	0.047061	0.016780
1.80	0.014391	0.104575	0.096467	0.048055	0.015958
1.85	0.088278	0.142473	0.105376	0.048680	0.014985
1.90	0.167627	0.181466	0.113738	0.048915	0.013863
1.95	0.252088	0.221248	0.121446	0.048739	0.012594
2.00	0.341209	0.261472	0.128391	0.048134	0.011184
2.05	0.434418	0.301751	0.134463	0.047085	0.009636
2.10	0.531020	0.341653	0.139550	0.045577	0.007959
2.15	0.630181	0.380709	0.143541	0.043601	0.006159
2.20	0.730927	0.418405	0.146326	0.041150	0.004246
2.25	0.832130	0.454186	0.147799	0.038220	0.002230
2.30	0.932502	0.487458	0.147856	0.034811	0.000124
2.35	1.03059	0.517591	0.146398	0.030927	−0.002059
2.40	1.12477	0.543919	0.143332	0.026576	−0.004305
2.45	1.21325	0.565744	0.138574	0.021773	−0.006597
2.50	1.29406	0.582344	0.132050	0.016535	−0.008919
2.55	1.36505	0.592975	0.123697	0.010884	−0.011253
2.60	1.42392	0.596880	0.113462	0.004849	−0.013578
2.65	1.46821	0.593295	0.101311	−0.001537	−0.015875
2.70	1.49531	0.581461	0.087225	−0.008234	−0.018122
2.75	1.50248	0.560631	0.071202	−0.015199	−0.020297
2.80	1.48688	0.530086	0.053262	−0.022379	−0.022378

Table 3.7 (continued)

			t		
v	1.5	2.0	3.0	4.0	5.0
2.85	1.44560	0.489144	0.033448	−0.029721	−0.024340
2.90	1.37568	0.437177	0.011826	−0.037163	−0.026161
2.95	1.27417	0.373626	−0.011514	−0.044637	−0.027817
3.00	1.13816	0.298022	−0.036451	−0.052071	−0.029283
3.05	0.964845	0.209998	−0.062836	−0.059389	−0.030536
3.10	0.751604	0.109319	−0.090491	−0.066509	−0.031554
3.15	0.496052	−0.004107	−0.119204	−0.073346	−0.032313
3.20	0.196139	−0.130199	−0.148733	−0.079808	−0.032792
3.25	−0.149763	−0.268681	−0.178799	−0.085804	−0.032972
3.30	−0.542754	−0.419060	−0.209091	−0.091238	−0.032832
3.35	−0.983296	−0.580597	−0.239265	−0.096014	−0.032357
3.40	−1.47109	−0.752286	−0.268942	−0.100032	−0.031531
3.45	−2.00494	−0.932827	−0.297709	−0.103195	−0.030341
3.50	−2.58263	−1.12060	−0.325124	−0.105407	−0.028778
3.55	−3.20076	−1.31365	−0.350715	−0.106572	−0.026835
3.60	−3.85459	−1.50966	−0.373981	−0.106599	−0.024507
3.65	−4.53791	−1.70593	−0.394398	−0.105402	−0.021794
3.70	−5.24287	−1.89935	−0.411422	−0.102899	−0.018700
3.75	−5.95981	−2.08642	−0.424494	−0.099018	−0.015232
3.80	−6.67714	−2.26322	−0.433043	−0.093694	−0.011402
3.85	−7.38114	−2.42542	−0.436495	−0.086874	−0.007227
3.90	−8.05593	−2.56825	−0.434279	−0.078516	−0.002726
3.95	−8.68324	−2.68659	−0.425835	−0.068593	0.002074
4.00	−9.24244	−2.77490	−0.410623	−0.057094	0.007143
4.05	−9.71041	−2.82733	−0.388130	−0.044022	0.012444
4.10	−10.0615	−2.83770	−0.357886	−0.029402	0.017937
4.15	−10.2678	−2.79963	−0.319470	−0.013279	0.023575
4.20	−10.2988	−2.70654	−0.272526	0.004282	0.029309
4.25	−10.1221	−2.55178	−0.216773	0.023191	0.035080
4.30	−9.70299	−2.32872	−0.152022	0.043334	0.040830
4.35	−9.00555	−2.03089	−0.078185	0.064571	0.046492
4.40	−7.99251	−1.65208	0.004703	0.086735	0.051997
4.45	−6.62601	−1.18654	0.096477	0.109632	0.057272
4.50	−4.86832	−0.629143	0.196820	0.133040	0.062239
4.55	−2.68263	0.024397	0.305255	0.156709	0.066820
4.60	−0.034055	0.777326	0.421123	0.180360	0.070933
4.65	3.10921	1.63159	0.543575	0.203687	0.074495
4.70	6.77453	2.58756	0.671556	0.226358	0.077422
4.75	10.983	3.64379	0.803792	0.248015	0.079631
4.80	15.749	4.79666	0.938782	0.268277	0.081040
4.85	21.077	6.04011	1.07479	0.286740	0.081568
4.90	26.957	7.36527	1.20984	0.302985	0.081139
4.95	33.370	8.76017	1.34170	0.316575	0.079682
5.00	40.276	10.2094	1.46791	0.327065	0.077130

Table 3.8: First derivatives with respect to the order of the Kelvin functions $\partial \mathrm{ker}_v(t)/\partial v$.

			t		
v	**0.05**	**0.10**	**0.15**	**0.20**	**0.25**
0.00	−1.22966	−1.22027	−1.20707	−1.19086	−1.17221
0.05	−1.04036	−1.22825	−1.27598	−1.28316	−1.27300
0.10	−0.957871	−1.30279	−1.39199	−1.41074	−1.40109
0.15	−0.998241	−1.45264	−1.56049	−1.57714	−1.55884
0.20	−1.18288	−1.68834	−1.78762	−1.78618	−1.74872
0.25	−1.53935	−2.02239	−2.08031	−2.04201	−1.97328
0.30	−2.10229	−2.46939	−2.44627	−2.34906	−2.23514
0.35	−2.91448	−3.04615	−2.89396	−2.71198	−2.53688
0.40	−4.02797	−3.77185	−3.43256	−3.13554	−2.88103
0.45	−5.50534	−4.66805	−4.07185	−3.62458	−3.26992
0.50	−7.42102	−5.75870	−4.82210	−4.18381	−3.70559
0.55	−9.86255	−7.07003	−5.69383	−4.81764	−4.18961
0.60	−12.9318	−8.63032	−6.69754	−5.52996	−4.72294
0.65	−16.7459	−10.4695	−7.84327	−6.32383	−5.30569
0.70	−21.4378	−12.6183	−9.14012	−7.20110	−5.93685
0.75	−27.1564	−15.1076	−10.5955	−8.16194	−6.61398
0.80	−34.0650	−17.9666	−12.2143	−9.20431	−7.33285
0.85	−42.3393	−21.2215	−13.9979	−10.32330	−8.08699
0.90	−52.1622	−24.8921	−15.9424	−11.5101	−8.86720
0.95	−63.7170	−28.9893	−18.0373	−12.7515	−9.66098
1.00	−77.1757	−33.5102	−20.2634	−14.0284	−10.4518
1.05	−92.6826	−38.4325	−22.5899	−15.3144	−11.2186
1.10	−110.331	−43.7077	−24.9719	−16.5747	−11.9345
1.15	−130.133	−49.2520	−27.3463	−17.7639	−12.5663
1.20	−151.971	−54.9354	−29.6279	−18.8244	−13.0732
1.25	−175.547	−60.5685	−31.7042	−19.6838	−13.4057
1.30	−200.300	−65.8857	−33.4298	−20.2527	−13.5047
1.35	−225.311	−70.5258	−34.6196	−20.4217	−13.2998
1.40	−249.172	−74.0083	−35.0416	−20.0587	−12.7081
1.45	−269.826	−75.7061	−34.4080	−19.0054	−11.6332
1.50	−284.360	−74.8122	−32.3665	−17.0742	−9.96342
1.55	−288.752	−70.3030	−28.4896	−14.0447	−7.57106
1.60	−277.550	−60.8950	−22.2641	−9.66040	−4.31111
1.65	−243.489	−44.9967	−13.0798	−3.62559	−0.020655
1.70	−177.022	−20.6552	−0.218019	4.39731	5.48151
1.75	−65.7611	14.5012	17.1590	14.7897	12.3946
1.80	106.179	63.3240	40.0182	27.9776	20.9346
1.85	358.946	129.206	69.4628	44.4308	31.3328
1.90	717.817	216.141	106.737	64.6602	43.8316
1.95	1,214.17	328.774	153.223	89.2116	58.6799
2.00	1,886.51	472.443	210.435	118.656	76.1262
2.05	2,781.57	653.190	279.999	153.575	96.4082
2.10	3,955.33	877.744	363.616	194.540	119.740

Table 3.8 (continued)

			t		
v	0.05	0.10	0.15	0.20	0.25
2.15	5,473.95	1,153.44	463.017	242.083	146.297
2.20	7,414.33	1,488.10	579.884	296.659	176.192
2.25	9,864.25	1,889.74	715.744	358.594	209.450
2.30	12,921.5	2,366.21	871.827	428.022	245.974
2.35	16,691.9	2,924.67	1,048.87	504.802	285.507
2.40	21,285.3	3,570.70	1,246.87	588.414	327.581
2.45	26,808.4	4,307.28	1,464.76	677.835	371.459
2.50	33,354.7	5,133.26	1,699.99	771.385	416.066
2.55	40,988.2	6,041.41	1,948.03	866.535	459.906
2.60	49,720.1	7,015.78	2,201.70	959.689	500.971
2.65	59,476.5	8,028.47	2,450.42	1,045.92	536.631
2.70	70,052.5	9,035.32	2,679.25	1,118.66	563.509
2.75	81,050.6	9,970.63	2,867.73	1,169.34	577.350
2.80	91,797.7	10,740.5	2,988.56	1,186.99	572.862
2.85	1,01,235	11,214.9	3,006.00	1,157.78	543.557
2.90	1,07,773	11,217.4	2,874.04	1,064.50	481.572
2.95	1,09,107	10,513.6	2,534.31	886.004	377.492
3.00	1,01,979	8,796.99	1,913.71	596.645	220.165
3.05	8,18,75.9	5,672.30	921.852	165.632	−3.46934
3.10	4,26,53.7	636.664	−551.743	−443.565	−308.536
3.15	−23,922.0	−6,941.95	−2,640.57	−1,273.79	−712.340
3.20	−1,28,740	−17,848.5	−5,504.05	−2,374.69	−1,234.43
3.25	−2,86,012	−33,046.4	−9,330.30	−3,803.08	−1,896.58
3.30	−5,14,068	−53,704.1	−14,338.40	−5,623.02	−2,722.66
3.35	−8,36,254	−81,220.6	−20,780.20	−7,905.62	−3,738.32
3.40	−1.282E+6	−1,17,249	−28,940.10	−10,728.3	−4,970.52
3.45	−1.888	−1,63,713	−39,134.20	−14,173.5	−6,446.75
3.50	−2.698	−2,22,814	−51,705.60	−18,326.3	−8,193.91
3.55	−3.767	−2,97,019	−67,016.20	−23,271.2	−10,236.8
3.60	−5.159	−3,89,026	−85,434.20	−29,087.1	−12,595.8
3.65	−6.949	−5,01,687	−1,07v313	−35,840.4	−15,284.5
3.70	−9.220	−6,37,886	−1,32,965	−43,576.2	−18,305.9
3.75	−1.207E+7	−8,00,339	−1,62,615	−52,305.8	−21,647.8
3.80	−1.558	−9,91,295	−1,96,354	−61,991.3	−25,277.0
3.85	−1.987	−1.212E+6	−2,34,056	−72,525.2	−29,132.7
3.90	−2.500	−1.463	−2,75,288	−83,705.3	−33,117.4
3.95	−3.104	−1.741	−3,19,182	−95,203.0	−37,086.4
4.00	−3.797	−2.040	−3,64,275	−1,06,525	−40,835.9
4.05	−4.572	−2.350	−4,08,308	−1,16,965	−44,087.5
4.10	−5.403	−2.654	−4,47,978	−1,25,548	−46,471.4
4.15	−6.247	−2.925	−4,78,629	−1,30,963	−47,506.5
4.20	−7.028E+7	−3.124	−4,93,878	−1,31,484	−46,577.4
4.25	−7.624	−3.196	−4,85,171	−1,24,880	−42,909.9
4.30	−7.854	−3.066	−4,41,247	−1,08,308	−35,542.7

Table 3.8 (continued)

v	0.05	0.10	0.15	0.20	0.25
			t		
4.35	−7.450	−2.631	−3,47,524	−78,198.8	−23,297.3
4.40	−6.030	−1.754	−1,85,387	−30,130.5	−4,747.7
4.45	−3.056	−257802	6,86,17.5	41,310.4	21,812.1
4.50	−2.209E+7	2.090E+6	4,43,664	1,42,669	58,397.4
4.55	1.077E+8	5.585	9,75,598	2,81,764	1,07,359
4.60	2.397	1.060E+7	1.708E+6	4,67,812	1,71,404
4.65	4.357	1.759	2.693	7,11,536	2,53,602
4.70	7.189	2.713	3.992	1.025E+6	357377
4.75	1.119E+9	3.991	5.676	1.423	4,86v478
4.80	1.674	5.675	7.829	1.919	6,44,908
4.85	2.431	7.862	1.054E+7	2.532	8,36,821
4.90	3.448	1.066E+8	1.392	3.277	1.066E+6
4.95	4.797	1.420	1.806	4.172	1.337
5.00	6.561	1.862	2.308	5.234	1.653

v	0.30	0.35	0.40	0.45	0.50
			t		
0.00	−1.15155	−1.129230	−1.10553	−1.08069	−1.05492
0.05	−1.25386	−1.229500	−1.20189	−1.17216	−1.14102
0.10	−1.377710	−1.346990	−1.31210	−1.27480	−1.23613
0.15	−1.524690	−1.482730	−1.43682	−1.38901	−1.34046
0.20	−1.696390	−1.637730	−1.57665	−1.51510	−1.45409
0.25	−1.894380	−1.812920	−1.73209	−1.65326	−1.57704
0.30	−2.120180	−2.009120	−1.90349	−1.80356	−1.70914
0.35	−2.375170	−2.227000	−2.09106	−1.96588	−1.85009
0.40	−2.660600	−2.466990	−2.29476	−2.13989	−1.99939
0.45	−2.977430	−2.729290	−2.51431	−2.32504	−2.15629
0.50	−3.326260	−3.013710	−2.74907	−2.52044	−2.31981
0.55	−3.707260	−3.319630	−2.99803	−2.72487	−2.48862
0.60	−4.119940	−3.645890	−3.25970	−2.93669	−2.66106
0.65	−4.563080	−3.990660	−3.53199	−3.15379	−2.83508
0.70	−5.034500	−4.351310	−3.81217	−3.37350	−3.00813
0.75	−5.530830	−4.724240	−4.09673	−3.59250	−3.17717
0.80	−6.047290	−5.104740	−4.38125	−3.80677	−3.33858
0.85	−6.577360	−5.486750	−4.66029	−4.01147	−3.48808
0.90	−7.11251	−5.86268	−4.92721	−4.20084	−3.62070
0.95	−7.64181	−6.22318	−5.17405	−4.36812	−3.73067
1.00	−8.15150	−6.55684	−5.39134	−4.50542	−3.81139
1.05	−8.62459	−6.84995	−5.56794	−4.60361	−3.85533
1.10	−9.04034	−7.08623	−5.69087	−4.65224	−3.85401
1.15	−9.37373	−7.24644	−5.74513	−4.63942	−3.79790
1.20	−9.59493	−7.30817	−5.71349	−4.55172	−3.67639
1.25	−9.66868	−7.24545	−5.57638	−4.37412	−3.47779

Table 3.8 (continued)

			t		
v	0.30	0.35	0.40	0.45	0.50
1.30	−9.55370	−7.02847	−5.31170	−4.08990	−3.18927
1.35	−9.20206	−6.62329	−4.89471	−3.68064	−2.79689
1.40	−8.55863	−5.99158	−4.29796	−3.12623	−2.28566
1.45	−7.56049	−5.09046	−3.49122	−2.40487	−1.63959
1.50	−6.13649	−3.87232	−2.44157	−1.49319	−0.841826
1.55	−4.20688	−2.28483	−1.11344	−0.366464	0.125165
1.60	−1.68315	−0.271096	0.531065	1.00121	1.27935
1.65	1.53194	2.23007	2.53185	2.63649	2.63880
1.70	5.54398	5.28357	4.93028	4.56633	4.22129
1.75	10.4661	8.95724	7.76841	6.81732	6.04373
1.80	16.4176	13.3206	11.0880	9.41490	8.12157
1.85	23.5223	18.4432	14.9293	12.3823	10.4680
1.90	31.9052	24.3925	19.3291	15.7395	13.0928
1.95	41.6891	31.2314	24.3193	19.5014	16.0017
2.00	52.9897	39.0139	29.9237	23.6760	19.1945
2.05	65.9085	47.7818	36.1555	28.2629	22.6638
2.10	80.5254	57.5583	43.0134	33.2498	26.3932
2.15	96.8878	68.3423	50.4773	38.6103	30.3552
2.20	114.999	80.1001	58.5034	44.3003	34.5090
2.25	134.801	92.7562	67.0181	50.2539	38.7975
2.30	156.158	106.183	75.9113	56.3794	43.1450
2.35	178.835	120.186	85.0287	62.5543	47.4538
2.40	202.469	134.492	94.1631	68.6196	51.6004
2.45	226.542	148.731	103.045	74.3744	55.4321
2.50	250.344	162.418	111.330	79.5690	58.7634
2.55	272.934	174.930	118.590	83.8981	61.3710
2.60	293.097	185.485	124.298	86.9942	62.9907
2.65	309.290	193.116	127.817	88.4193	63.3124
2.70	319.590	196.644	128.382	87.6583	61.9766
2.75	321.634	194.650	125.093	84.1115	58.5708
2.80	312.551	185.446	116.892	77.0881	52.6262
2.85	288.901	167.043	102.559	65.8006	43.6160
2.90	246.603	137.128	80.6947	49.3610	30.9543
2.95	180.873	93.040	49.7155	26.7786	13.9967
3.00	86.1588	31.743	7.84701	−3.03896	−7.95641
3.05	−43.9056	−50.176	−46.8736	−41.2803	−35.6543
3.10	−216.538	−156.515	−116.590	−89.2205	−69.8866
3.15	−439.819	−291.440	−203.607	−148.205	−111.470
3.20	−722.667	−459.447	−310.362	−219.625	−161.231
3.25	−1,074.77	−665.309	−439.377	−304.882	−219.977
3.30	−1,506.47	−913.974	−593.192	−405.347	−288.468
3.35	−2,028.53	−1,210.43	−774.278	−522.291	−367.376
3.40	−2,651.85	−1,559.53	−984.916	−656.820	−457.230
3.45	−3,387.03	−1,965.72	−1,227.05	−809.769	−558.357

Table 3.8 (continued)

			t		
v	0.30	0.35	0.40	0.45	0.50
3.50	−4,243.78	−2,432.71	−1,502.07	−981.588	−670.802
3.55	−5,230.14	−2,963.08	−1,810.62	−1,172.20	−794.244
3.60	−6,351.47	−3,557.74	−2,152.25	−1,380.80	−927.879
3.65	−7,609.18	−4,215.26	−2,525.09	−1,605.71	−1,070.30
3.70	−8,999.13	−4,931.08	−2,925.40	−1,844.06	−1,219.36
3.75	−10,509.6	−5,696.56	−3,347.09	−2,091.55	−1,371.98
3.80	−12,118.9	−6,497.80	−3,781.09	−2,342.10	−1,523.99
3.85	−13,792.4	−7,314.28	−4,214.66	−2,587.50	−1,669.89
3.90	−15,479.1	−8,117.28	−4,630.66	−2,816.98	−1,802.65
3.95	−17,107.5	−8,868.03	−5,006.56	−3,016.73	−1,913.40
4.00	−18,580.6	−9,515.61	−5,313.51	−3,169.40	−1,991.23
4.05	−19,770.5	−9,994.58	−5,515.21	−3,253.58	−2,022.85
4.10	−20,512.2	−10,222.4	−5,566.72	−3,243.15	−1,992.31
4.15	−20,596.3	−10,096.4	−5,413.20	−3,106.78	−1,880.71
4.20	−19,761.2	−9,490.92	−4,988.55	−2,807.22	−1,665.90
4.25	−17,684.9	−8,254.04	−4,214.14	−2,300.83	−1,322.25
4.30	−13,975.6	−6,204.21	−2,997.51	−1,536.96	−820.423
4.35	−8,162.83	−3,127.11	−1,231.18	−457.634	−127.264
4.40	311.806	1,227.21	1,208.24	1,002.81	794.259
4.45	12,103.1	7,146.75	4,460.71	2,917.70	1,984.93
4.50	27,969.3	14,959.1	8,682.75	5,367.80	3,488.96
4.55	48,777.1	25,031.8	14,046.9	8,440.63	5,353.50
4.60	75,503.0	37,770.7	20,740.2	12,229.2	7,627.72
4.65	1,09,229	53,616.1	28,961.2	16,830.4	10,361.6
4.70	1,51,134	73,035.6	38,915.6	22,341.7	13,604.2
4.75	2,02,470	96,512.6	50,809.8	28,858.0	17,401.5
4.80	2,64,533	1,24,529	64,841.7	36,466.2	21,793.0
4.85	3,38,610	1,57,544	81,188.8	45,238.3	26,808.5
4.90	4,25,914	1,95,956	99,992.0	55,223.6	32,462.8
4.95	5,27,481	2,40,069	121335	66,437.4	38,750.2
5.00	6,44,050	2,90,028	145218	78,847.9	45,637.0

			t		
v	0.55	0.60	0.65	0.70	0.75
0.00	−1.02840	−1.00130	−0.973759	−0.945895	−0.917825
0.05	−1.10895	−1.07629	−1.04330	−1.01015	−0.977009
0.10	−1.19675	−1.15707	−1.11740	−1.07796	−1.03889
0.15	−1.29184	−1.24358	−1.19595	−1.14913	−1.10323
0.20	−1.39418	−1.33567	−1.27871	−1.22339	−1.16973
0.25	−1.50364	−1.43310	−1.36537	−1.30037	−1.23800
0.30	−1.61991	−1.53548	−1.45549	−1.37962	−1.30755
0.35	−1.74256	−1.64230	−1.54852	−1.46054	−1.37780
0.40	−1.87097	−1.75288	−1.64373	−1.54242	−1.44803

Table 3.8 (continued)

			t		
v	0.55	0.60	0.65	0.70	0.75
0.45	−2.00433	−1.86637	−1.74028	−1.62441	−1.51743
0.50	−2.14158	−1.98169	−1.83711	−1.70550	−1.58504
0.55	−2.28139	−2.09756	−1.93297	−1.78450	−1.64973
0.60	−2.42214	−2.21240	−2.02640	−1.86006	−1.71025
0.65	−2.56185	−2.32439	−2.11569	−1.93059	−1.76516
0.70	−2.69818	−2.43135	−2.19887	−1.99432	−1.81284
0.75	−2.82835	−2.53079	−2.27369	−2.04920	−1.85148
0.80	−2.94913	−2.61981	−2.33757	−2.09299	−1.87907
0.85	−3.05675	−2.69513	−2.38764	−2.12314	−1.89340
0.90	−3.14692	−2.75301	−2.42067	−2.13687	−1.89203
0.95	−3.21471	−2.78926	−2.43306	−2.13108	−1.87234
1.00	−3.25457	−2.79919	−2.42085	−2.10241	−1.83145
1.05	−3.26027	−2.77760	−2.37969	−2.04722	−1.76630
1.10	−3.22486	−2.71877	−2.30484	−1.96158	−1.67363
1.15	−3.14063	−2.61644	−2.19119	−1.84128	−1.54997
1.20	−2.99914	−2.46382	−2.03325	−1.68189	−1.39172
1.25	−2.79117	−2.25360	−1.82518	−1.47872	−1.19511
1.30	−2.50678	−1.97799	−1.56086	−1.22692	−0.956308
1.35	−2.13528	−1.62875	−1.23388	−0.921501	−0.671420
1.40	−1.66537	−1.19727	−0.837650	−0.557418	−0.336576
1.45	−1.08517	−0.674647	−0.365486	−0.129633	0.052002
1.50	−0.382376	−0.051847	0.189283	0.366763	0.497905
1.55	0.455567	0.680178	0.833173	0.936437	1.00443
1.60	1.44125	1.53027	1.57237	1.58366	1.57445
1.65	2.58700	2.50683	2.41252	2.31213	2.21028
1.70	3.90455	3.61751	3.35849	3.12477	2.91347
1.75	5.40457	4.86886	4.41407	4.02350	3.68465
1.80	7.09619	6.26595	5.58163	5.00894	4.52330
1.85	8.98635	7.81180	6.86177	6.08011	5.42747
1.90	11.0790	9.50687	8.25280	7.23410	6.39356
1.95	13.3745	11.3484	9.75032	8.46564	7.41595
2.00	15.8683	13.3297	11.3466	9.76672	8.48672
2.05	18.5499	15.4391	13.0299	11.1261	9.59526
2.10	21.4017	17.6593	14.7840	12.5286	10.7279
2.15	24.3974	19.9662	16.5871	13.9550	11.8674
2.20	27.5006	22.3278	18.4112	15.3810	12.9928
2.25	30.6626	24.7028	20.2213	16.7766	14.0786
2.30	33.8210	27.0393	21.9744	18.1060	15.0945
2.35	36.8969	29.2735	23.6182	19.3262	16.0049
2.40	39.7935	31.3284	25.0906	20.3867	16.7686
2.45	42.3929	33.1115	26.3186	21.2290	17.3382
2.50	44.5541	34.5142	27.2171	21.7857	17.6596
2.55	46.1106	35.4098	27.6882	21.9800	17.6723

Table 3.8 (continued)

			t		
v	0.55	0.60	0.65	0.70	0.75
2.60	46.8676	35.6519	27.6200	21.7256	17.3084
2.65	46.6000	35.0737	26.8867	20.9258	16.4932
2.70	45.0501	33.4862	25.3474	19.4741	15.1446
2.75	41.9261	30.6783	22.8462	17.2536	13.1742
2.80	36.9006	26.4160	19.2125	14.1379	10.4869
2.85	29.6105	20.4427	14.2617	9.99189	6.98240
2.90	19.6572	12.4806	7.79608	4.67259	2.55614
2.95	6.60848	2.23259	-0.392819	-1.96857	-2.89925
3.00	-9.99802	-10.6147	-10.5215	-10.0836	-9.49172
3.05	-30.6496	-26.3861	-22.8107	-19.8245	-17.3272
3.10	-55.8508	-45.4116	-37.4800	-31.3397	-26.5064
3.15	-86.1127	-68.0180	-54.7419	-44.7680	-37.1206
3.20	-121.939	-94.5185	-74.7929	-60.2332	-49.2469
3.25	-163.807	-125.199	-97.8041	-77.8362	-62.9424
3.30	-212.145	-160.300	-123.908	-97.6457	-78.2376
3.35	-267.306	-200.001	-153.186	-119.688	-95.1280
3.40	-329.529	-244.389	-185.649	-143.933	-113.565
3.45	-398.898	-293.437	-221.217	-170.284	-133.448
3.50	-475.296	-346.965	-259.697	-198.554	-154.606
3.55	-558.339	-404.604	-300.759	-228.456	-176.794
3.60	-647.314	-465.748	-343.898	-259.576	-199.669
3.65	-741.096	-529.506	-388.410	-291.350	-222.783
3.70	-838.060	-594.642	-433.344	-323.045	-245.560
3.75	-935.977	-659.514	-477.472	-353.723	-267.278
3.80	-1,031.90	-722.000	-519.234	-382.222	-287.054
3.85	-1,122.06	-779.426	-556.698	-407.117	-303.822
3.90	-1,201.68	-828.484	-587.508	-426.696	-316.314
3.95	-1,264.88	-865.145	-608.833	-438.923	-323.040
4.00	-1,304.52	-884.575	-617.316	-441.416	-322.272
4.05	-1,312.02	-881.047	-609.026	-431.410	-312.029
4.10	-1,277.20	-847.859	-579.414	-405.740	-290.061
4.15	-1,188.18	-777.256	-523.273	-360.823	-253.847
4.20	-1,031.20	-660.370	-434.712	-292.642	-200.590
4.25	-790.506	-487.172	-307.141	-196.754	-127.227
4.30	-448.319	-246.460	-133.282	-68.3042	-30.4435
4.35	15.2300	74.1278	94.8040	97.9433	93.2896
4.40	622.029	488.04	385.646	307.544	247.673
4.45	1,395.86	1,009.69	748.261	566.289	436.510
4.50	2,362.12	1,654.23	1,191.99	880.086	663.613
4.55	3,547.44	2,437.30	1,726.29	1,254.79	932.690
4.60	4,979.04	3,374.54	2,360.41	1,696.02	1,247.19
4.65	6,683.97	4,481.09	3,103.06	2,208.85	1,610.10
4.70	8,687.94	5,770.83	3,961.87	2,797.50	2,023.76

Table 3.8 (continued)

			t		
v	0.55	0.60	0.65	0.70	0.75
4.75	11,013.9	7,255.49	4,942.81	3,464.94	2,489.51
4.80	13,680.4	8,943.49	6,049.48	4,212.39	3,007.42
4.85	16,698.8	10,838.5	7,282.13	5,038.71	3,575.87
4.90	20,071.2	12,937.8	8,636.67	5,939.73	4,191.10
4.95	23,786.4	15,230.3	10,103.4	6,907.44	4,846.66
5.00	27,816.4	17,693.8	11,665.4	7,929.06	5,532.83

			t		
v	0.80	0.85	0.90	0.95	1.0
0.00	−0.889648	−0.861455	−0.833325	−0.805330	−0.777536
0.05	−0.943993	−0.911203	−0.878722	−0.846622	−0.814963
0.10	−1.00031	−0.962310	−0.924962	−0.888318	−0.852422
0.15	−1.05834	−1.01450	−0.971756	−0.930125	−0.889623
0.20	−1.11775	−1.06744	−1.01876	−0.971701	−0.926227
0.25	−1.17815	−1.12070	−1.06557	−1.01265	−0.961847
0.30	−1.23903	−1.17382	−1.11170	−1.05250	−0.996039
0.35	−1.29983	−1.22622	−1.15662	−1.09074	−1.02831
0.40	−1.35984	−1.27724	−1.19969	−1.12677	−1.05810
0.45	−1.41830	−1.32613	−1.24021	−1.15994	−1.08479
0.50	−1.47428	−1.37205	−1.27739	−1.18950	−1.10770
0.55	−1.52675	−1.41404	−1.31034	−1.21464	−1.12609
0.60	−1.57455	−1.45101	−1.33808	−1.23448	−1.13915
0.65	−1.61636	−1.48179	−1.35953	−1.24803	−1.14599
0.70	−1.65072	−1.50506	−1.37352	−1.25424	−1.14568
0.75	−1.67603	−1.51938	−1.37876	−1.25197	−1.13720
0.80	−1.69050	−1.52318	−1.37387	−1.24000	−1.11948
0.85	−1.69219	−1.51476	−1.35737	−1.21704	−1.09140
0.90	−1.67901	−1.49231	−1.32767	−1.18172	−1.05176
0.95	−1.64867	−1.45387	−1.28311	−1.13261	−0.999338
1.00	−1.59876	−1.39737	−1.22192	−1.06820	−0.932872
1.05	−1.52668	−1.32064	−1.14227	−0.986963	−0.851070
1.10	−1.42972	−1.22141	−1.04228	−0.887331	−0.752632
1.15	−1.30502	−1.09734	−0.920020	−0.767719	−0.636264
1.20	−1.14966	−0.946038	−0.773542	−0.626556	−0.500702
1.25	−0.960607	−0.765078	−0.600915	−0.462303	−0.344732
1.30	−0.734823	−0.552060	−0.400249	−0.273489	−0.167223
1.35	−0.469279	−0.304636	−0.169742	−0.058744	0.032846
1.40	−0.161021	−0.020569	0.092281	0.183157	0.256345
1.45	0.192765	0.302209	0.387308	0.453249	0.503955
1.50	0.594672	0.665526	0.716579	0.752324	0.776117
1.55	1.04697	1.07089	1.08102	1.08086	1.07298
1.60	1.55152	1.51939	1.48116	1.43898	1.39436

Table 3.8 (continued)

			t		
v	0.80	0.85	0.90	0.95	1.0
1.65	2.10960	2.01159	1.91704	1.82634	1.73963
1.70	2.72183	2.54743	2.38812	2.24208	2.10772
1.75	3.38797	3.12607	2.89317	2.68470	2.49698
1.80	4.10672	3.74574	3.43014	3.15201	2.90515
1.85	4.87557	4.40362	3.99605	3.64098	3.32924
1.90	5.69057	5.09562	4.58681	4.14764	3.76544
1.95	6.54605	5.81620	5.19715	4.66700	4.20906
2.00	7.43441	6.55820	5.82037	5.19287	4.65443
2.05	8.34582	7.31259	6.44825	5.71777	5.09476
2.10	9.26794	8.06830	7.07085	6.23281	5.52210
2.15	10.1856	8.81190	7.67634	6.72755	5.92723
2.20	11.0804	9.52748	8.25084	7.18987	6.29954
2.25	11.9307	10.1963	8.77825	7.60587	6.62703
2.30	12.7109	10.7968	9.24013	7.95979	6.89615
2.35	13.3912	11.3040	9.61549	8.23387	7.09181
2.40	13.9378	11.6897	9.88076	8.40832	7.19735
2.45	14.3118	11.9220	10.0096	8.46129	7.19450
2.50	14.4698	11.9656	9.97312	8.36884	7.06345
2.55	14.3631	11.7813	9.73940	8.10504	6.78288
2.60	13.9383	11.3263	9.27406	7.64199	6.33013
2.65	13.1365	10.5542	8.54014	6.95008	5.68128
2.70	11.8943	9.41535	7.49841	5.99814	4.81149
2.75	10.1437	7.85692	6.10771	4.75385	3.69524
2.80	7.81259	5.82377	4.32543	3.18414	2.30672
2.85	4.82569	3.25892	2.10817	1.25574	0.620385
2.90	1.10555	0.104548	−0.587515	−1.06410	−1.38853
2.95	−3.42609	−3.69685	−3.80406	−3.80677	−3.74322
3.00	−8.84661	−8.20101	−7.58172	−7.00130	−6.46451
3.05	−15.2303	−13.4601	−11.9571	−10.6731	−9.56983
3.10	−22.6456	−19.5206	−16.9613	−14.8427	−13.0721
3.15	−31.1521	−26.4208	−22.6182	−19.5240	−16.9783
3.20	−40.7968	−34.1878	−28.9416	−24.7222	−21.2882
3.25	−51.6094	−42.8339	−35.9333	−30.4321	−25.9925
3.30	−63.5972	−52.3529	−43.5790	−36.6355	−31.0708
3.35	−76.7393	−62.7156	−51.8458	−43.2985	−36.4902
3.40	−90.9798	−73.8643	−60.6780	−50.3686	−42.2022
3.45	−106.220	−85.7078	−69.9926	−57.7719	−48.1410
3.50	−122.311	−98.1147	−79.6754	−65.4089	−54.2204
3.55	−139.041	−110.907	−89.5754	−73.1515	−60.3316
3.60	−156.131	−123.852	−99.4998	−80.8391	−66.3400
3.65	−173.217	−136.657	−109.209	−88.2744	−72.0824
3.70	−189.843	−148.956	−118.409	−95.2194	−77.3644

Table 3.8 (continued)

			t		
v	0.80	0.85	0.90	0.95	1.0
3.75	−205.450	−160.309	−126.748	−101.392	−81.9574
3.80	−219.355	−170.184	−133.810	−106.460	−85.5962
3.85	−230.750	−177.959	−139.108	−110.042	−87.9765
3.90	−238.680	−182.907	−142.082	−111.700	−88.7530
3.95	−242.037	−184.189	−142.090	−110.937	−87.5385
4.00	−239.546	−180.852	−138.410	−107.199	−83.9031
4.05	−229.761	−171.823	−130.234	−99.8730	−77.3746
4.10	−211.055	−155.907	−116.672	−88.2856	−67.4406
4.15	−181.620	−131.787	−96.7515	−71.7103	−53.5513
4.20	−139.471	−98.0306	−69.4242	−49.3702	−35.1252
4.25	−82.4547	−53.0998	−33.5754	−20.4478	−11.5558
4.30	−8.26809	4.63386	11.9624	15.9044	17.7782
4.35	85.5161	76.8661	68.3905	60.5473	53.5001
4.40	201.409	165.329	136.919	114.336	96.2186
4.45	341.954	271.747	218.733	178.092	146.508
4.50	509.656	397.790	314.955	252.574	204.884
4.55	706.894	544.999	426.590	338.440	271.771
4.60	935.809	714.711	554.468	436.197	347.473
4.65	1,198.17	907.954	699.170	546.152	432.126
4.70	1,495.21	1,125.34	860.942	668.344	525.657
4.75	1,827.42	1,366.90	1,039.59	802.474	627.723
4.80	2,194.35	1,631.96	1,234.39	947.820	737.658
4.85	2,594.29	1,918.94	1,443.90	1,103.14	854.399
4.90	3,024.01	2,225.10	1,665.87	1,266.59	976.411
4.95	3,478.38	2,546.37	1,897.06	1,435.55	1,101.61
5.00	3,949.97	2,877.04	2,133.03	1,606.57	1,227.27

			t		
v	1.5	2.0	3.0	4.0	5.0
0.00	−0.520555	−0.317929	−0.08030	0.003453	0.017573
0.05	−0.530716	−0.315514	−0.07304	0.007641	0.018945
0.10	−0.539387	−0.311721	−0.06527	0.011859	0.020235
0.15	−0.546378	−0.306470	−0.05701	0.016089	0.021436
0.20	−0.551487	−0.299679	−0.048272	0.020312	0.022538
0.25	−0.554499	−0.291267	−0.039077	0.024508	0.023536
0.30	−0.555189	−0.281154	−0.029442	0.028656	0.024419
0.35	−0.553320	−0.269266	−0.019393	0.032735	0.025182
0.40	−0.548646	−0.255528	−0.008959	0.036720	0.025817
0.45	−0.540914	−0.239874	0.001829	0.040589	0.026316
0.50	−0.529862	−0.222243	0.012933	0.044319	0.026673
0.55	−0.515226	−0.202580	0.024311	0.047883	0.026883
0.60	−0.496737	−0.180842	0.035918	0.051257	0.026939
0.65	−0.474127	−0.156993	0.047702	0.054415	0.026836

Table 3.8 (continued)

			t		
v	1.5	2.0	3.0	4.0	5.0
0.70	−0.447133	−0.131011	0.059608	0.057331	0.026571
0.75	−0.415495	−0.102889	0.071574	0.059980	0.026138
0.80	−0.378966	−0.072633	0.083535	0.062334	0.025536
0.85	−0.337313	−0.040270	0.095420	0.064370	0.024761
0.90	−0.290321	−0.005843	0.107153	0.066060	0.023813
0.95	−0.237800	0.030582	0.118654	0.067380	0.022692
1.00	−0.179591	0.068913	0.129838	0.068306	0.021397
1.05	−0.115570	0.109035	0.140613	0.068815	0.019930
1.10	−0.045658	0.150805	0.150886	0.068884	0.018293
1.15	0.030174	0.194049	0.160560	0.068492	0.016492
1.20	0.111897	0.238560	0.169530	0.067622	0.014530
1.25	0.199413	0.284098	0.177694	0.066255	0.012414
1.30	0.292551	0.330388	0.184942	0.064376	0.010151
1.35	0.391053	0.377115	0.191166	0.061974	0.007751
1.40	0.494566	0.423925	0.196253	0.059038	0.005222
1.45	0.602631	0.470424	0.200092	0.055561	0.002577
1.50	0.714676	0.516175	0.202572	0.051540	−0.000171
1.55	0.830003	0.560700	0.203583	0.046974	−0.003009
1.60	0.947776	0.603475	0.203016	0.041869	−0.005921
1.65	1.06701	0.643936	0.200768	0.036231	−0.008890
1.70	1.18658	0.681474	0.196738	0.030072	−0.011898
1.75	1.30517	0.715441	0.190833	0.023411	−0.014926
1.80	1.42132	0.745147	0.182968	0.016268	−0.017953
1.85	1.53335	0.769868	0.173065	0.008671	−0.020957
1.90	1.63943	0.788844	0.161059	0.000652	−0.023915
1.95	1.73753	0.801287	0.146895	−0.007752	−0.026804
2.00	1.82541	0.806386	0.130535	−0.016496	−0.029599
2.05	1.90064	0.803312	0.111954	−0.025532	−0.032275
2.10	1.96063	0.791227	0.091148	−0.034804	−0.034805
2.15	2.00258	0.769292	0.068131	−0.044252	−0.037163
2.20	2.02351	0.736679	0.042940	−0.053809	−0.039323
2.25	2.02031	0.692580	0.015635	−0.063402	−0.041256
2.30	1.98973	0.636221	−0.013699	−0.072952	−0.042938
2.35	1.92839	0.566880	−0.044947	−0.082377	−0.044340
2.40	1.83287	0.483897	−0.077966	−0.091587	−0.045438
2.45	1.69969	0.386699	−0.112582	−0.100487	−0.046205
2.50	1.52540	0.274815	−0.148587	−0.108980	−0.046618
2.55	1.30661	0.147897	−0.185739	−0.116962	−0.046654
2.60	1.04008	0.005745	−0.223762	−0.124328	−0.046292
2.65	0.722776	−0.151671	−0.262340	−0.130968	−0.045512
2.70	0.351970	−0.324183	−0.301124	−0.136773	−0.044298
2.75	−0.074665	−0.511400	−0.339724	−0.141628	−0.042634
2.80	−0.558952	−0.712683	−0.377714	−0.145423	−0.040509

Table 3.8 (continued)

v	1.5	2.0	3.0	4.0	5.0
2.85	−1.10208	−0.927109	−0.414630	−0.148045	−0.037914
2.90	−1.70448	−1.15345	−0.449973	−0.149384	−0.034843
2.95	−2.36569	−1.39014	−0.483207	−0.149334	−0.031296
3.00	−3.08416	−1.63526	−0.513766	−0.147792	−0.027274
3.05	−3.85715	−1.88647	−0.541052	−0.144662	−0.022785
3.10	−4.68049	−2.14103	−0.564441	−0.139855	−0.017840
3.15	−5.54845	−2.39576	−0.583286	−0.133290	−0.012455
3.20	−6.45353	−2.64701	−0.596924	−0.124899	−0.006652
3.25	−7.38624	−2.89064	−0.604679	−0.114623	−0.000458
3.30	−8.33498	−3.12205	−0.605871	−0.102421	0.006095
3.35	−9.28575	−3.33609	−0.599822	−0.088264	0.012969
3.40	−10.2221	−3.52715	−0.585865	−0.072145	0.020120
3.45	−11.1247	−3.68911	−0.563357	−0.054072	0.027498
3.50	−11.9716	−3.81539	−0.531682	−0.034079	0.035048
3.55	−12.7377	−3.89895	−0.490272	−0.012220	0.042708
3.60	−13.3949	−3.93236	−0.438612	0.011424	0.050410
3.65	−13.9119	−3.90783	−0.376256	0.036746	0.058083
3.70	−14.2543	−3.81730	−0.302845	0.063611	0.065649
3.75	−14.3847	−3.65251	−0.218117	0.091855	0.073023
3.80	−14.2626	−3.40513	−0.121925	0.121281	0.080118
3.85	−13.8448	−3.06684	−0.014256	0.151661	0.086843
3.90	−13.0858	−2.62955	0.104755	0.182736	0.093102
3.95	−11.9381	−2.08548	0.234806	0.214212	0.098796
4.00	−10.3528	−1.42744	0.375409	0.245762	0.103824
4.05	−8.28035	−0.648983	0.525876	0.277027	0.108086
4.10	−5.67140	0.255300	0.685297	0.307616	0.111477
4.15	−2.47792	1.28955	0.852529	0.337104	0.113897
4.20	1.34557	2.45630	1.02617	0.365039	0.115245
4.25	5.84048	3.75622	1.20457	0.390942	0.115424
4.30	11.0424	5.18767	1.38578	0.414307	0.114341
4.35	16.9793	6.74644	1.56757	0.434609	0.111909
4.40	23.6690	8.42527	1.74741	0.451305	0.108046
4.45	31.1170	10.2135	1.92249	0.463839	0.102683
4.50	39.3136	12.0966	2.08967	0.471649	0.095756
4.55	48.2306	14.0559	2.24553	0.474170	0.087216
4.60	57.8188	16.0678	2.38637	0.470847	0.077028
4.65	68.0034	18.1040	2.50820	0.461135	0.065169
4.70	78.6814	20.1303	2.60678	0.444514	0.051637
4.75	89.7168	22.1069	2.67766	0.420493	0.036445
4.80	100.937	23.9878	2.71618	0.388625	0.019628
4.85	112.129	25.7205	2.71757	0.348513	0.001243
4.90	123.034	27.2461	2.67694	0.299825	−0.018630
4.95	133.346	28.4988	2.58937	0.242304	−0.039887
5.00	142.706	29.4062	2.44999	0.175778	−0.062398

Table 3.9: Second derivatives with respect to the order of the Kelvin functions $\partial^2 \text{ker}_\nu(t)/\partial \nu^2$.

			t		
ν	0.05	0.10	0.15	0.20	0.25
0.00	4.77108	0.455957	−0.93970	−1.51479	−1.75756
0.05	2.76323	−0.798829	−1.83251	−2.18769	−2.28119
0.10	0.48315	−2.21177	−2.82582	−2.92753	−2.85042
0.15	−2.16904	−3.81718	−3.93447	−3.74097	−3.46789
0.20	−5.30805	−5.65249	−5.17383	−4.63440	−4.13576
0.25	−9.06582	−7.75820	−6.55952	−5.61371	−4.85546
0.30	−13.5941	−10.1777	−8.10704	−6.68392	−5.62744
0.35	−19.0665	−12.9568	−9.83118	−7.84873	−6.45082
0.40	−25.6812	−16.1431	−11.7453	−9.11004	−7.32303
0.45	−33.6615	−19.7847	−13.8606	−10.467300	−8.23937
0.50	−43.2572	−23.9287	−16.1846	−11.9165	−9.19246
0.55	−54.7431	−28.6190	−18.7199	−13.4497	−10.1716
0.60	−68.4165	−33.8926	−21.4621	−15.0535	−11.1622
0.65	−84.5904	−39.7761	−24.3977	−16.7077	−12.1446
0.70	−103.584	−46.2796	−27.5011	−18.3838	−13.0938
0.75	−125.705	−53.3898	−30.7312	−20.0433	−13.9775
0.80	−151.227	−61.0603	−34.0271	−21.6352	−14.7558
0.85	−180.355	−69.2010	−37.3030	−23.0939	−15.3793
0.90	−213.177	−77.6625	−40.4425	−24.3360	−15.7877
0.95	−249.595	−86.2186	−43.2915	−25.2574	−15.9087
1.00	−289.241	−94.5443	−45.6500	−25.7295	−15.6556
1.05	−331.358	−102.189	−47.2627	−25.5954	−14.9259
1.10	−374.641	−108.543	−47.8079	−24.6653	−13.5994
1.15	−417.044	−112.800	−46.8859	−22.7122	−11.5362
1.20	−455.517	−113.912	−44.0046	−19.4669	−8.57492
1.25	−485.681	−110.533	−38.5649	−14.6125	−4.53100
1.30	−501.421	−100.958	−29.8441	−7.77988	0.804929
1.35	−494.370	−83.0524	−16.9783	1.45776	7.66810
1.40	−453.282	−54.1697	1.05559	13.5889	16.3201
1.45	−363.259	−11.0626	25.4580	29.1673	27.0489
1.50	−204.824	50.2154	57.6242	48.8146	40.1673
1.55	47.1924	134.414	99.1595	73.2198	56.0102
1.60	424.953	247.193	151.891	103.137	74.9305
1.65	969.131	395.230	217.873	139.380	97.2919
1.70	1,730.630	586.315	299.386	182.807	123.459
1.75	2,772.500	829.426	398.916	234.308	153.784
1.80	4,171.970	1,134.78	519.127	294.774	188.590
1.85	6,022.640	1,513.80	662.805	365.061	228.146
1.90	8,436.460	1,979.06	832.767	445.938	272.637
1.95	11,545.6	2,544.05	1,031.74	538.023	322.129
2.00	15,503.5	3,222.85	1,262.18	641.695	376.519
2.05	20,485.2	4,029.49	1,526.04	756.978	435.482
2.10	26,685.6	4,977.17	1,824.44	883.407	498.397

Table 3.9 (continued)

			t		
v	0.05	0.10	0.15	0.20	0.25
2.15	34,315.3	6,076.97	2,157.25	1,019.85	564.267
2.20	43,592.8	7,336.12	2,522.57	1,164.30	631.617
2.25	54,730.6	8,755.70	2,916.00	1,313.58	698.377
2.30	67,914.3	10,327.5	3,329.84	1,463.10	761.748
2.35	83,270.7	12,029.7	3,751.99	1,606.39	818.039
2.40	1,00,822	13,822.0	4,164.62	1,734.73	862.491
2.45	1,20,419	1,56,38.2	4,542.63	1,836.59	889.069
2.50	1,41,654	1,73,77.9	4,851.71	1,897.04	890.238
2.55	1,63,733	18,895.7	5,046.08	1,897.06	856.708
2.60	1,85,313	19,987.6	5,065.79	1,812.80	777.171
2.65	2,04,280	20,374.5	4,833.66	1,614.68	638.009
2.70	2,17,463	19,682.7	4,251.69	1,266.48	423.005
2.75	2,20,256	17,419.7	3,197.07	724.358	113.049
2.80	2,06,145	12,946.5	1,517.69	−64.2116	−314.135
2.85	1,66,101	5,444.58	−972.680	−1,161.44	−884.228
2.90	8,78,20.7	−6,121.54	−4,499.5	−2,640.20	−1,626.49
2.95	−4,52,04.3	−23,046.4	−9,332.7	−4,584.87	−2,573.82
3.00	−2,54,831	−46,929.9	−15,791.7	−7,091.93	−3,762.73
3.05	−5,69,634	−79,726.9	−24,249.8	−10,270.2	−5,233.03
3.10	−1.027E+6	−123797	−35,138.0	−14,240.3	−7,027.30
3.15	−1.673	−181955	−48,945.6	−19,134.0	−9,190.03
3.20	−2.567	−257509	−66,219.9	−25,091.5	−11,766.2
3.25	−3.784	−354293	−87,559.8	−32,257.7	−14,799.5
3.30	−5.415	−476669	−113605	−40,777.1	−18,329.2
3.35	−7.570	−629497	−145014	−50,785.3	−22,387.3
3.40	−1.038E+7	−818051	−182438	−62,397.8	−26,992.8
3.45	−1.400	−1.048E+6	−226473	−75,694.9	−32,146.7
3.50	−1.860	−1.324	−277592	−90,701.2	−37,823.0
3.55	−2.438	−1.653	−336067	−1,07,360	−43,960.1
3.60	−3.155	−2.037	−401841	−1,25,499	−50,447.8
3.65	−4.031	−2.479	−474377	−1,44,788	−57,113.5
3.70	−5.085	−2.978	−552451	−1,64,687	−63,703.4
3.75	−6.331	−3.528	−633895	−1,84,378	−69,862.1
3.80	−7.772	−4.115	−715264	−2,02,691	−75,107.1
3.85	−9.395	−4.717	−791420	−2,18,001	−78,799.2
3.90	−1.116E+8	−5.297	−855026	−2,28,120	−80,109.2
3.95	−1.299	−5.799	−895916	−2,30,160	−77,979.0
4.00	−1.473	−6.142	−900342	−2,20,376	−71,078.9
4.05	−1.617	−6.211	−850074	−1,93,990	−57,759.3
4.10	−1.695	−5.851	−721335	−1,44,988	−36,000.4
4.15	−1.656	−4.849	−483575	−6,59,00	−3,357.9
4.20	−1.426	−2.926	−98,062.0	5,24,44	43,091.5
4.25	−8.995E+7	2,84,903	4,83,690	2,21,163	1,06,802

Table 3.9 (continued)

v	0.05	0.10	0.15	0.20	0.25
4.30	6.786E+6	5.258E+6	1.322E+6	4,53,583	1,91,806
4.35	1.674E+8	1.259E+7	2.489	7,65,473	3,02,751
4.40	4.185	2.305	4.074	1.175E+6	4,44,912
4.45	7.953	3.758	6.183	1.704	6,24,188
4.50	1.344E+9	5.732	8.938	2.376	8,47,051
4.55	2.125	8.369	1.248E+7	3.218	1.120E+6
4.60	3.213	1.183E+8	1.698	4.259	1.452
4.65	4.706	1.633	2.262	5.531	1.848
4.70	6.722	2.207	2.958	7.063	2.316
4.75	9.409	2.932	3.808	8.888	2.862
4.80	1.294E+10	3.836	4.833	1.103E+7	3.489
4.85	1.753	4.948	6.052	1.352	4.201
4.90	2.340	6.299	7.480	1.635	4.993
4.95	3.083	7.914	9.128	1.952	5.858
5.00	4.009	9.815	1.099E+8	2.300	6.780

v	0.30	0.35	0.40	0.45	0.50
0.00	−1.84091	−1.84039	−1.79356	−1.72075	−1.63365
0.05	−2.25637	−2.17402	−2.06338	−1.93969	−1.81134
0.10	−2.70306	−2.52883	−2.34717	−2.16733	−1.99384
0.15	−3.18151	−2.90421	−2.64369	−2.40212	−2.17944
0.20	−3.69171	−3.29901	−2.95122	−2.64206	−2.36607
0.25	−4.23294	−3.71143	−3.26746	−2.88466	−2.55123
0.30	−4.80350	−4.13883	−3.58939	−3.12684	−2.73190
0.35	−5.40056	−4.57758	−3.91317	−3.36480	−2.90452
0.40	−6.01983	−5.02285	−4.23396	−3.59398	−3.06486
0.45	−6.65527	−5.46840	−4.54580	−3.80889	−3.20799
0.50	−7.29870	−5.90632	−4.84144	−4.00303	−3.32817
0.55	−7.93945	−6.32678	−5.11213	−4.16877	−3.41877
0.60	−8.56385	−6.71770	−5.34745	−4.29718	−3.47223
0.65	−9.15474	−7.06446	−5.53507	−4.37795	−3.47989
0.70	−9.69091	−7.34952	−5.66058	−4.39921	−3.43200
0.75	−10.1464	−7.55207	−5.70721	−4.34744	−3.31761
0.80	−10.4901	−7.64760	−5.65565	−4.20730	−3.12447
0.85	−10.6845	−7.60752	−5.48378	−3.96154	−2.83905
0.90	−10.6854	−7.39873	−5.16649	−3.59091	−2.44645
0.95	−10.4408	−6.98321	−4.67546	−3.07403	−1.93039
1.00	−9.89038	−6.31761	−3.97901	−2.38740	−1.27328
1.05	−8.96424	−5.35291	−3.04194	−1.50538	−0.456229
1.10	−7.58243	−4.03409	−1.82549	−0.40023	0.540846
1.15	−5.65404	−2.29991	−0.28738	0.95779	1.73902
1.20	−3.07667	−0.082819	1.61813	2.60013	3.16017

Table 3.9 (continued)

			t		
v	0.30	0.35	0.40	0.45	0.50
1.25	0.26410	2.69101	3.93995	4.55972	4.82662
1.30	4.49486	6.10138	6.72974	6.87049	6.76072
1.35	9.75429	10.2336	10.0413	9.56675	8.98435
1.40	16.1925	15.1774	13.9295	12.6824	11.5182
1.45	23.9700	21.0261	18.4497	16.2500	14.3809
1.50	33.2556	27.8741	23.6552	20.2994	17.5882
1.55	44.2237	35.8151	29.5963	24.8562	21.1514
1.60	57.0499	44.9384	36.3171	29.9398	25.0763
1.65	71.9061	55.3248	43.8526	35.5614	29.3612
1.70	88.9528	67.0419	52.2251	41.7209	33.9949
1.75	108.330	80.1368	61.4397	48.4038	38.9546
1.80	130.145	94.6292	71.4787	55.5775	44.2032
1.85	154.459	110.501	82.2952	63.1871	49.6860
1.90	181.268	127.684	93.8058	71.1502	55.3277
1.95	210.478	146.051	105.882	79.3511	61.0284
2.00	241.885	165.392	118.340	87.6348	66.6595
2.05	275.137	185.402	130.929	95.7997	72.0589
2.10	309.700	205.658	143.321	103.590	77.0265
2.15	344.814	225.593	155.092	110.685	81.3183
2.20	379.438	244.470	165.709	116.695	84.6414
2.25	412.200	261.351	174.514	121.145	86.6480
2.30	441.324	275.061	180.703	123.469	86.9293
2.35	464.557	284.154	183.307	122.998	85.0106
2.40	479.090	286.870	181.174	118.948	80.3448
2.45	481.466	281.093	172.945	110.413	72.3088
2.50	467.482	264.311	157.035	96.3540	60.1987
2.55	432.092	233.567	131.615	75.5899	43.2280
2.60	369.296	185.420	94.5942	46.7949	20.5267
2.65	272.041	115.906	43.6092	8.49460	−8.85656
2.70	132.118	20.5043	−23.9859	−40.9314	−45.9493
2.75	−59.9218	−105.881	−111.117	−103.242	−91.8469
2.80	−314.845	−268.930	−220.977	−180.318	−147.693
2.85	−644.756	−474.892	−357.005	−274.139	−214.668
2.90	−1,063.09	−730.544	−522.838	−386.747	−293.951
2.95	−1,584.61	−1,043.11	−722.245	−520.194	−386.687
3.00	−2,225.16	−1,420.13	−959.038	−676.473	−493.937
3.05	−3,001.47	−1,869.30	−1,236.94	−857.435	−616.614
3.10	−3,930.78	−2,398.17	−1,559.41	−1,064.67	−755.411
3.15	−5,030.21	−3,013.82	−1,929.42	−1,299.36	−910.700
3.20	−6,316.06	−3,722.42	−2,349.21	−1,562.12	−1,082.42
3.25	−7,802.70	−4,528.60	−2,819.88	−1,852.77	−1,269.96
3.30	−9,501.22	−5,434.71	−3,341.00	−2,170.07	−1,471.96
3.35	−11,417.6	−6,439.89	−3,910.09	−2,511.43	−1,686.15

Table 3.9 (continued)

			t		
v	0.30	0.35	0.40	0.45	0.50
3.40	−13,550.7	−7,538.96	−4,521.96	−2,872.55	−1,909.13
3.45	−15,889.1	−8,720.94	−5,167.98	−3,246.96	−2,136.12
3.50	−18,407.9	−9,967.46	−5,835.23	−3,625.59	−2,360.67
3.55	−21,064.6	−11,250.7	−6,505.41	−3,996.16	−2,574.36
3.60	−23,793.9	−12,531.1	−7,153.73	−4,342.59	−2,766.43
3.65	−26,501.6	−13,754.7	−7,747.55	−4,644.30	−2,923.42
3.70	−29,058.0	−14,849.8	−8,244.86	−4,875.41	−3,028.75
3.75	−31,289.4	−15,723.5	−8,592.63	−5,003.95	−3,062.26
3.80	−32,968.4	−16,258.1	−8,724.99	−4,990.92	−2,999.82
3.85	−33,803.8	−16,306.0	−8,561.24	−4,789.46	−2,812.84
3.90	−33,428.2	−15,685.5	−8,003.86	−4,343.80	−2,467.83
3.95	−31,384.8	−14,175.6	−6,936.34	−3,588.45	−1,926.04
4.00	−27,113.5	−11,510.3	−5,221.10	−2,447.28	−1,143.07
4.05	−19,935.3	−7,374.12	−2,697.58	−832.871	−68.7088
4.10	−9,037.27	−1,396.17	819.51	1,354.07	1,353.22
4.15	6,542.66	6,853.85	5,541.47	4,224.86	3,184.70
4.20	27,930.0	17,869.5	11,706.2	7,902.69	5,493.14
4.25	56,425.2	32,209.4	19,577.9	12,521.8	8,350.64
4.30	93,511.7	50,497.5	29,445.3	18,226.0	11,832.8
4.35	140857	73,418.9	41,618.9	25,166.1	16,017.0
4.40	200308	101713	56,424.6	33,496.2	20,979.9
4.45	273869	136160	74,195.9	43,368.9	26,794.3
4.50	363668	177560	95,261.7	54,927.3	33,524.5
4.55	471901	226702	119930	68,296.6	41,221.4
4.60	600745	284324	148465	83,571.5	49,914.7
4.65	752244	351055	181062	100801	59,605.3
4.70	928137	427342	217805	119970	70,254.0
4.75	1.130E+6	513353	258626	140974	81,769.3
4.80	1.357	608856	303248	163594	93,992.2
4.85	1.610	713066	351113	187459	106678
4.90	1.886	824460	401301	212008	119477
4.95	2.180	940546	452428	236442	131909
5.00	2.484	1.058E+6	502530	259667	143335

			t		
v	0.55	0.60	0.65	0.70	0.75
0.00	−1.53928	−1.44201	−1.34456	−1.24871	−1.15554
0.05	−1.68310	−1.55771	−1.43677	−1.32115	−1.21129
0.10	−1.82881	−1.67317	−1.52712	−1.39055	−1.26313
0.15	−1.97470	−1.78666	−1.61397	−1.45535	−1.30962
0.20	−2.11871	−1.89621	−1.69545	−1.51383	−1.34916
0.25	−2.25840	−1.99954	−1.76945	−1.56403	−1.37997
0.30	−2.39095	−2.09401	−1.83354	−1.60376	−1.40009

Table 3.9 (continued)

			t		
v	0.55	0.60	0.65	0.70	0.75
0.35	−2.51305	−2.17661	−1.88502	−1.63059	−1.40734
0.40	−2.62087	−2.24391	−1.92083	−1.64181	−1.39933
0.45	−2.71002	−2.29204	−1.93757	−1.63444	−1.37348
0.50	−2.77548	−2.31661	−1.93142	−1.60521	−1.32694
0.55	−2.81154	−2.31273	−1.89822	−1.55052	−1.25667
0.60	−2.81176	−2.27495	−1.83333	−1.46649	−1.15938
0.65	−2.76894	−2.19723	−1.73173	−1.34890	−1.03157
0.70	−2.67502	−2.07295	−1.58794	−1.19325	−0.869520
0.75	−2.52111	−1.89485	−1.39606	−0.994737	−0.669326
0.80	−2.29740	−1.65509	−1.14977	−0.748276	−0.426918
0.85	−1.99322	−1.34518	−0.84237	−0.448554	−0.138093
0.90	−1.59697	−0.956097	−0.466781	−0.090059	0.201433
0.95	−1.09619	−0.478257	−0.015650	0.332858	0.595970
1.00	−0.477615	0.098374	0.518609	0.825907	1.04978
1.05	0.272761	0.784193	1.14370	1.39476	1.56700
1.10	1.16954	1.58983	1.86734	2.04495	2.15150
1.15	2.22774	2.52597	2.69706	2.78170	2.80681
1.20	3.46260	3.60316	3.64004	3.60977	3.53592
1.25	4.88921	4.83151	4.70289	4.53326	4.34115
1.30	6.52223	6.22039	5.89134	5.55538	5.22391
1.35	8.37534	7.77803	7.20992	6.67812	6.18451
1.40	10.4607	9.51106	8.66159	7.90203	7.22187
1.45	12.7884	11.4239	10.2473	9.225750	8.33325
1.50	15.3653	13.5183	11.9653	10.6457	9.51390
1.55	18.1945	15.7923	13.8110	12.1556	10.7567
1.60	21.2740	18.2395	15.7758	13.7457	12.0519
1.65	24.5953	20.8484	17.8466	15.4029	13.3864
1.70	28.1423	23.6007	20.0049	17.1093	14.7434
1.75	31.8893	26.4706	22.2263	18.8420	16.1022
1.80	35.7990	29.4231	24.4785	20.5723	17.4373
1.85	39.8209	32.4126	26.7215	22.2649	18.7177
1.90	43.8883	35.3814	28.9053	23.8772	19.9071
1.95	47.9165	38.2577	30.9697	25.3581	20.9622
2.00	51.7991	40.9540	32.8421	26.6476	21.8333
2.05	55.4058	43.3648	34.4369	27.6758	22.4628
2.10	58.5789	45.3648	35.6537	28.3618	22.7853
2.15	61.1298	46.8067	36.3766	28.6133	22.7267
2.20	62.8361	47.5196	36.4723	28.3258	22.2043
2.25	63.4378	47.3061	35.7896	27.3816	21.1260
2.30	62.6345	45.9416	34.1580	25.6504	19.3909
2.35	60.0817	43.1722	31.3874	22.9881	16.8888
2.40	55.3888	38.7134	27.2680	19.2377	13.5011
2.45	48.1170	32.2503	21.5697	14.2293	9.10115

Table 3.9 (continued)

			t		
ν	0.55	0.60	0.65	0.70	0.75
2.50	37.7777	23.4370	14.0440	7.78182	3.55553
2.55	23.8332	11.8981	4.42450	−0.296002	−3.27432
2.60	5.69747	−2.76906	−7.56996	−10.2029	−11.5301
2.65	−17.2597	−20.9907	−22.2320	−22.1420	−21.3537
2.70	−45.7073	−43.2110	−39.8608	−36.3166	−32.8838
2.75	−80.3440	−69.8842	−60.7554	−52.9248	−46.2516
2.80	−121.886	−101.465	−85.2066	−72.1527	−61.5751
2.85	−171.048	−138.394	−113.486	−94.1664	−78.9528
2.90	−228.527	−181.081	−145.835	−119.102	−98.4561
2.95	−294.965	−229.886	−182.444	−147.052	−120.120
3.00	−370.922	−285.090	−223.440	−178.056	−143.933
3.05	−456.827	−346.867	−268.860	−212.078	−169.824
3.10	−552.930	−415.245	−318.625	−248.992	−197.649
3.15	−659.230	−490.062	−372.509	−288.556	−227.177
3.20	−775.411	−570.915	−430.106	−330.393	−258.068
3.25	−900.743	−657.104	−490.788	−373.959	−289.858
3.30	−1,033.99	−747.558	−553.658	−418.511	−321.937
3.35	−1,173.26	−840.766	−617.504	−463.078	−353.522
3.40	−1,315.94	−934.686	−680.738	−506.419	−383.639
3.45	−1,458.46	−1,026.65	−741.341	−546.987	−411.090
3.50	−1,596.18	−1,113.26	−796.792	−582.886	−434.430
3.55	−1,723.19	−1,190.28	−844.002	−611.827	−451.940
3.60	−1,832.09	−1,252.52	−879.240	−631.085	−461.597
3.65	−1,913.81	−1,293.69	−898.063	−637.455	−461.052
3.70	−1,957.35	−1,306.30	−895.239	−627.210	−447.603
3.75	−1,949.57	−1,281.52	−864.679	−596.065	−418.178
3.80	−1,874.95	−1,209.04	−799.377	−539.145	−369.320
3.85	−1,715.36	−1,077.01	−691.355	−450.961	−297.182
3.90	−1,449.87	−871.897	−531.627	−325.406	−197.529
3.95	−1,054.57	−578.451	−310.190	−155.762	−65.760
4.00	−502.472	−179.682	−16.039	65.2714	103.065
4.05	236.546	343.113	362.778	345.513	314.162
4.10	1,195.67	1,010.24	839.048	693.206	572.950
4.15	2,410.94	1,843.42	1,426.23	1,116.88	884.933
4.20	3,920.91	2,865.45	2,138.19	1,625.17	1,255.560
4.25	5,766.01	4,099.80	2,988.91	2,226.53	1,690.040
4.30	7,987.69	5,569.92	3,991.95	2,928.95	2,193.070
4.35	10,627.2	7,298.45	5,159.95	3,739.50	2,768.610
4.40	13,724.1	9,306.13	6,503.85	4,663.87	3,419.440
4.45	17,313.9	11,610.4	8,031.95	5,705.68	4,146.780
4.50	21,425.6	14,223.7	9,748.84	6,865.78	4,949.770
4.55	26,078.1	17,151.3	11,654.0	8,141.33	5,824.86
4.60	31,276.6	20,389.0	13,740.1	9,524.75	6,765.10
4.65	37,006.7	23,919.5	15,991.4	11,002.5	7,759.38

Table 3.9 (continued)

v	0.55	0.60	0.65	0.70	0.75
4.70	43,229.3	27,709.7	18,381.1	12,553.7	8,791.44
4.75	49,872.5	31,705.4	20,869.0	14,148.4	9,838.95
4.80	56,824.1	35,827.4	23,398.7	15,746.0	10,872.3
4.85	63,921.1	39,965.2	25,894.1	17,293.1	11,853.4
4.90	70,939.0	43,971.3	28,255.8	18,721.3	12,734.3
4.95	77,578.8	47,653.7	30,357.0	19,944.9	13,455.8
5.00	83,453.2	50,768.3	32,039.4	20,858.5	13,946.0

v	0.80	0.85	0.90	0.95	1.0
0.00	−1.06577	−0.97981	−0.89789	−0.82011	−0.74649
0.05	−1.10736	−1.00938	−0.91723	−0.83076	−0.74976
0.10	−1.14444	−1.03401	−0.93139	−0.83609	−0.74765
0.15	−1.17566	−1.05251	−0.939244	−0.835073	−0.73926
0.20	−1.19958	−1.06351	−0.939593	−0.826638	−0.723616
0.25	−1.21457	−1.06559	−0.931135	−0.809621	−0.699681
0.30	−1.21888	−1.05716	−0.912471	−0.782782	−0.666366
0.35	−1.21057	−1.03651	−0.882104	−0.744810	−0.622525
0.40	−1.18755	−1.00183	−0.838438	−0.694319	−0.566960
0.45	−1.14757	−0.951157	−0.779784	−0.629855	−0.498426
0.50	−1.08820	−0.882400	−0.704361	−0.549905	−0.415640
0.55	−1.00682	−0.793358	−0.610305	−0.452898	−0.317287
0.60	−0.900688	−0.681713	−0.495682	−0.337225	−0.202039
0.65	−0.766882	−0.545053	−0.358498	−0.201252	−0.068562
0.70	−0.602354	−0.380879	−0.196720	−0.043336	0.084458
0.75	−0.403937	−0.186637	−0.008302	0.138153	0.258301
0.80	−0.168378	0.040257	0.208795	0.344798	0.454180
0.85	0.107626	0.302382	0.456560	0.578113	0.673218
0.90	0.427386	0.602267	0.736895	0.839496	0.916409
0.95	0.794159	0.942338	1.05157	1.13019	1.18458
1.00	1.21109	1.32486	1.40216	1.45124	1.47835
1.05	1.68111	1.75186	1.78999	1.80341	1.79807
1.10	2.20687	2.22504	2.21604	2.18717	2.14379
1.15	2.79061	2.74570	2.68091	2.60255	2.51515
1.20	3.43405	3.31459	3.18466	3.04913	2.91139
1.25	4.13821	3.93183	3.72674	3.52591	3.33117
1.30	4.90326	4.59674	4.30590	4.03123	3.77260
1.35	5.72836	5.30768	4.92000	4.56266	4.23307
1.40	6.61138	6.06191	5.56590	5.11686	4.70919
1.45	7.54875	6.85534	6.23932	5.68950	5.19669
1.50	8.53511	7.68240	6.93465	6.27510	5.69030
1.55	9.56311	8.53576	7.64480	6.86691	6.18368
1.60	10.6230	9.40610	8.36098	7.45674	6.66924

Table 3.9 (continued)

			t		
v	0.80	0.85	0.90	0.95	1.0
1.65	11.7025	10.2819	9.07252	8.03482	7.13813
1.70	12.7861	11.1490	9.76670	8.58969	7.58002
1.75	13.8550	11.9907	10.4285	9.10797	7.98306
1.80	14.8867	12.7870	11.0404	9.57432	8.33376
1.85	15.8544	13.5146	11.5822	9.97118	8.61692
1.90	16.7267	14.1468	12.0309	10.2788	8.81550
1.95	17.4673	14.6526	12.3603	10.4749	8.91061
2.00	18.0345	14.9974	12.5412	10.5349	8.88145
2.05	18.3809	15.1417	12.5410	10.4316	8.70532
2.10	18.4532	15.0418	12.3236	10.1353	8.35765
2.15	18.1916	14.6494	11.8497	9.61383	7.81208
2.20	17.5301	13.9116	11.0766	8.83268	7.04058
2.25	16.3963	12.7707	9.95851	7.75510	6.01368
2.30	14.7112	11.1650	8.44654	6.34248	4.70072
2.35	12.3901	9.028520	6.48940	4.55467	3.07022
2.40	9.34256	6.292090	4.03375	2.35049	1.09038
2.45	5.47332	2.883690	1.02495	−0.311601	−1.27039
2.50	0.683540	−1.270350	−2.59201	−3.47271	−4.04272
2.55	−5.12790	−6.244080	−6.87143	−7.17253	−7.25545
2.60	−12.0625	−12.1103	−11.8654	−11.4481	−10.9346
2.65	−20.2197	−18.9386	−17.6223	−16.3325	−15.1021
2.70	−29.6940	−26.7929	−24.1844	−21.8529	−19.7746
2.75	−40.5714	−35.7283	−31.5859	−28.0290	−24.9617
2.80	−52.9249	−45.7880	−39.8501	−34.8706	−30.6642
2.85	−66.8098	−56.9992	−48.9856	−42.3748	−36.8720
2.90	−82.2572	−69.3679	−58.9836	−50.5236	−43.5620
2.95	−99.2673	−82.8744	−69.8129	−59.2804	−50.6953
3.00	−117.802	−97.4665	−81.4159	−68.5862	−58.2147
3.05	−137.774	−113.053	−93.7027	−78.3561	−66.0416
3.10	−159.039	−129.496	−106.546	−88.4750	−74.0725
3.15	−181.382	−146.604	−119.776	−98.7924	−82.1761
3.20	−204.507	−164.117	−133.170	−109.118	−90.1894
3.25	−228.022	−181.707	−146.449	−119.218	−97.9138
3.30	−251.422	−198.958	−159.271	−128.806	−105.112
3.35	−274.078	−215.357	−171.218	−137.541	−111.503
3.40	−295.216	−230.289	−181.794	−145.021	−116.759
3.45	−313.901	−243.015	−190.414	−150.779	−120.505
3.50	−329.020	−252.671	−196.397	−154.275	−122.310
3.55	−339.264	−258.248	−198.962	−154.897	−121.690
3.60	−343.112	−258.590	−197.217	−151.951	−118.102
3.65	−338.817	−252.378	−190.158	−144.668	−110.948
3.70	−324.390	−238.131	−176.669	−132.193	−99.5721
3.75	−297.593	−214.197	−155.514	−113.597	−83.2662

Table 3.9 (continued)

			t		
v	0.80	0.85	0.90	0.95	1.0
3.80	−255.937	−178.757	−125.348	−87.8734	−61.2714
3.85	−196.678	−129.826	−84.7207	−53.9492	−32.7877
3.90	−116.833	−65.2720	−32.0869	−10.6957	3.01756
3.95	−13.1938	17.1711	34.1726	43.0565	46.9950
4.00	117.635	119.871	115.728	108.496	99.9964
4.05	279.192	245.260	214.267	186.803	162.852
4.10	475.091	395.778	331.453	279.117	236.343
4.15	708.936	573.806	468.867	386.490	321.167
4.20	984.207	781.581	627.947	509.841	417.902
4.25	1,304.12	1,021.09	809.903	649.887	526.951
4.30	1,671.45	1,293.92	1,015.62	807.076	648.493
4.35	2,088.31	1,601.15	1,245.55	981.498	782.418
4.40	2,555.92	1,943.12	1,499.55	1,172.79	928.248
4.45	3,074.26	2,319.22	1,776.78	1,380.00	1,085.06
4.50	3,641.78	2,727.66	2,075.46	1,601.51	1,251.39
4.55	4,254.91	3,165.19	2,392.73	1,834.83	1,425.13
4.60	4,907.64	3,626.73	2,724.36	2,076.50	1,603.41
4.65	5,590.97	4,105.07	3,064.58	2,321.85	1,782.49
4.70	6,292.32	4,590.40	3,405.71	2,564.86	1,957.61
4.75	6,994.84	5,069.94	3,737.96	2,797.96	2,122.85
4.80	7,676.71	5,527.39	4,049.04	3,011.80	2,271.03
4.85	8,310.33	5,942.50	4,323.88	3,195.01	2,393.51
4.90	8,861.50	6,290.47	4,544.26	3,334.05	2,480.07
4.95	9,288.55	6,541.46	4,688.48	3,412.92	2,518.79
5.00	9,541.43	6,660.04	4,731.05	3,413.01	2,495.92

			t		
v	1.5	2.0	3.0	4.0	5.0
0.00	−0.216923	0.03508	0.14026	0.08334	0.028190
0.05	−0.188930	0.06181	0.15038	0.08412	0.026644
0.10	−0.157272	0.09016	0.16031	0.08454	0.024934
0.15	−0.121693	0.12015	0.17000	0.08460	0.023060
0.20	−0.081946	0.151770	0.17938	0.08426	0.021026
0.25	−0.037793	0.184989	0.18838	0.08351	0.01884
0.30	0.010985	0.219764	0.196923	0.08234	0.01649
0.35	0.064591	0.256029	0.204937	0.08071	0.01400
0.40	0.123200	0.293697	0.212340	0.07863	0.01136
0.45	0.186964	0.332658	0.219047	0.07607	0.00859
0.50	0.255994	0.372774	0.224968	0.07302	0.00569
0.55	0.330363	0.413880	0.230012	0.069466	0.00267
0.60	0.410097	0.455783	0.234085	0.065406	−0.00045
0.65	0.495165	0.498255	0.237090	0.060831	−0.00367

Table 3.9 (continued)

ν	1.5	2.0	3.0	4.0	5.0
0.70	0.585473	0.541036	0.238928	0.055736	−0.00697
0.75	0.680856	0.583831	0.239500	0.050121	−0.010339
0.80	0.781070	0.626306	0.238706	0.043987	−0.013762
0.85	0.885778	0.668090	0.236446	0.037339	−0.017220
0.90	0.994545	0.708771	0.232621	0.030186	−0.020697
0.95	1.10683	0.747895	0.227135	0.022541	−0.024173
1.00	1.22196	0.784968	0.219894	0.014421	−0.027628
1.05	1.33914	0.819452	0.210810	0.005846	−0.031041
1.10	1.45744	0.850766	0.199799	−0.003157	−0.034390
1.15	1.57575	0.878289	0.186785	−0.012558	−0.037652
1.20	1.69284	0.901356	0.171698	−0.022322	−0.040802
1.25	1.80726	0.919265	0.154481	−0.032409	−0.043815
1.30	1.91741	0.931276	0.135086	−0.042771	−0.046665
1.35	2.02149	0.936613	0.113478	−0.053358	−0.049328
1.40	2.11749	0.934474	0.089639	−0.064113	−0.051774
1.45	2.20322	0.924029	0.063565	−0.074973	−0.053979
1.50	2.27627	0.904429	0.035271	−0.085870	−0.055914
1.55	2.33401	0.874813	0.004793	−0.096729	−0.057553
1.60	2.37364	0.834319	−0.027811	−0.107472	−0.058868
1.65	2.39211	0.782089	−0.062459	−0.118013	−0.059834
1.70	2.38622	0.717282	−0.099043	−0.128262	−0.060425
1.75	2.35257	0.639089	−0.137425	−0.138123	−0.060615
1.80	2.28760	0.546745	−0.177441	−0.147497	−0.060382
1.85	2.18764	0.439543	−0.218891	−0.156279	−0.059702
1.90	2.04888	0.316858	−0.261545	−0.164361	−0.058556
1.95	1.86748	0.178159	−0.305136	−0.171631	−0.056925
2.00	1.63956	0.023035	−0.349361	−0.177975	−0.054793
2.05	1.36128	−0.148785	−0.393879	−0.183276	−0.052145
2.10	1.02892	−0.337406	−0.438312	−0.187417	−0.048972
2.15	0.638922	−0.542741	−0.482242	−0.190280	−0.045265
2.20	0.187999	−0.764483	−0.525211	−0.191748	−0.041021
2.25	−0.326766	−1.00208	−0.566724	−0.191706	−0.036240
2.30	−0.907818	−1.25469	−0.606244	−0.190042	−0.030924
2.35	−1.55700	−1.52117	−0.643201	−0.186647	−0.025084
2.40	−2.27541	−1.80005	−0.676985	−0.181420	−0.018731
2.45	−3.06327	−2.08944	−0.706954	−0.174265	−0.011884
2.50	−3.91972	−2.38710	−0.732436	−0.165097	−0.004567
2.55	−4.84271	−2.69031	−0.752731	−0.153841	0.003192
2.60	−5.82871	−2.99589	−0.767117	−0.140432	0.011357
2.65	−6.87260	−3.30018	−0.774856	−0.124821	0.019889
2.70	−7.96742	−3.59899	−0.775195	−0.106976	0.028739
2.75	−9.10411	−3.88758	−0.767383	−0.086880	0.037856
2.80	−10.2714	−4.16065	−0.750668	−0.064537	0.047180

Table 3.9 (continued)

			t		
v	1.5	2.0	3.0	4.0	5.0
2.85	−11.4553	−4.41232	−0.724314	−0.039974	0.056645
2.90	−12.6393	−4.63616	−0.687609	−0.013241	0.066179
2.95	−13.8038	−4.82513	−0.639876	0.015588	0.075705
3.00	−14.9260	−4.97164	−0.580485	0.046409	0.085138
3.05	−15.9796	−5.06758	−0.508867	0.079091	0.094389
3.10	−16.9350	−5.10430	−0.424529	0.113470	0.103363
3.15	−17.7585	−5.07275	−0.327070	0.149350	0.111958
3.20	−18.4130	−4.96344	−0.216195	0.186502	0.120070
3.25	−18.8570	−4.76663	−0.091737	0.224662	0.127591
3.30	−19.0457	−4.47236	0.046328	0.263527	0.134407
3.35	−18.9302	−4.07060	0.197862	0.302761	0.140404
3.40	−18.4582	−3.55143	0.362543	0.341989	0.145465
3.45	−17.5738	−2.90515	0.539848	0.380800	0.149473
3.50	−16.2187	−2.12256	0.729030	0.418744	0.152310
3.55	−14.3321	−1.19512	0.929099	0.455339	0.153860
3.60	−11.8515	−0.115260	1.13881	0.490067	0.154010
3.65	−8.71425	1.12334	1.35662	0.522376	0.152651
3.70	−4.85778	2.52545	1.58071	0.551689	0.149678
3.75	−0.221632	4.09391	1.80892	0.577398	0.144995
3.80	5.25119	5.82923	2.03880	0.598876	0.138514
3.85	11.6123	7.72924	2.26752	0.615477	0.130157
3.90	18.9056	9.78857	2.49192	0.626541	0.119858
3.95	27.1652	11.9982	2.70849	0.631406	0.107565
4.00	36.4120	14.3450	2.91335	0.629408	0.093242
4.05	46.6510	16.8113	3.10229	0.619892	0.076871
4.10	57.8676	19.3740	3.27076	0.602222	0.058453
4.15	70.0239	22.0045	3.41386	0.575790	0.038010
4.20	83.0543	24.6681	3.52642	0.540023	0.015590
4.25	96.8613	27.3232	3.60298	0.494399	−0.008738
4.30	111.310	29.9211	3.63788	0.438459	−0.034875
4.35	126.225	32.4055	3.62526	0.371814	−0.062694
4.40	141.382	34.7122	3.55913	0.294166	−0.092038
4.45	156.506	36.7686	3.43349	0.205317	−0.122721
4.50	171.263	38.4939	3.24235	0.105187	−0.154521
4.55	185.256	39.7986	2.97984	−0.006174	−0.187185
4.60	198.022	40.5851	2.64034	−0.128570	−0.220427
4.65	209.022	40.7474	2.21858	−0.261641	−0.253925
4.70	217.645	40.1721	1.70980	−0.404852	−0.287324
4.75	223.196	38.7390	1.10982	−0.557475	−0.320237
4.80	224.903	36.3221	0.415319	−0.718577	−0.352242
4.85	221.910	32.7910	−0.376110	−0.887004	−0.382888
4.90	213.282	28.0126	−1.26570	−1.06137	−0.411695
4.95	198.006	21.8531	−2.25336	−1.24006	−0.438157
5.00	175.002	14.1808	−3.33743	−1.42119	−0.461743

3.4 The Kelvin Functions kei$_\nu(t)$, and Their First and Second Derivatives with Respect to the Order

The Kelvin functions kei$_\nu(t)$ (Tables 3.10–3.12) exist in MATHEMATICA program. Their first and second derivatives with respect to the order were determined directly using the central-difference formulas of order $O(h^4)$ with $h = 10^{-3}$. The functional form of these functions resembles those of ker$_\nu(t)$ functions (Figures 3.25–3.30):

$$\frac{\partial \text{kei}_\nu(t)}{\partial \nu} = \frac{-\text{kei}_{\nu+2h}(t) + 8\,\text{kei}_{\nu+h}(t) - 8\,\text{kei}_{\nu-h}(t) + \text{kei}_{\nu-2h}(t)}{12h} \tag{3.4.1}$$

$$\frac{\partial^2 \text{kei}_\nu(t)}{\partial \nu^2} = \frac{-\text{kei}_{\nu+2h}(t) + 16\,\text{kei}_{\nu+h}(t) - 30\,\text{kei}_\nu(t) + 16\,\text{kei}_{\nu-h}(t) - \text{kei}_{\nu-2h}(t)}{12h^2} \tag{3.4.2}$$

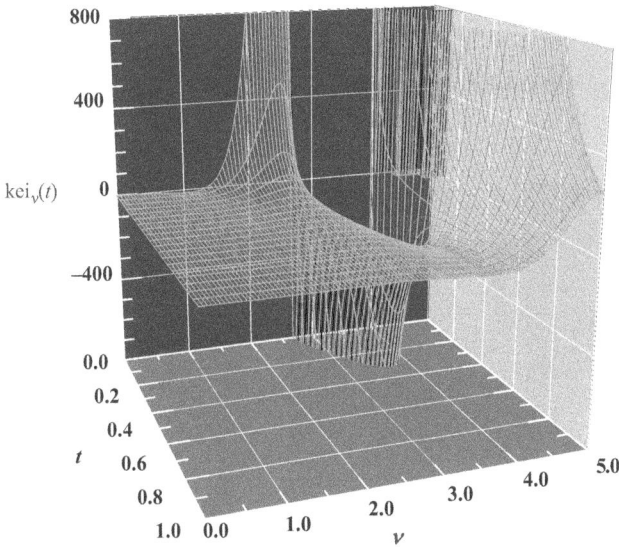

Figure 3.25: The Kelvin functions kei$_\nu(t)$ as a function of ν and t.

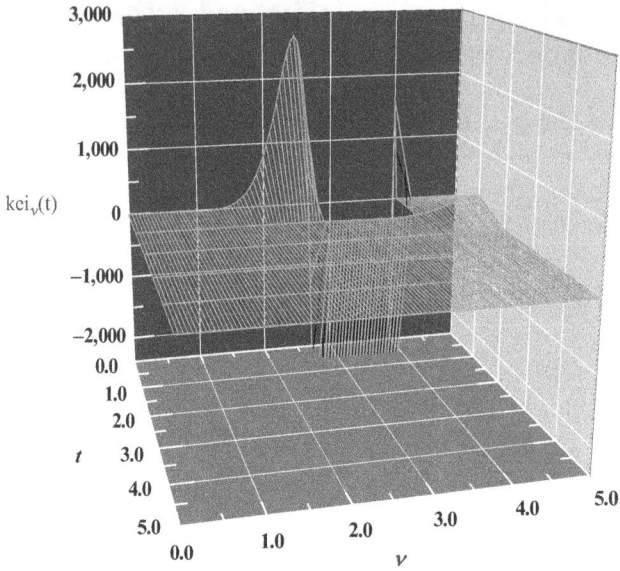

Figure 3.26: The Kelvin functions $\mathrm{kei}_\nu(t)$ as a function of ν and t.

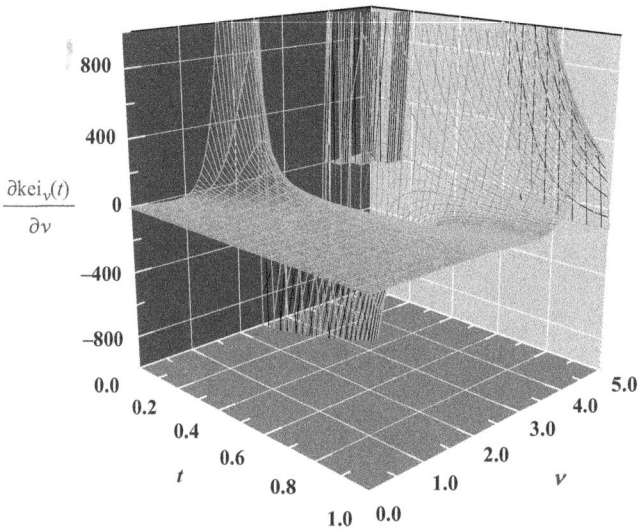

Figure 3.27: First derivatives with respect to the order of the Kelvin functions $\partial \mathrm{kei}_\nu(t)/\partial \nu$ as a function of ν and t.

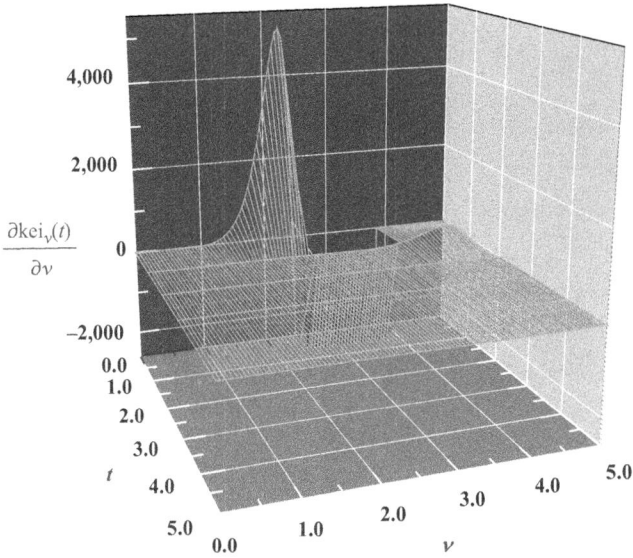

Figure 3.28: First derivatives with respect to the order of the Kelvin functions ∂kei$_v$(t)/∂v as a function of v and t.

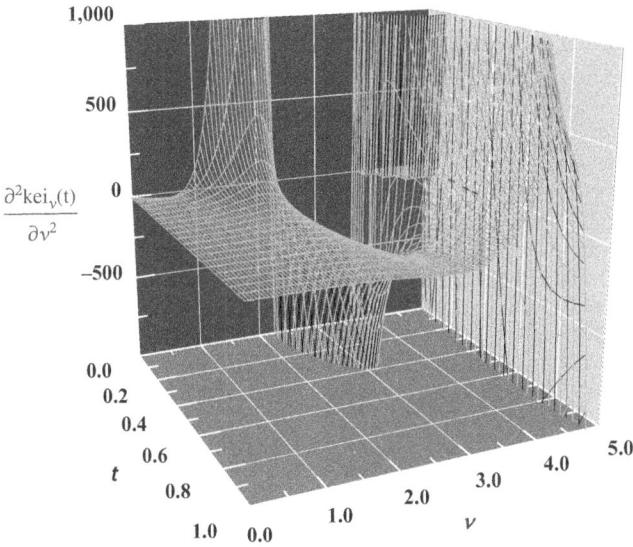

Figure 3.29: Second derivatives with respect to the order of the Kelvin functions ∂^2kei$_v$(t)/∂v^2 as a function of v and t.

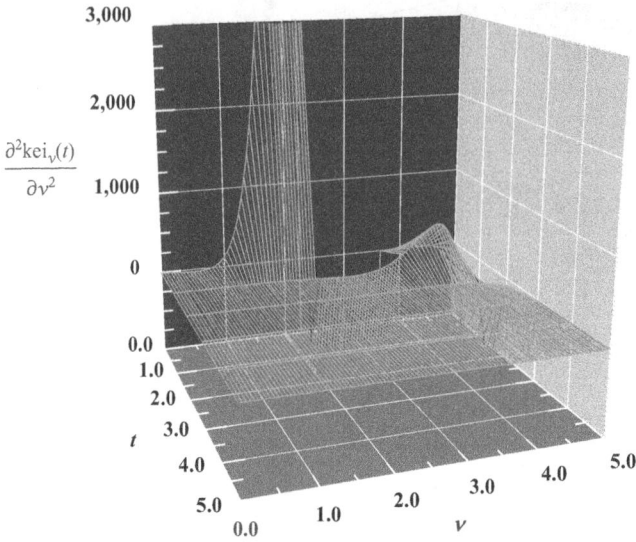

Figure 3.30: Second derivatives with respect to the order of the Kelvin functions $\partial^2 \mathrm{kei}_v(t)/\partial v^2$ as a function of v and t.

Table 3.10: The Kelvin functions $\mathrm{kei}_v(t)$.

			t		
v	0.05	0.10	0.15	0.20	0.25
0.00	−0.782828	−0.776851	−0.768444	−0.758125	−0.746253
0.05	−1.03671	−0.972081	−0.930035	−0.896216	−0.866382
0.10	−1.31340	−1.17924	−1.09877	−1.03870	−0.989120
0.15	−1.61881	−1.40072	−1.27579	−1.18613	−1.11469
0.20	−1.95898	−1.63876	−1.46208	−1.33888	−1.24316
0.25	−2.34002	−1.89539	−1.65835	−1.49711	−1.37441
0.30	−2.76813	−2.17235	−1.86507	−1.66076	−1.50811
0.35	−3.24943	−2.47103	−2.08233	−1.82944	−1.64370
0.40	−3.78982	−2.79233	−2.30984	−2.00245	−1.78033
0.45	−4.39469	−3.13651	−2.54676	−2.17867	−1.91684
0.50	−5.06852	−3.50298	−2.79164	−2.35651	−2.05170
0.55	−5.81433	−3.89005	−3.04225	−2.53380	−2.18294
0.60	−6.63296	−4.29464	−3.29540	−2.70774	−2.30814
0.65	−7.52207	−4.71186	−3.54680	−2.87472	−2.42431
0.70	−8.47488	−5.13458	−3.79078	−3.03028	−2.52785
0.75	−9.47849	−5.55286	−4.02009	−3.16890	−2.61445
0.80	−10.5118	−5.95329	−4.22556	−3.28390	−2.67902
0.85	−11.5429	−6.31825	−4.39578	−3.36726	−2.71563
0.90	−12.5255	−6.62494	−4.51674	−3.40944	−2.71738
0.95	−13.3953	−6.84439	−4.57146	−3.39919	−2.67632

Table 3.10 (continued)

ν	0.05	0.10	0.15	0.20	0.25
1.00	−14.0644	−6.94024	−4.53949	−3.32342	−2.58340
1.05	−14.4154	−6.86736	−4.39646	−3.16692	−2.42837
1.10	−14.2940	−6.57035	−4.11362	−2.91228	−2.19971
1.15	−13.4999	−5.98185	−3.65728	−2.53961	−1.88460
1.20	−11.7765	−5.02071	−2.98837	−2.02652	−1.46891
1.25	−8.79852	−3.59009	−2.06194	−1.34792	−0.937185
1.30	−4.15803	−1.57545	−0.82674	−0.47605	−0.272735
1.35	2.65166	1.15741	0.77506	0.61952	0.542244
1.40	12.2536	4.76415	2.80799	1.97168	1.52647
1.45	25.4063	9.42347	5.34324	3.61554	2.69928
1.50	43.0239	15.3381	8.45821	5.58786	4.08020
1.55	66.1958	22.7349	12.2357	7.926350	5.68838
1.60	96.2046	31.8643	16.7625	10.6686	7.54189
1.65	134.5410	42.9969	22.1271	13.8507	9.65670
1.70	182.9090	56.4195	28.4169	17.5054	12.0456
1.75	243.2230	72.4260	35.7137	21.6594	14.7164
1.80	317.5840	91.3059	44.0878	26.3306	17.6707
1.85	408.2260	113.327	53.5909	31.5241	20.9011
1.90	517.4350	138.709	64.2459	37.2273	24.3885
1.95	647.3990	167.593	76.0343	43.4040	28.0996
2.00	799.9990	199.996	88.8811	49.9876	31.9824
2.05	976.4910	235.754	102.635	56.8720	35.9623
2.10	1,177.07	274.445	117.044	63.9017	39.9369
2.15	1,400.24	315.298	131.726	70.8602	43.7704
2.20	1,641.99	357.072	146.138	77.4558	47.2870
2.25	1,894.69	397.906	159.528	83.3057	50.2637
2.30	2,145.55	435.142	170.890	87.9180	52.4222
2.35	2,374.77	465.098	178.908	90.6713	53.4207
2.40	2,552.97	482.804	181.888	90.7924	52.8439
2.45	2,638.10	481.682	177.685	87.3315	50.1940
2.50	2,571.33	453.177	163.620	79.1365	44.8806
2.55	2,272.19	386.321	136.388	64.8253	36.2112
2.60	1,632.33	267.236	91.9626	42.7587	23.3824
2.65	508.104	78.587	25.4900	11.0140	5.47212
2.70	−1,288.37	−201.036	−68.8123	−32.6387	−18.5653
2.75	−3,999.75	−597.762	−197.755	−90.7550	−49.9010
2.80	−7,935.70	−1,143.11	−369.261	−166.227	−89.8289
2.85	−13,486.7	−1,874.65	−592.436	−262.279	−139.756
2.90	−21,139.0	−2,836.58	−877.585	−382.454	−201.185
2.95	−31,491.1	−4,080.17	−1,236.19	−530.565	−275.685
3.00	−45,269.0	−5,663.92	−1,680.81	−710.624	−364.845
3.05	−63,340.8	−7,653.40	−2,224.82	−926.729	−470.216
3.10	−86,726.9	−10,120.4	−2,882.09	−1,182.89	−593.226

Table 3.10 (continued)

			t		
v	0.05	0.10	0.15	0.20	0.25
3.15	−1,16,603	−13,141.3	−3,666.39	−1,482.80	−735.065
3.20	−1,54,289	−16,794.2	−4,590.55	−1,829.51	−896.544
3.25	−2,01,222	−21,154.4	−5,665.29	−2,225.02	−1,077.91
3.30	−2,58,897	−26,287.9	−6,897.63	−2,669.70	−1,278.59
3.35	−3,28,764	−32,241.9	−8,288.77	−3,161.61	−1,496.96
3.40	−4,12,067	−39,031.2	−9,831.29	−3,695.63	−1,729.91
3.45	−5,09,598	−46,620.9	−11,505.6	−4,262.32	−1,972.50
3.50	−6,21,336	−54,901.8	−13,275.2	−4,846.58	−2,217.36
3.55	−7,45,927	−63,659.1	−15,081.2	−5,426.03	−2,454.18
3.60	−8,79,960	−72,531.6	−16,835.2	−5,969.01	−2,668.91
3.65	−1.017E+6	−80,959.1	−18,410.2	−6,432.25	−2,843.01
3.70	−1.146	−88,116.6	−19,630.4	−6,758.13	−2,952.48
3.75	−1.250	−92,831.6	−20,258.7	−6,871.48	−2,966.83
3.80	−1.304	−93,482.0	−19,981.4	−6,676.01	−2,847.90
3.85	−1.269	−87,872.6	−18,391.0	−6,050.17	−2,548.63
3.90	−1.094	−73,085.4	−14,966.3	−4,842.72	−2,011.68
3.95	−7,03,900	−45,302.6	−9,049.59	−2,867.82	−1,168.12
4.00	1,600.00	400.000	177.777	99.9992	63.9987
4.05	1.158E+6	70,290.3	13,722.8	4,331.02	1,780.88
4.10	2.946	1,72,246	32,808.6	10,145.3	4,094.06
4.15	5.602	3,16,044	58,902.3	17,916.7	7,131.10
4.20	9.430	5,13,670	93,741.2	28,076.7	11,035.8
4.25	1.482E+7	7,79,620	1,39,354	41,114.8	15,967.8
4.30	2.227	1.131E+6	1,98,071	57,577.8	22,100.4
4.35	3.238	1.589	2,72,528	78,063.6	29,618.2
4.40	4.591	2.176	3,65,637	1,03,210	38,711.1
4.45	6.376	2.918	4,80,537	1,33,677	49,567.7
4.50	8.700	3.846	6,20,501	1,70,114	62,363.6
4.55	1.169E+8	4.990	7,88,774	2,13,123	77,246.8
4.60	1.547	6.381	9,88,354	2,63,194	94,318.0
4.65	2.021	8.049	1.222E+6	3,20,628	1,13,604
4.70	2.604	1.002E+7	1.490	3,85,425	1,35,023
4.75	3.312	1.231	1.793	4,57,150	1,58,345
4.80	4.153	1.491	2.128	5,34,744	1,83,133
4.85	5.132	1.779	2.489	6,16,300	2,08,681
4.90	6.238	2.089	2.863	6,98,774	2,33,932
4.95	7.444	2.407	3.233	7,77,629	2,57,381
5.00	8.688	2.714	3.571	8,46,403	2,76,957

			t		
v	0.30	0.35	0.40	0.45	0.50
0.00	−0.733102	−0.718890	−0.703800	−0.687986	−0.671582
0.05	−0.838773	−0.812531	−0.787196	−0.762511	−0.738326

Table 3.10 (continued)

			t		
v	0.30	0.35	0.40	0.45	0.50
0.10	-0.945817	-0.906649	-0.870403	-0.836341	-0.803990
0.15	-1.05426	-1.00115	-0.953254	-0.909265	-0.868333
0.20	-1.16400	-1.09583	-1.03548	-0.980978	-0.931037
0.25	-1.27475	-1.19033	-1.11669	-1.05108	-0.991701
0.30	-1.38609	-1.28416	-1.19637	-1.11906	-1.04983
0.35	-1.49736	-1.37665	-1.27387	-1.18430	-1.10484
0.40	-1.60769	-1.46697	-1.34840	-1.24604	-1.15603
0.45	-1.71596	-1.55405	-1.41898	-1.30341	-1.20261
0.50	-1.82076	-1.63662	-1.48445	-1.35536	-1.24364
0.55	-1.92035	-1.71315	-1.54349	-1.40072	-1.27809
0.60	-2.01264	-1.78183	-1.59451	-1.43813	-1.30477
0.65	-2.09513	-1.84055	-1.63573	-1.46606	-1.32237
0.70	-2.16486	-1.88687	-1.66511	-1.48279	-1.32944
0.75	-2.21840	-1.91800	-1.68035	-1.48641	-1.32438
0.80	-2.25178	-1.93077	-1.67887	-1.47481	-1.30544
0.85	-2.26044	-1.92158	-1.65783	-1.44569	-1.27074
0.90	-2.23918	-1.88645	-1.61407	-1.39652	-1.21826
0.95	-2.18217	-1.82093	-1.54414	-1.32460	-1.14583
1.00	-2.08283	-1.72011	-1.44431	-1.22702	-1.05118
1.05	-1.93391	-1.57867	-1.31056	-1.10073	-0.931950
1.10	-1.72737	-1.39081	-1.13861	-0.942501	-0.785683
1.15	-1.45446	-1.15034	-0.923962	-0.749002	-0.609892
1.20	-1.10573	-0.850668	-0.661920	-0.516834	-0.402087
1.25	-0.671061	-0.484919	-0.347675	-0.242582	-0.159827
1.30	-0.139777	-0.045981	0.023623	0.077108	0.119217
1.35	0.499230	0.473351	0.456758	0.445439	0.437183
1.40	1.25731	1.08022	0.956311	0.865358	0.795934
1.45	2.14581	1.78152	1.52650	1.33942	1.19695
1.50	3.17573	2.58359	2.17099	1.86962	1.64119
1.55	4.35732	3.49197	2.89261	2.45724	2.12896
1.60	5.69954	4.51095	3.69314	3.10260	2.65977
1.65	7.20939	5.64314	4.57290	3.80482	3.23210
1.70	8.89115	6.88890	5.53039	4.56152	3.84320
1.75	10.7454	8.24571	6.56183	5.36852	4.48887
1.80	12.7677	9.70736	7.66063	6.21944	5.16318
1.85	14.9477	11.2631	8.81682	7.10531	5.85818
1.90	17.2669	12.8967	10.0163	8.01411	6.56360
1.95	19.6973	14.5850	11.2403	8.93026	7.26647
2.00	22.1989	16.2971	12.4641	9.83408	7.95078
2.05	24.7176	17.9924	13.6569	10.70120	8.59711
2.10	27.1822	19.6194	14.7799	11.5021	9.18222
2.15	29.5015	21.1137	15.7860	12.2010	9.67864
2.20	31.5610	22.3964	16.6184	12.7559	10.0543

Table 3.10 (continued)

v	0.30	0.35	0.40	0.45	0.50
2.25	33.2189	23.3717	17.2096	13.1175	10.2722
2.30	34.3025	23.9253	17.4802	13.2285	10.2898
2.35	34.6042	23.9221	17.3378	13.0234	10.0593
2.40	33.8768	23.2043	16.6762	12.4276	9.52681
2.45	31.8299	21.5893	15.3742	11.3576	8.63260
2.50	28.1255	18.8681	13.2954	9.72021	7.31112
2.55	22.3744	14.8043	10.2873	7.41307	5.49117
2.60	14.1337	9.13297	6.18223	4.32489	3.09638
2.65	2.90448	1.56102	0.797095	0.336220	0.04601
2.70	−11.8679	−8.23185	−6.06461	−4.67921	−3.74395
2.75	−30.7922	−20.5905	−14.6108	−10.8526	−8.35924
2.80	−54.5267	−35.8804	−25.0572	−18.3174	−13.8853
2.85	−83.7705	−54.4806	−37.6228	−27.2058	−20.4044
2.90	−119.250	−76.7751	−52.5231	−37.6439	−27.9927
2.95	−161.702	−103.140	−69.9617	−49.7463	−36.7157
3.00	−211.843	−133.927	−90.1196	−63.6081	−46.6230
3.05	−270.338	−169.441	−113.141	−79.2963	−57.7429
3.10	−337.754	−209.916	−139.118	−96.8390	−70.0740
3.15	−414.495	−255.477	−168.068	−116.212	−83.5777
3.20	−500.737	−306.100	−199.910	−137.323	−98.1671
3.25	−596.328	−361.565	−234.434	−159.995	−113.696
3.30	−700.676	−421.388	−271.269	−183.944	−129.947
3.35	−812.614	−484.757	−309.841	−208.754	−146.612
3.40	−930.237	−550.445	−349.325	−233.853	−163.283
3.45	−1,050.70	−616.709	−388.595	−258.481	−179.427
3.50	−1,170.01	−681.185	−426.164	−281.655	−194.369
3.55	−1,282.74	−740.757	−460.116	−302.134	−207.272
3.60	−1,381.75	−791.416	−488.034	−318.380	−217.113
3.65	−1,457.82	−828.106	−506.926	−328.515	−222.658
3.70	−1,499.33	−844.553	−513.136	−330.280	−222.446
3.75	−1,491.79	−833.088	−502.264	−320.990	−214.756
3.80	−1,417.44	−784.453	−469.076	−297.496	−197.598
3.85	−1,254.74	−687.622	−407.424	−256.144	−168.686
3.90	−977.988	−529.608	−310.170	−192.743	−125.432
3.95	−556.759	−295.306	−169.124	−102.548	−64.9392
4.00	44.4426	32.6505	24.9967	19.7489	15.9949
4.05	866.560	473.974	282.578	180.009	120.860
4.10	1,956.20	1,050.74	615.077	384.606	253.400
4.15	3,365.69	1,787.31	1,034.930	640.354	417.557
4.20	5,152.84	2,710.10	1,555.410	954.393	617.393
4.25	7,380.40	3,847.23	2,190.320	1,334.02	856.975
4.30	10,115.0	5,227.85	2,953.690	1,786.46	1,140.23
4.35	13,425.5	6,881.26	3,859.200	2,318.54	1,470.73

Table 3.10 (continued)

			t		
v	0.30	0.35	0.40	0.45	0.50
4.40	17,380.7	8,835.66	4,919.49	2,936.30	1,851.49
4.45	22,045.4	11,116.4	6,145.26	3,644.44	2,284.57
4.50	27,476.5	13,743.8	7,544.08	4,445.69	2,770.79
4.55	33,716.0	16,730.1	9,118.88	5,339.98	3,309.17
4.60	40,783.6	20,075.9	10,866.2	6,323.46	3,896.42
4.65	48,666.0	23,765.8	12,773.7	7,387.32	4,526.27
4.70	57,304.3	27,762.1	14,818.0	8,516.35	5,188.65
4.75	66,577.6	31,998.9	16,960.6	9,687.29	5,868.83
4.80	76,284.2	36,373.0	19,145.0	10,866.9	6,546.36
4.85	86,117.2	40,734.8	21,291.3	12,009.8	7,193.91
4.90	95,637.1	44,876.8	23,291.9	13,056.0	7,775.96
4.95	1,04,238	48,520.3	25,005.3	13,927.7	8,247.35
5.00	1,11,109	51,300.4	26,249.6	14,527.0	8,551.76

			t		
v	0.55	0.60	0.65	0.70	0.75
0.00	−0.654702	−0.637449	−0.619914	−0.602175	−0.584305
0.05	−0.714557	−0.691156	−0.668102	−0.645384	−0.623005
0.10	−0.773033	−0.743257	−0.714509	−0.686681	−0.659697
0.15	−0.829878	−0.793491	−0.758877	−0.725813	−0.694135
0.20	−0.884769	−0.841543	−0.800897	−0.762484	−0.726039
0.25	−0.937313	−0.887032	−0.840208	−0.796352	−0.755087
0.30	−0.987041	−0.929516	−0.876396	−0.827030	−0.780920
0.35	−1.03340	−0.968482	−0.908985	−0.854082	−0.803137
0.40	−1.07576	−1.00335	−0.937444	−0.877023	−0.821299
0.45	−1.11338	−1.03345	−0.961179	−0.895320	−0.834926
0.50	−1.14545	−1.05807	−0.979535	−0.908387	−0.843497
0.55	−1.17103	−1.07637	−0.991793	−0.915590	−0.846453
0.60	−1.18911	−1.08747	−0.997170	−0.916247	−0.843196
0.65	−1.19856	−1.09039	−0.994824	−0.909626	−0.833095
0.70	−1.19813	−1.08407	−0.983849	−0.894954	−0.815486
0.75	−1.18648	−1.06738	−0.963284	−0.871414	−0.789678
0.80	−1.16216	−1.03910	−0.932116	−0.838158	−0.754957
0.85	−1.12362	−0.997968	−0.889284	−0.794306	−0.710597
0.90	−1.06922	−0.942628	−0.833693	−0.738961	−0.655862
0.95	−0.997235	−0.871694	−0.764220	−0.671218	−0.590024
1.00	−0.905859	−0.783739	−0.679731	−0.590175	−0.512369
1.05	−0.793249	−0.677320	−0.579097	−0.494952	−0.422215
1.10	−0.657533	−0.550999	−0.461215	−0.384708	−0.318928
1.15	−0.496839	−0.403369	−0.325033	−0.258663	−0.201941
1.20	−0.309337	−0.233090	−0.169577	−0.116124	−0.070777
1.25	−0.093284	−0.038924	0.006013	0.043485	0.074926

Table 3.10 (continued)

			t		
v	0.55	0.60	0.65	0.70	0.75
1.30	0.152928	0.180215	0.202442	0.220590	0.235386
1.35	0.430700	0.425208	0.420221	0.415432	0.410645
1.40	0.741158	0.696672	0.659610	0.628018	0.600535
1.45	1.08506	0.994890	0.920562	0.858080	0.804635
1.50	1.46272	1.31972	1.20266	1.10502	1.02223
1.55	1.87384	1.67053	1.50504	1.36785	1.25228
1.60	2.31745	2.04607	1.82632	1.64512	1.49333
1.65	2.79170	2.44436	2.16451	1.93490	1.74350
1.70	3.29377	2.86261	2.51692	2.23462	2.00041
1.75	3.81961	3.29704	2.88004	2.54108	2.26112
1.80	4.36381	3.74274	3.24947	2.85031	2.52206
1.85	4.91937	4.19356	3.61977	3.15751	2.77900
1.90	5.47747	4.64189	3.98438	3.45700	3.02696
1.95	6.02724	5.07855	4.33547	3.74204	3.26017
2.00	6.55550	5.49258	4.66384	4.00487	3.47198
2.05	7.04651	5.87107	4.95880	4.23653	3.65486
2.10	7.48174	6.19901	5.20804	4.42686	3.80032
2.15	7.83958	6.45914	5.39759	4.56440	3.89888
2.20	8.09510	6.63177	5.51167	4.63635	3.94009
2.25	8.21987	6.69467	5.53266	4.62859	3.91245
2.30	8.18177	6.62304	5.44109	4.52561	3.80354
2.35	7.94482	6.38937	5.21561	4.31064	3.59997
2.40	7.46916	5.96354	4.83306	3.96561	3.28753
2.45	6.71109	5.31289	4.26861	3.47138	2.85128
2.50	5.62317	4.40236	3.49592	2.80788	2.27573
2.55	4.15454	3.19485	2.48741	1.95432	1.54505
2.60	2.25138	1.65157	1.21468	0.889603	0.643386
2.65	−0.142407	−0.267293	−0.351039	−0.407328	−0.444828
2.70	−3.08427	−2.60185	−2.23821	−1.95692	−1.73440
2.75	−6.63174	−5.39151	−4.47421	−3.77837	−3.23887
2.80	−10.8409	−8.673690	−7.084330	−5.88887	−4.96988
2.85	−15.7642	−12.4823	−10.0906	−8.30262	−6.93646
2.90	−21.4481	−16.8460	−13.5103	−11.0298	−9.14412
2.95	−27.9303	−21.7857	−17.3545	−14.0753	−11.5939
3.00	−35.2355	−27.3120	−21.6258	−17.4371	−14.2814
3.05	−43.3714	−33.4224	−26.3163	−21.1048	−17.1953
3.10	−52.3241	−40.0973	−31.4048	−25.0577	−20.3159
3.15	−62.0513	−47.2962	−36.8544	−29.2626	−23.6139
3.20	−72.4763	−54.9530	−42.6084	−33.6715	−27.0485
3.25	−83.4802	−62.9707	−48.5877	−38.2191	−30.5656
3.30	−94.8931	−71.2158	−54.6860	−42.8199	−34.0959

Table 3.10 (continued)

v	0.55	0.60	0.65	0.70	0.75
			t		
3.35	−106.485	−79.5118	−60.7661	−47.3658	−37.5527
3.40	−117.954	−87.6328	−66.6549	−51.7221	−40.8300
3.45	−128.917	−95.2957	−72.1392	−55.7251	−43.8002
3.50	−138.897	−102.153	−76.9603	−59.1788	−46.3121
3.55	−147.308	−107.784	−80.8092	−61.8512	−48.1886
3.60	−153.445	−111.688	−83.3217	−63.4719	−49.2253
3.65	−156.469	−113.277	−84.0739	−63.7292	−49.1885
3.70	−155.395	−111.867	−82.5785	−62.2675	−47.8143
3.75	−149.080	−106.671	−78.2806	−58.6866	−44.8076
3.80	−136.211	−96.7989	−70.5566	−52.5399	−39.8428
3.85	−115.303	−81.2478	−58.7134	−43.3362	−32.5642
3.90	−84.6871	−58.9075	−41.9893	−30.5406	−22.5888
3.95	−42.5190	−28.5616	−19.5591	−13.5796	−9.50932
4.00	13.2170	11.1039	9.45900	8.15352	7.09999
4.05	84.6908	61.4865	45.9919	35.2884	27.6761
4.10	174.199	124.046	90.9949	68.4657	52.6571
4.15	284.124	200.274	145.429	108.320	82.4699
4.20	416.880	291.655	210.234	155.457	117.514
4.25	574.831	399.611	286.285	210.430	158.140
4.30	760.196	525.435	374.354	273.702	204.631
4.35	974.922	670.207	475.043	345.609	257.165
4.40	1,220.52	834.687	588.722	426.309	315.789
4.45	1,497.88	1,019.19	715.437	515.726	380.376
4.50	1,807.01	1,223.45	854.818	613.484	450.577
4.55	2,146.79	1,446.39	1,005.96	718.830	525.773
4.60	2,514.59	1,686.00	1,167.29	830.542	605.013
4.65	2,905.93	1,939.02	1,336.42	946.839	686.949
4.70	3,313.96	2,200.68	1,509.95	1,065.26	769.763
4.75	3,729.02	2,464.43	1,683.33	1,182.55	851.087
4.80	4,137.98	2,721.55	1,850.59	1,294.53	927.920
4.85	4,523.62	2,960.80	2,004.15	1,395.95	996.539
4.90	4,863.95	3,168.00	2,134.56	1,480.33	1,052.40
4.95	5,131.37	3,325.59	2,230.27	1,539.83	1,090.06
5.00	5,291.87	3,412.16	2,277.32	1,565.11	1,103.07

v	0.80	0.85	0.90	0.95	1.0
			t		
0.00	−0.566368	−0.548419	−0.530511	−0.512689	−0.494995
0.05	−0.600970	−0.579290	−0.557976	−0.537041	−0.516497
0.10	−0.633497	−0.608042	−0.583298	−0.559241	−0.535852
0.15	−0.663715	−0.634451	−0.606264	−0.579089	−0.552873
0.20	−0.691354	−0.658265	−0.626639	−0.596366	−0.567357

Table 3.10 (continued)

			t		
v	0.80	0.85	0.90	0.95	1.0
0.25	-0.716117	-0.679207	-0.644165	-0.610835	-0.579084
0.30	-0.737671	-0.696970	-0.658562	-0.622237	-0.587819
0.35	-0.755650	-0.711221	-0.669526	-0.630298	-0.593314
0.40	-0.769655	-0.721599	-0.676732	-0.634725	-0.595306
0.45	-0.779256	-0.727718	-0.679834	-0.635211	-0.593522
0.50	-0.783989	-0.729167	-0.678468	-0.631432	-0.587677
0.55	-0.783363	-0.725513	-0.672254	-0.623056	-0.577479
0.60	-0.776854	-0.716300	-0.660795	-0.609738	-0.562630
0.65	-0.763917	-0.701057	-0.643685	-0.591128	-0.542832
0.70	-0.743984	-0.679297	-0.620509	-0.566875	-0.517786
0.75	-0.716468	-0.650525	-0.590849	-0.536628	-0.487200
0.80	-0.680772	-0.614243	-0.554291	-0.500044	-0.450792
0.85	-0.636295	-0.569954	-0.510427	-0.456791	-0.408296
0.90	-0.582439	-0.517173	-0.458867	-0.406559	-0.359468
0.95	-0.518617	-0.455434	-0.399245	-0.349064	-0.304094
1.00	-0.444270	-0.384302	-0.331227	-0.284058	-0.241996
1.05	-0.358875	-0.303382	-0.254525	-0.211338	-0.173041
1.10	-0.261962	-0.212338	-0.168907	-0.130757	-0.097152
1.15	-0.153131	-0.110901	-0.074211	-0.042237	-0.014315
1.20	-0.032074	0.001108	0.029641	0.054220	0.075403
1.25	0.101410	0.123761	0.142626	0.158517	0.171851
1.30	0.247380	0.257004	0.264601	0.270449	0.274775
1.35	0.405737	0.400632	0.395287	0.389680	0.383805
1.40	0.576186	0.554263	0.534244	0.515735	0.498441
1.45	0.758209	0.717317	0.680852	0.647975	0.618039
1.50	0.951026	0.888978	0.834287	0.785578	0.741794
1.55	1.15356	1.06818	0.993495	0.927524	0.868723
1.60	1.36440	1.25354	1.15717	1.07257	0.997652
1.65	1.58175	1.44338	1.32372	1.21923	1.12719
1.70	1.80340	1.63564	1.49125	1.36576	1.25574
1.75	2.02666	1.82788	1.65753	1.51012	1.38145
1.80	2.24832	2.01722	1.81997	1.64998	1.50220
1.85	2.46465	2.20032	1.97561	1.78270	1.61564
1.90	2.67127	2.37336	2.12106	1.90529	1.71912
1.95	2.86318	2.53197	2.25253	2.01442	1.80971
2.00	3.03471	2.67127	2.36577	2.10640	1.88420
2.05	3.17943	2.78578	2.45610	2.17719	1.93910
2.10	3.29023	2.86947	2.51837	2.22241	1.97062
2.15	3.35919	2.91571	2.54700	2.23729	1.97473
2.20	3.37769	2.91730	2.53597	2.21679	1.94716
2.25	3.33633	2.86652	2.47884	2.15554	1.88340
2.30	3.22501	2.75511	2.36883	2.04792	1.77878
2.35	3.03299	2.57435	2.19885	1.88811	1.62852

Table 3.10 (continued)

			t		
v	0.80	0.85	0.90	0.95	1.0
2.40	2.74895	2.31518	1.96156	1.67017	1.42777
2.45	2.36113	1.96827	1.64951	1.38811	1.17170
2.50	1.85748	1.52413	1.25523	1.03600	0.855596
2.55	1.22581	0.973364	0.771379	0.608117	0.474989
2.60	0.454110	0.306793	0.190950	0.099089	0.025759
2.65	-0.469199	-0.484237	-0.492541	-0.495930	-0.495696
2.70	-1.55490	-1.40758	-1.28479	-1.18104	-1.09228
2.75	-2.81254	-2.46992	-2.19043	-1.95937	-1.76602
2.80	-4.24992	-3.67639	-3.21269	-2.83279	-2.51780
2.85	-5.87255	-5.03007	-4.35299	-3.80159	-3.34717
2.90	-7.68293	-6.53151	-5.61059	-4.86421	-4.25202
2.95	-9.67982	-8.17805	-6.98205	-6.01676	-5.22830
3.00	-11.8574	-9.96324	-8.46071	-7.25270	-6.26971
3.05	-14.2042	-11.8761	-10.0362	-8.56236	-7.36732
3.10	-16.7022	-13.9001	-11.6937	-9.93245	-8.50921
3.15	-19.3257	-16.0128	-13.4134	-11.3456	-9.68013
3.20	-22.0399	-18.1844	-15.1699	-12.7798	-10.8610
3.25	-24.7996	-20.3770	-16.9310	-14.2078	-12.0287
3.30	-27.5479	-22.5436	-18.6576	-15.5968	-13.1554
3.35	-30.2150	-24.6270	-20.3026	-16.9078	-14.2084
3.40	-32.7160	-26.5589	-21.8103	-18.0947	-15.1496
3.45	-34.9502	-28.2587	-23.1156	-19.1046	-15.9356
3.50	-36.7992	-29.6327	-24.1435	-19.8768	-16.5166
3.55	-38.1261	-30.5731	-24.8083	-20.3427	-16.8372
3.60	-38.7736	-30.9578	-25.0139	-20.4254	-16.8357
3.65	-38.5639	-30.6493	-24.6526	-20.0400	-16.4442
3.70	-37.2975	-29.4948	-23.6060	-19.0937	-15.5891
3.75	-34.7536	-27.3266	-21.7449	-17.4857	-14.1913
3.80	-30.6903	-23.9621	-18.9300	-15.1086	-12.1669
3.85	-24.8459	-19.2060	-15.0133	-11.8489	-9.42816
3.90	-16.9409	-12.8512	-9.83947	-7.58869	-5.88468
3.95	-6.68122	-4.68200	-3.24821	-2.20761	-1.44507
4.00	6.23744	5.52225	4.92261	4.41482	3.98098
4.05	22.1235	17.9825	14.8332	12.3960	10.4802
4.10	41.2817	32.9138	26.6371	21.8464	18.1327
4.15	64.0039	50.5179	40.4750	32.8650	27.0081
4.20	90.5561	70.9740	56.4682	45.5337	37.1617
4.25	121.164	94.4275	74.7095	59.9109	48.6287
4.30	155.998	120.977	95.2541	76.0241	61.4193
4.35	195.145	150.660	118.108	93.8609	75.5114
4.40	238.594	183.432	143.213	113.360	90.8436
4.45	286.200	219.151	170.436	134.399	107.307
4.50	337.656	257.550	199.549	156.784	124.736

Table 3.10 (continued)

			t		
v	0.80	0.85	0.90	0.95	1.0
4.55	392.455	298.214	230.210	180.233	142.899
4.60	449.848	340.551	261.948	204.367	161.486
4.65	508.805	383.761	294.134	228.689	180.100
4.70	567.957	426.802	325.964	252.569	198.245
4.75	625.554	468.354	356.432	275.228	215.311
4.80	679.398	506.785	384.302	295.724	230.567
4.85	726.796	540.110	408.090	312.925	243.144
4.90	764.494	565.952	426.032	325.506	252.029
4.95	788.621	581.509	436.063	331.922	256.052
5.00	794.636	583.518	435.799	330.403	253.882

			t		
v	1.5	2.0	3.0	4.0	5.0
0.00	−0.331396	−0.202400	−0.051122	0.002198	0.011188
0.05	−0.334720	−0.198571	−0.045701	0.005038	0.012060
0.10	−0.336325	−0.193616	−0.039981	0.007864	0.012866
0.15	−0.336134	−0.187521	−0.033982	0.010663	0.013599
0.20	−0.334069	−0.180273	−0.027727	0.013419	0.014256
0.25	−0.330052	−0.171865	−0.021241	0.016117	0.014830
0.30	−0.324007	−0.162292	−0.014550	0.018741	0.015319
0.35	−0.315862	−0.151558	−0.007685	0.021276	0.015717
0.40	−0.305547	−0.139668	−0.000677	0.023706	0.016022
0.45	−0.292998	−0.126637	0.006441	0.026016	0.016228
0.50	−0.278159	−0.112484	0.013633	0.028189	0.016335
0.55	−0.260978	−0.097239	0.020860	0.030210	0.016338
0.60	−0.241417	−0.080936	0.028083	0.032063	0.016237
0.65	−0.219445	−0.063621	0.035260	0.033734	0.016029
0.70	−0.195048	−0.045348	0.042345	0.035208	0.015713
0.75	−0.168223	−0.026180	0.049294	0.036471	0.015289
0.80	−0.138989	−0.006193	0.056058	0.037508	0.014758
0.85	−0.107379	0.014527	0.062589	0.038308	0.014119
0.90	−0.073452	0.035883	0.068836	0.038858	0.013375
0.95	−0.037287	0.057765	0.074747	0.039148	0.012527
1.00	0.001009	0.080049	0.080270	0.039166	0.011578
1.05	0.041299	0.102601	0.085352	0.038905	0.010532
1.10	0.083418	0.125271	0.089940	0.038357	0.009392
1.15	0.127164	0.147896	0.093979	0.037516	0.008165
1.20	0.172302	0.170299	0.097418	0.036378	0.006855
1.25	0.218558	0.192291	0.100204	0.034940	0.005468
1.30	0.265615	0.213668	0.102287	0.033201	0.004013
1.35	0.313116	0.234213	0.103617	0.031162	0.002496
1.40	0.360659	0.253696	0.104148	0.028827	0.000926
1.45	0.407794	0.271877	0.103836	0.026201	−0.000687

Table 3.10 (continued)

ν	1.5	2.0	3.0	4.0	5.0
1.50	0.454026	0.288502	0.102640	0.023291	−0.002333
1.55	0.498810	0.303310	0.100524	0.020108	−0.004003
1.60	0.541553	0.316029	0.097455	0.016664	−0.005684
1.65	0.581613	0.326383	0.093407	0.012973	−0.007366
1.70	0.618301	0.334089	0.088359	0.009054	−0.009036
1.75	0.650882	0.338863	0.082296	0.004927	−0.010680
1.80	0.678577	0.340421	0.075212	0.000615	−0.012287
1.85	0.700568	0.338482	0.067106	−0.003858	−0.013843
1.90	0.716001	0.332773	0.057989	−0.008461	−0.015334
1.95	0.723993	0.323028	0.047879	−0.013166	−0.016746
2.00	0.723638	0.309001	0.036804	−0.017938	−0.018065
2.05	0.714018	0.290460	0.024805	−0.022741	−0.019278
2.10	0.694210	0.267201	0.011931	−0.027536	−0.020370
2.15	0.663302	0.239047	−0.001756	−0.032284	−0.021328
2.20	0.620402	0.205856	−0.016180	−0.036942	−0.022140
2.25	0.564659	0.167529	−0.031255	−0.041464	−0.022791
2.30	0.495276	0.124014	−0.046879	−0.045805	−0.023271
2.35	0.411535	0.075311	−0.062940	−0.049917	−0.023567
2.40	0.312813	0.021485	−0.079310	−0.053751	−0.023669
2.45	0.198612	−0.037334	−0.095848	−0.057257	−0.023567
2.50	0.068583	−0.100940	−0.112400	−0.060385	−0.023254
2.55	−0.077444	−0.169042	−0.128797	−0.063086	−0.022721
2.60	−0.239432	−0.241262	−0.144859	−0.065309	−0.021964
2.65	−0.417100	−0.317124	−0.160393	−0.067007	−0.020977
2.70	−0.609891	−0.396046	−0.175195	−0.068131	−0.019759
2.75	−0.816937	−0.477338	−0.189050	−0.068637	−0.018308
2.80	−1.03702	−0.560194	−0.201732	−0.068482	−0.016626
2.85	−1.26852	−0.643687	−0.213011	−0.067626	−0.014716
2.90	−1.50942	−0.726764	−0.222647	−0.066033	−0.012584
2.95	−1.75722	−0.808245	−0.230397	−0.063673	−0.010237
3.00	−2.00890	−0.886821	−0.236018	−0.060518	−0.007685
3.05	−2.26095	−0.961052	−0.239265	−0.056549	−0.004941
3.10	−2.50925	−1.02937	−0.239897	−0.051752	−0.002020
3.15	−2.74910	−1.09010	−0.237681	−0.046118	0.001060
3.20	−2.97520	−1.14142	−0.232392	−0.039651	0.004280
3.25	−3.18159	−1.18143	−0.223819	−0.032359	0.007617
3.30	−3.36170	−1.20813	−0.211769	−0.024260	0.011045
3.35	−3.50829	−1.21944	−0.196071	−0.015384	0.014537
3.40	−3.61354	−1.21322	−0.176580	−0.005769	0.018063
3.45	−3.66903	−1.18730	−0.153182	0.004535	0.021590
3.50	−3.66580	−1.13951	−0.125799	0.015468	0.025083

Table 3.10 (continued)

			t		
v	1.5	2.0	3.0	4.0	5.0
3.55	−3.59442	−1.06771	−0.094393	0.026956	0.028507
3.60	−3.44510	−0.969823	−0.058973	0.038914	0.031822
3.65	−3.20775	−0.843882	−0.019600	0.051247	0.034988
3.70	−2.87220	−0.688082	0.023609	0.063844	0.037964
3.75	−2.42830	−0.500831	0.070477	0.076584	0.040708
3.80	−1.86615	−0.280810	0.120759	0.089333	0.043177
3.85	−1.17637	−0.027038	0.174137	0.101947	0.045327
3.90	−0.350334	0.261059	0.230219	0.114268	0.047114
3.95	0.619496	0.583572	0.288533	0.126131	0.048497
4.00	1.73920	0.940031	0.348525	0.137357	0.049433
4.05	3.01301	1.32932	0.409551	0.147761	0.049883
4.10	4.44287	1.74959	0.470884	0.157152	0.049808
4.15	6.02792	2.19817	0.531703	0.165331	0.049173
4.20	7.76400	2.67149	0.591104	0.172094	0.047946
4.25	9.64305	3.16498	0.648090	0.177239	0.046098
4.30	11.6525	3.67296	0.701584	0.180560	0.043606
4.35	13.7747	4.18863	0.750425	0.181857	0.040451
4.40	15.9862	4.70391	0.793381	0.180932	0.036621
4.45	18.2571	5.20942	0.829153	0.177598	0.032107
4.50	20.5502	5.69444	0.856385	0.171680	0.026912
4.55	22.8207	6.14682	0.873678	0.163014	0.021042
4.60	25.0154	6.55300	0.879606	0.151458	0.014515
4.65	27.0719	6.89802	0.872729	0.136892	0.007356
4.70	28.9185	7.16552	0.851617	0.119221	−0.000402
4.75	30.4734	7.33782	0.814868	0.098381	−0.008713
4.80	31.6448	7.39602	0.761138	0.074340	−0.017525
4.85	32.3307	7.32015	0.689167	0.047108	−0.026773
4.90	32.4189	7.08936	0.597807	0.016736	−0.036380
4.95	31.7874	6.68213	0.486060	−0.016680	−0.046263
5.00	30.3054	6.07662	0.353111	−0.052991	−0.056323

Table 3.11: First derivatives with respect to the order of the Kelvin functions $\partial \mathrm{kei}_v(t)/\partial v$.

			t		
v	0.05	0.10	0.15	0.20	0.25
0.00	−4.88856	−3.80207	−3.16899	−2.72241	−2.37868
0.05	−5.28610	−4.01562	−3.29910	−2.80361	−2.42767
0.10	−5.80113	−4.27866	−3.45405	−2.89755	−2.48261
0.15	−6.43552	−4.58814	−3.63026	−3.00087	−2.54049
0.20	−7.19153	−4.94035	−3.82355	−3.10970	−2.59791

Table 3.11 (continued)

ν	t				
	0.05	0.10	0.15	0.20	0.25
0.25	−8.07091	−5.33056	−4.02888	−3.21952	−2.65099
0.30	−9.07390	−5.75249	−4.24007	−3.32503	−2.69524
0.35	−10.1979	−6.197750	−4.449570	−3.419920	−2.725520
0.40	−11.4358	−6.655190	−4.648060	−3.496770	−2.735860
0.45	−12.7740	−7.110080	−4.824110	−3.546760	−2.719400
0.50	−14.1894	−7.543190	−4.963710	−3.559530	−2.668280
0.55	−15.6462	−7.929670	−5.049840	−3.522870	−2.573450
0.60	−17.0917	−8.237790	−5.061930	−3.422550	−2.424640
0.65	−18.4508	−8.427430	−4.975220	−3.241990	−2.210180
0.70	−19.6193	−8.448350	−4.760180	−2.962030	−1.916910
0.75	−20.4557	−8.238270	−4.381810	−2.560660	−1.530060
0.80	−20.7715	−7.720590	−3.798890	−2.012750	−1.033210
0.85	−20.3185	−6.801970	−2.963250	−1.289800	−0.408198
0.90	−18.7747	−5.369520	−1.819000	−0.359759	0.364879
0.95	−15.7265	−3.28791	−0.301786	0.813136	1.30763
1.00	−10.6487	−0.396191	1.66184	2.26842	2.44325
1.05	−2.88055	3.49546	4.15501	4.04936	3.79637
1.10	8.40149	8.60891	7.27086	6.20274	5.39270
1.15	24.2143	15.2011	11.1126	8.77846	7.25863
1.20	45.8019	23.5666	15.7933	11.8288	9.42056
1.25	74.6715	34.0394	21.4348	15.4077	11.9042
1.30	112.631	46.9930	28.1660	19.5687	14.7333
1.35	161.827	62.8396	36.1209	24.3637	17.9284
1.40	224.772	82.0253	45.4338	29.8401	21.5052
1.45	304.377	105.023	56.2345	36.0370	25.4723
1.50	403.952	132.318	68.6410	42.9818	29.8285
1.55	527.192	164.391	82.7487	50.6837	34.5595
1.60	678.126	201.687	98.6185	59.1279	39.6344
1.65	861.003	244.578	116.259	68.2666	45.0011
1.70	1,080.11	293.308	135.608	78.0094	50.5804
1.75	1,339.48	347.923	156.500	88.2109	56.2602
1.80	1,642.43	408.173	178.641	98.6561	61.8880
1.85	1,990.95	473.398	201.563	109.044	67.2627
1.90	2,384.78	542.365	224.578	118.966	72.1249
1.95	2,820.16	613.078	246.716	127.885	76.1464
2.00	3,288.15	682.530	266.659	135.107	78.9182
2.05	3,772.37	746.406	282.657	139.752	79.9381
2.10	4,245.98	798.711	292.434	140.720	78.5962
2.15	4,667.82	831.327	293.076	136.654	74.1607
2.20	4,977.39	833.481	280.907	125.900	65.7624
2.25	5,088.43	791.113	251.351	106.462	52.3800
2.30	4,880.90	686.145	198.775	75.9643	32.8255
2.35	4,190.94	495.629	116.324	31.5985	5.73133

Table 3.11 (continued)

			t		
v	0.05	0.10	0.15	0.20	0.25
2.40	2,798.44	190.792	−4.25160	−29.9123	−30.4598
2.45	412.084	−264.031	−172.775	−112.353	−77.5008
2.50	−3,348.8	−912.558	−400.816	−220.045	−137.340
2.55	−8,975.5	−1,807.78	−701.838	−357.858	−212.111
2.60	−17,094.8	−3,013.18	−1,091.320	−531.206	−304.114
2.65	−28,497.0	−4,603.93	−1,586.820	−746.008	−415.782
2.70	−44,166.7	−6,667.85	−2,207.940	−1,008.62	−549.619
2.75	−65,316.2	−9,306.04	−2,976.140	−1,325.68	−708.126
2.80	−93,418.3	−12,633.0	−3,914.470	−1,703.94	−893.689
2.85	−130236	−16,775.8	−5,046.880	−2,149.98	−1,108.42
2.90	−177847	−21,872.2	−6,397.410	−2,669.75	−1,353.99
2.95	−238651	−28,066.2	−7,988.780	−3,268.10	−1,631.31
3.00	−315354	−35,502.6	−9,840.510	−3,948.00	−1,940.28
3.05	−410918	−44,316.4	−11,966.4	−4,709.64	−2,279.36
3.10	−528447	−54,619.1	−14,370.8	−5,549.23	−2,645.05
3.15	−670997	−66,478.4	−17,044.4	−6,457.50	−3,031.29
3.20	−841262	−79,890.2	−19,958.1	−7,417.81	−3,428.75
3.25	−1.041 +6	−94,741.7	−23,055.5	−8,403.91	−3,823.94
3.30	−1.271	−1,10,761	−26,243.1	−9,377.07	−4,198.16
3.35	−1.353	−1,27,452	−29,379.0	−10,282.8	−4,526.31
3.40	−1.807	−1,44,012	−32,257.9	−11,046.9	−4,775.52
3.45	−2.095	−1,59,222	−34,593.5	−11,570.8	−4,903.53
3.50	−2.371	−1,71,311	−35,997.2	−11,725.8	−4,856.92
3.55	−2.602	−1,77,795	−35,952.6	−11,347.5	−4,569.19
3.60	−2.738	−1,75,260	−33,785.9	−10,228.4	−3,958.57
3.65	−2.707	−1,59,119	−28,631.6	−8,110.16	−2,925.80
3.70	−2.402	−1,23,307	−19,393.2	−4,675.49	−1,351.79
3.75	−1.680	−59,925	−4,699.37	460.96	904.640
3.80	−3,38,835	41,176	17,143.6	7,761.55	4,008.94
3.85	1.891E+6	1,92,867	48,202.8	17,774.4	8,152.62
3.90	5.368	4,11,278	90,970.8	31,141.4	13,554.4
3.95	1.057E+7	7,16,369	1,48,416	48,604.3	20,460.7
4.00	1.810	1.133E+6	2,24,026	71,007.8	29,144.5
4.05	2.876	1.689	3,21,842	99,298.4	39,902.4
4.10	4.354	2.421	4,46,471	1,34,516	53,049.7
4.15	6.369	3.370	6,03,062	1,77,779	68,911.7
4.20	9.074	4.583	7,97,250	2,30,251	87,810.4
4.25	1.265E+8	6.113	1.035E+6	2,93,098	1,10,047
4.30	1.733	8.018	1.322	3,67,419	1,35,875
4.35	2.337	1.036E+7	1.666	4,54,147	1,65,470
4.40	3.105	1.320	2.069	5,53,925	1,98,880
4.45	4.071	1.660	2.538	6,66,921	2,35,973
4.50	5.266	2.061	3.072	7,92,599	2,76,362
4.55	6.725	2.525	3.669	9,29,420	3,19,313

Table 3.11 (continued)

v	0.05	0.10	0.15	0.20	0.25
			t		
4.60	8.472	3.050	4.322	1.074E+6	3,63,634
4.65	1.052E+9	3.632	5.015	1.223	4,07,540
4.70	1.287	4.254	5.720	1.368	4,48,481
4.75	1.546	4.891	6.397	1.498	4,82,948
4.80	1.821	5.499	6.984	1.600	5,06,242
4.85	2.091	6.013	7.397	1.653	5,12,198
4.90	2.326	6.337	7.518	1.631	4,92,877
4.95	2.476	6.333	7.186	1.502	4,38,216
5.00	2.466	5.815	6.191	1.220	3,35,636

v	0.30	0.35	0.40	0.45	0.50
			t		
0.00	−2.10050	−1.86796	−1.66917	−1.49642	−1.34445
0.05	−2.12684	−1.87771	−1.66642	−1.48412	−1.32476
0.10	−2.15503	−1.88673	−1.66131	−1.46840	−1.30099
0.15	−2.18243	−1.89272	−1.65183	−1.44753	−1.27167
0.20	−2.20609	−1.89315	−1.63581	−1.41965	−1.23517
0.25	−2.22270	−1.88521	−1.61084	−1.38270	−1.18974
0.30	−2.22852	−1.86574	−1.57429	−1.33446	−1.13353
0.35	−2.219300	−1.831250	−1.52325	−1.27252	−1.06452
0.40	−2.190250	−1.777850	−1.45455	−1.19428	−0.980559
0.45	−2.135940	−1.701210	−1.36471	−1.09692	−0.879364
0.50	−2.050290	−1.596550	−1.24993	−0.977412	−0.758506
0.55	−1.926450	−1.458600	−1.10610	−0.832515	−0.615427
0.60	−1.756780	−1.281560	−0.928783	−0.658793	−0.447452
0.65	−1.532790	−1.059120	−0.713207	−0.452618	−0.251802
0.70	−1.245110	−0.784445	−0.454302	−0.210195	−0.025629
0.75	−0.883432	−0.450166	−0.146714	0.072406	0.233961
0.80	−0.436560	−0.048426	0.215154	0.399204	0.529856
0.85	0.107619	0.429076	0.637084	0.774252	0.864888
0.90	0.762058	0.991006	1.12496	1.20156	1.24176
0.95	1.54046	1.64629	1.68467	1.68499	1.66297
1.00	2.45712	2.40397	2.32196	2.22816	2.13071
1.05	3.52673	3.27301	3.04228	2.83429	2.64674
1.10	4.76409	4.26201	3.85058	3.50604	3.21225
1.15	6.18369	5.37897	4.75102	4.24527	3.82770
1.20	7.79926	6.63080	5.74670	5.05287	4.49266
1.25	9.62311	8.02292	6.83930	5.92842	5.20553
1.30	11.6653	9.55866	8.02862	6.86988	5.96336
1.35	13.9329	11.2385	9.31209	7.87323	6.76153
1.40	16.4283	13.0594	10.6842	8.93206	7.59347
1.45	19.1485	15.0137	12.1357	10.0370	8.45028

Table 3.11 (continued)

			t		
v	0.30	0.35	0.40	0.45	0.50
1.50	22.0827	17.0879	13.6532	11.1755	9.32039
1.55	25.2110	19.2617	15.2178	12.3306	10.1891
1.60	28.5013	21.5063	16.8045	13.4810	11.0383
1.65	31.9074	23.7827	18.3809	14.5999	11.8456
1.70	35.3652	26.0397	19.9062	15.6543	12.5844
1.75	38.7897	28.2122	21.3297	16.6045	13.2227
1.80	42.0705	30.2184	22.5895	17.4026	13.7233
1.85	45.0678	31.9578	23.6113	17.9924	14.0427
1.90	47.6073	33.3080	24.3063	18.3079	14.1309
1.95	49.4745	34.1223	24.5703	18.2727	13.9306
2.00	50.4099	34.2262	24.2814	17.7993	13.3773
2.05	50.1017	33.4152	23.2993	16.7880	12.3987
2.10	48.1806	31.4510	21.4633	15.1269	10.9143
2.15	44.2129	28.0595	18.5918	12.6909	8.83585
2.20	37.6948	22.9280	14.4807	9.34215	6.06738
2.25	28.0465	15.7034	8.90379	4.92996	2.50562
2.30	14.6082	5.99143	1.61281	−0.708217	−1.95900
2.35	−3.36386	−6.64366	−7.66153	−7.74545	−7.44104
2.40	−26.6977	−22.6771	−19.2072	−16.3632	−14.0582
2.45	−56.3040	−42.6202	−33.3273	−26.7483	−21.9286
2.50	−93.170	−67.0143	−50.3350	−39.0887	−31.1680
2.55	−138.349	−96.4209	−70.5459	−53.5679	−41.8850
2.60	−192.943	−131.409	−94.2687	−70.3580	−54.1761
2.65	−258.074	−172.537	−121.792	−89.6104	−68.1192
2.70	−334.851	−220.328	−153.370	−111.445	−83.7658
2.75	−424.319	−275.239	−189.198	−135.937	−101.132
2.80	−527.396	−337.621	−229.391	−163.098	−120.185
2.85	−644.784	−407.670	−273.954	−192.860	−140.836
2.90	−776.872	−485.368	−322.740	−225.050	−162.916
2.95	−923.596	−570.406	−375.413	−259.364	−186.169
3.00	−1,084.28	−662.097	−431.392	−295.334	−210.222
3.05	−1,257.44	−759.272	−489.792	−332.294	−234.571
3.10	−1,440.55	−860.154	−549.355	−369.338	−258.550
3.15	−1,629.73	−962.213	−608.374	−405.278	−281.310
3.20	−1,819.48	−1,062.01	−664.598	−438.588	−301.785
3.25	−2,002.21	−1,154.98	−715.141	−467.354	−318.660
3.30	−2,167.87	−1,235.27	−756.366	−489.215	−330.341
3.35	−2,303.39	−1,295.45	−783.770	−501.297	−334.921
3.40	−2,392.13	−1,326.28	−791.860	−500.153	−330.141
3.45	−2,413.28	−1,316.45	−774.018	−481.695	−313.365
3.50	−2,341.13	−1,252.25	−722.373	−441.132	−281.540
3.55	−2,144.36	−1,117.31	−627.666	−372.914	−231.180
3.60	−1,785.29	−892.258	−479.130	−270.676	−158.342

Table 3.11 (continued)

			t		
v	0.30	0.35	0.40	0.45	0.50
3.65	−1,219.11	−554.494	−264.389	−127.212	−58.6196
3.70	−393.148	−77.9016	30.6215	65.5465	72.8504
3.75	753.821	567.321	421.69	316.52	241.362
3.80	2,292.06	1,414.76	926.30	635.42	452.591
3.85	4,301.22	2,501.73	1,563.55	1,032.66	712.514
3.90	6,870.27	3,869.11	2,353.90	1,519.19	1,027.29
3.95	10,097.1	5,560.83	3,318.88	2,106.28	1,403.10
4.00	14,087.4	7,623.14	4,480.58	2,805.15	1,845.91
4.05	18,952.7	10,103.4	5,860.89	3,626.53	2,361.20
4.10	24,807.5	13,048.5	7,480.58	4,580.10	2,953.57
4.15	31,764.9	16,502.4	9,357.94	5,673.67	3,626.29
4.20	39,930.6	20,503.3	11,507.2	6,912.29	4,380.75
4.25	49,394.5	25,079.3	13,936.3	8,297.01	5,215.69
4.30	60,220.3	30,243.5	16,644.4	9,823.47	6,126.45
4.35	72,431.2	35,987.6	19,618.4	11,480.1	7,103.89
4.40	85,992.4	42,273.7	22,829.3	13,246.1	8,133.29
4.45	1,00,789	49,024.8	26,227.2	15,088.8	9,192.95
4.50	1,16,597	56,112.4	29,735.7	16,961.2	10,252.6
4.55	1,33,053	63,343.0	33,245.4	18,798.0	11,271.8
4.60	1,49,611	70,441.1	36,606.4	20,512.4	12,197.7
4.65	1,65,494	77,029.5	39,619.1	21,991.7	12,962.9
4.70	1,79,639	82,606.9	42,024.6	23,092.4	13,483.5
4.75	1,90,630	86,522.5	43,493.9	23,635.3	13,656.0
4.80	1,96,620	87,946.6	43,615.0	23,399.7	13,354.7
4.85	1,95,248	85,839.6	41,880.9	22,117.9	12,429.4
4.90	1,83,541	78,916.8	37,674.9	19,469.4	10,702.4
4.95	1,57,809	65,612.2	30,257.2	15,074.7	7,966.0
5.00	1,13,533	44,040.7	18,750.7	8,490.11	3,980.65

			t		
v	0.55	0.60	0.65	0.70	0.75
0.00	−1.20950	−1.08875	−0.980080	−0.881811	−0.792608
0.05	−1.18405	−1.05884	−0.946727	−0.845831	−0.754668
0.10	−1.15417	−1.02431	−0.908692	−0.805207	−0.712185
0.15	−1.11854	−0.984020	−0.864996	−0.759091	−0.664430
0.20	−1.07579	−0.936779	−0.814613	−0.706603	−0.610650
0.25	−1.02442	−0.881323	−0.756467	−0.646831	−0.550069
0.30	−0.962860	−0.816323	−0.689437	−0.578829	−0.481897
0.35	−0.889431	−0.740378	−0.612357	−0.501629	−0.405330
0.40	−0.802364	−0.652030	−0.524024	−0.414241	−0.319560
0.45	−0.699795	−0.549758	−0.423202	−0.315667	−0.223779
0.50	−0.579772	−0.431997	−0.308638	−0.204903	−0.117196
0.55	−0.440272	−0.297145	−0.179065	−0.080959	0.000959

Table 3.11 (continued)

			t		
v	0.55	0.60	0.65	0.70	0.75
0.60	−0.279207	−0.143577	−0.033229	0.057128	0.131413
0.65	−0.094453	0.030328	0.130100	0.210277	0.274838
0.70	0.116133	0.226169	0.312101	0.379339	0.431830
0.75	0.354670	0.445488	0.513873	0.565074	0.602884
0.80	0.623223	0.689731	0.736402	0.768118	0.788370
0.85	0.923742	0.960210	0.980523	0.988951	0.988504
0.90	1.25801	1.25804	1.24687	1.22786	1.20331
0.95	1.62757	1.58409	1.53583	1.48488	1.43259
1.00	2.03363	1.93891	1.84749	1.75978	1.67588
1.05	2.47700	2.32263	2.18155	2.05196	1.93239
1.10	2.95791	2.73491	2.53724	2.36042	2.20098
1.15	3.47597	3.17477	2.91331	2.68370	2.48009
1.20	4.02993	3.64055	3.30782	3.01978	2.76768
1.25	4.61759	4.12972	3.71813	3.36601	3.06117
1.30	5.23556	4.63877	4.14077	3.71903	3.35738
1.35	5.87909	5.16303	4.57128	4.07468	3.65244
1.40	6.54181	5.69652	5.00413	4.42789	3.94176
1.45	7.21548	6.23174	5.43257	4.77260	4.21991
1.50	7.88977	6.75951	5.84848	5.10165	4.48058
1.55	8.55188	7.26873	6.24221	5.40667	4.71648
1.60	9.18633	7.74619	6.60249	5.67798	4.91930
1.65	9.77456	8.17635	6.91622	5.90452	5.07964
1.70	10.2947	8.54111	7.16837	6.07370	5.18693
1.75	10.7210	8.81960	7.34181	6.17139	5.22943
1.80	11.0239	8.98801	7.41724	6.18181	5.19416
1.85	11.1693	9.01937	7.37305	6.08753	5.06693
1.90	11.1186	8.88342	7.18531	5.86939	4.83232
1.95	10.8282	8.54652	6.82768	5.50659	4.47377
2.00	10.2497	7.97157	6.27148	4.97673	3.97363
2.05	9.32922	7.11807	5.48576	4.25593	3.31331
2.10	8.00803	5.94225	4.43751	3.31901	2.47346
2.15	6.22235	4.397280	3.091870	2.139760	1.434230
2.20	3.90384	2.43367	1.41254	0.691285	0.175561
2.25	0.980150	−0.000132	−0.637706	−1.05351	−1.32239
2.30	−2.62421	−2.95702	−3.09633	−3.12138	−3.07876
2.35	−6.98671	−6.49017	−6.00020	−5.53797	−5.11139
2.40	−12.1852	−10.6517	−9.38449	−8.32697	−7.43605
2.45	−18.2959	−15.4913	−13.2814	−11.5090	−10.0657
2.50	−25.3907	−21.0536	−17.7184	−15.1006	−13.0093
2.55	−33.5342	−27.3765	−22.7167	−19.1122	−16.2708
2.60	−42.7795	−34.4876	−28.2887	−23.5470	−19.8479
2.65	−53.1639	−42.4010	−34.4353	−28.3987	−23.7302
2.70	−64.7028	−51.1132	−41.1432	−33.6492	−27.8980

Table 3.11 (continued)

			t		
ν	0.55	0.60	0.65	0.70	0.75
2.75	−77.3835	−60.5983	−48.3815	−39.2665	−32.3199
2.80	−91.1572	−70.8028	−56.0972	−45.2011	−36.9510
2.85	−105.931	−81.6390	−64.2117	−51.3835	−41.7306
2.90	−121.555	−92.9785	−72.6153	−57.7207	−46.5799
2.95	−137.816	−104.644	−81.1620	−64.0921	−51.3987
3.00	−154.418	−116.402	−89.6639	−70.3460	−56.0637
3.05	−170.976	−127.950	−97.8851	−76.2949	−60.4246
3.10	−186.991	−138.913	−105.534	−81.7116	−64.3017
3.15	−201.843	−148.825	−112.260	−86.3246	−67.4832
3.20	−214.765	−157.125	−117.640	−89.8136	−69.7219
3.25	−224.828	−163.140	−121.179	−91.8054	−70.7326
3.30	−230.924	−166.077	−122.298	−91.8696	−70.1906
3.35	−231.743	−165.014	−120.330	−89.5160	−67.7290
3.40	−225.754	−158.884	−114.515	−84.1915	−62.9384
3.45	−211.195	−146.473	−103.997	−75.2786	−55.3664
3.50	−186.051	−126.410	−87.8178	−62.0959	−44.5187
3.55	−148.046	−97.1651	−64.9237	−43.9001	−29.8619
3.60	−94.643	−57.0516	−34.1644	−19.8901	−10.8280
3.65	−23.040	−4.23288	5.69685	10.7849	13.1793
3.70	69.810	63.2642	55.9708	49.0147	42.7745
3.75	187.179	147.525	118.023	95.7120	78.5783
3.80	332.515	250.716	193.246	151.791	121.200
3.85	509.388	375.031	283.025	218.142	171.220
3.90	721.395	522.637	388.691	295.594	229.157
3.95	972.051	695.583	511.459	384.874	295.444
4.00	1,264.63	895.702	652.360	486.553	370.388
4.05	1,601.99	1,124.48	812.147	600.984	454.120
4.10	1,986.30	1,382.89	991.187	728.229	546.550
4.15	2,418.79	1,671.21	1,189.34	867.965	647.300
4.20	2,899.35	1,988.79	1,405.78	1,019.39	755.634
4.25	3,426.13	2,333.78	1,638.87	1,181.09	870.38
4.30	3,995.03	2,702.81	1,885.91	1,350.93	989.84
4.35	4,599.11	3,090.63	2,142.92	1,525.89	1,111.69
4.40	5,227.87	3,489.69	2,404.40	1,701.88	1,232.87
4.45	5,866.54	3,889.68	2,663.01	1,873.58	1,349.45
4.50	6,495.10	4,277.01	2,909.28	2,034.26	1,456.55
4.55	7,087.39	4,634.23	3,131.24	2,175.51	1,548.15
4.60	7,609.95	4,939.38	3,314.08	2,287.08	1,616.99
4.65	8,020.87	5,165.39	3,439.74	2,356.62	1,654.42
4.70	8,268.51	5,279.32	3,486.54	2,369.44	1,650.26
4.75	8,290.21	5,241.72	3,428.79	2,308.34	1,592.72
4.80	8,010.93	5,005.86	3,236.40	2,153.37	1,468.23
4.85	7,341.93	4,517.13	2,874.57	1,881.69	1,261.41
4.90	6,179.52	3,712.43	2,303.50	1,467.44	955.026

Table 3.11 (continued)

			t		
ν	0.55	0.60	0.65	0.70	0.75
4.95	4,404.01	2,519.71	1,478.23	881.686	529.965
5.00	1,878.86	857.701	348.599	92.5002	−34.6481

			t		
ν	0.80	0.85	0.90	0.95	1.0
0.00	−0.711386	−0.637246	−0.569437	−0.507323	−0.450357
0.05	−0.672030	−0.596928	−0.528533	−0.466146	−0.409171
0.10	−0.628283	−0.552399	−0.483619	−0.421175	−0.364413
0.15	−0.579513	−0.503118	−0.434235	−0.372017	−0.315750
0.20	−0.525080	−0.448540	−0.379917	−0.318284	−0.262860
0.25	−0.464329	−0.388115	−0.320207	−0.259595	−0.205429
0.30	−0.396598	−0.321293	−0.254652	−0.195579	−0.143158
0.35	−0.321225	−0.247532	−0.182812	−0.125884	−0.075771
0.40	−0.237553	−0.166302	−0.104264	−0.050180	−0.003013
0.45	−0.144938	−0.077095	−0.018614	0.031832	0.075334
0.50	−0.042758	0.020569	0.074499	0.120416	0.159451
0.55	0.069573	0.127126	0.175390	0.215786	0.249471
0.60	0.192593	0.242962	0.284319	0.318101	0.345470
0.65	0.326778	0.368392	0.401476	0.427449	0.447456
0.70	0.472519	0.503653	0.526969	0.543841	0.555360
0.75	0.630110	0.648875	0.660809	0.667189	0.669023
0.80	0.799717	0.804071	0.802889	0.797299	0.788185
0.85	0.981356	0.969109	0.952968	0.933850	0.912466
0.90	1.17487	1.14369	1.11065	1.07638	1.04136
0.95	1.37987	1.32732	1.27536	1.22425	1.17420
1.00	1.59576	1.51929	1.44632	1.37667	1.31018
1.05	1.82162	1.71862	1.62253	1.53262	1.44829
1.10	2.05624	1.92406	1.80273	1.69087	1.58733
1.15	2.29802	2.13402	1.98538	1.84993	1.72590
1.20	2.54496	2.34658	2.16863	2.00804	1.86234
1.25	2.79458	2.55938	2.35029	2.16318	1.99475
1.30	3.04391	2.76966	2.52779	2.31297	2.12097
1.35	3.28939	2.97418	2.69818	2.45471	2.23856
1.40	3.52686	3.16915	2.85804	2.58536	2.34475
1.45	3.75147	3.35028	3.00354	2.70149	2.43650
1.50	3.95765	3.51263	3.13035	2.79926	2.51042
1.55	4.13907	3.65069	3.23362	2.87445	2.56283
1.60	4.28857	3.75827	3.30802	2.92243	2.58969
1.65	4.39811	3.82851	3.34766	2.93814	2.58666
1.70	4.45880	3.85387	3.34613	2.91614	2.54909
1.75	4.46082	3.82612	3.29652	2.85057	2.47204
1.80	4.39344	3.73634	3.19139	2.73524	2.35031
1.85	4.24504	3.57499	3.02285	2.56360	2.17847

Table 3.11 (continued)

| v | t | | | | |
---	0.80	0.85	0.90	0.95	1.0
1.90	4.00315	3.33190	2.78260	2.32884	1.95093
1.95	3.65447	2.99636	2.46198	2.02396	1.66200
2.00	3.18502	2.55724	2.05209	1.64182	1.30597
2.05	2.58025	2.00309	1.54389	1.17529	0.877205
2.10	1.82519	1.32229	0.928358	0.617343	0.370256
2.15	0.904715	0.503291	0.196671	-0.038741	-0.219972
2.20	-0.196180	-0.465151	-0.659587	-0.799218	-0.898079
2.25	-1.49205	-1.59369	-1.64816	-1.66964	-1.66797
2.30	-2.99657	-2.89202	-2.77581	-2.65462	-2.53265
2.35	-4.72201	-4.36844	-4.04798	-3.75752	-3.49394
2.40	-6.67869	-6.02940	-5.46838	-4.98010	-4.55228
2.45	-8.87422	-7.87891	-7.03854	-6.32221	-5.70631
2.50	-11.3128	-9.91790	-8.75729	-7.78127	-6.95264
2.55	-13.9941	-12.1435	-10.6201	-9.35192	-8.28541
2.60	-16.9126	-14.5483	-12.6187	-11.0254	-9.69588
2.65	-20.0559	-17.1192	-14.7400	-12.7891	-11.1721
2.70	-23.4041	-19.8368	-16.9656	-14.6259	-12.6982
2.75	-26.9276	-22.6741	-19.2709	-16.5136	-14.2543
2.80	-30.5863	-25.5951	-21.6242	-18.4241	-15.8156
2.85	-34.3274	-28.5543	-23.9857	-20.3230	-17.3522
2.90	-38.0839	-31.4944	-26.3069	-22.1687	-18.8284
2.95	-41.7725	-34.3460	-28.5292	-23.9117	-20.2021
3.00	-45.2919	-37.0252	-30.5830	-25.4939	-21.4246
3.05	-48.5208	-39.4333	-32.3872	-26.8482	-22.4401
3.10	-51.3155	-41.4544	-33.8476	-27.8975	-23.1850
3.15	-53.5088	-42.9551	-34.8569	-28.5547	-23.5882
3.20	-54.9075	-43.7827	-35.2933	-28.7219	-23.5703
3.25	-55.2911	-43.7649	-35.0205	-28.2904	-23.0441
3.30	-54.4108	-42.7087	-33.8875	-27.1409	-21.9144
3.35	-51.9884	-40.4005	-31.7284	-25.1434	-20.0785
3.40	-47.7162	-36.6063	-28.3632	-22.1581	-17.4271
3.45	-41.2577	-31.0724	-23.5987	-18.0364	-13.8448
3.50	-32.2491	-23.5274	-17.2303	-12.6225	-9.21209
3.55	-20.3017	-13.6845	-9.04371	-5.75500	-3.40652
3.60	-5.00634	-1.24532	1.18096	2.72962	3.69450
3.65	14.0609	14.0955	13.6657	12.9941	12.2110
3.70	37.3308	32.6434	28.6292	25.1960	22.2567
3.75	65.2310	54.6960	46.2799	39.4819	33.9349
3.80	98.1729	80.5328	66.8085	55.9816	47.3334
3.85	136.536	110.403	90.3778	74.8002	62.5181
3.90	180.647	144.510	117.111	96.0092	79.5266
3.95	230.758	182.995	147.081	119.636	98.3596
4.00	287.019	225.914	180.288	145.654	118.973

Table 3.11 (continued)

			t		
v	0.80	0.85	0.90	0.95	1.0
4.05	349.442	273.215	216.650	173.963	141.265
4.10	417.865	324.713	255.976	204.385	165.068
4.15	491.912	380.053	297.948	236.634	190.135
4.20	570.936	438.679	342.092	270.311	216.126
4.25	653.969	499.795	387.751	304.872	242.592
4.30	739.661	562.320	434.057	339.616	268.960
4.35	826.207	624.844	479.896	373.656	294.521
4.40	911.282	685.577	523.877	405.899	318.408
4.45	991.953	742.294	564.291	435.019	339.580
4.50	1,064.61	792.289	599.084	459.436	356.811
4.55	1,124.86	832.311	625.811	477.290	368.667
4.60	1,167.47	858.517	641.611	486.420	373.499
4.65	1,186.25	866.417	643.168	484.347	369.428
4.70	1,174.03	850.829	626.693	468.257	354.339
4.75	1,122.52	805.845	587.897	434.993	325.878
4.80	1,022.30	724.802	521.984	381.052	281.454
4.85	862.794	600.273	423.657	302.599	218.251
4.90	632.237	424.081	287.132	195.480	133.248
4.95	317.729	187.338	106.181	55.2660	23.2533
5.00	−94.6935	−119.477	−125.803	−122.697	−115.055

			t		
v	1.5	2.0	3.0	4.0	5.0
0.00	−0.083150	0.065447	0.10529	0.056830	0.018083
0.05	−0.049555	0.087780	0.11147	0.056709	0.016798
0.10	−0.014407	0.110452	0.11727	0.056299	0.015408
0.15	0.022304	0.133398	0.12262	0.055594	0.013916
0.20	0.060571	0.156544	0.127501	0.054586	0.012328
0.25	0.100371	0.179805	0.131860	0.053272	0.010648
0.30	0.141662	0.203086	0.135656	0.051645	0.008882
0.35	0.184377	0.226279	0.138844	0.049705	0.007038
0.40	0.228428	0.249265	0.141380	0.047449	0.005121
0.45	0.273701	0.271915	0.143220	0.044878	0.003141
0.50	0.320052	0.294083	0.144321	0.041993	0.001106
0.55	0.367306	0.315615	0.144640	0.038798	−0.000975
0.60	0.415257	0.336342	0.144135	0.035297	−0.003092
0.65	0.463661	0.356082	0.142767	0.031499	−0.005236
0.70	0.512240	0.374641	0.140498	0.027411	−0.007393
0.75	0.560671	0.391813	0.137293	0.023046	−0.009554
0.80	0.608593	0.407379	0.133119	0.018416	−0.011706
0.85	0.655599	0.421110	0.127948	0.013537	−0.013836
0.90	0.701237	0.432764	0.121755	0.008428	−0.015931

Table 3.11 (continued)

			t		
ν	1.5	2.0	3.0	4.0	5.0
0.95	0.745006	0.442094	0.114520	0.003108	−0.017978
1.00	0.786359	0.448839	0.106229	−0.002399	−0.019962
1.05	0.824698	0.452735	0.096872	−0.008068	−0.021870
1.10	0.859376	0.453512	0.086447	−0.013871	−0.023686
1.15	0.889697	0.450896	0.074959	−0.019779	−0.025397
1.20	0.914918	0.444612	0.062421	−0.025757	−0.026988
1.25	0.934247	0.434390	0.048854	−0.031771	−0.028444
1.30	0.946852	0.419961	0.034288	−0.037782	−0.029750
1.35	0.951857	0.401068	0.018763	−0.043751	−0.030893
1.40	0.948355	0.377462	0.002329	−0.049635	−0.031857
1.45	0.935408	0.348915	−0.014952	−0.055389	−0.032630
1.50	0.912056	0.315216	−0.033007	−0.060968	−0.033198
1.55	0.877328	0.276184	−0.051750	−0.066322	−0.033549
1.60	0.830249	0.231665	−0.071085	−0.071402	−0.033671
1.65	0.769857	0.181544	−0.090901	−0.076156	−0.033554
1.70	0.695215	0.125752	−0.111072	−0.080532	−0.033187
1.75	0.605424	0.064265	−0.131461	−0.084477	−0.032562
1.80	0.499649	−0.002881	−0.151917	−0.087938	−0.031671
1.85	0.377135	−0.075586	−0.172273	−0.090860	−0.030510
1.90	0.237230	−0.153681	−0.192352	−0.093191	−0.029073
1.95	0.079415	−0.236914	−0.211959	−0.094878	−0.027359
2.00	−0.096675	−0.324951	−0.230889	−0.095871	−0.025366
2.05	−0.291210	−0.417359	−0.248924	−0.096119	−0.023096
2.10	−0.504142	−0.513605	−0.265834	−0.095578	−0.020552
2.15	−0.735163	−0.613045	−0.281376	−0.094201	−0.017740
2.20	−0.983668	−0.714916	−0.295302	−0.091951	−0.014669
2.25	−1.24872	−0.818334	−0.307351	−0.088790	−0.011349
2.30	−1.52899	−0.922281	−0.317260	−0.084687	−0.007792
2.35	−1.82275	−1.02561	−0.324757	−0.079619	−0.004015
2.40	−2.12780	−1.12701	−0.329571	−0.073564	−0.000037
2.45	−2.44141	−1.22507	−0.331430	−0.066512	0.004123
2.50	−2.76032	−1.31819	−0.330066	−0.058457	0.008439
2.55	−3.08066	−1.40465	−0.325216	−0.049406	0.012884
2.60	−3.39793	−1.48257	−0.316628	−0.039370	0.017431
2.65	−3.70695	−1.54995	−0.304062	−0.028373	0.022045
2.70	−4.00183	−1.60464	−0.287297	−0.016449	0.026692
2.75	−4.27594	−1.64437	−0.266133	−0.003643	0.031335
2.80	−4.52191	−1.66679	−0.240397	0.009987	0.035933
2.85	−4.73159	−1.66942	−0.209948	0.024373	0.040444
2.90	−4.89607	−1.64973	−0.174680	0.039433	0.044823
2.95	−5.00567	−1.60514	−0.134531	0.055071	0.049025
3.00	−5.05003	−1.53308	−0.089487	0.071178	0.053001
3.05	−5.01810	−1.43096	−0.039586	0.087632	0.056702

Table 3.11 (continued)

			t		
v	1.5	2.0	3.0	4.0	5.0
3.10	−4.89824	−1.29628	0.015074	0.104296	0.060077
3.15	−4.67830	−1.12667	0.074331	0.121022	0.063076
3.20	−4.34578	−0.919887	0.137949	0.137643	0.065646
3.25	−3.88795	−0.673935	0.205618	0.153985	0.067737
3.30	−3.29206	−0.387105	0.276946	0.169858	0.069299
3.35	−2.54556	−0.058047	0.351453	0.185059	0.070282
3.40	−1.63639	0.314143	0.428567	0.199377	0.070638
3.45	−0.553255	0.729846	0.507618	0.212590	0.070324
3.50	0.713955	1.18882	0.587841	0.224466	0.069295
3.55	2.17380	1.69012	0.668365	0.234769	0.067515
3.60	3.83280	2.23198	0.748216	0.243255	0.064949
3.65	5.69493	2.81172	0.826314	0.249679	0.061567
3.70	7.76103	3.42562	0.901479	0.253796	0.057346
3.75	10.0282	4.06884	0.972425	0.255361	0.052270
3.80	12.4890	4.73529	1.037770	0.254136	0.046328
3.85	15.1310	5.41753	1.096040	0.249892	0.039518
3.90	17.9357	6.10668	1.145680	0.242410	0.031846
3.95	20.8779	6.79230	1.185050	0.231486	0.023329
4.00	23.9246	7.46234	1.212460	0.216939	0.013992
4.05	27.0346	8.10304	1.226170	0.198608	0.003871
4.10	30.1571	8.69888	1.224400	0.176362	−0.006986
4.15	33.2313	9.23257	1.205370	0.150102	−0.018520
4.20	36.1852	9.68505	1.167330	0.119767	−0.030661
4.25	38.9351	10.0355	1.108530	0.085338	−0.043325
4.30	41.3850	10.2613	1.027340	0.046842	−0.056417
4.35	43.4256	10.3385	0.922201	0.004360	−0.069826
4.40	44.9345	10.2414	0.791706	−0.041974	−0.083433
4.45	45.7756	9.94341	0.634632	−0.091960	−0.097102
4.50	45.7997	9.41676	0.449976	−0.145336	−0.110688
4.55	44.8443	8.63319	0.237006	−0.201769	−0.124031
4.60	42.7350	7.56428	−0.004691	−0.260853	−0.136962
4.65	39.2864	6.18193	−0.275159	−0.322103	−0.149301
4.70	34.3040	4.45898	−0.574021	−0.384958	−0.160857
4.75	27.5870	2.36989	−0.900423	−0.448772	−0.171433
4.80	18.9306	−0.108450	−1.25298	−0.512819	−0.180824
4.85	8.13081	−2.99585	−1.62972	−0.576289	−0.188819
4.90	−5.01146	−6.30778	−2.02802	−0.638289	−0.195206
4.95	−20.6843	−10.0543	−2.44458	−0.697847	−0.199770
5.00	−39.0583	−14.2388	−2.87536	−0.753913	−0.202297

Table 3.12: Second derivatives with respect to the order of the Kelvin functions $\partial^2 kei_\nu(t)/\partial \nu^2$.

			t		
ν	0.05	0.10	0.15	0.20	0.25
0.00	−6.78385	−3.75729	−2.33211	−1.47623	−0.902181
0.05	−9.12070	−4.77542	−2.86165	−1.76187	−1.04872
0.10	−11.4868	−5.73619	−3.32447	−1.98450	−1.13876
0.15	−13.8968	−6.63068	−3.71032	−2.13532	−1.16510
0.20	−16.3498	−7.44227	−4.00463	−2.20281	−1.11886
0.25	−18.8267	−8.14537	−4.18784	−2.17252	−0.989300
0.30	−21.2849	−8.70395	−4.23484	−2.02678	−0.763738
0.35	−23.6525	−9.06959	−4.11423	−1.74437	−0.427408
0.40	−25.8201	−9.17925	−3.78742	−1.30021	0.036654
0.45	−27.6306	−8.95267	−3.20781	−0.665021	0.647711
0.50	−28.8661	−8.28927	−2.31971	0.195028	1.42741
0.55	−29.2325	−7.06470	−1.05734	1.31860	2.39980
0.60	−28.3400	−5.12689	0.656280	2.74947	3.59137
0.65	−25.6789	−2.29173	2.91038	4.53672	5.03089
0.70	−20.5923	1.66168	5.80752	6.73483	6.74930
0.75	−12.2416	6.99607	9.46441	9.40359	8.77935
0.80	0.432401	14.0212	14.0126	12.6078	11.1552
0.85	18.7551	23.0991	19.5990	16.4170	13.9117
0.90	44.3706	34.6489	26.3854	20.9039	17.0835
0.95	79.2998	49.1508	34.5477	26.1437	20.7039
1.00	126.005	67.1496	44.2743	32.2120	24.8033
1.05	187.456	89.2548	55.7629	39.1817	29.4073
1.10	267.204	116.139	69.2156	47.1200	34.5341
1.15	369.443	148.531	84.8324	56.0834	40.1912
1.20	499.071	187.203	102.802	66.1117	46.3722
1.25	661.729	232.953	123.287	77.2210	53.0519
1.30	863.802	286.568	146.411	89.3936	60.1807
1.35	1,112.38	348.785	172.232	102.567	67.6781
1.40	1,415.14	420.224	200.718	116.619	75.4251
1.45	1,780.10	501.302	231.709	131.352	83.2552
1.50	2,215.23	592.117	264.871	146.471	90.9436
1.55	2,727.85	692.297	299.647	161.559	98.1953
1.60	3,323.72	800.803	335.183	176.052	104.631
1.65	4,005.75	915.674	370.250	189.200	109.774
1.70	4,772.12	1,033.71	403.146	200.033	113.027
1.75	5,613.80	1,150.08	431.578	207.316	113.660
1.80	6,511.13	1,257.80	452.527	209.498	110.786
1.85	7,429.21	1,347.16	462.088	204.658	103.341
1.90	8,311.97	1,404.95	455.285	190.447	90.0565
1.95	9,074.29	1,413.63	425.863	164.015	69.4428
2.00	9,591.91	1,350.19	366.057	121.949	39.7615
2.05	9,688.65	1,185.00	266.326	60.1997	−0.993881
2.10	9,120.18	880.312	115.082	−25.9912	−55.1163
2.15	7,553.86	388.635	−101.612	−142.156	−125.198

Table 3.12 (continued)

			t		
v	0.05	0.10	0.15	0.20	0.25
2.20	4,543.88	−349.100	−400.341	−294.660	−214.133
2.25	−499.043	−1,405.59	−800.630	−490.740	−325.117
2.30	−8,344.05	−2,869.33	−1,325.22	−738.525	−461.618
2.35	−19,983.2	−4,846.88	−2,000.26	−1,047.02	−627.339
2.40	−36,681.6	−7,465.09	−2,855.45	−1,426.02	−826.150
2.45	−60,034.5	−10,873.1	−3,924.01	−1,886.01	−1,061.98
2.50	−92,030.4	−15,244.1	−5,242.45	−2,437.90	−1,338.69
2.55	−1,35,118	−20,775.9	−6,850.03	−3,092.69	−1,659.83
2.60	−1,92,277	−27,690.5	−8,787.89	−3,860.99	−2,028.44
2.65	−2,67,079	−36,232.1	−11,097.5	−4,752.32	−2,446.64
2.70	−3,63,739	−46,661.5	−13,818.7	−5,774.19	−2,915.22
2.75	−4,87,141	−59,248.1	−16,986.5	−6,930.87	−3,433.08
2.80	−6,42,811	−74,255.0	−20,626.6	−8,221.81	−3,996.54
2.85	−8,36,829	−91,918.3	−24,750.5	−9,639.64	−4,598.46
2.90	−1.076E+6	−1,12,417	−29,347.1	−11,167.6	−5,227.22
2.95	−1.366	−1,35,830	−34,373.6	−12,776.5	−5,865.46
3.00	−1.712	−1,62,080	−39,742.8	−14,420.5	−6,488.61
3.05	−2.121	−1,90,851	−45,307.4	−16,033.0	−7,063.10
3.10	−2.591	−2,21,492	−50,840.4	−17,520.4	−7,544.37
3.15	−3.120	−2,52,877	−56,010.5	−18,755.6	−7,874.46
3.20	−3.697	−2,83,236	−60,352.8	−19,570.3	−7,979.39
3.25	−4.298	−3,09,938	−63,232.8	−19,745.5	−7,766.14
3.30	−4.883	−3,29,210	−63,803.4	−19,000.9	−7,119.35
3.35	−5.389	−3,35,803	−60,954.7	−16,983.6	−5,897.75
3.40	−5.718	−3,22,571	−53,254.9	−13,254.0	−3,930.34
3.45	−5.726	−2,79,961	−38,883.0	−7,272.3	−1,012.55
3.50	−5.209	−1,95,412	−15,552.6	1,617.6	3,097.67
3.55	−3.877	−52,634.2	19,572.0	14,203.3	8,683.37
3.60	−1.337	1,69,222	69,967.4	31,419.8	16,071.6
3.65	2.951	4,96,512	1,39,845	54,363.6	25,635.5
3.70	9.702	9,62,159	2,34,247	84,304.4	37,794.6
3.75	1.986E+7	1.607E+6	3,59,134	1,22,692	53,013.5
3.80	3.467	2.480	5,21,456	1,71,156	71,796.6
3.85	5.570	3.641	7,29,203	2,31,496	94,679.5
3.90	8.498	5.161	9,91,396	3,05,656	1,22,214
3.95	1.250E+8	7.123	1.318E+6	3,95,675	1,54,948
4.00	1.790	9.622	1.720	5,03,621	1,93,392
4.05	2.507	1.276E+07	2.208	6,31,476	2,37,978
4.10	3.447	1.666	2.794	7,80,982	2,89,004
4.15	4.663	2.145	3.488	9,53,429	3,46,553
4.20	6.218	2.724	4.299	1.149E+6	4,10,406
4.25	8.180	3.415	5.232	1.368	4,79,909
4.30	1.062E+9	4.226	6.286	1.608	5,53,826
4.35	1.362	5.164	7.453	1.864	6,30,145

Table 3.12 (continued)

			t		
ν	0.05	0.10	0.15	0.20	0.25
4.40	1.723	6.224	8.712	2.128	7,05,845
4.45	2.150	7.393	1.003E+7	2.390	7,76,616
4.50	2.644	8.639	1.133	2.632	8,36,518
4.55	3.198	9.907	1.254	2.831	8,77,590
4.60	3.795	1.111E+8	1.352	2.955	8,89,380
4.65	4.403	1.211	1.409	2.960	8,58,416
4.70	4.964	1.271	1.399	2.791	7,67,598
4.75	5.386	1.263	1.289	2.376	5,95,529
4.80	5.528	1.148	1.034	1.622	3,15,779
4.85	5.183	8.751E+7	5.788E+6	4,16,547	−1,03,895
4.90	4.048	3.734	−1.509	−1.384E+6	−7,02,358
4.95	1.698	−4.487	−1.247E+7	−3.951	−1.526E+6
5.00	−2.459	−1.710E+8	−2.822	−7.495	−2.629

			t		
ν	0.30	0.35	0.40	0.45	0.50
0.00	−0.492335	−0.188107	0.043504	0.222734	0.362803
0.05	−0.553605	−0.194917	0.072203	0.274681	0.429769
0.10	−0.565075	−0.158130	0.138964	0.359821	0.525663
0.15	−0.520997	−0.073185	0.247390	0.480982	0.652693
0.20	−0.414510	0.065211	0.401559	0.641300	0.813254
0.25	−0.237584	0.263096	0.606023	0.844199	1.009900
0.30	0.019016	0.527258	0.865787	1.09336	1.24533
0.35	0.365773	0.865225	1.18628	1.39269	1.52231
0.40	0.814439	1.28524	1.57331	1.74627	1.84368
0.45	1.37803	1.79624	2.03302	2.15828	2.21222
0.50	2.07082	2.40774	2.57179	2.63294	2.63064
0.55	2.90824	3.12979	3.19612	3.17439	3.10144
0.60	3.90682	3.97282	3.91255	3.78658	3.62685
0.65	5.08397	4.94742	4.72740	4.47308	4.20863
0.70	6.45775	6.06416	5.64664	5.23696	4.84797
0.75	8.04658	7.33323	6.67556	6.08050	5.54528
0.80	9.86870	8.76407	7.81848	7.00497	6.29999
0.85	11.9417	10.3649	9.07836	8.01031	7.11033
0.90	14.2816	12.1421	10.4564	9.09478	7.97298
0.95	16.9021	14.0995	11.9512	10.2545	8.88283
1.00	19.8134	16.2377	13.5587	11.4831	9.83262
1.05	23.0206	18.5525	15.2708	12.7711	10.8125
1.10	26.5222	21.0345	17.0750	14.1051	11.8095
1.15	30.3079	23.6669	18.9530	15.4674	12.8074
1.20	34.3563	26.4246	20.8801	16.8353	13.7857
1.25	38.6316	29.2717	22.8235	18.1799	14.7194
1.30	43.0809	32.1596	24.7409	19.4652	15.5784
1.35	47.6298	35.0247	26.5795	20.6477	16.3266

Table 3.12 (continued)

			t		
v	0.30	0.35	0.40	0.45	0.50
1.40	52.1779	37.7857	28.2738	21.6745	16.9213
1.45	56.5939	40.3402	29.7438	22.4828	17.3127
1.50	60.7098	42.5618	30.8933	22.9985	17.4431
1.55	64.3144	44.2964	31.6078	23.1349	17.2461
1.60	67.1460	45.3582	31.7520	22.7918	16.6462
1.65	68.8852	45.5258	31.1682	21.8543	15.5580
1.70	69.1459	44.5380	29.6739	20.1914	13.8860
1.75	67.4666	42.0894	27.0596	17.6558	11.5241
1.80	63.3016	37.8263	23.0873	14.0825	8.35566
1.85	56.0113	31.3429	17.4889	9.28915	4.25364
1.90	44.8533	22.1779	9.96544	3.07569	−0.918811
1.95	28.9742	9.81171	0.186438	−4.77474	−7.30736
2.00	7.40201	−6.33479	−12.2090	−14.4944	−15.0649
2.05	−20.9587	−26.9001	−27.6105	−26.3287	−24.3494
2.10	−57.3293	−52.5799	−46.4327	−40.5327	−35.3206
2.15	−103.055	−84.1215	−69.1099	−57.3661	−48.1363
2.20	−159.600	−122.315	−96.0884	−77.0869	−62.9470
2.25	−228.528	−167.982	−127.815	−99.9431	−79.8893
2.30	−311.484	−221.951	−164.723	−126.161	−99.0781
2.35	−410.158	−285.038	−207.212	−155.934	−120.596
2.40	−526.232	−358.007	−255.627	−189.401	−144.484
2.45	−661.3150	−441.526	−310.224	−226.632	−170.722
2.50	−816.8490	−536.112	−371.138	−267.598	−199.218
2.55	−993.9990	−642.058	−438.332	−312.147	−229.785
2.60	−1,193.50	−759.345	−511.550	−359.964	−262.120
2.65	−1,415.45	−887.534	−590.247	−410.537	−295.778
2.70	−1,659.14	−1,025.64	−673.513	−463.103	−330.142
2.75	−1,922.68	−1,171.98	−759.992	−516.605	−364.390
2.80	−2,202.75	−1,323.97	−847.769	−569.620	−397.459
2.85	−2,494.13	−1,477.97	−934.263	−620.304	−428.008
2.90	−2,789.23	−1,628.95	−1,016.09	−666.308	−454.367
2.95	−3,077.54	−1,770.31	−1,088.92	−704.702	−474.500
3.00	−3,344.97	−1,893.48	−1,147.31	−731.888	−485.950
3.05	−3,573.10	−1,987.63	−1,184.52	−743.502	−485.789
3.10	−3,738.32	−2,039.25	−1,192.35	−734.326	−470.573
3.15	−3,810.88	−2,031.72	−1,160.93	−698.183	−436.289
3.20	−3,753.89	−1,944.89	−1,078.47	−627.847	−378.315
3.25	−3,522.11	−1,754.61	−931.158	−514.953	−291.377
3.30	−3,060.84	−1,432.24	−702.864	−349.921	−169.530
3.35	−2,304.62	−944.190	−375.035	−121.900	−6.14085
3.40	−1,176.02	−251.539	73.4583	181.262	206.097
3.45	415.534	690.354	666.402	573.003	475.130
3.50	2,574.77	1,932.06	1,430.23	1,067.99	809.467
3.55	5,421.39	3,530.17	2,393.92	1,681.96	1,218.06

Table 3.12 (continued)

			t		
v	0.30	0.35	0.40	0.45	0.50
3.60	9,090.41	5,547.06	3,588.73	2,431.55	1,710.15
3.65	13,732.0	8,050.47	5,047.77	3,333.90	2,294.97
3.70	19,510.5	11,112.5	6,805.30	4,406.20	2,981.49
3.75	26,602.4	14,808.3	8,895.77	5,665.06	3,777.92
3.80	35,192.8	19,213.6	11,352.4	7,125.64	4,691.20
3.85	45,470.1	24,401.8	14,205.6	8,800.58	5,726.32
3.90	57,618.6	30,440.0	17,480.3	10,698.6	6,885.44
3.95	71,807.3	37,383.3	21,193.0	12,822.6	8,166.89
4.00	88,176.5	45,267.4	25,348.3	15,168.0	9,563.88
4.05	1,06,818	54,100.5	29,933.9	17,719.7	11,063.1
4.10	1,27,753	63,851.1	34,915.1	20,449.2	12,642.8
4.15	1,50,899	74,434.4	40,227.7	23,310.9	14,271.1
4.20	1,76,032	85,695.6	45,769.9	26,237.9	15,903.3
4.25	2,02,742	97,389.2	51,392.7	29,137.1	17,479.3
4.30	2,30,373	1,09,154	56,888.4	31,883.4	18,920.9
4.35	2,57,955	1,20,487	61,977.5	34,313.4	20,128.2
4.40	2,84,118	1,30,705	66,293.9	36,218.3	20,976.0
4.45	3,07,001	1,38,913	69,368.3	37,335.9	21,309.9
4.50	3,24,134	1,43,953	70,609.4	37,342.1	20,942.6
4.55	3,32,308	1,44,361	69,283.5	35,842.2	19,648.9
4.60	3,27,433	1,38,313	64,493.3	32,361.1	17,162.2
4.65	3,04,367	1,23,560	55,154.7	26,334.1	13,169.9
4.70	2,56,738	97,377	39,974.3	17,097.8	7,310.37
4.75	1,76,756	56,490	17,426.6	3,882.11	−830.522
4.80	55,004.7	−2,978.31	−14,265.9	−14,196.5	−11,720.8
4.85	−1,19,755	−85,572.1	−57,149.7	−38,138.9	−25,882.4
4.90	−3,60,781	−1,96,548	−1,13,551	−69,063.8	−43,889.6
4.95	−6,83,515	−3,41,908	−1,86,076	−1,08,203	−66,363.4
5.00	−1.106E+06	−5,28,410	−2,77,604	−1,56,889	−93,962.5

			t		
v	0.55	0.60	0.65	0.70	0.75
0.00	0.472769	0.559087	0.626503	0.678597	0.71813
0.05	0.549082	0.640762	0.710701	0.763282	0.801840
0.10	0.650594	0.744370	0.813972	0.864528	0.899900
0.15	0.779010	0.871214	0.937288	0.983042	1.01280
0.20	0.936141	1.02264	1.08163	1.11951	1.14100
0.25	1.12388	1.20001	1.24794	1.27455	1.28486
0.30	1.34414	1.40468	1.43715	1.44875	1.44470
0.35	1.59889	1.63796	1.65006	1.64255	1.62068
0.40	1.89001	1.90105	1.88739	1.85628	1.81284
0.45	2.21929	2.19503	2.14965	2.09006	2.02103
0.50	2.58837	2.52074	2.43716	2.34381	2.24488

Table 3.12 (continued)

			t		
v	0.55	0.60	0.65	0.70	0.75
0.55	2.99861	2.87878	2.74993	2.61716	2.48374
0.60	3.45106	3.26934	3.08763	2.90939	2.73667
0.65	3.94627	3.69221	3.44949	3.21941	3.00233
0.70	4.48423	4.14659	3.83423	3.54562	3.27899
0.75	5.06417	4.63099	4.23995	3.88590	3.56441
0.80	5.68444	5.14313	4.66401	4.23746	3.85578
0.85	6.34229	5.67975	5.10294	4.59681	4.14968
0.90	7.03362	6.23644	5.55230	4.95960	4.44197
0.95	7.75283	6.80750	6.00653	5.32056	4.72773
1.00	8.49246	7.38570	6.45880	5.67336	5.00117
1.05	9.24295	7.96207	6.90086	6.01051	5.25552
1.10	9.99231	8.52566	7.32288	6.32323	5.48299
1.15	10.72580	9.06335	7.71324	6.60130	5.67464
1.20	11.4254	9.55950	8.05839	6.83301	5.82034
1.25	12.0698	9.99575	8.34264	7.00496	5.90864
1.30	12.6334	10.3507	8.54802	7.10199	5.92676
1.35	13.0866	10.5998	8.65405	7.10708	5.86050
1.40	13.3947	10.7147	8.63763	7.00123	5.69421
1.45	13.5179	10.6635	8.47292	6.76346	5.41076
1.50	13.4109	10.4102	8.13116	6.37069	4.99160
1.55	13.0222	9.91460	7.58070	5.79784	4.41673
1.60	12.2940	9.13213	6.78690	5.01781	3.66484
1.65	11.1623	8.01396	5.71225	4.00160	2.71346
1.70	9.55613	6.50689	4.31648	2.71854	1.53911
1.75	7.39797	4.55360	2.55683	1.13652	0.117588
1.80	4.60389	2.09297	0.388365	−0.777643	−1.57565
1.85	1.08398	−0.939451	−2.23544	−3.05769	−3.56507
1.90	−3.25699	−4.61065	−5.36204	−5.73729	−5.87450
1.95	−8.51844	−8.98919	−9.03892	−8.84928	−8.52644
2.00	−14.8025	−14.1438	−13.3124	−12.4246	−11.5412
2.05	−22.2121	−20.1419	−18.2263	−16.4914	−14.9361
2.10	−30.8483	−27.0471	−23.8201	−21.0730	−18.7240
2.15	−40.8070	−34.9169	−30.1268	−26.1869	−22.9122
2.20	−52.1749	−43.7994	−37.1707	−31.8425	−27.5009
2.25	−65.0246	−53.7294	−44.9641	−38.0388	−32.4812
2.30	−79.4085	−64.7238	−53.5041	−44.7617	−37.8334
2.35	−95.3516	−76.7766	−62.7684	−51.9812	−43.5244
2.40	−112.843	−89.8524	−72.7109	−59.6481	−49.5056
2.45	−131.825	−103.879	−83.2568	−67.6899	−55.7096
2.50	−152.184	−118.741	−94.2962	−76.0070	−62.0477
2.55	−173.735	−134.267	−105.678	−84.4678	−68.4062
2.60	−196.206	−150.224	−117.204	−92.9038	−74.6434
2.65	−219.224	−166.301	−128.617	−101.105	−80.5852

Table 3.12 (continued)

			t		
v	0.55	0.60	0.65	0.70	0.75
2.70	−242.293	−182.102	−139.599	−108.812	−86.0220
2.75	−264.775	−197.127	−149.756	−115.714	−90.7042
2.80	−285.867	−210.762	−158.613	−121.439	−94.3388
2.85	−304.575	−222.261	−165.603	−125.550	−96.5850
2.90	−319.690	−230.731	−170.055	−127.537	−97.0515
2.95	−329.761	−235.117	−171.189	−126.815	−95.2925
3.00	−333.065	−234.185	−168.104	−122.715	−90.8059
3.05	−327.583	−226.508	−159.773	−114.485	−83.0315
3.10	−310.972	−210.452	−145.032	−101.281	−71.3504
3.15	−280.541	−184.167	−122.580	−82.1735	−55.0864
3.20	−233.234	−145.577	−90.9770	−56.1438	−33.5092
3.25	−165.616	−92.3808	−48.6466	−22.0920	−5.83938
3.30	−73.8641	−22.0547	6.11641	21.1552	28.7427
3.35	46.2173	68.1334	75.1311	74.8312	71.0850
3.40	199.181	181.101	160.308	140.210	122.048
3.45	389.894	319.911	263.619	218.579	182.483
3.50	623.470	487.720	387.050	311.207	253.205
3.55	905.177	687.705	532.549	419.298	334.961
3.60	1,240.30	922.960	701.949	543.940	428.387
3.65	1,633.98	1,196.37	896.877	686.033	533.956
3.70	2,090.95	1,510.47	1,118.64	846.210	651.927
3.75	2,615.30	1,867.21	1,368.09	1,024.74	782.263
3.80	3,210.08	2,267.74	1,645.46	1,221.41	924.561
3.85	3,876.86	2,712.13	1,950.16	1,435.40	1,077.95
3.90	4,615.20	3,198.97	2,280.54	1,665.11	1,240.99
3.95	5,422.01	3,725.03	2,633.66	1,908.00	1,411.55
4.00	6,290.81	4,284.74	3,004.93	2,160.36	1,586.65
4.05	7,210.80	4,869.64	3,387.80	2,417.13	1,762.36
4.10	8,165.86	5,467.79	3,773.37	2,671.61	1,933.60
4.15	9,133.41	6,063.02	4,149.91	2,915.19	2,093.96
4.20	10,083.0	6,634.18	4,502.46	3,137.08	2,235.53
4.25	10,974.9	7,154.26	4,812.23	3,323.96	2,348.71
4.30	11,758.3	7,589.47	5,056.12	3,459.69	2,422.00
4.35	12,370.0	7,898.20	5,206.11	3,524.95	2,441.80
4.40	12,731.6	8,030.04	5,228.69	3,496.92	2,392.23
4.45	12,748.5	7,924.64	5,084.28	3,348.93	2,254.95
4.50	12,307.2	7,510.69	4,726.67	3,050.26	2,009.02
4.55	11,273.3	6,704.92	4,102.51	2,565.79	1,630.80
4.60	9,489.79	5,411.15	3,150.96	1,855.93	1,093.92
4.65	6,775.17	3,519.62	1,803.39	876.544	369.285
4.70	2,922.06	906.503	−16.7080	−420.988	−574.751
4.75	−2,303.67	−2,566.16	−2,393.45	−2,089.46	−1,772.14
4.80	−9,162.92	−7,050.02	−5,418.07	−4,185.36	−3,258.69
4.85	−17,942.5	−12,709.3	−9,187.90	−6,768.04	−5,071.44

Table 3.12 (continued)

			t		
v	0.55	0.60	0.65	0.70	0.75
4.90	−28,952.5	−19,718.8	−13,804.9	−9,898.53	−7,247.78
4.95	−42,522.8	−28,261.0	−19,373.2	−13,638.0	−9,824.31
5.00	−58,995.9	−38,521.1	−25,996.3	−18,045.5	−12,835.3

			t		
v	0.80	0.85	0.90	0.95	1.0
0.00	0.74728	0.76778	0.78103	0.78819	0.79021
0.05	0.82896	0.84669	0.85665	0.86017	0.85833
0.10	0.923049	0.93629	0.94144	0.93999	0.93312
0.15	1.02986	1.03675	1.03548	1.02762	1.01446
0.20	1.14965	1.14820	1.13875	1.12297	1.10219
0.25	1.28261	1.27065	1.25116	1.22584	1.19605
0.30	1.42883	1.40404	1.37252	1.33596	1.29569
0.35	1.58827	1.54817	1.50252	1.45293	1.40066
0.40	1.76076	1.70272	1.64071	1.57622	1.51037
0.45	1.94593	1.86717	1.78650	1.70517	1.62412
0.50	2.14322	2.04083	1.93907	1.83891	1.74102
0.55	2.35181	2.22275	2.09744	1.97643	1.86004
0.60	2.57058	2.41173	2.26035	2.11646	1.97994
0.65	2.79810	2.60629	2.42630	2.25751	2.09927
0.70	3.03254	2.80457	2.59349	2.39785	2.21637
0.75	3.27165	3.00437	2.75976	2.53543	2.32931
0.80	3.51268	3.20305	2.92263	2.66791	2.43589
0.85	3.75235	3.39751	3.07920	2.79259	2.53365
0.90	3.98679	3.58414	3.22614	2.90642	2.61980
0.95	4.21143	3.75878	3.35965	3.00598	2.69124
1.00	4.42102	3.91666	3.47546	3.08742	2.74453
1.05	4.60949	4.05239	3.56875	3.14646	2.77590
1.10	4.76995	4.15986	3.63416	3.17840	2.78120
1.15	4.89459	4.23227	3.66575	3.17807	2.75595
1.20	4.97465	4.26206	3.65701	3.13985	2.69529
1.25	5.00036	4.24089	3.60080	3.05765	2.59405
1.30	4.96093	4.15963	3.48942	2.92494	2.44670
1.35	4.84449	4.00835	3.31454	2.73477	2.24742
1.40	4.63810	3.77634	3.06730	2.47980	1.99013
1.45	4.32779	3.45217	2.73832	2.15233	1.66853
1.50	3.89853	3.02367	2.31776	1.74440	1.27620
1.55	3.33438	2.47809	1.79541	1.24784	0.80662
1.60	2.61854	1.80216	1.16081	0.65442	0.25334
1.65	1.73351	0.982280	0.403360	−0.044054	−0.38997
1.70	0.661302	0.004651	−0.487453	−0.855578	−1.12928
1.75	−0.616316	−1.14443	−1.52188	−1.78773	−1.97013

Table 3.12 (continued)

			t		
v	0.80	0.85	0.90	0.95	1.0
1.80	−2.11749	−2.47831	−2.70963	−2.84747	−2.91737
1.85	−3.85989	−4.00961	−4.05958	−4.04084	−3.97492
1.90	−5.86021	−5.74982	−5.57940	−5.37270	−5.14555
1.95	−8.13359	−7.70881	−7.27512	−6.84633	−6.43056
2.00	−10.6929	−9.89423	−9.15070	−8.46306	−7.82945
2.05	−13.5481	−12.3108	−11.2074	−10.2218	−9.33951
2.10	−16.7049	−14.9598	−13.4433	−12.1185	−10.9555
2.15	−20.1641	−17.8377	−15.8523	−14.1458	−12.6690
2.20	−23.9203	−20.9356	−18.4239	−16.2919	−14.4683
2.25	−27.9603	−24.2383	−21.1416	−18.5406	−16.3374
2.30	−32.2615	−27.7225	−23.9827	−20.8699	−18.2558
2.35	−36.7908	−31.3560	−26.9167	−23.2519	−20.1976
2.40	−41.5018	−35.0960	−29.9048	−25.6511	−22.1314
2.45	−46.3336	−38.8877	−32.8981	−28.0246	−24.0189
2.50	−51.2078	−42.6627	−35.8370	−30.3203	−25.8149
2.55	−56.0269	−46.3373	−38.6495	−32.4765	−27.4666
2.60	−60.6712	−49.8104	−41.2506	−34.4211	−28.9125
2.65	−64.9966	−52.9623	−43.5400	−36.0702	−30.0824
2.70	−68.8320	−55.6526	−45.4021	−37.3281	−30.8964
2.75	−71.9764	−57.7183	−46.7040	−38.0857	−31.2651
2.80	−74.1967	−58.9726	−47.2950	−38.2210	−31.0885
2.85	−75.2250	−59.2032	−47.0057	−37.5976	−30.2566
2.90	−74.7571	−58.1715	−45.6473	−36.0653	−28.6493
2.95	−72.4503	−55.6115	−43.0117	−33.4602	−26.1364
3.00	−67.9227	−51.2298	−38.8716	−29.6047	−22.5786
3.05	−60.7529	−44.7061	−32.9814	−24.3093	−17.8285
3.10	−50.4803	−35.6942	−25.0784	−17.3734	−11.7321
3.15	−36.6077	−23.8244	−14.8858	−8.5878	−4.13061
3.20	−18.6043	−8.7071	−2.1149	2.2624	5.13679
3.25	4.08883	10.0625	13.5296	15.3939	16.2296
3.30	32.0510	32.8975	32.3443	31.0204	29.3019
3.35	65.8717	60.2111	54.6197	49.3466	44.4975
3.40	106.137	92.4049	80.6316	70.5608	61.9439
3.45	153.414	129.856	110.630	94.8263	81.7450
3.50	208.228	172.900	144.824	122.271	103.973
3.55	271.035	221.811	183.372	152.974	128.659
3.60	342.195	276.778	226.356	186.955	155.783
3.65	421.933	337.876	273.765	224.153	185.257
3.70	510.294	405.036	325.470	264.412	216.917
3.75	607.092	478.004	381.197	307.459	250.507
3.80	711.854	556.305	440.493	352.882	285.655
3.85	823.755	639.187	502.696	400.104	321.862

Table 3.12 (continued)

			t		
v	**0.80**	**0.85**	**0.90**	**0.95**	**1.0**
3.90	941.536	725.575	566.894	448.354	358.482
3.95	1,063.43	814.009	631.885	496.643	394.695
4.00	1,187.05	902.582	696.132	543.725	429.489
4.05	1,309.34	988.871	757.716	588.073	461.638
4.10	1,426.38	1,069.86	814.288	627.838	489.679
4.15	1,533.35	1,141.89	863.020	660.821	511.885
4.20	1,624.40	1,200.52	900.548	684.437	526.251
4.25	1,692.46	1,240.53	922.933	695.686	530.470
4.30	1,729.21	1,255.80	925.607	691.124	521.917
4.35	1,724.92	1,239.22	903.339	666.844	497.642
4.40	1,668.31	1,182.68	850.201	618.457	454.359
4.45	1,546.52	1,077.02	759.548	541.090	388.453
4.50	1,345.02	911.957	624.017	429.393	295.993
4.55	1,047.56	676.136	435.542	277.563	172.754
4.60	636.212	357.141	185.399	79.3888	14.2613
4.65	91.401	−58.417	−135.718	−171.683	−184.150
4.70	−607.903	−584.770	−537.580	−482.448	−427.255
4.75	−1,483.98	−1,236.79	−1,030.26	−859.809	−719.833
4.80	−2,559.99	−2,029.71	−1,623.91	−1,310.61	−1,066.53
4.85	−3,859.50	−2,978.74	−2,328.47	−1,841.39	−1,471.67
4.90	−5,405.81	−4,098.60	−3,153.32	−2,458.14	−1,939.08
4.95	−7,221.12	−5,402.87	−4,106.78	−3,165.95	−2,471.81
5.00	−9,325.46	−6,903.27	−5,195.60	−3,968.59	−3,071.83

			t		
v	**1.5**	**2.0**	**3.0**	**4.0**	**5.0**
0.00	0.656516	0.44291	0.12736	0.00042	−0.024614
0.05	0.687374	0.45026	0.11990	−0.00528	−0.026761
0.10	0.718576	0.45641	0.11162	−0.01113	−0.028833
0.15	0.749825	0.46119	0.10250	−0.01711	−0.030818
0.20	0.780786	0.464370	0.09253	−0.02321	−0.032703
0.25	0.811086	0.465753	0.08170	−0.02940	−0.03448
0.30	0.840311	0.465111	0.069982	−0.03566	−0.03612
0.35	0.868005	0.462212	0.057386	−0.04196	−0.03763
0.40	0.893670	0.456816	0.043907	−0.04827	−0.03899
0.45	0.916759	0.448679	0.029553	−0.05457	−0.04018
0.50	0.936681	0.437548	0.014336	−0.06082	−0.04120
0.55	0.952799	0.423171	−0.001723	−0.066979	−0.04202
0.60	0.964426	0.405292	−0.018596	−0.073019	−0.04264
0.65	0.970831	0.383657	−0.036247	−0.078896	−0.04305
0.70	0.971235	0.358016	−0.054630	−0.084571	−0.04323

Table 3.12 (continued)

			t		
v	1.5	2.0	3.0	4.0	5.0
0.75	0.964816	0.328126	−0.073689	−0.090001	−0.043168
0.80	0.950712	0.293751	−0.093360	−0.095140	−0.042861
0.85	0.928023	0.254673	−0.113566	−0.099942	−0.042295
0.90	0.895816	0.210689	−0.134220	−0.104361	−0.041463
0.95	0.853132	0.161618	−0.155223	−0.108347	−0.040355
1.00	0.798992	0.107308	−0.176464	−0.111851	−0.038967
1.05	0.732408	0.047636	−0.197819	−0.114823	−0.037292
1.10	0.652394	−0.017481	−0.219151	−0.117212	−0.035326
1.15	0.557976	−0.088081	−0.240311	−0.118968	−0.033067
1.20	0.448206	−0.164154	−0.261136	−0.120040	−0.030515
1.25	0.322185	−0.245631	−0.281450	−0.120381	−0.027670
1.30	0.179073	−0.332379	−0.301064	−0.119941	−0.024535
1.35	0.018118	−0.424195	−0.319775	−0.118675	−0.021116
1.40	−0.161326	−0.520792	−0.337370	−0.116537	−0.017419
1.45	−0.359768	−0.621800	−0.353621	−0.113489	−0.013454
1.50	−0.577553	−0.726750	−0.368292	−0.109490	−0.009232
1.55	−0.814832	−0.835070	−0.381135	−0.104508	−0.004768
1.60	−1.07152	−0.946078	−0.391892	−0.098512	−0.000079
1.65	−1.34727	−1.05897	−0.400300	−0.091479	0.004816
1.70	−1.64143	−1.17281	−0.406087	−0.083390	0.009894
1.75	−1.95299	−1.28654	−0.408979	−0.074234	0.015132
1.80	−2.28053	−1.39896	−0.408699	−0.064005	0.020501
1.85	−2.62222	−1.50869	−0.404971	−0.052708	0.025970
1.90	−2.97571	−1.61426	−0.397522	−0.040354	0.031508
1.95	−3.33814	−1.71398	−0.386084	−0.026966	0.037077
2.00	−3.70602	−1.80605	−0.370401	−0.012574	0.042639
2.05	−4.07525	−1.88850	−0.350229	0.002780	0.048153
2.10	−4.44102	−1.95919	−0.325341	0.019040	0.053575
2.15	−4.79781	−2.01585	−0.295532	0.036141	0.058859
2.20	−5.13928	−2.05606	−0.260624	0.054004	0.063956
2.25	−5.45831	−2.07726	−0.220470	0.072538	0.068817
2.30	−5.74691	−2.07679	−0.174958	0.091638	0.073390
2.35	−5.99621	−2.05188	−0.124021	0.111184	0.077621
2.40	−6.19647	−1.99967	−0.067638	0.131042	0.081455
2.45	−6.33705	−1.91728	−0.005840	0.151065	0.084838
2.50	−6.40642	−1.80177	0.061278	0.171090	0.087712
2.55	−6.39224	−1.65026	0.133559	0.190940	0.090023
2.60	−6.28133	−1.45992	0.210771	0.210424	0.091715
2.65	−6.05985	−1.22802	0.292605	0.229338	0.092734
2.70	−5.71330	−0.952022	0.378667	0.247463	0.093028
2.75	−5.22675	−0.629613	0.468472	0.264570	0.092544
2.80	−4.58492	−0.258788	0.561440	0.280417	0.091236

Table 3.12 (continued)

			t		
ν	1.5	2.0	3.0	4.0	5.0
2.85	−3.77247	0.162072	0.656889	0.294752	0.089060
2.90	−2.77422	0.634113	0.754029	0.307314	0.085973
2.95	−1.57542	1.15792	0.851962	0.317835	0.081942
3.00	−0.16216	1.73341	0.949677	0.326044	0.076935
3.05	1.47825	2.35973	1.04605	0.331664	0.070929
3.10	3.35684	3.03514	1.13982	0.334419	0.063907
3.15	5.48239	3.75689	1.22965	0.334037	0.055859
3.20	7.86090	4.52111	1.31405	0.330248	0.046786
3.25	10.4949	5.32267	1.39143	0.322794	0.036695
3.30	13.3827	6.15507	1.46010	0.311428	0.025607
3.35	16.5176	7.01028	1.51826	0.295921	0.013549
3.40	19.8871	7.87863	1.56404	0.276063	0.000564
3.45	23.4720	8.74872	1.59548	0.251671	−0.013295
3.50	27.2451	9.60725	1.61054	0.222589	−0.027963
3.55	31.1706	10.4390	1.60716	0.188700	−0.043360
3.60	35.2029	11.2265	1.58326	0.149924	−0.059393
3.65	39.2853	11.9505	1.53675	0.106227	−0.075955
3.70	43.3491	12.5893	1.46555	0.057625	−0.092925
3.75	47.3125	13.1192	1.36768	0.004193	−0.110167
3.80	51.0797	13.5141	1.24123	−0.053937	−0.127532
3.85	54.5394	13.7461	1.08441	−0.116563	−0.144854
3.90	57.5647	13.7852	0.895649	−0.183413	−0.161955
3.95	60.0117	13.5996	0.673557	−0.254135	−0.178644
4.00	61.7197	13.1561	0.417037	−0.328298	−0.194716
4.05	62.5103	12.4200	0.125312	−0.405385	−0.209954
4.10	62.1878	11.3560	−0.202011	−0.484789	−0.224131
4.15	60.5399	9.92824	−0.564883	−0.565810	−0.237011
4.20	57.3381	8.10129	−0.962754	−0.647652	−0.248349
4.25	52.3399	5.84050	−1.39450	−0.729424	−0.257894
4.30	45.2904	3.11295	−1.85838	−0.810139	−0.265393
4.35	35.9257	−0.11161	−2.35193	−0.888710	−0.270589
4.40	23.9763	−3.85976	−2.87195	−0.963959	−0.273227
4.45	9.17229	−8.15318	−3.41438	−1.03462	−0.273057
4.50	−8.75119	−13.0074	−3.97431	−1.09932	−0.269835
4.55	−30.0465	−18.4301	−4.54587	−1.15665	−0.263327
4.60	−54.9454	−24.4201	−5.12217	−1.20508	−0.253315
4.65	−83.6487	−30.9647	−5.69533	−1.24306	−0.239598
4.70	−116.316	−38.0388	−6.25634	−1.26898	−0.221997
4.75	−153.050	−45.6022	−6.79514	−1.28118	−0.200361
4.80	−193.886	−53.5979	−7.30054	−1.27800	−0.174569
4.85	−238.771	−61.9500	−7.76025	−1.25780	−0.144536
4.90	−287.549	−70.5613	−8.16092	−1.21895	−0.110217
4.95	−339.937	−79.3110	−8.48814	−1.15986	−0.071614
5.00	−395.507	−88.0529	−8.72658	−1.07906	−0.028778

4 Numerical Results – Tabulation of Functions, First and Second Derivatives with Respect to the Order of the Integral Bessel Functions

The Bessel and Related Functions of the Same Argument and Order

As already mentioned, the integral Bessel functions are tabulated in the Abramowitz–Stegun handbook [9], but only those Bessel functions that have zero order. Their integrands are $(1 - J_0(t))/t$, $(1 - I_0(t))/t$, $Y_0(t)/t$ and $K_0(t)/t$, and the integration limits vary in $0 \le t \le 5$ region in 0.1 intervals.

The integral Bessel functions $Ji_v(t)$, $Yi_v(t)$ and $Ki_v(t)$ were calculated using the integration program of MATHEMATICA. They are presented here in $0 \le v \le 5$ range, with 0.05 intervals, and arguments t are in $0.05 \le t \le 5$ range, with 0.05 intervals for $t \le 1.0$ and with 1.0 intervals for $t > 1.0$. The first and second derivatives with respect to the order of the integral Bessel functions were always calculated by using the central-difference formulas of order $O(h^4)$ with $h = 0.001$.

4.1 The Integral Bessel Functions of the First Kind and Their First and Second Derivatives with Respect to the Order

Integral Bessel functions $Ji_v(t)$ are defined by

$$Ji_v(z) = \int_z^\infty \frac{J_v(t)}{t} dt \tag{4.1.1}$$

They were calculated by applying MATHEMATICA program, and their derivatives with respect to the order of the integral Bessel function of the first kind were determined using the central – difference formulas:

$$\frac{\partial Ji_v(t)}{\partial v} = \frac{-Ji_{v+2h}(t) + 8Ji_{v+h}(t) - 8Ji_{v-h}(t) + Ji_{v-2h}(t)}{12h} \tag{4.1.2}$$

$$\frac{\partial^2 Ji_v(t)}{\partial v^2} = \frac{-Ji_{v+2h}(t) + 16Ji_{v+h}(t) - 30Ji_v(t) + 16Ji_{v-h}(t) - Ji_{v-2h}(t)}{12h^2} \tag{4.1.3}$$

Changes in values of the integral Bessel functions and the first and second derivatives are important only for small values of the order v (see Figures 4.1–4.9; Tables 4.1–4.3).

https://doi.org/10.1515/9783110682472-004

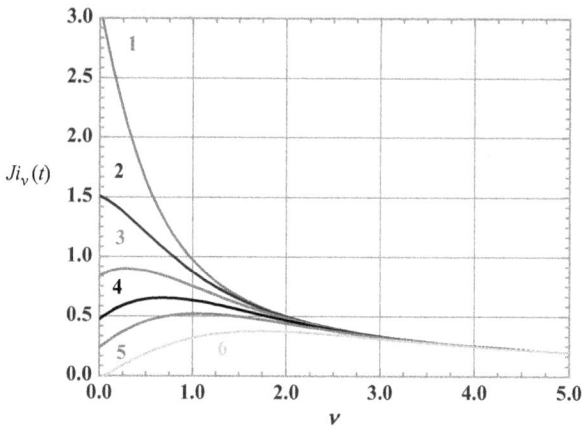

Figure 4.1: The integral Bessel functions of the first kind $Ji_\nu(t)$ plotted as a function of the order ν, at constant values of argument t.

$1 - t = 0.05;\ 2 - t = 0.25;\ 3 - t = 0.50;\ 4 - t = 0.75;\ 5 - t = 1.0;\ 6 - t = 1.5.$

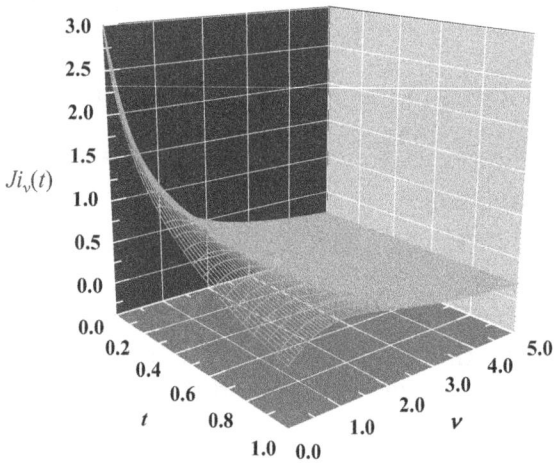

Figure 4.2: The integral Bessel functions of the first kind $Ji_\nu(t)$ as a function of ν and t.

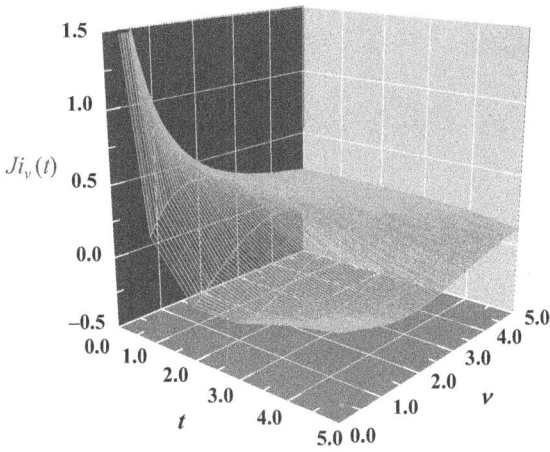

Figure 4.3: The integral Bessel functions of the first kind $Ji_v(t)$ as a function of v and t.

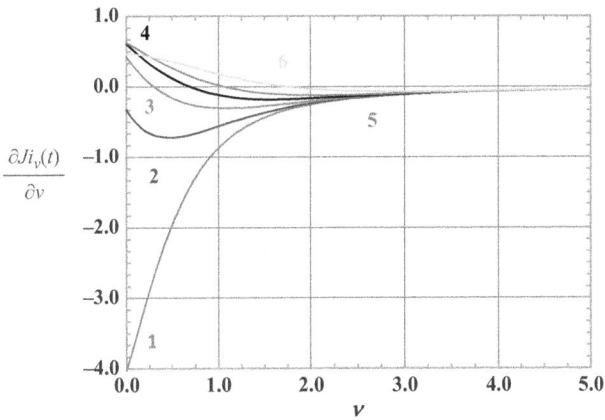

Figure 4.4: First derivatives of the integral Bessel functions of the first kind with respect to the order $\partial Ji_v(t)/\partial v$ plotted as a function of the order v, at constant values of argument t.

$1 - t = 0.05; 2 - t = 0.25; 3 - t = 0.50; 4 - t = 0.75; 5 - t = 1.0; 6 - t = 1.5.$

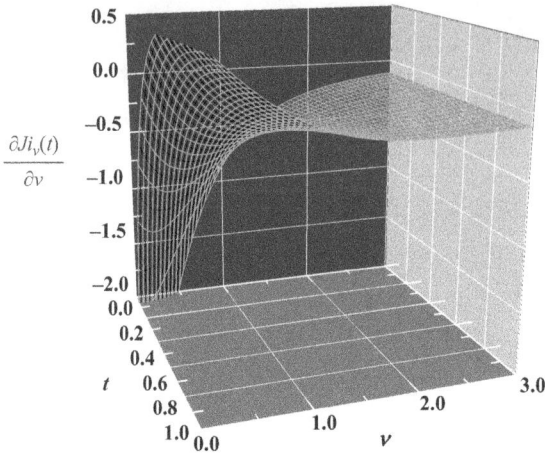

Figure 4.5: First derivatives of the integral Bessel functions of the first kind with respect to the order $\partial Ji_v(t)/\partial v$ as a function of v and t.

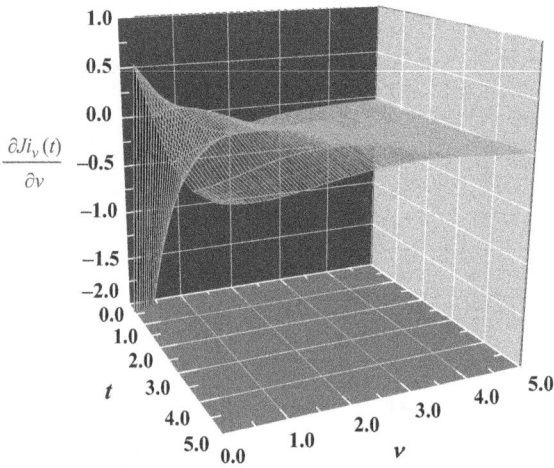

Figure 4.6: First derivatives of the integral Bessel functions of the first kind with respect to the order $\partial Ji_v(t)/\partial v$ as a function of v and t.

$$\frac{\partial^2 Ji_\nu(t)}{\partial \nu^2}$$

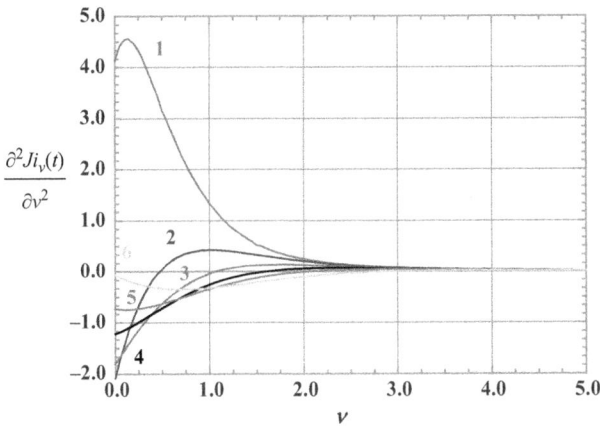

Figure 4.7: Second derivatives of the integral Bessel functions of the first kind with respect to the order $\partial^2 Ji_\nu(t)/\partial \nu^2$ plotted as a function of the order ν, at constant values of argument t.
1 – $t = 0.05$; 2 – $t = 0.25$; 3 – $t = 0.50$; 4 – $t = 0.75$; 5 – $t = 1.0$; 6 – $t = 1.5$.

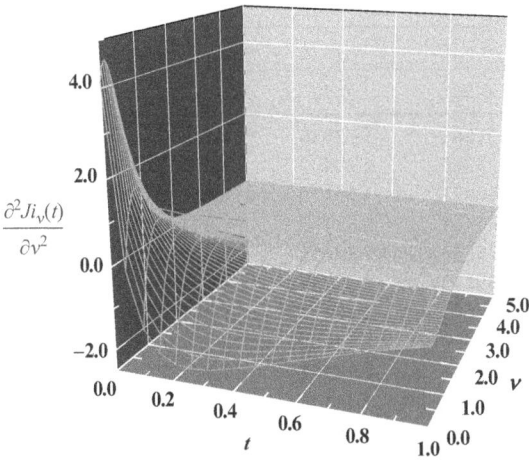

$$\frac{\partial^2 Ji_\nu(t)}{\partial \nu^2}$$

Figure 4.8: Second derivatives of the integral Bessel functions of the first kind with respect to the order $\partial^2 Ji_\nu(t)/\partial \nu^2$ as a function of ν and t.

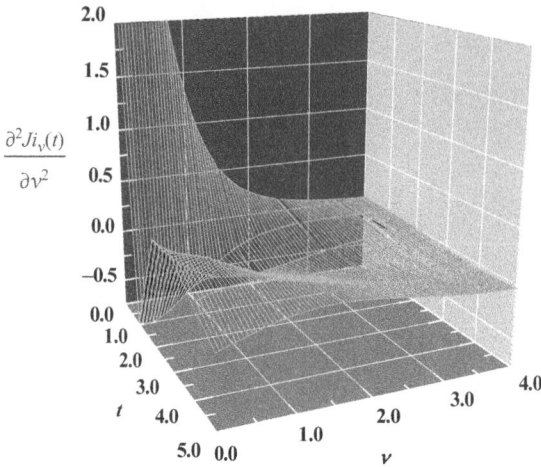

$$\frac{\partial^2 Ji_\nu(t)}{\partial \nu^2}$$

Figure 4.9: Second derivatives of the integral Bessel functions of the first kind with respect to the order $\partial^2 Ji_\nu(t)/\partial \nu^2$ as a function of ν and t.

Table 4.1: The integral Bessel functions of the first kind $Ji_\nu(t)$.

			t			
ν	0.05	0.10	0.15	0.20	0.25	0.30
0.00	3.11198	2.41977	2.01586	1.73036	1.51002	1.33112
0.05	2.91626	2.31458	1.95369	1.69409	1.49111	1.32458
0.10	2.73155	2.21050	1.88927	1.65413	1.46788	1.31347
0.15	2.55817	2.10853	1.82361	1.61139	1.44110	1.29845
0.20	2.39616	2.00939	1.75754	1.56664	1.41147	1.28014
0.25	2.24531	1.91366	1.69173	1.52056	1.37962	1.25909
0.30	2.10529	1.82172	1.62673	1.47372	1.34609	1.23581
0.35	1.97563	1.73383	1.56300	1.42660	1.31135	1.21074
0.40	1.85578	1.65014	1.50087	1.37961	1.27582	1.18428
0.45	1.74518	1.57072	1.44062	1.33308	1.23986	1.15679
0.50	1.64320	1.49554	1.38242	1.28730	1.20377	1.12858
0.55	1.54926	1.42456	1.32643	1.24248	1.16782	1.09991
0.60	1.46274	1.35766	1.27273	1.19880	1.13221	1.07101
0.65	1.38308	1.29470	1.22135	1.15639	1.09712	1.04210
0.70	1.30971	1.23554	1.17231	1.11535	1.06271	1.01334
0.75	1.24213	1.17999	1.12560	1.07575	1.02909	0.984884
0.80	1.17984	1.12788	1.08117	1.03763	0.996351	0.956840
0.85	1.12239	1.07901	1.03898	1.00101	0.964562	0.929314
0.90	1.06935	1.03320	0.998949	0.965907	0.933774	0.902386
0.95	1.02034	0.990266	0.961008	0.932298	0.904022	0.876119
1.00	0.975003	0.950021	0.925070	0.900167	0.875325	0.850561
1.05	0.933013	0.912292	0.891049	0.869481	0.847692	0.825750

Table 4.1 (continued)

			t			
v	0.05	0.10	0.15	0.20	0.25	0.30
1.10	0.894075	0.876913	0.858853	0.840203	0.821122	0.801711
1.15	0.857916	0.843723	0.828393	0.812290	0.795605	0.778459
1.20	0.824293	0.812570	0.799576	0.785693	0.771125	0.756001
1.25	0.792982	0.783313	0.772314	0.760362	0.747660	0.734340
1.30	0.763781	0.755816	0.746519	0.736243	0.725184	0.713469
1.35	0.736509	0.729956	0.722107	0.713285	0.703669	0.693379
1.40	0.710999	0.705615	0.698997	0.691433	0.683082	0.674056
1.45	0.687103	0.682684	0.677112	0.670633	0.663392	0.655485
1.50	0.664684	0.661062	0.656376	0.650835	0.644562	0.637644
1.55	0.643622	0.640656	0.636720	0.631986	0.626560	0.620514
1.60	0.623805	0.621379	0.618076	0.614037	0.609349	0.604072
1.65	0.605133	0.603151	0.600383	0.596941	0.592894	0.588293
1.70	0.587516	0.585898	0.583581	0.580650	0.577161	0.573155
1.75	0.570870	0.569551	0.567613	0.565121	0.562117	0.558632
1.80	0.555122	0.554048	0.552429	0.550312	0.547728	0.544700
1.85	0.540205	0.539330	0.537980	0.536183	0.533963	0.531335
1.90	0.526055	0.525345	0.524219	0.522696	0.520790	0.518511
1.95	0.512619	0.512041	0.511104	0.509814	0.508180	0.506206
2.00	0.499844	0.499375	0.498595	0.497504	0.496104	0.494396
2.05	0.487684	0.487304	0.486656	0.485734	0.484535	0.483059
2.10	0.476097	0.475789	0.475251	0.474472	0.473448	0.472173
2.15	0.465044	0.464795	0.464348	0.463691	0.462816	0.461716
2.20	0.454489	0.454288	0.453918	0.453364	0.452617	0.451669
2.25	0.444401	0.444238	0.443932	0.443466	0.442829	0.442012
2.30	0.434749	0.434618	0.434364	0.433972	0.433429	0.432727
2.35	0.425506	0.425400	0.425190	0.424861	0.424399	0.423795
2.40	0.416647	0.416561	0.416388	0.416111	0.415718	0.415200
2.45	0.408148	0.408079	0.407936	0.407704	0.407370	0.406925
2.50	0.399988	0.399933	0.399815	0.399620	0.399337	0.398955
2.55	0.392148	0.392103	0.392006	0.391843	0.391602	0.391275
2.60	0.384608	0.384573	0.384492	0.384356	0.384152	0.383872
2.65	0.377353	0.377324	0.377258	0.377144	0.376972	0.376732
2.70	0.370366	0.370343	0.370289	0.370193	0.370047	0.369843
2.75	0.363633	0.363615	0.363570	0.363490	0.363367	0.363192
2.80	0.357140	0.357126	0.357089	0.357022	0.356918	0.356769
2.85	0.350875	0.350863	0.350833	0.350778	0.350690	0.350563
2.90	0.344826	0.344817	0.344792	0.344746	0.344671	0.344563
2.95	0.338982	0.338974	0.338954	0.338916	0.338853	0.338761
3.00	0.333332	0.333326	0.333310	0.333278	0.333225	0.333146
3.05	0.327868	0.327863	0.327850	0.327823	0.327779	0.327712
3.10	0.322580	0.322576	0.322565	0.322543	0.322506	0.322449
3.15	0.317460	0.317457	0.317448	0.317429	0.317398	0.317350
3.20	0.312500	0.312497	0.312490	0.312475	0.312448	0.312407

Table 4.1 (continued)

			t			
v	0.05	0.10	0.15	0.20	0.25	0.30
3.25	0.307692	0.307690	0.307684	0.307671	0.307649	0.307615
3.30	0.303030	0.303029	0.303024	0.303013	0.302995	0.302965
3.35	0.298507	0.298506	0.298502	0.298493	0.298478	0.298453
3.40	0.294118	0.294117	0.294113	0.294106	0.294093	0.294072
3.45	0.289855	0.289854	0.289852	0.289846	0.289835	0.289817
3.50	0.285714	0.285714	0.285711	0.285707	0.285697	0.285682
3.55	0.281690	0.281690	0.281688	0.281684	0.281676	0.281663
3.60	0.277778	0.277777	0.277776	0.277773	0.277766	0.277755
3.65	0.273973	0.273972	0.273971	0.273968	0.273963	0.273954
3.70	0.270270	0.270270	0.270269	0.270267	0.270262	0.270255
3.75	0.266667	0.266666	0.266666	0.266664	0.266660	0.266654
3.80	0.263158	0.263158	0.263157	0.263156	0.263152	0.263147
3.85	0.259740	0.259740	0.259740	0.259738	0.259736	0.259731
3.90	0.256410	0.256410	0.256410	0.256409	0.256407	0.256403
3.95	0.253165	0.253164	0.253164	0.253163	0.253161	0.253158
4.00	0.250000	0.250000	0.250000	0.249999	0.249997	0.249995
4.05	0.246914	0.246914	0.246913	0.246913	0.246911	0.246909
4.10	0.243902	0.243902	0.243902	0.243902	0.243901	0.243899
4.15	0.240964	0.240964	0.240964	0.240963	0.240962	0.240961
4.20	0.238095	0.238095	0.238095	0.238095	0.238094	0.238093
4.25	0.235294	0.235294	0.235294	0.235294	0.235293	0.235292
4.30	0.232558	0.232558	0.232558	0.232558	0.232557	0.232556
4.35	0.229885	0.229885	0.229885	0.229885	0.229884	0.229884
4.40	0.227273	0.227273	0.227273	0.227273	0.227272	0.227272
4.45	0.224719	0.224719	0.224719	0.224719	0.224719	0.224718
4.50	0.222222	0.222222	0.222222	0.222222	0.222222	0.222221
4.55	0.219780	0.219780	0.219780	0.219780	0.219780	0.219780
4.60	0.217391	0.217391	0.217391	0.217391	0.217391	0.217391
4.65	0.215054	0.215054	0.215054	0.215054	0.215054	0.215053
4.70	0.212766	0.212766	0.212766	0.212766	0.212766	0.212766
4.75	0.210526	0.210526	0.210526	0.210526	0.210526	0.210526
4.80	0.208333	0.208333	0.208333	0.208333	0.208333	0.208333
4.85	0.206186	0.206186	0.206186	0.206186	0.206185	0.206185
4.90	0.204082	0.204082	0.204082	0.204082	0.204082	0.204081
4.95	0.202020	0.202020	0.202020	0.202020	0.202020	0.202020
5.00	0.200000	0.200000	0.200000	0.200000	0.200000	0.200000

			t			
v	0.35	0.40	0.45	0.50	0.55	0.60
0.00	1.18101	1.05212	0.939592	0.840086	0.751226	0.671254
0.05	1.18362	1.06166	0.954454	0.859071	0.773409	0.695909
0.10	1.18161	1.06667	0.964949	0.873891	0.791656	0.716866

Table 4.1 (continued)

			t			
v	0.35	0.40	0.45	0.50	0.55	0.60
0.15	1.17557	1.06764	0.971484	0.884886	0.806244	0.734352
0.20	1.16602	1.06503	0.974450	0.892386	0.817450	0.748597
0.25	1.15345	1.05926	0.974219	0.896711	0.825548	0.759830
0.30	1.13832	1.05074	0.971144	0.898169	0.830803	0.768280
0.35	1.12103	1.03983	0.965557	0.897052	0.833472	0.774168
0.40	1.10195	1.02689	0.957767	0.893640	0.833803	0.777714
0.45	1.08143	1.01222	0.948063	0.888194	0.832031	0.779127
0.50	1.05977	0.996106	0.936711	0.880959	0.828379	0.778608
0.55	1.03723	0.978813	0.923956	0.872164	0.823059	0.776351
0.60	1.01406	0.960572	0.910024	0.862021	0.816268	0.772537
0.65	0.990453	0.941594	0.895117	0.850725	0.808191	0.767339
0.70	0.966605	0.922064	0.879422	0.838456	0.798998	0.760918
0.75	0.942671	0.902150	0.863105	0.825378	0.788849	0.753426
0.80	0.918791	0.881998	0.846316	0.811639	0.777889	0.745004
0.85	0.895081	0.861735	0.829188	0.797376	0.766252	0.735781
0.90	0.871641	0.841474	0.811840	0.782709	0.754059	0.725877
0.95	0.848555	0.821311	0.794376	0.767747	0.741422	0.715403
1.00	0.825890	0.801328	0.776889	0.752588	0.728440	0.704460
1.05	0.803705	0.781596	0.759457	0.737318	0.715203	0.693139
1.10	0.782042	0.762174	0.742150	0.722012	0.701792	0.681523
1.15	0.760938	0.743109	0.725027	0.706737	0.688279	0.669688
1.20	0.740417	0.724443	0.708138	0.691552	0.674727	0.657701
1.25	0.720497	0.706205	0.691525	0.676506	0.661192	0.645622
1.30	0.701191	0.688422	0.675222	0.661642	0.647723	0.633505
1.35	0.682503	0.671111	0.659259	0.646996	0.634363	0.621398
1.40	0.664436	0.654285	0.643658	0.632599	0.621148	0.609341
1.45	0.646985	0.637953	0.628436	0.618476	0.608110	0.597372
1.50	0.630145	0.622117	0.613605	0.604647	0.595276	0.585523
1.55	0.613905	0.606780	0.599176	0.591128	0.582667	0.573819
1.60	0.598255	0.591937	0.585153	0.577932	0.570301	0.562285
1.65	0.583179	0.577585	0.571540	0.565068	0.558194	0.550940
1.70	0.568664	0.563716	0.558335	0.552542	0.546357	0.539799
1.75	0.554693	0.550321	0.545536	0.540357	0.534798	0.528878
1.80	0.541249	0.537390	0.533141	0.528514	0.523525	0.518185
1.85	0.528313	0.524912	0.521142	0.517014	0.512540	0.507729
1.90	0.515870	0.512874	0.509533	0.505854	0.501846	0.497516
1.95	0.503899	0.501263	0.498305	0.495030	0.491443	0.487551
2.00	0.492383	0.490066	0.487450	0.484537	0.481331	0.477835
2.05	0.481304	0.479270	0.476958	0.474370	0.471506	0.468370
2.10	0.470644	0.468860	0.466820	0.464522	0.461967	0.459156
2.15	0.460386	0.458823	0.457023	0.454986	0.452709	0.450192
2.20	0.450513	0.449144	0.447559	0.445754	0.443726	0.441474
2.25	0.441008	0.439811	0.438416	0.436818	0.435014	0.433001

Table 4.1 (continued)

				t		
v	0.35	0.40	0.45	0.50	0.55	0.60
2.30	0.431856	0.430809	0.429582	0.428169	0.426565	0.424768
2.35	0.423040	0.422126	0.421048	0.419799	0.418375	0.416772
2.40	0.414546	0.413749	0.412802	0.411700	0.410436	0.409007
2.45	0.406359	0.405664	0.404834	0.403861	0.402741	0.401468
2.50	0.398465	0.397861	0.397133	0.396276	0.395284	0.394151
2.55	0.390853	0.390326	0.389689	0.388934	0.388056	0.387049
2.60	0.383507	0.383049	0.382492	0.381828	0.381051	0.380157
2.65	0.376417	0.376020	0.375532	0.374948	0.374262	0.373468
2.70	0.369571	0.369226	0.368800	0.368288	0.367682	0.366977
2.75	0.362958	0.362659	0.362287	0.361837	0.361303	0.360678
2.80	0.356568	0.356308	0.355984	0.355589	0.355118	0.354565
2.85	0.350390	0.350165	0.349882	0.349536	0.349121	0.348632
2.90	0.344414	0.344220	0.343974	0.343671	0.343305	0.342873
2.95	0.338633	0.338465	0.338251	0.337985	0.337664	0.337282
3.00	0.333037	0.332892	0.332705	0.332473	0.332191	0.331853
3.05	0.327618	0.327492	0.327331	0.327128	0.326880	0.326582
3.10	0.322368	0.322260	0.322120	0.321943	0.321725	0.321462
3.15	0.317281	0.317187	0.317065	0.316911	0.316720	0.316489
3.20	0.312348	0.312268	0.312162	0.312027	0.311860	0.311656
3.25	0.307564	0.307495	0.307403	0.307286	0.307139	0.306960
3.30	0.302922	0.302862	0.302783	0.302681	0.302552	0.302395
3.35	0.298416	0.298365	0.298296	0.298207	0.298095	0.297956
3.40	0.294041	0.293996	0.293937	0.293860	0.293761	0.293640
3.45	0.289790	0.289752	0.289701	0.289633	0.289548	0.289441
3.50	0.285659	0.285627	0.285583	0.285524	0.285449	0.285356
3.55	0.281644	0.281616	0.281578	0.281527	0.281462	0.281380
3.60	0.277739	0.277715	0.277682	0.277638	0.277581	0.277509
3.65	0.273940	0.273919	0.273891	0.273853	0.273803	0.273740
3.70	0.270243	0.270225	0.270201	0.270167	0.270124	0.270069
3.75	0.266643	0.266628	0.266607	0.266579	0.266541	0.266493
3.80	0.263138	0.263126	0.263107	0.263083	0.263050	0.263008
3.85	0.259724	0.259713	0.259697	0.259676	0.259647	0.259611
3.90	0.256396	0.256387	0.256374	0.256355	0.256330	0.256298
3.95	0.253153	0.253145	0.253133	0.253117	0.253096	0.253068
4.00	0.249990	0.249983	0.249973	0.249960	0.249941	0.249917
4.05	0.246905	0.246900	0.246891	0.246879	0.246863	0.246842
4.10	0.243896	0.243891	0.243883	0.243873	0.243859	0.243840
4.15	0.240958	0.240954	0.240948	0.240939	0.240927	0.240910
4.20	0.238090	0.238087	0.238081	0.238074	0.238063	0.238049
4.25	0.235290	0.235287	0.235282	0.235276	0.235267	0.235255
4.30	0.232555	0.232552	0.232548	0.232543	0.232535	0.232524
4.35	0.229882	0.229880	0.229877	0.229872	0.229865	0.229856
4.40	0.227270	0.227268	0.227266	0.227261	0.227256	0.227248

Table 4.1 (continued)

			t			
v	0.35	0.40	0.45	0.50	0.55	0.60
4.45	0.224717	0.224716	0.224713	0.224709	0.224704	0.224697
4.50	0.222221	0.222219	0.222217	0.222214	0.222210	0.222204
4.55	0.219779	0.219778	0.219776	0.219773	0.219769	0.219764
4.60	0.217390	0.217389	0.217388	0.217385	0.217382	0.217378
4.65	0.215053	0.215052	0.215051	0.215049	0.215046	0.215042
4.70	0.212765	0.212764	0.212763	0.212762	0.212759	0.212756
4.75	0.210526	0.210525	0.210524	0.210523	0.210521	0.210518
4.80	0.208333	0.208332	0.208331	0.208330	0.208328	0.208326
4.85	0.206185	0.206185	0.206184	0.206183	0.206181	0.206179
4.90	0.204081	0.204081	0.204080	0.204079	0.204078	0.204076
4.95	0.202020	0.202020	0.202019	0.202018	0.202017	0.202016
5.00	0.200000	0.199999	0.199999	0.199998	0.199997	0.199996

			t			
	0.65	0.70	0.75	0.80	0.85	0.90
0.00	0.598835	0.532927	0.472703	0.417494	0.366751	0.320017
0.05	0.625377	0.560882	0.501680	0.447169	0.396851	0.350312
0.10	0.648467	0.585631	0.527695	0.474118	0.424457	0.378337
0.15	0.668286	0.607314	0.550851	0.498417	0.449615	0.404114
0.20	0.685020	0.626081	0.571265	0.520150	0.472386	0.427679
0.25	0.698862	0.642088	0.589062	0.539415	0.492841	0.449083
0.30	0.710003	0.655496	0.604373	0.556318	0.511064	0.468386
0.35	0.718635	0.666468	0.617337	0.570971	0.527142	0.485659
0.40	0.724949	0.675167	0.628091	0.583491	0.541174	0.500977
0.45	0.729130	0.681757	0.636777	0.593998	0.553260	0.514424
0.50	0.731358	0.686396	0.643534	0.602614	0.563503	0.526088
0.55	0.731807	0.689242	0.648502	0.609460	0.572009	0.536059
0.60	0.730644	0.690444	0.651815	0.614657	0.578886	0.544429
0.65	0.728029	0.690149	0.653607	0.618325	0.584239	0.551293
0.70	0.724112	0.688498	0.654005	0.620579	0.588172	0.556745
0.75	0.719037	0.685622	0.653134	0.621534	0.590791	0.560876
0.80	0.712936	0.681648	0.651111	0.621299	0.592195	0.563781
0.85	0.705936	0.676697	0.648049	0.619980	0.592481	0.565547
0.90	0.698152	0.670879	0.644054	0.617676	0.591746	0.566264
0.95	0.689694	0.664301	0.639229	0.614485	0.590078	0.566015
1.00	0.680661	0.657059	0.633666	0.610498	0.587566	0.564884
1.05	0.671146	0.649245	0.627457	0.605799	0.584290	0.562947
1.10	0.661231	0.640943	0.620682	0.600471	0.580330	0.560281
1.15	0.650995	0.632230	0.613419	0.594588	0.575759	0.556956
1.20	0.640508	0.623178	0.605740	0.588222	0.570647	0.553040
1.25	0.629831	0.613851	0.597711	0.581438	0.565058	0.548596
1.30	0.619023	0.604309	0.589391	0.574297	0.559054	0.543686
1.35	0.608135	0.594604	0.580836	0.566856	0.552691	0.538364

Table 4.1 (continued)

			t			
	0.65	0.70	0.75	0.80	0.85	0.90
1.40	0.597211	0.584787	0.572096	0.559166	0.546021	0.532683
1.45	0.586293	0.574899	0.563218	0.551275	0.539092	0.526693
1.50	0.575416	0.564981	0.554243	0.543225	0.531950	0.520438
1.55	0.564611	0.555066	0.545208	0.535057	0.524635	0.513961
1.60	0.553906	0.545186	0.536147	0.526806	0.517185	0.507300
1.65	0.543324	0.535368	0.527089	0.518505	0.509633	0.500490
1.70	0.532886	0.525635	0.518061	0.510182	0.502011	0.493565
1.75	0.522609	0.516008	0.509088	0.501863	0.494347	0.486553
1.80	0.512507	0.506504	0.500188	0.493571	0.486665	0.479482
1.85	0.502592	0.497139	0.491381	0.485328	0.478989	0.472375
1.90	0.492873	0.487926	0.482682	0.477150	0.471338	0.465256
1.95	0.483359	0.478875	0.474104	0.469053	0.463731	0.458143
2.00	0.474055	0.469994	0.465658	0.461053	0.456183	0.451055
2.05	0.464964	0.461291	0.457354	0.453159	0.448708	0.444007
2.10	0.456090	0.452771	0.449201	0.445382	0.441318	0.437012
2.15	0.447434	0.444438	0.441203	0.437731	0.434024	0.430084
2.20	0.438997	0.436294	0.433366	0.430212	0.426834	0.423233
2.25	0.430778	0.428342	0.425693	0.422831	0.419756	0.416467
2.30	0.422774	0.420582	0.418189	0.415594	0.412796	0.409796
2.35	0.414985	0.413013	0.410853	0.408502	0.405960	0.403225
2.40	0.407408	0.405635	0.403687	0.401559	0.399251	0.396760
2.45	0.400038	0.398447	0.396691	0.394767	0.392673	0.390407
2.50	0.392873	0.391445	0.389864	0.388126	0.386228	0.384168
2.55	0.385908	0.384628	0.383206	0.381637	0.379918	0.378046
2.60	0.379139	0.377992	0.376714	0.375298	0.373743	0.372044
2.65	0.372561	0.371535	0.370386	0.369111	0.367705	0.366164
2.70	0.366169	0.365252	0.364221	0.363072	0.361802	0.360406
2.75	0.359959	0.359139	0.358215	0.357182	0.356035	0.354771
2.80	0.353925	0.353194	0.352365	0.351436	0.350402	0.349258
2.85	0.348063	0.347411	0.346669	0.345834	0.344902	0.343868
2.90	0.342368	0.341786	0.341122	0.340373	0.339533	0.338600
2.95	0.336834	0.336315	0.335722	0.335050	0.334294	0.333451
3.00	0.331456	0.330995	0.330465	0.329862	0.329183	0.328422
3.05	0.326230	0.325820	0.325347	0.324807	0.324196	0.323511
3.10	0.321151	0.320786	0.320364	0.319880	0.319332	0.318715
3.15	0.316213	0.315889	0.315513	0.315080	0.314588	0.314033
3.20	0.311412	0.311125	0.310790	0.310403	0.309962	0.309462
3.25	0.306744	0.306489	0.306191	0.305846	0.305450	0.305001
3.30	0.302205	0.301978	0.301713	0.301405	0.301051	0.300647
3.35	0.297788	0.297588	0.297352	0.297077	0.296760	0.296398
3.40	0.293492	0.293314	0.293105	0.292860	0.292576	0.292251
3.45	0.289311	0.289154	0.288967	0.288749	0.288496	0.288205
3.50	0.285241	0.285102	0.284937	0.284742	0.284516	0.284255

Table 4.1 (continued)

	0.65	0.70	0.75	0.80	0.85	0.90
3.55	0.281278	0.281156	0.281009	0.280836	0.280634	0.280401
3.60	0.277420	0.277312	0.277182	0.277028	0.276848	0.276639
3.65	0.273662	0.273566	0.273451	0.273315	0.273154	0.272967
3.70	0.270000	0.269916	0.269814	0.269693	0.269550	0.269383
3.75	0.266433	0.266358	0.266268	0.266160	0.266033	0.265884
3.80	0.262955	0.262889	0.262810	0.262714	0.262601	0.262468
3.85	0.259564	0.259506	0.259436	0.259351	0.259250	0.259132
3.90	0.256258	0.256207	0.256145	0.256070	0.255980	0.255874
3.95	0.253032	0.252988	0.252933	0.252866	0.252787	0.252692
4.00	0.249885	0.249846	0.249798	0.249739	0.249668	0.249584
4.05	0.246814	0.246780	0.246737	0.246685	0.246622	0.246548
4.10	0.243817	0.243786	0.243749	0.243703	0.243647	0.243581
4.15	0.240890	0.240863	0.240830	0.240789	0.240740	0.240681
4.20	0.238031	0.238008	0.237979	0.237943	0.237899	0.237846
4.25	0.235239	0.235218	0.235193	0.235161	0.235122	0.235076
4.30	0.232510	0.232492	0.232470	0.232442	0.232408	0.232366
4.35	0.229844	0.229828	0.229808	0.229784	0.229753	0.229716
4.40	0.227237	0.227223	0.227206	0.227184	0.227157	0.227125
4.45	0.224688	0.224676	0.224661	0.224642	0.224618	0.224589
4.50	0.222196	0.222185	0.222172	0.222155	0.222134	0.222108
4.55	0.219757	0.219748	0.219736	0.219722	0.219703	0.219680
4.60	0.217371	0.217364	0.217353	0.217340	0.217324	0.217304
4.65	0.215037	0.215030	0.215021	0.215009	0.214995	0.214977
4.70	0.212751	0.212745	0.212737	0.212727	0.212715	0.212699
4.75	0.210514	0.210508	0.210501	0.210493	0.210481	0.210468
4.80	0.208322	0.208318	0.208312	0.208304	0.208294	0.208282
4.85	0.206176	0.206172	0.206167	0.206160	0.206151	0.206141
4.90	0.204074	0.204070	0.204065	0.204059	0.204052	0.204042
4.95	0.202013	0.202010	0.202006	0.202001	0.201994	0.201986
5.00	0.199994	0.199991	0.199988	0.199983	0.199977	0.199970

v	0.95	1.0	2.0	3.0	4.0	5.0
0.00	0.276908	0.237097	−0.135296	−0.126534	−0.023126	0.046841
0.05	0.307204	0.267227	−0.121916	−0.128495	−0.029793	0.043075
0.10	0.335444	0.295508	−0.107993	−0.129694	−0.036240	0.039073
0.15	0.361631	0.321923	−0.093616	−0.130142	−0.042436	0.034861
0.20	0.385779	0.346471	−0.078873	−0.129857	−0.048350	0.030467
0.25	0.407920	0.369163	−0.063850	−0.128859	−0.053957	0.025919
0.30	0.428095	0.390023	−0.048625	−0.127170	−0.059231	0.021245
0.35	0.446354	0.409087	−0.033275	−0.124818	−0.064151	0.016472
0.40	0.462758	0.426397	−0.017873	−0.121829	−0.068699	0.011629

Table 4.1 (continued)

			t			
v	0.95	1.0	2.0	3.0	4.0	5.0
0.45	0.477373	0.442006	−0.002486	−0.118234	−0.072857	0.006742
0.50	0.490271	0.455970	0.012823	−0.114065	−0.076613	0.001838
0.55	0.501529	0.468352	0.027995	−0.109355	−0.079955	−0.003057
0.60	0.511225	0.479220	0.042976	−0.104140	−0.082875	−0.007918
0.65	0.519442	0.488644	0.057716	−0.098453	−0.085367	−0.012721
0.70	0.526262	0.496695	0.072172	−0.092332	−0.087428	−0.017441
0.75	0.531769	0.503448	0.086302	−0.085813	−0.089054	−0.022058
0.80	0.536045	0.508976	0.100071	−0.078931	−0.090248	−0.026550
0.85	0.539173	0.513354	0.113446	−0.071724	−0.091012	−0.030897
0.90	0.541233	0.516656	0.126400	−0.064228	−0.091350	−0.035080
0.95	0.542305	0.518954	0.138910	−0.056477	−0.091269	−0.039084
1.00	0.542464	0.520320	0.150955	−0.048508	−0.090778	−0.042891
1.05	0.541786	0.520823	0.162518	−0.040355	−0.089884	−0.046488
1.10	0.540341	0.520530	0.173587	−0.032050	−0.088601	−0.049861
1.15	0.538198	0.519507	0.184152	−0.023625	−0.086939	−0.052998
1.20	0.535422	0.517815	0.194206	−0.015113	−0.084913	−0.055890
1.25	0.532075	0.515516	0.203745	−0.006543	−0.082537	−0.058528
1.30	0.528215	0.512665	0.212766	0.002057	−0.079826	−0.060904
1.35	0.523898	0.509316	0.221272	0.010660	−0.076797	−0.063013
1.40	0.519177	0.505522	0.229264	0.019239	−0.073465	−0.064848
1.45	0.514099	0.501331	0.236748	0.027771	−0.069850	−0.066408
1.50	0.508711	0.496788	0.243731	0.036232	−0.065968	−0.067690
1.55	0.503055	0.491935	0.250220	0.044602	−0.061837	−0.068692
1.60	0.497170	0.486812	0.256225	0.052860	−0.057477	−0.069415
1.65	0.491093	0.481458	0.261758	0.060988	−0.052904	−0.069861
1.70	0.484858	0.475905	0.266830	0.068970	−0.048139	−0.070031
1.75	0.478495	0.470186	0.271454	0.076789	−0.043199	−0.069929
1.80	0.472033	0.464331	0.275644	0.084433	−0.038103	−0.069559
1.85	0.465497	0.458366	0.279415	0.091888	−0.032868	−0.068926
1.90	0.458912	0.452316	0.282782	0.099144	−0.027513	−0.068037
1.95	0.452298	0.446204	0.285759	0.106191	−0.022055	−0.066898
2.00	0.445675	0.440051	0.288362	0.113020	−0.016511	−0.065516
2.05	0.439061	0.433874	0.290608	0.119623	−0.010898	−0.063899
2.10	0.432469	0.427692	0.292513	0.125996	−0.005232	−0.062057
2.15	0.425915	0.421520	0.294091	0.132132	0.000472	−0.059998
2.20	0.419411	0.415371	0.295360	0.138029	0.006198	−0.057731
2.25	0.412967	0.409256	0.296335	0.143683	0.011932	−0.055266
2.30	0.406593	0.403188	0.297032	0.149092	0.017661	−0.052615
2.35	0.400297	0.397176	0.297466	0.154255	0.023370	−0.049787
2.40	0.394086	0.391228	0.297652	0.159173	0.029048	−0.046792
2.45	0.387966	0.385351	0.297606	0.163846	0.034683	−0.043643
2.50	0.381943	0.379551	0.297341	0.168274	0.040263	−0.040349
2.55	0.376019	0.373835	0.296871	0.172461	0.045779	−0.036922

Table 4.1 (continued)

			t			
v	0.95	1.0	2.0	3.0	4.0	5.0
2.60	0.370200	0.368206	0.296211	0.176409	0.051220	−0.033373
2.65	0.364486	0.362668	0.295373	0.180121	0.056577	−0.029712
2.70	0.358881	0.357225	0.294370	0.183602	0.061842	−0.025952
2.75	0.353386	0.351878	0.293215	0.186854	0.067008	−0.022102
2.80	0.348002	0.346630	0.291918	0.189883	0.072067	−0.018173
2.85	0.342729	0.341482	0.290492	0.192694	0.077013	−0.014177
2.90	0.337568	0.336435	0.288947	0.195292	0.081840	−0.010122
2.95	0.332518	0.331489	0.287293	0.197682	0.086543	−0.006020
3.00	0.327578	0.326645	0.285541	0.199871	0.091117	−0.001880
3.05	0.322747	0.321902	0.283699	0.201865	0.095559	0.002289
3.10	0.318025	0.317259	0.281776	0.203669	0.099865	0.006477
3.15	0.313410	0.312717	0.279781	0.205290	0.104032	0.010675
3.20	0.308901	0.308274	0.277722	0.206734	0.108059	0.014874
3.25	0.304495	0.303928	0.275606	0.208009	0.111943	0.019067
3.30	0.300191	0.299680	0.273440	0.209120	0.115682	0.023246
3.35	0.295988	0.295526	0.271230	0.210073	0.119277	0.027402
3.40	0.291882	0.291466	0.268984	0.210876	0.122726	0.031529
3.45	0.287873	0.287497	0.266706	0.211535	0.126029	0.035621
3.50	0.283957	0.283619	0.264402	0.212057	0.129187	0.039669
3.55	0.280133	0.279828	0.262077	0.212447	0.132200	0.043670
3.60	0.276399	0.276125	0.259737	0.212711	0.135070	0.047616
3.65	0.272752	0.272505	0.257384	0.212857	0.137797	0.051504
3.70	0.269190	0.268968	0.255024	0.212890	0.140383	0.055327
3.75	0.265711	0.265512	0.252660	0.212817	0.142831	0.059081
3.80	0.262313	0.262134	0.250295	0.212642	0.145141	0.062763
3.85	0.258993	0.258833	0.247932	0.212371	0.147316	0.066368
3.90	0.255750	0.255607	0.245575	0.212011	0.149359	0.069893
3.95	0.252582	0.252453	0.243227	0.211566	0.151272	0.073335
4.00	0.249485	0.249370	0.240889	0.211041	0.153059	0.076692
4.05	0.246460	0.246356	0.238563	0.210442	0.154721	0.079960
4.10	0.243502	0.243410	0.236253	0.209774	0.156263	0.083137
4.15	0.240611	0.240528	0.233959	0.209040	0.157687	0.086223
4.20	0.237784	0.237710	0.231683	0.208246	0.158996	0.089215
4.25	0.235020	0.234954	0.229427	0.207396	0.160193	0.092111
4.30	0.232317	0.232258	0.227192	0.206494	0.161283	0.094913
4.35	0.229672	0.229620	0.224978	0.205544	0.162269	0.097617
4.40	0.227086	0.227039	0.222788	0.204550	0.163153	0.100225
4.45	0.224554	0.224513	0.220622	0.203515	0.163939	0.102735
4.50	0.222077	0.222040	0.218481	0.202443	0.164632	0.105148
4.55	0.219653	0.219620	0.216365	0.201337	0.165234	0.107465
4.60	0.217279	0.217250	0.214274	0.200201	0.165748	0.109684
4.65	0.214956	0.214929	0.212211	0.199038	0.166178	0.111808
4.70	0.212680	0.212657	0.210173	0.197849	0.166528	0.113836

Table 4.1 (continued)

v	0.95	1.0	2.0	3.0	4.0	5.0
4.75	0.210451	0.210430	0.208163	0.196639	0.166801	0.115770
4.80	0.208267	0.208249	0.206180	0.195409	0.167000	0.117611
4.85	0.206127	0.206111	0.204224	0.194162	0.167128	0.119359
4.90	0.204031	0.204016	0.202296	0.192900	0.167189	0.121017
4.95	0.201975	0.201963	0.200395	0.191626	0.167186	0.122586
5.00	0.199961	0.199949	0.198521	0.190341	0.167122	0.124067

Table 4.2: First derivatives with respect to the order of the integral Bessel functions of the first kind $\partial Ji_v(t)/\partial v$.

v	0.05	0.10	0.15	0.20	0.25	0.30
0.00	−4.02020	−2.10704	−1.21359	−0.682088	−0.329280	−0.080234
0.05	−3.80590	−2.09632	−1.26949	−0.765505	−0.424196	−0.178880
0.10	−3.58146	−2.06338	−1.30374	−0.829751	−0.502608	−0.263490
0.15	−3.35367	−2.01326	−1.31985	−0.877283	−0.566236	−0.335253
0.20	−3.12767	−1.95021	−1.32093	−0.910355	−0.616709	−0.395336
0.25	−2.90726	−1.87780	−1.30971	−0.931020	−0.655555	−0.444863
0.30	−2.69515	−1.79896	−1.28858	−0.941130	−0.684194	−0.484911
0.35	−2.49318	−1.71612	−1.25962	−0.942351	−0.703936	−0.516498
0.40	−2.30251	−1.63123	−1.22459	−0.936164	−0.715982	−0.540580
0.45	−2.12377	−1.54588	−1.18502	−0.923882	−0.721421	−0.558046
0.50	−1.95719	−1.46130	−1.14220	−0.906657	−0.721242	−0.569718
0.55	−1.80268	−1.37844	−1.09720	−0.885501	−0.716330	−0.576355
0.60	−1.65993	−1.29803	−1.05093	−0.861287	−0.707476	−0.578645
0.65	−1.52848	−1.22059	−1.00413	−0.834771	−0.695383	−0.577216
0.70	−1.40778	−1.14650	−0.957402	−0.806600	−0.680670	−0.572634
0.75	−1.29717	−1.07599	−0.911229	−0.777323	−0.663881	−0.565407
0.80	−1.19599	−1.00919	−0.865993	−0.747402	−0.645491	−0.555989
0.85	−1.10357	−0.946153	−0.821990	−0.717224	−0.625910	−0.544783
0.90	−1.01923	−0.886853	−0.779440	−0.687109	−0.605492	−0.532145
0.95	−0.942323	−0.831225	−0.738505	−0.657316	−0.584538	−0.518387
1.00	−0.872222	−0.779165	−0.699292	−0.628056	−0.563306	−0.503784
1.05	−0.808336	−0.730532	−0.661866	−0.599493	−0.542010	−0.488571
1.10	−0.750114	−0.685179	−0.626260	−0.571756	−0.520829	−0.472954
1.15	−0.697040	−0.642939	−0.592476	−0.544939	−0.499910	−0.457108
1.20	−0.648640	−0.603639	−0.560496	−0.519113	−0.479374	−0.441182
1.25	−0.604477	−0.567105	−0.530282	−0.494320	−0.459315	−0.425300
1.30	−0.564151	−0.533162	−0.501786	−0.470589	−0.439808	−0.409569
1.35	−0.527298	−0.501640	−0.474948	−0.447929	−0.420908	−0.394073

Table 4.2 (continued)

				t		
v	0.05	0.10	0.15	0.20	0.25	0.30
1.40	−0.493586	−0.472374	−0.449701	−0.426338	−0.402657	−0.378883
1.45	−0.462717	−0.445203	−0.425974	−0.405804	−0.385083	−0.364054
1.50	−0.434417	−0.419977	−0.403692	−0.386304	−0.368202	−0.349632
1.55	−0.408443	−0.396553	−0.382781	−0.367813	−0.352022	−0.335648
1.60	−0.384573	−0.374795	−0.363164	−0.350298	−0.336543	−0.322126
1.65	−0.362608	−0.354577	−0.344767	−0.333723	−0.321759	−0.309083
1.70	−0.342369	−0.335780	−0.327517	−0.318050	−0.307657	−0.296529
1.75	−0.323694	−0.318296	−0.311344	−0.303239	−0.294223	−0.284467
1.80	−0.306440	−0.302021	−0.296180	−0.289250	−0.281439	−0.272896
1.85	−0.290474	−0.286862	−0.281960	−0.276042	−0.269283	−0.261813
1.90	−0.275681	−0.272731	−0.268622	−0.263574	−0.257733	−0.251209
1.95	−0.261955	−0.259548	−0.256108	−0.251807	−0.246766	−0.241074
2.00	−0.249201	−0.247239	−0.244362	−0.240702	−0.236356	−0.231397
2.05	−0.237334	−0.235737	−0.233333	−0.230222	−0.226479	−0.222164
2.10	−0.226277	−0.224978	−0.222972	−0.220330	−0.217111	−0.213359
2.15	−0.215961	−0.214905	−0.213233	−0.210992	−0.208226	−0.204969
2.20	−0.206323	−0.205466	−0.204073	−0.202175	−0.199800	−0.196975
2.25	−0.197308	−0.196613	−0.195453	−0.193847	−0.191810	−0.189363
2.30	−0.188863	−0.188300	−0.187336	−0.185978	−0.184234	−0.182115
2.35	−0.180944	−0.180488	−0.179688	−0.178540	−0.177047	−0.175215
2.40	−0.173508	−0.173139	−0.172475	−0.171506	−0.170230	−0.168648
2.45	−0.166517	−0.166219	−0.165669	−0.164852	−0.163763	−0.162397
2.50	−0.159938	−0.159697	−0.159242	−0.158554	−0.157624	−0.156447
2.55	−0.153739	−0.153545	−0.153168	−0.152589	−0.151796	−0.150782
2.60	−0.147892	−0.147735	−0.147424	−0.146937	−0.146262	−0.145389
2.65	−0.142371	−0.142245	−0.141987	−0.141578	−0.141004	−0.140254
2.70	−0.137152	−0.137051	−0.136838	−0.136495	−0.136006	−0.135362
2.75	−0.132214	−0.132133	−0.131957	−0.131670	−0.131255	−0.130702
2.80	−0.127538	−0.127472	−0.127328	−0.127087	−0.126734	−0.126260
2.85	−0.123105	−0.123052	−0.122933	−0.122731	−0.122432	−0.122026
2.90	−0.118898	−0.118856	−0.118758	−0.118589	−0.118336	−0.117988
2.95	−0.114904	−0.114870	−0.114789	−0.114648	−0.114434	−0.114136
3.00	−0.111107	−0.111079	−0.111013	−0.110895	−0.110714	−0.110460
3.05	−0.107494	−0.107473	−0.107418	−0.107320	−0.107167	−0.106950
3.10	−0.104055	−0.104038	−0.103994	−0.103911	−0.103782	−0.103597
3.15	−0.100779	−0.100765	−0.100729	−0.100660	−0.100551	−0.100393
3.20	−0.097655	−0.097644	−0.097614	−0.097557	−0.097465	−0.097330
3.25	−0.094673	−0.094664	−0.094640	−0.094593	−0.094515	−0.094400
3.30	−0.091826	−0.091819	−0.091799	−0.091760	−0.091694	−0.091597
3.35	−0.089106	−0.089100	−0.089084	−0.089051	−0.088996	−0.088913
3.40	−0.086505	−0.086500	−0.086487	−0.086460	−0.086413	−0.086343
3.45	−0.084016	−0.084012	−0.084001	−0.083978	−0.083940	−0.083880
3.50	−0.081632	−0.081629	−0.081621	−0.081602	−0.081569	−0.081518

Table 4.2 (continued)

				t		
v	0.05	0.10	0.15	0.20	0.25	0.30
3.55	−0.079349	−0.079347	−0.079340	−0.079324	−0.079297	−0.079254
3.60	−0.077160	−0.077159	−0.077153	−0.077140	−0.077117	−0.077080
3.65	−0.075061	−0.075059	−0.075055	−0.075044	−0.075025	−0.074994
3.70	−0.073046	−0.073045	−0.073041	−0.073032	−0.073016	−0.072990
3.75	−0.071111	−0.071110	−0.071107	−0.071100	−0.071086	−0.071064
3.80	−0.069252	−0.069251	−0.069249	−0.069243	−0.069231	−0.069213
3.85	−0.067465	−0.067464	−0.067462	−0.067457	−0.067448	−0.067432
3.90	−0.065746	−0.065746	−0.065744	−0.065740	−0.065732	−0.065719
3.95	−0.064092	−0.064092	−0.064091	−0.064087	−0.064081	−0.064069
4.00	−0.062500	−0.062500	−0.062499	−0.062496	−0.062490	−0.062481
4.05	−0.060966	−0.060966	−0.060965	−0.060963	−0.060958	−0.060950
4.10	−0.059488	−0.059488	−0.059488	−0.059486	−0.059482	−0.059475
4.15	−0.058064	−0.058063	−0.058063	−0.058061	−0.058058	−0.058052
4.20	−0.056689	−0.056689	−0.056689	−0.056688	−0.056685	−0.056680
4.25	−0.055363	−0.055363	−0.055363	−0.055362	−0.055360	−0.055356
4.30	−0.054083	−0.054083	−0.054083	−0.054082	−0.054080	−0.054077
4.35	−0.052847	−0.052847	−0.052847	−0.052846	−0.052845	−0.052842
4.40	−0.051653	−0.051653	−0.051653	−0.051652	−0.051651	−0.051648
4.45	−0.050499	−0.050499	−0.050499	−0.050498	−0.050497	−0.050495
4.50	−0.049383	−0.049383	−0.049383	−0.049382	−0.049381	−0.049380
4.55	−0.048303	−0.048303	−0.048303	−0.048303	−0.048302	−0.048301
4.60	−0.047259	−0.047259	−0.047259	−0.047259	−0.047258	−0.047257
4.65	−0.046248	−0.046248	−0.046248	−0.046248	−0.046247	−0.046246
4.70	−0.045269	−0.045269	−0.045269	−0.045269	−0.045269	−0.045268
4.75	−0.044321	−0.044321	−0.044321	−0.044321	−0.044321	−0.044320
4.80	−0.043403	−0.043403	−0.043403	−0.043403	−0.043402	−0.043402
4.85	−0.042513	−0.042513	−0.042513	−0.042512	−0.042512	−0.042512
4.90	−0.041649	−0.041649	−0.041649	−0.041649	−0.041649	−0.041649
4.95	−0.040812	−0.040812	−0.040812	−0.040812	−0.040812	−0.040812
5.00	−0.040000	−0.040000	−0.040000	−0.040000	−0.040000	−0.040000

				t		
v	0.35	0.40	0.45	0.50	0.55	0.60
0.00	0.102327	0.239359	0.343677	0.423626	0.484901	0.531545
0.05	0.004084	0.143831	0.252187	0.336918	0.403373	0.455373
0.10	−0.082310	0.058242	0.168972	0.257039	0.327414	0.383674
0.15	−0.157656	−0.017920	0.093739	0.183862	0.257024	0.316543
0.20	−0.222770	−0.085206	0.026144	0.117201	0.192142	0.254016
0.25	−0.278468	−0.144188	−0.034197	0.056824	0.132654	0.196071
0.30	−0.325552	−0.195451	−0.087694	0.002460	0.078402	0.142643
0.35	−0.364801	−0.239580	−0.134772	−0.046186	0.029197	0.093630
0.40	−0.396965	−0.277153	−0.175867	−0.089431	−0.015178	0.048896
0.45	−0.422760	−0.308734	−0.211414	−0.127602	−0.054959	0.008284

Table 4.2 (continued)

				t		
v	0.35	0.40	0.45	0.50	0.55	0.60
0.50	−0.442858	−0.334867	−0.241846	−0.161034	−0.090396	−0.028386
0.55	−0.457895	−0.356075	−0.267585	−0.190061	−0.121748	−0.061305
0.60	−0.468459	−0.372853	−0.289040	−0.215015	−0.149278	−0.090675
0.65	−0.475097	−0.385666	−0.306603	−0.236220	−0.173250	−0.116706
0.70	−0.478310	−0.394954	−0.320649	−0.253994	−0.193924	−0.139609
0.75	−0.478560	−0.401123	−0.331534	−0.268639	−0.211557	−0.159595
0.80	−0.476265	−0.404551	−0.339590	−0.280445	−0.226397	−0.176874
0.85	−0.471803	−0.405584	−0.345129	−0.289689	−0.238685	−0.191652
0.90	−0.465517	−0.404541	−0.348441	−0.296630	−0.248649	−0.204127
0.95	−0.457712	−0.401712	−0.349796	−0.301513	−0.256509	−0.214494
1.00	−0.448663	−0.397360	−0.349439	−0.304566	−0.262471	−0.222936
1.05	−0.438612	−0.391723	−0.347599	−0.305999	−0.266729	−0.229630
1.10	−0.427772	−0.385016	−0.344482	−0.306010	−0.269467	−0.234744
1.15	−0.416332	−0.377428	−0.340276	−0.304778	−0.270854	−0.238433
1.20	−0.404456	−0.369131	−0.335151	−0.302469	−0.271047	−0.240848
1.25	−0.392287	−0.360275	−0.329259	−0.299234	−0.270192	−0.242124
1.30	−0.379947	−0.350993	−0.322739	−0.295209	−0.268423	−0.242392
1.35	−0.367543	−0.341402	−0.315712	−0.290520	−0.265862	−0.241769
1.40	−0.355163	−0.331604	−0.308288	−0.285277	−0.262623	−0.240367
1.45	−0.342882	−0.321687	−0.300562	−0.279581	−0.258806	−0.238286
1.50	−0.330764	−0.311726	−0.292619	−0.273522	−0.254503	−0.235618
1.55	−0.318860	−0.301788	−0.284533	−0.267181	−0.249799	−0.232448
1.60	−0.307211	−0.291925	−0.276369	−0.260626	−0.244767	−0.228853
1.65	−0.295852	−0.282186	−0.268182	−0.253922	−0.239475	−0.224903
1.70	−0.284808	−0.272608	−0.260020	−0.247122	−0.233982	−0.220659
1.75	−0.274098	−0.263222	−0.251923	−0.240275	−0.228341	−0.216178
1.80	−0.263736	−0.254053	−0.243926	−0.233421	−0.222598	−0.211511
1.85	−0.253731	−0.245122	−0.236057	−0.226596	−0.216795	−0.206703
1.90	−0.244088	−0.236444	−0.228340	−0.219831	−0.210967	−0.201793
1.95	−0.234808	−0.228029	−0.220793	−0.213150	−0.205144	−0.196817
2.00	−0.225889	−0.219885	−0.213433	−0.206576	−0.199355	−0.191805
2.05	−0.217328	−0.212016	−0.206270	−0.200127	−0.193621	−0.186785
2.10	−0.209119	−0.204425	−0.199314	−0.193816	−0.187962	−0.181781
2.15	−0.201254	−0.197111	−0.192570	−0.187656	−0.182395	−0.176811
2.20	−0.193725	−0.190073	−0.186042	−0.181654	−0.176932	−0.171895
2.25	−0.186522	−0.183306	−0.179732	−0.175819	−0.171585	−0.167046
2.30	−0.179634	−0.176805	−0.173640	−0.170155	−0.166362	−0.162276
2.35	−0.173052	−0.170565	−0.167766	−0.164664	−0.161270	−0.157597
2.40	−0.166763	−0.164580	−0.162106	−0.159348	−0.156315	−0.153016
2.45	−0.160756	−0.158841	−0.156657	−0.154208	−0.151501	−0.148540
2.50	−0.155020	−0.153342	−0.151416	−0.149243	−0.146828	−0.144175
2.55	−0.149542	−0.148074	−0.146377	−0.144451	−0.142299	−0.139924
2.60	−0.144313	−0.143029	−0.141535	−0.139830	−0.137914	−0.135789

Table 4.2 (continued)

				t		
v	0.35	0.40	0.45	0.50	0.55	0.60
2.65	−0.139320	−0.138199	−0.136885	−0.135376	−0.133673	−0.131774
2.70	−0.134553	−0.133574	−0.132420	−0.131087	−0.129573	−0.127878
2.75	−0.130002	−0.129148	−0.128134	−0.126957	−0.125613	−0.124101
2.80	−0.125655	−0.124911	−0.124022	−0.122983	−0.121792	−0.120444
2.85	−0.121502	−0.120855	−0.120076	−0.119161	−0.118105	−0.116905
2.90	−0.117537	−0.116973	−0.116291	−0.115485	−0.114550	−0.113482
2.95	−0.113746	−0.113256	−0.112660	−0.111950	−0.111124	−0.110175
3.00	−0.110124	−0.109698	−0.109177	−0.108553	−0.107822	−0.106980
3.05	−0.106660	−0.106291	−0.105835	−0.105288	−0.104642	−0.103894
3.10	−0.103348	−0.103027	−0.102630	−0.102149	−0.101579	−0.100917
3.15	−0.100179	−0.099901	−0.099555	−0.099133	−0.098631	−0.098043
3.20	−0.097146	−0.096906	−0.096604	−0.096234	−0.095791	−0.095272
3.25	−0.094242	−0.094034	−0.093771	−0.093448	−0.093058	−0.092598
3.30	−0.091461	−0.091282	−0.091053	−0.090770	−0.090427	−0.090020
3.35	−0.088797	−0.088642	−0.088443	−0.088195	−0.087894	−0.087535
3.40	−0.086243	−0.086109	−0.085936	−0.085720	−0.085455	−0.085139
3.45	−0.083794	−0.083679	−0.083529	−0.083340	−0.083108	−0.082828
3.50	−0.081445	−0.081346	−0.081216	−0.081051	−0.080847	−0.080601
3.55	−0.079191	−0.079105	−0.078992	−0.078849	−0.078670	−0.078453
3.60	−0.077027	−0.076953	−0.076855	−0.076730	−0.076574	−0.076383
3.65	−0.074948	−0.074885	−0.074800	−0.074691	−0.074554	−0.074386
3.70	−0.072951	−0.072896	−0.072823	−0.072728	−0.072608	−0.072461
3.75	−0.071031	−0.070984	−0.070921	−0.070838	−0.070733	−0.070604
3.80	−0.069184	−0.069144	−0.069089	−0.069018	−0.068926	−0.068813
3.85	−0.067408	−0.067373	−0.067326	−0.067264	−0.067184	−0.067084
3.90	−0.065698	−0.065669	−0.065628	−0.065574	−0.065504	−0.065417
3.95	−0.064052	−0.064027	−0.063991	−0.063944	−0.063884	−0.063807
4.00	−0.062466	−0.062444	−0.062414	−0.062373	−0.062320	−0.062253
4.05	−0.060938	−0.060919	−0.060893	−0.060858	−0.060812	−0.060753
4.10	−0.059464	−0.059448	−0.059426	−0.059395	−0.059355	−0.059304
4.15	−0.058043	−0.058030	−0.058010	−0.057984	−0.057949	−0.057904
4.20	−0.056672	−0.056661	−0.056644	−0.056621	−0.056591	−0.056552
4.25	−0.055349	−0.055339	−0.055325	−0.055305	−0.055279	−0.055245
4.30	−0.054071	−0.054063	−0.054051	−0.054034	−0.054011	−0.053981
4.35	−0.052837	−0.052830	−0.052819	−0.052805	−0.052785	−0.052759
4.40	−0.051645	−0.051638	−0.051629	−0.051617	−0.051599	−0.051577
4.45	−0.050492	−0.050486	−0.050479	−0.050468	−0.050453	−0.050433
4.50	−0.049377	−0.049372	−0.049366	−0.049356	−0.049343	−0.049326
4.55	−0.048298	−0.048295	−0.048289	−0.048281	−0.048270	−0.048255
4.60	−0.047255	−0.047252	−0.047247	−0.047240	−0.047230	−0.047217
4.65	−0.046245	−0.046242	−0.046238	−0.046232	−0.046223	−0.046212
4.70	−0.045266	−0.045264	−0.045261	−0.045255	−0.045248	−0.045238
4.75	−0.044319	−0.044317	−0.044314	−0.044309	−0.044303	−0.044295

Table 4.2 (continued)

			t			
v	0.35	0.40	0.45	0.50	0.55	0.60
4.80	-0.043401	-0.043399	-0.043396	-0.043393	-0.043387	-0.043380
4.85	-0.042511	-0.042509	-0.042507	-0.042504	-0.042499	-0.042493
4.90	-0.041648	-0.041647	-0.041645	-0.041642	-0.041638	-0.041632
4.95	-0.040811	-0.040810	-0.040808	-0.040806	-0.040802	-0.040798
5.00	-0.039999	-0.039998	-0.039997	-0.039995	-0.039992	-0.039988

			t			
v	0.65	0.70	0.75	0.80	0.85	0.90
0.00	0.566514	0.592030	0.609798	0.621154	0.627158	0.628669
0.05	0.495737	0.526594	0.549592	0.566025	0.576932	0.583154
0.10	0.428475	0.463839	0.491337	0.512212	0.527466	0.537916
0.15	0.364899	0.403988	0.435296	0.460005	0.479070	0.493275
0.20	0.305115	0.347206	0.381678	0.409642	0.432004	0.449508
0.25	0.249178	0.293604	0.330638	0.361317	0.386488	0.406852
0.30	0.197096	0.243251	0.282292	0.315179	0.342699	0.365506
0.35	0.148837	0.196170	0.236711	0.271338	0.300777	0.325635
0.40	0.104334	0.152354	0.193933	0.229871	0.260831	0.287369
0.45	0.063496	0.111764	0.153966	0.190823	0.222936	0.250812
0.50	0.026204	0.074334	0.116789	0.154211	0.187142	0.216040
0.55	-0.007677	0.039982	0.082360	0.120030	0.153475	0.183103
0.60	-0.038294	0.008602	0.050617	0.088254	0.121938	0.152034
0.65	-0.065807	-0.019919	0.021483	0.058838	0.092517	0.122842
0.70	-0.090384	-0.045709	-0.005132	0.031724	0.065183	0.095523
0.75	-0.112196	-0.068901	-0.029330	0.006842	0.039891	0.070057
0.80	-0.131414	-0.089635	-0.051218	-0.015889	0.016586	0.046412
0.85	-0.148213	-0.108054	-0.070911	-0.036557	-0.004796	0.024544
0.90	-0.162762	-0.124300	-0.088527	-0.055256	-0.024327	0.004403
0.95	-0.175230	-0.138518	-0.104185	-0.072084	-0.042084	-0.014069
1.00	-0.185780	-0.150848	-0.118007	-0.087142	-0.058149	-0.030939
1.05	-0.194568	-0.161429	-0.130112	-0.100531	-0.072608	-0.046274
1.10	-0.201747	-0.170395	-0.140619	-0.112355	-0.085548	-0.060146
1.15	-0.207459	-0.177877	-0.149643	-0.122715	-0.097055	-0.072629
1.20	-0.211841	-0.183999	-0.157297	-0.131711	-0.107218	-0.083800
1.25	-0.215023	-0.188881	-0.163690	-0.139440	-0.116124	-0.093734
1.30	-0.217127	-0.192636	-0.168926	-0.145998	-0.123858	-0.102506
1.35	-0.218266	-0.195372	-0.173104	-0.151477	-0.130503	-0.110192
1.40	-0.218546	-0.197188	-0.176321	-0.155965	-0.136141	-0.116864
1.45	-0.218065	-0.198181	-0.178665	-0.159546	-0.140849	-0.122595
1.50	-0.216916	-0.198438	-0.180221	-0.162300	-0.144702	-0.127454
1.55	-0.215181	-0.198041	-0.181070	-0.164303	-0.147773	-0.131508
1.60	-0.212938	-0.197067	-0.181284	-0.165627	-0.150129	-0.134821
1.65	-0.210257	-0.195587	-0.180935	-0.166339	-0.151835	-0.137455
1.70	-0.207204	-0.193666	-0.180085	-0.166503	-0.152953	-0.139469

Table 4.2 (continued)

				t		
v	**0.65**	**0.70**	**0.75**	**0.80**	**0.85**	**0.90**
1.75	−0.203837	−0.191362	−0.178796	−0.166177	−0.153539	−0.140917
1.80	−0.200208	−0.188731	−0.177121	−0.165416	−0.153649	−0.141853
1.85	−0.196365	−0.185822	−0.175112	−0.164271	−0.153332	−0.142326
1.90	−0.192351	−0.182680	−0.172815	−0.162789	−0.152635	−0.142382
1.95	−0.188206	−0.179346	−0.170272	−0.161014	−0.151602	−0.142064
2.00	−0.183962	−0.175857	−0.167522	−0.158984	−0.150273	−0.141414
2.05	−0.179651	−0.172246	−0.164600	−0.156738	−0.148686	−0.140469
2.10	−0.175299	−0.168543	−0.161537	−0.154306	−0.146874	−0.139263
2.15	−0.170930	−0.164773	−0.158363	−0.151721	−0.144870	−0.137830
2.20	−0.166564	−0.160960	−0.155102	−0.149010	−0.142702	−0.136198
2.25	−0.162220	−0.157125	−0.151778	−0.146196	−0.140396	−0.134394
2.30	−0.157913	−0.153286	−0.148411	−0.143303	−0.137976	−0.132445
2.35	−0.153656	−0.149459	−0.145020	−0.140350	−0.135463	−0.130372
2.40	−0.149460	−0.145658	−0.141619	−0.137356	−0.132878	−0.128197
2.45	−0.145336	−0.141894	−0.138224	−0.134335	−0.130237	−0.125939
2.50	−0.141290	−0.138178	−0.134847	−0.131304	−0.127557	−0.123614
2.55	−0.137328	−0.134518	−0.131497	−0.128273	−0.124850	−0.121237
2.60	−0.133457	−0.130921	−0.128185	−0.125253	−0.122131	−0.118824
2.65	−0.129680	−0.127394	−0.124918	−0.122255	−0.119410	−0.116385
2.70	−0.126000	−0.123942	−0.121703	−0.119287	−0.116696	−0.113933
2.75	−0.122419	−0.120567	−0.118545	−0.116355	−0.113998	−0.111476
2.80	−0.118938	−0.117273	−0.115449	−0.113465	−0.111323	−0.109024
2.85	−0.115558	−0.114063	−0.112418	−0.110623	−0.108679	−0.106585
2.90	−0.112279	−0.110938	−0.109456	−0.107834	−0.106070	−0.104164
2.95	−0.109100	−0.107898	−0.106565	−0.105099	−0.103500	−0.101768
3.00	−0.106021	−0.104944	−0.103745	−0.102423	−0.100975	−0.099401
3.05	−0.103040	−0.102076	−0.100999	−0.099807	−0.098497	−0.097069
3.10	−0.100156	−0.099294	−0.098327	−0.097253	−0.096069	−0.094774
3.15	−0.097366	−0.096596	−0.095729	−0.094762	−0.093693	−0.092520
3.20	−0.094670	−0.093982	−0.093205	−0.092336	−0.091371	−0.090308
3.25	−0.092064	−0.091450	−0.090755	−0.089973	−0.089103	−0.088142
3.30	−0.089546	−0.088999	−0.088377	−0.087675	−0.086891	−0.086022
3.35	−0.087114	−0.086627	−0.086070	−0.085441	−0.084735	−0.083951
3.40	−0.084765	−0.084332	−0.083835	−0.083270	−0.082635	−0.081927
3.45	−0.082498	−0.082112	−0.081668	−0.081163	−0.080592	−0.079954
3.50	−0.080308	−0.079966	−0.079570	−0.079117	−0.078604	−0.078029
3.55	−0.078195	−0.077890	−0.077537	−0.077132	−0.076672	−0.076154
3.60	−0.076154	−0.075884	−0.075570	−0.075208	−0.074795	−0.074329
3.65	−0.074184	−0.073945	−0.073665	−0.073342	−0.072972	−0.072553
3.70	−0.072283	−0.072070	−0.071821	−0.071533	−0.071201	−0.070825
3.75	−0.070447	−0.070259	−0.070037	−0.069780	−0.069483	−0.069145
3.80	−0.068674	−0.068508	−0.068311	−0.068081	−0.067816	−0.067513
3.85	−0.066962	−0.066815	−0.066641	−0.066436	−0.066199	−0.065927

Table 4.2 (continued)

			t			
v	0.65	0.70	0.75	0.80	0.85	0.90
3.90	−0.065309	−0.065179	−0.065024	−0.064842	−0.064631	−0.064387
3.95	−0.063713	−0.063598	−0.063460	−0.063298	−0.063109	−0.062891
4.00	−0.062170	−0.062069	−0.061947	−0.061803	−0.061635	−0.061439
4.05	−0.060680	−0.060591	−0.060483	−0.060355	−0.060205	−0.060030
4.10	−0.059240	−0.059161	−0.059066	−0.058952	−0.058818	−0.058662
4.15	−0.057848	−0.057779	−0.057694	−0.057593	−0.057474	−0.057335
4.20	−0.056502	−0.056441	−0.056367	−0.056277	−0.056171	−0.056047
4.25	−0.055201	−0.055148	−0.055082	−0.055002	−0.054908	−0.054797
4.30	−0.053943	−0.053896	−0.053838	−0.053767	−0.053683	−0.053584
4.35	−0.052726	−0.052684	−0.052633	−0.052571	−0.052496	−0.052408
4.40	−0.051548	−0.051511	−0.051466	−0.051411	−0.051345	−0.051266
4.45	−0.050408	−0.050376	−0.050336	−0.050287	−0.050228	−0.050158
4.50	−0.049304	−0.049276	−0.049241	−0.049198	−0.049146	−0.049083
4.55	−0.048235	−0.048211	−0.048180	−0.048142	−0.048096	−0.048040
4.60	−0.047200	−0.047179	−0.047151	−0.047118	−0.047077	−0.047028
4.65	−0.046197	−0.046178	−0.046154	−0.046125	−0.046089	−0.046045
4.70	−0.045225	−0.045209	−0.045188	−0.045162	−0.045130	−0.045091
4.75	−0.044283	−0.044269	−0.044250	−0.044227	−0.044199	−0.044165
4.80	−0.043370	−0.043357	−0.043341	−0.043321	−0.043296	−0.043265
4.85	−0.042484	−0.042473	−0.042459	−0.042441	−0.042419	−0.042392
4.90	−0.041625	−0.041615	−0.041603	−0.041587	−0.041568	−0.041544
4.95	−0.040791	−0.040783	−0.040772	−0.040758	−0.040741	−0.040720
5.00	−0.039982	−0.039975	−0.039965	−0.039953	−0.039938	−0.039919

			t			
v	0.95	1.00	2.00	3.00	4.00	5.00
0.00	0.626387	0.620893	0.261540	−0.046924	−0.135304	−0.072763
0.05	0.585384	0.584197	0.273342	−0.031559	−0.131252	−0.077767
0.10	0.544237	0.546994	0.283303	−0.016416	−0.126533	−0.082233
0.15	0.503273	0.509609	0.291480	−0.001571	−0.121196	−0.086151
0.20	0.462777	0.472334	0.297938	0.012905	−0.115295	−0.089514
0.25	0.422999	0.435426	0.302749	0.026948	−0.108883	−0.092318
0.30	0.384154	0.399110	0.305992	0.040501	−0.102012	−0.094563
0.35	0.346423	0.363580	0.307750	0.053514	−0.094736	−0.096253
0.40	0.309959	0.329005	0.308109	0.065941	−0.087110	−0.097395
0.45	0.274884	0.295523	0.307160	0.077747	−0.079186	−0.097998
0.50	0.241296	0.263250	0.304993	0.088898	−0.071015	−0.098075
0.55	0.209269	0.232278	0.301700	0.099369	−0.062649	−0.097640
0.60	0.178856	0.202679	0.297374	0.109140	−0.054138	−0.096710
0.65	0.150090	0.174506	0.292105	0.118196	−0.045529	−0.095305
0.70	0.122988	0.147793	0.285984	0.126529	−0.036868	−0.093445
0.75	0.097552	0.122561	0.279100	0.134132	−0.028199	−0.091152

Table 4.2 (continued)

			t			
v	0.95	1.00	2.00	3.00	4.00	5.00
0.80	0.073769	0.098818	0.271538	0.141006	-0.019564	-0.088452
0.85	0.051616	0.076556	0.263382	0.147154	-0.011002	-0.085368
0.90	0.031061	0.055760	0.254712	0.152585	-0.002551	-0.081928
0.95	0.012062	0.036404	0.245604	0.157309	0.005755	-0.078157
1.00	-0.005428	0.018458	0.236132	0.161341	0.013883	-0.074084
1.05	-0.021464	0.001880	0.226366	0.164697	0.021805	-0.069736
1.10	-0.036102	-0.013374	0.216370	0.167398	0.029493	-0.065142
1.15	-0.049405	-0.027352	0.206207	0.169466	0.036923	-0.060330
1.20	-0.061436	-0.040107	0.195934	0.170924	0.044074	-0.055327
1.25	-0.072260	-0.051694	0.185603	0.171798	0.050924	-0.050162
1.30	-0.081943	-0.062168	0.175266	0.172114	0.057459	-0.044862
1.35	-0.090551	-0.071588	0.164965	0.171901	0.063664	-0.039454
1.40	-0.098150	-0.080011	0.154744	0.171188	0.069526	-0.033964
1.45	-0.104805	-0.087496	0.144639	0.170003	0.075037	-0.028417
1.50	-0.110580	-0.094100	0.134684	0.168377	0.080187	-0.022839
1.55	-0.115536	-0.099879	0.124910	0.166340	0.084973	-0.017253
1.60	-0.119733	-0.104889	0.115342	0.163922	0.089390	-0.011681
1.65	-0.123229	-0.109185	0.106005	0.161152	0.093438	-0.006146
1.70	-0.126081	-0.112817	0.096919	0.158061	0.097115	-0.000668
1.75	-0.128340	-0.115836	0.088100	0.154678	0.100425	0.004734
1.80	-0.130058	-0.118291	0.079564	0.151031	0.103370	0.010042
1.85	-0.131282	-0.120228	0.071321	0.147147	0.105956	0.015239
1.90	-0.132058	-0.121690	0.063383	0.143055	0.108189	0.020309
1.95	-0.132428	-0.122719	0.055755	0.138780	0.110076	0.025237
2.00	-0.132433	-0.123355	0.048443	0.134347	0.111625	0.030011
2.05	-0.132111	-0.123634	0.041449	0.129782	0.112846	0.034619
2.10	-0.131496	-0.123593	0.034776	0.125106	0.113750	0.039050
2.15	-0.130620	-0.123262	0.028423	0.120342	0.114347	0.043294
2.20	-0.129515	-0.122673	0.022387	0.115511	0.114650	0.047344
2.25	-0.128209	-0.121855	0.016667	0.110634	0.114670	0.051193
2.30	-0.126726	-0.120833	0.011257	0.105727	0.114420	0.054834
2.35	-0.125091	-0.119633	0.006153	0.100810	0.113914	0.058263
2.40	-0.123326	-0.118276	0.001349	0.095898	0.113165	0.061476
2.45	-0.121451	-0.116784	-0.003164	0.091007	0.112186	0.064471
2.50	-0.119484	-0.115176	-0.007391	0.086150	0.110992	0.067245
2.55	-0.117441	-0.113468	-0.011342	0.081340	0.109596	0.069799
2.60	-0.115337	-0.111679	-0.015025	0.076590	0.108012	0.072132
2.65	-0.113187	-0.109821	-0.018449	0.071910	0.106253	0.074245
2.70	-0.111002	-0.107908	-0.021624	0.067309	0.104334	0.076140
2.75	-0.108793	-0.105952	-0.024559	0.062797	0.102268	0.077820
2.80	-0.106570	-0.103964	-0.027263	0.058381	0.100067	0.079287
2.85	-0.104343	-0.101954	-0.029747	0.054069	0.097745	0.080546
2.90	-0.102117	-0.099931	-0.032021	0.049865	0.095315	0.081601

Table 4.2 (continued)

ν	0.95	1.00	2.00	3.00	4.00	5.00
			t			
2.95	−0.099901	−0.097901	−0.034093	0.045776	0.092788	0.082457
3.00	−0.097701	−0.095873	−0.035975	0.041805	0.090176	0.083119
3.05	−0.095521	−0.093852	−0.037675	0.037956	0.087492	0.083594
3.10	−0.093366	−0.091843	−0.039203	0.034232	0.084745	0.083886
3.15	−0.091240	−0.089852	−0.040568	0.030636	0.081947	0.084003
3.20	−0.089146	−0.087882	−0.041780	0.027168	0.079107	0.083952
3.25	−0.087088	−0.085938	−0.042846	0.023830	0.076236	0.083740
3.30	−0.085066	−0.084021	−0.043777	0.020623	0.073343	0.083373
3.35	−0.083085	−0.082135	−0.044580	0.017546	0.070436	0.082860
3.40	−0.081144	−0.080281	−0.045263	0.014599	0.067523	0.082207
3.45	−0.079245	−0.078463	−0.045833	0.011781	0.064612	0.081422
3.50	−0.077388	−0.076680	−0.046300	0.009091	0.061711	0.080514
3.55	−0.075576	−0.074934	−0.046669	0.006527	0.058825	0.079489
3.60	−0.073807	−0.073226	−0.046947	0.004088	0.055963	0.078355
3.65	−0.072082	−0.071556	−0.047142	0.001770	0.053129	0.077119
3.70	−0.070401	−0.069926	−0.047258	−0.000427	0.050328	0.075790
3.75	−0.068763	−0.068334	−0.047304	−0.002507	0.047566	0.074375
3.80	−0.067169	−0.066782	−0.047283	−0.004473	0.044847	0.072880
3.85	−0.065618	−0.065269	−0.047201	−0.006327	0.042176	0.071313
3.90	−0.064109	−0.063794	−0.047064	−0.008073	0.039555	0.069682
3.95	−0.062642	−0.062358	−0.046876	−0.009713	0.036988	0.067992
4.00	−0.061215	−0.060960	−0.046642	−0.011252	0.034479	0.066250
4.05	−0.059829	−0.059600	−0.046365	−0.012692	0.032029	0.064463
4.10	−0.058482	−0.058276	−0.046051	−0.014037	0.029641	0.062637
4.15	−0.057173	−0.056988	−0.045703	−0.015290	0.027317	0.060778
4.20	−0.055902	−0.055736	−0.045324	−0.016455	0.025058	0.058892
4.25	−0.054668	−0.054519	−0.044918	−0.017536	0.022866	0.056984
4.30	−0.053469	−0.053335	−0.044487	−0.018534	0.020741	0.055059
4.35	−0.052305	−0.052185	−0.044035	−0.019455	0.018685	0.053123
4.40	−0.051174	−0.051067	−0.043565	−0.020302	0.016697	0.051181
4.45	−0.050076	−0.049980	−0.043078	−0.021077	0.014778	0.049236
4.50	−0.049010	−0.048924	−0.042577	−0.021784	0.012929	0.047293
4.55	−0.047975	−0.047898	−0.042065	−0.022427	0.011148	0.045356
4.60	−0.046969	−0.046901	−0.041542	−0.023008	0.009436	0.043429
4.65	−0.045993	−0.045932	−0.041012	−0.023530	0.007792	0.041516
4.70	−0.045045	−0.044990	−0.040475	−0.023997	0.006216	0.039620
4.75	−0.044124	−0.044075	−0.039934	−0.024412	0.004706	0.037743
4.80	−0.043229	−0.043185	−0.039389	−0.024776	0.003262	0.035889
4.85	−0.042360	−0.042321	−0.038842	−0.025094	0.001883	0.034060
4.90	−0.041515	−0.041480	−0.038295	−0.025368	0.000567	0.032259
4.95	−0.040694	−0.040663	−0.037747	−0.025600	−0.000686	0.030488
5.00	−0.039896	−0.039869	−0.037200	−0.025793	−0.001878	0.028749

Table 4.3: Second derivatives with respect to the order of the integral Bessel functions of the first kind $\partial^2 Ji_\nu(t)/\partial\nu^2$.

				t		
ν	0.05	0.10	0.15	0.20	0.25	0.30
0.00	4.12951	−0.045498	−1.35992	−1.87744	−2.07524	−2.12131
0.05	4.41311	0.454824	−0.889065	−1.46811	−1.72741	−1.82856
0.10	4.54200	0.845979	−0.492553	−1.10992	−1.41481	−1.55979
0.15	4.55262	1.14458	−0.162151	−0.79886	−1.13574	−1.31463
0.20	4.47479	1.36541	0.109905	−0.530858	−0.888273	−1.09242
0.25	4.33269	1.52145	0.330846	−0.301897	−0.670297	−0.89226
0.30	4.14572	1.62406	0.507331	−0.108082	−0.479631	−0.713061
0.35	3.92928	1.68310	0.645429	0.054312	−0.314060	−0.553608
0.40	3.69540	1.70700	0.750629	0.188800	−0.171393	−0.412616
0.45	3.45338	1.70300	0.827851	0.298664	−0.049490	−0.288755
0.50	3.15998	1.70377	0.918378	0.350176	0.034514	−0.188153
0.55	2.97128	1.63474	0.915363	0.456416	0.140146	−0.087073
0.60	2.74016	1.57991	0.932917	0.509624	0.211672	−0.006629
0.65	2.51951	1.51623	0.937094	0.548854	0.269997	0.061896
0.70	2.31100	1.44661	0.930457	0.576165	0.316707	0.119690
0.75	2.11567	1.37338	0.915212	0.593389	0.353260	0.167876
0.80	1.93371	1.29844	0.893244	0.602154	0.380984	0.207506
0.85	1.76538	1.22325	0.866153	0.603892	0.401083	0.239557
0.90	1.61033	1.14898	0.835288	0.599861	0.414645	0.264932
0.95	1.46806	1.07650	0.801774	0.591153	0.422642	0.284459
1.00	1.33797	1.00647	0.766548	0.578717	0.425942	0.298897
1.05	1.21930	0.939305	0.730371	0.563368	0.425313	0.308925
1.10	1.11131	0.875363	0.693871	0.545804	0.421435	0.315168
1.15	1.01322	0.814816	0.657546	0.526618	0.414902	0.318181
1.20	0.924241	0.757754	0.621789	0.506309	0.406233	0.318460
1.25	0.843622	0.704190	0.586906	0.485295	0.395879	0.316449
1.30	0.770635	0.654083	0.553127	0.463925	0.384229	0.312537
1.35	0.704595	0.607344	0.520619	0.442481	0.371615	0.307069
1.40	0.644858	0.563854	0.489498	0.421197	0.358322	0.300348
1.45	0.590830	0.523471	0.459840	0.400255	0.344590	0.292636
1.50	0.516417	0.675374	0.635482	0.597445	0.331725	0.528096
1.55	0.497743	0.451389	0.405037	0.359949	0.316581	0.275122
1.60	0.457721	0.419355	0.379895	0.340779	0.302609	0.265685
1.65	0.421476	0.389768	0.356228	0.322352	0.288817	0.255995
1.70	0.388602	0.362459	0.333997	0.304706	0.275295	0.246172
1.75	0.358835	0.337267	0.313150	0.287865	0.262115	0.236319
1.80	0.331789	0.314037	0.293631	0.271839	0.249329	0.226520
1.85	0.307213	0.292621	0.275380	0.256625	0.236980	0.216844
1.90	0.284858	0.272878	0.258331	0.242213	0.225094	0.207346
1.95	0.264563	0.254677	0.242419	0.228588	0.213691	0.198072
2.00	0.246475	0.237788	0.227519	0.215725	0.202782	0.189057
2.05	0.228998	0.222415	0.213745	0.203601	0.192370	0.180326

Table 4.3 (continued)

v	0.05	0.10	0.15	0.20	0.25	0.30
2.10	0.213513	0.208133	0.200856	0.192185	0.182453	0.171899
2.15	0.199342	0.194950	0.188848	0.181447	0.173024	0.163789
2.20	0.186356	0.182775	0.177665	0.171354	0.164075	0.156004
2.25	0.174441	0.171524	0.167249	0.161875	0.155591	0.148547
2.30	0.163494	0.161120	0.157548	0.152977	0.147559	0.141418
2.35	0.153422	0.151492	0.148511	0.144628	0.139962	0.134615
2.40	0.144144	0.142577	0.140091	0.136795	0.132782	0.128132
2.45	0.135586	0.134314	0.132244	0.129450	0.126001	0.121963
2.50	0.114872	0.196441	0.197176	0.197225	0.116516	0.195236
2.55	0.120371	0.119536	0.118104	0.116102	0.113564	0.110527
2.60	0.113601	0.112925	0.111736	0.110044	0.107870	0.105241
2.65	0.107324	0.106777	0.105791	0.104362	0.102503	0.100229
2.70	0.101497	0.101055	0.100237	0.099033	0.097443	0.095478
2.75	0.096080	0.095724	0.095047	0.094031	0.092674	0.090979
2.80	0.091040	0.090753	0.090192	0.089337	0.088180	0.086717
2.85	0.086344	0.086112	0.085649	0.084930	0.083943	0.082684
2.90	0.081964	0.081777	0.081395	0.080627	0.079950	0.078866
2.95	0.077874	0.077724	0.077408	0.076901	0.076186	0.075254
3.00	0.074050	0.073929	0.073669	0.073243	0.072636	0.071835
3.05	0.070472	0.070375	0.070161	0.069804	0.069288	0.068601
3.10	0.067631	0.067042	0.066866	0.066567	0.066129	0.065540
3.15	0.063977	0.063915	0.063769	0.063519	0.063148	0.062643
3.20	0.061027	0.060977	0.060857	0.060648	0.060334	0.059902
3.25	0.058255	0.058215	0.058117	0.057942	0.057676	0.057306
3.30	0.055648	0.055616	0.055535	0.055390	0.055165	0.054849
3.35	0.053194	0.053169	0.053103	0.052981	0.052791	0.052521
3.40	0.050883	0.050862	0.050808	0.050706	0.050546	0.050316
3.45	0.048703	0.048686	0.048642	0.048557	0.048422	0.048226
3.50	0.049877	0.049745	0.050156	-0.167607	-0.181221	0.050544
3.55	0.044703	0.044692	0.044662	0.044604	0.044508	0.044365
3.60	0.042866	0.042858	0.042833	0.042785	0.042704	0.042583
3.65	0.041129	0.041122	0.041102	0.041062	0.040994	0.040891
3.70	0.039484	0.039478	0.039462	0.039429	0.039371	0.039284
3.75	0.037926	0.037921	0.037908	0.037880	0.037832	0.037758
3.80	0.036448	0.036445	0.036434	0.036411	0.036371	0.036308
3.85	0.035047	0.035044	0.035035	0.035016	0.034982	0.034929
3.90	0.033716	0.033714	0.033707	0.033691	0.033662	0.033617
3.95	0.032452	0.032450	0.032444	0.032431	0.032407	0.032369
4.00	0.031250	0.031249	0.031244	0.031233	0.031213	0.031181
4.05	0.030107	0.030106	0.030102	0.030093	0.030076	0.030049
4.10	0.029019	0.029018	0.029015	0.029007	0.028993	0.028970
4.15	0.027982	0.027982	0.027979	0.027973	0.027961	0.027942
4.20	0.026995	0.026994	0.026992	0.026987	0.026978	0.026961

Table 4.3 (continued)

			t			
v	0.05	0.10	0.15	0.20	0.25	0.30
4.25	0.026053	0.026053	0.026051	0.026047	0.026039	0.026025
4.30	0.025155	0.025155	0.025153	0.025150	0.025143	0.025131
4.35	0.024298	0.024297	0.024296	0.024293	0.024288	0.024278
4.40	0.023479	0.023478	0.023478	0.023475	0.023470	0.023462
4.45	0.022696	0.022696	0.022695	0.022693	0.022689	0.022682
4.50	0.010306	0.017539	0.017801	0.017996	0.018600	-0.056657
4.55	0.021232	0.021232	0.021232	0.021230	0.021228	0.021223
4.60	0.020547	0.020547	0.020547	0.020546	0.020544	0.020540
4.65	0.019892	0.019892	0.019891	0.019890	0.019889	0.019885
4.70	0.019264	0.019264	0.019263	0.019263	0.019261	0.019258
4.75	0.018662	0.018662	0.018661	0.018661	0.018660	0.018657
4.80	0.018085	0.018085	0.018084	0.018084	0.018083	0.018081
4.85	0.017531	0.017531	0.017531	0.017530	0.017530	0.017528
4.90	0.017000	0.017000	0.017000	0.016999	0.016999	0.016997
4.95	0.016490	0.016490	0.016490	0.016489	0.016489	0.016488
5.00	0.016000	0.016000	0.016000	0.016000	0.015999	0.015998

			t			
v	0.35	0.40	0.45	0.50	0.55	0.60
0.00	-2.08859	-2.01310	-1.91420	-1.80294	-1.68592	-1.56718
0.05	-1.84374	-1.80953	-1.74617	-1.66555	-1.57499	-1.47915
0.10	-1.61469	-1.61573	-1.58340	-1.53003	-1.46337	-1.38851
0.15	-1.40187	-1.43260	-1.42706	-1.39767	-1.35241	-1.29660
0.20	-1.20542	-1.26074	-1.27804	-1.26954	-1.24325	-1.20458
0.25	-1.02517	-1.10050	-1.13697	-1.14646	-1.13682	-1.11344
0.30	-0.860776	-0.951975	-1.00430	-1.02907	-1.03388	-1.02401
0.35	-0.711694	-0.815111	-0.880273	-0.917835	-0.935039	-0.936978
0.40	-0.577269	-0.689687	-0.764974	-0.813053	-0.840744	-0.852895
0.45	-0.456748	-0.575366	-0.658371	-0.714909	-0.751331	-0.772199
0.50	-0.355880	-0.480270	-0.570251	-0.633741	-0.677020	-0.704729
0.55	-0.254113	-0.378247	-0.470595	-0.538710	-0.587942	-0.622208
0.60	-0.170270	-0.294404	-0.388899	-0.460529	-0.514141	-0.553313
0.65	-0.096911	-0.219608	-0.314883	-0.388762	-0.445596	-0.488629
0.70	-0.033172	-0.153261	-0.248162	-0.323193	-0.382230	-0.428185
0.75	0.021786	-0.094757	-0.188325	-0.263570	-0.323916	-0.371960
0.80	0.068772	-0.043496	-0.134944	-0.209610	-0.270494	-0.319890
0.85	0.108557	0.001113	-0.087584	-0.161009	-0.221770	-0.271876
0.90	0.141869	0.039642	-0.045810	-0.117453	-0.177531	-0.227792
0.95	0.169393	0.072641	-0.009194	-0.078618	-0.137547	-0.187486
1.00	0.191768	0.100634	0.022684	-0.044180	-0.101578	-0.150794
1.05	0.209583	0.124117	0.050229	-0.013818	-0.069378	-0.117535
1.10	0.223383	0.143556	0.073829	0.012785	-0.040698	-0.087521
1.15	0.233668	0.159384	0.093854	0.035933	-0.015291	-0.060562

Table 4.3 (continued)

			t			
ν	0.35	0.40	0.45	0.50	0.55	0.60
1.20	0.240891	0.172006	0.110652	0.055922	0.007086	−0.036463
1.25	0.245466	0.181794	0.124550	0.073034	0.026668	−0.015030
1.30	0.247765	0.189089	0.135856	0.087535	0.043685	0.003928
1.35	0.248122	0.194205	0.144853	0.099679	0.058356	0.020598
1.40	0.246837	0.197426	0.151803	0.109703	0.070890	0.035158
1.45	0.244177	0.199010	0.156950	0.117827	0.081485	0.047785
1.50	0.235897	0.194552	0.155729	0.119311	0.080745	0.050069
1.55	0.235647	0.198176	0.162698	0.129183	0.097595	0.067891
1.60	0.230170	0.196155	0.163684	0.132781	0.103448	0.075678
1.65	0.224107	0.193292	0.163641	0.135211	0.108038	0.082145
1.70	0.217595	0.189736	0.162716	0.136620	0.111507	0.087422
1.75	0.210756	0.185618	0.161045	0.137141	0.113985	0.091634
1.80	0.203693	0.181052	0.158748	0.136897	0.115588	0.094894
1.85	0.196495	0.176139	0.155932	0.135996	0.116428	0.097307
1.90	0.189237	0.170967	0.152692	0.134536	0.116601	0.098970
1.95	0.181982	0.165611	0.149111	0.132607	0.116199	0.099973
2.00	0.174781	0.160136	0.145264	0.130287	0.115303	0.100398
2.05	0.167680	0.154598	0.141215	0.127646	0.113986	0.100319
2.10	0.160713	0.149045	0.137020	0.124746	0.112314	0.099803
2.15	0.153908	0.143516	0.132728	0.121642	0.110345	0.098912
2.20	0.147287	0.138045	0.128380	0.118383	0.108134	0.097702
2.25	0.140868	0.132659	0.124012	0.115010	0.105725	0.096222
2.30	0.134662	0.127380	0.119656	0.111561	0.103161	0.094517
2.35	0.128678	0.122227	0.115335	0.108066	0.100478	0.092626
2.40	0.122921	0.117214	0.111073	0.104554	0.097708	0.090586
2.45	0.117394	0.112351	0.106886	0.101046	0.094879	0.088428
2.50	0.193012	0.190189	0.186774	0.096794	0.178280	0.173264
2.55	0.107028	0.103104	0.098791	0.094124	0.089137	0.083864
2.60	0.102184	0.098729	0.094904	0.090739	0.086263	0.081505
2.65	0.097561	0.094521	0.091133	0.087421	0.083408	0.079120
2.70	0.093152	0.090482	0.087484	0.084178	0.080585	0.076725
2.75	0.088953	0.086608	0.083959	0.081019	0.077805	0.074334
2.80	0.084955	0.082899	0.080560	0.077948	0.075076	0.071958
2.85	0.082023	0.079351	0.077288	0.074969	0.072406	0.069609
2.90	0.077000	0.075960	0.074142	0.072087	0.069801	0.067295
2.95	0.074100	0.072722	0.071121	0.069301	0.067266	0.065021
3.00	0.070835	0.069632	0.068224	0.066613	0.064802	0.062795
3.05	0.067735	0.066684	0.065448	0.064024	0.062414	0.060621
3.10	0.064791	0.063875	0.062789	0.061532	0.060102	0.058501
3.15	0.061996	0.061198	0.060246	0.059137	0.057868	0.056440
3.20	0.059342	0.058648	0.057814	0.056836	0.055711	0.054440
3.25	0.056824	0.056220	0.055490	0.054628	0.053632	0.052500
3.30	0.054432	0.053908	0.053269	0.052511	0.051630	0.050623

Table 4.3 (continued)

			t			
v	0.35	0.40	0.45	0.50	0.55	0.60
3.35	0.052162	0.051707	0.051149	0.050482	0.049703	0.048809
3.40	0.050007	0.049612	0.049125	0.048539	0.047851	0.047057
3.45	0.047960	0.047618	0.047193	0.046679	0.046071	0.045367
3.50	0.050493	0.050360	0.050139	0.049825	0.049413	0.048900
3.55	0.044169	0.043913	0.043590	0.043195	0.042723	0.042170
3.60	0.042414	0.042193	0.041911	0.041566	0.041150	0.040661
3.65	0.040746	0.040555	0.040310	0.040007	0.039642	0.039210
3.70	0.039160	0.038995	0.038782	0.038518	0.038196	0.037814
3.75	0.037652	0.037509	0.037325	0.037094	0.036811	0.036474
3.80	0.036217	0.036094	0.035934	0.035732	0.035484	0.035187
3.85	0.034851	0.034745	0.034606	0.034430	0.034213	0.033951
3.90	0.033551	0.033460	0.033339	0.033186	0.032995	0.032764
3.95	0.032312	0.032234	0.032130	0.031996	0.031829	0.031626
4.00	0.031133	0.031065	0.030975	0.030858	0.030712	0.030534
4.05	0.030008	0.029950	0.029872	0.029770	0.029643	0.029486
4.10	0.028935	0.028885	0.028818	0.028730	0.028618	0.028480
4.15	0.027912	0.027869	0.027811	0.027735	0.027637	0.027516
4.20	0.026936	0.026899	0.026849	0.026782	0.026697	0.026591
4.25	0.026003	0.025972	0.025929	0.025871	0.025797	0.025704
4.30	0.025113	0.025086	0.025049	0.024999	0.024934	0.024852
4.35	0.024262	0.024239	0.024207	0.024164	0.024107	0.024036
4.40	0.023449	0.023429	0.023401	0.023364	0.023315	0.023252
4.45	0.022671	0.022654	0.022630	0.022598	0.022555	0.022501
4.50	−0.058571	−0.060290	−0.061811	0.023131	−0.064247	−0.065159
4.55	0.021215	0.021202	0.021185	0.021160	0.021128	0.021087
4.60	0.020533	0.020522	0.020507	0.020486	0.020458	0.020422
4.65	0.019879	0.019870	0.019857	0.019839	0.019815	0.019783
4.70	0.019253	0.019246	0.019234	0.019219	0.019198	0.019169
4.75	0.018653	0.018646	0.018637	0.018623	0.018605	0.018581
4.80	0.018077	0.018072	0.018063	0.018052	0.018036	0.018015
4.85	0.017525	0.017520	0.017513	0.017503	0.017489	0.017471
4.90	0.016995	0.016991	0.016985	0.016976	0.016964	0.016948
4.95	0.016485	0.016482	0.016477	0.016470	0.016459	0.016445
5.00	0.015996	0.015994	0.015989	0.015983	0.015974	0.015962

			t			
v	0.65	0.70	0.75	0.80	0.85	0.90
0.00	−1.44925	−1.33372	−1.22160	−1.11350	−1.00981	−0.910698
0.05	−1.38106	−1.28276	−1.18559	−1.09047	−0.998029	−0.908666
0.10	−1.30884	−1.22670	−1.14374	−1.06109	−0.979576	−0.899784
0.15	−1.23387	−1.16679	−1.09720	−1.02642	−0.955421	−0.884935
0.20	−1.15731	−1.10413	−1.04703	−0.987442	−0.926475	−0.864954
0.25	−1.08014	−1.03970	−0.994162	−0.945064	−0.893581	−0.840622

Table 4.3 (continued)

			t			
v	0.65	0.70	0.75	0.80	0.85	0.90
0.30	−1.00325	−0.974369	−0.939457	−0.900097	−0.857515	−0.812668
0.35	−0.927346	−0.908882	−0.883659	−0.853276	−0.818981	−0.781762
0.40	−0.853057	−0.843890	−0.827430	−0.805257	−0.778622	−0.748521
0.45	−0.780887	−0.779944	−0.771340	−0.756620	−0.737011	−0.713506
0.50	−0.720252	−0.726062	−0.724039	−0.715649	−0.702062	−0.684236
0.55	−0.644456	−0.656955	−0.661482	−0.659457	−0.652026	−0.640128
0.60	−0.580763	−0.598597	−0.608481	−0.611752	−0.609500	−0.602623
0.65	−0.520342	−0.542674	−0.557169	−0.565079	−0.567427	−0.565064
0.70	−0.463306	−0.489364	−0.507778	−0.519705	−0.526101	−0.527763
0.75	−0.409719	−0.438796	−0.460485	−0.475849	−0.485769	−0.490988
0.80	−0.359597	−0.391052	−0.415426	−0.433685	−0.446637	−0.454968
0.85	−0.312917	−0.346173	−0.372692	−0.393347	−0.408873	−0.419896
0.90	−0.269627	−0.304166	−0.332342	−0.354936	−0.372609	−0.385930
0.95	−0.229644	−0.265008	−0.294401	−0.318517	−0.337947	−0.353200
1.00	−0.192866	−0.228652	−0.258871	−0.284132	−0.304958	−0.321806
1.05	−0.159173	−0.195030	−0.225729	−0.251794	−0.273693	−0.291824
1.10	−0.128433	−0.164057	−0.194927	−0.221500	−0.244176	−0.263307
1.15	−0.100502	−0.135635	−0.166412	−0.193225	−0.216416	−0.236292
1.20	−0.075232	−0.109656	−0.140112	−0.166931	−0.190403	−0.210790
1.25	−0.052469	−0.086004	−0.115945	−0.142567	−0.166115	−0.186808
1.30	−0.032058	−0.064557	−0.093821	−0.120073	−0.143516	−0.164333
1.35	−0.013845	−0.045193	−0.073644	−0.099379	−0.122561	−0.143341
1.40	0.002323	−0.027785	−0.055316	−0.080410	−0.103198	−0.123800
1.45	0.016594	−0.012207	−0.038732	−0.063086	−0.085368	−0.105672
1.50	0.021574	−0.004830	−0.029220	−0.051675	−0.072269	−0.091075
1.55	0.040024	0.013946	−0.010391	−0.033039	−0.054045	−0.073461
1.60	0.049456	0.024763	0.001572	−0.020144	−0.040415	−0.059272
1.65	0.057539	0.034225	0.012197	−0.008555	−0.028043	−0.046284
1.70	0.064394	0.042443	0.021581	0.001814	−0.016858	−0.034438
1.75	0.070135	0.049521	0.029819	0.011045	−0.006788	−0.023672
1.80	0.074869	0.055559	0.037000	0.019219	0.002239	−0.013926
1.85	0.078697	0.060651	0.043212	0.026416	0.010292	−0.005138
1.90	0.081712	0.064885	0.048540	0.032711	0.017439	0.002752
1.95	0.084001	0.068343	0.053053	0.038175	0.023747	0.009804
2.00	0.085644	0.071105	0.056833	0.042878	0.029280	0.016075
2.05	0.086716	0.073240	0.059947	0.046885	0.034097	0.021621
2.10	0.087284	0.074818	0.062460	0.050258	0.038258	0.026497
2.15	0.087410	0.075899	0.064431	0.053056	0.041817	0.030753
2.20	0.087152	0.076540	0.065917	0.055367	0.044827	0.034440
2.25	0.086560	0.076793	0.066971	0.057138	0.047337	0.037605
2.30	0.085682	0.076708	0.067640	0.058522	0.049394	0.040291
2.35	0.084560	0.076327	0.067969	0.059527	0.051042	0.042542
2.40	0.083232	0.075689	0.067998	0.060195	0.052317	0.044397

Table 4.3 (continued)

			t			
v	0.65	0.70	0.75	0.80	0.85	0.90
2.45	0.081733	0.074833	0.067765	0.060563	0.053262	0.045893
2.50	0.167779	0.161862	0.066542	0.148851	0.141877	0.025111
2.55	0.078337	0.072587	0.066644	0.060537	0.054294	0.047943
2.60	0.076493	0.071254	0.065815	0.060203	0.054443	0.048561
2.65	0.074580	0.069813	0.064842	0.059692	0.054385	0.048944
2.70	0.072618	0.068285	0.063749	0.059028	0.054145	0.049120
2.75	0.070623	0.066690	0.062555	0.058234	0.053746	0.049111
2.80	0.068609	0.065044	0.061278	0.057328	0.053209	0.048939
2.85	0.066590	0.063361	0.059936	0.056329	0.052554	0.048625
2.90	0.064576	0.061655	0.058544	0.055254	0.051797	0.048186
2.95	0.062575	0.059936	0.057113	0.054116	0.050954	0.047640
3.00	0.060597	0.058215	0.055656	0.052928	0.050040	0.047002
3.05	0.058647	0.056499	0.054182	0.051702	0.049067	0.046285
3.10	0.056732	0.054797	0.052701	0.050449	0.048047	0.045502
3.15	0.054855	0.053114	0.051220	0.049177	0.046990	0.044664
3.20	0.053020	0.051455	0.049745	0.047894	0.045905	0.043782
3.25	0.051231	0.049825	0.048283	0.046607	0.044800	0.042864
3.30	0.049489	0.048228	0.046838	0.045322	0.043681	0.041918
3.35	0.047797	0.046666	0.045415	0.044045	0.042557	0.040952
3.40	0.046154	0.045141	0.044016	0.042779	0.041431	0.039972
3.45	0.044563	0.043656	0.042645	0.041530	0.040309	0.038984
3.50	0.048281	0.047555	0.046719	0.045772	0.044712	0.043539
3.55	0.041533	0.040809	0.039995	0.039090	0.038092	0.037001
3.60	0.040095	0.039448	0.038719	0.037904	0.037004	0.036016
3.65	0.038707	0.038130	0.037477	0.036745	0.035932	0.035038
3.70	0.037368	0.036854	0.036270	0.035612	0.034880	0.034071
3.75	0.036078	0.035621	0.035098	0.034508	0.033848	0.033118
3.80	0.034836	0.034429	0.033962	0.033433	0.032839	0.032179
3.85	0.033640	0.033278	0.032861	0.032387	0.031854	0.031258
3.90	0.032490	0.032168	0.031796	0.031372	0.030892	0.030356
3.95	0.031383	0.031097	0.030766	0.030386	0.029956	0.029472
4.00	0.030319	0.030065	0.029770	0.029431	0.029045	0.028610
4.05	0.029296	0.029071	0.028809	0.028505	0.028159	0.027768
4.10	0.028313	0.028114	0.027880	0.027610	0.027300	0.026948
4.15	0.027369	0.027192	0.026985	0.026743	0.026466	0.026150
4.20	0.026461	0.026305	0.026121	0.025905	0.025657	0.025374
4.25	0.025589	0.025451	0.025288	0.025096	0.024874	0.024620
4.30	0.024752	0.024630	0.024485	0.024314	0.024116	0.023888
4.35	0.023947	0.023840	0.023711	0.023559	0.023382	0.023178
4.40	0.023175	0.023080	0.022966	0.022831	0.022673	0.022490
4.45	0.022432	0.022349	0.022248	0.022128	0.021987	0.021824
4.50	−0.065866	−0.066367	−0.066665	0.043099	0.043172	0.043170
4.55	0.021034	0.020969	0.020890	0.020796	0.020684	0.020554

Table 4.3 (continued)

			t			
v	0.65	0.70	0.75	0.80	0.85	0.90
4.60	0.020376	0.020318	0.020249	0.020165	0.020066	0.019949
4.65	0.019743	0.019692	0.019631	0.019557	0.019468	0.019364
4.70	0.019135	0.019090	0.019036	0.018970	0.018892	0.018799
4.75	0.018550	0.018511	0.018463	0.018405	0.018335	0.018253
4.80	0.017988	0.017954	0.017912	0.017860	0.017798	0.017725
4.85	0.017447	0.017418	0.017380	0.017335	0.017280	0.017214
4.90	0.016928	0.016901	0.016869	0.016828	0.016780	0.016721
4.95	0.016427	0.016404	0.016376	0.016340	0.016297	0.016245
5.00	0.015946	0.015926	0.015901	0.015869	0.015831	0.015785

			t			
v	0.95	1.00	2.00	3.00	4.00	5.00
0.00	−0.816266	−0.726515	0.254791	0.308959	0.074032	−0.105376
0.05	−0.822648	−0.740137	0.217463	0.305349	0.087879	−0.094751
0.10	−0.822128	−0.746900	0.181177	0.300130	0.100728	−0.083873
0.15	−0.815497	−0.747506	0.146116	0.293433	0.112547	−0.072821
0.20	−0.803512	−0.742637	0.112437	0.285392	0.123315	−0.061668
0.25	−0.786892	−0.732947	0.080270	0.276141	0.133015	−0.050484
0.30	−0.766314	−0.719058	0.049722	0.265813	0.141641	−0.039339
0.35	−0.742408	−0.701558	0.020877	0.254538	0.149195	−0.028296
0.40	−0.715761	−0.681000	−0.006202	0.242450	0.155683	−0.017416
0.45	−0.686913	−0.657896	−0.031471	0.229657	0.161121	−0.006754
0.50	−0.662961	−0.638896	−0.050832	0.225860	0.168298	0.001378
0.55	−0.624540	−0.605913	−0.076498	0.202481	0.168932	0.013709
0.60	−0.591869	−0.577870	−0.096253	0.188317	0.171362	0.023420
0.65	−0.558704	−0.548952	−0.114189	0.173912	0.172852	0.032729
0.70	−0.525366	−0.519484	−0.130339	0.159364	0.173443	0.041603
0.75	−0.492137	−0.489756	−0.144744	0.144767	0.173175	0.050011
0.80	−0.459262	−0.460023	−0.157454	0.130205	0.172094	0.057928
0.85	−0.426953	−0.430511	−0.168528	0.115760	0.170247	0.065330
0.90	−0.395389	−0.401414	−0.178031	0.101504	0.167681	0.072202
0.95	−0.364721	−0.372899	−0.186032	0.087504	0.164449	0.078528
1.00	−0.335072	−0.345108	−0.192606	0.073820	0.160599	0.084300
1.05	−0.306541	−0.318160	−0.197831	0.060506	0.156185	0.089512
1.10	−0.279205	−0.292148	−0.201785	0.047610	0.151258	0.094159
1.15	−0.253121	−0.267151	−0.204550	0.035172	0.145869	0.098244
1.20	−0.228327	−0.243227	−0.206208	0.023229	0.140070	0.101769
1.25	−0.204848	−0.220416	−0.206839	0.011810	0.133910	0.104742
1.30	−0.182692	−0.198748	−0.206526	0.000940	0.127439	0.107171
1.35	−0.161857	−0.178238	−0.205347	−0.009362	0.120705	0.109069
1.40	−0.142329	−0.158890	−0.203381	−0.019080	0.113754	0.110449
1.45	−0.124087	−0.140697	−0.200703	−0.028204	0.106631	0.111327
1.50	−0.108164	−0.123605	−0.281113	−0.036388	0.104911	0.111530

Table 4.3 (continued)

			t			
v	0.95	1.00	2.00	3.00	4.00	5.00
1.55	−0.091335	−0.107715	−0.193503	−0.044652	0.092039	0.111651
1.60	−0.076748	−0.092878	−0.189119	−0.051975	0.084649	0.111138
1.65	−0.063297	−0.079102	−0.184300	−0.058702	0.077246	0.110202
1.70	−0.050932	−0.066352	−0.179105	−0.064841	0.069865	0.108868
1.75	−0.039605	−0.054586	−0.173594	−0.070403	0.062538	0.107159
1.80	−0.029263	−0.043765	−0.167821	−0.075400	0.055295	0.105099
1.85	−0.019855	−0.033844	−0.161835	−0.079847	0.048162	0.102714
1.90	−0.011327	−0.024778	−0.155685	−0.083760	0.041166	0.100028
1.95	−0.003628	−0.016524	−0.149414	−0.087159	0.034330	0.097067
2.00	0.003294	−0.009034	−0.143063	−0.090062	0.027673	0.093856
2.05	0.009490	−0.002265	−0.136668	−0.092491	0.021215	0.090419
2.10	0.015010	0.003829	−0.130263	−0.094466	0.014972	0.086783
2.15	0.019900	0.009290	−0.123880	−0.096011	0.008957	0.082970
2.20	0.024208	0.014162	−0.117545	−0.097148	0.003184	0.079006
2.25	0.027977	0.018486	−0.111283	−0.097901	−0.002338	0.074913
2.30	0.031250	0.022301	−0.105117	−0.098292	−0.007601	0.070714
2.35	0.034067	0.025645	−0.099066	−0.098345	−0.012598	0.066431
2.40	0.036466	0.028555	−0.093146	−0.098083	−0.017323	0.062085
2.45	0.038485	0.031066	−0.087372	−0.097529	−0.021775	0.057695
2.50	0.018105	0.033125	−0.086483	−0.096001	−0.026498	0.053573
2.55	0.041510	0.035020	−0.076311	−0.095635	−0.029849	0.048863
2.60	0.042580	0.036524	−0.071042	−0.094339	−0.033472	0.044456
2.65	0.043392	0.037750	−0.065957	−0.092838	−0.036822	0.040076
2.70	0.043972	0.038723	−0.061062	−0.091153	−0.039902	0.035740
2.75	0.044345	0.039469	−0.056360	−0.089303	−0.042715	0.031461
2.80	0.044533	0.040009	−0.051852	−0.087307	−0.045268	0.027252
2.85	0.044557	0.040365	−0.047542	−0.085182	−0.047566	0.023126
2.90	0.044435	0.040556	−0.043427	−0.082947	−0.049615	0.019092
2.95	0.044185	0.040600	−0.039508	−0.080617	−0.051423	0.015162
3.00	0.043823	0.040514	−0.035783	−0.078208	−0.052998	0.011344
3.05	0.043364	0.040313	−0.032248	−0.075734	−0.054348	0.007646
3.10	0.042821	0.040011	−0.028901	−0.073209	−0.055482	0.004076
3.15	0.042206	0.039622	−0.025738	−0.070646	−0.056408	0.000639
3.20	0.041530	0.039156	−0.022754	−0.068057	−0.057136	−0.002660
3.25	0.040804	0.038624	−0.019945	−0.065452	−0.057676	−0.005817
3.30	0.040035	0.038036	−0.017306	−0.062843	−0.058037	−0.008827
3.35	0.039233	0.037402	−0.014830	−0.060238	−0.058229	−0.011688
3.40	0.038404	0.036728	−0.012512	−0.057646	−0.058261	−0.014399
3.45	0.037554	0.036023	−0.010348	−0.055075	−0.058144	−0.016957
3.50	0.042255	0.040858	0.075217	−0.072687	−0.063405	−0.019171
3.55	0.035818	0.034541	−0.006451	−0.050025	−0.057498	−0.021616
3.60	0.034940	0.033776	−0.004708	−0.047558	−0.056988	−0.023718
3.65	0.034061	0.033001	−0.003093	−0.045136	−0.056366	−0.025669

Table 4.3 (continued)

			t			
v	0.95	1.00	2.00	3.00	4.00	5.00
3.70	0.033185	0.032220	−0.001600	−0.042765	−0.055641	−0.027471
3.75	0.032314	0.031437	−0.000223	−0.040448	−0.054822	−0.029126
3.80	0.031452	0.030655	0.001043	−0.038188	−0.053917	−0.030637
3.85	0.030600	0.029876	0.002205	−0.035989	−0.052935	−0.032007
3.90	0.029760	0.029104	0.003267	−0.033853	−0.051883	−0.033240
3.95	0.028934	0.028339	0.004236	−0.031782	−0.050769	−0.034338
4.00	0.028124	0.027585	0.005117	−0.029777	−0.049600	−0.035306
4.05	0.027330	0.026842	0.005916	−0.027840	−0.048384	−0.036148
4.10	0.026553	0.026112	0.006636	−0.025972	−0.047127	−0.036868
4.15	0.025794	0.025395	0.007284	−0.024172	−0.045836	−0.037470
4.20	0.025053	0.024694	0.007863	−0.022441	−0.044516	−0.037960
4.25	0.024332	0.024007	0.008378	−0.020780	−0.043174	−0.038343
4.30	0.023629	0.023336	0.008834	−0.019187	−0.041814	−0.038622
4.35	0.022945	0.022682	0.009234	−0.017662	−0.040442	−0.038803
4.40	0.022281	0.022044	0.009582	−0.016204	−0.039063	−0.038892
4.45	0.021636	0.021423	0.009883	−0.014813	−0.037681	−0.038892
4.50	0.043091	0.021665	0.014872	−0.014179	−0.035588	0.200319
4.55	0.020403	0.020230	0.010355	−0.012224	−0.034923	−0.038647
4.60	0.019814	0.019659	0.010532	−0.011024	−0.033555	−0.038412
4.65	0.019244	0.019105	0.010675	−0.009885	−0.032199	−0.038109
4.70	0.018691	0.018567	0.010785	−0.008806	−0.030858	−0.037742
4.75	0.018156	0.018045	0.010866	−0.007784	−0.029533	−0.037316
4.80	0.017639	0.017538	0.010921	−0.006819	−0.028229	−0.036835
4.85	0.017138	0.017048	0.010950	−0.005907	−0.026946	−0.036304
4.90	0.016653	0.016573	0.010958	−0.005048	−0.025687	−0.035727
4.95	0.016184	0.016112	0.010944	−0.004240	−0.024454	−0.035108
5.00	0.015731	0.015667	0.010913	−0.003480	−0.023247	−0.034452

4.2 Integral Bessel Functions of the Second Kind and Their First and Second Derivatives with Respect to the Order

Integral Bessel functions $Yi_v(t)$ were calculated by applying MATHEMATICA program from their definition:

$$Yi_v(z) = \int_z^\infty \frac{Yi_v(t)}{t}\,dt \qquad (4.2.1)$$

Their derivatives with respect to the order of the integral Bessel function of the second kind were determined using the following expressions (Figures 4.10–4.12; Tables 4.4–4.6):

$$\frac{\partial Yi_v(t)}{\partial v} = \frac{-Yi_{v+2h}(t) + 8\,Yi_{v+h}(t) - 8\,Yi_{v-h}(t) + Yi_{v-2h}(t)}{12h} \tag{4.2.2}$$

$$\frac{\partial^2 Yi_v(t)}{\partial v^2} = \frac{-Yi_{v+2h}(t) + 16\,Yi_{v+h}(t) - 30\,Yi_v(t) + 16\,Yi_{v-h}(t) - Yi_{v-2h}(t)}{12h^2} \tag{4.2.3}$$

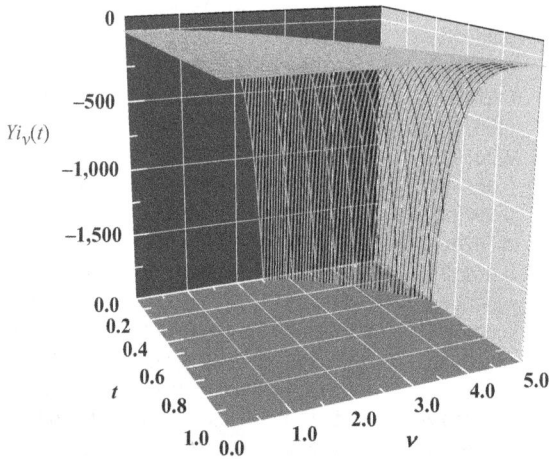

Figure 4.10: The integral Bessel functions of the second kind $Yi_v(t)$ as a function of v and t.

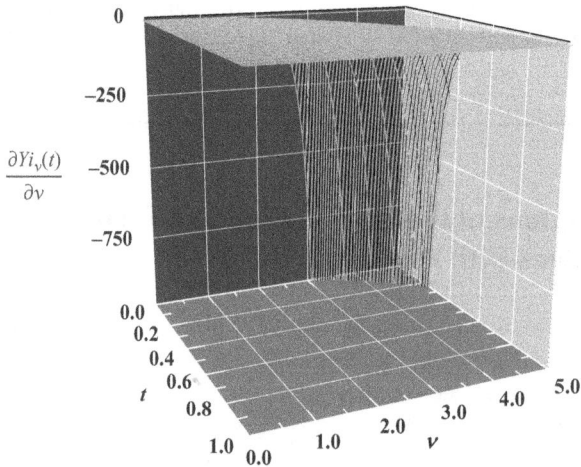

Figure 4.11: First derivatives of the integral Bessel functions of the second kind with respect to the order $\partial Yi_v(t)/\partial v$ as a function of v and t.

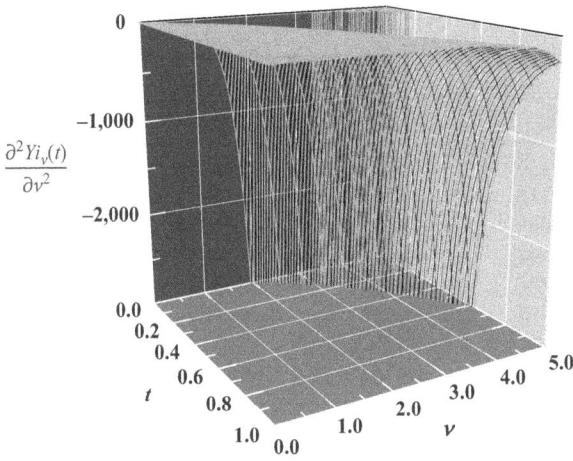

Figure 4.12: Second derivatives of the integral Bessel functions of the second kind with respect to the order $\partial^2 Yi_\nu(t)/\partial \nu^2$ as a function of ν and t.

Table 4.4: The integral Bessel functions of the second kind $Yi_\nu(t)$.

			t		
ν	0.05	0.10	0.15	0.20	0.25
0.00	−2.55934	−1.34138	−0.772598	−0.434231	−0.209626
0.05	−2.79740	−1.52608	−0.926819	−0.567024	−0.325894
0.10	−3.02627	−1.70163	−1.07363	−0.694019	−0.437704
0.15	−3.25108	−1.87033	−1.21425	−0.815888	−0.545407
0.20	−3.47684	−2.03447	−1.34995	−0.933362	−0.649415
0.25	−3.70852	−2.19635	−1.48203	−1.04722	−0.750191
0.30	−3.95114	−2.35830	−1.61180	−1.15827	−0.848245
0.35	−4.20988	−2.52264	−1.74064	−1.26737	−0.944123
0.40	−4.49019	−2.69180	−1.86993	−1.37540	−1.03841
0.45	−4.79792	−2.86827	−2.00112	−1.48329	−1.13172
0.50	−5.13947	−3.05467	−2.13570	−1.59200	−1.22471
0.55	−5.52197	−3.25381	−2.27523	−1.70253	−1.31805
0.60	−5.95347	−3.46870	−2.42138	−1.81594	−1.41247
0.65	−6.44314	−3.70263	−2.57590	−1.93332	−1.50872
0.70	−7.00156	−3.95923	−2.74070	−2.05588	−1.60761
0.75	−7.64106	−4.24252	−2.91784	−2.18486	−1.70997
0.80	−8.37601	−4.55705	−3.10957	−2.32162	−1.81672
0.85	−9.22335	−4.90793	−3.31838	−2.46764	−1.92882
0.90	−10.2031	−5.30102	−3.54703	−2.62453	−2.04734
0.95	−11.3389	−5.74299	−3.79860	−2.79403	−2.17338
1.00	−12.6590	−6.24152	−4.07656	−2.97812	−2.30821
1.05	−14.1970	−6.80548	−4.38482	−3.17893	−2.45316
1.10	−15.9930	−7.44513	−4.72783	−3.39889	−2.60973

Table 4.4 (continued)

			t		
v	0.05	0.10	0.15	0.20	0.25
1.15	−18.0951	−8.17240	−5.11063	−3.64070	−2.77958
1.20	−20.5611	−9.00118	−5.53900	−3.90739	−2.96451
1.25	−23.4603	−9.94765	−6.01953	−4.20238	−3.16659
1.30	−26.8761	−11.0308	−6.55981	−4.52956	−3.38808
1.35	−30.9095	−12.2727	−7.16855	−4.89331	−3.63154
1.40	−35.6820	−13.6995	−7.85581	−5.29864	−3.89984
1.45	−41.3410	−15.3417	−8.63320	−5.75125	−4.19623
1.50	−48.0649	−17.2354	−9.51412	−6.25764	−4.52435
1.55	−56.0705	−19.4229	−10.5141	−6.82524	−4.88836
1.60	−65.6211	−21.9543	−11.6513	−7.46260	−5.29295
1.65	−77.0371	−24.8888	−12.9465	−8.17948	−5.74345
1.70	−90.7094	−28.2965	−14.4242	−8.98711	−6.24593
1.75	−107.115	−32.2601	−16.1127	−9.89843	−6.80730
1.80	−126.836	−36.8782	−18.0452	−10.9283	−7.43544
1.85	−150.587	−42.2675	−20.2601	−12.0939	−8.13936
1.90	−179.241	−48.5670	−22.8028	−13.4151	−8.92936
1.95	−213.872	−55.9420	−25.7258	−14.9146	−9.81721
2.00	−255.797	−64.5895	−29.0912	−16.6191	−10.8164
2.05	−306.638	−74.7448	−32.9714	−18.5592	−11.9425
2.10	−368.393	−86.6885	−37.4514	−20.7704	−13.2132
2.15	−443.523	−100.756	−42.6311	−23.2940	−14.6489
2.20	−535.069	−117.351	−48.6282	−26.1780	−16.2733
2.25	−646.790	−136.952	−55.5809	−29.4779	−18.1134
2.30	−783.337	−160.140	−63.6522	−33.2587	−20.2003
2.35	−950.471	−187.606	−73.0345	−37.5959	−22.5701
2.40	−1,155.34	−220.185	−83.9545	−42.5775	−25.2645
2.45	−1,406.82	−258.879	−96.6806	−48.3062	−28.3313
2.50	−1,715.93	−304.897	−111.530	−54.9019	−31.8263
2.55	−2,096.40	−359.695	−128.878	−62.5050	−35.8137
2.60	−2,565.32	−425.030	−149.169	−71.2794	−40.3680
2.65	−3,144.01	−503.024	−172.932	−81.4173	−45.5758
2.70	−3,859.07	−596.244	−200.793	−93.1437	−51.5373
2.75	−4,743.73	−707.795	−233.496	−106.7230	−58.3689
2.80	−5,839.56	−841.436	−271.925	−122.4650	−66.2062
2.85	−7,198.58	−1,001.73	−317.135	−140.7330	−75.2065
2.90	−8,886.00	−1,194.20	−370.379	−161.9570	−85.5533
2.95	−10,983.6	−1,425.57	−433.153	−186.6390	−97.4601
3.00	−13,594.0	−1,704.01	−507.240	−215.3740	−111.176
3.05	−16,846.1	−2,039.44	−594.772	−248.8600	−126.992
3.10	−20,902.2	−2,443.97	−698.295	−287.9220	−145.247
3.15	−25,966.4	−2,932.32	−820.852	−333.5350	−166.338
3.20	−32,296.0	−3,522.47	−966.088	−386.8480	−190.728
3.25	−40,215.2	−4,236.35	−1,138.37	−449.2210	−218.961

Table 4.4 (continued)

v	0.05	0.10	0.15	0.20	0.25
3.30	−50,133.3	−5,100.75	−1,342.93	−522.2630	−251.672
3.35	−62,567.3	−6,148.44	−1,586.04	−607.8780	−289.605
3.40	−78,170.5	−7,419.47	−1,875.24	−708.3220	−333.635
3.45	−97,769.7	−8,962.91	−2,219.60	−826.2700	−384.787
3.50	−122412	−10,838.9	−2,629.98	−964.8980	−444.265
3.55	−153422	−13,121.1	−3,119.50	−1,127.97	−513.485
3.60	−192484	−15,900.0	−3,703.92	−1,319.97	−594.109
3.65	−241731	−19,286.7	−4,402.25	−1,546.22	−688.097
3.70	−303874	−23,417.7	−5,237.40	−1,813.04	−797.754
3.75	−382359	−28,460.9	−6,237.03	−2,127.99	−925.799
3.80	−481569	−34,622.8	−7,434.51	−2,500.04	−1,075.43
3.85	−607083	−42,158.1	−8,870.18	−2,939.90	−1,250.44
3.90	−766007	−51,380.1	−10,592.8	−3,460.35	−1,455.28
3.95	−967401	−62,675.8	−12,661.4	−4,076.63	−1,695.22
4.00	−1.223E+6	−76,522.3	−15,147.4	−4,806.95	−1,976.51
4.05	−1.547	−93,508.9	−18,137.2	−5,673.08	−2,306.50
4.10	−1.959	−114364	−21,735.9	−6,701.06	−2,693.92
4.15	−2.482	−139988	−26,070.6	−7,922.02	−3,149.11
4.20	−3.148	−171495	−31,295.7	−9,373.27	−3,684.30
4.25	−3.996	−210266	−37,598.9	−11,099.5	−4,314.01
4.30	−5.077	−258008	−45,207.9	−13,154.3	−5,055.47
4.35	−6.455	−316840	−54,399.9	−15,601.8	−5,929.11
4.40	−8.213	−389392	−65,512.1	−18,519.3	−6,959.20
4.45	−1.046E+7	−478925	−78,954.9	−21,999.4	−8,174.61
4.50	−1.333	−589490	−95,228.5	−26,153.4	−9,609.63
4.55	−1.699	−726122	−114942	−31,115.1	−11,305.1
4.60	−2.169	−895081	−138839	−37,045.5	−13,309.6
4.65	−2.770	−1.104E+6	−167827	−44,138.4	−15,681.0
4.70	−3.540	−1.363	−203012	−52,627.2	−18,488.2
4.75	−4.527	−1.684	−245748	−62,793.1	−21,813.5
4.80	−5.794	−2.082	−297688	−74,975.3	−25,755.1
4.85	−7.420	−2.575	−360855	−89,582.8	−30,429.9
4.90	−9.509	−3.188	−437723	−107109	−35,977.9
4.95	−1.220E+8	−3.949	−531322	−128151	−42,566.1
5.00	−1.565	−4.894	−645364	−153429	−50,394.3

v	0.30	0.35	0.40	0.45	0.50
0.00	−0.051078	0.065144	0.152380	0.218792	0.269689
0.05	−0.153913	−0.026396	0.070562	0.145483	0.203924
0.10	−0.253388	−0.115479	−0.009546	0.073264	0.138731
0.15	−0.349652	−0.202120	−0.087870	0.002268	0.074284

Table 4.4 (continued)

			t		
v	0.30	0.35	0.40	0.45	0.50
0.20	−0.442914	−0.286394	−0.164390	−0.067419	0.010711
0.25	−0.533434	−0.368423	−0.239134	−0.135759	−0.051901
0.30	−0.621521	−0.448378	−0.312176	−0.202754	−0.113501
0.35	−0.707524	−0.526468	−0.383624	−0.268442	−0.174075
0.40	−0.791825	−0.602941	−0.453627	−0.332896	−0.233640
0.45	−0.874842	−0.678073	−0.522361	−0.396218	−0.292242
0.50	−0.957024	−0.752173	−0.590033	−0.458540	−0.349955
0.55	−1.03885	−0.825576	−0.656876	−0.520016	−0.406874
0.60	−1.12082	−0.898644	−0.723147	−0.580826	−0.463120
0.65	−1.20347	−0.971766	−0.789127	−0.641173	−0.518833
0.70	−1.28738	−1.04535	−0.855121	−0.701278	−0.574173
0.75	−1.37315	−1.11985	−0.921455	−0.761385	−0.629319
0.80	−1.46141	−1.19572	−0.988479	−0.821757	−0.684470
0.85	−1.55285	−1.27347	−1.05657	−0.882676	−0.739839
0.90	−1.64819	−1.35362	−1.12612	−0.944447	−0.795659
0.95	−1.74824	−1.43676	−1.19756	−1.00740	−0.852182
1.00	−1.85382	−1.52348	−1.27135	−1.07187	−0.909677
1.05	−1.96587	−1.61446	−1.34797	−1.13824	−0.968434
1.10	−2.08538	−1.71040	−1.42796	−1.20692	−1.02876
1.15	−2.21346	−1.81207	−1.51188	−1.27832	−1.09100
1.20	−2.35129	−1.92032	−1.60035	−1.35293	−1.15550
1.25	−2.50022	−2.03606	−1.69404	−1.43122	−1.22265
1.30	−2.66171	−2.16031	−1.79366	−1.51376	−1.29287
1.35	−2.83739	−2.29418	−1.90002	−1.60113	−1.36660
1.40	−3.02907	−2.43889	−2.01399	−1.69396	−1.44434
1.45	−3.23878	−2.59580	−2.13652	−1.79297	−1.52660
1.50	−3.46879	−2.76641	−2.26867	−1.89891	−1.61398
1.55	−3.72164	−2.95240	−2.41159	−2.01263	−1.70709
1.60	−4.00020	−3.15565	−2.56659	−2.13506	−1.80663
1.65	−4.30767	−3.37823	−2.73508	−2.26721	−1.91335
1.70	−4.64770	−3.62249	−2.91865	−2.41022	−2.02807
1.75	−5.02438	−3.89104	−3.11909	−2.56533	−2.15171
1.80	−5.44236	−4.18685	−3.33836	−2.73393	−2.28528
1.85	−5.90690	−4.51321	−3.57868	−2.91755	−2.42989
1.90	−6.42397	−4.87386	−3.84252	−3.11792	−2.58676
1.95	−7.00033	−5.27301	−4.13265	−3.33695	−2.75728
2.00	−7.64368	−5.71542	−4.45218	−3.57677	−2.94294
2.05	−8.36281	−6.20645	−4.80462	−3.83975	−3.14545
2.10	−9.16769	−6.75220	−5.19389	−4.12858	−3.36668
2.15	−10.0697	−7.35956	−5.62443	−4.44624	−3.60872
2.20	−11.0819	−8.03633	−6.10124	−4.79608	−3.87391
2.25	−12.2190	−8.79139	−6.62995	−5.18186	−4.16486
2.30	−13.4982	−9.63481	−7.21694	−5.60782	−4.48449

Table 4.4 (continued)

v	0.30	0.35	0.40	0.45	0.50
			t		
2.35	−14.9388	−10.5781	−7.86940	−6.07871	−4.83608
2.40	−16.5632	−11.6342	−8.59549	−6.59989	−5.22329
2.45	−18.3968	−12.8180	−9.40443	−7.17739	−5.65022
2.50	−20.4690	−14.1465	−10.3067	−7.81802	−6.12151
2.55	−22.8134	−15.6390	−11.3141	−8.52947	−6.64232
2.60	−25.4688	−17.3175	−12.4402	−9.32040	−7.21850
2.65	−28.4798	−19.2072	−13.7003	−10.2006	−7.85659
2.70	−31.8974	−21.3371	−15.1117	−11.1812	−8.56400
2.75	−35.7809	−23.7400	−16.6943	−12.2748	−9.34905
2.80	−40.1984	−26.4538	−18.4706	−13.4956	−10.2211
2.85	−45.2283	−29.5218	−20.4664	−14.8597	−11.1908
2.90	−50.9616	−32.9938	−22.7110	−16.3855	−12.2702
2.95	−57.5030	−36.9266	−25.2377	−18.0937	−13.4726
3.00	−64.9739	−41.3860	−28.0849	−20.0080	−14.8135
3.05	−73.5148	−46.4472	−31.2962	−22.1552	−16.3102
3.10	−83.2882	−52.1969	−34.9215	−24.5659	−17.9823
3.15	−94.4827	−58.7349	−39.0180	−27.2749	−19.8521
3.20	−107.317	−66.1760	−43.6511	−30.3219	−21.9446
3.25	−122.045	−74.6529	−48.8959	−33.7520	−24.2887
3.30	−138.961	−84.3183	−54.8384	−37.6168	−26.9167
3.35	−158.408	−95.3486	−61.5773	−41.9752	−29.8657
3.40	−180.785	−107.948	−69.2259	−46.8944	−33.1775
3.45	−206.555	−122.351	−77.9144	−52.4513	−36.9000
3.50	−236.259	−138.832	−87.7927	−58.7338	−41.0876
3.55	−270.527	−157.704	−99.0329	−65.8424	−45.8021
3.60	−310.093	−179.334	−111.833	−73.8924	−51.1142
3.65	−355.813	−204.144	−126.423	−83.0158	−57.1043
3.70	−408.690	−232.626	−143.064	−93.3638	−63.8645
3.75	−469.893	−265.349	−162.060	−105.110	−71.4995
3.80	−540.789	−302.974	−183.763	−118.454	−80.1293
3.85	−622.977	−346.270	−208.577	−133.625	−89.8908
3.90	−718.332	−396.128	−236.969	−150.885	−100.941
3.95	−829.048	−453.589	−269.480	−170.537	−113.459
4.00	−957.696	−519.861	−306.735	−192.930	−127.650
4.05	−1,107.29	−596.352	−349.459	−218.464	−143.750
4.10	−1,281.38	−684.702	−398.488	−247.600	−162.028
4.15	−1,484.11	−786.824	−454.795	−280.871	−182.794
4.20	−1,720.38	−904.950	−519.507	−318.891	−206.403
4.25	−1,995.91	−1,041.68	−593.928	−362.367	−233.263
4.30	−2,317.47	−1,200.07	−679.576	−412.117	−263.843
4.35	−2,693.01	−1,383.66	−778.213	−469.084	−298.681
4.40	−3,131.89	−1,596.61	−891.885	−534.361	−338.397
4.45	−3,645.14	−1,843.78	−1,022.97	−609.209	−383.704

Table 4.4 (continued)

			t		
v	0.30	0.35	0.40	0.45	0.50
4.50	−4,245.77	−2,130.87	−1,174.24	−695.088	−435.422
4.55	−4,949.12	−2,464.55	−1,348.91	−793.687	−494.498
4.60	−5,773.29	−2,852.61	−1,550.75	−906.965	−562.021
4.65	−6,739.67	−3,304.23	−1,784.10	−1,037.19	−639.249
4.70	−7,873.52	−3,830.13	−2,054.09	−1,186.99	−727.631
4.75	−9,204.71	−4,442.94	−2,366.64	−1,359.41	−828.841
4.80	−10,768.6	−5,157.45	−2,728.70	−1,558.00	−944.812
4.85	−12,606.9	−5,991.06	−3,148.36	−1,786.87	−1,077.78
4.90	−14,769.2	−6,964.21	−3,635.08	−2,050.79	−1,230.32
4.95	−17,314.1	−8,100.94	−4,199.92	−2,355.31	−1,405.43
5.00	−20,311.1	−9,429.55	−4,855.81	−2,706.88	−1,606.56

			t		
v	0.55	0.60	0.65	0.70	0.75
0.00	0.308698	0.338392	0.360654	0.376898	0.388210
0.05	0.249688	0.285477	0.313273	0.334565	0.350501
0.10	0.190816	0.232336	0.265359	0.291444	0.311790
0.15	0.132281	0.179183	0.217136	0.247760	0.272302
0.20	0.074243	0.126198	0.168796	0.203711	0.232235
0.25	0.016822	0.073525	0.120499	0.159469	0.191767
0.30	−0.039894	0.021280	0.072379	0.115180	0.151053
0.35	−0.095851	−0.030455	0.024541	0.070964	0.110223
0.40	−0.151023	−0.081623	−0.022936	0.026919	0.069388
0.45	−0.205415	−0.132193	−0.069994	−0.016879	0.028639
0.50	−0.259053	−0.182160	−0.116600	−0.060375	−0.011953
0.55	−0.311990	−0.231537	−0.162741	−0.103533	−0.052335
0.60	−0.364297	−0.280361	−0.208423	−0.146337	−0.092471
0.65	−0.416068	−0.328686	−0.253670	−0.188786	−0.132341
0.70	−0.467414	−0.376586	−0.298524	−0.230897	−0.171941
0.75	−0.518465	−0.424149	−0.343042	−0.272700	−0.211280
0.80	−0.569365	−0.471481	−0.387297	−0.314241	−0.250381
0.85	−0.620278	−0.518702	−0.431375	−0.355578	−0.289280
0.90	−0.671381	−0.565947	−0.475376	−0.396784	−0.328027
0.95	−0.722869	−0.613365	−0.519413	−0.437943	−0.366681
1.00	−0.774951	−0.661120	−0.563613	−0.479150	−0.405313
1.05	−0.827854	−0.709389	−0.608115	−0.520514	−0.444005
1.10	−0.881823	−0.758365	−0.653070	−0.562154	−0.482849
1.15	−0.937117	−0.808255	−0.698644	−0.604200	−0.521950
1.20	−0.994020	−0.859283	−0.745015	−0.646795	−0.561419
1.25	−1.05283	−0.911688	−0.792375	−0.690094	−0.601383
1.30	−1.11388	−0.965730	−0.840930	−0.734263	−0.641975
1.35	−1.17752	−1.02168	−0.890905	−0.779483	−0.683341
1.40	−1.24412	−1.07985	−0.942539	−0.825947	−0.725640

Table 4.4 (continued)

			t		
v	0.55	0.60	0.65	0.70	0.75
1.45	−1.31409	−1.14056	−0.996089	−0.873865	−0.769041
1.50	−1.38789	−1.20415	−1.05183	−0.923461	−0.813727
1.55	−1.46598	−1.27100	−1.11008	−0.974977	−0.859897
1.60	−1.54890	−1.34152	−1.17114	−1.02867	−0.907762
1.65	−1.63721	−1.41615	−1.23537	−1.08484	−0.957551
1.70	−1.73155	−1.49539	−1.30315	−1.14376	−1.00951
1.75	−1.83260	−1.57975	−1.37490	−1.20579	−1.06391
1.80	−1.94110	−1.66981	−1.45107	−1.27127	−1.12104
1.85	−2.05790	−1.76620	−1.53214	−1.34060	−1.18120
1.90	−2.18389	−1.86962	−1.61866	−1.41420	−1.24474
1.95	−2.32009	−1.98083	−1.71120	−1.49252	−1.31203
2.00	−2.46761	−2.10065	−1.81042	−1.57607	−1.38346
2.05	−2.62767	−2.23002	−1.91701	−1.66541	−1.45947
2.10	−2.80165	−2.36994	−2.03175	−1.76112	−1.54054
2.15	−2.99106	−2.52155	−2.15550	−1.86388	−1.62719
2.20	−3.19757	−2.68609	−2.28920	−1.97441	−1.71998
2.25	−3.42307	−2.86492	−2.43388	−2.09351	−1.81955
2.30	−3.66963	−3.05959	−2.59068	−2.22205	−1.92656
2.35	−3.93957	−3.27179	−2.76089	−2.36100	−2.04177
2.40	−4.23550	−3.50340	−2.94589	−2.51142	−2.16602
2.45	−4.56031	−3.75653	−3.14726	−2.67450	−2.30020
2.50	−4.91723	−4.03350	−3.36670	−2.85154	−2.44532
2.55	−5.30989	−4.33692	−3.60615	−3.04398	−2.60248
2.60	−5.74234	−4.66969	−3.86773	−3.25342	−2.77290
2.65	−6.21912	−5.03507	−4.15381	−3.48162	−2.95794
2.70	−6.74534	−5.43666	−4.46704	−3.73056	−3.15908
2.75	−7.32671	−5.87852	−4.81035	−4.00242	−3.37797
2.80	−7.96967	−6.36518	−5.18702	−4.29962	−3.61644
2.85	−8.68144	−6.90172	−5.60071	−4.62487	−3.87652
2.90	−9.47014	−7.49383	−6.05552	−4.98115	−4.16046
2.95	−10.3449	−8.14788	−6.55600	−5.37183	−4.47077
3.00	−11.3161	−8.87103	−7.10726	−5.80063	−4.81021
3.05	−12.3952	−9.67131	−7.71503	−6.27170	−5.18188
3.10	−13.5955	−10.5578	−8.38568	−6.78969	−5.58921
3.15	−14.9316	−11.5405	−9.12641	−7.35979	−6.03603
3.20	−16.4203	−12.6311	−9.94524	−7.98778	−6.52661
3.25	−18.0804	−13.8422	−10.8512	−8.68016	−7.06570
3.30	−19.9333	−15.1884	−11.8544	−9.44416	−7.65860
3.35	−22.0031	−16.6861	−12.9662	−10.2879	−8.31124
3.40	−24.3172	−18.3536	−14.1995	−11.2205	−9.03022
3.45	−26.9065	−20.2117	−15.5685	−12.2522	−9.82295
3.50	−29.8061	−22.2840	−17.0895	−13.3943	−10.6977
3.55	−33.0558	−24.5969	−18.7806	−14.6597	−11.6637

Table 4.4 (continued)

			t		
v	0.55	0.60	0.65	0.70	0.75
3.60	−36.7009	−27.1804	−20.6625	−16.0629	−12.7313
3.65	−40.7927	−30.0685	−22.7582	−17.6200	−13.9121
3.70	−45.3894	−33.2997	−25.0939	−19.3492	−15.2192
3.75	−50.5573	−36.9172	−27.6989	−21.2711	−16.6671
3.80	−56.3721	−40.9706	−30.6066	−23.4086	−18.2721
3.85	−62.9193	−45.5157	−33.8545	−25.7878	−20.0527
3.90	−70.2970	−50.6159	−37.4852	−28.4379	−22.0294
3.95	−78.6165	−56.3431	−41.5466	−31.3919	−24.2256
4.00	−88.0049	−62.7793	−46.0931	−34.6869	−26.6672
4.05	−98.6072	−70.0171	−51.1863	−38.3651	−29.3836
4.10	−110.589	−78.1624	−56.8960	−42.4737	−32.4078
4.15	−124.139	−87.3353	−63.3013	−47.0664	−35.7771
4.20	−139.473	−97.6726	−70.4917	−52.2036	−39.5335
4.25	−156.839	−109.3300	−78.5692	−57.9540	−43.7240
4.30	−176.518	−122.4850	−87.6491	−64.3949	−48.4022
4.35	−198.835	−137.3410	−97.8629	−71.6141	−53.6281
4.40	−224.159	−154.1270	−109.360	−79.7109	−59.4698
4.45	−252.914	−173.1070	−122.309	−88.7979	−66.0040
4.50	−285.587	−194.5830	−136.903	−99.0029	−73.3175
4.55	−322.736	−218.8970	−153.363	−110.471	−81.5084
4.60	−365.000	−246.4430	−171.938	−123.365	−90.6878
4.65	−413.115	−277.6700	−192.912	−137.874	−100.981
4.70	−467.925	−313.0900	−216.612	−154.208	−112.531
4.75	−530.398	−353.2940	−243.407	−172.609	−125.499
4.80	−601.652	−398.953	−273.719	−193.350	−140.067
4.85	−682.970	−450.840	−308.032	−216.745	−156.443
4.90	−775.826	−509.840	−346.897	−243.146	−174.862
4.95	−881.924	−576.966	−390.942	−272.959	−195.591
5.00	−1,003.22	−653.384	−440.888	−306.644	−218.933

			t		
v	0.80	0.85	0.90	0.95	1.0
0.00	0.395439	0.399261	0.400223	0.398770	0.395273
0.05	0.361980	0.369718	0.374294	0.376184	0.375780
0.10	0.327341	0.338848	0.346920	0.352056	0.354670
0.15	0.291742	0.306865	0.318304	0.326582	0.332130
0.20	0.255382	0.273964	0.288638	0.299945	0.308334
0.25	0.218440	0.240323	0.258096	0.272315	0.283445
0.30	0.181074	0.206102	0.226836	0.243849	0.257618
0.35	0.143422	0.171444	0.195004	0.214691	0.230992
0.40	0.105605	0.136474	0.162729	0.184970	0.203699
0.45	0.067723	0.101301	0.130123	0.154804	0.175855
0.50	0.029861	0.066018	0.097287	0.124297	0.147569

Table 4.4 (continued)

			t		
v	0.80	0.85	0.90	0.95	1.0
0.55	−0.007916	0.030702	0.064306	0.093539	0.118935
0.60	−0.045555	−0.004583	0.031252	0.062610	0.090036
0.65	−0.083022	−0.039791	−0.001817	0.031575	0.060945
0.70	−0.120294	−0.074884	−0.034855	0.000491	0.031723
0.75	−0.157364	−0.109842	−0.067827	−0.030601	0.002423
0.80	−0.194237	−0.144654	−0.100710	−0.061666	−0.026914
0.85	−0.230931	−0.179321	−0.133495	−0.092682	−0.056257
0.90	−0.267473	−0.213856	−0.166178	−0.123637	−0.085584
0.95	−0.303905	−0.248283	−0.198769	−0.154527	−0.114881
1.00	−0.340275	−0.282633	−0.231286	−0.185359	−0.144143
1.05	−0.376646	−0.316951	−0.263757	−0.216146	−0.173375
1.10	−0.413087	−0.351287	−0.296218	−0.246911	−0.202587
1.15	−0.449679	−0.385702	−0.328714	−0.277687	−0.231799
1.20	−0.486511	−0.420267	−0.361297	−0.308510	−0.261037
1.25	−0.523683	−0.455060	−0.394029	−0.339428	−0.290335
1.30	−0.561303	−0.490168	−0.426979	−0.370495	−0.319734
1.35	−0.599490	−0.525688	−0.460225	−0.401771	−0.349281
1.40	−0.638372	−0.561724	−0.493850	−0.433326	−0.379031
1.45	−0.678090	−0.598389	−0.527949	−0.465234	−0.409044
1.50	−0.718793	−0.635809	−0.562623	−0.497580	−0.439389
1.55	−0.760642	−0.674115	−0.597982	−0.530454	−0.470141
1.60	−0.803813	−0.713453	−0.634145	−0.563955	−0.501380
1.65	−0.848493	−0.753976	−0.671242	−0.598190	−0.533197
1.70	−0.894884	−0.795852	−0.709410	−0.633275	−0.565687
1.75	−0.943204	−0.839261	−0.748799	−0.669333	−0.598954
1.80	−0.993689	−0.884396	−0.789569	−0.706500	−0.633112
1.85	−1.04659	−0.931467	−0.831895	−0.744919	−0.668281
1.90	−1.10219	−0.980699	−0.875961	−0.784747	−0.704592
1.95	−1.16078	−1.03233	−0.921971	−0.826151	−0.742186
2.00	−1.22268	−1.08664	−0.970141	−0.869312	−0.781213
2.05	−1.28825	−1.14390	−1.02070	−0.914424	−0.821836
2.10	−1.35786	−1.20442	−1.07392	−0.961697	−0.864232
2.15	−1.43193	−1.26853	−1.13005	−1.01136	−0.908590
2.20	−1.51092	−1.33662	−1.18941	−1.06366	−0.955113
2.25	−1.59531	−1.40906	−1.25232	−1.11886	−1.00402
2.30	−1.68566	−1.48630	−1.31912	−1.17724	−1.05556
2.35	−1.78254	−1.56880	−1.39020	−1.23913	−1.10998
2.40	−1.88661	−1.65709	−1.46597	−1.30487	−1.16757
2.45	−1.99858	−1.75174	−1.54690	−1.37482	−1.22862
2.50	−2.11924	−1.85335	−1.63347	−1.44938	−1.29348
2.55	−2.24944	−1.96261	−1.72625	−1.52900	−1.36250
2.60	−2.39013	−2.08026	−1.82581	−1.61416	−1.43607
2.65	−2.54235	−2.20713	−1.93281	−1.70538	−1.51461

Table 4.4 (continued)

			t		
v	0.80	0.85	0.90	0.95	1.0
2.70	−2.70726	−2.34412	−2.04796	−1.80323	−1.59860
2.75	−2.88611	−2.49221	−2.17204	−1.90835	−1.68855
2.80	−3.08033	−2.65250	−2.30593	−2.02142	−1.78500
2.85	−3.29145	−2.82619	−2.45057	−2.14319	−1.88857
2.90	−3.52119	−3.01461	−2.60699	−2.27450	−1.99992
2.95	−3.77146	−3.21923	−2.77636	−2.41625	−2.11978
3.00	−4.04436	−3.44166	−2.95992	−2.56945	−2.24896
3.05	−4.34221	−3.68371	−3.15909	−2.73520	−2.38832
3.10	−4.66763	−3.94736	−3.37540	−2.91471	−2.53885
3.15	−5.02347	−4.23480	−3.61056	−3.10931	−2.70159
3.20	−5.41294	−4.54847	−3.86644	−3.32049	−2.87771
3.25	−5.83960	−4.89106	−4.14512	−3.54985	−3.06850
3.30	−6.30738	−5.26558	−4.44892	−3.79921	−3.27538
3.35	−6.82071	−5.67535	−4.78038	−4.07054	−3.49990
3.40	−7.38447	−6.12406	−5.14232	−4.36603	−3.74378
3.45	−8.00413	−6.61582	−5.53789	−4.68810	−4.00893
3.50	−8.68578	−7.15519	−5.97056	−5.03945	−4.29744
3.55	−9.43622	−7.74727	−6.44418	−5.42305	−4.61162
3.60	−10.2630	−8.39771	−6.96304	−5.84217	−4.95403
3.65	−11.1747	−9.11281	−7.53192	−6.30048	−5.32752
3.70	−12.1807	−9.89961	−8.15610	−6.80203	−5.73521
3.75	−13.2917	−10.7659	−8.84148	−7.35131	−6.18058
3.80	−14.5194	−11.7205	−9.59461	−7.95330	−6.66747
3.85	−15.8772	−12.7732	−10.4228	−8.61356	−7.20015
3.90	−17.3799	−13.9348	−11.3342	−9.33825	−7.78335
3.95	−19.0442	−15.2176	−12.3379	−10.1342	−8.42232
4.00	−20.8888	−16.6352	−13.4440	−11.0091	−9.12288
4.05	−22.9346	−18.2028	−14.6638	−11.9714	−9.89152
4.10	−25.2052	−19.9376	−16.0099	−13.0306	−10.7354
4.15	−27.7270	−21.8586	−17.4965	−14.1972	−11.6626
4.20	−30.5297	−23.9874	−19.1392	−15.4830	−12.6820
4.25	−33.6466	−26.3479	−20.9558	−16.9011	−13.8034
4.30	−37.1154	−28.9672	−22.9658	−18.4661	−15.0380
4.35	−40.9782	−31.8754	−25.1914	−20.1945	−16.3979
4.40	−45.2826	−35.1066	−27.6574	−22.1044	−17.8970
4.45	−50.0823	−38.6989	−30.3913	−24.2163	−19.5505
4.50	−55.4375	−42.6953	−33.4243	−26.5531	−21.3755
4.55	−61.4165	−47.1440	−36.7912	−29.1403	−23.3910
4.60	−68.0959	−52.0993	−40.5310	−32.0065	−25.6182
4.65	−75.5626	−57.6222	−44.6876	−35.1837	−28.0810
4.70	−83.9143	−63.7816	−49.3103	−38.7079	−30.8058
4.75	−93.2618	−70.6549	−54.4545	−42.6193	−33.8225
4.80	−103.730	−78.3294	−60.1823	−46.9629	−37.1641

Table 4.4 (continued)

	t				
v	0.80	0.85	0.90	0.95	1.0
4.85	−115.460	−86.9038	−66.5638	−51.7895	−40.8679
4.90	−128.612	−96.4891	−73.6778	−57.1559	−44.9756
4.95	−143.367	−107.211	−81.6129	−63.1258	−49.5338
5.00	−159.929	−119.210	−90.4690	−69.7709	−54.5949

	t				
v	1.5	2.0	3.0	4.0	5.0
0.00	0.297697	0.166501	−0.029873	−0.086137	−0.046322
0.05	0.298992	0.176582	−0.019898	−0.084071	−0.049847
0.10	0.298643	0.185562	−0.009887	−0.081522	−0.053056
0.15	0.296721	0.193435	0.000103	−0.078513	−0.055934
0.20	0.293303	0.200199	0.010019	−0.075070	−0.058471
0.25	0.288469	0.205860	0.019811	−0.071222	−0.060659
0.30	0.282299	0.210428	0.029432	−0.066997	−0.062490
0.35	0.274875	0.213921	0.038837	−0.062426	−0.063961
0.40	0.266280	0.216359	0.047984	−0.057538	−0.065070
0.45	0.256595	0.217768	0.056836	−0.052367	−0.065816
0.50	0.245900	0.218176	0.065358	−0.046943	−0.066201
0.55	0.234275	0.217615	0.073517	−0.041299	−0.066230
0.60	0.221797	0.216119	0.081287	−0.035466	−0.065908
0.65	0.208539	0.213726	0.088641	−0.029477	−0.065242
0.70	0.194572	0.210472	0.095558	−0.023363	−0.064240
0.75	0.179966	0.206399	0.102019	−0.017156	−0.062914
0.80	0.164783	0.201546	0.108007	−0.010884	−0.061274
0.85	0.149085	0.195955	0.113510	−0.004580	−0.059335
0.90	0.132930	0.189666	0.118517	0.001730	−0.057108
0.95	0.116368	0.182721	0.123021	0.008018	−0.054610
1.00	0.099450	0.175160	0.127016	0.014256	−0.051856
1.05	0.082220	0.167025	0.130500	0.020420	−0.048863
1.10	0.064719	0.158354	0.133472	0.026485	−0.045648
1.15	0.046982	0.149185	0.135934	0.032428	−0.042228
1.20	0.029041	0.139555	0.137888	0.038228	−0.038622
1.25	0.010925	0.129501	0.139341	0.043865	−0.034848
1.30	−0.007343	0.119055	0.140299	0.049321	−0.030924
1.35	−0.025743	0.108252	0.140770	0.054577	−0.026870
1.40	−0.044259	0.097120	0.140764	0.059618	−0.022703
1.45	−0.062879	0.085690	0.140292	0.064429	−0.018443
1.50	−0.081595	0.073988	0.139366	0.068999	−0.014108
1.55	−0.100403	0.062038	0.137998	0.073314	−0.009716
1.60	−0.119303	0.049865	0.136203	0.077366	−0.005285
1.65	−0.138297	0.037489	0.133994	0.081144	−0.000833
1.70	−0.157391	0.024928	0.131386	0.084643	0.003624

Table 4.4 (continued)

			t		
v	1.5	2.0	3.0	4.0	5.0
1.75	−0.176596	0.012199	0.128394	0.087854	0.008068
1.80	−0.195926	−0.000682	0.125034	0.090775	0.012485
1.85	−0.215396	−0.013704	0.121322	0.093400	0.016858
1.90	−0.235028	−0.026856	0.117272	0.095727	0.021173
1.95	−0.254844	−0.040128	0.112901	0.097754	0.025416
2.00	−0.274872	−0.053516	0.108225	0.099481	0.029573
2.05	−0.295143	−0.067015	0.103258	0.100909	0.033631
2.10	−0.315689	−0.080621	0.098017	0.102039	0.037580
2.15	−0.336549	−0.094335	0.092515	0.102872	0.041407
2.20	−0.357764	−0.108159	0.086767	0.103412	0.045102
2.25	−0.379377	−0.122096	0.080788	0.103663	0.048656
2.30	−0.401438	−0.136151	0.074591	0.103628	0.052059
2.35	−0.423998	−0.150333	0.068189	0.103313	0.055305
2.40	−0.447115	−0.164650	0.061595	0.102724	0.058384
2.45	−0.470850	−0.179114	0.054820	0.101866	0.061292
2.50	−0.495268	−0.193739	0.047875	0.100745	0.064022
2.55	−0.520440	−0.208539	0.040771	0.099369	0.066568
2.60	−0.546441	−0.223533	0.033518	0.097744	0.068928
2.65	−0.573354	−0.238739	0.026125	0.095879	0.071096
2.70	−0.601264	−0.254180	0.018600	0.093780	0.073071
2.75	−0.630267	−0.269878	0.010951	0.091455	0.074849
2.80	−0.660463	−0.285860	0.003184	0.088912	0.076430
2.85	−0.691960	−0.302153	−0.004694	0.086159	0.077811
2.90	−0.724875	−0.318788	−0.012679	0.083204	0.078993
2.95	−0.759332	−0.335798	−0.020766	0.080055	0.079975
3.00	−0.795467	−0.353219	−0.028950	0.076720	0.080758
3.05	−0.833423	−0.371087	−0.037231	0.073206	0.081343
3.10	−0.873358	−0.389445	−0.045605	0.069522	0.081732
3.15	−0.915440	−0.408335	−0.054071	0.065675	0.081925
3.20	−0.959851	−0.427805	−0.062630	0.061673	0.081926
3.25	−1.00679	−0.447905	−0.071281	0.057521	0.081738
3.30	−1.05646	−0.468689	−0.080026	0.053229	0.081362
3.35	−1.10909	−0.490215	−0.088868	0.048801	0.080803
3.40	−1.16495	−0.512543	−0.097809	0.044246	0.080064
3.45	−1.22428	−0.535740	−0.106853	0.039568	0.079148
3.50	−1.28739	−0.559877	−0.116005	0.034774	0.078060
3.55	−1.35458	−0.585029	−0.125271	0.029869	0.076804
3.60	−1.42621	−0.611278	−0.134657	0.024859	0.075383
3.65	−1.50264	−0.638709	−0.144171	0.019748	0.073804
3.70	−1.58428	−0.667417	−0.153821	0.014541	0.072069
3.75	−1.67156	−0.697500	−0.163615	0.009242	0.070184
3.80	−1.76497	−0.729067	−0.173565	0.003854	0.068153
3.85	−1.86503	−0.762231	−0.183682	−0.001618	0.065981

Table 4.4 (continued)

v	1.5	2.0	3.0	4.0	5.0
			t		
3.90	−1.97229	−0.797118	−0.193977	−0.007173	0.063673
3.95	−2.08738	−0.833858	−0.204465	−0.012806	0.061233
4.00	−2.21098	−0.872596	−0.215158	−0.018518	0.058667
4.05	−2.34381	−0.913485	−0.226074	−0.024304	0.055978
4.10	−2.48667	−0.956690	−0.237227	−0.030165	0.053171
4.15	−2.64045	−1.00239	−0.248637	−0.036100	0.050251
4.20	−2.80610	−1.05078	−0.260321	−0.042107	0.047223
4.25	−2.98466	−1.10206	−0.272301	−0.048188	0.044090
4.30	−3.17728	−1.15646	−0.284598	−0.054341	0.040856
4.35	−3.38521	−1.21421	−0.297235	−0.060569	0.037526
4.40	−3.60981	−1.27559	−0.310237	−0.066873	0.034104
4.45	−3.85259	−1.34087	−0.323630	−0.073254	0.030593
4.50	−4.11519	−1.41035	−0.337442	−0.079715	0.026996
4.55	−4.39941	−1.48438	−0.351704	−0.086258	0.023317
4.60	−4.70724	−1.56329	−0.366447	−0.092886	0.019560
4.65	−5.04082	−1.64749	−0.381704	−0.099604	0.015726
4.70	−5.40256	−1.73738	−0.397513	−0.106415	0.011820
4.75	−5.79505	−1.83342	−0.413911	−0.113325	0.007842
4.80	−6.22118	−1.93611	−0.430939	−0.120337	0.003797
4.85	−6.68410	−2.04598	−0.448641	−0.127459	−0.000315
4.90	−7.18727	−2.16360	−0.467063	−0.134696	−0.004491
4.95	−7.73452	−2.28961	−0.486254	−0.142055	−0.008730
5.00	−8.33005	−2.42469	−0.506268	−0.149543	−0.013031

Table 4.5: First derivatives with respect to the order of the integral Bessel functions of the second kind $\partial Yi_v(t)/\partial v$.

v	0.05	0.10	0.15	0.20	0.25
			t		
0.00	−0.058659	−0.045612	−0.037998	−0.032617	−0.028463
0.05	−0.055823	−0.043137	−0.036075	−0.031149	−0.027357
0.10	−0.054239	−0.041217	−0.034442	−0.029835	−0.026326
0.15	−0.053870	−0.039849	−0.033108	−0.028691	−0.025388
0.20	−0.054694	−0.039031	−0.032081	−0.027728	−0.024555
0.25	−0.056714	−0.038766	−0.031369	−0.026955	−0.023839
0.30	−0.059952	−0.039060	−0.030979	−0.026383	−0.023249
0.35	−0.064461	−0.039922	−0.030919	−0.026020	−0.022796
0.40	−0.070320	−0.041373	−0.031199	−0.025874	−0.022487
0.45	−0.077645	−0.043437	−0.031830	−0.025953	−0.022329
0.50	−0.086585	−0.046150	−0.032829	−0.026268	−0.022332

Table 4.5 (continued)

			t		
v	0.05	0.10	0.15	0.20	0.25
0.55	−0.097338	−0.049558	−0.034213	−0.026829	−0.022502
0.60	−0.110147	−0.053721	−0.036008	−0.027649	−0.022849
0.65	−0.125316	−0.058710	−0.038240	−0.028744	−0.023383
0.70	−0.143219	−0.064616	−0.040947	−0.030132	−0.024115
0.75	−0.164311	−0.071546	−0.044170	−0.031833	−0.025056
0.80	−0.189141	−0.079632	−0.047960	−0.033873	−0.026222
0.85	−0.218380	−0.089028	−0.052378	−0.036281	−0.027631
0.90	−0.252836	−0.099921	−0.057495	−0.039093	−0.029300
0.95	−0.293483	−0.112530	−0.063397	−0.042349	−0.031253
1.00	−0.341505	−0.127118	−0.070181	−0.046097	−0.033517
1.05	−0.398327	−0.143992	−0.077964	−0.050391	−0.036121
1.10	−0.465678	−0.163518	−0.086882	−0.055298	−0.039100
1.15	−0.545650	−0.186127	−0.097094	−0.060891	−0.042497
1.20	−0.640778	−0.212330	−0.108785	−0.067256	−0.046356
1.25	−0.754143	−0.242732	−0.122173	−0.074495	−0.050731
1.30	−0.889486	−0.278047	−0.137509	−0.082724	−0.055686
1.35	−1.05136	−0.319120	−0.155089	−0.092077	−0.061290
1.40	−1.24531	−0.366957	−0.175259	−0.102709	−0.067625
1.45	−1.47811	−0.422746	−0.198421	−0.114802	−0.074786
1.50	−1.75802	−0.487901	−0.225046	−0.128562	−0.082880
1.55	−2.09515	−0.564102	−0.255685	−0.144231	−0.092032
1.60	−2.50188	−0.653348	−0.290985	−0.162089	−0.102383
1.65	−2.99338	−0.758018	−0.331701	−0.182461	−0.114099
1.70	−3.58829	−0.880949	−0.378721	−0.205721	−0.127367
1.75	−4.30949	−1.02553	−0.433087	−0.232306	−0.142405
1.80	−5.18516	−1.19580	−0.496021	−0.262722	−0.159463
1.85	−6.25000	−1.39661	−0.568963	−0.297560	−0.178830
1.90	−7.54681	−1.63373	−0.653605	−0.337503	−0.200838
1.95	−9.12845	−1.91412	−0.751943	−0.383350	−0.225870
2.00	−11.0603	−2.24608	−0.866329	−0.436029	−0.254370
2.05	−13.4231	−2.63963	−0.999537	−0.496626	−0.286849
2.10	−16.3171	−3.10675	−1.15485	−0.566405	−0.323901
2.15	−19.8664	−3.66190	−1.33613	−0.646843	−0.366209
2.20	−24.2255	−4.32248	−1.54798	−0.739668	−0.414569
2.25	−29.5858	−5.10944	−1.79582	−0.846900	−0.469899
2.30	−36.1860	−6.04807	−2.08609	−0.970904	−0.533268
2.35	−44.3230	−7.16891	−2.42643	−1.11445	−0.605914
2.40	−54.3670	−8.50887	−2.82589	−1.28080	−0.689277
2.45	−66.7802	−10.1126	−3.29527	−1.47375	−0.785028
2.50	−82.1396	−12.0341	−3.84736	−1.69780	−0.895116
2.55	−101.167	−14.3389	−4.49742	−1.95821	−1.02181
2.60	−124.765	−17.1065	−5.26361	−2.26117	−1.16774
2.65	−154.065	−20.4332	−6.16760	−2.61399	−1.33599

Table 4.5 (continued)

			t		
v	0.05	0.10	0.15	0.20	0.25
2.70	−190.485	−24.4362	−7.23522	−3.02525	−1.53016
2.75	−235.806	−29.2579	−8.49734	−3.50509	−1.75444
2.80	−292.262	−35.0717	−9.99082	−4.06547	−2.01372
2.85	−362.664	−42.0887	−11.7598	−4.72051	−2.31375
2.90	−450.549	−50.5662	−13.8570	−5.48688	−2.66123
2.95	−560.371	−60.8180	−16.3456	−6.38433	−3.06401
3.00	−697.741	−73.2275	−19.3015	−7.43618	−3.53129
3.05	−869.742	−88.2627	−22.8154	−8.67008	−4.07386
3.10	−1,085.31	−106.496	−26.9966	−10.1188	−4.70438
3.15	−1,355.75	−128.628	−31.9761	−11.8212	−5.43771
3.20	−1,695.34	−155.517	−37.9114	−13.8233	−6.29132
3.25	−2,122.16	−188.213	−44.9920	−16.1799	−7.28574
3.30	−2,659.10	−228.006	−53.4461	−18.9560	−8.44511
3.35	−3,335.19	−276.478	−63.5485	−22.2290	−9.79788
3.40	−4,187.25	−335.572	−75.6305	−26.0909	−11.3775
3.45	−5,262.00	−407.675	−90.0918	−30.6511	−13.2235
3.50	−6,618.83	−495.726	−107.415	−36.0403	−15.3824
3.55	−8,333.23	−603.338	−128.182	−42.4139	−17.9092
3.60	−10,501.2	−734.963	−153.098	−49.9577	−20.8687
3.65	−13,245.1	−896.086	−183.014	−58.8930	−24.3375
3.70	−16,720.7	−1,093.48	−218.962	−69.4845	−28.4064
3.75	−21,126.5	−1,335.48	−262.189	−82.0485	−33.1826
3.80	−26,716.1	−1,632.40	−314.209	−96.9629	−38.7929
3.85	−33,813.1	−1,997.00	−376.856	−114.680	−45.3878
3.90	−42,830.7	−2,445.00	−452.356	−135.742	−53.1453
3.95	−54,297.7	−2,995.92	−543.410	−160.798	−62.2767
4.00	−68,890.0	−3,673.86	−653.301	−190.625	−73.0327
4.05	−87,473.5	−4,508.73	−786.018	−226.156	−85.7108
4.10	−111157	−5,537.58	−946.414	−268.512	−100.664
4.15	−141361	−6,806.35	−1,140.39	−319.036	−118.314
4.20	−179910	−8,372.09	−1,375.15	−379.344	−139.158
4.25	−229143	−10,305.6	−1,659.43	−451.378	−163.792
4.30	−292064	−12,694.9	−2,003.93	−537.472	−192.922
4.35	−372535	−15,649.4	−2,421.67	−640.439	−227.392
4.40	−475520	−19,305.2	−2,928.54	−763.661	−268.205
4.45	−607407	−23,831.7	−3,543.95	−911.217	−316.558
4.50	−776418	−29,439.9	−4,291.63	−1,088.02	−373.881
4.55	−993145	−36,392.7	−5,200.56	−1,300.00	−441.877
4.60	−1.271E+6	−45,017.9	−6,306.20	−1,554.31	−522.583
4.65	−1.628	−55,724.4	−7,651.94	−1,859.58	−618.429
4.70	−2.087	−69,022.7	−9,290.93	−2,226.25	−732.325
4.75	−2.677	−85,550.2	−11,288.2	−2,666.92	−867.748
4.80	−3.435	−106103	−13,723.7	−3,196.84	−1,028.86

Table 4.5 (continued)

			t		
v	0.05	0.10	0.15	0.20	0.25
4.85	−4.412	−131678	−16,695.0	−3,834.44	−1,220.64
4.90	−5.670	−163520	−20,322.3	−4,602.06	−1,449.05
4.95	−7.290	−203187	−24,753.0	−5,526.71	−1,721.25
5.00	−9.380	−252633	−30,168.0	−6,641.16	−2,045.82

			t		
v	0.30	0.35	0.40	0.45	0.50
0.00	−0.025091	−0.022262	−0.019832	−0.017711	−0.015835
0.05	−0.024273	−0.021675	−0.019435	−0.017470	−0.015723
0.10	−0.023482	−0.021085	−0.019014	−0.017190	−0.015563
0.15	−0.022734	−0.020506	−0.018581	−0.016884	−0.015367
0.20	−0.022042	−0.019950	−0.018150	−0.016564	−0.015145
0.25	−0.021420	−0.019431	−0.017731	−0.016239	−0.014907
0.30	−0.020876	−0.018956	−0.017334	−0.015920	−0.014661
0.35	−0.020420	−0.018537	−0.016968	−0.015614	−0.014415
0.40	−0.020061	−0.018181	−0.016641	−0.015328	−0.014178
0.45	−0.019805	−0.017895	−0.016360	−0.015072	−0.013954
0.50	−0.019661	−0.017686	−0.016132	−0.014849	−0.013751
0.55	−0.019635	−0.017562	−0.015963	−0.014667	−0.013575
0.60	−0.019733	−0.017527	−0.015859	−0.014530	−0.013429
0.65	−0.019964	−0.017588	−0.015824	−0.014445	−0.013320
0.70	−0.020336	−0.017752	−0.015866	−0.014415	−0.013251
0.75	−0.020856	−0.018025	−0.015989	−0.014447	−0.013227
0.80	−0.021536	−0.018414	−0.016198	−0.014543	−0.013253
0.85	−0.022384	−0.018926	−0.016500	−0.014710	−0.013333
0.90	−0.023414	−0.019571	−0.016901	−0.014953	−0.013471
0.95	−0.024640	−0.020358	−0.017409	−0.015276	−0.013671
1.00	−0.026078	−0.021297	−0.018029	−0.015686	−0.013938
1.05	−0.027746	−0.022400	−0.018772	−0.016189	−0.014278
1.10	−0.029665	−0.023681	−0.019646	−0.016792	−0.014694
1.15	−0.031859	−0.025155	−0.020661	−0.017502	−0.015194
1.20	−0.034356	−0.026841	−0.021831	−0.018327	−0.015782
1.25	−0.037188	−0.028757	−0.023167	−0.019278	−0.016467
1.30	−0.040391	−0.030928	−0.024685	−0.020364	−0.017255
1.35	−0.044006	−0.033378	−0.026403	−0.021597	−0.018156
1.40	−0.048082	−0.036137	−0.028340	−0.022992	−0.019178
1.45	−0.052671	−0.039239	−0.030517	−0.024562	−0.020333
1.50	−0.057836	−0.042722	−0.032960	−0.026325	−0.021632
1.55	−0.063648	−0.046630	−0.035696	−0.028299	−0.023089
1.60	−0.070187	−0.051011	−0.038758	−0.030507	−0.024719
1.65	−0.077546	−0.055922	−0.042181	−0.032971	−0.026538
1.70	−0.085831	−0.061427	−0.046007	−0.035721	−0.028566

Table 4.5 (continued)

ν			t		
	0.30	**0.35**	**0.40**	**0.45**	**0.50**
1.75	−0.095163	−0.067598	−0.050282	−0.038787	−0.030823
1.80	−0.105679	−0.074519	−0.055057	−0.042202	−0.033334
1.85	−0.117538	−0.082283	−0.060394	−0.046008	−0.036126
1.90	−0.130922	−0.090998	−0.066359	−0.050247	−0.039229
1.95	−0.146038	−0.100785	−0.073028	−0.054971	−0.042678
2.00	−0.163124	−0.111785	−0.080488	−0.060236	−0.046510
2.05	−0.182454	−0.124156	−0.088838	−0.066105	−0.050769
2.10	−0.204342	−0.138079	−0.098190	−0.072652	−0.055503
2.15	−0.229148	−0.153762	−0.108670	−0.079958	−0.060769
2.20	−0.257286	−0.171442	−0.120423	−0.088116	−0.066626
2.25	−0.289235	−0.191389	−0.133613	−0.097230	−0.073147
2.30	−0.325542	−0.213912	−0.148428	−0.107420	−0.080408
2.35	−0.366840	−0.239367	−0.165079	−0.118821	−0.088501
2.40	−0.413860	−0.268160	−0.183811	−0.131586	−0.097524
2.45	−0.467441	−0.300756	−0.204898	−0.145887	−0.107591
2.50	−0.528557	−0.337690	−0.228658	−0.161923	−0.118832
2.55	−0.598328	−0.379575	−0.255450	−0.179916	−0.131392
2.60	−0.678054	−0.427115	−0.285686	−0.200122	−0.145434
2.65	−0.769235	−0.481120	−0.319836	−0.222830	−0.161147
2.70	−0.873609	−0.542522	−0.358440	−0.248370	−0.178741
2.75	−0.993190	−0.612392	−0.402112	−0.277118	−0.198457
2.80	−1.13031	−0.691967	−0.451558	−0.309500	−0.220565
2.85	−1.28769	−0.782667	−0.507588	−0.346006	−0.245376
2.90	−1.46845	−0.886135	−0.571127	−0.387191	−0.273239
2.95	−1.67627	−1.00426	−0.643240	−0.433692	−0.304555
3.00	−1.91538	−1.13924	−0.725149	−0.486234	−0.339776
3.05	−2.19072	−1.29359	−0.818256	−0.545648	−0.379419
3.10	−2.50804	−1.47024	−0.924174	−0.612883	−0.424072
3.15	−2.87404	−1.67257	−1.04476	−0.689026	−0.474404
3.20	−3.29652	−1.90448	−1.18215	−0.775320	−0.531179
3.25	−3.78458	−2.17051	−1.33879	−0.873191	−0.595269
3.30	−4.34884	−2.47592	−1.51754	−0.984274	−0.667665
3.35	−5.00170	−2.82679	−1.72164	−1.11044	−0.749505
3.40	−5.75766	−3.23020	−1.95488	−1.25385	−0.842084
3.45	−6.63364	−3.69435	−2.22160	−1.41697	−0.946885
3.50	−7.64947	−4.22878	−2.52683	−1.60263	−1.06560
3.55	−8.82834	−4.84458	−2.87638	−1.81411	−1.20018
3.60	−10.1974	−5.55464	−3.27696	−2.05516	−1.35284
3.65	−11.7885	−6.37398	−3.73634	−2.33010	−1.52612
3.70	−13.6390	−7.32008	−4.26353	−2.64390	−1.72294
3.75	−15.7927	−8.41330	−4.86893	−3.00232	−1.94666
3.80	−18.3010	−9.67739	−5.56462	−3.41195	−2.20111

Table 4.5 (continued)

			t		
v	0.30	0.35	0.40	0.45	0.50
3.85	−21.2243	−11.1401	−6.36462	−3.88042	−2.49069
3.90	−24.6336	−12.8336	−7.28515	−4.41655	−2.82049
3.95	−28.6124	−14.7958	−8.34509	−5.03049	−3.19631
4.00	−33.2590	−17.0708	−9.56632	−5.73399	−3.62485
4.05	−38.6889	−19.7102	−10.9743	−6.54063	−4.11382
4.10	−45.0384	−22.7742	−12.5986	−7.46610	−4.67208
4.15	−52.4681	−26.3335	−14.4738	−8.52858	−5.30984
4.20	−61.1673	−30.4708	−16.6397	−9.74909	−6.03887
4.25	−71.3594	−35.2829	−19.1431	−11.152	−6.87273
4.30	−83.3080	−40.8834	−22.0384	−12.766	−7.82706
4.35	−97.3247	−47.4054	−25.3889	−14.623	−8.91992
4.40	−113.778	−55.0051	−29.2685	−16.761	−10.1721
4.45	−133.102	−63.8660	−33.7634	−19.225	−11.6078
4.50	−155.812	−74.2035	−38.9744	−22.065	−13.2548
4.55	−182.517	−86.2708	−45.0189	−25.341	−15.1451
4.60	−213.939	−100.366	−52.0344	−29.123	−17.3162
4.65	−250.932	−116.838	−60.1814	−33.490	−19.8109
4.70	−294.509	−136.100	−69.6480	−38.536	−22.6792
4.75	−345.872	−158.638	−80.6539	−44.370	−25.9788
4.80	−406.447	−185.022	−93.4566	−51.118	−29.7767
4.85	−477.924	−215.926	−108.358	−58.929	−34.1504
4.90	−562.315	−252.145	−125.710	−67.974	−39.1900
4.95	−662.005	−294.616	−145.929	−78.453	−44.9999
5.00	−779.835	−344.446	−169.500	−90.602	−51.7015

			t		
v	0.55	0.60	0.65	0.70	0.75
0.00	−0.014160	−0.012653	−0.011288	−0.010045	−0.008910
0.05	−0.014155	−0.012736	−0.011445	−0.010264	−0.009180
0.10	−0.014096	−0.012763	−0.011545	−0.010425	−0.009392
0.15	−0.013994	−0.012743	−0.011595	−0.010535	−0.009554
0.20	−0.013859	−0.012684	−0.011602	−0.010601	−0.009671
0.25	−0.013699	−0.012594	−0.011575	−0.010629	−0.009748
0.30	−0.013522	−0.012480	−0.011519	−0.010625	−0.009790
0.35	−0.013336	−0.012350	−0.011441	−0.010595	−0.009804
0.40	−0.013147	−0.012209	−0.011346	−0.010544	−0.009793
0.45	−0.012962	−0.012064	−0.011241	−0.010477	−0.009763
0.50	−0.012787	−0.011920	−0.011130	−0.010400	−0.009719
0.55	−0.012626	−0.011782	−0.011018	−0.010316	−0.009663
0.60	−0.012485	−0.011655	−0.010910	−0.010230	−0.009601
0.65	−0.012369	−0.011544	−0.010810	−0.010146	−0.009536

Table 4.5 (continued)

			t		
v	0.55	0.60	0.65	0.70	0.75
0.70	−0.012282	−0.011452	−0.010722	−0.010068	−0.009472
0.75	−0.012228	−0.011383	−0.010650	−0.009999	−0.009412
0.80	−0.012211	−0.011341	−0.010596	−0.009943	−0.009358
0.85	−0.012235	−0.011330	−0.010565	−0.009902	−0.009315
0.90	−0.012303	−0.011353	−0.010560	−0.009880	−0.009285
0.95	−0.012420	−0.011414	−0.010583	−0.009880	−0.009271
1.00	−0.012589	−0.011516	−0.010638	−0.009904	−0.009275
1.05	−0.012815	−0.011662	−0.010729	−0.009955	−0.009300
1.10	−0.013101	−0.011855	−0.010857	−0.010037	−0.009349
1.15	−0.013452	−0.012101	−0.011026	−0.010151	−0.009424
1.20	−0.013873	−0.012402	−0.011240	−0.010301	−0.009527
1.25	−0.014370	−0.012763	−0.011501	−0.010489	−0.009661
1.30	−0.014948	−0.013188	−0.011814	−0.010719	−0.009829
1.35	−0.015613	−0.013682	−0.012183	−0.010994	−0.010033
1.40	−0.016372	−0.014251	−0.012611	−0.011317	−0.010277
1.45	−0.017234	−0.014901	−0.013104	−0.011692	−0.010563
1.50	−0.018206	−0.015637	−0.013666	−0.012123	−0.010895
1.55	−0.019299	−0.016468	−0.014303	−0.012615	−0.011275
1.60	−0.020524	−0.017400	−0.015021	−0.013171	−0.011709
1.65	−0.021892	−0.018444	−0.015826	−0.013798	−0.012200
1.70	−0.023417	−0.019609	−0.016727	−0.014501	−0.012752
1.75	−0.025115	−0.020907	−0.017731	−0.015286	−0.013371
1.80	−0.027001	−0.022348	−0.018848	−0.016161	−0.014062
1.85	−0.029096	−0.023949	−0.020088	−0.017133	−0.014831
1.90	−0.031421	−0.025723	−0.021463	−0.018211	−0.015684
1.95	−0.033999	−0.027688	−0.022984	−0.019404	−0.016630
2.00	−0.036858	−0.029864	−0.024668	−0.020724	−0.017675
2.05	−0.040028	−0.032272	−0.026528	−0.022181	−0.018830
2.10	−0.043542	−0.034937	−0.028584	−0.023790	−0.020105
2.15	−0.047440	−0.037885	−0.030854	−0.025565	−0.021509
2.20	−0.051762	−0.041147	−0.033361	−0.027522	−0.023057
2.25	−0.056559	−0.044757	−0.036130	−0.029679	−0.024760
2.30	−0.061883	−0.048753	−0.039188	−0.032058	−0.026636
2.35	−0.067796	−0.053178	−0.042566	−0.034680	−0.028701
2.40	−0.074366	−0.058080	−0.046299	−0.037571	−0.030973
2.45	−0.081670	−0.063513	−0.050424	−0.040760	−0.033474
2.50	−0.089796	−0.069537	−0.054987	−0.044277	−0.036227
2.55	−0.098841	−0.076221	−0.060033	−0.048158	−0.039259
2.60	−0.108916	−0.083641	−0.065620	−0.052443	−0.042599
2.65	−0.120145	−0.091883	−0.071806	−0.057175	−0.046278
2.70	−0.132669	−0.101043	−0.078660	−0.062404	−0.050334
2.75	−0.146648	−0.111231	−0.086259	−0.068185	−0.054806

Table 4.5 (continued)

			t		
v	0.55	0.60	0.65	0.70	0.75
2.80	−0.162261	−0.122570	−0.094690	−0.074580	−0.059741
2.85	−0.179711	−0.135197	−0.104047	−0.081657	−0.065188
2.90	−0.199229	−0.149269	−0.114441	−0.089495	−0.071204
2.95	−0.221075	−0.164961	−0.125994	−0.098179	−0.077852
3.00	−0.245545	−0.182472	−0.138841	−0.107809	−0.085203
3.05	−0.272972	−0.202026	−0.153140	−0.118492	−0.093335
3.10	−0.303737	−0.223877	−0.169063	−0.130352	−0.102337
3.15	−0.338269	−0.248311	−0.186807	−0.143526	−0.112308
3.20	−0.377057	−0.275652	−0.206593	−0.158170	−0.123359
3.25	−0.420658	−0.306267	−0.228671	−0.174459	−0.135615
3.30	−0.469701	−0.340572	−0.253323	−0.192588	−0.149215
3.35	−0.524904	−0.379036	−0.280867	−0.212777	−0.164316
3.40	−0.587085	−0.422194	−0.311662	−0.235277	−0.181094
3.45	−0.657173	−0.470651	−0.346115	−0.260367	−0.199747
3.50	−0.736227	−0.525093	−0.384685	−0.288363	−0.220498
3.55	−0.825454	−0.586300	−0.427894	−0.319621	−0.243596
3.60	−0.926231	−0.655159	−0.476328	−0.354545	−0.269323
3.65	−1.04013	−0.732675	−0.530656	−0.393588	−0.297997
3.70	−1.16894	−0.819995	−0.591633	−0.437263	−0.329973
3.75	−1.31471	−0.918421	−0.660116	−0.486151	−0.365655
3.80	−1.47978	−1.02944	−0.737078	−0.540906	−0.405495
3.85	−1.66683	−1.15473	−0.823621	−0.602272	−0.450006
3.90	−1.87891	−1.29623	−0.921000	−0.671088	−0.499764
3.95	−2.11953	−1.45612	−1.03064	−0.748306	−0.555423
4.00	−2.39270	−1.63692	−1.15415	−0.835002	−0.617717
4.05	−2.70301	−1.84148	−1.29339	−0.932399	−0.687479
4.10	−3.05573	−2.07306	−1.45043	−1.04188	−0.765651
4.15	−3.45690	−2.33539	−1.62767	−1.16502	−0.853296
4.20	−3.91345	−2.63273	−1.82782	−1.30360	−0.951619
4.25	−4.43333	−2.96995	−2.05397	−1.45966	−1.06198
4.30	−5.02569	−3.35262	−2.30966	−1.63548	−1.18594
4.35	−5.70100	−3.78712	−2.59889	−1.83369	−1.32523
4.40	−6.47136	−4.28076	−2.92627	−2.05727	−1.48184
4.45	−7.35062	−4.84190	−3.29702	−2.30960	−1.65804
4.50	−8.35477	−5.48013	−3.71714	−2.59455	−1.85636
4.55	−9.50219	−6.20646	−4.19346	−2.91649	−2.07971
4.60	−10.8141	−7.03349	−4.73378	−3.28043	−2.33140
4.65	−12.3148	−7.97571	−5.34706	−3.69208	−2.61516
4.70	−14.0325	−9.04976	−6.04352	−4.15794	−2.93525
4.75	−15.9996	−10.2747	−6.83486	−4.68543	−3.29652
4.80	−18.2535	−11.6726	−7.73448	−5.28301	−3.70447
4.85	−20.8376	−13.2686	−8.75775	−5.96036	−4.16538

Table 4.5 (continued)

			t		
v	0.55	0.60	0.65	0.70	0.75
4.90	−23.8016	−15.0917	−9.92226	−6.72852	−4.68638
4.95	−27.2033	−17.1755	−11.2482	−7.60011	−5.27563
5.00	−31.1093	−19.5583	−12.7587	−8.58956	−5.94238

			t		
v	0.80	0.85	0.90	0.95	1.0
0.00	−0.007870	−0.006913	−0.006032	−0.005220	−0.004469
0.05	−0.008181	−0.007258	−0.006405	−0.005614	−0.004880
0.10	−0.008437	−0.007551	−0.006727	−0.005960	−0.005245
0.15	−0.008643	−0.007794	−0.007001	−0.006260	−0.005567
0.20	−0.008803	−0.007992	−0.007232	−0.006519	−0.005849
0.25	−0.008923	−0.008149	−0.007422	−0.006737	−0.006092
0.30	−0.009007	−0.008271	−0.007576	−0.006920	−0.006300
0.35	−0.009061	−0.008360	−0.007698	−0.007070	−0.006475
0.40	−0.009088	−0.008421	−0.007790	−0.007191	−0.006621
0.45	−0.009092	−0.008458	−0.007857	−0.007285	−0.006740
0.50	−0.009079	−0.008475	−0.007901	−0.007355	−0.006834
0.55	−0.009052	−0.008474	−0.007927	−0.007405	−0.006907
0.60	−0.009014	−0.008461	−0.007937	−0.007438	−0.006962
0.65	−0.008969	−0.008437	−0.007934	−0.007456	−0.007000
0.70	−0.008921	−0.008407	−0.007922	−0.007463	−0.007024
0.75	−0.008873	−0.008373	−0.007903	−0.007460	−0.007038
0.80	−0.008827	−0.008337	−0.007880	−0.007450	−0.007043
0.85	−0.008787	−0.008304	−0.007856	−0.007437	−0.007041
0.90	−0.008755	−0.008274	−0.007832	−0.007421	−0.007035
0.95	−0.008734	−0.008252	−0.007812	−0.007406	−0.007027
1.00	−0.008726	−0.008238	−0.007797	−0.007393	−0.007019
1.05	−0.008734	−0.008236	−0.007790	−0.007385	−0.007013
1.10	−0.008761	−0.008248	−0.007793	−0.007384	−0.007010
1.15	−0.008807	−0.008275	−0.007807	−0.007390	−0.007013
1.20	−0.008876	−0.008320	−0.007835	−0.007407	−0.007023
1.25	−0.008970	−0.008385	−0.007879	−0.007436	−0.007042
1.30	−0.009092	−0.008471	−0.007940	−0.007479	−0.007071
1.35	−0.009243	−0.008582	−0.008021	−0.007537	−0.007113
1.40	−0.009426	−0.008719	−0.008123	−0.007612	−0.007169
1.45	−0.009644	−0.008885	−0.008249	−0.007707	−0.007240
1.50	−0.009900	−0.009082	−0.008399	−0.007823	−0.007328
1.55	−0.010195	−0.009311	−0.008578	−0.007961	−0.007435
1.60	−0.010534	−0.009577	−0.008786	−0.008124	−0.007563
1.65	−0.010920	−0.009881	−0.009026	−0.008314	−0.007713
1.70	−0.011356	−0.010227	−0.009301	−0.008532	−0.007887
1.75	−0.011847	−0.010617	−0.009612	−0.008781	−0.008086

Table 4.5 (continued)

			t		
v	0.80	0.85	0.90	0.95	1.0
1.80	−0.012396	−0.011056	−0.009964	−0.009064	−0.008314
1.85	−0.013009	−0.011547	−0.010359	−0.009383	−0.008572
1.90	−0.013690	−0.012094	−0.010801	−0.009741	−0.008863
1.95	−0.014445	−0.012702	−0.011293	−0.010140	−0.009188
2.00	−0.015282	−0.013375	−0.011838	−0.010585	−0.009551
2.05	−0.016206	−0.014120	−0.012443	−0.011078	−0.009955
2.10	−0.017225	−0.014943	−0.013111	−0.011623	−0.010403
2.15	−0.018348	−0.015849	−0.013847	−0.012225	−0.010897
2.20	−0.019585	−0.016846	−0.014658	−0.012888	−0.011443
2.25	−0.020946	−0.017944	−0.015550	−0.013618	−0.012044
2.30	−0.022442	−0.019150	−0.016530	−0.014420	−0.012704
2.35	−0.024088	−0.020474	−0.017605	−0.015301	−0.013429
2.40	−0.025896	−0.021928	−0.018786	−0.016266	−0.014224
2.45	−0.027883	−0.023525	−0.020080	−0.017325	−0.015096
2.50	−0.030067	−0.025276	−0.021499	−0.018484	−0.016050
2.55	−0.032467	−0.027199	−0.023055	−0.019753	−0.017094
2.60	−0.035106	−0.029309	−0.024759	−0.021143	−0.018236
2.65	−0.038007	−0.031625	−0.026628	−0.022665	−0.019485
2.70	−0.041198	−0.034167	−0.028676	−0.024330	−0.020851
2.75	−0.044710	−0.036960	−0.030921	−0.026153	−0.022344
2.80	−0.048575	−0.040027	−0.033383	−0.028150	−0.023977
2.85	−0.052832	−0.043398	−0.036084	−0.030335	−0.025763
2.90	−0.057521	−0.047103	−0.039046	−0.032730	−0.027715
2.95	−0.062691	−0.051179	−0.042299	−0.035353	−0.029851
3.00	−0.068392	−0.055663	−0.045870	−0.038228	−0.032188
3.05	−0.074684	−0.060600	−0.049793	−0.041380	−0.034746
3.10	−0.081630	−0.066038	−0.054104	−0.044837	−0.037547
3.15	−0.089303	−0.072030	−0.058845	−0.048631	−0.040615
3.20	−0.097785	−0.078638	−0.064061	−0.052796	−0.043976
3.25	−0.107166	−0.085928	−0.069801	−0.057371	−0.047661
3.30	−0.117548	−0.093974	−0.076123	−0.062399	−0.051702
3.35	−0.129043	−0.102862	−0.083090	−0.067926	−0.056136
3.40	−0.141780	−0.112683	−0.090770	−0.074006	−0.061004
3.45	−0.155900	−0.123544	−0.099241	−0.080699	−0.066350
3.50	−0.171564	−0.135560	−0.108592	−0.088068	−0.072225
3.55	−0.188951	−0.148863	−0.118918	−0.096188	−0.078684
3.60	−0.208262	−0.163599	−0.130328	−0.105139	−0.085789
3.65	−0.229723	−0.179931	−0.142943	−0.115013	−0.093609
3.70	−0.253587	−0.198044	−0.156898	−0.125909	−0.102221
3.75	−0.280139	−0.218143	−0.172345	−0.137942	−0.111709
3.80	−0.309701	−0.240460	−0.189452	−0.151237	−0.122169
3.85	−0.342632	−0.265252	−0.208409	−0.165934	−0.133706
3.90	−0.379338	−0.292812	−0.229428	−0.182191	−0.146439

Table 4.5 (continued)

			t		
v	0.80	0.85	0.90	0.95	1.0
3.95	−0.420276	−0.323465	−0.252747	−0.200182	−0.160499
4.00	−0.465962	−0.357579	−0.278631	−0.220105	−0.176032
4.05	−0.516973	−0.395566	−0.307379	−0.242179	−0.193202
4.10	−0.573966	−0.437889	−0.339326	−0.266649	−0.212193
4.15	−0.637677	−0.485070	−0.374848	−0.293790	−0.233209
4.20	−0.708938	−0.537697	−0.414367	−0.323912	−0.256477
4.25	−0.788691	−0.596431	−0.458357	−0.357358	−0.282254
4.30	−0.877995	−0.662017	−0.507349	−0.394516	−0.310825
4.35	−0.978052	−0.735294	−0.561944	−0.435821	−0.342511
4.40	−1.09022	−0.817210	−0.622814	−0.481760	−0.377668
4.45	−1.21603	−0.908832	−0.690718	−0.532880	−0.416698
4.50	−1.35722	−1.01137	−0.766508	−0.589795	−0.460050
4.55	−1.51575	−1.12617	−0.851146	−0.653195	−0.508228
4.60	−1.69385	−1.25479	−0.945714	−0.723856	−0.561797
4.65	−1.89405	−1.39896	−1.05143	−0.802652	−0.621391
4.70	−2.11920	−1.56063	−1.16968	−0.890562	−0.687721
4.75	−2.37255	−1.74204	−1.30200	−0.988693	−0.761585
4.80	−2.65778	−1.94570	−1.45016	−1.09829	−0.843883
4.85	−2.97905	−2.17444	−1.61612	−1.22075	−0.935621
4.90	−3.34112	−2.43150	−1.80213	−1.35765	−1.03793
4.95	−3.74937	−2.72052	−2.01071	−1.51077	−1.15210
5.00	−4.20991	−3.04565	−2.24471	−1.68213	−1.27954

			t		
v	1.5	2.0	3.0	4.0	5.0
0.00	0.000514	0.002550	0.00239	0.000436	−0.000883
0.05	0.000111	0.002288	0.00240	0.000555	−0.000809
0.10	−0.000276	0.002023	0.00240	0.000668	−0.000731
0.15	−0.000644	0.001756	0.00239	0.000775	−0.000650
0.20	−0.000993	0.001491	0.002367	0.000876	−0.000567
0.25	−0.001324	0.001227	0.002331	0.000970	−0.000482
0.30	−0.001635	0.000967	0.002285	0.001057	−0.000396
0.35	−0.001926	0.000711	0.002228	0.001136	−0.000310
0.40	−0.002197	0.000461	0.002161	0.001208	−0.000223
0.45	−0.002449	0.000217	0.002086	0.001273	−0.000136
0.50	−0.002681	−0.000020	0.002003	0.001329	−0.000050
0.55	−0.002895	−0.000248	0.001913	0.001379	0.000035
0.60	−0.003091	−0.000468	0.001816	0.001420	0.000119
0.65	−0.003270	−0.000679	0.001713	0.001454	0.000201
0.70	−0.003431	−0.000881	0.001606	0.001480	0.000280
0.75	−0.003577	−0.001073	0.001494	0.001499	0.000356
0.80	−0.003708	−0.001255	0.001379	0.001510	0.000430

Table 4.5 (continued)

			t		
v	1.5	2.0	3.0	4.0	5.0
0.85	−0.003825	−0.001427	0.001262	0.001515	0.000501
0.90	−0.003928	−0.001590	0.001142	0.001513	0.000568
0.95	−0.004019	−0.001742	0.001020	0.001504	0.000631
1.00	−0.004099	−0.001885	0.000898	0.001489	0.000690
1.05	−0.004169	−0.002018	0.000775	0.001468	0.000746
1.10	−0.004230	−0.002142	0.000652	0.001442	0.000797
1.15	−0.004283	−0.002257	0.000530	0.001410	0.000844
1.20	−0.004328	−0.002363	0.000409	0.001373	0.000886
1.25	−0.004367	−0.002461	0.000289	0.001332	0.000924
1.30	−0.004401	−0.002551	0.000171	0.001286	0.000958
1.35	−0.004430	−0.002633	0.000055	0.001236	0.000987
1.40	−0.004457	−0.002709	−0.000058	0.001183	0.001012
1.45	−0.004481	−0.002777	−0.000168	0.001126	0.001032
1.50	−0.004503	−0.002839	−0.000276	0.001067	0.001048
1.55	−0.004525	−0.002896	−0.000380	0.001004	0.001059
1.60	−0.004547	−0.002947	−0.000481	0.000940	0.001067
1.65	−0.004570	−0.002993	−0.000579	0.000874	0.001070
1.70	−0.004595	−0.003035	−0.000673	0.000805	0.001069
1.75	−0.004624	−0.003074	−0.000763	0.000736	0.001064
1.80	−0.004655	−0.003109	−0.000849	0.000666	0.001055
1.85	−0.004691	−0.003141	−0.000932	0.000594	0.001043
1.90	−0.004733	−0.003171	−0.001011	0.000523	0.001027
1.95	−0.004780	−0.003199	−0.001086	0.000451	0.001008
2.00	−0.004835	−0.003226	−0.001158	0.000379	0.000986
2.05	−0.004897	−0.003253	−0.001226	0.000307	0.000961
2.10	−0.004967	−0.003278	−0.001290	0.000235	0.000933
2.15	−0.005047	−0.003304	−0.001350	0.000165	0.000903
2.20	−0.005137	−0.003331	−0.001408	0.000095	0.000870
2.25	−0.005239	−0.003359	−0.001462	0.000026	0.000835
2.30	−0.005352	−0.003388	−0.001512	−0.000042	0.000798
2.35	−0.005479	−0.003419	−0.001560	−0.000109	0.000759
2.40	−0.005620	−0.003453	−0.001605	−0.000174	0.000719
2.45	−0.005776	−0.003490	−0.001647	−0.000238	0.000677
2.50	−0.005948	−0.003530	−0.001686	−0.000300	0.000633
2.55	−0.006138	−0.003575	−0.001723	−0.000360	0.000589
2.60	−0.006346	−0.003623	−0.001758	−0.000419	0.000543
2.65	−0.006575	−0.003677	−0.001791	−0.000476	0.000497
2.70	−0.006826	−0.003736	−0.001821	−0.000531	0.000450
2.75	−0.007100	−0.003801	−0.001850	−0.000584	0.000403
2.80	−0.007399	−0.003872	−0.001878	−0.000636	0.000355
2.85	−0.007725	−0.003950	−0.001904	−0.000685	0.000308
2.90	−0.008079	−0.004036	−0.001929	−0.000733	0.000260

Table 4.5 (continued)

v	1.5	2.0	3.0	4.0	5.0
			t		
2.95	−0.008465	−0.004130	−0.001953	−0.000778	0.000212
3.00	−0.008885	−0.004233	−0.001976	−0.000822	0.000164
3.05	−0.009341	−0.004345	−0.001999	−0.000864	0.000117
3.10	−0.009835	−0.004468	−0.002021	−0.000904	0.000070
3.15	−0.010372	−0.004601	−0.002043	−0.000942	0.000023
3.20	−0.010953	−0.004746	−0.002065	−0.000979	−0.000023
3.25	−0.011584	−0.004904	−0.002088	−0.001014	−0.000068
3.30	−0.012268	−0.005075	−0.002110	−0.001047	−0.000112
3.35	−0.013008	−0.005260	−0.002134	−0.001078	−0.000156
3.40	−0.013811	−0.005460	−0.002158	−0.001108	−0.000199
3.45	−0.014681	−0.005677	−0.002183	−0.001137	−0.000241
3.50	−0.015623	−0.005912	−0.002210	−0.001164	−0.000282
3.55	−0.016645	−0.006165	−0.002238	−0.001190	−0.000321
3.60	−0.017752	−0.006438	−0.002268	−0.001215	−0.000360
3.65	−0.018951	−0.006733	−0.002299	−0.001238	−0.000398
3.70	−0.020252	−0.007051	−0.002333	−0.001261	−0.000435
3.75	−0.021663	−0.007394	−0.002369	−0.001283	−0.000470
3.80	−0.023194	−0.007763	−0.002408	−0.001303	−0.000505
3.85	−0.024855	−0.008161	−0.002449	−0.001323	−0.000538
3.90	−0.026657	−0.008590	−0.002493	−0.001343	−0.000570
3.95	−0.028614	−0.009051	−0.002541	−0.001361	−0.000601
4.00	−0.030740	−0.009549	−0.002592	−0.001380	−0.000631
4.05	−0.033050	−0.010084	−0.002648	−0.001398	−0.000660
4.10	−0.035561	−0.010661	−0.002707	−0.001415	−0.000687
4.15	−0.038291	−0.011282	−0.002770	−0.001433	−0.000714
4.20	−0.041261	−0.011952	−0.002839	−0.001451	−0.000740
4.25	−0.044494	−0.012672	−0.002912	−0.001468	−0.000764
4.30	−0.048013	−0.013449	−0.002991	−0.001486	−0.000788
4.35	−0.051846	−0.014285	−0.003076	−0.001504	−0.000810
4.40	−0.056023	−0.015187	−0.003166	−0.001522	−0.000832
4.45	−0.060577	−0.016159	−0.003264	−0.001541	−0.000853
4.50	−0.065544	−0.017207	−0.003368	−0.001560	−0.000873
4.55	−0.070963	−0.018338	−0.003479	−0.001580	−0.000892
4.60	−0.076878	−0.019557	−0.003599	−0.001601	−0.000911
4.65	−0.083339	−0.020873	−0.003726	−0.001623	−0.000929
4.70	−0.090398	−0.022294	−0.003863	−0.001646	−0.000946
4.75	−0.098114	−0.023827	−0.004009	−0.001670	−0.000963
4.80	−0.106553	−0.025484	−0.004166	−0.001696	−0.000979
4.85	−0.115785	−0.027275	−0.004333	−0.001723	−0.000995
4.90	−0.125891	−0.029210	−0.004512	−0.001751	−0.001010
4.95	−0.136957	−0.031302	−0.004702	−0.001781	−0.001025
5.00	−0.149081	−0.033565	−0.004906	−0.001813	−0.001039

Table 4.6: Second derivatives with respect to the order of the integral Bessel functions of the second kind $\partial^2 Yi_\nu(t)/\partial\nu^2$.

			t		
ν	0.05	0.10	0.15	0.20	0.25
0.00	5.80129	4.58710	3.43908	2.56561	1.89790
0.05	3.67040	3.66140	2.96683	2.32207	1.78566
0.10	1.61891	2.73969	2.47558	2.05182	1.64482
0.15	-0.382261	1.82134	1.96946	1.75962	1.47974
0.20	-2.36644	0.902982	1.45081	1.44905	1.29400
0.25	-4.37107	-0.021214	0.920350	1.12260	1.09044
0.30	-6.43789	-0.959319	0.377368	0.781772	0.871193
0.35	-8.61344	-1.92153	-0.180214	0.427129	0.637762
0.40	-10.9499	-2.92015	-0.755734	0.058382	0.391025
0.45	-13.5064	-3.96964	-1.35376	-0.325580	0.131280
0.50	-16.5434	-5.03818	-2.15586	-0.663314	-0.127709
0.55	-19.5586	-6.29081	-2.64164	-1.14752	-0.428810
0.60	-23.2222	-7.60401	-3.34682	-1.59161	-0.731281
0.65	-27.4466	-9.05187	-4.10530	-2.06319	-1.05100
0.70	-32.3564	-10.6638	-4.92833	-2.56732	-1.39035
0.75	-38.1001	-12.4737	-5.82889	-3.11000	-1.75229
0.80	-44.8546	-14.5211	-6.82191	-3.69815	-2.14033
0.85	-52.8322	-16.8517	-7.92460	-4.33983	-2.55863
0.90	-62.2884	-19.5190	-9.15677	-5.04426	-3.01202
0.95	-73.5316	-22.5855	-10.5413	-5.82207	-3.50608
1.00	-86.9350	-26.1247	-12.1045	-6.68544	-4.04721
1.05	-102.951	-30.2230	-13.8769	-7.64836	-4.64278
1.10	-122.131	-34.9825	-15.8941	-8.72688	-5.30118
1.15	-145.143	-40.5235	-18.1972	-9.93950	-6.03203
1.20	-172.805	-46.9889	-20.8341	-11.3075	-6.84635
1.25	-206.114	-54.5479	-23.8607	-12.8553	-7.75670
1.30	-246.290	-63.4015	-27.3422	-14.6112	-8.77752
1.35	-294.827	-73.7890	-31.3551	-16.6080	-9.92531
1.40	-353.556	-85.9953	-35.9885	-18.8834	-11.21900
1.45	-424.727	-100.360	-41.3475	-21.4814	-12.68040
1.50	-509.834	-115.994	-46.2366	-23.1149	-14.59580
1.55	-616.088	-137.268	-54.7556	-27.8564	-16.21010
1.60	-743.871	-160.876	-63.1195	-31.7612	-18.34040
1.65	-899.622	-188.809	-72.8472	-36.2471	-20.76370
1.70	-1,089.72	-221.901	-84.1751	-41.4071	-23.52400
1.75	-1,322.07	-261.152	-97.3819	-47.3498	-26.67250
1.80	-1,606.42	-307.765	-112.797	-54.2021	-30.26820
1.85	-1,954.88	-363.185	-130.811	-62.1121	-34.37950
1.90	-2,382.45	-429.154	-151.884	-71.2529	-39.08540
1.95	-2,907.75	-507.768	-176.561	-81.8273	-44.47790
2.00	-3,553.94	-601.557	-205.492	-94.0727	-50.66340
2.05	-4,349.79	-713.574	-239.443	-108.267	-57.76580

Table 4.6 (continued)

			t		
v	0.05	0.10	0.15	0.20	0.25
2.10	−5,331.18	−847.507	−279.327	−124.738	−65.92880
2.15	−6,542.77	−1,007.82	−326.228	−143.868	−75.31990
2.20	−8,040.32	−1,199.90	−381.435	−166.108	−86.13380
2.25	−9,893.45	−1,430.30	−446.482	−191.987	−98.59750
2.30	−12,189.2	−1,706.93	−523.199	−222.129	−112.976
2.35	−15,036.4	−2,039.42	−613.764	−257.269	−129.577
2.40	−18,571.5	−2,439.45	−720.778	−298.271	−148.761
2.45	−22,965.3	−2,921.21	−847.346	−346.156	−170.950
2.50	−28,437.4	−3,507.22	−1,002.49	−407.485	−202.024
2.55	−35,241.4	−4,202.73	−1,174.72	−467.613	−226.395
2.60	−43,731.4	−5,049.11	−1,385.27	−544.289	−260.900
2.65	−54,327.5	−6,072.32	−1,635.19	−634.145	−300.942
2.70	−67,565.5	−7,310.44	−1,932.10	−739.537	−347.447
2.75	−84,120.5	−8,809.99	−2,285.16	−863.251	−401.500
2.80	−104843	−10,627.8	−2,705.32	−1,008.59	−464.377
2.85	−130808	−12,833.4	−3,205.77	−1,179.48	−537.576
2.90	−163372	−15,511.7	−3,802.35	−1,380.56	−622.855
2.95	−204249	−18,767.1	−4,514.10	−1,617.37	−722.285
3.00	−255609	−22,727.2	−5,363.95	−1,896.46	−838.302
3.05	−320197	−27,548.5	−6,379.52	−2,225.63	−973.774
3.10	−401494	−33,423.3	−7,594.08	−2,614.18	−1,132.08
3.15	−503910	−40,587.7	−9,047.75	−3,073.16	−1,317.21
3.20	−633044	−49,331.7	−10,789.0	−3,615.74	−1,533.85
3.25	−796004	−60,012.3	−12,876.2	−4,257.63	−1,787.56
3.30	−1.002E+6	−73,068.6	−15,380.2	−5,017.55	−2,084.89
3.35	−1.262	−89,041.7	−18,386.2	−5,917.87	−2,433.59
3.40	−1.591	−108598	−21,997.8	−6,985.28	−2,842.81
3.45	−2.008	−132560	−26,340.0	−8,251.70	−3,323.40
3.50	−2.536	−161991	−31,616.7	−9,810.07	−3,945.30
3.55	−3.205	−197996	−37,854.8	−11,541.6	−4,552.32
3.60	−4.055	−242271	−45,434.0	−13,665.5	−5,333.86
3.65	−5.134	−296680	−54,572.4	−16,192.2	−6,254.16
3.70	−6.506	−363591	−65,598.6	−19,200.4	−7,338.56
3.75	−8.250	−445936	−78,911.4	−22,784.0	−8,617.1
3.80	−1.047E+7	−547343	−94,996.2	−27,056.0	−10,125.6
3.85	−1.330	−672315	−114443	−32,151.9	−11,906.5
3.90	−1.691	−826431	−137970	−38,234.5	−14,010.3
3.95	−2.150	−1.017E+6	−166452	−45,499.5	−16,497.2
4.00	−2.738	−1.251	−200955	−54,182.3	−19,438.5
4.05	−3.488	−1.542	−242779	−64,566.1	−22,919.7
4.10	−4.446	−1.901	−293508	−76,991.6	−27,042.1
4.15	−5.672	−2.345	−355079	−91,869.5	−31,926.9
4.20	−7.242	−2.894	−429854	−109695	−37,718.6

Table 4.6 (continued)

			t		
v	0.05	0.10	0.15	0.20	0.25
4.25	−9.252	−3.575	−520722	−131064	−44,589.3
4.30	−1.183E+8	−4.419	−631211	−156696	−52,745.1
4.35	−1.513	−5.467	−765641	−187462	−62,431.8
4.40	−1.937	−6.766	−929296	−224409	−73,943.3
4.45	−2.482	−8.380	−1.129E+6	−268805	−87,631.2
4.50	−3.181	−1.039E+7	−1.372	−322178	−103911
4.55	−4.081	−1.288	−1.668	−386397	−123301
4.60	−5.238	−1.598	−2.030	−463689	−146390
4.65	−6.727	−1.985	−2.471	−556774	−173905
4.70	−8.645	−2.466	−3.010	−668942	−206711
4.75	−1.112E+9	−3.065	−3.669	−804178	−245849
4.80	−1.431	−3.813	−4.476	−967313	−292564
4.85	−1.842	−4.746	−5.462	−1.164E+6	−348352
4.90	−2.373	−5.911	−6.670	−1.402	−415012
4.95	−3.059	−7.366	−8.149	−1.689	−494702
5.00	−3.945	−9.184	−9.962	−2.037	−590018

			t		
v	0.30	0.35	0.40	0.45	0.50
0.00	1.37742	0.964508	0.632351	0.362365	0.141282
0.05	1.34596	0.984639	0.685965	0.437646	0.230233
0.10	1.28650	0.978472	0.715448	0.491114	0.299679
0.15	1.20275	0.949063	0.723258	0.524705	0.351113
0.20	1.09791	0.899132	0.711646	0.540242	0.385978
0.25	0.974680	0.831070	0.682649	0.539419	0.405654
0.30	0.835270	0.746932	0.638084	0.523788	0.411434
0.35	0.681433	0.648450	0.579538	0.494749	0.404519
0.40	0.514481	0.537036	0.508375	0.453544	0.386001
0.45	0.335299	0.413792	0.425732	0.401257	0.356866
0.50	0.137613	0.272422	0.327935	0.335247	0.314253
0.55	−0.058228	0.134728	0.229444	0.266947	0.270092
0.60	−0.272785	−0.020363	0.116972	0.186273	0.213824
0.65	−0.499980	−0.185812	−0.004626	0.097213	0.149680
0.70	−0.740865	−0.361958	−0.135297	0.000036	0.078037
0.75	−0.996871	−0.549417	−0.275197	−0.105153	−0.000853
0.80	−1.26982	−0.749086	−0.424692	−0.218409	−0.086863
0.85	−1.56195	−0.962150	−0.584363	−0.339951	−0.179994
0.90	−1.87592	−1.19009	−0.755008	−0.470164	−0.280375
0.95	−2.21485	−1.43471	−0.937651	−0.609602	−0.388266
1.00	−2.58237	−1.69813	−1.13355	−0.758993	−0.504059
1.05	−2.98266	−1.98284	−1.34422	−0.919249	−0.628288
1.10	−3.42053	−2.29172	−1.57143	−1.09147	−0.761628
1.15	−3.90144	−2.62808	−1.81725	−1.27697	−0.904907

Table 4.6 (continued)

			t		
v	0.30	0.35	0.40	0.45	0.50
1.20	−4.43167	−2.99571	−2.08406	−1.47728	−1.05911
1.25	−5.01835	−3.39891	−2.37457	−1.69415	−1.22541
1.30	−5.66960	−3.84260	−2.69190	−1.92962	−1.40513
1.35	−6.39469	−4.33236	−3.03958	−2.18600	−1.59984
1.40	−7.20421	−4.87454	−3.42160	−2.46591	−1.81129
1.45	−8.11019	−5.47633	−3.84253	−2.77233	−2.04149
1.50	−9.37582	−6.06500	−4.22492	−3.02460	−2.20745
1.55	−10.2686	−6.89264	−4.82235	−3.47863	−2.56755
1.60	−11.5548	−7.72705	−5.39369	−3.88664	−2.86890
1.65	−13.0056	−8.66121	−6.02899	−4.33751	−3.20002
1.70	−14.6445	−9.70885	−6.73674	−4.83676	−3.56463
1.75	−16.4987	−10.8856	−7.52658	−5.39060	−3.96689
1.80	−18.5994	−12.2094	−8.40944	−6.00605	−4.41148
1.85	−20.9821	−13.7005	−9.39773	−6.69106	−4.90369
1.90	−23.6880	−15.3823	−10.5056	−7.45462	−5.44951
1.95	−26.7644	−17.2814	−11.7490	−8.30693	−6.05565
2.00	−30.2657	−19.4284	−13.1463	−9.25953	−6.72971
2.05	−34.2548	−21.8581	−14.7184	−10.3255	−7.48028
2.10	−38.8040	−24.6107	−16.4889	−11.5197	−8.31706
2.15	−43.9970	−27.7321	−18.4850	−12.8590	−9.25102
2.20	−49.9305	−31.2750	−20.7375	−14.3626	−10.2946
2.25	−56.7161	−35.3000	−23.2819	−16.0523	−11.4617
2.30	−64.4830	−39.8769	−26.1586	−17.9529	−12.7685
2.35	−73.3810	−45.0858	−29.4138	−20.0925	−14.2329
2.40	−83.5834	−51.0190	−33.1004	−22.5035	−15.8753
2.45	−95.2913	−57.7828	−37.2792	−25.2225	−17.7192
2.50	−114.144	−70.9108	−42.2860	−28.5688	−20.0786
2.55	−124.194	−74.3112	−47.4012	−31.7577	−22.1201
2.60	−141.974	−84.3803	−53.5158	−35.6762	−24.7414
2.65	−162.444	−95.8956	−60.4685	−40.1093	−27.6934
2.70	−186.028	−109.075	−68.3801	−45.1283	−31.0204
2.75	−213.223	−124.170	−77.3894	−50.8148	−34.7727
2.80	−244.605	−141.472	−87.6565	−57.2621	−39.0079
2.85	−280.845	−161.318	−99.3652	−64.5774	−43.7914
2.90	−322.727	−184.100	−112.728	−72.8833	−49.1980
2.95	−371.167	−210.270	−127.988	−82.3206	−55.3130
3.00	−427.230	−240.355	−145.429	−93.0507	−62.2339
3.05	−492.165	−274.962	−165.374	−105.259	−70.0724
3.10	−567.429	−314.802	−188.201	−119.159	−78.9558
3.15	−654.728	−360.696	−214.341	−134.995	−89.0301
3.20	−756.056	−413.602	−244.298	−153.048	−100.462
3.25	−873.752	−474.631	−278.650	−173.643	−113.444
3.30	−1,010.55	−545.078	−318.068	−197.153	−128.194

Table 4.6 (continued)

			t		
v	0.30	0.35	0.40	0.45	0.50
3.35	−1,169.67	−626.453	−363.330	−224.007	−144.964
3.40	−1,354.87	−720.510	−415.336	−254.702	−164.044
3.45	−1,570.57	−829.300	−475.127	−289.807	−185.763
3.50	−1,823.32	−956.583	−545.295	−331.366	−211.888
3.55	−2,115.14	−1,101.03	−623.102	−375.987	−238.701
3.60	−2,457.28	−1,270.01	−714.318	−428.702	−270.859
3.65	−2,856.81	−1,465.97	−819.455	−489.140	−307.556
3.70	−3,323.67	−1,693.33	−940.711	−558.476	−349.457
3.75	−3,869.53	−1,957.32	−1,080.64	−638.067	−397.328
3.80	−4,508.17	−2,264.00	−1,242.23	−729.484	−452.051
3.85	−5,255.83	−2,620.50	−1,428.92	−834.546	−514.645
3.90	−6,131.65	−3,035.17	−1,644.76	−955.359	−586.281
3.95	−7,158.23	−3,517.79	−1,894.43	−1,094.37	−668.313
4.00	−8,362.24	−4,079.83	−2,183.41	−1,254.40	−762.302
4.05	−9,775.20	−4,734.73	−2,518.09	−1,438.74	−870.051
4.10	−11,434.4	−5,498.28	−2,905.90	−1,651.20	−993.643
4.15	−13,383.7	−6,389.02	−3,355.55	−1,896.22	−1,135.49
4.20	−15,675.5	−7,428.75	−3,877.18	−2,178.92	−1,298.36
4.25	−18,371.2	−8,643.05	−4,482.66	−2,505.29	−1,485.50
4.30	−21,544.0	−10,062.0	−5,185.84	−2,882.29	−1,700.61
4.35	−25,280.3	−11,721.2	−6,002.95	−3,317.99	−1,948.03
4.40	−29,682.8	−13,662.1	−6,952.95	−3,821.80	−2,232.75
4.45	−34,873.1	−15,934.0	−8,058.07	−4,404.70	−2,560.55
4.50	−40,990.2	−18,589.2	−9,338.88	−5,074.02	−2,932.77
4.55	−48,221.6	−21,712.3	−10,842.1	−5,860.90	−3,373.40
4.60	−56,754.6	−25,367.4	−12,587.3	−6,766.44	−3,875.28
4.65	−66,836.4	−29,655.0	−14,621.7	−7,816.29	−4,454.31
4.70	−78,754.5	−34,686.9	−16,994.4	−9,034.07	−5,122.69
4.75	−92,850.9	−40,595.6	−19,763.2	−10,447.4	−5,894.59
4.80	−109532	−47,537.4	−22,995.9	−12,088.4	−6,786.49
4.85	−129283	−55,697.0	−26,772.0	−13,994.7	−7,817.55
4.90	−152679	−65,293.0	−31,185.1	−16,210.4	−9,010.05
4.95	−180409	−76,584.0	−36,345.3	−18,787.0	−10,390.0
5.00	−213291	−89,875.8	−42,381.9	−21,784.6	−11,987.5

			t		
v	0.55	0.60	0.65	0.70	0.75
0.00	−0.040628	−0.190676	−0.314467	−0.416379	−0.499891
0.05	0.056435	−0.089441	−0.211890	−0.314512	−0.400222
0.10	0.136208	−0.003410	−0.122581	−0.224127	−0.310401
0.15	0.199813	0.068224	−0.045991	−0.144893	−0.230273
0.20	0.248366	0.126300	0.018480	−0.076408	−0.159606
0.25	0.282962	0.171665	0.071465	−0.018225	−0.098109

Table 4.6 (continued)

			t		
ν	0.55	0.60	0.65	0.70	0.75
0.30	0.304658	0.205161	0.113616	0.030142	−0.045443
0.35	0.314458	0.227608	0.145586	0.069195	−0.001237
0.40	0.313300	0.239793	0.168019	0.099448	0.034903
0.45	0.302050	0.242459	0.181538	0.121409	0.063381
0.50	0.277159	0.231315	0.181185	0.129537	0.078129
0.55	0.252336	0.221948	0.184177	0.142434	0.098977
0.60	0.215183	0.199976	0.174371	0.142431	0.106885
0.65	0.170555	0.170885	0.157789	0.135992	0.108703
0.70	0.118877	0.135109	0.134847	0.123504	0.104778
0.75	0.060476	0.093005	0.105908	0.105314	0.095432
0.80	−0.004418	0.044857	0.071273	0.081726	0.080956
0.85	−0.075676	−0.009130	0.031188	0.052996	0.061605
0.90	−0.153267	−0.068831	−0.014167	0.019336	0.037600
0.95	−0.237265	−0.134201	−0.064675	−0.019095	0.009120
1.00	−0.327848	−0.205279	−0.120284	−0.062192	−0.023693
1.05	−0.425301	−0.282187	−0.181011	−0.109899	−0.060745
1.10	−0.530022	−0.365136	−0.246941	−0.162222	−0.101983
1.15	−0.642522	−0.454429	−0.318234	−0.219220	−0.147403
1.20	−0.763436	−0.550461	−0.395122	−0.281014	−0.197046
1.25	−0.893527	−0.653727	−0.477917	−0.347788	−0.251007
1.30	−1.03370	−0.764827	−0.567011	−0.419788	−0.309432
1.35	−1.18499	−0.884472	−0.662883	−0.497331	−0.372520
1.40	−1.34861	−1.01349	−0.766104	−0.580806	−0.440529
1.45	−1.52594	−1.15285	−0.877343	−0.670676	−0.513777
1.50	−1.63226	−1.21652	−0.909673	−0.679414	−0.504421
1.55	−1.92822	−1.46710	−1.12708	−0.871871	−0.677590
1.60	−2.15696	−1.64466	−1.26747	−0.984557	−0.769130
1.65	−2.40703	−1.83792	−1.41971	−1.10637	−0.867868
1.70	−2.68099	−2.04868	−1.58507	−1.23826	−0.974495
1.75	−2.98171	−2.27897	−1.76501	−1.38128	−1.08979
1.80	−3.31242	−2.53106	−1.96119	−1.53662	−1.21463
1.85	−3.67676	−2.80753	−2.17542	−1.70563	−1.35002
1.90	−4.07881	−3.11123	−2.40979	−1.88983	−1.49706
1.95	−4.52317	−3.44539	−2.66659	−2.09088	−1.65701
2.00	−5.01501	−3.81364	−2.94842	−2.31068	−1.83126
2.05	−5.56015	−4.22003	−3.25818	−2.55134	−2.02137
2.10	−6.16515	−4.66912	−3.59911	−2.81522	−2.22909
2.15	−6.83740	−5.16604	−3.97485	−3.10497	−2.45637
2.20	−7.58523	−5.71654	−4.38949	−3.42352	−2.70536
2.25	−8.41802	−6.32710	−4.84760	−3.77419	−2.97852
2.30	−9.34639	−7.00499	−5.35430	−4.16066	−3.27854
2.35	−10.3823	−7.75842	−5.91536	−4.58707	−3.60844
2.40	−11.5393	−8.59662	−6.53723	−5.05805	−3.97161

Table 4.6 (continued)

			t		
v	0.55	0.60	0.65	0.70	0.75
2.45	−12.8328	−9.52998	−7.22719	−5.57878	−4.37183
2.50	−14.5773	−10.8765	−8.30771	−6.47671	−5.14145
2.55	−15.9003	−11.7306	−8.84499	−6.79346	−5.30079
2.60	−17.7163	−13.0261	−9.79233	−7.50123	−5.83954
2.65	−19.7530	−14.4734	−10.8471	−8.28660	−6.43547
2.70	−22.0390	−16.0918	−12.0222	−9.15878	−7.09520
2.75	−24.6067	−17.9028	−13.3326	−10.1281	−7.82617
2.80	−27.4928	−19.9306	−14.7949	−11.2063	−8.63670
2.85	−30.7392	−22.2030	−16.4278	−12.4064	−9.53611
2.90	−34.3933	−24.7513	−18.2525	−13.7431	−10.5349
2.95	−38.5092	−27.6108	−20.2930	−15.2330	−11.6448
3.00	−43.1485	−30.8217	−22.5763	−16.8949	−12.8790
3.05	−48.3810	−34.4296	−25.1331	−18.7498	−14.2525
3.10	−54.2866	−38.4862	−27.9979	−20.8215	−15.7818
3.15	−60.9561	−43.0503	−31.2098	−23.1368	−17.4858
3.20	−68.4932	−48.1886	−34.8133	−25.7259	−19.3857
3.25	−77.0162	−53.9770	−38.8586	−28.6231	−21.5053
3.30	−86.6599	−60.5017	−43.4025	−31.8670	−23.8713
3.35	−97.5787	−67.8610	−48.5098	−35.5013	−26.5141
3.40	−109.949	−76.1666	−54.2537	−39.5753	−29.4679
3.45	−123.971	−85.5459	−60.7174	−44.1451	−32.7710
3.50	−141.259	−95.9733	−67.8402	−49.1345	−36.3439
3.55	−157.929	−108.126	−76.1949	−55.0333	−40.6051
3.60	−178.429	−121.681	−85.4382	−61.5048	−45.2406
3.65	−201.724	−137.025	−95.8640	−68.7803	−50.4363
3.70	−228.210	−154.402	−107.630	−76.9645	−56.2632
3.75	−258.340	−174.095	−120.917	−86.1758	−62.8014
3.80	−292.637	−196.424	−135.928	−96.5490	−70.1419
3.85	−331.699	−221.756	−152.898	−108.237	−78.3876
3.90	−376.212	−250.512	−172.093	−121.414	−87.6551
3.95	−426.967	−283.171	−193.815	−136.276	−98.0764
4.00	−484.870	−320.284	−218.412	−153.050	−109.802
4.05	−550.964	−362.482	−246.278	−171.990	−123.001
4.10	−626.449	−410.487	−277.864	−193.388	−137.866
4.15	−712.708	−465.127	−313.687	−217.575	−154.618
4.20	−811.329	−527.353	−354.337	−244.930	−173.505
4.25	−924.146	−598.255	−400.486	−275.882	−194.809
4.30	−1,053.27	−679.085	−452.907	−310.923	−218.853
4.35	−1,201.14	−771.281	−512.482	−350.612	−246.002
4.40	−1,370.55	−876.495	−580.223	−395.590	−276.673
4.45	−1,564.76	−996.628	−657.286	−446.587	−311.340
4.50	−1,782.13	−1,128.53	−739.723	−499.228	−345.415
4.55	−2,043.10	−1,290.71	−844.886	−570.090	−394.893

Table 4.6 (continued)

			t		
ν	0.55	0.60	0.65	0.70	0.75
4.60	−2,336.55	−1,470.08	−958.689	−644.640	−445.096
4.65	−2,673.63	−1,675.28	−1,088.41	−729.331	−501.950
4.70	−3,061.01	−1,910.16	−1,236.35	−825.590	−566.366
4.75	−3,506.42	−2,179.15	−1,405.15	−935.048	−639.384
4.80	−4,018.80	−2,487.35	−1,597.85	−1,059.58	−722.192
4.85	−4,608.52	−2,840.64	−1,817.92	−1,201.31	−816.147
4.90	−5,287.57	−3,245.81	−2,069.38	−1,362.71	−922.799
4.95	−6,069.87	−3,710.70	−2,356.84	−1,546.59	−1,043.92
5.00	−6,971.52	−4,244.37	−2,685.61	−1,756.17	−1,181.53

			t		
ν	0.80	0.85	0.90	0.95	1.0
0.00	−0.567801	−0.622393	−0.665550	−0.698841	−0.723585
0.05	−0.471408	−0.530050	−0.577800	−0.616054	−0.645996
0.10	−0.383373	−0.444710	−0.495826	−0.537933	−0.572074
0.15	−0.303686	−0.366482	−0.419832	−0.464759	−0.502157
0.20	−0.232256	−0.295391	−0.349939	−0.396730	−0.436507
0.25	−0.168925	−0.231391	−0.286194	−0.333972	−0.375315
0.30	−0.113478	−0.174377	−0.228585	−0.276549	−0.318708
0.35	−0.065662	−0.124194	−0.177043	−0.224469	−0.266758
0.40	−0.025189	−0.080650	−0.131460	−0.177694	−0.219491
0.45	0.008250	−0.043522	−0.091689	−0.136148	−0.176889
0.50	0.028094	−0.019835	−0.065197	−0.107708	−0.147217
0.55	0.055318	0.012472	−0.028878	−0.068281	−0.105447
0.60	0.069600	0.031861	−0.005437	−0.041674	−0.076423
0.65	0.078141	0.045865	0.012978	−0.019734	−0.051707
0.70	0.081249	0.054746	0.026584	−0.002284	−0.031166
0.75	0.079212	0.058759	0.035600	0.010854	−0.014655
0.80	0.072298	0.058145	0.040237	0.019860	−0.002026
0.85	0.060753	0.053128	0.040696	0.024912	0.006872
0.90	0.044793	0.043916	0.037167	0.026180	0.012186
0.95	0.024606	0.030694	0.029826	0.023824	0.014060
1.00	0.000347	0.013621	0.018831	0.017994	0.012632
1.05	−0.027861	−0.007168	0.004319	0.008825	0.008028
1.10	−0.059933	−0.031566	−0.013593	−0.003566	0.000365
1.15	−0.095819	−0.059497	−0.034813	−0.019076	−0.010256
1.20	−0.135509	−0.090915	−0.059273	−0.037626	−0.023747
1.25	−0.179035	−0.125808	−0.086930	−0.059154	−0.040037
1.30	−0.226469	−0.164198	−0.117772	−0.083623	−0.059073
1.35	−0.277929	−0.206142	−0.151815	−0.111020	−0.080822
1.40	−0.333578	−0.251738	−0.189105	−0.141357	−0.105269
1.45	−0.393630	−0.301120	−0.229722	−0.174671	−0.132425
1.50	−0.370188	−0.266590	−0.186411	−0.124386	−0.184432

Table 4.6 (continued)

			t		
v	**0.80**	**0.85**	**0.90**	**0.95**	**1.0**
1.55	−0.528050	−0.411995	−0.321423	−0.250525	−0.195009
1.60	−0.603113	−0.473978	−0.372845	−0.293288	−0.230576
1.65	−0.683976	−0.540732	−0.428270	−0.339477	−0.269128
1.70	−0.771144	−0.612625	−0.487969	−0.389286	−0.310802
1.75	−0.865193	−0.690087	−0.552256	−0.442947	−0.355767
1.80	−0.966780	−0.773606	−0.621498	−0.500731	−0.404223
1.85	−1.07664	−0.863736	−0.696112	−0.562951	−0.456404
1.90	−1.19562	−0.961106	−0.776571	−0.629968	−0.512581
1.95	−1.32464	−1.06642	−0.863411	−0.702189	−0.573067
2.00	−1.46475	−1.18047	−0.957236	−0.780075	−0.638214
2.05	−1.61711	−1.30414	−1.05872	−0.864146	−0.708424
2.10	−1.78305	−1.43842	−1.16862	−0.954985	−0.784146
2.15	−1.96401	−1.58442	−1.28779	−1.05324	−0.865885
2.20	−2.16162	−1.74336	−1.41715	−1.15964	−0.954205
2.25	−2.37770	−1.91662	−1.55777	−1.27499	−1.04974
2.30	−2.61425	−2.10572	−1.71080	−1.40020	−1.15318
2.35	−2.87353	−2.31236	−1.87755	−1.53626	−1.26531
2.40	−3.15805	−2.53843	−2.05945	−1.68429	−1.38699
2.45	−3.47061	−2.78604	−2.25811	−1.84551	−1.51919
2.50	−3.81000	−3.39623	−14.95440	−2.02021	−1.66298
2.55	−4.19269	−3.35548	−2.71303	−2.21319	−1.81950
2.60	−4.60958	−3.68283	−2.97346	−2.42286	−1.99011
2.65	−5.06934	−4.04280	−3.25907	−2.65217	−2.17623
2.70	−5.57681	−4.43901	−3.57256	−2.90322	−2.37948
2.75	−6.13742	−4.87548	−3.91698	−3.17832	−2.60164
2.80	−6.75724	−5.35669	−4.29569	−3.48004	−2.84470
2.85	−7.44303	−5.88766	−4.71246	−3.81122	−3.11084
2.90	−8.20239	−6.47398	−5.17146	−4.17506	−3.40251
2.95	−9.04384	−7.12191	−5.67737	−4.57507	−3.72241
3.00	−9.97687	−7.83842	−6.23540	−5.01519	−4.07354
3.05	−11.0122	−8.63134	−6.85134	−5.49980	−4.45926
3.10	−12.1618	−9.50941	−7.53168	−6.03377	−4.88327
3.15	−13.4390	−10.4824	−8.28366	−6.62254	−5.34969
3.20	−14.8591	−11.5613	−9.11537	−7.27215	−5.86313
3.25	−16.4389	−12.7583	−10.0358	−7.98937	−6.42869
3.30	−18.1974	−14.0873	−11.0552	−8.78172	−7.05207
3.35	−20.1562	−15.5636	−12.1847	−9.65763	−7.73961
3.40	−22.3392	−17.2046	−13.4371	−10.6265	−8.49837
3.45	−24.7735	−19.0298	−14.8265	−11.6988	−9.33625
3.50	−28.8076	−22.3556	−16.2959	−12.8305	−10.2236
3.55	−30.5223	−23.3223	−18.0818	−14.2020	−11.2855
3.60	−33.9098	−25.8419	−19.9855	−15.6608	−12.4176
3.65	−37.6959	−28.6504	−22.1022	−17.2790	−13.6707

Table 4.6 (continued)

v	0.80	0.85	0.90	0.95	1.0
3.70	−41.9299	−31.7829	−24.4571	−19.0750	−15.0583
3.75	−46.6673	−35.2785	−27.0785	−21.0694	−16.5957
3.80	−51.9710	−39.1814	−29.9979	−23.2854	−18.3000
3.85	−57.9116	−43.5415	−33.2511	−25.7488	−20.1904
3.90	−64.5694	−48.4148	−36.8781	−28.4887	−22.2883
3.95	−72.0347	−53.8646	−40.9239	−31.5378	−24.6175
4.00	−80.4100	−59.9622	−45.4392	−34.9326	−27.2050
4.05	−89.8110	−66.7882	−50.4812	−38.7142	−30.0808
4.10	−100.369	−74.4335	−56.1139	−42.9289	−33.2786
4.15	−112.232	−83.0008	−62.4099	−47.6285	−36.8363
4.20	−125.568	−92.6060	−69.4507	−52.8715	−40.7964
4.25	−140.568	−103.380	−77.3284	−58.7236	−45.2063
4.30	−157.448	−115.472	−86.1466	−65.2586	−50.1198
4.35	−176.452	−129.049	−96.0227	−72.5599	−55.5967
4.40	−197.859	−144.300	−107.089	−80.7211	−61.7048
4.45	−221.985	−161.442	−119.494	−89.8479	−68.5199
4.50	−244.157	−165.058	−128.604	−95.3882	−71.6003
4.55	−279.874	−202.400	−149.019	−111.490	−84.6233
4.60	−314.507	−226.806	−166.546	−124.291	−94.1160
4.65	−353.614	−254.288	−186.231	−138.633	−104.727
4.70	−397.792	−285.249	−208.350	−154.709	−116.594
4.75	−447.722	−320.145	−233.216	−172.738	−129.871
4.80	−504.180	−359.495	−261.183	−192.965	−144.733
4.85	−568.050	−403.887	−292.651	−215.669	−161.375
4.90	−640.337	−453.990	−328.076	−241.165	−180.021
4.95	−722.187	−510.565	−367.972	−269.808	−200.919
5.00	−814.908	−574.476	−412.924	−302.002	−224.353

v	1.5	2.0	3.0	4.0	5.0
0.00	−0.685234	−0.434234	0.037419	0.202516	0.120647
0.05	−0.658084	−0.440277	0.014110	0.193593	0.126858
0.10	−0.629002	−0.443355	−0.008241	0.183852	0.132144
0.15	−0.598418	−0.443688	−0.029547	0.173392	0.136508
0.20	−0.566734	−0.441497	−0.049734	0.162314	0.139961
0.25	−0.534317	−0.437005	−0.068741	0.150715	0.142519
0.30	−0.501509	−0.430434	−0.086522	0.138691	0.144203
0.35	−0.468618	−0.422004	−0.103041	0.126333	0.145038
0.40	−0.435923	−0.411929	−0.118280	0.113732	0.145053
0.45	−0.403675	−0.400418	−0.132211	0.100973	0.144282
0.50	−0.382215	−0.396796	−0.145740	0.097867	0.151210
0.55	−0.341381	−0.373885	−0.156186	0.075302	0.140527
0.60	−0.311700	−0.359240	−0.166247	0.062540	0.137624

Table 4.6 (continued)

			t		
v	1.5	2.0	3.0	4.0	5.0
0.65	−0.283199	−0.343914	−0.175050	0.049918	0.134095
0.70	−0.256002	−0.328070	−0.182626	0.037500	0.129985
0.75	−0.230212	−0.311863	−0.189010	0.025342	0.125340
0.80	−0.205911	−0.295438	−0.194242	0.013498	0.120206
0.85	−0.183165	−0.278925	−0.198368	0.002014	0.114632
0.90	−0.162025	−0.262449	−0.201438	−0.009066	0.108665
0.95	−0.142525	−0.246122	−0.203504	−0.019706	0.102352
1.00	−0.124689	−0.230044	−0.204621	−0.029872	0.095741
1.05	−0.108528	−0.214307	−0.204847	−0.039538	0.088877
1.10	−0.094043	−0.198993	−0.204241	−0.048681	0.081806
1.15	−0.081230	−0.184176	−0.202862	−0.057281	0.074571
1.20	−0.070073	−0.169919	−0.200770	−0.065325	0.067216
1.25	−0.060555	−0.156278	−0.198027	−0.072802	0.059782
1.30	−0.052652	−0.143300	−0.194691	−0.079707	0.052308
1.35	−0.046338	−0.131027	−0.190823	−0.086036	0.044831
1.40	−0.041583	−0.119492	−0.186479	−0.091790	0.037388
1.45	−0.038358	−0.108722	−0.181718	−0.096974	0.030011
1.50	−0.035484	−0.097436	−0.175900	−0.102153	0.023023
1.55	−0.036378	−0.089564	−0.171160	−0.105659	0.015579
1.60	−0.037567	−0.081205	−0.165468	−0.109181	0.008581
1.65	−0.040172	−0.073671	−0.159568	−0.112176	0.001762
1.70	−0.044173	−0.066968	−0.153506	−0.114658	−0.004856
1.75	−0.049550	−0.061098	−0.147327	−0.116646	−0.011252
1.80	−0.056289	−0.056059	−0.141073	−0.118159	−0.017409
1.85	−0.064381	−0.051849	−0.134784	−0.119218	−0.023311
1.90	−0.073822	−0.048465	−0.128498	−0.119844	−0.028944
1.95	−0.084616	−0.045900	−0.122249	−0.120060	−0.034296
2.00	−0.096771	−0.044149	−0.116070	−0.119889	−0.039358
2.05	−0.110304	−0.043204	−0.109990	−0.119356	−0.044122
2.10	−0.125241	−0.043061	−0.104038	−0.118485	−0.048583
2.15	−0.141614	−0.043712	−0.098238	−0.117299	−0.052737
2.20	−0.159467	−0.045153	−0.092613	−0.115823	−0.056581
2.25	−0.178850	−0.047379	−0.087185	−0.114082	−0.060115
2.30	−0.199828	−0.050387	−0.081973	−0.112100	−0.063339
2.35	−0.222474	−0.054178	−0.076992	−0.109901	−0.066256
2.40	−0.246873	−0.058751	−0.072259	−0.107507	−0.068870
2.45	−0.273125	−0.064111	−0.067786	−0.104943	−0.071185
2.50	−0.311956	0.013388	−0.063927	−0.107762	−0.073028
2.55	−0.331651	−0.077216	−0.059667	−0.099391	−0.074943
2.60	−0.364195	−0.084983	−0.056039	−0.096446	−0.076402
2.65	−0.399135	−0.093580	−0.052710	−0.093415	−0.077591
2.70	−0.436649	−0.103025	−0.049687	−0.090318	−0.078520
2.75	−0.476935	−0.113344	−0.046973	−0.087174	−0.079200
2.80	−0.520214	−0.124563	−0.044574	−0.084000	−0.079641

Table 4.6 (continued)

ν	1.5	2.0	3.0	4.0	5.0
			t		
2.85	−0.566728	−0.136716	−0.042494	−0.080812	−0.079853
2.90	−0.616746	−0.149842	−0.040736	−0.077628	−0.079849
2.95	−0.670566	−0.163983	−0.039301	−0.074462	−0.079641
3.00	−0.728512	−0.179188	−0.038192	−0.071328	−0.079239
3.05	−0.790945	−0.195515	−0.037412	−0.068239	−0.078657
3.10	−0.858260	−0.213025	−0.036961	−0.065209	−0.077907
3.15	−0.930889	−0.231789	−0.036841	−0.062248	−0.077001
3.20	−1.00931	−0.251884	−0.037054	−0.059368	−0.075952
3.25	−1.09405	−0.273397	−0.037601	−0.056577	−0.074770
3.30	−1.18568	−0.296423	−0.038484	−0.053886	−0.073470
3.35	−1.28483	−0.321066	−0.039706	−0.051303	−0.072062
3.40	−1.39220	−0.347442	−0.041268	−0.048835	−0.070558
3.45	−1.50855	−0.375679	−0.043175	−0.046490	−0.068970
3.50	−1.76800	−0.401191	−0.046135	−0.043726	−0.066017
3.55	−1.77162	−0.438302	−0.048037	−0.042192	−0.065586
3.60	−1.92027	−0.473009	−0.051003	−0.040250	−0.063811
3.65	−2.08178	−0.510216	−0.054333	−0.038452	−0.061995
3.70	−2.25736	−0.550124	−0.058035	−0.036804	−0.060147
3.75	−2.44837	−0.592950	−0.062117	−0.035307	−0.058277
3.80	−2.65628	−0.638932	−0.066589	−0.033967	−0.056394
3.85	−2.88271	−0.688330	−0.071463	−0.032786	−0.054507
3.90	−3.12946	−0.741425	−0.076750	−0.031766	−0.052624
3.95	−3.39850	−0.798528	−0.082465	−0.030911	−0.050752
4.00	−3.69199	−0.859977	−0.088625	−0.030222	−0.048901
4.05	−4.01233	−0.926135	−0.095245	−0.029701	−0.047074
4.10	−4.36216	−0.997407	−0.102347	−0.029351	−0.045281
4.15	−4.74438	−1.07423	−0.109952	−0.029174	−0.043527
4.20	−5.16221	−1.15708	−0.118083	−0.029171	−0.041819
4.25	−5.61918	−1.24648	−0.126768	−0.029345	−0.040160
4.30	−6.11920	−1.34300	−0.136036	−0.029697	−0.038558
4.35	−6.66658	−1.44726	−0.145918	−0.030229	−0.037016
4.40	−7.26610	−1.55993	−0.156449	−0.030944	−0.035538
4.45	−7.92302	−1.68177	−0.167667	−0.031843	−0.034130
4.50	−9.81850	−1.89696	−0.159434	−0.027412	−0.032986
4.55	−9.43297	−1.95623	−0.192333	−0.034207	−0.031536
4.60	−10.2996	−2.11071	−0.205875	−0.035678	−0.030356
4.65	−11.2508	−2.27807	−0.220293	−0.037345	−0.029259
4.70	−12.2955	−2.45946	−0.235644	−0.039214	−0.028247
4.75	−13.4432	−2.65616	−0.251991	−0.041287	−0.027324
4.80	−14.7048	−2.86955	−0.269402	−0.043571	−0.026490
4.85	−16.0919	−3.10115	−0.287949	−0.046070	−0.025749
4.90	−17.6179	−3.35262	−0.307713	−0.048790	−0.025102
4.95	−19.2972	−3.62578	−0.328780	−0.051738	−0.024551
5.00	−21.1461	−3.92263	−0.351244	−0.054922	−0.024099

4.3 The Integral Modified Bessel Functions of the Second Kind and Their First and Second Derivatives with Respect to the Order

Integral modified Bessel functions $Ki_v(t)$ are defined by

$$Ki_v(z) = \int\limits_z^\infty \frac{Ki_v(t)}{t}\, dt \tag{4.3.1}$$

and they were calculated directly by applying MATHEMATICA program (Figures 4.13– Figures 4.15; Tables 4.7–4.9).

First and second derivatives with respect to the order of the integral modified Bessel function of the second kind were determined using the following expressions:

$$\frac{\partial Ki_v(t)}{\partial v} = \frac{-Ki_{v+2h}(t) + 8\,Ki_{v+h}(t) - 8\,Ki_{v-h}(t) + Ki_{v-2h}(t)}{12h} \tag{4.3.2}$$

$$\frac{\partial^2 Ki_v(t)}{\partial v^2} = \frac{-Ki_{v+2h}(t) + 16\,Ki_{v+h}(t) - 30\,Ki_v(t) + 16\,Ki_{v-h}(t) - Ki_{v-2h}(t)}{12h^2} \tag{4.3.3}$$

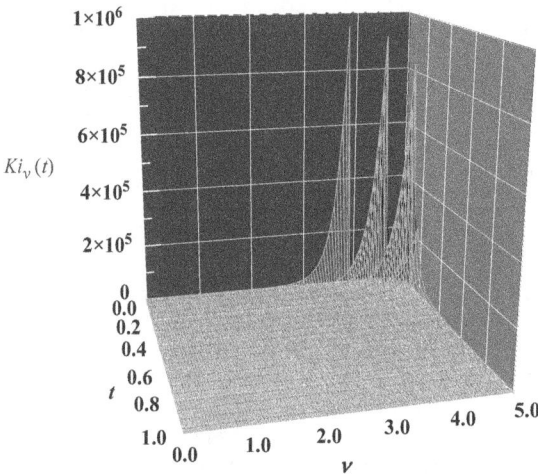

Figure 4.13: The integral modified Bessel functions of the second kind $Ki_v(t)$ as a function of v and t.

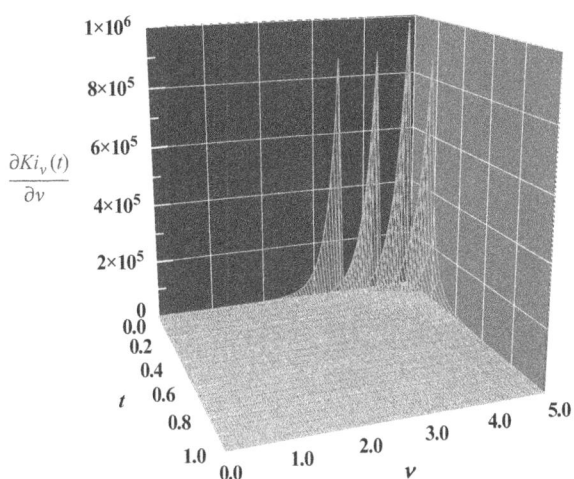

Figure 4.14: First derivatives of the integral modified Bessel functions of the second kind with respect to the order $\partial Ki_v(t)/\partial v$ as a function of v and t.

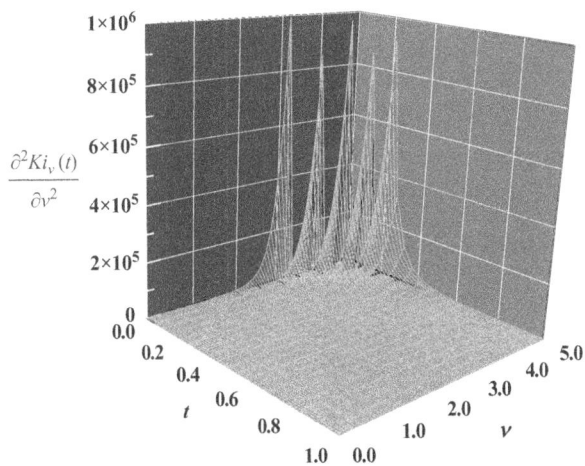

Figure 4.15: Second derivatives of the integral modified Bessel functions of the second kind with respect to the order $\partial^2 Ki_v(t)/\partial v^2$ as a function of v and t.

Table 4.7: The integral modified Bessel functions of the second kind $Ki_v(t)$.

v	\multicolumn{5}{c}{t}				
	0.05	0.10	0.15	0.20	0.25
0.00	5.25102	3.33095	2.42753	1.88353	1.51607
0.05	5.26886	3.33929	2.43254	1.88689	1.51847
0.10	5.32269	3.36442	2.44761	1.89699	1.52569
0.15	5.41349	3.40669	2.47291	1.91393	1.53780
0.20	5.54290	3.46666	2.50873	1.93788	1.55490
0.25	5.71330	3.54514	2.55546	1.96907	1.57713
0.30	5.92781	3.64322	2.61365	2.00781	1.60470
0.35	6.19047	3.76225	2.68395	2.05448	1.63784
0.40	6.50627	3.90390	2.76717	2.10954	1.67686
0.45	6.88131	4.07018	2.86429	2.17356	1.72210
0.50	7.32297	4.26349	2.97646	2.24718	1.77397
0.55	7.84016	4.48666	3.10502	2.33118	1.83294
0.60	8.44350	4.74303	3.25155	2.42643	1.89958
0.65	9.14569	5.03650	3.41787	2.53395	1.97450
0.70	9.96186	5.37164	3.60607	2.65490	2.05843
0.75	10.9100	5.75375	3.81859	2.79062	2.15218
0.80	12.0117	6.18904	4.05821	2.94264	2.25669
0.85	13.2924	6.68471	4.32814	3.11269	2.37300
0.90	14.7826	7.24916	4.63209	3.30276	2.50230
0.95	16.5188	7.89220	4.97428	3.51510	2.64596
1.00	18.5445	8.62521	5.35960	3.75229	2.80549
1.05	20.9118	9.46153	5.79368	4.01727	2.98262
1.10	23.6830	10.4167	6.28296	4.31337	3.17933
1.15	26.9331	11.5089	6.83491	4.64443	3.39781
1.20	30.7520	12.7593	7.45811	5.01478	3.64060
1.25	35.2479	14.1929	8.16247	5.42942	3.91052
1.30	40.5515	15.8388	8.95942	5.89403	4.21083
1.35	46.8201	17.7314	9.86221	6.41512	4.54518
1.40	54.2442	19.9110	10.8861	7.00013	4.91774
1.45	63.0541	22.4248	12.0490	7.65761	5.33325
1.50	73.5294	25.3289	13.3712	8.39736	5.79709
1.55	86.0092	28.6892	14.8768	9.23061	6.31540
1.60	100.906	32.5834	16.5934	10.1703	6.89517
1.65	118.721	37.1036	18.5534	11.2313	7.54437
1.70	140.067	42.3588	20.7943	12.4306	8.27209
1.75	165.692	48.4782	23.3599	13.7882	9.08875
1.80	196.507	55.6150	26.3014	15.3266	10.0062
1.85	233.632	63.9514	29.6785	17.0721	11.0381
1.90	278.436	73.7039	33.5610	19.0552	12.2001
1.95	332.601	85.1304	38.0307	21.3109	13.5099
2.00	398.193	98.5384	43.1833	23.8799	14.9881
2.05	477.755	114.295	49.1313	26.8093	16.6583
2.10	574.416	132.838	56.0066	30.1539	18.5475

Table 4.7 (continued)

			t		
v	0.05	0.10	0.15	0.20	0.25
2.15	692.039	154.692	63.9641	33.9772	20.6870
2.20	835.392	180.484	73.1864	38.3528	23.1126
2.25	1,010.37	210.967	83.8882	43.3668	25.8656
2.30	1,224.26	247.043	96.3227	49.1191	28.9938
2.35	1,486.11	289.796	110.789	55.7261	32.5523
2.40	1,807.12	340.528	127.639	63.3237	36.6047
2.45	2,201.21	400.807	147.290	72.0706	41.2246
2.50	2,685.68	472.521	170.236	82.1522	46.4971
2.55	3,282.06	557.948	197.060	93.7851	52.5211
2.60	4,017.16	659.835	228.455	107.223	59.4108
2.65	4,924.43	781.501	265.243	122.763	67.2990
2.70	6,045.59	926.959	308.398	140.754	76.3398
2.75	7,432.80	1,101.07	359.079	161.604	86.7122
2.80	9,151.26	1,309.71	418.665	185.794	98.6244
2.85	11,282.6	1,560.01	488.796	213.887	112.319
2.90	13,929.2	1,860.64	571.427	246.547	128.078
2.95	17,219.2	2,222.10	668.889	284.556	146.230
3.00	21,313.9	2,657.18	783.962	328.832	167.159
3.05	26,415.4	3,181.41	919.968	380.462	191.314
3.10	32,778.5	3,813.74	1,080.88	440.724	219.217
3.15	40,723.5	4,577.22	1,271.44	511.130	251.480
3.20	50,654.0	5,500.01	1,497.33	593.465	288.819
3.25	63,079.0	6,616.43	1,765.38	689.840	332.071
3.30	78,640.9	7,968.44	2,083.73	802.754	382.219
3.35	98,151.0	9,607.35	2,462.19	935.164	440.413
3.40	122635	11,595.9	2,912.52	1,090.58	508.002
3.45	153390	14,010.9	3,448.86	1,273.15	586.572
3.50	192060	16,946.6	4,088.20	1,487.82	677.984
3.55	240725	20,518.4	4,851.00	1,740.44	784.427
3.60	302026	24,868.0	5,761.88	2,037.98	908.476
3.65	379314	30,169.4	6,850.53	2,388.71	1,053.16
3.70	476844	36,636.5	8,152.73	2,802.49	1,222.05
3.75	600024	44,532.5	9,711.68	3,291.04	1,419.35
3.80	755735	54,180.8	11,579.5	3,868.36	1,650.03
3.85	952734	65,980.5	13,819.3	4,551.10	1,919.93
3.90	1.20218E+6	80,422.7	16,507.2	5,359.14	2,235.98
3.95	1.518	98,113.5	19,735.4	6,316.24	2,606.35
4.00	1.919	119801	23,615.6	7,450.74	3,040.69
4.05	2.428	146409	28,283.0	8,796.56	3,550.43
4.10	3.075	179078	33,901.6	10,394.2	4,149.11
4.15	3.896	219220	40,670.2	12,292.3	4,852.75
4.20	4.942	268582	48,830.3	14,548.8	5,680.35
4.25	6.273	329326	58,675.2	17,233.5	6,654.42

Table 4.7 (continued)

			t		
v	0.05	0.10	0.15	0.20	0.25
4.30	7.969	404131	70,561.1	20,429.8	7,801.69
4.35	1.013E+7	496318	84,921.5	24,237.9	9,153.90
4.40	1.289	610008	102284	28,778.0	10,748.7
4.45	1.642	750315	123290	34,194.6	12,631.0
4.50	2.092	923590	148721	40,661.2	14,854.0
4.55	2.668	1.138E+6	179533	48,386.5	17,481.1
4.60	3.404	1.403	216886	57,621.6	20,587.9
4.65	4.348	1.730	262200	68,668.8	24,264.1
4.70	5.557	2.136	317207	81,892.2	28,617.2
4.75	7.107	2.639	384025	97,730.5	33,774.8
4.80	9.096	3.262	465241	116713	39,889.4
4.85	1.165E+8	4.036	564018	139478	47,143.2
4.90	1.493	4.996	684231	166795	55,753.7
4.95	1.915	6.189	830620	199596	65,980.7
5.00	2.457	7.672	1.009E+06	239005	78,134.9

			t		
v	0.30	0.35	0.40	0.45	0.50
0.00	1.25049	1.04973	0.893051	0.767789	0.665751
0.05	1.25228	1.05111	0.894138	0.768661	0.666461
0.10	1.25767	1.05526	0.897407	0.771283	0.668596
0.15	1.26670	1.06221	0.902879	0.775672	0.672167
0.20	1.27945	1.07201	0.910592	0.781854	0.677197
0.25	1.29600	1.08473	0.920596	0.789869	0.683715
0.30	1.31651	1.10047	0.932961	0.799769	0.691760
0.35	1.34112	1.11933	0.947770	0.811614	0.701380
0.40	1.37003	1.14146	0.965125	0.825483	0.712633
0.45	1.40349	1.16703	0.985146	0.841464	0.725587
0.50	1.44177	1.19623	1.00797	0.859661	0.740322
0.55	1.48518	1.22927	1.03377	0.880195	0.756928
0.60	1.53410	1.26642	1.06271	0.903200	0.775510
0.65	1.58893	1.30796	1.09501	0.928833	0.796183
0.70	1.65015	1.35422	1.13091	0.957268	0.819081
0.75	1.71830	1.40558	1.17067	0.988699	0.844350
0.80	1.79398	1.46245	1.21459	1.02335	0.872155
0.85	1.87789	1.52530	1.26300	1.06146	0.902681
0.90	1.97079	1.59465	1.31628	1.10330	0.936132
0.95	2.07355	1.67110	1.37484	1.14918	0.972737
1.00	2.18716	1.75532	1.43915	1.19944	1.01275
1.05	2.31271	1.84804	1.50974	1.25446	1.05644
1.10	2.45146	1.95009	1.58718	1.31466	1.10414
1.15	2.60478	2.06242	1.67213	1.38050	1.15618
1.20	2.77428	2.18606	1.76531	1.45250	1.21294

Table 4.7 (continued)

			t		
v	0.30	0.35	0.40	0.45	0.50
1.25	2.96170	2.32219	1.86753	1.53125	1.27486
1.30	3.16907	2.47213	1.97969	1.61739	1.34240
1.35	3.39862	2.63735	2.10281	1.71163	1.41609
1.40	3.65292	2.81950	2.23802	1.81478	1.49651
1.45	3.93483	3.02046	2.38658	1.92772	1.58429
1.50	4.24761	3.24231	2.54991	2.05144	1.68017
1.55	4.59494	3.48741	2.72958	2.18706	1.78493
1.60	4.98098	3.75841	2.92738	2.33580	1.89946
1.65	5.41043	4.05830	3.14529	2.49905	2.02475
1.70	5.88866	4.39045	3.38555	2.67834	2.16189
1.75	6.42173	4.75865	3.65066	2.87540	2.31210
1.80	7.01652	5.16719	3.94344	3.09216	2.47675
1.85	7.68086	5.62092	4.26706	3.33077	2.65736
1.90	8.42367	6.12533	4.62508	3.59365	2.85562
1.95	9.25508	6.68660	5.02152	3.88350	3.07342
2.00	10.1866	7.31178	5.46089	4.20338	3.31288
2.05	11.2315	8.00882	5.94829	4.55669	3.57638
2.10	12.4048	8.78675	6.48949	4.94727	3.86655
2.15	13.7235	9.65584	7.09097	5.37943	4.18636
2.20	15.2075	10.6277	7.76008	5.85802	4.53914
2.25	16.8790	11.7156	8.50514	6.38851	4.92862
2.30	18.7638	12.9347	9.33553	6.97703	5.35898
2.35	20.8915	14.3020	10.2619	7.63054	5.83491
2.40	23.2956	15.8371	11.2964	8.35686	6.36169
2.45	26.0151	17.5624	12.4526	9.16483	6.94525
2.50	29.0944	19.5032	13.7461	10.0644	7.59228
2.55	32.5848	21.6888	15.1945	11.0669	8.31028
2.60	36.5451	24.1522	16.8180	12.1851	9.10772
2.65	41.0431	26.9315	18.6393	13.4335	9.99415
2.70	46.1569	30.0702	20.6845	14.8283	10.9803
2.75	51.9765	33.6182	22.9832	16.3883	12.0784
2.80	58.6059	37.6326	25.5690	18.1344	13.3022
2.85	66.1649	42.1790	28.4806	20.0906	14.6671
2.90	74.7924	47.3325	31.7618	22.2840	16.1907
2.95	84.6484	53.1796	35.4629	24.7456	17.8929
3.00	95.9187	59.8197	39.6413	27.5103	19.7962
3.05	108.818	67.3671	44.3626	30.6183	21.9261
3.10	123.595	75.9534	49.7019	34.1150	24.3115
3.15	140.539	85.7302	55.7453	38.0523	26.9852
3.20	159.985	96.8721	62.5914	42.4894	29.9844
3.25	182.321	109.581	70.3534	47.4938	33.3515
3.30	208.001	124.089	79.1611	53.1426	37.1345
3.35	237.549	140.665	89.1635	59.5238	41.3881

Table 4.7 (continued)

			t		
v	0.30	0.35	0.40	0.45	0.50
3.40	271.579	159.619	100.532	66.7381	46.1746
3.45	310.801	181.312	113.463	74.9008	51.5648
3.50	356.047	206.158	128.184	84.1436	57.6397
3.55	408.284	234.638	144.955	94.6176	64.4911
3.60	468.642	267.311	164.077	106.496	72.2242
3.65	538.439	304.823	185.896	119.976	80.9589
3.70	619.215	347.924	210.812	135.287	90.8322
3.75	712.770	397.484	239.284	152.690	102.000
3.80	821.211	454.517	271.847	172.484	114.643
3.85	947.002	520.196	309.113	195.015	128.964
3.90	1,093.03	595.890	351.795	220.680	145.198
3.95	1,262.67	683.189	400.714	249.934	163.613
4.00	1,459.90	783.945	456.822	283.305	184.517
4.05	1,689.37	900.318	521.221	321.397	208.263
4.10	1,956.53	1,034.82	595.187	364.908	235.255
4.15	2,267.81	1,190.39	680.201	414.643	265.958
4.20	2,630.74	1,370.45	777.980	471.530	300.904
4.25	3,054.18	1,579.00	890.516	536.642	340.707
4.30	3,548.58	1,820.72	1,020.12	611.218	386.070
4.35	4,126.21	2,101.05	1,169.49	696.687	437.805
4.40	4,801.54	2,426.39	1,341.74	794.705	496.843
4.45	5,591.62	2,804.21	1,540.52	907.188	564.259
4.50	6,516.55	3,243.27	1,770.04	1,036.35	641.288
4.55	7,600.07	3,753.82	2,035.25	1,184.76	729.358
4.60	8,870.17	4,347.87	2,341.87	1,355.39	830.112
4.65	10,359.9	5,039.52	2,696.59	1,551.68	945.446
4.70	12,108.4	5,845.31	3,107.22	1,777.64	1,077.55
4.75	14,161.9	6,784.65	3,582.86	2,037.91	1,228.95
4.80	16,575.1	7,880.34	4,134.12	2,337.87	1,402.58
4.85	19,412.7	9,159.18	4,773.42	2,683.79	1,601.80
4.90	22,751.3	10,652.7	5,515.26	3,082.93	1,830.53
4.95	26,681.7	12,397.9	6,376.58	3,543.76	2,093.29
5.00	31,311.7	14,438.4	7,377.20	4,076.10	2,395.31

			t		
v	0.55	0.60	0.65	0.70	0.75
0.00	0.581372	0.510730	0.450974	0.399984	0.356145
0.05	0.581957	0.511217	0.451384	0.400330	0.356440
0.10	0.583716	0.512681	0.452613	0.401371	0.357326
0.15	0.586659	0.515131	0.454670	0.403111	0.358809
0.20	0.590801	0.518578	0.457564	0.405559	0.360893
0.25	0.596167	0.523041	0.461310	0.408727	0.363590
0.30	0.602787	0.528545	0.465926	0.412630	0.366912

Table 4.7 (continued)

v	0.55	0.60	0.65	0.70	0.75
0.35	0.610697	0.535118	0.471438	0.417287	0.370873
0.40	0.619943	0.542796	0.477872	0.422721	0.375494
0.45	0.630578	0.551621	0.485262	0.428959	0.380796
0.50	0.642664	0.561642	0.493648	0.436031	0.386803
0.55	0.656271	0.572913	0.503071	0.443974	0.393545
0.60	0.671478	0.585497	0.513584	0.452828	0.401055
0.65	0.688376	0.599466	0.525241	0.462637	0.409368
0.70	0.707067	0.614898	0.538106	0.473452	0.418527
0.75	0.727664	0.631882	0.552250	0.485330	0.428576
0.80	0.750294	0.650517	0.567749	0.498332	0.439565
0.85	0.775098	0.670912	0.584691	0.512529	0.451552
0.90	0.802231	0.693190	0.603171	0.527995	0.464596
0.95	0.831868	0.717484	0.623295	0.544816	0.478767
1.00	0.864202	0.743944	0.645181	0.563085	0.494139
1.05	0.899444	0.772735	0.668957	0.582904	0.510793
1.10	0.937832	0.804036	0.694764	0.604384	0.528820
1.15	0.979625	0.838050	0.722760	0.627651	0.548319
1.20	1.02511	0.874997	0.753117	0.652839	0.569398
1.25	1.07461	0.915121	0.786023	0.680098	0.592176
1.30	1.12848	0.958692	0.821689	0.709593	0.616784
1.35	1.18710	1.00601	0.860343	0.741502	0.643364
1.40	1.25091	1.05739	0.902240	0.776025	0.672074
1.45	1.32039	1.11321	0.947657	0.813379	0.703086
1.50	1.39607	1.17387	0.996902	0.853804	0.736588
1.55	1.47853	1.23980	1.05031	0.897562	0.772788
1.60	1.56843	1.31150	1.10826	0.944944	0.811913
1.65	1.66649	1.38950	1.17117	0.996268	0.854214
1.70	1.77351	1.47440	1.23948	1.05189	0.899965
1.75	1.89037	1.56688	1.31370	1.11218	0.949467
1.80	2.01808	1.66765	1.39438	1.17758	1.00305
1.85	2.15772	1.77753	1.48214	1.24856	1.06109
1.90	2.31052	1.89743	1.57765	1.32563	1.12397
1.95	2.47784	2.02835	1.68167	1.40937	1.19215
2.00	2.66120	2.17139	1.79502	1.50041	1.26611
2.05	2.86229	2.32779	1.91863	1.59944	1.34638
2.10	3.08298	2.49893	2.05352	1.70723	1.43357
2.15	3.32538	2.68633	2.20081	1.82465	1.52832
2.20	3.59185	2.89168	2.36177	1.95264	1.63136
2.25	3.88499	3.11688	2.53778	2.09224	1.74349
2.30	4.20774	3.36405	2.73040	2.24461	1.86558
2.35	4.56339	3.63552	2.94136	2.41105	1.99862
2.40	4.95560	3.93393	3.17256	2.59297	2.14368
2.45	5.38849	4.26220	3.42614	2.79196	2.30196

Table 4.7 (continued)

			t		
v	0.55	0.60	0.65	0.70	0.75
2.50	5.86667	4.62361	3.70448	3.00978	2.47478
2.55	6.39532	5.02182	4.01023	3.24838	2.66362
2.60	6.98023	5.46091	4.34633	3.50995	2.87010
2.65	7.62793	5.94548	4.71610	3.79690	3.09603
2.70	8.34575	6.48065	5.12319	4.11192	3.34342
2.75	9.14193	7.07216	5.57174	4.45802	3.61450
2.80	10.0257	7.72648	6.06633	4.83854	3.91175
2.85	11.0076	8.45083	6.61210	5.25721	4.23792
2.90	12.0993	9.25334	7.21482	5.71820	4.59609
2.95	13.3141	10.1431	7.88091	6.22616	4.98967
3.00	14.6670	11.1305	8.61759	6.78627	5.42248
3.05	16.1748	12.2269	9.43298	7.40434	5.89876
3.10	17.8567	13.4454	10.3361	8.08688	6.42324
3.15	19.7341	14.8005	11.3372	8.84113	7.00122
3.20	21.8314	16.3089	12.4478	9.67525	7.63859
3.25	24.1763	17.9890	13.6806	10.5984	8.34195
3.30	26.7999	19.8617	15.0501	11.6206	9.11867
3.35	29.7376	21.9509	16.5726	12.7536	9.97699
3.40	33.0296	24.2831	18.2664	14.0101	10.9261
3.45	36.7212	26.8885	20.1522	15.4045	11.9765
3.50	40.8641	29.8013	22.2531	16.9531	13.1395
3.55	45.5169	33.0601	24.5954	18.6742	14.4283
3.60	50.7462	36.7086	27.2087	20.5882	15.8573
3.65	56.6276	40.7963	30.1263	22.7182	17.4428
3.70	63.2472	45.3793	33.3859	25.0902	19.2033
3.75	70.7029	50.5211	37.0300	27.7335	21.1592
3.80	79.1063	56.2940	41.1069	30.6810	23.3338
3.85	88.5845	62.7798	45.6709	33.9700	25.7529
3.90	99.2823	70.0715	50.7837	37.6424	28.4459
3.95	111.365	78.2748	56.5151	41.7456	31.4457
4.00	125.021	87.5098	62.9442	46.3330	34.7893
4.05	140.466	97.9131	70.1605	51.4652	38.5185
4.10	157.945	109.640	78.2657	57.2104	42.6803
4.15	177.740	122.868	87.3751	63.6459	47.3280
4.20	200.172	137.799	97.6197	70.8592	52.5212
4.25	225.609	154.662	109.148	78.9494	58.3278
4.30	254.472	173.720	122.129	88.0285	64.8239
4.35	287.243	195.272	136.755	98.2236	72.0959
4.40	324.475	219.659	153.245	109.679	80.2413
4.45	366.801	247.271	171.847	122.558	89.3705
4.50	414.949	278.555	192.844	137.046	99.6081
4.55	469.752	314.019	216.560	153.355	111.096
4.60	532.168	354.246	243.361	171.722	123.993

Table 4.7 (continued)

			t		
v	0.55	0.60	0.65	0.70	0.75
4.65	603.298	399.903	273.667	192.421	138.481
4.70	684.407	451.754	307.957	215.760	154.766
4.75	776.948	510.673	346.776	242.092	173.081
4.80	882.595	577.662	390.747	271.817	193.690
4.85	1,003.27	653.871	440.584	305.390	216.894
4.90	1,141.20	740.617	497.099	343.332	243.034
4.95	1,298.94	839.412	561.224	386.234	272.496
5.00	1,479.42	951.995	634.024	434.772	305.722

			t		
v	0.80	0.85	0.90	0.95	1.0
0.00	0.318206	0.285186	0.256299	0.230915	0.208518
0.05	0.318459	0.285403	0.256488	0.231079	0.208661
0.10	0.319219	0.286058	0.257054	0.231571	0.209091
0.15	0.320489	0.287152	0.258001	0.232394	0.209808
0.20	0.322274	0.288689	0.259331	0.233550	0.210816
0.25	0.324584	0.290678	0.261051	0.235044	0.212119
0.30	0.327428	0.293125	0.263168	0.236882	0.213722
0.35	0.330819	0.296043	0.265690	0.239072	0.215631
0.40	0.334771	0.299443	0.268628	0.241623	0.217854
0.45	0.339305	0.303340	0.271995	0.244544	0.220398
0.50	0.344439	0.307751	0.275804	0.247848	0.223275
0.55	0.350197	0.312697	0.280072	0.251548	0.226496
0.60	0.356607	0.318198	0.284818	0.255659	0.230073
0.65	0.363697	0.324280	0.290061	0.260199	0.234020
0.70	0.371503	0.330970	0.295824	0.265188	0.238355
0.75	0.380060	0.338299	0.302134	0.270645	0.243095
0.80	0.389410	0.346301	0.309017	0.276594	0.248259
0.85	0.399598	0.355012	0.316505	0.283061	0.253869
0.90	0.410675	0.364474	0.324632	0.290075	0.259948
0.95	0.422695	0.374733	0.333435	0.297666	0.266522
1.00	0.435720	0.385838	0.342955	0.305868	0.273621
1.05	0.449815	0.397842	0.353236	0.314718	0.281274
1.10	0.465053	0.410806	0.364328	0.324257	0.289515
1.15	0.481515	0.424794	0.376284	0.334528	0.298381
1.20	0.499288	0.439879	0.389162	0.345581	0.307913
1.25	0.518467	0.456136	0.403026	0.357467	0.318153
1.30	0.539157	0.473652	0.417945	0.370245	0.329150
1.35	0.561474	0.492520	0.433996	0.383975	0.340954
1.40	0.585542	0.512841	0.451261	0.398727	0.353623
1.45	0.611500	0.534725	0.469831	0.414575	0.367217
1.50	0.639497	0.558295	0.489804	0.431598	0.381804

Table 4.7 (continued)

			t		
v	0.80	0.85	0.90	0.95	1.0
1.55	0.669700	0.583684	0.511287	0.449886	0.397455
1.60	0.702288	0.611035	0.534399	0.469535	0.414250
1.65	0.737461	0.640509	0.559268	0.490648	0.432275
1.70	0.775435	0.672279	0.586035	0.513341	0.451623
1.75	0.816449	0.706535	0.614852	0.537737	0.472396
1.80	0.860765	0.743486	0.645887	0.563974	0.494706
1.85	0.908669	0.783360	0.679324	0.592199	0.518673
1.90	0.960477	0.826407	0.715363	0.622575	0.544431
1.95	1.01654	0.872903	0.754224	0.655279	0.572123
2.00	1.07723	0.923148	0.796148	0.690505	0.601907
2.05	1.14297	0.977474	0.841399	0.728465	0.633955
2.10	1.21422	1.03624	0.890265	0.769391	0.668455
2.15	1.29150	1.09985	0.943063	0.813537	0.705611
2.20	1.37535	1.16875	1.00014	0.861182	0.745649
2.25	1.46641	1.24341	1.06188	0.912630	0.788815
2.30	1.56534	1.32436	1.12870	0.968216	0.835376
2.35	1.67290	1.41219	1.20107	1.02831	0.885628
2.40	1.78992	1.50755	1.27948	1.09330	0.939893
2.45	1.91731	1.61115	1.36450	1.16365	0.998524
2.50	2.05608	1.72377	1.45674	1.23983	1.06191
2.55	2.20737	1.84627	1.55688	1.32238	1.13047
2.60	2.37239	1.97961	1.66565	1.41188	1.20468
2.65	2.55254	2.12484	1.78389	1.50898	1.28505
2.70	2.74932	2.28313	1.91250	1.61439	1.37215
2.75	2.96442	2.45577	2.05247	1.72890	1.46658
2.80	3.19971	2.64419	2.20490	1.85336	1.56903
2.85	3.45725	2.84996	2.37103	1.98873	1.68026
2.90	3.73936	3.07483	2.55219	2.13605	1.80108
2.95	4.04858	3.32075	2.74987	2.29649	1.93241
3.00	4.38776	3.58985	2.96572	2.47132	2.07524
3.05	4.76006	3.88454	3.20157	2.66194	2.23069
3.10	5.16898	4.20744	3.45943	2.86993	2.39997
3.15	5.61844	4.56150	3.74154	3.09701	2.58442
3.20	6.11279	4.94999	4.05039	3.34509	2.78552
3.25	6.65690	5.37653	4.38872	3.61627	3.00493
3.30	7.25617	5.84517	4.75959	3.91290	3.24445
3.35	7.91665	6.36040	5.16640	4.23758	3.50608
3.40	8.64507	6.92722	5.61291	4.59317	3.79204
3.45	9.44896	7.55121	6.10332	4.98287	4.10480
3.50	10.3367	8.23859	6.64227	5.41021	4.44705
3.55	11.3178	8.99627	7.23495	5.87912	4.82183
3.60	12.4026	9.83199	7.88714	6.39396	5.23247
3.65	13.6031	10.7544	8.60525	6.95959	5.68266

Table 4.7 (continued)

			t		
v	0.80	0.85	0.90	0.95	1.0
3.70	14.9322	11.7731	9.39644	7.58139	6.17653
3.75	16.4049	12.8989	10.2687	8.26535	6.71862
3.80	18.0376	14.1438	11.2309	9.01815	7.31401
3.85	19.8489	15.5213	12.2930	9.84723	7.96832
3.90	21.8595	17.0464	13.4662	10.7608	8.68781
3.95	24.0930	18.7361	14.7626	11.7682	9.47944
4.00	26.5754	20.6092	16.1963	12.8797	10.3509
4.05	29.3362	22.6868	17.7827	14.1067	11.3109
4.10	32.4086	24.9929	19.5391	15.4620	12.3690
4.15	35.8298	27.5539	21.4849	16.9599	13.5359
4.20	39.6418	30.3999	23.6417	18.6164	14.8235
4.25	43.8917	33.5642	26.0339	20.4494	16.2451
4.30	48.6326	37.0848	28.6886	22.4788	17.8155
4.35	53.9244	41.0038	31.6364	24.7269	19.5514
4.40	59.8347	45.3691	34.9115	27.2187	21.4711
4.45	66.4395	50.2341	38.5524	29.9821	23.5953
4.50	73.8247	55.6593	42.6021	33.0486	25.9471
4.55	82.0874	61.7124	47.1091	36.4532	28.5524
4.60	91.3370	68.4701	52.1278	40.2352	31.4399
4.65	101.697	76.0186	57.7194	44.4388	34.6421
4.70	113.308	84.4550	63.9527	49.1135	38.1950
4.75	126.328	93.8889	70.9052	54.3149	42.1392
4.80	140.935	104.444	78.6640	60.1055	46.5202
4.85	157.333	116.261	87.3272	66.5553	51.3887
4.90	175.750	129.496	97.0054	73.7433	56.8020
4.95	196.447	144.328	107.823	81.7581	62.8240
5.00	219.719	160.958	119.921	90.6995	69.5267

			t		
v	1.5	2.0	3.0	4.0	5.0
0.00	0.082372	0.036178	0.008236	0.002120	0.000586
0.05	0.082414	0.036193	0.008239	0.002121	0.000586
0.10	0.082542	0.036238	0.008246	0.002122	0.000587
0.15	0.082756	0.036314	0.008259	0.002125	0.000587
0.20	0.083056	0.036421	0.008277	0.002129	0.000588
0.25	0.083443	0.036558	0.008300	0.002133	0.000589
0.30	0.083918	0.036727	0.008328	0.002139	0.000591
0.35	0.084483	0.036927	0.008361	0.002146	0.000592
0.40	0.085140	0.037160	0.008399	0.002154	0.000594
0.45	0.085890	0.037425	0.008443	0.002163	0.000596
0.50	0.086736	0.037723	0.008493	0.002173	0.000598
0.55	0.087679	0.038056	0.008547	0.002184	0.000601

Table 4.7 (continued)

			t		
v	1.5	2.0	3.0	4.0	5.0
0.60	0.088724	0.038423	0.008608	0.002196	0.000604
0.65	0.089873	0.038826	0.008674	0.002209	0.000607
0.70	0.091130	0.039266	0.008746	0.002224	0.000610
0.75	0.092498	0.039744	0.008824	0.002240	0.000614
0.80	0.093982	0.040260	0.008908	0.002257	0.000617
0.85	0.095586	0.040817	0.008998	0.002275	0.000622
0.90	0.097316	0.041416	0.009095	0.002294	0.000626
0.95	0.099176	0.042058	0.009198	0.002315	0.000631
1.00	0.101174	0.042745	0.009308	0.002337	0.000636
1.05	0.103315	0.043479	0.009425	0.002360	0.000641
1.10	0.105606	0.044261	0.009550	0.002385	0.000647
1.15	0.108055	0.045095	0.009681	0.002411	0.000652
1.20	0.110671	0.045981	0.009821	0.002439	0.000659
1.25	0.113461	0.046922	0.009969	0.002468	0.000665
1.30	0.116435	0.047921	0.010124	0.002498	0.000672
1.35	0.119604	0.048981	0.010289	0.002530	0.000679
1.40	0.122979	0.050104	0.010462	0.002564	0.000687
1.45	0.126571	0.051294	0.010645	0.002600	0.000695
1.50	0.130394	0.052554	0.010837	0.002637	0.000703
1.55	0.134461	0.053887	0.011039	0.002676	0.000712
1.60	0.138787	0.055298	0.011251	0.002717	0.000721
1.65	0.143388	0.056791	0.011475	0.002760	0.000730
1.70	0.148281	0.058369	0.011710	0.002805	0.000740
1.75	0.153485	0.060038	0.011956	0.002852	0.000750
1.80	0.159019	0.061803	0.012215	0.002901	0.000761
1.85	0.164907	0.063669	0.012487	0.002952	0.000772
1.90	0.171170	0.065642	0.012772	0.003006	0.000784
1.95	0.177834	0.067728	0.013071	0.003062	0.000796
2.00	0.184925	0.069933	0.013386	0.003121	0.000809
2.05	0.192474	0.072265	0.013715	0.003182	0.000822
2.10	0.200511	0.074731	0.014061	0.003246	0.000836
2.15	0.209070	0.077340	0.014424	0.003313	0.000850
2.20	0.218189	0.080100	0.014804	0.003382	0.000865
2.25	0.227906	0.083020	0.015204	0.003455	0.000881
2.30	0.238265	0.086110	0.015622	0.003531	0.000897
2.35	0.249312	0.089382	0.016062	0.003610	0.000914
2.40	0.261097	0.092845	0.016523	0.003693	0.000931
2.45	0.273673	0.096513	0.017007	0.003779	0.000949
2.50	0.287100	0.100399	0.017515	0.003870	0.000968
2.55	0.301440	0.104517	0.018049	0.003964	0.000988
2.60	0.316763	0.108882	0.018609	0.004062	0.001008
2.65	0.333142	0.113510	0.019197	0.004164	0.001029
2.70	0.350659	0.118419	0.019814	0.004272	0.001051

Table 4.7 (continued)

			t		
v	1.5	2.0	3.0	4.0	5.0
2.75	0.369400	0.123628	0.020462	0.004383	0.001074
2.80	0.389460	0.129156	0.021144	0.004500	0.001098
2.85	0.410942	0.135026	0.021860	0.004622	0.001123
2.90	0.433957	0.141260	0.022612	0.004749	0.001149
2.95	0.458626	0.147884	0.023403	0.004883	0.001176
3.00	0.485081	0.154925	0.024235	0.005022	0.001204
3.05	0.513464	0.162411	0.025110	0.005167	0.001233
3.10	0.543931	0.170375	0.026031	0.005319	0.001263
3.15	0.576651	0.178849	0.026999	0.005477	0.001295
3.20	0.611807	0.187870	0.028019	0.005643	0.001328
3.25	0.649600	0.197477	0.029092	0.005817	0.001362
3.30	0.690247	0.207713	0.030222	0.005998	0.001397
3.35	0.733985	0.218622	0.031413	0.006188	0.001434
3.40	0.781073	0.230255	0.032668	0.006387	0.001473
3.45	0.831793	0.242663	0.033990	0.006594	0.001513
3.50	0.886452	0.255904	0.035385	0.006812	0.001555
3.55	0.945386	0.270040	0.036855	0.007040	0.001598
3.60	1.00896	0.285138	0.038406	0.007278	0.001644
3.65	1.07757	0.301270	0.040043	0.007528	0.001691
3.70	1.15167	0.318513	0.041771	0.007789	0.001740
3.75	1.23171	0.336953	0.043596	0.008063	0.001791
3.80	1.31824	0.356681	0.045523	0.008351	0.001845
3.85	1.41180	0.377795	0.047559	0.008652	0.001901
3.90	1.51305	0.400403	0.049711	0.008968	0.001959
3.95	1.62264	0.424620	0.051986	0.009299	0.002020
4.00	1.74134	0.450573	0.054392	0.009646	0.002083
4.05	1.86996	0.478397	0.056938	0.010011	0.002149
4.10	2.00941	0.508240	0.059632	0.010394	0.002218
4.15	2.16066	0.540262	0.062483	0.010796	0.002290
4.20	2.32480	0.574638	0.065503	0.011217	0.002365
4.25	2.50301	0.611555	0.068702	0.011661	0.002444
4.30	2.69660	0.651219	0.072091	0.012126	0.002526
4.35	2.90699	0.693852	0.075684	0.012616	0.002612
4.40	3.13576	0.739696	0.079493	0.013130	0.002701
4.45	3.38462	0.789013	0.083534	0.013671	0.002795
4.50	3.65548	0.842090	0.087821	0.014240	0.002893
4.55	3.95042	0.899237	0.092372	0.014838	0.002995
4.60	4.27173	0.960792	0.097203	0.015468	0.003103
4.65	4.62194	1.02712	0.102335	0.016131	0.003215
4.70	5.00384	1.09863	0.107787	0.016829	0.003332
4.75	5.42047	1.17575	0.113581	0.017564	0.003455
4.80	5.87522	1.25896	0.119742	0.018339	0.003583
4.85	6.37181	1.34878	0.126293	0.019155	0.003718

Table 4.7 (continued)

			t		
v	1.5	2.0	3.0	4.0	5.0
4.90	6.91433	1.44577	0.133264	0.020015	0.003858
4.95	7.50731	1.55054	0.140682	0.020922	0.004006
5.00	8.15575	1.66378	0.148579	0.021879	0.004161

Table 4.8: First derivatives with respect to the order of the integral modified Bessel functions of the second kind $\partial Ki_v(t)/\partial v$.

			t		
v	0.05	0.10	0.15	0.20	0.25
0.00	0.000000	0.000000	0.000000	0.000000	0.000000
0.05	0.714575	0.334016	0.200366	0.134345	0.096096
0.10	1.44201	0.672514	0.402963	0.269990	0.193021
0.15	2.19549	1.02007	0.610057	0.408253	0.291617
0.20	2.98891	1.38143	0.823990	0.550489	0.392745
0.25	3.83723	1.76162	1.04722	0.698116	0.497302
0.30	4.75685	2.16607	1.28236	0.852625	0.606227
0.35	5.76615	2.60066	1.53223	1.01561	0.720521
0.40	6.88594	3.07193	1.79990	1.18880	0.841252
0.45	8.14008	3.58716	2.08873	1.37407	0.969579
0.50	9.55622	4.15455	2.40247	1.57348	1.10676
0.55	11.1666	4.78342	2.74532	1.78929	1.25418
0.60	13.0092	5.48440	3.12195	2.02405	1.41336
0.65	15.1287	6.26970	3.53769	2.28058	1.58598
0.70	17.5780	7.15339	3.99856	2.56204	1.77392
0.75	20.4201	8.15174	4.51142	2.87200	1.97928
0.80	23.7304	9.28365	5.08411	3.21451	2.20440
0.85	27.5985	10.5711	5.72559	3.59413	2.45192
0.90	32.1322	12.0397	6.44618	4.01605	2.72481
0.95	37.4609	13.7195	7.25774	4.48617	3.02642
1.00	43.7402	15.6455	8.17392	5.01123	3.36054
1.05	51.1578	17.8588	9.21050	5.59891	3.73145
1.10	59.9402	20.4079	10.3857	6.25801	4.14402
1.15	70.3614	23.3494	11.7208	6.99861	4.60378
1.20	82.7534	26.7506	13.2401	7.83226	5.11702
1.25	97.5186	30.6903	14.9722	8.77222	5.69089
1.30	115.146	35.2619	16.9501	9.83377	6.33356
1.35	136.231	40.5759	19.2124	11.0345	7.05434
1.40	161.497	46.7629	21.8039	12.3945	7.86388
1.45	191.830	53.9780	24.7769	13.9373	8.77436
1.50	228.308	62.4051	28.1927	15.6898	9.79971

Table 4.8 (continued)

			t		
ν	0.05	0.10	0.15	0.20	0.25
1.55	272.251	72.2628	32.1226	17.6832	10.9559
1.60	325.278	83.8111	36.6503	19.9535	12.2613
1.65	389.369	97.3598	41.8738	22.5424	13.7369
1.70	466.960	113.278	47.9079	25.4984	15.4069
1.75	561.040	132.007	54.8874	28.8777	17.2990
1.80	675.291	154.072	62.9708	32.7454	19.4454
1.85	814.247	180.103	72.3444	37.1774	21.8827
1.90	983.502	210.854	83.2274	42.2618	24.6537
1.95	1,189.96	247.229	95.8781	48.1014	27.8073
2.00	1,442.16	290.311	110.601	54.8158	31.4001
2.05	1,750.68	341.401	127.757	62.5445	35.4977
2.10	2,128.59	402.063	147.769	71.4506	40.1758
2.15	2,592.15	474.178	171.140	81.7244	45.5221
2.20	3,161.51	560.014	198.465	93.5885	51.6382
2.25	3,861.72	662.301	230.449	107.303	58.6418
2.30	4,723.97	784.335	267.925	123.174	66.6698
2.35	5,787.08	930.095	311.885	141.558	75.8807
2.40	7,099.47	1,104.39	363.505	162.875	86.4592
2.45	8,721.53	1,313.05	424.184	187.618	98.6197
2.50	10,728.7	1,563.10	495.585	216.364	112.612
2.55	13,215.4	1,863.10	579.687	249.795	128.727
2.60	16,299.7	2,223.40	678.851	288.711	147.303
2.65	20,129.4	2,656.57	795.890	334.054	168.737
2.70	24,890.1	3,177.89	934.160	386.936	193.489
2.75	30,814.4	3,805.95	1,097.67	448.667	222.099
2.80	38,194.8	4,563.35	1,291.22	520.792	255.196
2.85	47,398.8	5,477.64	1,520.53	605.140	293.519
2.90	58,888.8	6,582.38	1,792.46	703.867	337.929
2.95	73,247.4	7,918.52	2,115.24	819.527	389.438
3.00	91,208.6	9,536.09	2,498.72	955.142	449.229
3.05	113699	11,496.2	2,954.71	1,114.29	518.693
3.10	141887	13,873.4	3,497.42	1,301.22	599.460
3.15	177250	16,759.4	4,143.88	1,520.95	693.445
3.20	221658	20,265.9	4,914.59	1,779.46	802.898
3.25	277473	24,530.3	5,834.23	2,083.85	930.467
3.30	347693	29,720.7	6,932.46	2,442.53	1,079.27
3.35	436111	36,043.8	8,245.07	2,865.55	1,252.96
3.40	547546	43,753.1	9,815.17	3,364.81	1,455.88
3.45	688109	53,160.6	11,694.8	3,954.53	1,693.11
3.50	865568	64,649.5	13,946.8	4,651.64	1,970.66
3.55	1.090E+6	78,692.0	16,646.9	5,476.32	2,295.62
3.60	1.373	95,869.3	19,886.9	6,452.64	2,676.39
3.65	1.732	116898	23,777.8	7,609.34	3,122.85

Table 4.8 (continued)

			t		
v	0.05	0.10	0.15	0.20	0.25
3.70	2.187	142661	28,453.8	8,980.77	3,646.71
3.75	2.763	174249	34,077.5	10,608.0	4,261.85
3.80	3.494	213010	40,846.0	12,540.0	4,984.65
3.85	4.423	260605	48,998.2	14,835.6	5,834.56
3.90	5.602	319094	58,824.1	17,565.2	6,834.61
3.95	7.102	391022	70,675.8	20,812.9	8,012.12
4.00	9.011	479540	84,980.9	24,679.9	9,399.51
4.05	1.144E+7	588553	102259	29,287.3	11,035.3
4.10	1.454	722901	123144	34,780.6	12,965.2
4.15	1.849	888587	148403	41,334.4	15,243.5
4.20	2.353	1.093E+6	178975	49,158.6	17,935.0
4.25	2.998	1.346	216002	58,505.7	21,116.6
4.30	3.821	1.658	260874	69,679.0	24,879.8
4.35	4.873	2.044	315292	83,044.0	29,333.8
4.40	6.221	2.521	381326	99,040.6	34,608.6
4.45	7.946	3.112	461508	118199	40,859.3
4.50	1.016E+8	3.845	558931	141158	48,271.0
4.55	1.299	4.753	677374	168688	57,064.6
4.60	1.663	5.880	821461	201720	67,503.8
4.65	2.130	7.279	996853	241377	79,903.9
4.70	2.730	9.016	1.210E+06	289016	94,641.9
4.75	3.502	1.118E+7	1.471	346276	112168
4.80	4.495	1.386	1.788	415140	133023
4.85	5.772	1.720	2.176	498008	157852
4.90	7.418	2.136	2.649	597784	187429
4.95	9.538	2.655	3.226	717986	222682
5.00	1.227E+9	3.301	3.932	862875	264722

			t		
v	0.30	0.35	0.40	0.45	0.50
0.00	0.000000	0.000000	0.000000	0.000000	0.000000
0.05	0.071742	0.055227	0.043509	0.034907	0.028419
0.10	0.144046	0.110852	0.087310	0.070032	0.057007
0.15	0.217483	0.167279	0.131697	0.105598	0.085930
0.20	0.292634	0.224919	0.176971	0.141828	0.115363
0.25	0.370105	0.284197	0.223441	0.178954	0.145482
0.30	0.450525	0.345556	0.271429	0.217217	0.176470
0.35	0.534562	0.409464	0.321273	0.256868	0.208517
0.40	0.622927	0.476414	0.373331	0.298172	0.241826
0.45	0.716384	0.546937	0.427981	0.341410	0.276608
0.50	0.815759	0.621602	0.485633	0.386882	0.313089
0.55	0.921953	0.701026	0.546726	0.434910	0.351512

Table 4.8 (continued)

			t		
v	0.30	0.35	0.40	0.45	0.50
0.60	1.03595	0.785882	0.611736	0.485842	0.392137
0.65	1.15884	0.876906	0.681182	0.540057	0.435246
0.70	1.29181	0.974905	0.755632	0.597965	0.481144
0.75	1.43621	1.08077	0.835708	0.660015	0.530164
0.80	1.59349	1.19549	0.922096	0.726699	0.582668
0.85	1.76533	1.32015	1.01555	0.798560	0.639054
0.90	1.95355	1.45598	1.11691	0.876192	0.699758
0.95	2.16024	1.60433	1.22711	0.960256	0.765261
1.00	2.38771	1.76670	1.34717	1.05148	0.836091
1.05	2.63857	1.94480	1.47825	1.15067	0.912833
1.10	2.91577	2.14053	1.62162	1.25872	0.996133
1.15	3.22263	2.35600	1.77872	1.37663	1.08671
1.20	3.56291	2.59360	1.95114	1.50551	1.18535
1.25	3.94084	2.85604	2.14067	1.64660	1.29294
1.30	4.36123	3.14633	2.34931	1.80127	1.41046
1.35	4.82953	3.46788	2.57932	1.97108	1.53899
1.40	5.35192	3.82455	2.83322	2.15773	1.67976
1.45	5.93544	4.22069	3.11385	2.36315	1.83412
1.50	6.58805	4.66120	3.42439	2.58952	2.00357
1.55	7.31886	5.15166	3.76845	2.83925	2.18981
1.60	8.13821	5.69836	4.15007	3.11505	2.39472
1.65	9.05790	6.30844	4.57382	3.41998	2.62040
1.70	10.0914	6.99000	5.04485	3.75747	2.86923
1.75	11.2541	7.75224	5.56900	4.13138	3.14385
1.80	12.5636	8.60560	6.15284	4.54604	3.44721
1.85	14.0399	9.56197	6.80382	5.00634	3.78266
1.90	15.7061	10.6349	7.53038	5.51780	4.15393
1.95	17.5886	11.8397	8.34207	6.08663	4.56522
2.00	19.7176	13.1939	9.24972	6.71984	5.02124
2.05	22.1278	14.7177	10.2656	7.42537	5.52731
2.10	24.8590	16.4338	11.4038	8.21217	6.08941
2.15	27.9571	18.3682	12.6800	9.09037	6.71426
2.20	31.4746	20.5510	14.1123	10.0714	7.40947
2.25	35.4722	23.0160	15.7212	11.1684	8.18359
2.30	40.0196	25.8024	17.5301	12.3959	9.04628
2.35	45.1974	28.9549	19.5655	13.7706	10.0085
2.40	51.0982	32.5248	21.8579	15.3116	11.0825
2.45	57.8293	36.5707	24.4419	17.0403	12.2823
2.50	65.5143	41.1604	27.3569	18.9811	13.6237
2.55	74.2963	46.3711	30.6481	21.1619	15.1245
2.60	84.3406	52.2920	34.3671	23.6142	16.8050
2.65	95.8386	59.0256	38.5729	26.3741	18.6881
2.70	109.012	66.6898	43.3331	29.4825	20.7999

Table 4.8 (continued)

			t		
v	0.30	0.35	0.40	0.45	0.50
2.75	124.118	75.4203	48.7250	32.9861	23.1698
2.80	141.455	85.3736	54.8373	36.9382	25.8316
2.85	161.369	96.7301	61.7716	41.3997	28.8232
2.90	184.261	109.698	69.6447	46.4400	32.1881
2.95	210.599	124.518	78.5904	52.1383	35.9755
3.00	240.926	141.467	88.7626	58.5853	40.2415
3.05	275.873	160.867	100.338	65.8849	45.0502
3.10	316.177	183.089	113.521	74.1557	50.4742
3.15	362.695	208.562	128.544	83.5337	56.5968
3.20	416.427	237.785	145.678	94.1747	63.5126
3.25	478.540	271.335	165.234	106.257	71.3298
3.30	550.394	309.879	187.569	119.986	80.1719
3.35	633.580	354.196	213.096	135.597	90.1801
3.40	729.958	405.185	242.294	153.360	101.516
3.45	841.700	463.894	275.712	173.585	114.364
3.50	971.351	531.539	313.988	196.630	128.935
3.55	1,121.89	609.535	357.857	222.904	145.472
3.60	1,296.80	699.529	408.172	252.881	164.252
3.65	1,500.19	803.437	465.917	287.105	185.592
3.70	1,736.84	923.494	532.234	326.202	209.858
3.75	2,012.38	1,062.30	608.447	370.896	237.468
3.80	2,333.44	1,222.90	696.090	422.021	268.903
3.85	2,707.78	1,408.82	796.941	480.540	304.714
3.90	3,144.54	1,624.21	913.066	547.563	345.537
3.95	3,654.45	1,873.91	1,046.86	624.375	392.101
4.00	4,250.18	2,163.55	1,201.12	712.459	445.245
4.05	4,946.60	2,499.74	1,379.08	813.533	505.937
4.10	5,761.26	2,890.21	1,584.50	929.581	575.289
4.15	6,714.86	3,344.01	1,821.77	1,062.90	654.585
4.20	7,831.78	3,871.73	2,096.00	1,216.16	745.302
4.25	9,140.81	4,485.80	2,413.14	1,392.44	849.148
4.30	10,675.9	5,200.77	2,780.11	1,595.33	968.089
4.35	12,477.4	6,033.74	3,204.99	1,828.96	1,104.40
4.40	14,592.5	7,004.73	3,697.23	2,098.15	1,260.70
4.45	17,077.5	8,137.31	4,267.82	2,408.50	1,440.03
4.50	19,998.7	9,459.13	4,929.61	2,766.50	1,645.90
4.55	23,434.9	11,002.7	5,697.63	3,179.70	1,882.36
4.60	27,479.0	12,806.3	6,589.42	3,656.88	2,154.10
4.65	32,241.4	14,914.8	7,625.52	4,208.24	2,466.56
4.70	37,853.0	17,381.3	8,829.94	4,845.66	2,826.04
4.75	44,468.8	20,268.1	10,230.8	5,582.99	3,239.84
4.80	52,272.9	23,648.7	11,861.0	6,436.34	3,716.40
4.85	61,484.0	27,609.7	13,759.2	7,424.48	4,265.56

Table 4.8 (continued)

			t		
v	0.30	0.35	0.40	0.45	0.50
4.90	72,361.6	32,253.4	15,970.5	8,569.32	4,898.68
4.95	85,214.2	37,700.3	18,548.1	9,896.40	5,628.98
5.00	100409	44,092.7	21,554.1	11,435.5	6,471.83

			t		
v	0.55	0.60	0.65	0.70	0.75
0.00	0.000000	0.000000	0.000000	0.000000	0.000000
0.05	0.023421	0.019500	0.016378	0.013860	0.011808
0.10	0.046972	0.039102	0.032838	0.027787	0.023670
0.15	0.070785	0.058912	0.049464	0.041848	0.035641
0.20	0.094995	0.079036	0.066340	0.056112	0.047778
0.25	0.119739	0.099581	0.083554	0.070647	0.060137
0.30	0.145159	0.120659	0.101194	0.085528	0.072777
0.35	0.171403	0.142387	0.119353	0.100827	0.085758
0.40	0.198625	0.164886	0.138127	0.116622	0.099143
0.45	0.226990	0.188285	0.157618	0.132996	0.112999
0.50	0.256671	0.212719	0.177934	0.150033	0.127394
0.55	0.287855	0.238332	0.199187	0.167825	0.142403
0.60	0.320738	0.265278	0.221500	0.186468	0.158102
0.65	0.355537	0.293724	0.245003	0.206066	0.174576
0.70	0.392482	0.323848	0.269836	0.226731	0.191913
0.75	0.431824	0.355844	0.296149	0.248580	0.210209
0.80	0.473838	0.389920	0.324107	0.271744	0.229565
0.85	0.518821	0.426306	0.353885	0.296361	0.250095
0.90	0.567100	0.465250	0.385677	0.322583	0.271917
0.95	0.619033	0.507024	0.419693	0.350575	0.295162
1.00	0.675013	0.551926	0.456162	0.380515	0.319971
1.05	0.735473	0.600283	0.495336	0.412598	0.346499
1.10	0.800891	0.652454	0.537489	0.447040	0.374914
1.15	0.871791	0.708834	0.582923	0.484072	0.405399
1.20	0.948757	0.769858	0.631969	0.523953	0.438156
1.25	1.03243	0.836008	0.684993	0.566962	0.473403
1.30	1.12353	0.907814	0.742396	0.613410	0.511381
1.35	1.22284	0.985862	0.804621	0.663635	0.552355
1.40	1.33125	1.07080	0.872158	0.718012	0.596615
1.45	1.44972	1.16335	0.945546	0.776952	0.644480
1.50	1.57934	1.26431	1.02538	0.840911	0.696299
1.55	1.72133	1.37456	1.11233	0.910390	0.752461
1.60	1.87702	1.49509	1.20711	0.985942	0.813390
1.65	2.04792	1.62699	1.31055	1.06818	0.879555
1.70	2.23570	1.77146	1.42353	1.15778	0.951475
1.75	2.44223	1.92987	1.54707	1.25550	1.02972

Table 4.8 (continued)

			t		
v	0.55	0.60	0.65	0.70	0.75
1.80	2.66960	2.10372	1.68226	1.36216	1.11492
1.85	2.92014	2.29469	1.83034	1.47868	1.20779
1.90	3.19646	2.50464	1.99268	1.60608	1.30908
1.95	3.50148	2.73568	2.17080	1.74551	1.41965
2.00	3.83849	2.99011	2.36640	1.89821	1.54046
2.05	4.21115	3.27057	2.58137	2.06558	1.67256
2.10	4.62358	3.57995	2.81781	2.24919	1.81710
2.15	5.08040	3.92152	3.07809	2.45075	1.97539
2.20	5.58681	4.29893	3.36482	2.67219	2.14885
2.25	6.14864	4.71626	3.68093	2.91566	2.33909
2.30	6.77246	5.17811	4.02971	3.18355	2.54788
2.35	7.46565	5.68960	4.41481	3.47852	2.77719
2.40	8.23653	6.25651	4.84033	3.80354	3.02922
2.45	9.09446	6.88531	5.31086	4.16194	3.30641
2.50	10.0500	7.58327	5.83155	4.55741	3.61148
2.55	11.1151	8.35859	6.40814	4.99411	3.94748
2.60	12.3031	9.22046	7.04709	5.47666	4.31777
2.65	13.6293	10.17920	7.75566	6.01024	4.72615
2.70	15.1109	11.24660	8.54198	6.60067	5.17684
2.75	16.7671	12.43560	9.41518	7.25443	5.67453
2.80	18.6200	13.76120	10.3855	7.97880	6.22450
2.85	20.6944	15.24000	11.4646	8.78196	6.83264
2.90	23.0184	16.89080	12.6653	9.67305	7.50552
2.95	25.6239	18.73510	14.0023	10.6624	8.25054
3.00	28.5471	20.79680	15.4922	11.7614	9.07593
3.05	31.8290	23.10310	17.1533	12.9832	9.99096
3.10	35.5160	25.68490	19.0067	14.3423	11.0060
3.15	39.6612	28.57680	21.0760	15.8551	12.1327
3.20	44.3245	31.81850	23.3877	17.5400	13.3841
3.25	49.5742	35.45440	25.9720	19.4179	14.7748
3.30	55.4882	39.53530	28.8629	21.5121	16.3214
3.35	62.1547	44.11860	32.0988	23.8489	18.0422
3.40	69.6746	49.26940	35.7230	26.4582	19.9583
3.45	78.1625	55.06170	39.7849	29.3734	22.0928
3.50	87.7493	61.57940	44.3400	32.6325	24.4723
3.55	98.5843	68.91820	49.4513	36.2781	27.1262
3.60	110.838	77.18650	55.1903	40.3585	30.0882
3.65	124.704	86.50790	61.6381	44.9284	33.3958
3.70	140.405	97.02290	68.8864	50.0494	37.0915
3.75	158.195	108.892	77.0396	55.7915	41.2232
3.80	178.364	122.296	86.2160	62.2336	45.8450
3.85	201.245	137.445	96.5502	69.4653	51.0180
3.90	227.217	154.575	108.195	77.5881	56.8112

Table 4.8 (continued)

			t		
v	0.55	0.60	0.65	0.70	0.75
3.95	256.716	173.956	121.324	86.7169	63.3026
4.00	290.241	195.898	136.136	96.9821	70.5804
4.05	328.364	220.753	152.855	108.532	78.7443
4.10	371.741	248.923	171.737	121.534	87.9072
4.15	421.125	280.870	193.075	136.179	98.1971
4.20	477.380	317.121	217.202	152.684	109.759
4.25	541.498	358.279	244.496	171.295	122.756
4.30	614.621	405.032	275.392	192.292	137.376
4.35	698.061	458.173	310.382	215.994	153.829
4.40	793.324	518.607	350.032	242.764	172.355
4.45	902.149	587.373	394.986	273.015	193.226
4.50	1,026.53	665.660	445.981	307.217	216.750
4.55	1,168.78	754.836	503.859	345.907	243.280
4.60	1,331.54	856.469	569.584	389.697	273.214
4.65	1,517.87	972.361	644.258	439.283	307.006
4.70	1,731.30	1,104.58	729.144	495.462	345.173
4.75	1,975.91	1,255.51	825.688	559.142	388.303
4.80	2,256.38	1,427.88	935.547	631.363	437.065
4.85	2,578.15	1,624.85	1,060.62	713.310	492.221
4.90	2,947.49	1,850.02	1,203.09	806.339	554.642
4.95	3,371.64	2,107.59	1,365.45	912.000	625.318
5.00	3,859.00	2,402.34	1,550.58	1,032.07	705.379

			t		
v	0.80	0.85	0.90	0.95	1.0
0.00	0.000000	0.000000	0.000000	0.000000	0.000000
0.05	0.010118	0.008716	0.007543	0.006556	0.005720
0.10	0.020281	0.017469	0.015117	0.013138	0.011461
0.15	0.030534	0.026297	0.022754	0.019772	0.017247
0.20	0.040924	0.035237	0.030484	0.026485	0.023099
0.25	0.051495	0.044329	0.038341	0.033304	0.029041
0.30	0.062298	0.053612	0.046357	0.040256	0.035095
0.35	0.073381	0.063127	0.054566	0.047371	0.041286
0.40	0.084796	0.072917	0.063005	0.054678	0.047638
0.45	0.096597	0.083026	0.071709	0.062207	0.054178
0.50	0.108841	0.093500	0.080717	0.069990	0.060933
0.55	0.121587	0.104390	0.090070	0.078062	0.067930
0.60	0.134899	0.115746	0.099811	0.086459	0.075199
0.65	0.148844	0.127624	0.109985	0.095217	0.082772
0.70	0.163494	0.140083	0.120641	0.104377	0.090683
0.75	0.178926	0.153185	0.131830	0.113982	0.098967

Table 4.8 (continued)

			t		
v	0.80	0.85	0.90	0.95	1.0
0.80	0.195224	0.166999	0.143607	0.124076	0.107661
0.85	0.212476	0.181596	0.156032	0.134711	0.116807
0.90	0.230779	0.197054	0.169169	0.145936	0.126448
0.95	0.250237	0.213458	0.183085	0.157810	0.136630
1.00	0.270964	0.230899	0.197857	0.170392	0.147404
1.05	0.293082	0.249477	0.213563	0.183749	0.158823
1.10	0.316725	0.269297	0.230291	0.197952	0.170947
1.15	0.342039	0.290479	0.248135	0.213077	0.183839
1.20	0.369182	0.313147	0.267199	0.229209	0.197566
1.25	0.398329	0.337442	0.287593	0.246437	0.212204
1.30	0.429669	0.363514	0.309439	0.264861	0.227833
1.35	0.463409	0.391528	0.332870	0.284588	0.244540
1.40	0.499779	0.421666	0.358031	0.305735	0.262420
1.45	0.539027	0.454124	0.385080	0.328430	0.281578
1.50	0.581427	0.489121	0.414189	0.352810	0.302126
1.55	0.627282	0.526892	0.445548	0.379030	0.324187
1.60	0.676921	0.567700	0.479364	0.407254	0.347896
1.65	0.730710	0.611829	0.515865	0.437664	0.373399
1.70	0.789050	0.659595	0.555298	0.470460	0.400857
1.75	0.852381	0.711343	0.597937	0.505858	0.430444
1.80	0.921193	0.767452	0.644082	0.544099	0.462353
1.85	0.996022	0.828343	0.694063	0.585444	0.496794
1.90	1.07746	0.894476	0.748241	0.630181	0.533996
1.95	1.16617	0.966360	0.807016	0.678624	0.574213
2.00	1.26287	1.04455	0.870826	0.731121	0.617719
2.05	1.36835	1.12968	0.940155	0.788052	0.664819
2.10	1.48352	1.22242	1.01553	0.849838	0.715846
2.15	1.60935	1.32352	1.09755	0.916940	0.771165
2.20	1.74693	1.43383	1.18685	0.989864	0.831180
2.25	1.89746	1.55426	1.28416	1.06917	0.896332
2.30	2.06228	1.68584	1.39025	1.15548	0.967107
2.35	2.24287	1.82969	1.50600	1.24947	1.04404
2.40	2.44088	1.98708	1.63238	1.35189	1.12773
2.45	2.65814	2.15938	1.77046	1.46356	1.21881
2.50	2.89668	2.34813	1.92140	1.58541	1.31801
2.55	3.15877	2.55505	2.08653	1.71845	1.42612
2.60	3.44691	2.78203	2.26728	1.86379	1.54401
2.65	3.76391	3.03118	2.46526	2.02267	1.67264
2.70	4.11289	3.30483	2.68225	2.19646	1.81307
2.75	4.49732	3.60558	2.92022	2.38667	1.96649
2.80	4.92107	3.93634	3.18137	2.59498	2.13419
2.85	5.38847	4.30031	3.46813	2.82325	2.31761

Table 4.8 (continued)

			t		
v	0.80	0.85	0.90	0.95	1.0
2.90	5.90433	4.70109	3.78319	3.07354	2.51833
2.95	6.47404	5.14267	4.12956	3.34815	2.73813
3.00	7.10363	5.62949	4.51058	3.64960	2.97895
3.05	7.79979	6.16653	4.92998	3.98072	3.24296
3.10	8.57007	6.75932	5.39187	4.34464	3.53255
3.15	9.42286	7.41403	5.90088	4.74483	3.85037
3.20	10.3676	8.13758	6.46214	5.18518	4.19939
3.25	11.4148	8.93766	7.08136	5.66997	4.58287
3.30	12.5762	9.82290	7.76493	6.20400	5.00446
3.35	13.8652	10.8029	8.51997	6.79260	5.46820
3.40	15.2966	11.8885	9.35443	7.44172	5.97857
3.45	16.8870	13.0918	10.2772	8.15797	6.54059
3.50	18.6552	14.4262	11.2981	8.94873	7.15981
3.55	20.6220	15.9069	12.4284	9.82223	7.84242
3.60	22.8113	17.5508	13.6804	10.7877	8.59531
3.65	25.2493	19.3771	15.0679	11.8553	9.42617
3.70	27.9661	21.4069	16.6065	13.0365	10.3436
3.75	30.9952	23.6644	18.3135	14.3442	11.3570
3.80	34.3744	26.1764	20.2086	15.7927	12.4772
3.85	38.1462	28.9731	22.3133	17.3979	13.7160
3.90	42.3586	32.0886	24.6524	19.1777	15.0867
3.95	47.0657	35.5609	27.2531	21.1522	16.6041
4.00	52.3284	39.4332	30.1464	23.3438	18.2848
4.05	58.2156	43.7537	33.3668	25.7777	20.1473
4.10	64.8048	48.5769	36.9532	28.4821	22.2122
4.15	72.1837	53.9641	40.9493	31.4884	24.5029
4.20	80.4515	59.9845	45.4041	34.8322	27.0451
4.25	89.7200	66.7160	50.3728	38.5532	29.8681
4.30	100.116	74.2464	55.9177	42.6961	33.0042
4.35	111.782	82.6749	62.1085	47.3109	36.4900
4.40	124.882	92.1136	69.0242	52.4541	40.3664
4.45	139.597	102.689	76.7534	58.1888	44.6792
4.50	156.136	114.543	85.3960	64.5864	49.4799
4.55	174.735	127.838	95.0649	71.7269	54.8262
4.60	195.661	142.757	105.887	79.7004	60.7831
4.65	219.216	159.505	118.006	88.6084	67.4233
4.70	245.744	178.316	131.585	98.5652	74.8289
4.75	275.635	199.455	146.804	109.700	83.0918
4.80	309.332	223.222	163.873	122.157	92.3157
4.85	347.338	249.956	183.024	136.100	102.617
4.90	390.225	280.041	204.521	151.715	114.127

Table 4.8 (continued)

			t		
v	0.80	0.85	0.90	0.95	1.0
4.95	438.643	313.914	228.663	169.209	126.993
5.00	493.331	352.070	255.788	188.818	141.383

			t		
v	1.5	2.0	3.0	4.0	5.0
0.00	0.000000	0.000000	0.00000	0.000000	0.000000
0.05	0.001704	0.000606	0.00010	0.000021	0.000005
0.10	0.003414	0.001214	0.00020	0.000041	0.000009
0.15	0.005133	0.001825	0.00030	0.000062	0.000014
0.20	0.006867	0.002440	0.000406	0.000083	0.000019
0.25	0.008622	0.003061	0.000509	0.000104	0.000024
0.30	0.010402	0.003689	0.000613	0.000125	0.000029
0.35	0.012212	0.004326	0.000717	0.000146	0.000034
0.40	0.014059	0.004974	0.000824	0.000168	0.000038
0.45	0.015948	0.005633	0.000931	0.000189	0.000043
0.50	0.017885	0.006306	0.001040	0.000211	0.000048
0.55	0.019875	0.006995	0.001151	0.000234	0.000053
0.60	0.021926	0.007701	0.001265	0.000256	0.000059
0.65	0.024043	0.008426	0.001380	0.000279	0.000064
0.70	0.026234	0.009171	0.001498	0.000302	0.000069
0.75	0.028505	0.009940	0.001619	0.000326	0.000074
0.80	0.030866	0.010733	0.001742	0.000351	0.000080
0.85	0.033323	0.011554	0.001869	0.000375	0.000085
0.90	0.035885	0.012404	0.002000	0.000401	0.000091
0.95	0.038562	0.013286	0.002134	0.000427	0.000097
1.00	0.041363	0.014202	0.002272	0.000453	0.000103
1.05	0.044298	0.015155	0.002414	0.000480	0.000109
1.10	0.047379	0.016148	0.002561	0.000508	0.000115
1.15	0.050617	0.017183	0.002713	0.000537	0.000121
1.20	0.054024	0.018265	0.002870	0.000567	0.000127
1.25	0.057614	0.019396	0.003032	0.000597	0.000134
1.30	0.061401	0.020579	0.003201	0.000628	0.000141
1.35	0.065401	0.021819	0.003375	0.000661	0.000148
1.40	0.069629	0.023120	0.003557	0.000694	0.000155
1.45	0.074105	0.024486	0.003745	0.000728	0.000162
1.50	0.078847	0.025921	0.003942	0.000764	0.000170
1.55	0.083875	0.027431	0.004146	0.000800	0.000177
1.60	0.089211	0.029020	0.004358	0.000838	0.000185
1.65	0.094880	0.030694	0.004580	0.000878	0.000193
1.70	0.100906	0.032460	0.004811	0.000919	0.000202
1.75	0.107319	0.034322	0.005053	0.000961	0.000211
1.80	0.114147	0.036289	0.005305	0.001005	0.000220
1.85	0.121423	0.038367	0.005568	0.001050	0.000229

Table 4.8 (continued)

v	1.5	2.0	3.0	4.0	5.0
			t		
1.90	0.129181	0.040565	0.005844	0.001098	0.000239
1.95	0.137461	0.042890	0.006132	0.001147	0.000249
2.00	0.146302	0.045351	0.006435	0.001198	0.000259
2.05	0.155748	0.047959	0.006751	0.001251	0.000270
2.10	0.165849	0.050723	0.007083	0.001306	0.000281
2.15	0.176655	0.053655	0.007431	0.001364	0.000292
2.20	0.188223	0.056766	0.007796	0.001424	0.000304
2.25	0.200615	0.060069	0.008179	0.001486	0.000316
2.30	0.213896	0.063579	0.008582	0.001551	0.000329
2.35	0.228138	0.067309	0.009005	0.001619	0.000342
2.40	0.243421	0.071276	0.009449	0.001690	0.000356
2.45	0.259828	0.075497	0.009917	0.001765	0.000370
2.50	0.277453	0.079990	0.010409	0.001842	0.000385
2.55	0.296395	0.084776	0.010928	0.001923	0.000400
2.60	0.316765	0.089876	0.011473	0.002007	0.000417
2.65	0.338681	0.095312	0.012048	0.002096	0.000433
2.70	0.362273	0.101111	0.012654	0.002188	0.000451
2.75	0.387682	0.107298	0.013293	0.002285	0.000469
2.80	0.415062	0.113903	0.013967	0.002386	0.000487
2.85	0.444580	0.120958	0.014679	0.002492	0.000507
2.90	0.476421	0.128496	0.015429	0.002604	0.000527
2.95	0.510784	0.136554	0.016222	0.002720	0.000549
3.00	0.547885	0.145171	0.017060	0.002842	0.000571
3.05	0.587965	0.154391	0.017945	0.002970	0.000594
3.10	0.631282	0.164260	0.018881	0.003104	0.000618
3.15	0.678121	0.174828	0.019871	0.003245	0.000643
3.20	0.728793	0.186149	0.020918	0.003393	0.000669
3.25	0.783637	0.198283	0.022026	0.003548	0.000697
3.30	0.843025	0.211292	0.023200	0.003711	0.000725
3.35	0.907365	0.225247	0.024442	0.003882	0.000755
3.40	0.977100	0.240221	0.025759	0.004062	0.000786
3.45	1.05272	0.256296	0.027154	0.004251	0.000819
3.50	1.13476	0.273560	0.028633	0.004450	0.000853
3.55	1.22381	0.292108	0.030201	0.004660	0.000889
3.60	1.32050	0.312045	0.031865	0.004880	0.000926
3.65	1.42555	0.333482	0.033631	0.005111	0.000965
3.70	1.53972	0.356542	0.035506	0.005355	0.001006
3.75	1.66387	0.381359	0.037497	0.005612	0.001048
3.80	1.79893	0.408075	0.039611	0.005882	0.001093
3.85	1.94593	0.436848	0.041858	0.006167	0.001140
3.90	2.10599	0.467849	0.044247	0.006467	0.001189
3.95	2.28034	0.501264	0.046786	0.006784	0.001241
4.00	2.47036	0.537294	0.049487	0.007118	0.001294

Table 4.8 (continued)

			t		
v	1.5	2.0	3.0	4.0	5.0
4.05	2.67754	0.576159	0.052361	0.007470	0.001351
4.10	2.90353	0.618100	0.055420	0.007841	0.001410
4.15	3.15015	0.663377	0.058677	0.008233	0.001472
4.20	3.41940	0.712274	0.062145	0.008647	0.001537
4.25	3.71347	0.765101	0.065840	0.009084	0.001606
4.30	4.03481	0.822197	0.069778	0.009545	0.001678
4.35	4.38609	0.883929	0.073976	0.010033	0.001753
4.40	4.77026	0.950701	0.078452	0.010548	0.001832
4.45	5.19059	1.02295	0.083226	0.011093	0.001915
4.50	5.65068	1.10116	0.088321	0.011669	0.002002
4.55	6.15449	1.18584	0.093759	0.012278	0.002094
4.60	6.70642	1.27758	0.099564	0.012922	0.002191
4.65	7.31133	1.37700	0.105765	0.013604	0.002292
4.70	7.97456	1.48477	0.112389	0.014325	0.002399
4.75	8.70205	1.60165	0.119468	0.015089	0.002511
4.80	9.50036	1.72846	0.127036	0.015898	0.002629
4.85	10.3767	1.86608	0.135129	0.016755	0.002753
4.90	11.3392	2.01550	0.143785	0.017663	0.002883
4.95	12.3967	2.17778	0.153047	0.018626	0.003021
5.00	13.5590	2.35410	0.162961	0.019646	0.003166

Table 4.9: Second derivatives with respect to the order of the integral modified Bessel functions of the second kind $\partial^2 Ki_v(t)/\partial v^2$.

			t		
v	0.05	0.10	0.15	0.20	0.25
0.00	14.2489	6.66544	3.99992	2.68259	1.91916
0.05	14.3769	6.71011	4.02216	2.69555	1.92744
0.10	14.7643	6.84503	4.08925	2.73463	1.95238
0.15	15.4213	7.07288	4.20230	2.80039	1.99432
0.20	16.3654	7.39822	4.36322	2.89380	2.05379
0.25	17.6218	7.82762	4.57471	3.01622	2.13160
0.30	19.2244	8.36977	4.84034	3.16947	2.22875
0.35	21.2168	9.03575	5.16463	3.35582	2.34655
0.40	23.6539	9.83929	5.55315	3.57806	2.48657
0.45	26.6036	10.7972	6.01263	3.83953	2.65070
0.50	30.1494	11.9295	6.55112	4.14423	2.84118
0.55	34.3931	13.2606	7.17815	4.49684	3.06061
0.60	39.4585	14.8192	7.90496	4.90285	3.31206

Table 4.9 (continued)

			t		
v	0.05	0.10	0.15	0.20	0.25
0.65	45.4955	16.6395	8.74473	5.36864	3.59904
0.70	52.6861	18.7621	9.71291	5.90165	3.92564
0.75	61.2507	21.2349	10.8275	6.51047	4.29654
0.80	71.4561	24.1147	12.1097	7.20506	4.71714
0.85	83.6258	27.4686	13.5840	7.99694	5.19363
0.90	98.1524	31.3761	15.2791	8.89941	5.73314
0.95	115.512	35.9311	17.2285	9.92785	6.34383
1.00	136.284	41.2449	19.4712	11.1000	7.03506
1.05	161.172	47.4494	22.0530	12.4365	7.81757
1.10	191.035	54.7010	25.0270	13.9611	8.70371
1.15	226.917	63.1853	28.4558	15.7012	9.70764
1.20	270.097	73.1228	32.4121	17.6888	10.8456
1.25	322.135	84.7758	36.9813	19.9607	12.1364
1.30	384.941	98.4564	42.2636	22.5596	13.6013
1.35	460.854	114.536	48.3764	25.5353	15.2653
1.40	552.744	133.459	55.4573	28.9453	17.1568
1.45	664.135	155.755	63.6683	32.8566	19.3086
1.50	799.361	182.055	73.1995	37.3471	21.7586
1.55	963.754	213.117	84.2751	42.5072	24.5505
1.60	1,163.88	249.846	97.1588	48.4426	27.7346
1.65	1,407.86	293.328	112.162	55.2762	31.3693
1.70	1,705.69	344.866	129.650	63.1514	35.5220
1.75	2,069.74	406.021	150.058	72.2358	40.2707
1.80	2,515.34	478.674	173.898	82.7248	45.7057
1.85	3,061.46	565.084	201.774	94.8474	51.9316
1.90	3,731.64	667.971	234.404	108.871	59.0699
1.95	4,555.08	790.616	272.639	125.110	67.2615
2.00	5,568.10	936.973	317.485	143.930	76.6700
2.05	6,815.87	1,111.82	370.138	165.763	87.4857
2.10	8,354.63	1,320.92	432.018	191.114	99.9297
2.15	10,254.5	1,571.25	504.815	220.577	114.259
2.20	12,602.9	1,871.26	590.535	254.851	130.775
2.25	15,509.1	2,231.17	691.572	294.756	149.826
2.30	19,109.7	2,663.40	810.773	341.260	171.820
2.35	23,575.4	3,182.99	951.537	395.500	197.233
2.40	29,120.3	3,808.22	1,117.92	458.820	226.620
2.45	36,012.5	4,561.30	1,314.76	532.803	260.631
2.50	44,588.4	5,469.27	1,547.85	619.319	300.027
2.55	55,270.6	6,565.02	1,824.10	720.576	345.696
2.60	68,589.9	7,888.64	2,151.81	839.186	398.678
2.65	85,214.2	9,488.99	2,540.89	978.239	460.196
2.70	105984	11,425.7	3,003.24	1,141.39	531.678
2.75	131959	13,771.7	3,553.13	1,332.97	614.805

Table 4.9 (continued)

			t		
v	0.05	0.10	0.15	0.20	0.25
2.80	164475	16,615.8	4,207.69	1,558.13	711.547
2.85	205217	20,067.0	4,987.48	1,822.94	824.221
2.90	256316	24,258.5	5,917.24	2,134.64	955.549
2.95	320461	29,353.4	7,026.70	2,501.82	1,108.73
3.00	401058	35,551.8	8,351.67	2,934.69	1,287.55
3.05	502418	43,099.0	9,935.28	3,445.39	1,496.43
3.10	630001	52,296.1	11,829.5	4,048.36	1,740.61
3.15	790733	63,512.8	14,097.0	4,760.82	2,026.27
3.20	993400	77,203.9	16,813.5	5,603.27	2,360.68
3.25	1.249E+6	93,928.4	20,070.3	6,600.14	2,752.44
3.30	1.572	114375	23,977.9	7,780.61	3,211.72
3.35	1.981	139390	28,669.7	9,179.48	3,750.51
3.40	2.497	170019	34,307.3	10,838.3	4,383.03
3.45	3.151	207551	41,086.3	12,806.9	5,126.07
3.50	3.980	253575	49,243.7	15,144.5	5,999.53
3.55	5.031	310054	59,066.7	17,922.5	7,026.99
3.60	6.365	379415	70,903.7	21,225.8	8,236.40
3.65	8.059	464657	85,177.5	25,156.5	9,660.91
3.70	1.021E+7	569492	102402	29,836.9	11,339.9
3.75	1.295	698513	123200	35,413.7	13,320.0
3.80	1.644	857412	148332	42,062.9	15,656.7
3.85	2.088	1.053E+6	178719	49,995.8	18,416.1
3.90	2.654	1.295	215486	59,466.2	21,676.6
3.95	3.376	1.593	260001	70,779.4	25,531.6
4.00	4.297	1.961	313931	84,302.3	30,092.2
4.05	5.475	2.416	379309	100477	35,490.8
4.10	6.980	2.978	458616	119834	41,885.3
4.15	8.905	3.674	554878	143016	49,464.0
4.20	1.137E+8	4.536	671794	170793	58,451.2
4.25	1.452	5.603	813883	204097	69,115.2
4.30	1.857	6.927	986667	244051	81,775.9
4.35	2.376	8.568	1.197E+6	292011	96,815.8
4.40	3.041	1.061E+7	1.453	349614	114692
4.45	3.896	1.314	1.765	418840	135952
4.50	4.994	1.628	2.145	502080	161250
4.55	6.407	2.019	2.608	602229	191369
4.60	8.223	2.506	3.174	722787	227247
4.65	1.056E+9	3.111	3.865	867995	270010
4.70	1.357	3.866	4.708	1.043E+6	321006
4.75	1.746	4.806	5.740	1.254	381851
4.80	2.246	5.979	7.001	1.509	454487
4.85	2.892	7.442	8.545	1.816	541244
4.90	3.726	9.268	1.044E+7	2.187	644920

Table 4.9 (continued)

			t		
v	0.05	0.10	0.15	0.20	0.25
4.95	4.803	1.155E+8	1.275	2.635	768879
5.00	6.195	1.440	1.559	3.177	917164

			t		
v	0.30	0.35	0.40	0.45	0.50
0.00	1.43296	1.10321	0.86922	0.697407	0.567831
0.05	1.43858	1.10719	0.87213	0.699591	0.569505
0.10	1.45549	1.11917	0.88089	0.706166	0.574541
0.15	1.48391	1.13927	0.89559	0.717194	0.582987
0.20	1.52417	1.16773	0.91638	0.732782	0.594917
0.25	1.57675	1.20486	0.94348	0.753080	0.610442
0.30	1.64228	1.25107	0.97717	0.778284	0.629702
0.35	1.72157	1.30688	1.01779	0.808642	0.652875
0.40	1.81557	1.37290	1.06577	0.844448	0.680174
0.45	1.92545	1.44990	1.12162	0.88606	0.711853
0.50	2.05255	1.53875	1.18592	0.93388	0.748208
0.55	2.19848	1.64046	1.25937	0.98840	0.789582
0.60	2.36506	1.75623	1.34276	1.05016	0.836366
0.65	2.55442	1.88740	1.43698	1.11979	0.889008
0.70	2.76901	2.03553	1.54308	1.19801	0.948017
0.75	3.01161	2.20239	1.66224	1.28563	1.01397
0.80	3.28542	2.38999	1.79579	1.38356	1.08751
0.85	3.59410	2.60065	1.94524	1.49284	1.16938
0.90	3.94182	2.83696	2.11232	1.61465	1.26039
0.95	4.33333	3.10189	2.29896	1.75030	1.36148
1.00	4.77408	3.39881	2.50735	1.90128	1.47368
1.05	5.27024	3.73154	2.73998	2.06927	1.59816
1.10	5.82888	4.10441	2.99965	2.25615	1.73623
1.15	6.45806	4.52235	3.28952	2.46404	1.88937
1.20	7.16699	4.99094	3.61319	2.69535	2.05922
1.25	7.96617	5.51654	3.97470	2.95276	2.24764
1.30	8.86761	6.10637	4.37863	3.23931	2.45672
1.35	9.88504	6.76863	4.83019	3.55844	2.68879
1.40	11.0341	7.51265	5.33523	3.91399	2.94649
1.45	12.3329	8.34908	5.90042	4.31034	3.23277
1.50	13.8019	9.29001	6.53330	4.75240	3.55096
1.55	15.4648	10.34920	7.24243	5.24572	3.90480
1.60	17.3485	11.5425	8.0375	5.7966	4.2985
1.65	19.4841	12.8877	8.9296	6.4121	4.7368
1.70	21.9073	14.4054	9.9312	7.1003	5.2251
1.75	24.6591	16.1189	11.0565	7.87030	5.76938
1.80	27.7865	18.0551	12.3218	8.73234	6.37645
1.85	31.3438	20.2445	13.7455	9.69813	7.05401

Table 4.9 (continued)

			t		
v	0.30	0.35	0.40	0.45	0.50
1.90	35.3933	22.7223	15.3486	10.7809	7.81073
1.95	40.0071	25.5285	17.1550	11.9956	8.65642
2.00	45.2681	28.7091	19.1921	13.3594	9.60217
2.05	51.2719	32.3170	21.4909	14.8915	10.6605
2.10	58.1291	36.4127	24.0869	16.6141	11.8457
2.15	65.9674	41.0657	27.0209	18.5520	13.1738
2.20	74.9344	46.3560	30.3393	20.7339	14.6630
2.25	85.2009	52.3755	34.0951	23.1920	16.3340
2.30	96.9646	59.2298	38.3491	25.9634	18.2103
2.35	110.455	67.0409	43.1710	29.0901	20.3185
2.40	125.937	75.9488	48.6404	32.6202	22.6889
2.45	143.718	86.1153	54.8489	36.6085	25.3558
2.50	164.158	97.7269	61.9014	41.1175	28.3584
2.55	187.670	110.999	69.9183	46.2188	31.7411
2.60	214.738	126.179	79.0379	51.9941	35.5546
2.65	245.922	143.556	89.4192	58.5368	39.8566
2.70	281.877	163.461	101.245	65.9541	44.7129
2.75	323.362	186.279	114.726	74.3684	50.1983
2.80	371.265	212.454	130.104	83.9201	56.3985
2.85	426.617	242.501	147.658	94.7701	63.4110
2.90	490.625	277.018	167.710	107.103	71.3474
2.95	564.694	316.697	190.630	121.131	80.3351
3.00	650.470	362.342	216.847	137.097	90.5198
3.05	749.870	414.886	246.854	155.281	102.068
3.10	865.141	475.412	281.223	176.003	115.171
3.15	998.910	545.181	320.612	199.634	130.046
3.20	1,154.25	625.657	365.785	226.599	146.945
3.25	1,334.77	718.546	417.624	257.387	166.154
3.30	1,544.69	825.831	477.150	292.562	188.001
3.35	1,788.95	949.826	545.549	332.774	212.865
3.40	2,073.37	1,093.22	624.189	378.773	241.178
3.45	2,404.78	1,259.17	714.663	431.423	273.439
3.50	2,791.17	1,451.33	818.815	491.722	310.220
3.55	3,241.97	1,673.99	938.787	560.824	352.178
3.60	3,768.25	1,932.14	1,077.06	640.059	400.069
3.65	4,383.04	2,231.63	1,236.54	730.969	454.765
3.70	5,101.67	2,579.29	1,420.57	835.333	517.268
3.75	5,942.21	2,983.12	1,633.06	955.213	588.733
3.80	6,925.95	3,452.47	1,878.56	1,092.99	670.491
3.85	8,077.99	3,998.31	2,162.36	1,251.44	764.076
3.90	9,427.94	4,633.46	2,490.62	1,433.76	871.259

Table 4.9 (continued)

			t		
v	0.30	0.35	0.40	0.45	0.50
3.95	11,010.7	5,372.98	2,870.54	1,643.65	994.084
4.00	12,867.7	6,234.53	3,310.49	1,885.44	1,134.91
4.05	15,047.5	7,238.80	3,820.25	2,164.11	1,296.46
4.10	17,607.9	8,410.13	4,411.21	2,485.48	1,481.89
4.15	20,617.0	9,777.08	5,096.71	2,856.28	1,694.85
4.20	24,155.5	11,373.2	5,892.29	3,284.35	1,939.53
4.25	28,318.8	13,238.0	6,816.14	3,778.81	2,220.83
4.30	33,220.2	15,417.8	7,889.53	4,350.24	2,544.39
4.35	38,993.7	17,967.3	9,137.34	5,010.99	2,916.75
4.40	45,798.2	20,950.9	10,588.7	5,775.40	3,345.49
4.45	53,822.2	24,444.2	12,277.6	6,660.22	3,839.41
4.50	63,289.4	28,536.6	14,244.1	7,684.93	4,408.69
4.55	74,465.6	33,333.3	16,535.0	8,872.26	5,065.17
4.60	87,666.2	38,958.6	19,205.0	10,248.7	5,822.59
4.65	103266	45,559.0	22,318.7	11,845.2	6,696.89
4.70	121711	53,307.5	25,951.6	13,697.9	7,706.62
4.75	143532	62,408.5	30,192.3	15,848.8	8,873.31
4.80	169360	73,103.5	35,145.0	18,347.4	10,222.0
4.85	199945	85,678.0	40,932.2	21,251.1	11,781.9
4.90	236184	100470	47,697.8	24,627.4	13,586.9
4.95	279142	117878	55,611.1	28,555.0	15,676.5
5.00	330091	138377	64,871.3	33,126.0	18,096.7

			t		
v	0.55	0.60	0.65	0.70	0.75
0.00	0.46798	0.38965	0.32728	0.27698	0.23597
0.05	0.469281	0.39068	0.32810	0.27765	0.23651
0.10	0.473206	0.393779	0.33058	0.27965	0.23815
0.15	0.479784	0.398978	0.334742	0.283016	0.240898
0.20	0.489074	0.406316	0.340611	0.287759	0.244767
0.25	0.501154	0.415853	0.348234	0.293918	0.249790
0.30	0.516129	0.427668	0.357672	0.301540	0.256003
0.35	0.534130	0.441858	0.369000	0.310681	0.263449
0.40	0.555314	0.458542	0.382306	0.321411	0.272185
0.45	0.579867	0.477859	0.397698	0.333812	0.282272
0.50	0.608005	0.499970	0.415299	0.347978	0.293786
0.55	0.639981	0.525063	0.435250	0.364020	0.306812
0.60	0.676080	0.553353	0.457713	0.382061	0.321446
0.65	0.716630	0.585081	0.482873	0.402243	0.337799
0.70	0.762000	0.620524	0.510937	0.424725	0.355994
0.75	0.812609	0.659990	0.542139	0.449686	0.376168

Table 4.9 (continued)

			t		
v	0.55	0.60	0.65	0.70	0.75
0.80	0.868927	0.703828	0.576740	0.477326	0.398479
0.85	0.931483	0.752429	0.615035	0.507868	0.423097
0.90	1.00087	0.806230	0.657351	0.541563	0.450217
0.95	1.07776	0.865721	0.704054	0.578689	0.480051
1.00	1.16289	0.931448	0.755554	0.619555	0.512839
1.05	1.25711	1.00402	0.812304	0.664507	0.548845
1.10	1.36134	1.08413	0.874813	0.713926	0.588362
1.15	1.47663	1.17253	0.943644	0.768238	0.631715
1.20	1.60416	1.27007	1.01943	0.827918	0.679266
1.25	1.74523	1.37770	1.10287	0.893489	0.731414
1.30	1.90132	1.49648	1.19473	0.965538	0.788603
1.35	2.07407	1.62760	1.29591	1.04471	0.851327
1.40	2.26533	1.77238	1.40735	1.13173	0.920130
1.45	2.47716	1.93230	1.53014	1.22741	0.995618
1.50	2.71187	2.10900	1.66549	1.33262	1.07846
1.55	2.97207	2.30434	1.81474	1.44837	1.16942
1.60	3.26066	2.52039	1.97937	1.57576	1.26930
1.65	3.58092	2.75945	2.16108	1.71603	1.37904
1.70	3.93653	3.02412	2.36172	1.87054	1.49965
1.75	4.33161	3.31731	2.58338	2.04082	1.63229
1.80	4.77082	3.64226	2.82839	2.22858	1.77820
1.85	5.25936	4.00263	3.09937	2.43571	1.93880
1.90	5.80314	4.40251	3.39922	2.66434	2.11589
1.95	6.40876	4.84650	3.73122	2.91684	2.31053
2.00	7.08369	5.33976	4.09903	3.19584	2.52535
2.05	7.83634	5.88810	4.50673	3.50432	2.76229
2.10	8.67622	6.49804	4.95894	3.84556	3.02377
2.15	9.61402	7.17692	5.46081	4.22328	3.31249
2.20	10.6618	7.93301	6.01812	4.64161	3.63147
2.25	11.8334	8.77562	6.63737	5.10518	3.98408
2.30	13.1440	9.71522	7.32585	5.61918	4.37408
2.35	14.6113	10.7636	8.09179	6.18944	4.80567
2.40	16.2550	11.9342	8.94439	6.82249	5.28357
2.45	18.0975	13.2420	9.89403	7.52564	5.81304
2.50	20.1641	14.7040	10.9524	8.30711	6.39998
2.55	22.4837	16.3393	12.1327	9.17614	7.05100
2.60	25.0888	18.1698	13.4497	10.1431	7.77349
2.65	28.0165	20.2199	14.9201	11.2197	8.57578
2.70	31.3088	22.5174	16.5629	12.4189	9.46716
2.75	35.0135	25.0937	18.3992	13.7557	10.4581
2.80	39.1848	27.9844	20.4532	15.2465	11.5603

Table 4.9 (continued)

			t		
v	0.55	0.60	0.65	0.70	0.75
2.85	43.8843	31.2299	22.7519	16.9102	12.7870
2.90	49.1824	34.8759	25.3261	18.7677	14.1530
2.95	55.1589	38.9743	28.2103	20.8430	15.6750
3.00	61.9049	43.5840	31.4440	23.1628	17.3717
3.05	69.5240	48.7719	35.0715	25.7574	19.2641
3.10	78.1345	54.6139	39.1431	28.6610	21.3762
3.15	87.8713	61.1964	43.7159	31.9121	23.7346
3.20	98.8882	68.6175	48.8543	35.5546	26.3694
3.25	111.361	76.9890	54.6317	39.6376	29.3148
3.30	125.491	86.4381	61.1313	44.2171	32.6091
3.35	141.506	97.1095	68.4473	49.3562	36.2955
3.40	159.670	109.168	76.6869	55.1263	40.4230
3.45	180.283	122.802	85.9719	61.6087	45.0469
3.50	203.688	138.227	96.4407	68.8950	50.2294
3.55	230.278	155.685	108.251	77.0894	56.0412
3.60	260.505	175.458	121.581	86.3099	62.5622
3.65	294.884	197.864	136.635	96.6906	69.8825
3.70	334.009	223.268	153.645	108.384	78.1046
3.75	378.559	252.087	172.876	121.562	87.3441
3.80	429.315	284.796	194.629	136.421	97.7323
3.85	487.172	321.775	219.248	153.185	109.418
3.90	553.161	364.149	247.124	172.108	122.570
3.95	628.463	412.131	278.717	193.478	137.379
4.00	714.441	466.708	314.504	217.624	154.063
4.05	812.660	528.819	355.102	244.920	172.868
4.10	924.922	599.539	401.168	275.794	194.075
4.15	1,053.30	680.105	453.463	310.731	218.002
4.20	1,200.19	771.934	512.862	350.285	245.011
4.25	1,368.35	876.653	580.361	395.089	275.513
4.30	1,560.94	996.133	657.103	445.864	309.978
4.35	1,781.64	1,132.52	744.399	503.434	348.938
4.40	2,034.68	1,288.29	843.746	568.741	393.001
4.45	2,324.94	1,466.28	956.865	642.858	442.858
4.50	2,658.07	1,669.76	1,085.73	727.015	499.298
4.55	3,040.57	1,902.50	1,232.59	822.618	563.220
4.60	3,479.99	2,168.83	1,400.06	931.272	635.650
4.65	3,985.04	2,473.74	1,591.10	1,054.82	717.757
4.70	4,565.81	2,823.00	1,809.15	1,195.36	810.877
4.75	5,233.96	3,223.24	2,058.13	1,355.32	916.535
4.80	6,003.03	3,682.11	2,342.57	1,537.45	1,036.47
4.85	6,888.66	4,208.47	2,667.66	1,744.93	1,172.68

Table 4.9 (continued)

			t		
v	0.55	0.60	0.65	0.70	0.75
4.90	7,909.01	4,812.50	3,039.38	1,981.38	1,327.44
4.95	9,085.11	5,506.00	3,464.63	2,250.98	1,503.35
5.00	10,441.4	6,302.57	3,951.31	2,558.51	1,703.40

			t		
v	0.80	0.85	0.90	0.95	1.0
0.00	0.20221	0.17419	0.15076	0.13103	0.114318
0.05	0.20266	0.17456	0.15107	0.13129	0.114538
0.10	0.20401	0.17568	0.15200	0.13207	0.115199
0.15	0.20627	0.17755	0.15356	0.13338	0.116304
0.20	0.209452	0.180192	0.15576	0.13523	0.117862
0.25	0.213582	0.183613	0.15862	0.13762	0.11988
0.30	0.218688	0.187841	0.162140	0.14058	0.12237
0.35	0.224806	0.192904	0.166358	0.14411	0.12535
0.40	0.231977	0.198836	0.171298	0.14825	0.12884
0.45	0.240253	0.205677	0.176991	0.15302	0.13285
0.50	0.249692	0.213474	0.183475	0.15844	0.13742
0.55	0.260361	0.222279	0.190793	0.164563	0.14256
0.60	0.272336	0.232154	0.198994	0.171415	0.14832
0.65	0.285703	0.243168	0.208131	0.179044	0.15473
0.70	0.300560	0.255396	0.218267	0.187501	0.16182
0.75	0.317015	0.268925	0.229471	0.196839	0.169651
0.80	0.335189	0.283851	0.241819	0.207121	0.178264
0.85	0.355218	0.300280	0.255396	0.218414	0.187715
0.90	0.377252	0.318332	0.270296	0.230796	0.198066
0.95	0.401457	0.338138	0.286624	0.244349	0.209385
1.00	0.428020	0.359843	0.304496	0.259167	0.221747
1.05	0.457146	0.383609	0.324040	0.275351	0.235233
1.10	0.489061	0.409614	0.345396	0.293014	0.249935
1.15	0.524018	0.438054	0.368721	0.312280	0.265953
1.20	0.562296	0.469149	0.394185	0.333286	0.283395
1.25	0.604203	0.503138	0.421980	0.356184	0.302384
1.30	0.650081	0.540288	0.452314	0.381138	0.323052
1.35	0.700308	0.580892	0.485417	0.408332	0.345545
1.40	0.755303	0.625275	0.521545	0.437967	0.370023
1.45	0.815528	0.673796	0.560978	0.470265	0.396664
1.50	0.881499	0.726852	0.604026	0.505471	0.425663
1.55	0.953782	0.784881	0.651032	0.543854	0.457232
1.60	1.03301	0.848369	0.702372	0.585710	0.491608
1.65	1.11988	0.917852	0.758464	0.631367	0.529049
1.70	1.21517	0.993925	0.819768	0.681186	0.569841

Table 4.9 (continued)

			t		
ν	0.80	0.85	0.90	0.95	1.0
1.75	1.31973	1.07725	0.886795	0.735565	0.614298
1.80	1.43453	1.16854	0.960107	0.794944	0.662767
1.85	1.56061	1.26862	1.04033	0.859808	0.715630
1.90	1.69917	1.37839	1.12814	0.930693	0.773306
1.95	1.85150	1.49882	1.22432	1.00819	0.836260
2.00	2.01906	1.63102	1.32970	1.09296	0.905005
2.05	2.20347	1.77623	1.44522	1.18571	0.980105
2.10	2.40652	1.93578	1.57191	1.28726	1.06219
2.15	2.63023	2.11120	1.71093	1.39848	1.15194
2.20	2.87683	2.30416	1.86355	1.52037	1.25012
2.25	3.14879	2.51652	2.03119	1.65399	1.35758
2.30	3.44891	2.75036	2.21541	1.80057	1.47524
2.35	3.78026	3.00798	2.41797	1.96143	1.60414
2.40	4.14631	3.29196	2.64079	2.13805	1.74542
2.45	4.55090	3.60516	2.88604	2.33208	1.90035
2.50	4.99835	3.95077	3.15612	2.54534	2.07033
2.55	5.49346	4.33234	3.45368	2.77985	2.25690
2.60	6.04161	4.75386	3.78171	3.03786	2.46180
2.65	6.64882	5.21973	4.14350	3.32187	2.68693
2.70	7.32183	5.73492	4.54274	3.63466	2.93441
2.75	8.06817	6.30495	4.98354	3.97933	3.20661
2.80	8.89628	6.93599	5.47047	4.35930	3.50612
2.85	9.81563	7.63494	6.00866	4.77843	3.83587
2.90	10.8368	8.40952	6.60379	5.24096	4.19908
2.95	11.9718	9.26838	7.26223	5.75166	4.59935
3.00	13.2338	10.2212	7.99110	6.31583	5.04068
3.05	14.6380	11.2788	8.79836	6.93938	5.52751
3.10	16.2010	12.4533	9.69287	7.62891	6.06480
3.15	17.9419	13.7583	10.6846	8.39178	6.65809
3.20	19.8820	15.2091	11.7847	9.23622	7.31351
3.25	22.0450	16.8229	13.0055	10.1714	8.03795
3.30	24.4580	18.6187	14.3611	11.2076	8.83904
3.35	27.1513	20.6184	15.8670	12.3563	9.72533
3.40	30.1590	22.8459	17.5409	13.6304	10.7064
3.45	33.5194	25.3288	19.4022	15.0441	11.7928
3.50	37.2760	28.0975	21.4732	16.6137	12.9965
3.55	41.4777	31.1866	23.7784	18.3571	14.3308
3.60	46.1794	34.6348	26.3458	20.2944	15.8106
3.65	51.4436	38.4859	29.2064	22.4484	17.4524
3.70	57.3403	42.7891	32.3954	24.8445	19.2750
3.75	63.9490	47.5998	35.9522	27.5110	21.2991
3.80	71.3594	52.9805	39.9211	30.4799	23.5482
3.85	79.6727	59.0018	44.3521	33.7871	26.0483

Table 4.9 (continued)

			t		
v	0.80	0.85	0.90	0.95	1.0
3.90	89.0038	65.7431	49.3012	37.4728	28.8287
3.95	99.4823	73.2942	54.8316	41.5824	31.9224
4.00	111.255	81.7565	61.0147	46.1667	35.3661
4.05	124.489	91.2444	67.9307	51.2829	39.2014
4.10	139.371	101.887	75.6701	56.9953	43.4745
4.15	156.116	113.832	84.3350	63.3765	48.2377
4.20	174.966	127.243	94.0405	70.5078	53.5496
4.25	196.196	142.308	104.917	78.4812	59.4760
4.30	220.116	159.239	117.110	87.4000	66.0910
4.35	247.082	178.287	130.787	97.3810	73.4778
4.40	277.494	199.690	146.135	108.556	81.7301
4.45	311.810	223.790	163.365	121.072	90.9534
4.50	350.548	250.924	182.717	135.098	101.266
4.55	394.298	281.489	204.462	150.821	112.803
4.60	443.731	315.933	228.907	168.457	125.714
4.65	499.611	354.767	256.400	188.244	140.169
4.70	562.807	398.570	287.333	210.457	156.359
4.75	634.310	447.999	322.153	235.402	174.502
4.80	715.246	503.801	361.365	263.428	194.840
4.85	806.901	566.825	405.542	294.929	217.650
4.90	910.740	638.038	455.334	330.350	243.242
4.95	1,028.43	718.537	511.479	370.196	271.967
5.00	1,161.89	809.573	574.814	415.040	304.222

			t		
v	1.5	2.0	3.0	4.0	5.0
0.00	0.034069	0.01212	0.00202	0.00041	0.000095
0.05	0.034119	0.01214	0.00202	0.00041	0.000095
0.10	0.034269	0.012180	0.00203	0.00041	0.000095
0.15	0.034520	0.012252	0.00204	0.00042	0.000095
0.20	0.034872	0.012354	0.00205	0.00042	0.000096
0.25	0.035327	0.012486	0.00207	0.00042	0.00010
0.30	0.035888	0.012648	0.00209	0.00042	0.00010
0.35	0.036558	0.012841	0.00211	0.00043	0.00010
0.40	0.037338	0.013066	0.00214	0.00043	0.00010
0.45	0.038234	0.013323	0.00217	0.00044	0.00010
0.50	0.039248	0.013613	0.00220	0.00044	0.00010
0.55	0.040386	0.013939	0.00224	0.00045	0.00010
0.60	0.041654	0.014300	0.00229	0.00046	0.00010
0.65	0.043057	0.014698	0.002333	0.00046	0.00010
0.70	0.044602	0.015135	0.002386	0.00047	0.00011
0.75	0.046296	0.015613	0.002443	0.00048	0.00011

Table 4.9 (continued)

			t		
v	1.5	2.0	3.0	4.0	5.0
0.80	0.048148	0.016133	0.002504	0.00049	0.00011
0.85	0.050167	0.016698	0.002571	0.00050	0.00011
0.90	0.052362	0.017309	0.002643	0.00051	0.00011
0.95	0.054744	0.017970	0.002720	0.00053	0.00012
1.00	0.057326	0.018682	0.002802	0.00054	0.00012
1.05	0.060121	0.019449	0.002891	0.00055	0.00012
1.10	0.063141	0.020274	0.002986	0.00057	0.00012
1.15	0.066403	0.021161	0.003087	0.00058	0.00013
1.20	0.069924	0.022112	0.003194	0.00060	0.00013
1.25	0.073722	0.023133	0.003309	0.000617	0.00013
1.30	0.077817	0.024226	0.003431	0.000635	0.00014
1.35	0.082230	0.025398	0.003561	0.000655	0.00014
1.40	0.086984	0.026652	0.003699	0.000676	0.00014
1.45	0.092107	0.027995	0.003845	0.000699	0.00015
1.50	0.097624	0.029432	0.004001	0.000722	0.00015
1.55	0.103568	0.030970	0.004166	0.000747	0.00016
1.60	0.109969	0.032615	0.004341	0.000773	0.00016
1.65	0.116865	0.034374	0.004527	0.000801	0.00017
1.70	0.124295	0.036257	0.004724	0.000830	0.00017
1.75	0.132300	0.038271	0.004933	0.000861	0.00018
1.80	0.140926	0.040425	0.005155	0.000893	0.00018
1.85	0.150225	0.042730	0.005390	0.000928	0.00019
1.90	0.160249	0.045196	0.005639	0.000964	0.00020
1.95	0.171060	0.047835	0.005903	0.001002	0.00020
2.00	0.182722	0.050659	0.006182	0.001042	0.00021
2.05	0.195305	0.053683	0.006479	0.001084	0.00022
2.10	0.208886	0.056920	0.006793	0.001129	0.00023
2.15	0.223549	0.060387	0.007127	0.001175	0.00023
2.20	0.239385	0.064102	0.007480	0.001225	0.00024
2.25	0.256495	0.068081	0.007855	0.001277	0.00025
2.30	0.274987	0.072346	0.008252	0.001332	0.00026
2.35	0.294980	0.076918	0.008674	0.001389	0.00027
2.40	0.316603	0.081821	0.009121	0.001450	0.00028
2.45	0.339999	0.087080	0.009595	0.001515	0.00029
2.50	0.365322	0.092722	0.010099	0.001582	0.00030
2.55	0.392741	0.098778	0.010633	0.001654	0.00031
2.60	0.422442	0.105280	0.011201	0.001729	0.00033
2.65	0.454627	0.112263	0.011803	0.001808	0.00034
2.70	0.489516	0.119764	0.012443	0.001892	0.00035
2.75	0.527354	0.127825	0.013123	0.001980	0.00037
2.80	0.568405	0.136490	0.013845	0.002073	0.00038
2.85	0.612959	0.145808	0.014613	0.002171	0.00040

Table 4.9 (continued)

			t		
v	1.5	2.0	3.0	4.0	5.0
2.90	0.661336	0.155832	0.015429	0.002275	0.00042
2.95	0.713884	0.166617	0.016297	0.002384	0.00043
3.00	0.770986	0.178227	0.017220	0.002499	0.00045
3.05	0.833063	0.190729	0.018201	0.002621	0.00047
3.10	0.900574	0.204194	0.019246	0.002750	0.000492
3.15	0.974026	0.218704	0.020358	0.002886	0.000513
3.20	1.05397	0.234344	0.021541	0.003029	0.000536
3.25	1.14103	0.251208	0.022802	0.003181	0.000560
3.30	1.23586	0.269398	0.024144	0.003341	0.000584
3.35	1.33920	0.289025	0.025575	0.003510	0.000610
3.40	1.45186	0.310210	0.027099	0.003689	0.000638
3.45	1.57473	0.333085	0.028724	0.003879	0.000667
3.50	1.70880	0.357793	0.030457	0.004079	0.000697
3.55	1.85514	0.384491	0.032305	0.004291	0.000729
3.60	2.01494	0.413349	0.034277	0.004515	0.000763
3.65	2.18951	0.444551	0.036381	0.004752	0.000798
3.70	2.38029	0.478302	0.038627	0.005003	0.000835
3.75	2.58886	0.514821	0.041025	0.005269	0.000874
3.80	2.81699	0.554349	0.043586	0.005551	0.000915
3.85	3.06660	0.597150	0.046323	0.005849	0.000959
3.90	3.33982	0.643510	0.049247	0.006165	0.001005
3.95	3.63901	0.693743	0.052372	0.006500	0.001053
4.00	3.96676	0.748192	0.055714	0.006855	0.001103
4.05	4.32595	0.807230	0.059288	0.007231	0.001157
4.10	4.71974	0.871266	0.063111	0.007630	0.001213
4.15	5.15163	0.940749	0.067203	0.008053	0.001272
4.20	5.62551	1.01617	0.071582	0.008503	0.001335
4.25	6.14565	1.09805	0.076271	0.008979	0.001401
4.30	6.71680	1.18700	0.081293	0.009485	0.001471
4.35	7.34419	1.28364	0.086673	0.010022	0.001544
4.40	8.03363	1.38868	0.092438	0.010592	0.001622
4.45	8.79155	1.50289	0.098617	0.011198	0.001703
4.50	9.62508	1.62711	0.105242	0.011841	0.001790
4.55	10.5421	1.76227	0.112348	0.012525	0.001881
4.60	11.5514	1.90937	0.119970	0.013252	0.001977
4.65	12.6626	2.06953	0.128150	0.014025	0.002079
4.70	13.8865	2.24397	0.136929	0.014847	0.002186
4.75	15.2350	2.43401	0.146355	0.015721	0.002299
4.80	16.7215	2.64114	0.156479	0.016650	0.002419
4.85	18.3605	2.86696	0.167348	0.017640	0.002546
4.90	20.1685	3.11324	0.179040	0.018693	0.002680
4.95	22.1636	3.38191	0.191600	0.019814	0.002822
5.00	24.3659	3.67513	0.205105	0.021008	0.002971

4.4 The Bessel and Related Functions with the Same Argument and Order

There are no tabulations of the Bessel and related functions with the same argument and order in the literature. Evidently, such values, but only in limited range, are available in tables of considered functions. Here, in systematic way, they are presented for the first time. The following functions $J_v(v)$, $Y_v(v)$, $I_v(v)$, $K_v(v)$, $H_v(v)$, $L_v(v)$, $\mathbf{J}_v(v)$, $\mathbf{E}_v(v)$, $\mathrm{ber}_v(v)$, $\mathrm{bei}_v(v)$, $\mathrm{ker}_v(v)$, $\mathrm{kei}_v(v)$, $Ji_v(v)$, $Yi_v(v)$ and $Ki_v(v)$.are tabulated and plotted in figures. Usually, these functions are monotonic descending or rising curves (Figures 4.16, 4.17 and 4.19), or they have an oscillatory character (Figures 4.17 and 4.18).

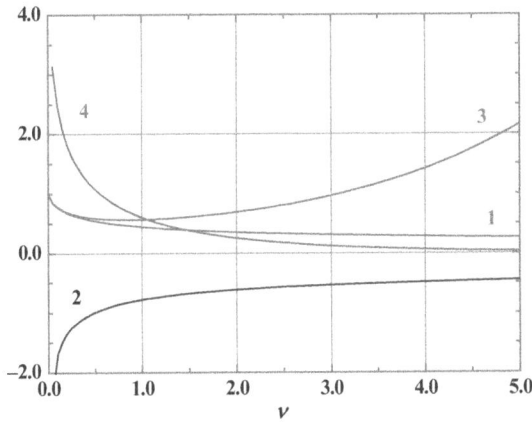

Figure 4.16: Functions with the same argument and order 1 – $J_v(v)$; 2 – $Y_v(v)$; 3 – $I_v(v)$; 4 – $K_v(v)$.

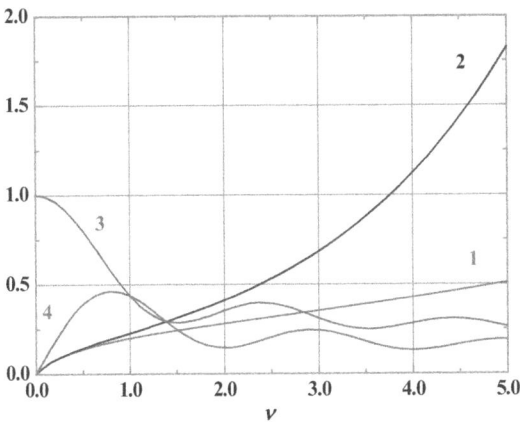

Figure 4.17: Functions with the same argument and order 1 – $H_v(v)$; 2 – $L_v(v)$; 3 – $\mathbf{J}_v(v)$; 4 – $\mathbf{E}_v(v)$.

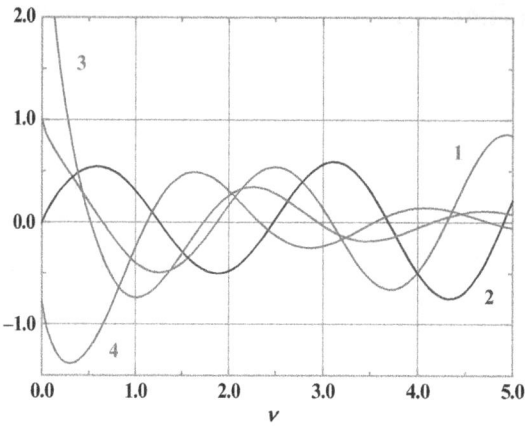

Figure 4.18: Functions with the same argument and order 1 – $ber_v(v)$; 2 – $bei_v(v)$; 3 – $ker_v(v)$; 4 – $kei_v(v)$.

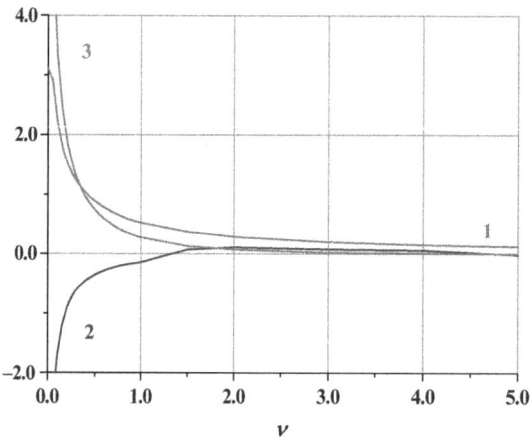

Figure 4.19: Functions with the same argument and order 1 – $Ji_v(v)$; 2 – $Yi_v(v)$; 3 – $Ki_v(v)$.

Table 4.10: The Bessel functions with the same argument and order $J_v(v)$, $Y_v(v)$, $I_v(v)$ and $K_v(v)$.

v	$J_v(v)$	$Y_v(v)$	$I_v(v)$	$K_v(v)$
0.00	1.000000		1.000000	
0.05	0.853691	−2.057930	0.854708	3.132250
0.10	0.777264	−1.682440	0.780805	2.467050
0.15	0.723156	−1.481550	0.730265	2.090610
0.20	0.681477	−1.349180	0.692931	1.831810
0.25	0.647832	−1.252810	0.664232	1.637010
0.30	0.619815	−1.178330	0.641646	1.482340
0.35	0.595953	−1.118380	0.623615	1.355120
0.40	0.575276	−1.068690	0.609108	1.247800
0.45	0.557111	−1.026570	0.597408	1.155540
0.50	0.540974	−0.990246	0.587993	1.075050
0.55	0.526502	−0.958463	0.580478	1.003980
0.60	0.513419	−0.930328	0.574567	0.940638
0.65	0.501510	−0.905175	0.570031	0.883712
0.70	0.490606	−0.882499	0.566688	0.832208
0.75	0.480568	−0.861905	0.564393	0.785335
0.80	0.471284	−0.843086	0.563027	0.742461
0.85	0.462662	−0.825792	0.562494	0.703071
0.90	0.454624	−0.809821	0.562714	0.666742
0.95	0.447105	−0.795007	0.563621	0.633120
1.00	0.440051	−0.781213	0.565159	0.601907
1.05	0.433413	−0.768322	0.567281	0.572851
1.10	0.427150	−0.756237	0.569949	0.545736
1.15	0.421229	−0.744874	0.573127	0.520375
1.20	0.415617	−0.734160	0.576788	0.496606
1.25	0.410288	−0.724036	0.580907	0.474287
1.30	0.405219	−0.714445	0.585463	0.453294
1.35	0.400387	−0.705341	0.590440	0.433516
1.40	0.395774	−0.696683	0.595821	0.414858
1.45	0.391364	−0.688433	0.601595	0.397231
1.50	0.387142	−0.680560	0.607750	0.380558
1.55	0.383094	−0.673035	0.614277	0.364770
1.60	0.379209	−0.665830	0.621170	0.349802
1.65	0.375474	−0.658925	0.628422	0.335599
1.70	0.371881	−0.652296	0.636027	0.322108
1.75	0.368420	−0.645926	0.643983	0.309283
1.80	0.365083	−0.639798	0.652287	0.297081
1.85	0.361862	−0.633896	0.660936	0.285461
1.90	0.358752	−0.628205	0.669929	0.274390
1.95	0.355744	−0.622713	0.679267	0.263833
2.00	0.352834	−0.617408	0.688948	0.253760
2.05	0.350016	−0.612279	0.698975	0.244143
2.10	0.347285	−0.607317	0.709349	0.234955
2.15	0.344637	−0.602511	0.720072	0.226174

Table 4.10 (continued)

v	$J_v(v)$	$Y_v(v)$	$I_v(v)$	$K_v(v)$
2.20	0.342068	−0.597855	0.731145	0.217776
2.25	0.339572	−0.593338	0.742573	0.209741
2.30	0.337148	−0.588955	0.754358	0.202050
2.35	0.334790	−0.584699	0.766505	0.194684
2.40	0.332497	−0.580564	0.779018	0.187627
2.45	0.330265	−0.576543	0.791901	0.180863
2.50	0.328091	−0.572631	0.805159	0.174377
2.55	0.325974	−0.568823	0.818799	0.168155
2.60	0.323909	−0.565114	0.832825	0.162186
2.65	0.321895	−0.561501	0.847244	0.156456
2.70	0.319931	−0.557978	0.862061	0.150953
2.75	0.318013	−0.554541	0.877285	0.145669
2.80	0.316140	−0.551188	0.892922	0.140591
2.85	0.314310	−0.547915	0.908980	0.135711
2.90	0.312522	−0.544718	0.925465	0.131020
2.95	0.310773	−0.541595	0.942387	0.126509
3.00	0.309063	−0.538542	0.959754	0.122170
3.05	0.307389	−0.535556	0.977574	0.117996
3.10	0.305751	−0.532636	0.995856	0.113979
3.15	0.304148	−0.529779	1.014610	0.110113
3.20	0.302577	−0.526981	1.033850	0.106390
3.25	0.301038	−0.524243	1.053570	0.102806
3.30	0.299530	−0.521560	1.073800	0.099354
3.35	0.298051	−0.518931	1.094540	0.096028
3.40	0.296601	−0.516355	1.115810	0.092824
3.45	0.295179	−0.513829	1.137610	0.089735
3.50	0.293783	−0.511351	1.159950	0.086759
3.55	0.292414	−0.508921	1.182860	0.083889
3.60	0.291069	−0.506537	1.206330	0.081122
3.65	0.289749	−0.504196	1.230390	0.078454
3.70	0.288453	−0.501898	1.255040	0.075881
3.75	0.287179	−0.499642	1.280310	0.073398
3.80	0.285927	−0.497425	1.306200	0.071003
3.85	0.284697	−0.495248	1.332720	0.068691
3.90	0.283488	−0.493108	1.359900	0.066461
3.95	0.282299	−0.491005	1.387750	0.064307
4.00	0.281129	−0.488937	1.416280	0.062229
4.05	0.279979	−0.486903	1.445500	0.060222
4.10	0.278847	−0.484903	1.475450	0.058284
4.15	0.277733	−0.482936	1.506120	0.056413
4.20	0.276637	−0.481000	1.537540	0.054606
4.25	0.275557	−0.479094	1.569730	0.052860
4.30	0.274495	−0.477219	1.602710	0.051173
4.35	0.273448	−0.475373	1.636480	0.049544

Table 4.10 (continued)

v	$J_v(v)$	$Y_v(v)$	$I_v(v)$	$K_v(v)$
4.40	0.272417	−0.473555	1.671080	0.047969
4.45	0.271402	−0.471764	1.706520	0.046448
4.50	0.270402	−0.470000	1.742820	0.044977
4.55	0.269416	−0.468262	1.780000	0.043556
4.60	0.268444	−0.466550	1.818090	0.042182
4.65	0.267486	−0.464863	1.857100	0.040854
4.70	0.266541	−0.463199	1.897050	0.039570
4.75	0.265610	−0.461560	1.937980	0.038328
4.80	0.264692	−0.459943	1.979900	0.037128
4.85	0.263786	−0.458349	2.022830	0.035966
4.90	0.262892	−0.456776	2.066800	0.034843
4.95	0.262011	−0.455225	2.111840	0.033757
5.00	0.261141	−0.453695	2.157970	0.032706

Table 4.11: The Struve, Anger and Weber functions with the same argument and order $H_v(v)$, $L_v(v)$, $J_v(v)$ and $E_v(v)$.

v	$H_v(v)$	$L_v(v)$	$J_v(v)$	$E_v(v)$
0.00	0.000000	0.000000	1.000000	0.000000
0.05	0.026384	0.026398	0.997765	0.046624
0.10	0.046749	0.046846	0.991093	0.092740
0.15	0.063605	0.063895	0.980076	0.137849
0.20	0.078048	0.078662	0.964869	0.181467
0.25	0.090711	0.091798	0.945684	0.223134
0.30	0.102005	0.103720	0.922788	0.262422
0.35	0.112213	0.114718	0.896498	0.298939
0.40	0.121541	0.125001	0.867174	0.332339
0.45	0.130142	0.134725	0.835216	0.362322
0.50	0.138133	0.144010	0.801052	0.388643
0.55	0.145609	0.152950	0.765138	0.411114
0.60	0.152642	0.161619	0.727944	0.429605
0.65	0.159293	0.170077	0.689949	0.444046
0.70	0.165612	0.178376	0.651634	0.454428
0.75	0.171640	0.186557	0.613470	0.460801
0.80	0.177411	0.194655	0.575917	0.463274
0.85	0.182955	0.202700	0.539409	0.462010
0.90	0.188297	0.210719	0.504354	0.457223
0.95	0.193458	0.218733	0.471124	0.449173
1.00	0.198457	0.226764	0.440051	0.438162
1.05	0.203311	0.234829	0.411420	0.424528
1.10	0.208034	0.242945	0.385470	0.408635

Table 4.11 (continued)

v	$H_v(v)$	$L_v(v)$	$J_v(v)$	$E_v(v)$
1.15	0.212638	0.251124	0.362387	0.390870
1.20	0.217135	0.259382	0.342304	0.371634
1.25	0.221534	0.267729	0.325299	0.351336
1.30	0.225846	0.276178	0.311396	0.330384
1.35	0.230076	0.284739	0.300568	0.309178
1.40	0.234234	0.293421	0.292736	0.288106
1.45	0.238325	0.302235	0.287772	0.267532
1.50	0.242356	0.311190	0.285506	0.247797
1.55	0.246331	0.320293	0.285728	0.229207
1.60	0.250256	0.329554	0.288193	0.212034
1.65	0.254135	0.338981	0.292629	0.196510
1.70	0.257972	0.348581	0.298740	0.182822
1.75	0.261772	0.358363	0.306215	0.171114
1.80	0.265537	0.368334	0.314733	0.161484
1.85	0.269272	0.378502	0.323969	0.153983
1.90	0.272978	0.388874	0.333604	0.148617
1.95	0.276659	0.399458	0.343324	0.145347
2.00	0.280318	0.410261	0.352834	0.144095
2.05	0.283957	0.421291	0.361858	0.144744
2.10	0.287578	0.432555	0.370144	0.147141
2.15	0.291183	0.444061	0.377471	0.151105
2.20	0.294775	0.455816	0.383650	0.156429
2.25	0.298355	0.467829	0.388527	0.162886
2.30	0.301926	0.480106	0.391985	0.170235
2.35	0.305488	0.492657	0.393946	0.178227
2.40	0.309044	0.505488	0.394370	0.186609
2.45	0.312595	0.518608	0.393254	0.195129
2.50	0.316143	0.532024	0.390631	0.203546
2.55	0.319688	0.545746	0.386570	0.211628
2.60	0.323233	0.559782	0.381170	0.219163
2.65	0.326779	0.574140	0.374558	0.225958
2.70	0.330326	0.588829	0.366887	0.231847
2.75	0.333876	0.603859	0.358329	0.236690
2.80	0.337431	0.619237	0.349070	0.240377
2.85	0.340991	0.634974	0.339310	0.242829
2.90	0.344557	0.651079	0.329252	0.244000
2.95	0.348130	0.667561	0.319103	0.243874
3.00	0.351711	0.684431	0.309063	0.242467
3.05	0.355302	0.701699	0.299326	0.239827
3.10	0.358903	0.719375	0.290075	0.236026
3.15	0.362514	0.737469	0.281475	0.231164
3.20	0.366138	0.755992	0.273671	0.225362
3.25	0.369775	0.774956	0.266788	0.218762
3.30	0.373425	0.794372	0.260926	0.211518

Table 4.11 (continued)

v	$H_v(v)$	$L_v(v)$	$J_v(v)$	$E_v(v)$
3.35	0.377089	0.814250	0.256157	0.203798
3.40	0.380769	0.834604	0.252529	0.195775
3.45	0.384464	0.855445	0.250060	0.187624
3.50	0.388176	0.876785	0.248743	0.179521
3.55	0.391906	0.898638	0.248543	0.171636
3.60	0.395654	0.921015	0.249403	0.164127
3.65	0.399420	0.943931	0.251239	0.157143
3.70	0.403207	0.967399	0.253951	0.150815
3.75	0.407013	0.991433	0.257419	0.145256
3.80	0.410841	1.016050	0.261510	0.140559
3.85	0.414690	1.041260	0.266079	0.136792
3.90	0.418562	1.067070	0.270975	0.134003
3.95	0.422456	1.093520	0.276043	0.132215
4.00	0.426374	1.120600	0.281129	0.131426
4.05	0.430317	1.148340	0.286082	0.131612
4.10	0.434284	1.176760	0.290758	0.132725
4.15	0.438277	1.205860	0.295024	0.134699
4.20	0.442296	1.235670	0.298760	0.137448
4.25	0.446342	1.266210	0.301863	0.140869
4.30	0.450415	1.297490	0.304246	0.144847
4.35	0.454517	1.329540	0.305843	0.149257
4.40	0.458647	1.362360	0.306607	0.153965
4.45	0.462807	1.395980	0.306514	0.158835
4.50	0.466997	1.430430	0.305561	0.163729
4.55	0.471217	1.465710	0.303764	0.168513
4.60	0.475468	1.501860	0.301163	0.173057
4.65	0.479751	1.538890	0.297813	0.177242
4.70	0.484067	1.576830	0.293788	0.180959
4.75	0.488416	1.615690	0.289177	0.184111
4.80	0.492798	1.655510	0.284079	0.186622
4.85	0.497215	1.696300	0.278608	0.188426
4.90	0.501667	1.738090	0.272879	0.189482
4.95	0.506154	1.780910	0.267016	0.189764
5.00	0.510678	1.824780	0.261141	0.189267

Table 4.12: The Kelvin functions with the same argument and order $ber_v(v)$, $bei_v(v)$, $ker_v(v)$ and $kei_v(v)$.

v	$ber_v(v)$	$bei_v(v)$	$ker_v(v)$	$kei_v(v)$
0.00	1.000000	0.000000		-0.785398
0.05	0.848218	0.100905	3.055820	-1.036710
0.10	0.757095	0.183583	2.296540	-1.179240
0.15	0.680554	0.254858	1.815480	-1.275790
0.20	0.609680	0.317075	1.445810	-1.338880
0.25	0.540868	0.371258	1.137240	-1.374410
0.30	0.472441	0.417858	0.868691	-1.386090
0.35	0.403643	0.457034	0.629896	-1.376650
0.40	0.334231	0.488808	0.415420	-1.348400
0.45	0.264280	0.513137	0.222259	-1.303410
0.50	0.194077	0.529967	0.048708	-1.243640
0.55	0.124053	0.539263	-0.106225	-1.171030
0.60	0.054745	0.541032	-0.243148	-1.087470
0.65	-0.013240	0.535338	-0.362479	-0.994824
0.70	-0.079247	0.522314	-0.464570	-0.894954
0.75	-0.142594	0.502168	-0.549779	-0.789678
0.80	-0.202591	0.475188	-0.618520	-0.680772
0.85	-0.258558	0.441745	-0.671292	-0.569954
0.90	-0.309833	0.402291	-0.708694	-0.458867
0.95	-0.355795	0.357359	-0.731432	-0.349064
1.00	-0.395868	0.307557	-0.740322	-0.241996
1.05	-0.429540	0.253564	-0.736282	-0.138996
1.10	-0.456366	0.196124	-0.720324	-0.041270
1.15	-0.475988	0.136035	-0.693545	0.050114
1.20	-0.488131	0.074143	-0.657110	0.134235
1.25	-0.492619	0.011327	-0.612240	0.210319
1.30	-0.489377	-0.051508	-0.560194	0.277752
1.35	-0.478435	-0.113447	-0.502253	0.336070
1.40	-0.459930	-0.173576	-0.439701	0.384968
1.45	-0.434110	-0.230994	-0.373813	0.424289
1.50	-0.401327	-0.284830	-0.305834	0.454026
1.55	-0.362040	-0.334253	-0.236968	0.474310
1.60	-0.316806	-0.378486	-0.168360	0.485402
1.65	-0.266279	-0.416818	-0.101085	0.487686
1.70	-0.211197	-0.448617	-0.036139	0.481655
1.75	-0.152376	-0.473341	0.025579	0.467900
1.80	-0.090699	-0.490547	0.083265	0.447095
1.85	-0.027104	-0.499899	0.136224	0.419984
1.90	0.037431	-0.501175	0.183872	0.387366
1.95	0.101899	-0.494276	0.225740	0.350083
2.00	0.165279	-0.479225	0.261472	0.309001
2.05	0.226556	-0.456170	0.290831	0.265000
2.10	0.284730	-0.425390	0.313688	0.218958
2.15	0.338836	-0.387282	0.330026	0.171737

Table 4.12 (continued)

v	ber$_v$(v)	bei$_v$(v)	ker$_v$(v)	kei$_v$(v)
2.20	0.387956	−0.342368	0.339932	0.124173
2.25	0.431241	−0.291283	0.343590	0.077063
2.30	0.467916	−0.234768	0.341271	0.031154
2.35	0.497301	−0.173663	0.333330	−0.012863
2.40	0.518820	−0.108893	0.320193	−0.054366
2.45	0.532013	−0.041456	0.302347	−0.092801
2.50	0.536544	0.027592	0.280328	−0.127695
2.55	0.532212	0.097148	0.254713	−0.158654
2.60	0.518952	0.166088	0.226110	−0.185369
2.65	0.496844	0.233273	0.195141	−0.207613
2.70	0.466109	0.297578	0.162438	−0.225244
2.75	0.427113	0.357901	0.128629	−0.238202
2.80	0.380361	0.413186	0.094327	−0.246504
2.85	0.326493	0.462441	0.060126	−0.250242
2.90	0.266278	0.504751	0.026588	−0.249581
2.95	0.200598	0.539299	−0.005764	−0.244745
3.00	0.130443	0.565381	−0.036451	−0.236018
3.05	0.056896	0.582415	−0.065043	−0.223734
3.10	−0.018888	0.589956	−0.091166	−0.208268
3.15	−0.095691	0.587709	−0.114505	−0.190030
3.20	−0.172258	0.575530	−0.134804	−0.169454
3.25	−0.247313	0.553437	−0.151869	−0.146996
3.30	−0.319578	0.521612	−0.165571	−0.123116
3.35	−0.387797	0.480400	−0.175837	−0.098281
3.40	−0.450757	0.430306	−0.182659	−0.072948
3.45	−0.507303	0.371995	−0.186084	−0.047563
3.50	−0.556368	0.306279	−0.186212	−0.022552
3.55	−0.596982	0.234110	−0.183194	0.001686
3.60	−0.628298	0.156566	−0.177225	0.024782
3.65	−0.649604	0.074836	−0.168540	0.046402
3.70	−0.660341	−0.009796	−0.157408	0.066254
3.75	−0.660110	−0.095969	−0.144127	0.084086
3.80	−0.648691	−0.182271	−0.129017	0.099692
3.85	−0.626041	−0.267255	−0.112412	0.112913
3.90	−0.592306	−0.349466	−0.094658	0.123634
3.95	−0.547818	−0.427464	−0.076103	0.131789
4.00	−0.493097	−0.499846	−0.057094	0.137357
4.05	−0.428844	−0.565274	−0.037967	0.140359
4.10	−0.355936	−0.622496	−0.019050	0.140861
4.15	−0.275414	−0.670370	−0.000648	0.138963
4.20	−0.188468	−0.707883	0.016952	0.134806
4.25	−0.096424	−0.734177	0.033492	0.128557
4.30	−0.000724	−0.748557	0.048741	0.120413
4.35	0.097099	−0.750516	0.062501	0.110594
4.40	0.195438	−0.739744	0.074606	0.099338

Table 4.12 (continued)

v	ber$_v$(v)	bei$_v$(v)	ker$_v$(v)	kei$_v$(v)
4.45	0.292644	−0.716135	0.084925	0.086896
4.50	0.387047	−0.679799	0.093364	0.073527
4.55	0.476985	−0.631061	0.099862	0.059497
4.60	0.560834	−0.570461	0.104395	0.045068
4.65	0.637035	−0.498754	0.106971	0.030501
4.70	0.704119	−0.416899	0.107632	0.016045
4.75	0.760738	−0.326046	0.106448	0.001938
4.80	0.805688	−0.227527	0.103519	−0.011599
4.85	0.837933	−0.122832	0.098968	−0.024363
4.90	0.856627	−0.013595	0.092941	−0.036172
4.95	0.861133	0.098440	0.085602	−0.046870
5.00	0.851038	0.211434	0.077130	−0.056323

Table 4.13: The integral Bessel functions with the same argument and order $Ji_v(v)$, $Yi_v(v)$ and $Ki_v(v)$.

v	$Ji_v(v)$	$Yi_v(v)$	$Ki_v(v)$
0.00	3.111980		
0.05	2.916260	−2.797400	5.251020
0.10	2.210500	−1.701630	3.339290
0.15	1.823610	−1.214250	2.472910
0.20	1.566640	−0.933362	1.937880
0.25	1.379620	−0.750191	1.577130
0.30	1.235810	−0.621521	1.316510
0.35	1.121030	−0.526468	1.119330
0.40	1.026890	−0.453627	0.965125
0.45	0.948063	−0.396218	0.841464
0.50	0.880959	−0.349955	0.740322
0.55	0.823059	−0.311990	0.656271
0.60	0.772537	−0.280361	0.585497
0.65	0.728029	−0.253670	0.525241
0.70	0.688498	−0.230897	0.473452
0.75	0.653134	−0.211280	0.428576
0.80	0.621299	−0.194237	0.389410
0.85	0.592481	−0.179321	0.355012
0.90	0.566264	−0.166178	0.324632
0.95	0.542305	−0.154527	0.297666
1.00	0.520320	−0.144143	0.273621
1.50	0.370784	0.073988	0.130394
2.00	0.288362	0.108225	0.069933
3.00	0.199871	0.076720	0.024235
4.00	0.153059	0.058667	0.009646
5.00	0.124067	−0.013031	0.004161

Index

https://doi.org/10.1515/9783110682472-005

www.ingramcontent.com/pod-product-compliance
Lightning Source LLC
Chambersburg PA
CBHW060948210326
41598CB00031B/4760